Advanced Biology

Third Edition

J. Simpkins ... Senior ... Biology
People's College of Further Education
...

... Lecturer and Head of Biology
People's College of Further Education
Nottingham

Series Editor: **M. K. Sands**

UNWIN HYMAN

Published by
UNWIN HYMAN LIMITED
15/17 Broadwick Street
London WIV 1FP

© J. Simpkins and J. I. Williams 1980, 1984, 1989
First published in 1980 by Mills & Boon Limited
Reprinted by Bell & Hyman Limited 1981, 1982
Second Edition 1984
Reprinted 1984, 1985, 1986
Reprinted by Unwin Hyman Limited 1988
Third edition 1989

British Library Cataloguing in Publication Data

Simpkins, J.
    Advanced biology—3rd ed.
    1. Organisms
    I. Title   II. Williams, J. I.   III. Simpkins, J.
    Biology of the cell, mammal and flowering plant

ISBN 0 04 448003 2

Typeset by MS Filmsetting Limited, Frome, Somerset
Printed and bound in Great Britain by
Butler & Tanner Ltd, Frome and London

Contents

Preface (v)

Acknowledgements (vii)

1 Water and aqueous solutions 1

2 Carbohydrates, lipoids and proteins 15

3 Enzymes 39

4 Nucleic acids and protein synthesis 53

5 Energy conversion in living cells 75

6 Cell structure 95

7 Cell division 119

8 Gas exchange in mammals 133

9 Circulation of blood and lymph 155

10 Blood functions and immune system 179

11 The renal system 199

12 Uptake and transport of water in flowering plants 219

13 Photosynthesis and translocation of photosynthetic products 239

14 Mineral nutrition of flowering plants 261

15 The mammalian alimentary system 277

16 Skeletal and muscle systems 305

17 Nervous control and co-ordination in mammals 1: Neurones and muscles 319

18 Nervous control and co-ordination in mammals 2: Nervous integration 341

19 The eye and ear 367

20 Thermoregulation 379

21 Endocrine system 393

22 Reproduction in flowering plants 413

23 Seed structure and germination 437

24 Growth and development of flowering plants 451

25 Mammalian reproduction 479

26 Genetics 495

27 The origin and evolution of life 525

28 Variety of life: Micro-organisms and plants 551

29 Variety of life: Non-chordate animals 583

30 Variety of life: Chordate animals 613

31 Ecology 1: Distribution of organisms; associations 631

32 Ecology 2: Populations, communities and ecosystems 655

33 Pollution and conservation 687

34 Biotechnology 715

Appendix 1 Units, Symbols and Quantities 731

Appendix 2 Chemical Nomenclature 731

Appendix 3 Practical Work 732

Index 751

Preface to the Third Edition

All examining boards include a Common Core in Biology at GCE A level. It includes reference to the applications of biology. Most examining boards also have an assessment strategy which incorporates practical coursework. These requirements were foremost in our minds in preparing this new edition of *Advanced Biology*.

The book also covers the syllabuses in Biochemistry, Cell Biology and Mammalian Physiology for BTEC National Certificate and Diploma courses in Science.

As in the second edition, we have endeavoured to write in a style and language which the average sixth former can readily understand. The chapters can be read in any order, but in our opinion the subject matter is likely to be more intelligible if the topics are studied in the order presented. The first seven chapters provide a foundation in biological, biochemical and physical knowledge and principles which should enable students to cope with the topics which follow.

Key words have been presented in bold type to emphasise the important facts and concepts. Applications of biological knowledge are highlighted throughout the text in boxed sections. Chapter Summaries have been included for ease of revision. The text has also been made more accessible by means of thought-provoking questions in the margins. We have followed the recommendations of the Institute of Biology regarding Nomenclature, Taxonomy and Units. These have been adopted by the examining boards. Students should no longer have to remember lists of alternative names.

Biology is a practical subject and so we have included suggestions for laboratory and field work. Second-hand evidence from experimental work is also quoted extensively in the text. It is hoped that this will encourage students to realise that factual evidence is essential if valid conclusions are to be drawn. Where we have written about controversial issues we have tried to offer a reasonable balance from the conflicting evidence available.

J. SIMPKINS & J. I WILLIAMS   Nottingham 1989

Acknowledgements

We gratefully acknowledge the assistance of:

Our editor, Margaret Sands, for constant encouragement and helpful guidance.

Christopher Blake and Pat Winter of Unwin Hyman Ltd., for their endless patience and thoughtful advice in ensuring the completion of the project.

Dr R. Jones, Dr P. M. Davies and Dr D. M. Holdich of the Department of Zoology, University of Nottingham; Dr I. Mackenzie and Dr A. F. Braithwaite of the Department of Botany, University of Nottingham; Dr A. Booth of the Department of Botany, University of Sheffield and Dr G. Triggs of Liverpool Polytechnic who devoted much time to reading the typescript. Their comments were immensely helpful.

Dr D. Brindley of the Queen's Medical Centre, University of Nottingham for his useful advice on biochemical nomenclature.

The Joint Matriculation Board, the University of London School Examinations Board, the Associated Examining Board, the University of Oxford Delegacy of Local Examinations, the Cambridge Local Examinations Syndicate and the Welsh Joint Education Committee, for their permission to use recent examination questions.

Mr J. Kugler of the Queen's Medical Centre, University of Nottingham and Mr D. Wark of King's Mill Hospital, Sutton-in-Ashfield, without whose expertise and hard work the task of obtaining most of the photographs would have been immensely more demanding. Nearly all of the photographs not acknowledged below or in the text were provided by Mr Kugler and Mr Wark.

Mrs M. Hollingsworth of the Department of Anatomy, University of Sheffield, Mrs A. Tomlinson of the Queen's Medical Centre, University of Nottingham, and Mr B. Case of the Department of Botany, University of Nottingham, who kindly gave their time and skills in preparing some of the photographs.

Other people who kindly provided photographs and drawings were:

Mr P. Bailey and Mr A. Bezear of the Queen's Medical Centre, University of Nottingham, Professor R. M. H. McMinn of the Department of Anatomy, the Royal College of Surgeons of England, Dr M. Davey of the Department of Botany, University of Nottingham, Dr J. Freer of the University of Glasgow, Dr M. A. Tribe of the School of Biological Sciences, University of Sussex, Mr D. Gould, of the Department of Dental Surgery, General Hospital, Nottingham, Mr R. Dainty of the Queen's Medical Centre, Nottingham, Mr S. Hyman of the University of Leicester, and Mr S. Wood (Philip Harris Biological) who also supplied microscopical preparations from which many of the photomicrographs were prepared.

The many other individuals, institutions and publishers who provided or who gave permission to use illustrative material. They have been credited in the text alongside the appropriate figures.

Any errors in the book are entirely the responsibility of the authors and the publishers.

Examination Questions

The publisher thanks the examination boards for permission to reproduce examination questions. Questions are acknowledged as follows:

 AEB Associated Examining Board
    C Cambridge Local Examinations Syndicate
 JMB Joint Matriculation Board
    L University of London School Examinations Board
    O University of Oxford Delegacy of Local Examinations
WJEC Welsh Joint Education Committee

1 Water and aqueous solutions

1.1 The water molecule   2

1.2 Properties of water   2

   1.2.1 Water as a solvent   2
   1.2.2 Thermal properties   4
   1.2.3 Density and viscosity   5
   1.2.4 Surface tension   5
   1.2.5 Penetration by light   6
   1.2.6 Dissociation of water   6

1.3 Some properties of aqueous solutions   7

   1.3.1 Acids, bases and buffers   7
   1.3.2 Osmotic properties   8

1.4 Gain and loss of water by living cells   8

   1.4.1 Osmosis in plant cells   9
   1.4.2 Osmosis in animal cells   11
   1.4.3 Osmoregulation   12

1.5 Non-osmotic movement of water   12

1.6 Water balance   13

Summary   13

Questions   13

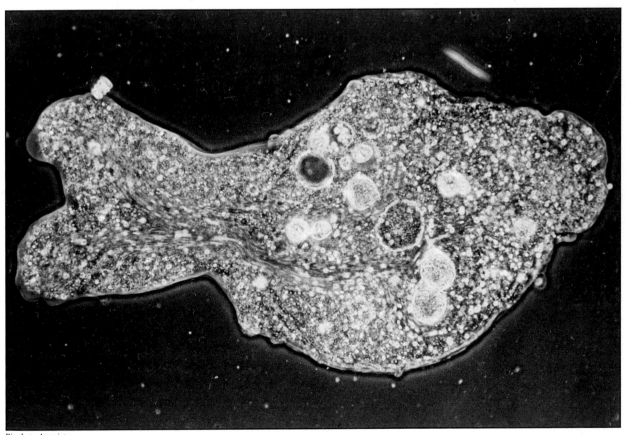

Biophoto Associates

1 Water and aqueous solutions

Life is thought to have originated in water, and many species of plants and animals living today have the sea or freshwater as their habitat. Whether a creature lives in water, or lives on land as humans do, water is vital for its life processes. No less than two thirds of the fresh body mass of a human is water, about half occurring inside our cells and the remainder in our body fluids (Table 1.1). Remarkably, as much as 85% of your brain is water! The bodies of all organisms contain a high percentage of water (Table 1.2).

It is in aqueous solution that all metabolic reactions take place. Water is a reactant in hydrolytic reactions (Chapter 3) and a raw material in photosynthesis (Chapter 5). However, there are many other ways in which water is indispensable to all living creatures. To appreciate why water is so vital it is first necessary to understand some of the properties of water.

Table 1.1 Distribution of water in a human (after Armstrong and Bennett 1979)

Site	Water in body/%
intracellular	55.0
extracellular:	
blood plasma	7.5
tissue fluid and lymph	22.5
connective tissue,	
cartilage and bone	15.0

Table 1.2 Water content of a variety of organisms

Specimen	Water content/ % fresh mass
lettuce leaf	93–95
carrot taproot	89–91
strawberry fruit	88–90
jellyfish	95–98
earthworm	82–84

1.1 The water molecule

A water molecule consists of two atoms of hydrogen and one atom of oxygen. The atoms are joined in such a way that the single electron of each of the hydrogen atoms is shared with electrons in the outer shell of the oxygen atom (Fig 1.1). The unshared pairs of electrons repel the shared pairs so that the hydrogen nuclei are pushed towards one another causing the molecule to be bent (Fig 1.2). In the region of the hydrogen nuclei the water molecule is positively charged, whilst the part of the oxygen atom away from the hydrogen nuclei is negatively charged.

Fig 1.1 Formation of a molecule of water

Fig 1.2 The non-linear water molecule

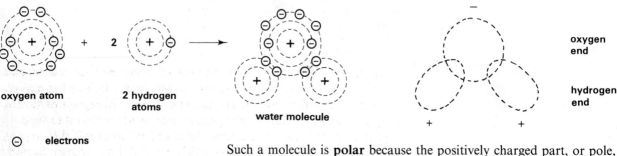

oxygen atom

2 hydrogen atoms

water molecule

oxygen end

hydrogen end

⊖ electrons

(+) atomic nuclei

Such a molecule is **polar** because the positively charged part, or pole, differs from the negatively charged pole in the way it reacts with ions and charged molecules. The polarity of water molecules accounts for many of the properties of water which are important to living organisms.

1.2 Properties of water

1.2.1 Water as a solvent

A wide range of inorganic and organic substances readily dissolve in water. The reason why water acts as a **solvent** for so many inorganic compounds can be understood by considering the way a simple salt such as sodium chloride reacts with water. In a crystal of sodium chloride the sodium and chloride ions are held together by electrovalent bonds. When placed in

Fig 1.3 Reaction between sodium chloride and water

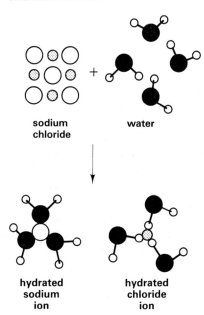

sodium
chloride

water

hydrated
sodium
ion

hydrated
chloride
ion

water the positively charged sodium ions (Na^+) are attracted to the negatively charged oxygen poles of water molecules. Conversely, the negatively charged chloride ions (Cl^-) are pulled towards the positively charged hydrogen poles of water molecules. Consequently the sodium and chloride ions become separated by clusters of water molecules and an **aqueous solution** of sodium chloride is formed (Fig 1.3). The polarity of water molecules is clearly a key factor in the ability of water to dissolve inorganic substances.

Many biologically important organic substances do not ionise and yet they also dissolve in water. This is because they form hydrogen bonds with water. For example, hydroxyl (—OH) groups of sugar molecules, imino ($>$NH) groups of proteins, and carbonyl ($>$C=O) groups of organic acids react with water in this way. Molecules of such organic compounds thus become surrounded by water molecules and go into solution or suspension.

1 True solutions

For the reason outlined above many inorganic and organic particles smaller than 10^{-5} cm in diameter dissolve quickly in water to form a **true solution**. The dissolving power of water is particularly important in the uptake and transportation of, substances inside the bodies of all living organisms. Metabolic reactions catalysed by enzymes also take place in aqueous solution (Chapter 3). Nevertheless, the inability of some substances to dissolve readily in water sometimes poses serious problems to living organisms. For example, if we compare the volumes of some common gases in air and in water in direct contact with air, we see some startling differences (Table 1.3). Whereas carbon dioxide dissolves in water with ease, oxygen and nitrogen do not.

Table 1.3 Amounts of some common gases in air and in water in direct contact with air (at 10 °C)

| | Volume (cm^3) of gas in 100 cm^3 of air or water | | |
	Oxygen	Nitrogen	Carbon dioxide
air	20.95	78.0	0.03
water	0.79	1.2	0.03

The difference in the oxygen content of water and air has several consequences. Aquatic organisms, such as fish, have relatively less oxygen available to them for respiration in water than terrestrial organisms have in air; the lower concentration of oxygen in water is not sufficient to support respiration in humans, for example. Another consequence is that oxygen transport in our blood would be inefficient if it did not contain haemoglobin (Chapter 8).

The rate at which gases diffuse through water is much slower than through air. For this reason, the rate at which carbon dioxide diffuses through the protoplasm of photosynthetic cells before reaching the chloroplasts is a factor which limits the productivity of plants (Chapter 13).

2 The colloidal state

Particles between 10^{-3} and 10^{-5} cm in diameter, such as polysaccharides and proteins of high relative molar mass, do not form true solutions with water. They attract water to form a **colloidal state**, with water acting as a **dispersion medium** in which the particles, known as the **disperse phase**, are permanently suspended. Substances which react with water in this way are called hydrophilic colloids. Cytoplasm is an example of a colloidal suspension. Blood plasma is another. The blood proteins which form the disperse phase play important roles in blood clotting, combating infections and in the formation of tissue fluid (Chapters 9 and 10). Hydrophilic colloids cling to water and thus help minimise evaporation of water from organisms which are exposed to air.

3 Suspensions

Particles larger than 10^{-3} cm in diameter can be temporarily dispersed in water to form a **suspension**. On standing, the disperse phase of a suspension will gradually separate from the dispersion medium unless an emulsifying agent is added. For example, bile emulsifies fats and oils in the mammalian gut (Chapter 15). The disperse phase of fats and oils then has a larger surface area on which fat-splitting enzymes can work.

1.2.2 Thermal properties

Compared with other compounds of about the same relative molar mass, water has higher melting and boiling points and is liquid over a wider temperature range (Table 1.4). This can be understood by looking at the way in which water molecules react with each other. Because particles of

Fig 1.4 Hydrogen bonding between water molecules

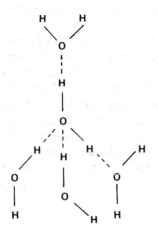

--- H = hydrogen bond

Table 1.4 Some thermal properties of water and compounds of comparable relative molar mass

Substance	Formula	Relative molar mass	Melting point/°C	Boiling point/°C
water	H_2O	18	0	100
ammonia	NH_3	17	-78	-33
methane	CH_4	16	-184	-161

opposite charge attract each other, the negative pole of a water molecule is drawn towards the positive pole of another. Weak **hydrogen bonds** are formed between linked water molecules. In this way a water molecule can bind with up to four others, two attracted to the oxygen atom and one to each of the hydrogen atoms (Fig 1.4). Water therefore exists as **molecular clusters** rather than as individual molecules.

Temperature has an effect on the extent of bonding between water molecules. At 0 °C, the freezing point of water, the molecules are arranged in a regular, hexagonal, crystalline network in which each water molecule is hydrogen-bonded to four others (Fig 1.5(a)). At temperatures just above 0 °C some of the bonds are broken, producing clusters of molecules which fit together more compactly than the evenly spaced water molecules in ice (Fig 1.5(b)). This is why liquid water occupies a smaller volume and is therefore denser than ice. Water is most dense at 4 °C.

Fig 1.5 (a) Crystalline structure of ice

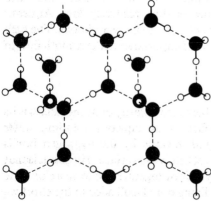

● oxygen atoms

○ hydrogen atoms

--- hydrogen bonds

Fig 1.5 (b) Molecules in liquid water

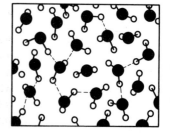

Changes in the density of water are important for the survival of many aquatic organisms. The temperature of a body of open water such as a pond or a lake is affected by the temperature of the air above the water. Very cold air causes the body of water to cool from the surface downwards. When the temperature of the upper layers falls to 4 °C the dense water sinks bringing warmer water to the surface. Convection currents of this sort delay the freezing of a body of water. Ultimately a point may be reached when the water near the surface freezes. The ice, being less dense than the water beneath, floats on the surface where it insulates the water from further heat loss. Consequently liquid water, in which aquatic life can remain active, remains beneath the ice.

Heat energy can break the bonds between water molecules. Because of the extensive hydrogen-bonding in ice, a relatively large amount of heat ($333.6 \, J \, g^{-1}$ at 0 °C), called the **latent heat of melting**, is required to convert ice to liquid water. For this reason water has a relatively high melting point. The formation of water vapour involves the breaking of hydrogen bonds between molecules of liquid water. Evaporation uses a lot of heat energy called the **latent heat of vaporisation** ($2411 \, J \, g^{-1}$ at 25 °C). This is why water has a relatively high boiling point. It also explains why a considerable cooling effect occurs when water evaporates from the bodies of living organisms. The evaporation of water from terrestrial organisms plays an important role in preventing them from becoming overheated (Chapters 12 and 20).

The amount of heat energy required to raise the temperature of 1 gram of a compound by 1 °C is called the **specific heat capacity**. Water has a much higher specific heat capacity ($4.2 \, J \, g^{-1}$ between 0 and 50 °C) than other compounds of similar relative molar mass. The significance of this to living organisms is that water does not quickly heat up or cool down as air does. It is for this reason that large bodies of water such as oceans and lakes have a steady low temperature which is suitable for cold-blooded organisms such as fish. The ability of water to act as a **thermal buffer** also helps to prevent rapid changes in body temperature of terrestrial organisms. Remember that land-dwelling creatures, like all forms of life, contain a large proportion of water.

1.2.3 Density and viscosity

A given mass of water is heavier than a similar mass of many other substances. The high **relative density** (specific gravity) of water explains why many living organisms, humans included, readily float in water. The buoyancy of water also helps the swimming of motile gametes and in the dispersal of fruits, seeds and spores.

Viscosity is a measure of the difficulty with which molecules slide over one another. The hydrogen bonds between water molecules are continually broken and re-formed so that the molecules slide over each other with relative ease. The viscosity of water is thus relatively low. Aqueous solutions and suspensions on the other hand can be very viscous. The high viscosity of blood plasma and lymph is of importance in the smooth flow of blood and lymph (Chapter 9).

1.2.4 Surface tension

Another property of water attributable to the polarity of its molecules is its high **surface tension**. At the surface of an aqueous solution, water molecules are pulled downwards and inwards by the hydrogen bonds which join them to water molecules just below the surface. The result is that the water molecules in the surface layer draw together and form a skin. The surface tension in the layer of water lining our alveoli adds to the effort we need to make to stretch our lungs when breathing in. Phospholipid molecules in the alveolar membrane lower the surface tension, thus reducing the work required (Chapters 2 and 8).

1.2.5 Penetration by light

Light rays penetrate water with relative ease because water is transparent. The depth to which light can penetrate is of course reduced if the water is turbid due to suspended particles. In clear water, red and yellow light can reach to a depth of 50 m while blue and violet rays can go down to 200 m (Chapter 31).

The ability of light to penetrate water enables photosynthetic organisms to inhabit the vast surface volumes of lakes and seas. It also means that light can easily penetrate the transparent water-filled epidermis of leaves and reach the underlying cells which contain the light-absorbing pigments (Chapter 13).

1.2.6 Dissociation of water

The nucleus of a hydrogen atom consists of a single positively charged particle called a **hydrogen ion** (H^+), also called a proton. In an aqueous solution small numbers of water molecules lose their hydrogen ions. The remainder of the water molecule is called a **hydroxyl ion** (OH^-). Such molecules are said to be **dissociated**. The process is usually written as:

$$H_2O \rightleftharpoons \quad H^+ \quad + \quad OH^-$$

water   hydrogen ion   hydroxyl ion

In pure water the concentration of hydrogen ions $[H^+]$ is $10^{-7}\,mol\,dm^{-3}$ at 298 K. The product of the concentrations of hydrogen and hydroxyl ions, which is called the **ionic-product of water** K_w, is always $10^{-14}\,mol\,dm^{-3}$ in any aqueous solution at 298 K:

$$K_w \quad = \quad [H^+] \quad \times \quad [OH^-]$$

ionic-product   concentration   concentration
of water   of H^+ of OH^-

Thus for pure water at 298 K: $K_w = 10^{-7} \times 10^{-7} = 10^{-14}\,mol\,dm^{-3}$.

Remember: add indices when multiplying numbers expressed in this way.

K_w is the basis of the **pH scale**, a means of indicating the concentration of hydrogen ions in an aqueous solution. The pH of a solution is defined as:

> **the negative logarithm to the base 10 of the hydrogen ion concentration of the solution: pH $= -\log_{10}[H^+]$**

A solution of pH 7.0 is a **neutral** solution. If the pH is less than 7.0 the solution is **acidic** but if the pH is greater than 7.0 the solution is **alkaline.** The pH of pure water at 298 K is:

$$-\log_{10} 10^{-7} = 7.0, \quad \text{so pure water is neutral.}$$

Table 1.5 gives the concentrations of hydrogen and hydroxyl ions in solutions ranging from pH 0 to pH 14, the full range of the pH scale. Note that the $[H^+]$ of a solution of pH 6.0 is ten times more than one of pH 7.0 and a hundred times more than a solution of pH 8.0.

Use the equation on the right to calculate the pH of a solution containing $1.5 \times 10^{-7}\,mol\,dm^{-3}$ of hydrogen ions.

Table 1.5 Concentrations ($mol\,dm^{-3}$) of H^+ and OH^- in solutions of pH ranging from 0 to 14

pH	0.0	1.0	2.0	3.0	4.0	5.0	6.0	7.0	8.0	9.0	10	11	12	13	14
H^+	1.0	10^{-1}	10^{-2}	10^{-3}	10^{-4}	10^{-5}	10^{-6}	10^{-7}	10^{-8}	10^{-9}	10^{-10}	10^{-11}	10^{-12}	10^{-13}	10^{-14}
OH^-	10^{-14}	10^{-13}	10^{-12}	10^{-11}	10^{-10}	10^{-9}	10^{-8}	10^{-7}	10^{-6}	10^{-5}	10^{-4}	10^{-3}	10^{-2}	10^{-1}	1.0

The pH of intracellular fluid is normally between 6.5 and 8.0. The body fluids also have a pH within this range. Even more significant, the hydrogen ion concentration in living organisms is kept fairly constant. Examples of the way in which this is achieved are described in section 1.3.1. One of the advantages of maintaining a relatively fixed pH inside living organisms is that enzymes can then catalyse metabolic reactions at their optimum efficiency. Should the pH of body fluids go outside the pH range of 6.5–8.0, most enzymes will not work (Chapter 3).

1.3 Some properties of aqueous solutions

Some of the properties of aqueous solutions have already been touched on. It is worthwhile exploring a number of these properties more fully. Two properties in particular are important to living organisms. Firstly, there are those properties governed by the nature of the solute which dissolves in water; of special importance in this context are **acids, bases** and **buffers**. Then there are the **osmotic properties** which are governed by the number of solute particles per unit volume of water, that is the concentration of an aqueous solution.

1.3.1 Acids, bases and buffers

An **acid** is a substance which dissociates in solution to release hydrogen ions (H^+). Most organic acids are called **weak acids** because few of their molecules dissociate. Conversely, many inorganic acids are **strong acids** because most of their molecules dissociate. Both weak and strong acids are found in living cells. Carboxylic acids are weak acids produced by all living organisms. Hydrochloric acid, produced by cells in the stomach lining, is a strong acid.

Substances which accept H^+ are called **bases**. A reaction between an acid and a base involves a **conjugate acid–base pair** which consists of a H^+ donor and a H^+ acceptor. Acid–base pairs are found in solutions containing weak acids and their salts. Let us consider a mixture of carbonic acid and sodium hydrogencarbonate. The acid, being a weak acid, dissociates to yield a small number of hydrogen ions:

$$H_2CO_3 \rightleftharpoons H^+ + HCO_3^-$$

carbonic acid   hydrogen ion   hydrogencarbonate ion

However, the salt dissociates to produce a large number of hydrogen-carbonate ions:

$$NaHCO_3 \rightleftharpoons HCO_3^- + Na^+$$

sodium salt   hydrogencarbonate ion   sodium ion

Because the hydrogencarbonate ions have a high affinity for H^+, they remove a large proportion of any further H^+ added to the mixture, so the pH of the solution remains almost constant. This effect is called the **buffering capacity** of the acid–base pair.

Blood and tissue fluid are buffered at about pH 7.2–7.4, by the carbonic acid–hydrogencarbonate acid–base pair (Chapters 8 and 10). The main buffer which prevents large fluctuations of pH in cells is the dihydrogen-phosphate and hydrogenphosphate acid–base pair. This acid–base pair ionises as follows at pH 7.2:

$$H_2PO_4^- \rightleftharpoons H^+ + HPO_4^-$$

The importance of buffers in living organisms cannot be overstressed. They help to stabilize the pH of cellular and body fluids, so helping to maintain suitable conditions for enzymes to efficiently catalyse metabolic reactions (Chapter 3).

1.3.2 Osmotic properties

When pure water is separated from an aqueous solution by a membrane which is permeable to water but not to dissolved solutes there is a net flow of water into the solution by a process called **osmosis** (Fig 1.6).

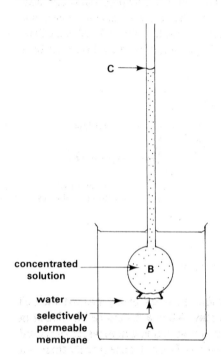

Fig 1.6 A simple osmometer. The net flow of water from A to B causes a rise in the height of the solution at C

concentrated solution

water

selectively permeable membrane

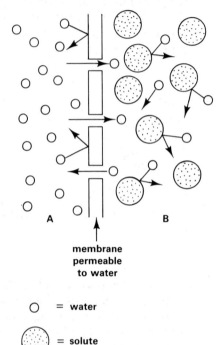

Fig 1.7 Osmosis. Note that water molecules pass randomly both ways through the membrane; however, there is a net osmotic flow from A to B

The arrows indicate the direction of movement of water molecules.

membrane permeable to water

○ = water

⬤ = solute

How many grammes of the following are required to make a 1.0 M solution of each? (i) Sucrose, (ii) sodium chloride.

In which direction would there by a net flow of water in Fig 1.6 if the water at A was replaced by an aqueous solution which is more concentrated than in B?
Explain in terms of a water potential gradient.

Free water molecules have kinetic energy and move about at random. The total amount of kinetic energy of the water molecules on either side of the membrane shown in Fig 1.7 is called **water potential** (WP). Pure water at STP has a water potential of zero whereas aqueous solutions have negative water potentials. The term **solute potential** denotes the lowering of the water potential of a solution or suspension caused by attraction of water to solute molecules. An aqueous solution of an organic solute such as sucrose of 1.0 M concentration has a water potential of $-2270\,kPa$. Water molecules moving near the membrane may hit the membrane and bounce off or they may slip through the pores. Hence water moves either way through the membrane from A to B or from B to A. Movement from A to B is impeded only by other water molecules whereas movement from B to A is also impeded by solute molecules. Hence the water potential of A is greater than B; **a water potential gradient** (ΔWP) exists and there is a net flow of water down such a gradient. One way of defining osmosis is therefore:

the net flow of water through a membrane permeable to water but not to solute down a water potential gradient

1.4 Gain and loss of water by living cells

Osmosis is one of several ways in which water can enter and leave cells. It is convenient to look at osmosis in plant and animal cells separately, despite the fact that there are common features.

1.4.1 Osmosis in plant cells

The main part of the body of many kinds of plants is a tissue called parenchyma (Chapter 24). Parenchyma cells have a wall made of a jelly-like matrix in which fibres of cellulose are enmeshed. The **cell wall** is porous and permeable to aqueous solutions. In contrast the outer boundary of the protoplast, the live matter of the cell, is a **cell membrane** which is more permeable to water than to many kinds of solutes. The protoplast contains **sap**, an aqueous solution of inorganic and organic substances. They cause the sap to have a negative water potential (solute potential). Much of the sap is often localised in one or more large **vacuoles**, each bounded by a tonoplast membrane (Fig 1.8).

Fig 1.8 A parenchyma cell

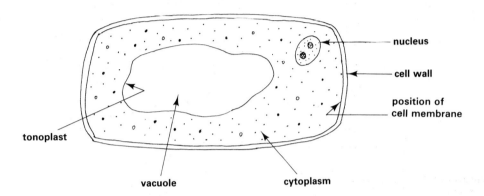

nucleus

cell wall

position of cell membrane

tonoplast

vacuole

cytoplasm

If the protoplast is sufficiently swollen to press firmly against the cell wall, the cell is **turgid**. If not it is **flaccid**. When a flaccid cell is placed in water the cell membrane separates the sap with a negative water potential from the water which has a water potential of zero. Consequently there is a net osmotic flow of water down the water potential gradient into the cell. The entry of water enlarges the protoplast which now presses firmly against the cell wall. Though able to stretch to accommodate the enlarged protoplast, the cell wall opposes the expansion causing the water in the cell to be subjected to a hydrostatic pressure called **pressure potential**. Hence two factors determine the magnitude of the water potential of a cell, the solute potential (SP) of its contents and its pressure potential (PP).

$$WP = SP + PP$$
$$(\psi = \psi_s + \psi_p)$$

When a cell is immersed in an aqueous solution, the direction of net osmotic flow of water is determined by the difference in water potential of the cell and the water potential of the solution. A non-turgid cell will absorb water by osmosis when placed in a solution which creates a negative water potential gradient. This happens when the solution is relatively dilute.

For example, if SP $= -1000$ kPa   and PP $= +500$ kPa
then   WP (cell) $= -1000$ kPa $+ 500 = -500$ kPa.
If   WP (soln.) $= -300$ kPa
then ΔWP $= -500 - (-300) = -500 + 300 = -200$ kPa.

In contrast, should ΔWP be a positive value there is a net osmotic withdrawal of water from the cell. This occurs when the solution in which the cell is immersed is relatively concentrated.

For example if WP (cell) $= -500$ kPa, and WP (soln.) $= -600$ kPa
then ΔWP $= -500 (-600) = -500 + 600 = +100$ kPa

As water is lost the protoplast gradually shrinks and eventually just ceases to press against the cell wall. This point is called **incipient plasmolysis** (Fig 1.9). Further withdrawal of water causes the protoplast to be reduced to a more or less spherical mass completely detached from the cell wall. This is **total plasmolysis**. The space between the protoplast and cell wall is occupied by the solution in which the cell was immersed.

Fig 1.9 Osmosis in epidermal cells of beetroot petiole, × 250

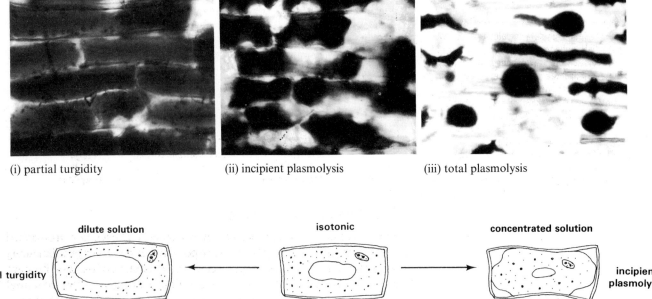

(i) partial turgidity   (ii) incipient plasmolysis   (iii) total plasmolysis

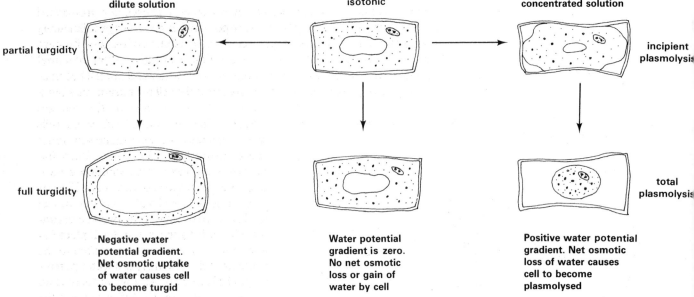

A third possibility is that the water potential of the cell and the external solution are the same. Here there is no net osmotic loss or gain of water by the cell. Fig 1.10 illustrates the changes in magnitude of the various factors which affect osmotic flow of water from a cell as it moves from turgidity to plasmolysis. Changes in the turgidity of guard cells are the cause of stomatal opening and closing (Chapter 13).

Fig 1.10 Changes in magnitude of factors affecting osmotic intake of water by a plant cell as it moves from incipient plasmolysis to full turgidity

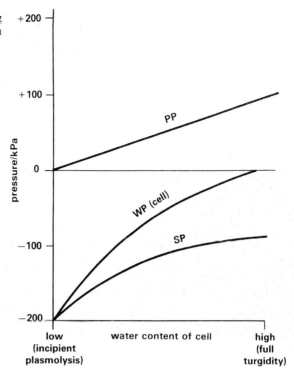

Use the following data to calculate the water potential gradient and hence the direction of net osmotic flow of water for a plant cell immersed in three solutions, each having a different water potential.

(i) SP = −1000 kPa, PP = +300 kPa,
WP (soln.) = −900 kPa

(ii) SP = −1000 kPa, PP = +300 kPa,
WP (soln.) = −400 kPa

(iii) SP = −1000 kPa, PP = +300 kPa,
WP (soln.) = −700 kPa

1.4.2 Osmosis in animal cells

Animal cells do not have a cell wall, hence unless enclosed in a pressurised system such as blood vessels, pressure potential is not a factor influencing their water potential. As with plant cells, three patterns of osmotic flow are observed. When placed in a dilute solution a negative water potential gradient exists so there is a net osmotic influx of water and the cell becomes turgid. If a large volume of water is absorbed the cell membrane may burst (lysis) and the contents of the cell are released. The bursting of red blood cells when immersed in a weak saline solution is called **haemolysis** (Fig 1.11). When immersed in a concentrated solution a positive water potential gradient exists. There is a net osmotic loss of water causing the cell to shrink. Red blood cells become **crenated** when this happens. Finally, if placed in a solution of similar water potential to the contents of the cell, there is a water potential gradient of zero. As a result there is no net osmotic loss or gain of water and the cell's water content remains steady.

Fig 1.11 Osmotic behaviour of a red blood cell following immersion in different solutions

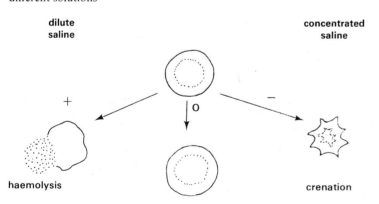

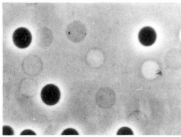

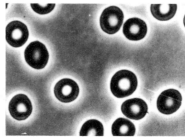

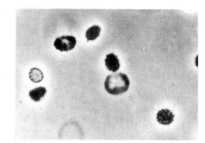

(i) haemolysis

(ii) red blood cells, × 800

(iii) crenation

11

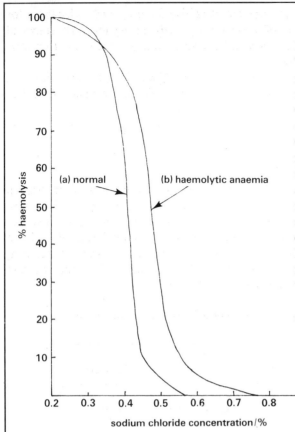

Fig 1.12 Osmotic fragility curves: (a) normal, (b), for a patient suffering from haemolytic anaemia

% haemolysis

(a) normal   (b) haemolytic anaemia

sodium chloride concentration/%

MEDICAL APPLICATION

The **osmotic fragility test** is designed to determine the resistance of red cells to haemolysis when subjected to a range of saline solutions of different solute potential. 0.05 cm^3 heparinised blood is added to 10 cm^3 saline solution in tubes over a range of concentration between 0.0 and 0.9 % NaCl. After 30 minutes the blood–saline mixtures are centrifuged and the absorbance of the supernatant fluids is determined with a colorimeter. The absorbance is dependent on the amount of haemoglobin released, which is proportional to the degree of haemolysis which has occurred. The percentage haemolysis is calculated for each mixture assuming that all the cells haemolyse in distilled water. An **osmotic fragility curve** is plotted from the results. Red cells which are more prone to haemolysis produce a curve to the right of normal (Fig 1.12). The results are of help in diagnosing blood diseases such as haemolytic anaemia.

1.4.3 Osmoregulation

Cells cannot function efficiently if there is a continual fluctuation of their water content. The amount of water a plant cell can take in is limited by the resistance offered by the cell wall which restricts excessive enlargement of the protoplast. A variety of osmoregulatory mechanisms have evolved in animals which maintain a relatively constant water content in their bodies (Chapters 29 and 30). One of the simplest devices is seen in fresh water protozoans such as *Amoeba* and *Paramecium* (Fig 1.13). Excess water taken in by osmosis collects in **contractile vacuoles** which periodically discharge their contents at the cell surface. The rate at which such a vacuole empties is proportional to the rate of osmotic influx of water and is affected by the magnitude of the water potential gradient between the cell and fresh water. Marine protozoa do not form contractile vacuoles in seawater where the water potential gradient is zero. In mammals the water content of the body is regulated by an elaborate process in which the kidney plays a key role (Chapter 11).

Fig 1.13   The protozoan *Paramecium* as seen with a phase-contrast microscope, × 700 (courtesy Philip Harris Biological Ltd.)

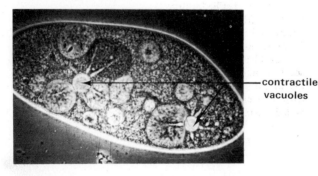

— contractile vacuoles

1.5 Non-osmotic movement of water

Water is taken in by all cells when pockets are formed in the cell membrane during **pinocytosis** and **phago-cytosis** (Chapter 6). Cells and cell walls also contain substances such as polysaccharides and proteins which chemically attract water. The term **matric**

potential is used to describe the attractive force by such substances for water. The absorption of water by dry seeds during the early stages of germination, a phenomenon called **imbibition**, is largely due to matric potential (Chapter 23).

Evaporation and subsequent **diffusion** of water vapour into the air is one of the main causes of water loss from terrestrial organisms (Chapters 11 and 12).

1.6 Water balance

Water is of such fundamental importance that it is essential to maintain a balance between gain and loss (Table 1.6). An adult human consumes on average about $1700 \, cm^3$ of water each day, about half in liquid drinks and the remainder in solid food. About $300 \, cm^3$ of water is formed daily in an adult human as a product of respiration (**oxidation water**). Water loss, mostly in urine but also in faeces and by evaporation from the skin and lungs, balances water intake and production.

Table 1.6 Daily water balance of an adult human (from Armstrong and Bennett 1979)

	Gain (cm^3)		Loss (cm^3)
in liquid drinks	900	in urine	1050
in solid food	800	from skin and lungs	850
oxidation water	300	in faeces	100
Total	$2 \, dm^3$	Total	$2 \, dm^3$

SUMMARY

Water is a vital and substantial component of all known forms of life. It has unusual physical properties which enable it to remain liquid over a wide temperature range and to resist temperature fluctuations. Because of its polarity water is able to dissolve many ionic compounds with ease. Its ability to form hydrogen bonds with neutral organic compounds such as sugars enables it to dissolve them.

Water molecules ionise to form hydrogen ions. The main providers of hydrogen ions in aqueous solutions are acids. The concentration of hydrogen ions in solution is the basis of the pH scale. Metabolic processes such as enzyme activity are sensitive to changes in pH; in living organisms buffers regulate the pH.

Water passes through cell membranes by osmosis, always moving down a water potential gradient.

QUESTIONS

1 $1 \, cm^3$ of defibrinated sheep's blood was added to $10 \, cm^3$ of isotonic saline ($0.17 \, M$). $1 \, cm^3$ of the diluted blood was added to $10 \, cm^3$ of each of the following:

distilled water; ammonium chloride solution ($0.17 \, M$); glycerol solution ($0.32 \, M$).

The times taken for haemolysis (bursting of red blood cells) to occur are given below.

Distilled water	Immediately
Ammonium chloride solution	45 seconds
Glycerol solution.	10 minutes

(a) Why did haemolysis occur when the diluted blood was added to the distilled water?

(b) Why did haemolysis occur when the diluted blood was added to the ammonium chloride solution?

(c) Why did haemolysis take place more quickly when the diluted blood was added to the ammonium chloride solution than when it was added to the glycerol solution?

(d) Briefly explain, with reasons, what you would expect to observe under the microscope if the red blood cells were added to a drop of concentrated sodium chloride solution.

(O)

2 The graph shows erythrocyte osmotic fragility curves for two breeds of pig. 0.02 cm³ of freshly collected blood was mixed with 10 cm³ of each concentration of saline. After 30 minutes the tubes were centrifuged at 2000 rpm for 10 minutes. 3.5 cm³ of supernatant liquid were transferred from each tube to a colorimeter and the absorbance measured at wavelength 540 nm. The percentage lysis for each sample was calculated assuming 100 % haemolysis in the tube containing distilled water.

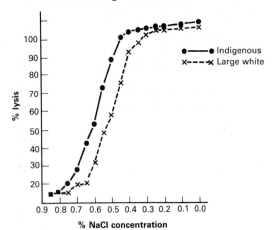

Erythrocyte osmotic fragility curves in the two breeds of piglets (from M. O. Makinde, *Animal Technology* 1986, p. 75).

(a) The saline solutions were prepared by diluting a 1.0 % stock solution of sodium chloride with distilled water. Calculate the volumes (in cm³) of stock solution and distilled water used to prepare the 0.3 % NaCl solution.

(b) Explain why erythrocytes lyse when immersed in dilute solutions.

(c) Discuss the probable reason why only a proportion of cells lysed at each NaCl concentration.

(d) Why was each sample centrifuged?

(e) Name the substance in the supernatant liquid, the absorbance of which was measured.

(f) Suggest a hypothesis to explain the difference in osmotic fragility shown by the two breeds.

(g) Outline another way in which comparable data could have been obtained by microscopy.

(h) What is the potential clinical significance of the results of osmotic fragility tests?

(from *Advanced Human Biology*)

3 The water potential (ψ_w) of a plant cell is determined by two factors, its pressure potential (ψ_p) and its solute potential (ψ_s). Thus $\psi_w = \psi_p + \psi_s$.

(a) What does the water potential of a cell measure?

(b) Consider the two cells **A** and **B**.

(i) State which cell has the higher water potential.

(ii) State the direction in which water will move by osmosis. Give the reason for your answer.

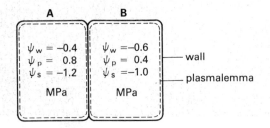

(c) If cell **A** was allowed to equilibrate by being placed in pure water, what would be its expected pressure potential? Show how you derived your answer.

(d) The following experiment was set up to determine the water potential of plant cells. Comparable discs, about 1 mm thick, of potato tuber were cut out and weighed. Replicate samples were taken and placed in a series of sucrose solutions ranging from 0.0 to 0.6 M in covered dishes at a constant temperature of 20 °C. After 1 hour the samples were blotted rapidly between sheets of filter paper and re-weighed. The results are shown in the table as the mean percentage change in mass.

Concentration of sucrose solution/M	0.0	0.1	0.2	0.3	0.4	0.5	0.6
Mean percentage in mass	+22	+17	+9	+3	−3	−10	−15

(i) Plot a graph of these figures.

(ii) Use your graph to work out which concentration of sucrose solution has a water potential equal to the mean water potential of the cells of the tissue.

(iii) What further information would you need to express the water potential of the tissue in units of MPa?

(iv) Why were the dishes containing the discs and sucrose solution covered during the experiment?

(v) Suggest the reason for expressing the change in mass as a percentage change. (C)

4 In an experiment, individuals of two different species A and B of *Amoeba* were transferred from their natural habitats to different dilutions of seawater. Each individual was given time to adjust to its new environment, and then the rate of contraction of its contractile vacuole was studied. The following results were obtained.

Concentration of sea water (normal sea water = 100 %)	Number of vacuolar contractions per hour	
	Species A	Species B
5 %	82	20
10 %	74	63
15 %	65	64
20 %	58	56
30 %	34	31
40 %	14	13
50 %	0	6
60 %	0	0

(a) Plot the results of the experiment as a graph.

(b) (i) What is the function of a contractile vacuole?

(ii) Explain how it carries out this function.

(c) Explain, by reference to the data, the difference in vacuolar contraction in the two species of *Amoeba* when placed in the higher concentrations of seawater.

(d) What information may be deduced about the natural habitats of the two species from the rates of vacuolar contractions? (L)

2 Carbohydrates, lipoids and proteins

2.1 **Carbohydrates** 16

   2.1.1 Monosaccharides 16
   2.1.2 Disaccharides 18
   2.1.3 Polysaccharides 20

2.2 **Lipoids** 22

   2.2.1 Simple lipids 22
   2.2.2 Complex lipids 25
   2.2.3 Waxes 26
   2.2.4 Steroids and sterols 26

2.3 **Proteins** 27

   2.3.1 Amino acids 27
   2.3.2 Polypeptides and proteins 28
   2.3.3 Glycoproteins and
        lipoproteins 34

2.4 **Biochemical techniques** 34

   2.4.1 Chromatography 34
   2.4.2 Electrophoresis 36

Summary 37

Questions 38

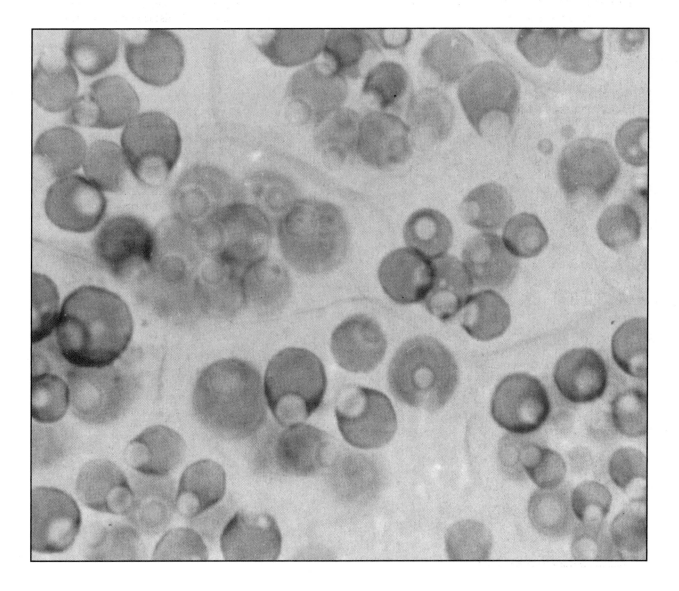

2 Carbohydrates, lipoids and proteins

Many of the compounds which occur in living organisms are **organic substances**. They include carbohydrates, lipoids and proteins. Their molecules contain carbon atoms. Carbon is **tetravalent**: its atoms have four sites to which other atoms such as hydrogen, oxygen, nitrogen, phosphorus and sulphur can bond. Consequently a variety of functional groups such as amino, carboxyl and phosphate are found in naturally occurring organic compounds.

Carbon atoms also bond to each other, forming carbon chains as in fatty acid molecules and rings as in sugars. Atoms which bond to carbon are arranged spatially in three dimensions. An enormous range of organic molecules of various shapes and sizes thus exists.

2.1 Carbohydrates

Carbohydrates have the elements of a molecule of water for each carbon atom. The term carbohydrate is thus derived from 'hydrated carbon'. It reflects the fact that all carbohydrates are made up of carbon, hydrogen and oxygen, usually in the proportion 1:2:1 respectively.

2.1.1 Monosaccharides

Monosaccharides are the simplest of carbohydrates, having the empirical formula $(CH_2O)_n$ where $n = 3$–7. They are sweet-tasting and dissolve in water. One way of classifying monosaccharides is according to the number of carbon atoms in each of their molecules (Table 2.1).

Fig 2.1 Reducing groups of monosaccharides

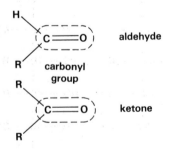

Table 2.1 Some monosaccharide sugars

	Molecular formula	Examples
trioses	$C_3H_6O_3$	glyceraldehyde
pentoses	$C_5H_{10}O_5$	ribose, ribulose
hexoses	$C_6H_{12}O_6$	glucose, fructose, galactose

The molecules of all monosaccharides contain a **carbonyl group**, either as part of an **aldehyde** group or as part of a **ketone** group (Fig 2.1). The carbonyl group readily donates electrons. It is this property which makes monosaccharides **reducing sugars**. When heated with Fehling's or with Benedict's reagent they reduce copper(II) ions to copper(I), forming a brick-red precipitate of copper(I) oxide:

$$2Cu(OH)_2 + R\!-\!\underset{H}{\overset{}{C}}\!=\!O \rightarrow Cu_2O + R.COOH + 2H_2O$$

copper(II) hydroxide   copper(I) oxide   sugar acid

Table 2.2 Classification of monosaccharides

Category	Example
aldotriose	glyceraldehyde
aldopentose	ribose, deoxyribose
aldohexose	glucose, galactose
ketohexose	fructose

Aldoses are monosaccharides whose molecules contain an aldehyde group, whereas **ketoses** have a ketone group instead. An alternative way of classifying monosaccharides is based on the nature of the reducing group and the number of carbon atoms per molecule (Table 2.2).

The molecules of all monosaccharides contain one or more **asymmetric carbon atoms**. They are identifiable because they have four different functional groups bonded to them. **Glyceraldehyde** has one asymmetric carbon atom to which the functional groups can be attached in two ways (Fig 2.2). Note that the hydroxyl group is on the right of the asymmetric carbon atom in D-glyceraldehyde and on the left in L-glyceraldehyde. The difference can be detected by using a polarimeter to pass a beam of polarised light through an aqueous solution of glyceraldehyde. The D-form rotates polarised light to the right (dextro-rotatory), the L-form to the left (laevo-rotatory). The two forms, called **stereoisomers,** cannot be superimposed. Imagine placing a molecule of D-glyceraldehyde in front of a mirror. The image you would see in the mirror is L-glyceraldehyde.

Fig 2.2 Stereoisomers of glyceraldehyde

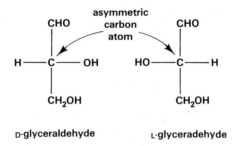

D-glyceraldehyde   L-glyceradehyde

Pentoses and hexoses have several asymmetric carbon atoms per molecule. Of particular importance is the one furthest from the reducing group which is also the last-but-one carbon atom in the carbon chain. In the stereochemical formulae of D-sugars, the hydroxyl group is on the right of this atom, as in D-glyceraldehyde, whereas in L-sugars it is on the left as in L-glyceraldehyde. Only D-isomers of sugars are commonly found in living organisms (Fig. 2.3).

Fig 2.3 Stereochemical formulae of some D-sugars

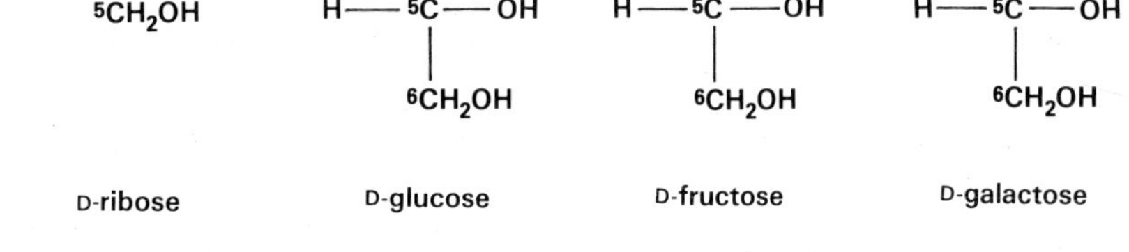

D-ribose   D-glucose   D-fructose   D-galactose

Which of the numbered carbon atoms in each molecule is asymmetrical?

The relatively long carbon chains of pentose and hexose sugars can bend, bringing the carbonyl group close enough to reduce one of the hydroxyl groups in the same molecule. In this way ring-shaped molecules are formed. **Glucose** forms six-sided rings where the hydroxyl group attached to carbon atom 5 is reduced. In the ring form, carbon atom 1 is asymmetric, having four different functional groups bonded to it. This is not so in the straight-chain form. The additional asymmetric carbon atom enables α- and β-forms of D-glucose rings to exist. In **α-D-glucose** the hydroxyl group is below carbon atom 1, whereas it is above it in **β-D-glucose** (Fig 2.4).

Fig 2.4 Ring forms of D-glucose

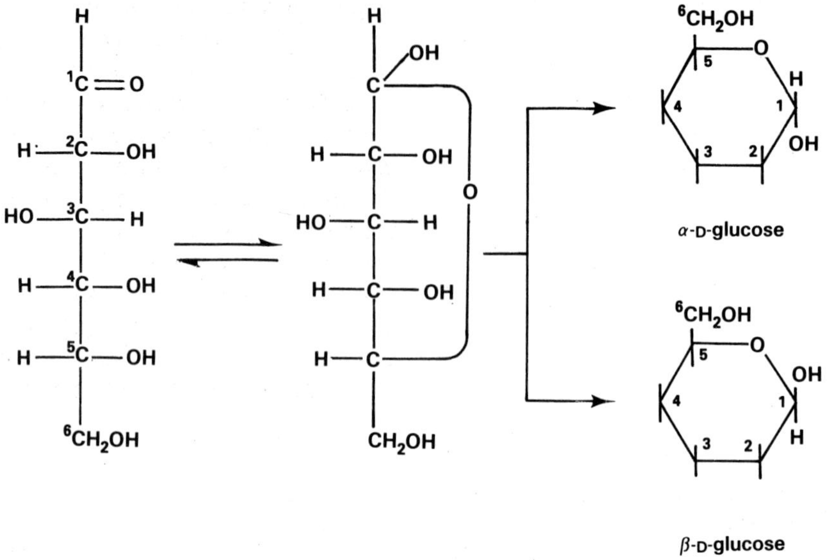

Fructose molecules can also exist as six-sided rings, but more often they occur as more stable five-sided rings in which the carbonyl group at carbon atom 2 reduces the hydroxyl group at carbon atom 5 (Fig 2.5).

Fig 2.5 Ring forms of D-fructose

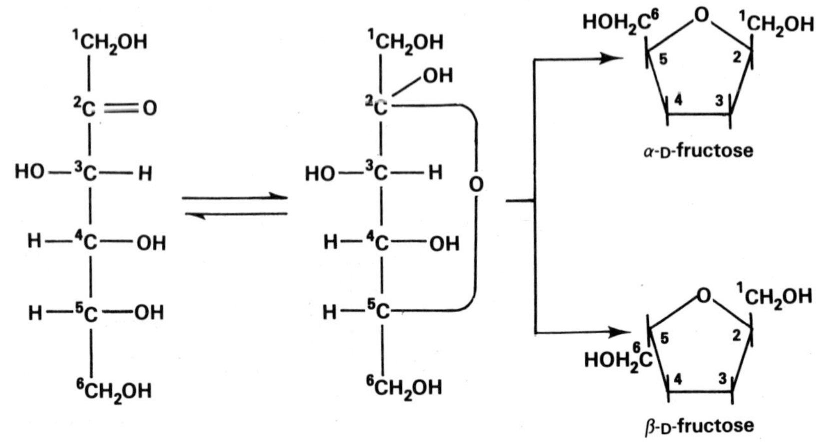

2.1.2 Disaccharides

Sucrose, lactose and maltose are important **disaccharides**. They are formed when two hexose molecules bond to each other in a condensation reaction:

$$2(C_6H_{12}O_6) \rightarrow C_{12}H_{22}O_{11} + H_2O$$

Sucrose is consumed in large amounts to sweeten many of the foods we eat. It is stored in some plants such as sugar cane and sugar beet which are cultivated on a large scale to supply us with sugar. A molecule of sucrose is formed when one of α-D-glucose and one of β-D-fructose join in condensation. A bond is created between carbon atom 1 of the α-glucose ring and

18

carbon atom 2 of the β-fructose. For this reason it is called an $\alpha(1 \rightarrow 2)$ **glycosidic linkage** (Fig 2.6(a)). It is at these carbon atoms that the carbonyl group occurs in both glucose and fructose. It means that there are no free carbonyl groups in sucrose which is therefore a **non-reducing sugar**. However, sucrose can be easily hydrolysed into the monosaccharides of which it is made, by boiling sucrose with dilute hydrochloric acid or by incubating it with the enzyme sucrase (invertase).

Lactose is often called milk sugar because it is the main carbohydrate in milk. Its molecules consist of a ring of α-D-glucose bonded by a $\beta(1 \rightarrow 4)$ glycosidic linkage to a ring of β-D-galactose (Fig 2.6(b)). The linkage uses the carbonyl group of the galactose ring, leaving that of glucose free. It is therefore a **reducing sugar**.

Maltose is an intermediate product of the hydrolysis of starch by the enzyme amylase (Chapters 15 and 23). A maltose molecule consists of two α-D-glucose rings joined together by an $\alpha(1 \rightarrow 4)$ glycosidic linkage (Fig 2.6(c)). The linkage involves the carbonyl group of only one of the glucose rings. Maltose thus has a free carbonyl group and is a **reducing sugar**.

Fig 2.6 Formation of some disaccharides

(a) Sucrose

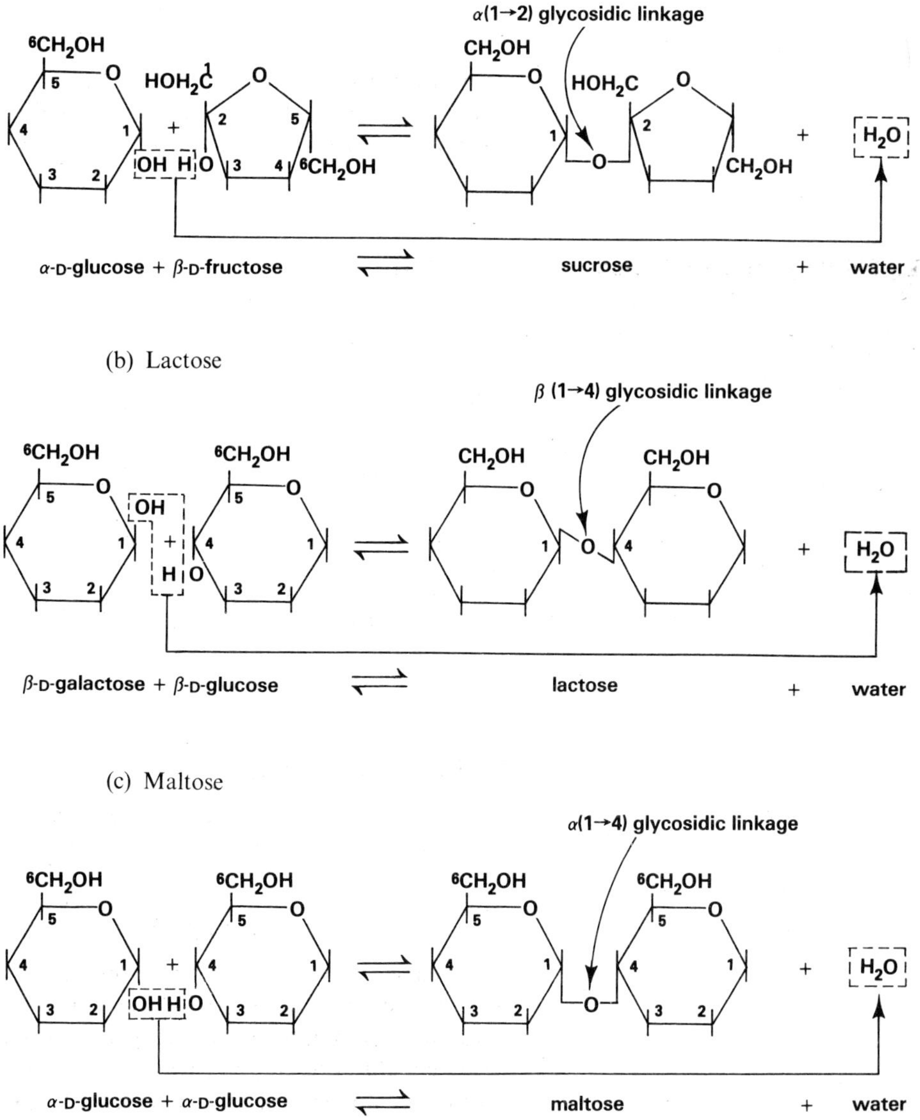

α-D-glucose + β-D-fructose ⇌ sucrose + water

(b) Lactose

β-D-galactose + β-D-glucose ⇌ lactose + water

(c) Maltose

α-D-glucose + α-D-glucose ⇌ maltose + water

2.1.3 Polysaccharides

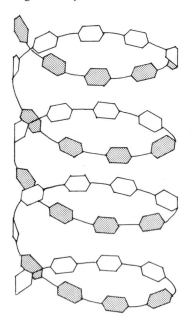

Fig 2.7 Amylose helix

A polymer is a substance of large relative molar mass formed by the joining together of a large number of basically similar smaller molecules (monomers). Polysaccharides are **polymers** formed by the condensation of many monosaccharide molecules. D-glucose is the main sugar involved, and the products serve as storage and structural materials in plants and animals.

Starch is the main storage material of green plants. Its molecules have two components, **amylose** and **amylopectin**. The proportion of each varies from one type of starch to another.

'Soluble' starches consist mainly of amylose. Amylose does not truly dissolve in water but forms a colloidal suspension. Amylose is a polymer of several hundred to a few thousand α-D-glucose rings joined by $\alpha(1\rightarrow4)$ glycosidic linkages. The resulting molecule is unbranched and wound into a helix (Fig 2.7). The bore of the helix is just big enough to trap iodine molecules. It is this reaction which produces the blue coloured complex when starch is mixed with iodine in potassium iodide solution. The iodine test is frequently used to investigate the presence of starch in materials of plant origin.

In contrast, amylopectin is a branched molecule, though again it is a polymer of α-D-glucose. The backbone of the molecule is held together by $\alpha(1\rightarrow4)$ glycosidic linkages as in amylose. Branches arise about every twenty-fifth glucose ring where $\alpha(1\rightarrow6)$ glycosidic bonds occur (Fig 2.8). The branches which consist of 15–20 glucose rings joined by $\alpha(1\rightarrow4)$ linkages may themselves be branched. Each molecule of amylopectin contains several thousand glucose rings. Like amylose, amylopectin forms a colloidal suspension when mixed with water. The suspension reacts with iodine in potassium iodide solution giving a red-violet colour.

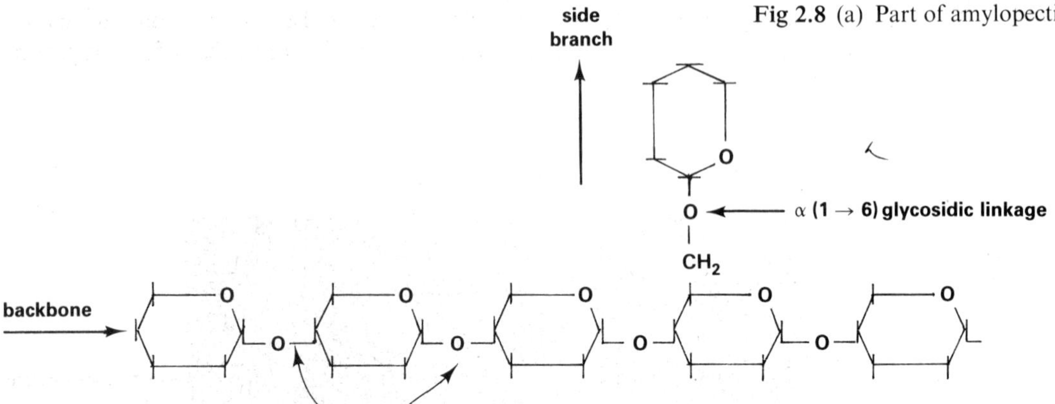

side branch

Fig 2.8 (a) Part of amylopectin molecule

α (1 → 6) glycosidic linkage

CH$_2$

backbone

α (1 → 4) glycosidic linkages

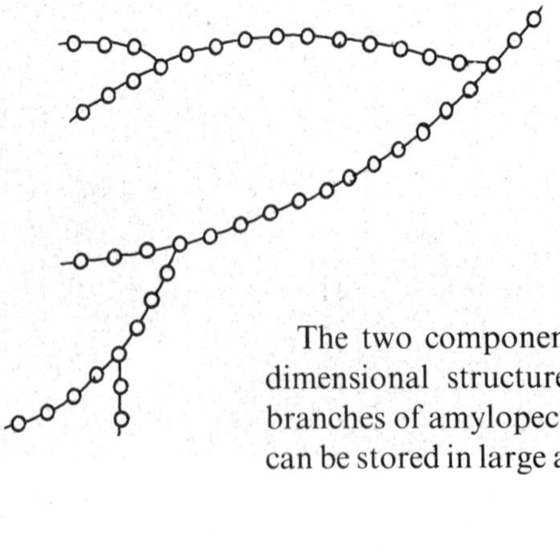

Fig 2.8 (b) Structure of amylopectin, simplified. Each circle represents an α-D glucose ring

The two components of starch fit together to form a complex three-dimensional structure in which amylose helices are entangled in the branches of amylopectin molecules. Because starch is insoluble in water, it can be stored in large amounts without having any great effect on the water potential of cells. It is usually found as grains in the tissues of storage organs such as potato tubers (Fig 2.9).

Fig 2.9 Starch grains in potato tuber

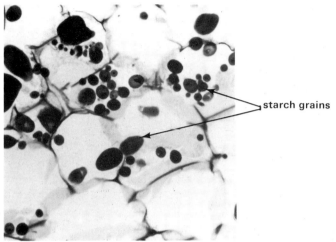

starch grains

Fig 2.10 Structure of glycogen, simplified. Each circle represents an α-D glucose ring

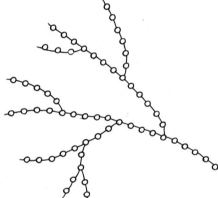

Glycogen is a storage polysaccharide found mainly in muscle and liver. In molecular structure it is very similar to amylopectin except that side branches occur more frequently, are somewhat longer and are more branched (Fig 2.10). Glycogen is insoluble in water and therefore is a useful storage material because it has little effect on the water potential of cellular fluid (Fig 2.11). Its reaction with iodine is similar to that of amylopectin.

Fig 2.11 Glycogen granules in a liver cell

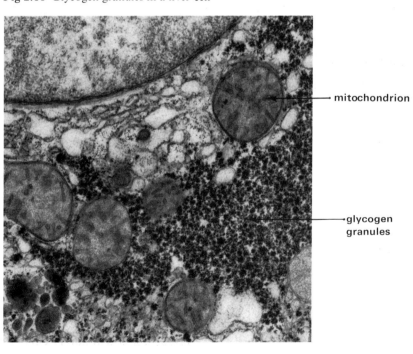

mitochondrion

glycogen granules

Cellulose is by far the most abundant structural polysaccharide in living organisms, although it is found only in plants. Unlike starch and glycogen, it is a polymer of β-D-glucose. Several hundred to a few thousand glucose rings are joined by $\beta(1\rightarrow4)$ glycosidic linkages in the long, unbranched molecules of cellulose (Fig 2.12). Attraction between hydroxyl (—OH) and hydrogen (—H) groups of the glucose rings of adjacent cellulose molecules results in the formation of **hydrogen bonds** which bind the molecules in a regular crystal-like lattice.

Fig 2.12 (a) Part of a cellulose molecule

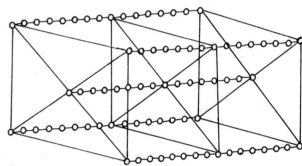

— $\beta(1\rightarrow4)$ **glycosidic linkages**

Fig 2.12 (b) Cellulose lattice, simplified. Each circle represents a β-D glucose ring

The crystalline property of cellulose has been known for a long time but the way in which cellulose is laid down in the walls of plant cells was not known until the electron microscope came into use. Electronmicrographs show that the plant cell walls consist of many fine **fibres** of cellulose which in some instances are laid down in parallel bundles but in others are distributed at random. Each fibre contains a large number of cellulose molecules. Other polysaccharides called hemicelluloses and pectic compounds cement the fibres together. In sclerenchyma fibres and in xylem elements the walls are thickened with extra cellulose and stiffened with deposits of an alcohol polymer called lignin. This gives the tissues considerable rigidity, enabling them to take on a supporting role.

Cellulose reacts with iodine to form a yellow–brown complex. Lignified cell walls stain a yellow–orange colour when stained with phenylamine (aniline) dyes and red with acidified phloroglucinol.

2.2 Lipoids

The term lipoid includes a range of organic compounds which can be extracted from plant and animal tissues using non-polar organic solvents such as benzene and ether. **Lipoids** do not readily dissolve in polar solvents such as water. They include fats, oils, phospho-, sphingo- and glycolipids, waxes, steroids and sterols.

2.2.1 Simple lipids

Fats and **oils** are sometimes called **simple lipids**. They are made of just two ingredients, **glycerol** and **fatty acids**. Glycerol is an alcohol derivative of glyceraldehyde, each of its molecules having three hydroxyl groups (Fig 2.13). Over seventy different fatty acids have been extracted from natural

Fig 2.13 Stereochemical formula of glycerol

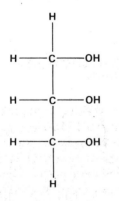

Molecular formula
$CH_2OH.CHOH.CH_2OH$
Empirical formula $C_3H_8O_3$

sources. All fatty acids have a long pleated hydrocarbon chain and a terminal carboxyl group. The length of the hydrocarbon chain differs from one fatty acid to another. Those most frequently found in living organisms have 15–17 carbon atoms, for example palmitic, stearic and oleic acids. In **saturated fatty acids**, each carbon atom in the chain is joined to the next by single covalent bonds. In contrast, the hydrocarbon chains of **unsaturated fatty acids** have one or more double covalent bonds. Pronounced bends occur in the hydrocarbon chains where double bonds exist (Fig 2.14).

Fig 2.14 Stereochemical formula of three fatty acids

(a) stearic and palmitic acids (saturated)

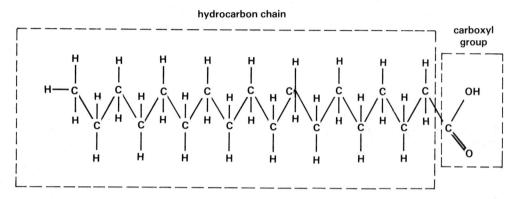

hydrocarbon chain

carboxyl group

stearic acid: molecular formula $CH_3 (CH_2)_{16} COOH$
empirical formula $C_{17} H_{35} COOH$

palmitic acid: molecular formula $CH_3 (CH_2)_{14} COOH$
empirical formula $C_{15} H_{33} COOH$

(b) oleic acid (unsaturated)

double covalent bond

oleic acid: molecular formula $CH_3(CH_2)_7CH : CH(CH_2)_7COOH$
empirical formula $C_{17} H_{33} COOH$

Carboxyl groups of fatty acids react with the hydroxyl groups of glycerol to form **acylglycerols** (glycerides) and water (Fig 2.15). The bonds which join the two components are called **ester linkages** and the process known as **esterification**. If only one hydroxyl group is esterified, the product is a monoacylglycerol (monoglyceride), if two a diacylglycerol (diglyceride), and if three a triacylglycerol (triglyceride). Where more than one hydroxyl group becomes esterified, the fatty acids involved may be similar or different. Many acylglycerols of animal origin are formed from saturated fatty acids. They are solid at room temperature and called **fats**. Butter and

23

lard are products which consist almost entirely of fats. In contrast, acylglycerols formed from unsaturated fatty acids are liquid at 15–20 °C and are called **oils**. Most vegetable oils such as olive oil and corn oil are of this kind.

Fig 2.15 Formation of an acylglycerol

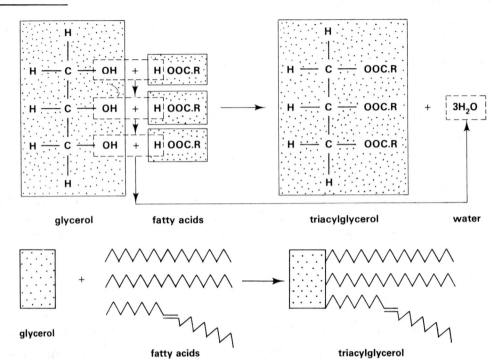

glycerol fatty acids triacylglycerol water

glycerol fatty acids triacylglycerol

Fats and oils are virtually insoluble in water. For this reason they can be stored in the body without affecting the water potential of cells. In mammals, fats are stored in adipose tissue beneath the dermis of the skin and around internal organs (Chapter 20). Oil is often stored in the seeds of flowering plants (Fig 2.16).

Fig 2.16 Oil stored in castor oil seed

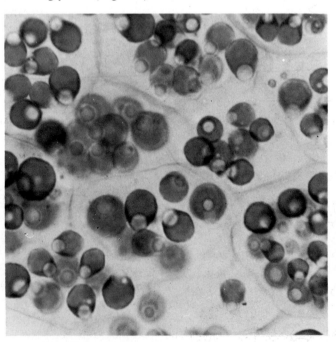

Fats and oils take on a pink colour when stained with the lipid-soluble dye Sudan Red.

2.2.2 Complex lipids

Phospholipids are triacylglycerols in which one of the hydroxyl groups of glycerol is esterified by **phosphoric acid** which in turn is esterified to an **amino alcohol. Choline** is an amino alcohol commonly used for this

Fig 2.17 Formation of a phospholipid

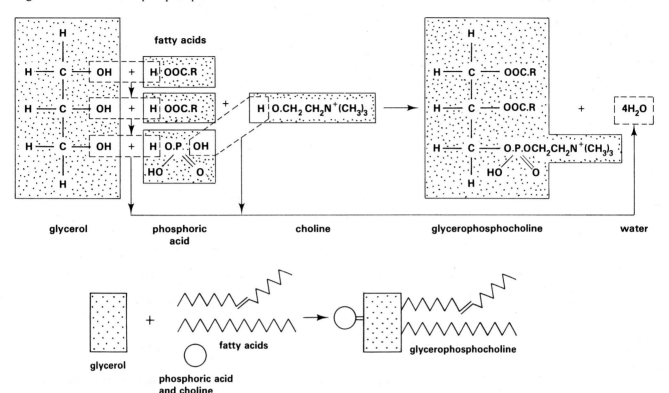

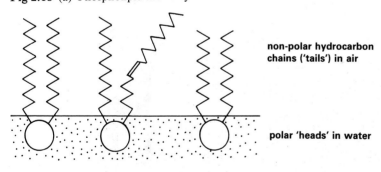

Fig 2.18 (a) Phospholipid monolayer

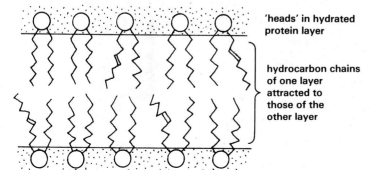

non-polar hydrocarbon chains ('tails') in air

polar 'heads' in water

Fig 2.18 (b) Phospholipid bilayer

'heads' in hydrated protein layer

hydrocarbon chains of one layer attracted to those of the other layer

purpose (Fig 2.17). **Lecithin** (glycerophosphocholine) is a phospholipid found in cell membranes (Chapter 6). The phosphate-containing 'head' of a phospholipid molecule attracts water whereas the hydrocarbon 'tails' of the fatty acids are hydrophobic. If lecithin is dissolved in ether and carefully poured on top of some water, the ether soon evaporates leaving a **monolayer** of the phospholipid on the surface of the water. The 'head' of each phospholipid molecule dissolves in the water whereas the 'tails' stand out of the water (Fig 2.18(a)).

If a second layer of phospholipid is carefully poured on top of the first, a **bilayer** is formed in which the 'tails' of the second layer are attracted to those of the first (Fig 2.18(b)). This arrangement is seen in the ultrastructure of cell membranes (Chapter 6).

The phospholipid monolayer greatly reduces the surface tension of the water, making the meniscus more easily stretched. Such an effect is important in the air sacs of our lungs because it enables the alveoli to become inflated with air and allows us to breathe without much effort. Some children are unable to produce phospholipids in their alveoli at birth. They suffer from **respiratory distress syndrome**. One of the symptoms of the condition is that a lot of effort is required when breathing to overcome the surface tension of the water layer on the alveolar membrane. Such children have to be kept in intensive care until they can synthesise lecithin.

Fig 2.19

(a) A sphingolipid, simplified (b) A glycolipid, simplified

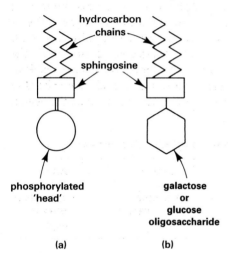

(a) (b)

Fig 2.20 (a) A steroid, simplified

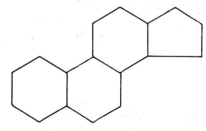

Fig 2.20 (b) Cholesterol

Sphingolipids are comparable to phospholipids except that **sphingosine** is present instead of glycerol. Sphinogosine has a hydrocarbon chain of its own and two hydroxyl groups. One of these is esterified by phosphoric acid which in turn is esterified by an amino alcohol such as choline. Sphingosine also has an amino group which is esterified by a fatty acid with a long hydrocarbon chain. Each sphingolipid molecule thus has a water-soluble 'head', and one long and one shorter hydrocarbon 'tail' (Fig 2.19(a)). Sphingolipids are common in cell membranes.

Glycolipids are similar to sphingolipids except that an oligosaccharide made usually of glucose or galactose is bonded to sphingosine instead of the phosphoric acid esterified amino alcohol (Fig 2.19(b)). The oligosaccharide is water soluble, hence glycolipids have similar physical properties to sphingolipids. Galactocerebroside is the main glycolipid in the myelin sheath of nerve cells.

2.2.3 Waxes

Waxes are esters of fatty acids with long hydrocarbon chains and alcohols of high relative molar mass. The alcohols in waxes have only one hydroxyl group compared with three in glycerol.

A waxy coating called the **cuticle** covers the epidermis of leaves and stems of many kinds of terrestrial plants. It prevents excessive evaporation of water from the shoot system. The sebaceous glands in our skin release **sebum** onto the epidermis. Sebum is a mixture of waxes and triacylglycerols. It keeps the epidermis supple and inhibits some species of bacteria which would otherwise grow on the skin. Sebum also reduces the loss of water by evaporation from the surface of our bodies.

2.2.4 Steroids and sterols

The properties of these compounds have little in common with the other lipids so far described, except that they are soluble in non-aqueous solvents. **Steroids** have a characteristic heterocyclic structure (Fig 2.20(a)). They include bile salts, hormones of the adrenal cortex and the sex hormones (Chapters 15, 21 and 25). **Sterols** are alcohol derivatives of steroids (Fig 2.20(b)). They include **cholesterol**, which is attracted to the hydrocarbon 'tails' of complex lipids in cell membranes (Chapter 6).

2.3 Proteins

Proteins are polymers of high relative molar mass. Their monomer components are **amino acids**. The smallest protein molecules contain at least fifty monomers, the largest over a thousand. A few proteins contain nearly twenty different amino acids; others are made of a much smaller variety. The number of ways in which amino acids can be put together in making proteins is infinite. It accounts for the enormous variety of proteins and their diverse functions.

2.3.1 Amino acids

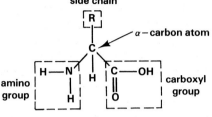

Fig 2.21 Stereochemical formula of an amino acid

All amino acids contain the elements carbon, hydrogen, oxygen and nitrogen. Amino means nitrogen-containing. A few amino acids contain sulphur too. Every amino acid molecule has at least one **amino group** and one **carboxyl group** (Fig 2.21). The structural formula for an amino acid is $NH_2.R.CH.COOH$. Chemists use Greek letters to denote the carbon atoms in molecules of this kind. The one next to the carboxyl group is the α-carbon atom. It is to this carbon atom that the amino group also is bonded. For this reason such compounds are called **α-amino acids**. With the exception of glycine, the α-carbon atom of all amino acids is asymmetrical. It means that D- and L-stereoisomers of most amino acids can exist (Fig 2.22). In contrast to carbohydrates it is the L-isomers of amino acids which are mainly found in living organisms.

Fig 2.22 Stereoisomers of alanine. Compare with Fig 2.2. Note that the amino group is on the right of the α-carbon atom in D-alanine and on the left in L-alanine

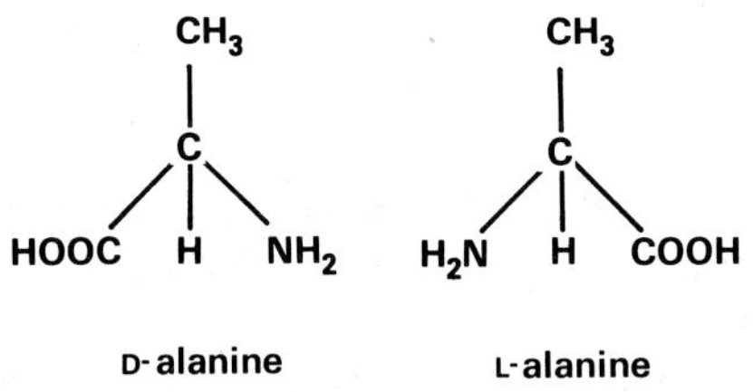

Table 2.3 α-amino acids commonly found in plant and animal proteins

alanine	glutamine	leucine	serine
arginine	glutamic acid	lysine	threonine
asparagine	glycine	methionine	tryptophan
aspartic acid	histidine	phenylalanine	tyrosine
cysteine	isoleucine	proline	valine

The twenty different amino acids isolated from plant and animal proteins are listed in Table 2.3. They differ according to the nature of the amino acid side-chain, R. In glycine, the simplest amino acid, R is a hydrogen atom. The amino acid side chain can be a short hydrocarbon chain as in alanine, a hydrocarbon ring as in tyrosine, or sulphur-containing as in cysteine. The R group of glutamic acid has a carboxyl group whereas in lysine it contains an amino group (Fig 2.23).

Fig 2.23 Stereochemical formulae of some amino acids

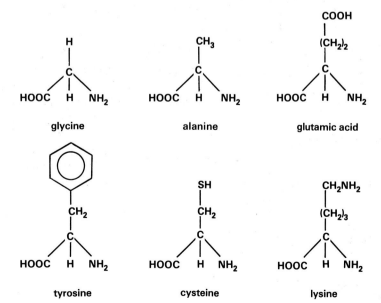

27

A knowledge of the nature of amino acid side chains is vital as a basis for understanding some of the techniques used to separate mixtures of amino acids and also proteins, and in appreciating some of the properties of proteins. Consider the simplest amino acid glycine. In aqueous solution its carboxyl group can ionise to release hydrogen ions. It is this property which gives amino acids their acidic characteristic. However, the amino group is basic and so attracts hydrogen ions.

Amino acids thus have acidic and basic properties and are described as **amphoteric**. In very acidic solutions ionisation of the carboxyl group is suppressed whilst the amino group attracts hydrogen ions and becomes positively charged. The effect is to give glycine an overall net electrostatic charge of $+1$. Conversely, in very alkaline solutions the carboxyl group ionises and the amino group is uncharged, giving glycine an overall net charge of -1. Between pH 6.5 and 7.5 both carboxyl and amino groups are charged. Such an ion with both positive and negative charges is called an **amphion** (Fig 2.24(a)). The pH at which amphions predominate is called the **isoelectric point**.

Fig 2.24 Effect of pH on the ionisation of amino acids:

(a) glycine

(b) lysine
(c) glutamic acid

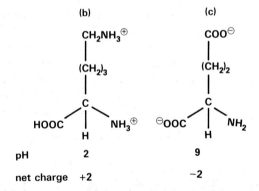

Where the amino acid side chain is also ionisable the situation is a little more complex. Nevertheless the same principle applies. If it contains an amino group this attracts hydrogen ions in acidic solutions giving an additional positive charge. Hence lysine has a net electrostatic charge of $+2$ at low pH, whereas glutamic acid has a net charge of -2 at high pH (Fig 2.24(b)). Thus at a given pH some amino acids in a mixture may be cationic, others anionic and others amphionic. Furthermore, some may be more strongly cationic or anionic than others. This is the basis on which mixtures of amino acids or proteins can be separated by electrophoresis (section 2.4.2).

2.3.2 Polypeptides and proteins

In suitable conditions amino acids polymerise. The α-amino group of one molecule joins to the carboxyl group of another in a condensation reaction which results in the formation of **peptide linkages** (Fig 2.25). The amino and carboxyl groups at each end of the **dipeptide** can form peptide linkages with other amino acid molecules, so building up a chain of amino acid residues. Some short-chain peptides are of biological importance. Antidiuretic hormone and oxytocin, two mammalian hormones (Chapter 21), are each made from just nine amino acid molecules. Long chains of amino acid residues are called **polypeptides**. One or more polypeptide chains occur in a protein molecule. The number of amino acid residues ranges from about fifty in the smallest proteins to several thousand in the largest. The relative molar mass of proteins ranges from 6000 to over 1 000 000.

Fig 2.25 Formation of a dipeptide

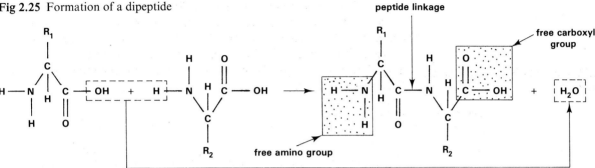

Chromatographic and other techniques have made it possible to identify the variety of amino acids in many proteins. However, the entire sequence of amino acids in any protein molecule was a mystery until the early 1950s when Frederick Sanger analysed the structure of the hormone insulin. Sanger showed that insulin consists of two polypeptide chains bonded to each other by disulphide bridges (Fig 2.26). One chain contains 21 amino acid residues, the other 30. In the late 1950s and during the 1960s, the amino acid sequences of other small protein molecules were determined. Adrenocorticotrophic hormone (ACTH) was found to contain 39 amino acid molecules, ribonuclease has 124 while the α-globin and β-globin chains of haemoglobin have 141 and 146 respectively. More recently proteins of much higher relative molar mass have been analysed and a great deal of information has been compiled on the composition and amino acid sequences of a range of proteins from many different sources.

Fig 2.26 Structure of insulin

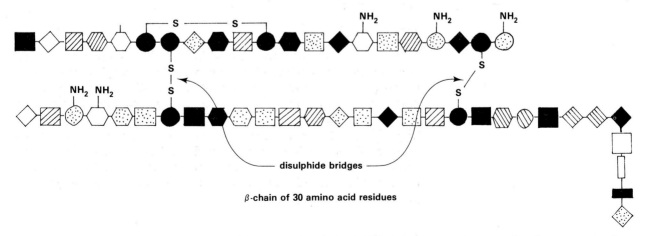

Such studies have shown that few proteins contain all twenty α-amino acids. Insulin, for example, has seventeen different amino acids. Some proteins consist of a small range of amino acids; in others a much wider range is found. However, the results of the sequence analyses have revealed two general principles. Firstly, the sequence of amino acids differs from one type of protein to another. Secondly, the sequence in any one type of protein is precisely fixed.

Without question, amino acid composition and sequence studies have yielded much valuable information about proteins. Yet proteins display many properties which cannot be accounted for from the results of such studies. In order to understand some of these properties it is necessary to know something about protein structure.

1 Protein structure

The **primary structure** is the number and sequence of amino acids in a polypeptide chain. The only form of bonding in a protein's primary structure is the **peptide linkage**.

X-ray diffraction analysis has been a very important tool in unravelling the spatial arrangement of the atoms in protein molecules. In this technique molecules are bombarded with a beam of X-rays. Some of the rays are deflected by atoms in the molecules and are then passed through a photographic film to give a diffraction pattern. From the pattern, which is characteristic for a particular compound, the shape of the molecules can be deduced. It all sounds very simple but the interpretation of the pattern necessitates the use of a computer. In 1939 Astbury obtained an X-ray diffraction pattern for keratin, a fibrous protein found in hair. The pattern indicated that the polypeptide chains in keratin were twisted or folded in a regular manner. Pauling and Corey in 1951 showed that the chains were twisted into a right handed helix. In this **α-helix** the peptide linkages form the backbone from which the R groups of the amino acids jut out in all directions. The helical structure is stabilised by **hydrogen bonds** which occur between the carbonyl ($>C=O$) and imino ($>N-H$) groups of every fourth peptide link (Fig 2.27). It is now generally agreed that many fibrous proteins have a **secondary structure** of this kind.

Fig 2.27 The α-helix, simplified

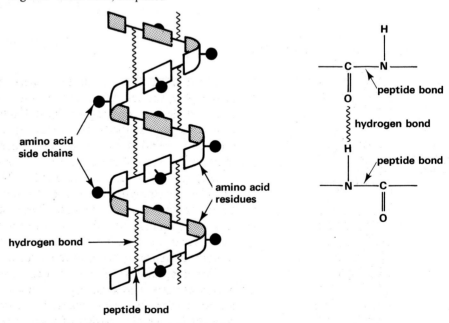

Another form of secondary structure also identified by Pauling and Corey is the **pleated sheet**. Here two or more pleated polypeptides are hydrogen bonded to each other. Once again the hydrogen bonds develop between adjacent carbonyl and imino groups. When the terminal amino groups are at the same end of each polypeptide a parallel pleated sheet is formed. If at opposite ends, the pleated sheet is **antiparallel** (Fig 2.28(a)). In some polypeptides a mixture of α-helix alternates with a pleated sheet secondary structure. Other parts of the same polypeptide have no regular arrangement and exist as a **random coil**.

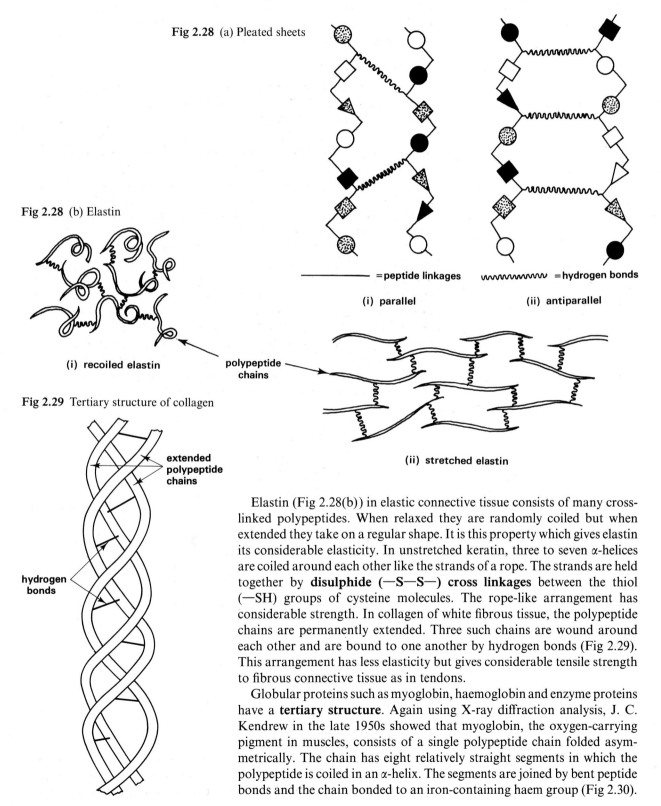

Fig 2.28 (a) Pleated sheets

———— = peptide linkages   〜〜〜〜 = hydrogen bonds

(i) parallel   (ii) antiparallel

Fig 2.28 (b) Elastin

(i) recoiled elastin

polypeptide chains

(ii) stretched elastin

Fig 2.29 Tertiary structure of collagen

extended polypeptide chains

hydrogen bonds

Elastin (Fig 2.28(b)) in elastic connective tissue consists of many cross-linked polypeptides. When relaxed they are randomly coiled but when extended they take on a regular shape. It is this property which gives elastin its considerable elasticity. In unstretched keratin, three to seven α-helices are coiled around each other like the strands of a rope. The strands are held together by **disulphide (—S—S—) cross linkages** between the thiol (—SH) groups of cysteine molecules. The rope-like arrangement has considerable strength. In collagen of white fibrous tissue, the polypeptide chains are permanently extended. Three such chains are wound around each other and are bound to one another by hydrogen bonds (Fig 2.29). This arrangement has less elasticity but gives considerable tensile strength to fibrous connective tissue as in tendons.

Globular proteins such as myoglobin, haemoglobin and enzyme proteins have a **tertiary structure**. Again using X-ray diffraction analysis, J. C. Kendrew in the late 1950s showed that myoglobin, the oxygen-carrying pigment in muscles, consists of a single polypeptide chain folded asymmetrically. The chain has eight relatively straight segments in which the polypeptide is coiled in an α-helix. The segments are joined by bent peptide bonds and the chain bonded to an iron-containing haem group (Fig 2.30).

31

Fig 2.30 Tertiary structure of myoglobin

Disulphide, hydrogen, hydrophobic and **electrovalent bonds** help to stabilise the three-dimensional shape of such molecules (Fig 2.31).

Fig 2.31 Some of the bonds which maintain tertiary structure

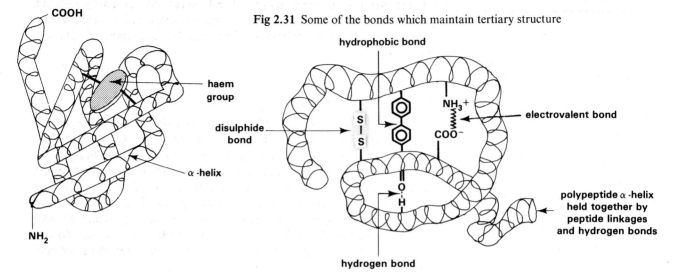

COOH

haem group

disulphide bond

α-helix

NH₂

hydrophobic bond

NH₃⁺ ── electrovalent bond

COO⁻

S
S

O
H

polypeptide α-helix held together by peptide linkages and hydrogen bonds

hydrogen bond

Fig 2.32 Quaternary structure of haemoglobin

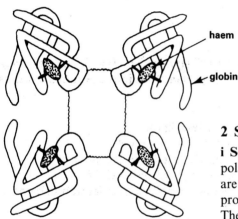

haem

globin

The sub-units are held more tightly together than illustrated here.

Most globular proteins with a relative molar mass of more than 50 000 consist of two or more polypeptide chains. The way in which they fit together is called the **quaternary structure** of the protein. Haemoglobin for example consists of four chains, two of α-globin and two of β-globin, each bonded to a haem group (Fig 2.32). The α- and β-chains are very similar in structure to myoglobin. Many enzymes have a quaternary structure, their molecules consisting of two to four globular units.

2 Some properties of proteins in aqueous solutions

i Solubility Globular proteins are generally **soluble in water** because the polar R groups jutting outwards from the molecule attract water. Enzymes are globular proteins which function in aqueous solutions. Some globular proteins of high relative molar mass form **colloidal suspensions** in water. They include some of the proteins in blood plasma. The **insolubility** of fibrous proteins in water is accounted for by the large number of non-polar hydrophobic R groups on the exterior of their molecules.

ii Buffering capacity The ability of a protein to accept or donate hydrogen ions (H⁺) depends on the number and charge of ionisable R groups in its molecules. The most important R groups in this respect are the amino (—NH₂) and carboxyl (—COOH) groups. Amino groups bind H⁺ in acidic solutions (pH < 7) whilst carboxyl groups donate H⁺ in alkaline solutions (pH > 7). In this way proteins act as **buffers** in stabilising the pH of their surroundings (Fig 2.33).

Fig 2.33 Buffering by a protein

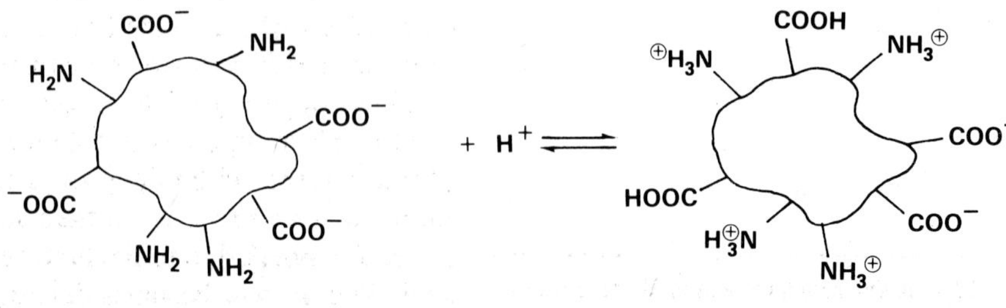

COO⁻   NH₂

H₂N

COO⁻

⁻OOC

NH₂ NH₂   COO⁻

$+ H^+ \rightleftharpoons$

COOH   NH₃⊕

⊕H₃N

COO⁻

HOOC   COO⁻

H₃⊕N   NH₃⊕

iii Isoelectric point At a precise pH in aqueous solution each protein exists as an amphion which carries no net electrostatic charge. This pH is

Proteins are least soluble in water at their isoelectric point. Why is this so?

called the **isoelectric point** of the protein. If the protein has a large number of amino groups in its side chains the isoelectric point is at a pH above 7, whilst proteins with a large number of carboxyl groups form amphions at a pH below 7 (Fig 2.34). A knowledge of the isoelectric points of proteins is exploited in separating mixtures of proteins by electrophoresis (section 2.4.2).

Fig 2.34 Iso-electric points of two proteins

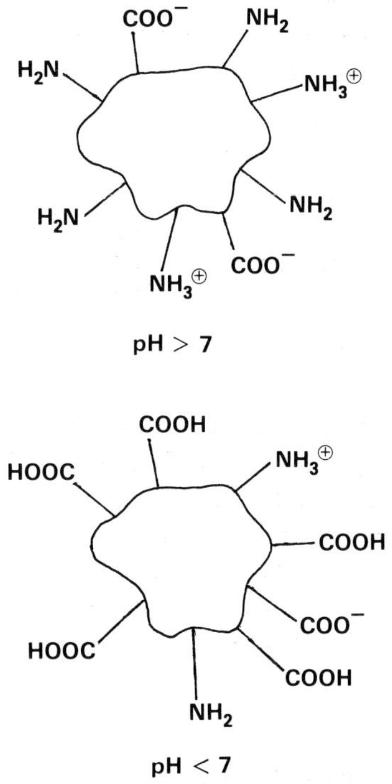

pH > 7

pH < 7

3 Protein denaturation

Most proteins function effectively only within a limited range of temperature and pH. At extremes of temperature and pH proteins undergo **denaturation**. Denaturation is a change in the physical shape of a protein molecule. It happens because some of the bonds which normally maintain the protein's three-dimensional structure are broken. Excessive heat can cause atoms in the molecule to vibrate so strongly that bonds between them are destroyed. Proteins also become denatured in very acidic or very alkaline solutions. At such extremes of pH electrovalent bonds do not develop. Denatured proteins are unable to carry out their normal functions.

Provided denaturation is not too extreme, protein molecules often spontaneously return to their original form when placed in ideal conditions of temperature and pH.

4 Functions of proteins

Proteins are used for an extraordinary range of activities in living organisms. Some of their more important functions are summarised as follows.

i	Enzymes	eg pepsin, amylase, lipase
ii	Contraction	eg actin and myosin in muscle
iii	Protection	eg antibodies, fibrinogen, prothrombin
iv	Hormones	eg insulin, growth hormone, ACTH
v	Transport	eg haemoglobin, transferrin (carries iron in blood plasma)
iv	Support	eg collagen, elastin
vii	Storage	eg myoglobin, ferritin (stores iron in the liver)

5 Biochemical tests for proteins and amino acids

One of the most common biochemical tests used to demonstrate the presence of amino acids in extracts from living organisms is the **ninhydrin reaction**. When heated with ninhydrin the α-amino group forms a blue coloured complex. Proline gives a yellow colour in this reaction.

The **biuret test** is frequently used to investigate the presence of compounds with peptide links in materials of living origin. Proteins and polypeptides form a violet complex when their peptide linkages react with alkaline copper(II) sulphate solution (biuret reagent).

Millon's reagent is a solution of mercury(II) nitrate in nitric acid. Mercury reacts with thiol groups of amino acids, such as cysteine which is found in many proteins. At the same time electrostatic forces in the protein molecules break due to the high concentration of hydrogen ions from the acid. The protein becomes denatured and a white precipitate is formed. When heated the precipitate takes on a red coloration if the protein contains the amino acid tyrosine.

What is the main limitation of Millon's test in identifying the presence of proteins in biological fluids?

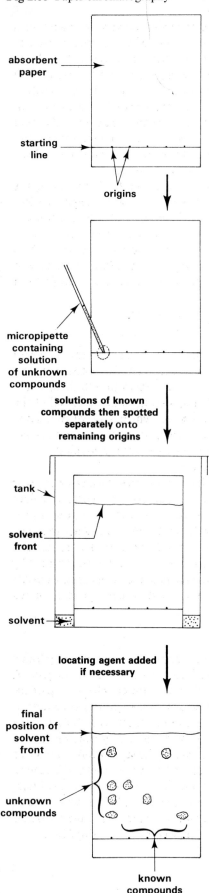

Fig 2.35 Paper chromatography

absorbent paper

starting line

origins

micropipette containing solution of unknown compounds

solutions of known compounds then spotted separately onto remaining origins

tank

solvent front

solvent

solvent

locating agent added if necessary

final position of solvent front

unknown compounds

known compounds

2.3.3 Glycoproteins and lipoproteins

These are conjugated proteins, the molecules of which contain an organic **prosthetic group** in addition to protein. In **glycoproteins** the additional component is usually a monosaccharide sugar such as galactose or mannose. Glycoproteins are part of the structure of cell membranes, assisting cell adhesion (Chapter 6). The α- and γ-immunoglobulins, prothrombin and fibrinogen are also glycoproteins (Chapter 10). In **lipoproteins** the prosthetic group is a simple lipid, phospholipid or cholesterol. The β-immunoglobulins are lipoproteins.

2.4 Biochemical techniques

Some technological developments used in biochemistry such as X-ray crystallography have already been mentioned earlier in this chapter. Let us now look at other techniques which have been used even more widely to provide a wealth of information about the organic compounds found in living organisms. Some of the techniques are simple and you can use them yourself. Scientific progress does not rely exclusively on sophisticated equipment and complicated technology. The techniques described here are mainly concerned with the separation and identification of biological compounds in mixtures and the analysis of complex molecules.

2.4.1 Chromatography

Chromatography is a technique used to separate the components of mixtures. It is based on the partitioning of compounds between a stationary phase and a moving phase. In **partition chromatography** the stationary phase is a liquid whilst in **adsorption chromatography** the stationary phase is a solid. Partition chromatography depends mainly on the relative solubility of substances in two or more solvents. When a solute is added to a mixture of equal volumes of two immiscible solvents, it may dissolve entirely in one or other of the solvents. If the solute dissolves in both solvents, it may not dissolve to the same extent in each solvent. The ratio:

$$\frac{\text{concentration of solute in solvent 1}}{\text{concentration of solute in solvent 2}}$$

at equilibrium is called the **partition coefficient**. In partition chromatography, use is mainly made of differences in the partition coefficient to separate components of mixtures.

In contrast, adsorption chromatography relies mainly on **electrostatic interaction** between solutes and the stationary phase. Some adsorption also occurs in partition chromatography. Paper, column and thin-layer are examples of the various forms of chromatography commonly used by biochemists.

1 Paper chromatography

Paper chromatography is used extensively in biological research yet it is a simple technique and is often practised in school laboratories (Fig 2.35). A **starting line** is drawn in pencil about 1.0 cm from one edge of a sheet of absorbent paper which is the **support medium**. On the line are marked a number of **origins** about 1.5 cm apart. A drop of the mixture to be analysed is spotted on to one of the origins. On each of the remaining origins is spotted a drop of a pure solution of substances suspected to be in the mixture. The technique is extremely sensitive and can be used to detect

Fig 2.36 (a) A column chromatogram showing separated photosynthetic pigments

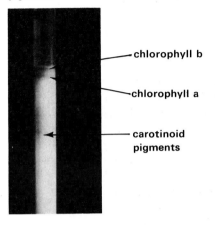

— chlorophyll b

— chlorophyll a

— carotinoid pigments

Fig 2.36 (b) Stages in the separation of the components of a mixture by column chromatography

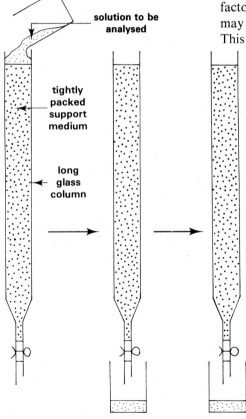

solution to be analysed

tightly packed support medium

long glass column

first fraction containing one component

second fraction containing another component

further fractions may be collected

minute quantities of unknown substances. For this reason a micropipette is used to spot the origins.

Before the paper is prepared, the **developing solvent** is placed in a chromatography tank and the lid replaced. Some of the solvent evaporates to saturate the air in the tank with solvent vapour. The paper is then placed vertically in the tank so that the starting line is just above the solvent which runs up (in **ascending chromatography**) or down (in **descending chromatography**) the support medium by capillarity. The known compounds and the components of the unknown mixture are carried different distances according to their partition coefficients between the solvent and water present in the paper. Adsorption to the support medium also affects movement of the solutes as the chromatogram develops.

When the solvent has travelled most of the length of the paper, the distance it has moved, the **solvent front**, is marked and the paper is dried. If the solutes are coloured, as in the case of leaf pigments, the positions of the components can be seen without further steps. The positions of colourless substances have to be detected using **locating agents** which react with the compounds under investigation to form coloured end products. Ninhydrin, for example, is used to locate amino acids (section 2.3.2). The ratio:

$$\frac{\text{distance travelled by a compound}}{\text{distance travelled by solvent front}}$$

is called the **retardation factor (R_f value)**. For a given solvent at a given temperature every compound has a characteristic R_f value. Thus by comparing the R_f values of the components of a mixture with those of the known compounds it is usually possible to identify some if not all of the substances in the mixture. Should the first solvent not produce a satisfactory separation of all the compounds in the mixture, the chromatogram may then be turned through 90° and a second solvent run along it. This is **two-dimensional chromatography**.

The amounts of substances which have been separated can also be determined. The first step is to cut out the coloured spots and place them separately in a standard volume of a solvent. This procedure is called **elution**. After a period of time the paper discs are removed and the intensity of colour in the solvent is measured using a colorimeter. The amount of substance in solution is determined by comparing its absorbance value read on the scale of the colorimeter with the absorbance of a standard solution containing a known concentration of the same substance.

2 Column chromatography

A long glass tube is packed with a support medium such as hydrated starch, silica gel or a powder of diatom shells called kieselguhr. A solution of the mixture to be analysed is poured into the top of the column. The components of the mixture move down the column at different rates. Differences in partition coefficients between the solvent and water in the support medium mainly account for the rates at which the solutes travel. Again, the separation of coloured compounds such as leaf pigments can be seen without further steps. When the solvent begins to run out at the bottom of the column, fractions collected at staggered intervals of time will contain different components of the mixture. The amounts of the separated components can be measured using a colorimeter (Fig 2.36).

3 Thin-layer chromatography

A thin glass or aluminium plate is covered with an aqueous slurry of a support medium, for example, silica gel or cellulose powder. The plate is dried and the mixture to be analysed, together with known compounds, spotted on to separate origins on the starting line. The plate is placed vertically in a tank so that the solvent almost touches the starting line. As in ascending paper chromatography the solvent rises by capillarity and the components of the mixture become separated according to their partition coefficients and adsorption to the support medium. When the solvent front has moved a sufficient distance to separate the components of the mixture, the plate is removed from the tank, dried, and if necessary a locating agent sprayed on to the support medium. The procedure can be made two-dimensional if the first solvent gives an incomplete separation. Once more the amounts of substances in the mixture can be determined. After scraping off the coloured spots and eluting the coloured complex, the intensity of colour can be determined with a colorimeter.

Sugars, amino acids, lipids and pigments in biological extracts can be separated using any of the methods of chromatography described above. Thin-layer chromatography is more rapid and is especially useful in analysing the acylglycerols in natural fats and oils.

2.4.2 Electrophoresis

An amino acid or protein molecule may be cationic, amphionic or anionic in aqueous solution depending on the nature of the amino acid side chains R and the pH of the solution. At a given pH some molecules may be more strongly cationic or anionic than others (section 2.3.1). When an electric current is passed through such a mixture, cations will move towards the cathode, anions to the anode, while amphions will fail to move either way. It is this difference which is exploited in separating mixtures of amino acids and of proteins by **electrophoresis**.

The pH of the mixture is first adjusted with a buffer so that each of the expected components has a different charge or a different strength of the same charge. In paper electrophoresis the mixture is applied to a starting line marked across the centre of a piece of paper or cellulose acetate support medium which has been soaked in the same buffer. Each end of the paper is then immersed in separated troughs of the buffer solution (see Fig 2.37). An electric current is then passed through the buffer. The current must be **direct** not alternate as on mains electricity, otherwise the components will migrate towards one pole them immediately back towards the other as the

Fig 2.37 Diagram of electrophoresis tank

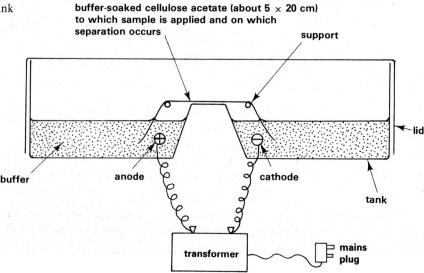

buffer-soaked cellulose acetate (about 5 × 20 cm) to which sample is applied and on which separation occurs

support

lid

buffer

anode

cathode

tank

transformer

mains plug

current alternates. A **power pack transformer** converts alternating mains electricity to the direct current required and also lowers the voltage (Fig 2.38). If the voltage is too high the support medium becomes overheated and may cause denaturation of proteins as well as evaporation of the buffer. After a period of time the power supply is disconnected and the support medium treated with a locating agent to find the positions of the separated components. Ninhydrin can be used to locate amino acids and a dye such as light green for proteins. The pattern of separation is compared with that of known mixtures run under identical conditions. Unknowns which migrate similar distances to knowns are probably identical.

Electrophoresis is used extensively to analyse plasma proteins such as immunoglobulins (Chapter 10).

Fig 2.38 Equipment for electrophoresis

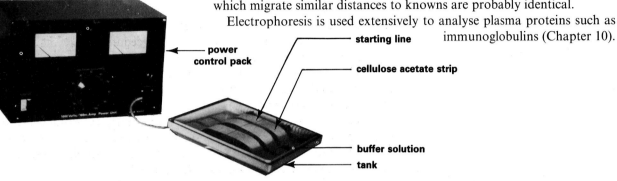

- power control pack
- starting line
- cellulose acetate strip
- buffer solution
- tank

SUMMARY

Carbon is a key element in the structure of all organic compounds. It is tetravalent and bonds to a variety of elements as well as to other carbon atoms forming a diverse range of molecules. When covalently bonded to four different functional groups, carbon atoms are asymmetric and give rise to stereoisomers.

Carbohydrates are the main products of photosynthesis and are important sources of energy for many living organisms. Each molecule of monosaccharide is a single reducing sugar unit, being either a ketose or an aldose. Triose monosaccharides have three, pentoses five and hexoses six carbon atoms per molecule. Disaccharides have two monosaccharide units in each of their molecules. Sucrose is a non-reducing disaccharide whereas maltose and lactose are reducing sugars.

Molecules of polysaccharides contain numerous monosaccharide units. Starch, the main storage carbohydrate of plants is made of two components, amylose which has a helical structure and amylopectin, a branched molecule. Both have α-D-glucose as their monomer. Glycogen is similar to amylopectin but more branched and is the main carbohydrate energy store in animal cells. Cellulose, an important structural ingredient in the walls of plant cells is an unbranched polymer made of β-D-glucose units.

Lipoids are a group of natural substances which are soluble in organic solvents. They include simple lipids, acylglycerols(glycerides) made when fatty acids esterify glycerol e.g. fats and oils. All the carbon atoms in the hydrocarbon chains of saturated fatty acids are joined to each other by single covalent bonds. In unsaturated fatty acids the hydrocarbon chain contains one or more double covalent bonds. Complex lipids include phospho-, glyco- and sphingolipids, all components of cell membranes. These are polar molecules unlike fats and oils which are neutral. Several mammalian hormones are steroids, whilst cholesterol is an example of a sterol.

Proteins are polymers of amino acids, joined to each other by peptide bonds. Varied in structure and function, protein molecules are made of one or more polypeptide chains which exist as α-helices, pleated sheets or random coils. The amino acid sequence in polypeptide chains determines the structure and hence the properties of proteins. Any change in structure is called denaturation and leads to loss of function. Like the amino acids of which they are formed proteins can buffer against pH changes in living cells.

Chromatography and electrophoresis are examples of biochemical techniques which can be used to separate and identify such compounds in cell extracts and biological fluids.

QUESTIONS

1 (a) What general type of molecule is shown in the diagram below?

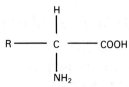

(b) What is the simplest form of **R**?

(c) Which part of the structure gives acidic properties to the molecule?

(d) Which part of the structure gives basic properties to the molecule?

(e) It may be said that because molecules of this type can show polymerisation they are very important biologically.
 (i) What is meant by *polymerisation*?
 (ii) If molecules of this type polymerise, what will be formed?

(f) With the aid of a diagram, illustrate the product when two of the units shown have joined.

(g) What general type of biochemical reaction is this?

(h) What name is given to the type of bond formed between the two units? (O)

2 Describe how the structure of:
(a) **named** polysaccharides,
(b) **named** proteins,
make them suitable for their functions in living cells. (C)

3 At its iso-electric point a protein such as casein will have no overall charge. It is at this point that there is the highest precipitation out of solution. The iso-electric point is usually reached at a particular pH and can be demonstrated in the laboratory as follows.

First a series of tubes containing solutions of known pH are prepared. Next, 1 cm³ of prepared casein solution is added to each tube and varying degrees of turbidity (cloudiness) observed in each tube. The turbidity is estimated visually and recorded by a simple + system: the more +'s the greater the turbidity (Table 1).

Table 1. Turbidity of casein solutions at different pH

Tube number	1	2	3	4	5	6	7
Initial pH reading	3.9	4.2	4.5	4.8	5.1	5.4	5.7
pH after addition of casein	4.1	4.4	4.7	4.9	5.3	5.6	5.9
Turbidity: after 30 seconds		++	+++	+++	++	+	
after 5 minutes	+	++	+++	++++	+	+	
after 30 minutes	+	+	++	++	+	+	

(a) (i) Which time is most likely to give an accurate assessment of the iso-electric point of casein? Explain your answer.
 (ii) Which tube gives the best indication of the iso-electric point?
 (iii) Explain why the pH rises on the addition of the casein to the tubes.

(b) The turbidity of the same series of tubes was also measured using an absorption meter and is recorded in Table 2.

Table 2. Turbidity of casein solutions at different pH using an absorption meter

Tube number	1	2	3	4	5	6	7
Percentage transmission (after 10 minutes)	74	24	5	7	49	72	91

(i) Plot a graph of turbidity against pH.
(ii) Determine the iso-electric point from your graph.
(iii) Comment on the possible physiological significance of the precipitation of proteins at their isoelectric point. (JMB)

4 Show how the properties of amino acids are important in the formation of proteins which have a diversity of roles. (C)

3 Enzymes

3.1	The structure of enzymes	40
3.2	Enzyme catalysis	41
	3.2.1 The lock and key mechanism	42
	3.2.2 Control of enzyme action	43
3.3	Factors affecting enzyme action	43
	3.3.1 Co-factors	43
	3.3.2 Inhibitors	44
	3.3.3 Temperature	46
	3.3.4 pH	46
	3.3.5 Enzyme concentration	47
	3.3.6 Substrate concentration	48

3.4	The classification of enzymes	49
	3.4.1 Hydrolases	49
	3.4.2 Oxido-reductases	50
	3.4.3 Transferases	50
	3.4.4 Isomerases	51
	3.4.5 Ligases	51
	3.4.6 Lyases	51
	Summary	51
	Questions	52

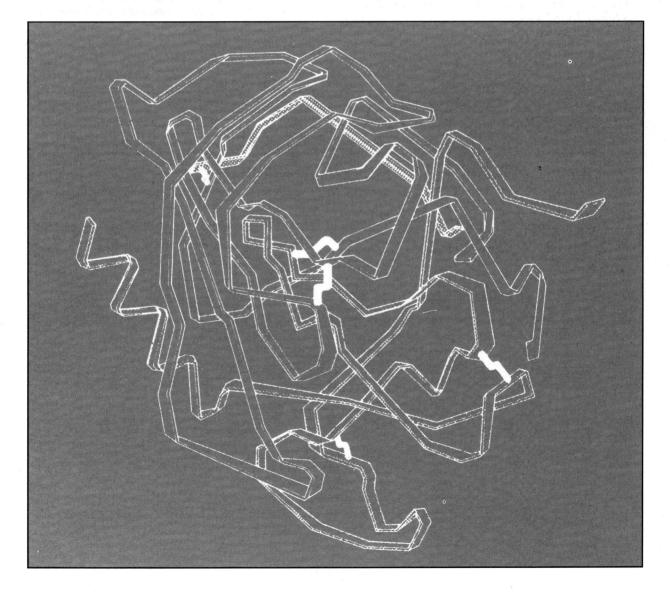

3 Enzymes

In living cells hundreds of different biochemical reactions take place rapidly and simultaneously. The reactions go on at relatively low temperatures and are normally controlled in such a way that useful products are made and wastes removed at rates which satisfy the metabolic needs of cells. How is it possible for there to be such orderliness in what must be a potentially chaotic situation? How can reactions take place so rapidly at such modest temperatures? The answers to these questions come from a study of enzymes.

The first enzyme to be discovered was amylase which catalyses the conversion of starch to maltose. Its presence in malt extract was detected in 1833 by two French chemists Payen and Persoz. However, it was not until 1876 that the term **enzyme** was proposed by Wilhelm Kühne, the distinguished German biochemist.

3.1 The structure of enzymes

To appreciate the structure of enzymes it is necessary to have a knowledge of protein structure (Chapter 2). Over 90 % of enzymes are simple **globular proteins**. The remainder are **conjugated proteins** which have a non-protein fraction called the **prosthetic group**. Many enzymes have relative molar masses of between 10 000 and 500 000.

Amino acid analysis, X-ray diffraction studies and other techniques have in recent years provided a lot of information about the structure of proteins (Chapter 2). The three-dimensional structure of a number of enzymes is now known in detail. Chymotrypsinogen, an inactive precursor of the pancreatic enzyme chymotrypsin, consists of three polypeptide chains containing 245 amino acid molecules (Fig 3.1).

Fig 3.1 Three-dimensional structure of chymotrypsinogen (after P B Singler et al *J. Mol. Biol.* **35,** 143, 1968)

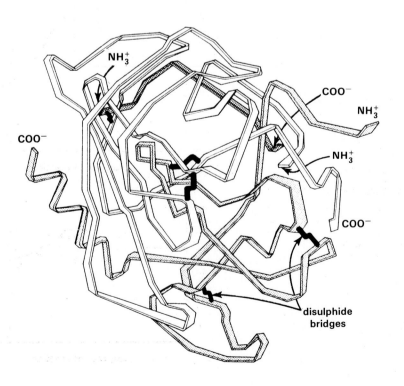

3.2 Enzyme catalysis

Enzymes are soluble in water and work in aqueous solution in living cells. They are sometimes described as **organic catalysts**. A catalyst is a substance which affects the rate of chemical reactions. Many of the biochemical reactions catalysed by enzymes are reversible reactions. An example of a reversible reaction is as follows:

$$A + B \rightleftharpoons C$$

reactants   product

At equilibrium the rate at which A and B are converted to C may or may not be equal to the rate at which C is converted back to A and B. The position of the equilibrium depends on the energy difference between the reactants and the product. Enzymes do not alter the direction of a reaction; they speed up the rate at which equilibrium is reached. In so doing they can catalyse reversible reactions in either direction provided it is energetically feasible.

Biochemical reactions involve the formation or the destruction of chemical bonds. When two or more reactants are joined, chemical bonds are formed. When a complex molecule is split into simpler components, chemical bonds are destroyed. In either case energy is required to bring about the changes. The energy required for a chemical reaction to proceed is called **activation energy** (Fig 3.2(a)). Heat can be used as a source of activation energy. Indeed, many chemical reactions do not proceed quickly unless the reactants are raised to relatively high temperatures. In living cells however, reactions take place rapidly at relatively low temperatures. Enzymes lower the amount of activation energy needed, making it possible for reactions to occur at temperatures which are otherwise energetically unfavourable (Fig 3.2(b)).

Fig 3.2 Energetics of biochemical reactions

(a) Uncatalysed

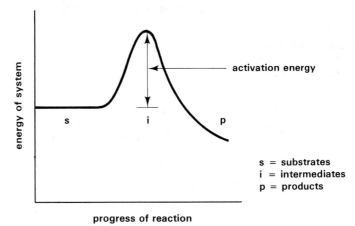

s = substrates
i = intermediates
p = products

(b) Catalysed

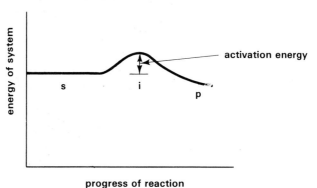

3.2.1 The lock and key mechanism

Emil Fischer in 1894 proposed the **lock and key mechanism** to explain how enzymes and **substrates** interact when mixed. Fischer suggested that enzyme and substrate molecules combine to form an **enzyme–substrate complex** before the products of the reaction are released (Fig 3.3).

Fig 3.3 The lock and key mechanism of enzyme action

(a) Bond formation

(b) Bond destruction

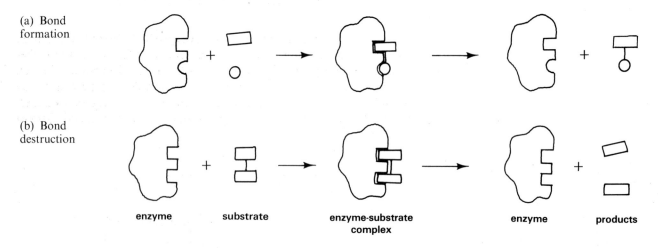

enzyme   substrate   enzyme-substrate complex   enzyme   products

---

One of the systems used to name enzymes is based on the substrates with which they react. For example, lipases react with lipids. What types of substrates do you think carbohydrases and peptidases react with?

The sites on enzyme molecules where substrates fit are called **active centres**. Bringing substrate molecules close to each other at the active centre increases the probability that they will react. Because enzyme and substrate molecules are three-dimensional structures, the formation of an enzyme–substrate complex requires that the shapes of the reactants and the active centres of enzymes are **complementary**. Otherwise the enzyme and substrate cannot unite (Fig 3.4(a)). It is comparable to the way in which a lock can only be opened by a key of a specific shape. The phenomenon whereby most enzymes work with only one or with a limited range of substrates is called **enzyme specificity**. The specificity of enzymes supports the lock and key hypothesis.

Fig 3.4 (a) Enzyme specificity

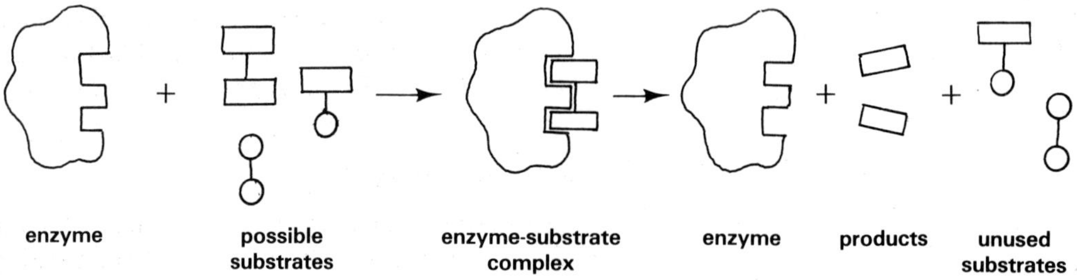

enzyme   possible substrates   enzyme-substrate complex   enzyme   products   unused substrates

Fig 3.4 (b) Induced fit mechanism

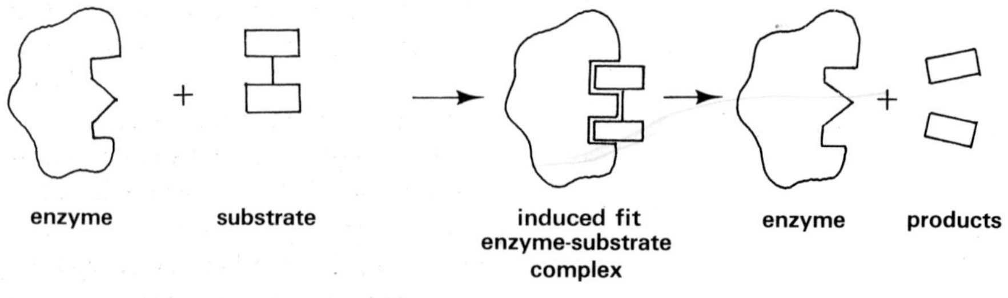

enzyme   substrate   induced fit enzyme-substrate complex   enzyme   products

Koshland has more recently proposed the **induced-fit mechanism** to explain the action of some enzymes. According to this theory, the shape of the active centre is changed when a substrate molecule binds to such enzymes (Fig 3.4(b)).

3.2.2 Control of enzyme action

Biochemical pathways usually involve a number of linked reactions. Each reaction is catalysed by a specific enzyme:

$$A \underset{}{\overset{\text{enzyme X}}{\rightleftharpoons}} B \underset{}{\overset{\text{enzyme Y}}{\rightleftharpoons}} C \underset{}{\overset{\text{enzyme Z}}{\rightleftharpoons}} D$$

In some instances accumulation of one of the products formed near the end of the chain inhibits the action of an enzyme used in one of the earlier reactions. In the hypothetical example shown above, the presence of a high concentration of product D for example may slow down the rate at which enzyme X converts A to B. This is an example of **negative feedback inhibition,** an important device in the co-ordination of the metabolism of cells.

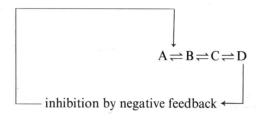

Negative feedback ensures that reactants are used efficiently and prevents the excess manufacture of end products. Like other homeostatic devices described elsewhere in this book, the control of enzyme action helps to maintain a stable environment in living organisms.

3.3 Factors affecting enzyme action

To function at all, some enzymes need the presence of **co-factors**. To function efficiently, enzymes need a suitable **temperature** and **pH** and there must be enough **substrate** available. Enzyme action can be stopped partially or completely if **inhibitors** are present in the reaction mixture.

3.3.1 Co-factors

A **prosthetic group** is an essential co-factor attached to the protein part of a conjugated enzyme. Some prosthetic groups are organic, others are atoms of metals such as copper. If the prosthetic group is removed the enzyme fails to function. Mineral ions must be mixed with the reactants before some enzymes will work. Zinc, iron and magnesium act as **enzyme activators** in this way. It is partly for these reasons that minerals are essential to living organisms (Chapters 14 and 15).

Other enzymes have **co-enzymes** as co-factors. Co-enzymes are organic, non-protein substances which are not bonded to enzyme molecules like prosthetic groups. Several important co-enzymes are vitamin derivatives. **Nicotinamide-adenine dinucleotide (NAD)** and its phosphate ester **nicotinamide-adenine dinucleotide phosphate (NADP)** for example, are derived from nicotinamide, a B-group vitamin. Riboflavin (vitamin B_2) forms part of **flavin-adenine dinucleotide (FAD)**. NAD, NADP and FAD function as hydrogen acceptors in reactions catalysed by dehydrogenase enzymes (section 3.4).

3.3.2 Inhibitors

Many substances inhibit the activity of enzymes. **Inhibitors** fall into two categories, **reversible** and **non-reversible**.

1 Reversible inhibitors

Reversible inhibitors are substances which prevent enzymes from combining with substrates. Activity of the enzyme is restored when the inhibitor is removed.

Competitive inhibitors affect enzyme action by becoming attached to active centres, so stopping the substrate from binding to the enzyme. A well-known example of this behaviour is inhibition of the enzyme succinate dehydrogenase by malonic acid. Succinate dehydrogenase catalyses the oxidation of succinate to fumarate in the Krebs cycle (Chapter 5). In the presence of malonate the reaction rate is slowed down. Malonate has a molecular structure which is very similar to that of succinate. Inhibition happens because the active centre of some of the dehydrogenase enzyme molecules becomes occupied by malonate rather than by the normal substrate succinate. In effect, malonate and succinate are competing for the active centre of the enzyme (Fig 3.5).

Fig 3.5 (a) Competitive inhibition

Fig 3.5 (b) Molecular structures of succinate and malonate

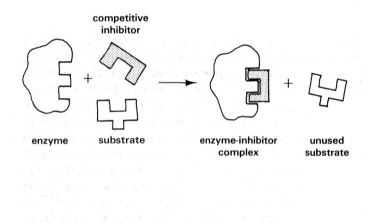

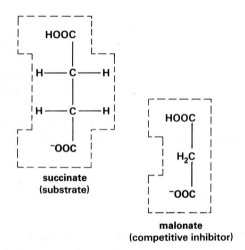

MEDICAL APPLICATION

Inhibitors of this kind can be used as drugs to reduce the rate at which undesirable reactions occur in the human body. Inhibiting such reactions is one of the ways of treating some forms of cancer. Another application of competitive inhibition is the use of sulphonamide drugs such as prontosil to combat bacterial infections. Prontosil is similar in molecular structure to *p*-aminobenzoic acid (see Fig 3.5) which bacteria use to synthesise folic acid. When unable to make folic acid, they cannot grow and multiply in the human body.

Fig 3.5 (c) Molecular structures of PABA and prontosil

The degree of inhibition by a competitive inhibitor is less if the ratio of substrate-to-inhibitor is high (Fig 3.6).

Fig 3.6 Effect of concentration of a competitive inhibitor on the rate of an enzyme catalysed reaction

(a) Low concentration

(b) High concentration

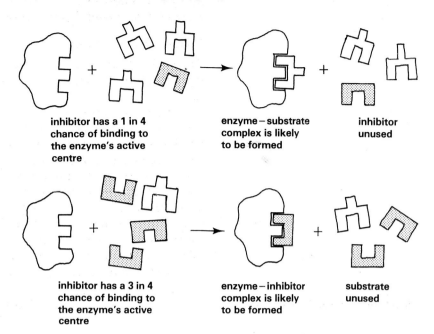

inhibitor has a 1 in 4 chance of binding to the enzyme's active centre

enzyme–substrate complex is likely to be formed

inhibitor unused

inhibitor has a 3 in 4 chance of binding to the enzyme's active centre

enzyme–inhibitor complex is likely to be formed

substrate unused

The degree of inhibition by **non-competitive inhibitors** cannot be reduced by increasing the number of substrate molecules (Fig 3.7). Here the inhibitor becomes attached to the enzyme at a position other than the active centre. Nevertheless the enzyme, the substrate, or possibly both, become changed so that enzyme activity stops. Disulphide bridges are important in maintaining the tertiary structure of enzyme molecules (Chapter 2). If the disulphide bridges are broken the three-dimensional shape of the enzyme changes. The ions of heavy metals such as mercury (Hg), silver (Ag) and copper (Cu) affect enzymes in this way. Hg^{2+}, Ag^+ and Cu^{2+} ions combine with thiol (—SH) groups in enzymes, so denaturing enzyme molecules and inhibiting enzyme activity. Cyanide is another non-competitive inhibitor. It blocks the action of some enzymes by combining with iron which may be present in a prosthetic group or which may be required as an enzyme activator. It is not surprising therefore that the salts of heavy metals and cyanide are potent poisons to living organisms. Nevertheless, non-competitive inhibitors do not bind strongly to enzymes and can be removed by dialysis. Enzyme activity is then restored.

Fig 3.7 Comparative effects of a non-competitive and competitive inhibitor on the rate of an enzyme-catalysed reaction

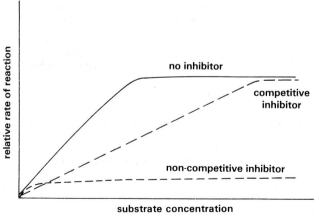

relative rate of reaction

no inhibitor

competitive inhibitor

non-competitive inhibitor

substrate concentration

2 Non-reversible inhibitors

Organophosphorus insecticides such as malathion are good examples of **non-reversible inhibitors**. They become firmly bound to active centres so that substrate molecules cannot bind to enzymes and activity of the enzyme is permanently stopped. Insecticides of this type inactivate the enzyme cholinesterase which is essential for the functioning of the nervous system (Chapter 17). Nerve gases such as sarin do the same.

Fig 3.8 Effect of temperature on the rate of an enzyme-catalysed reaction

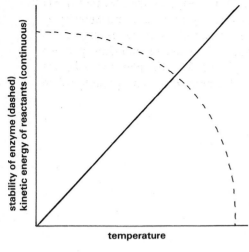

(a)

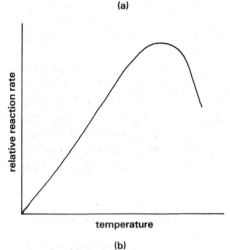

(b)

3.3.3 Temperature

Heat supplies kinetic energy to reacting molecules, causing them to move more rapidly. The chances of molecular collision taking place are thus increased at higher temperatures so it is more likely that enzyme–substrate complexes will be formed. However, heat energy also increases the vibration of the atoms which make up enzyme molecules. If the vibrations become too violent, chemical bonds in the enzyme break and the precise three-dimensional structure, so essential for enzyme activity, is lost. At high temperatures therefore enzymes become **denatured**.

When the effect of temperature on enzyme activity is investigated experimentally, a temperature usually called the **optimum temperature** is observed at which the reaction proceeds most rapidly. This temperature is not necessarily that at which the enzyme is most stable. It is the resultant of the contrary effects of temperature on the movement of reactants and of enzyme denaturation (Fig 3.8).

The term **temperature coefficient (Q_{10})** is used to express the effect of a $10\,°C$ rise in temperature on the rate of a chemical reaction.

$$Q_{10} = \frac{\text{rate of reaction at } t + 10\,°C}{\text{rate of reaction at } t\,°C}$$

Between $4\,°C$ and $37\,°C$, the optimum temperature for enzymes in the human body, the Q_{10} for enzyme-catalysed reactions is 2.

3.3.4 pH

The symbol pH refers to the concentration of hydrogen ions in solution (Chapter 1). The concentration of hydrogen ions $[H^+]$ affects the stability of the electrovalent bonds which help to maintain the tertiary structure of protein molecules (Chapter 2). Extremes of pH cause the bonds to break resulting in enzyme **denaturation**.

It must be remembered that even small changes in pH mean relatively large changes in $[H^+]$. A change of 1 on the pH scale involves a ten-fold increase or decrease in $[H^+]$ whilst a change in pH of 2 represents a hundred-fold change in $[H^+]$ (Chapter 1). Thus even small changes in pH can have a great effect on enzyme activity.

Apart from the effect in denaturing enzymes, changes in $[H^+]$ can alter the ionisation of the amino acid side chains at the active centres of enzymes. Ionisation of substrate molecules can also be affected. The formation of enzyme–substrate complexes sometimes depends on the active centres and substrate molecules having opposite electrostatic charges. If the charges are altered by changes in pH, some enzymes fail to function (Fig 3.9).

Fig 3.9 One way in which pH may affect an enzyme

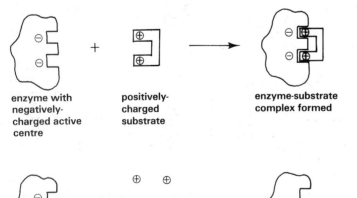

enzyme with negatively-charged active centre

positively-charged substrate

enzyme-substrate complex formed

enzyme

hydrogen ions

active centre neutralised

substrate repelled

At 35°C it took an enzyme-catalysed reaction 5 minutes to be completed compared with 10 minutes at 25°C. Calculate the Q_{10} for the reaction.

Fig 3.10 Effect of pH on the rates of enzyme-catalysed reactions

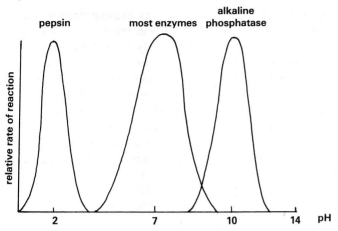

For every enzyme there is an **optimum pH** at which the reaction it catalyses proceeds most rapidly (Fig 3.10). Many enzymes work within a pH range of 5–9 and catalyse reactions most efficiently at pH 7. There are exceptions. For example, pepsin and rennin secreted in the mammalian stomach work best at pH 1.5–2.5 (Chapter 15). Alkaline phosphatase in the kidneys has an optimum pH of 10.

Fig 3.11 Effect of enzyme concentration on the rate of an enzyme-catalysed reaction in the presence of unlimited concentration of substrate

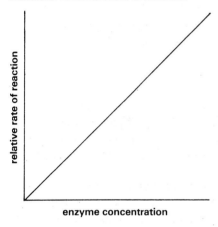

3.3.5 Enzyme concentration

Enzymes catalyse reactions rapidly at very low enzyme concentrations. This is because enzyme molecules form complexes with substrates only very briefly. The products of the reaction are quickly released and the enzyme is then available for further activity.

The rate at which enzymes use substrates is described as the **turnover number**. For some enzymes the turnover number is very high. A molecule of catalase for example can break down 40 000 molecules of hydrogen peroxide into water and oxygen every second! Even the slowest of enzymes have turnover numbers of about $100\,s^{-1}$. The larger the number of enzyme molecules present the greater is the amount of substrate used in a given period of time, provided that there is an excess of substrate available (Figs 3.11 and 3.12).

Fig 3.12 Formation of enzyme–substrate complexes at three concentrations of enzyme

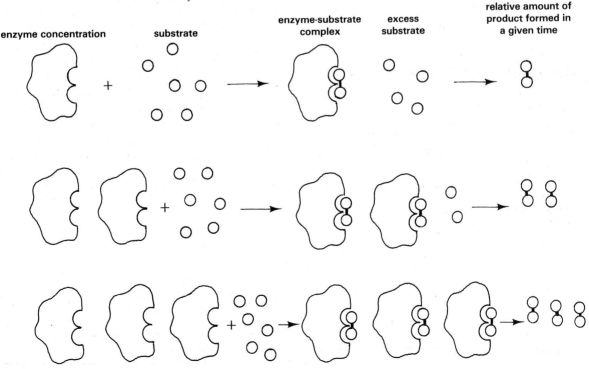

3.3.6 Substrate concentration

Fig 3.13 shows the effect of substrate concentration on the rate of a biochemical reaction when the amount of enzyme is limited. At low concentrations of substrate there is a linear relationship between the reaction rate and substrate concentration. In these conditions the ratio of enzyme to substrate molecules is high. Consequently some active centres are always free for substrate molecules to bind with the enzyme. However, a point is reached when a further increase in substrate concentration does not cause the reaction to go any faster. The enzyme-to-substrate ratio is then lower, and there are more substrate molecules present than there are free active centres with which to bind. Adding more substrate will not make the reaction go more quickly (Fig 3.14).

Fig 3.13 Effect of substrate concentration on the rate of an enzyme-catalysed reaction

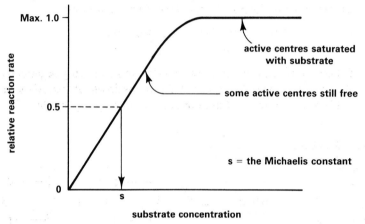

The concentration of substrate at which the reaction rate is half of its maximum is characteristic for each combination of enzyme and substrate.

It is called the **Michaelis constant (K_M)**. A high K_M value indicates that the enzyme has a low affinity for its substrate. Conversely a low K_M value indicates a high affinity.

Fig 3.14 Formation of enzyme-substrate complexes at four concentrations of substrate

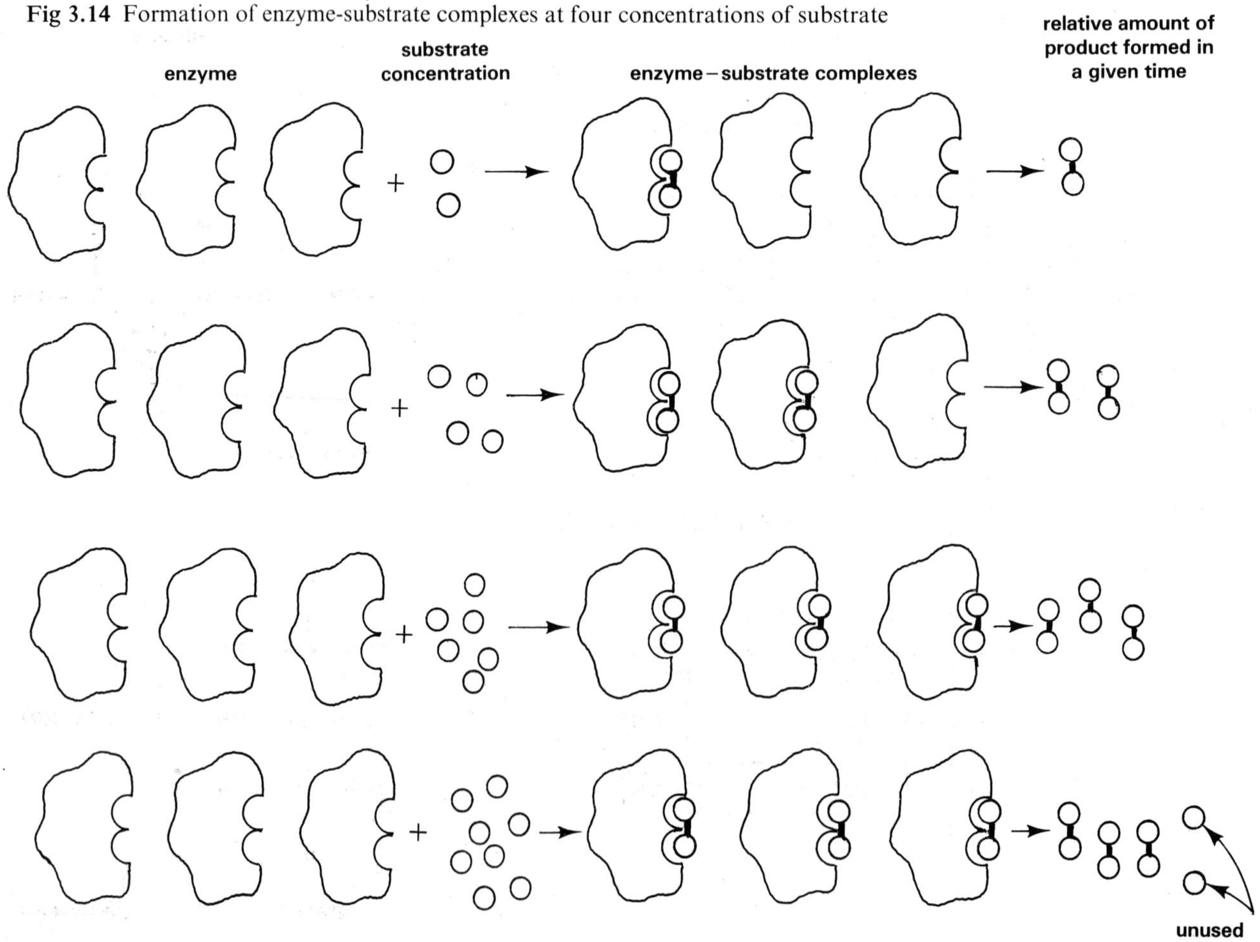

3.4 The classification of enzymes

In recent times many new enzymes have been discovered. Older methods of classifying enzymes became impractical and in the early 1960s the International Union of Biochemistry (IUB) recommended a new system for the classification and naming of enzymes. According to the IUB system there are six groups of enzymes.

3.4.1 Hydrolases

These enzymes catalyse reactions in which a substrate is hydrolysed into two simpler products. During **hydrolysis**, hydrogen atoms from water enter one of the products whilst the hydroxyl groups end up in the other product. The reaction is shown in a simplified form as follows:

$$AB + H.OH \rightleftharpoons AH + B.OH$$
substrate   water   products

Fig 3.15 shows how some important substrates found in living organisms are hydrolysed. Hydrolases bring about the breakdown of materials in lysosomes (Chapter 6) and the digestion of food in the gut (Chapter 15).

Fig 3.15 Hydrolysis of some important biological compounds

(a) Glycosidic linkage in a disaccharide

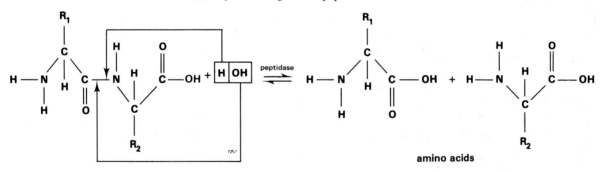

(b) Peptide linkage in a dipeptide

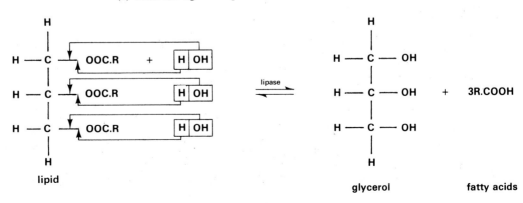

(c) Ester linkage in a lipid

49

3.4.2 Oxido-reductases

The oxidation of substrates is catalysed by this group of enzymes of which there are two kinds:

1 Oxidases

Oxidases catalyse the transfer of hydrogen to molecules of oxygen. A simplified way of expressing the reaction is as follows:

$$AH_2 \; + \; \tfrac{1}{2}O_2 \; \rightleftharpoons \; A \; + \; H_2O$$

substrate   oxygen   oxidised   water
substrate   (reduced oxygen)

An example is **cytochrome oxidase** which catalyses the oxidation of reduced cytochrome:

$$Cyt . H_2 \; + \; \tfrac{1}{2}O_2 \rightleftharpoons cytochrome \; + \; H_2O$$

reduced
cytochrome

Hydrogen peroxide is sometimes formed instead of water.

2 Dehydrogenases

Dehydrogenases catalyse the oxidation of substrates by transferring hydrogen to co-enzymes such as NAD and NADP.

$$AH_2 \; + \; 2NAD^+ \rightleftharpoons \; A \; + 2NADH$$

substrate   co-enzyme   oxidised   reduced
substrate   co-enzyme

For example, **alcohol dehydrogenase** controls the rate at which ethanol is oxidised to ethanal:

$$CH_3CH_2OH \; + 2NAD^+ \rightleftharpoons CH_3CHO \; + \; 2NADH$$

ethanol   ethanal

In reactions catalysed by **oxido-reductases**, substrates are oxidised whilst oxygen or co-enzymes are reduced. The importance of these reactions is described more fully in Chapter 5.

3.4.3 Transferases

Transferases catalyse the transfer of functional groups from one substrate to another:

$$AB + C \rightleftharpoons A + BC$$

Phosphotransferases control the transfer of phosphate groups in respiration (Chapter 5):

$$glucose + ATP \rightleftharpoons glucose\text{-}6\text{-}phosphate + ADP$$

Aminotransferases regulate the transfer of amino groups (Fig 3.16 and Chapter 15).

Fig 3.16 Action of an aminotransferase enzyme

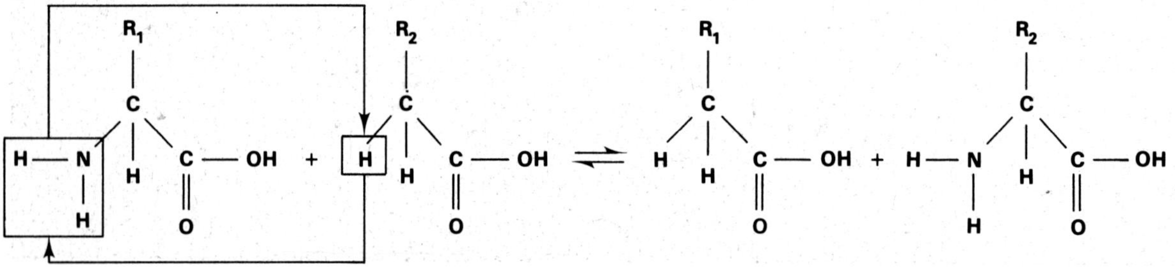

3.4.4 Isomerases

Isomerases control the conversion of one isomer of a compound to another isomer of the same compound:

$$ABC \rightleftharpoons ACB$$

The interconversion of sugar isomers in glycolysis is catalysed by isomerase enzymes such as **hexosephosphate isomerase** (Chapter 5):

$$glucose\text{-}6\text{-}phosphate \rightleftharpoons fructose\text{-}6\text{-}phosphate$$

3.4.5 Ligases

This group of enzymes catalyses reactions in which new chemical bonds are formed. ATP provides the energy to make the new chemical bonds. For example **DNA** and **RNA ligase** control the synthesis of macromolecules such as nucleic acid (Chapter 4).

3.4.6 Lyases

Lyases catalyse the breakdown of complex substrates into simpler products but, unlike the case of hydrolytic reactions, water is not used:

$$AB \rightleftharpoons A + B$$

Decarboxylases which regulate the release of carbon dioxide from respiratory substrates are examples of lyase enzymes (Chapter 5). **Deaminases** catalyse the release of ammonia from amino acids. Their role in the metabolism of unwanted amino acids absorbed from the gut is outlined in Chapter 15.

Much of what is known about enzymes has come from investigations carried out *in vitro* on purified enzyme extracts. *In vitro* means in glass, indicating that it is easier to obtain information from test-tube studies than it is *in vivo*, in living cells. The way in which enzymes control vast numbers of different reactions taking place simultaneously in microscopic packets of protoplasm is therefore all the more remarkable. Compartmentalisation of cells (Chapter 6) undoubtedly helps: each of the organelles of a living cell has a specific range of enzymes so that different kinds of metabolic reactions are separated. Even so, the way in which enzymes regulate the many, rapid and complex metabolic reactions in living cells surely gives cause for wonder. The commercial production of enzymes is one of the many branches of **biotechnology** (Chapter 34).

SUMMARY

Enzymes are either simple or conjugated proteins which are very efficient biological catalysts. By lowering the activation energy they enable reactions to proceed rapidly at temperatures and pressures suitable for life. A given enzyme catalyses a reaction using only one of a limited range of substrates. Such specificity is explained by the lock-and-key mechanism of enzyme action.

The activity of some enzymes requires the presence of a variety of co-factors which include prosthetic groups, enzyme activators and co-enzymes. A number of substances can act as enzyme inhibitors.

Some forms of life have the ability to maintain the optimum temperature and pH for enzyme action. Extremes of temperature and pH lead to enzyme denaturation.

The IUB system for enzyme classification has six groups—hydrolases, oxidoreductases, transferases, isomerases, ligases and lyases. ▶

Enzymes play major commercial roles in food processing industries and are vital ingredients of biological detergents. There is increasing demand for enzymes for biochemical analyses and for the treatment of a few kinds of life-threatening diseases such as lymphatic leukaemia. Immobilised enzymes are used on a large scale in the production of penicillin derivatives and fructose.

QUESTIONS

1 The graph below illustrates the relationship between time, temperature and the amount of product formed in an enzyme-catalysed reaction. In the experiment, the samples were incubated at different temperatures for periods of 1, 2 and 5 hours. The quantities of products formed were then determined.

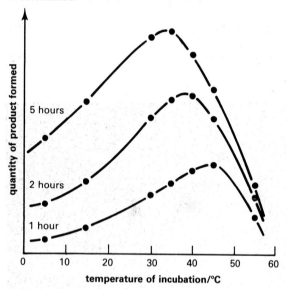

(a) (i) Explain the effect of increasing the incubation temperature on the quantity of product formed, as shown by any one of the curves presented in the graph above.
(ii) Explain why the *optimum temperature* is higher if the quantity of product formed is measured after 1 hour rather than after 5 hours.

The graph below illustrates the influence of substrate concentration on the rate of an enzyme-catalysed reaction, the enzyme concentration and the temperature being kept constant.

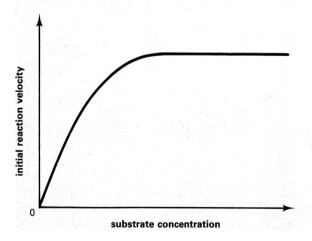

(b) (i) Explain why an increase in substrate concentration at low substrate concentrations increases the initial reaction velocity but an increase at high substrate concentrations does not do so.
(ii) What change in conditions would cause a marked increase in reaction velocity at high substrate concentrations? (C)

2 The enzyme polyphenol oxidase is involved in the browning of a cut surface of a potato tuber when it is exposed to the air; phenolic compounds, such as catechol, yield dark coloured melanic products which cause browning. The table below shows the time taken for a standard dark brown colour to develop when different amounts of a standard enzyme solution and water were added, at different concentrations, to tubes containing catechol solution. The tubes were kept at constant temperature and shaken during the experiment.

Tube	Enzyme /cm^3	Water /cm^3	Catechol sol^n /cm^3	Time for colour to develop minutes
1	2.0	2.0	0.0	(No colour)
2	2.0	0.0	2.0	10
3	1.5	0.5	2.0	13
4	1.0	1.0	2.0	21
5	0.5	1.5	2.0	40

(a) Draw a graph to show the **rate** of reaction against enzyme concentration.

(b) At what enzyme concentration is the standard colour obtained after 17 minutes?
(c) What was the purpose of tube 1?
(d) Why were the tubes shaken during the experiment?
(e) The rate of the reaction can be altered by the presence of enzyme inhibitors which may be competitive or non-competitive. State what is understood by (i) competitive and (ii) non-competitive inhibition. (O)

3 (a) Describe the characteristic structural features of an enzyme molecule.
(b) Explain how
 (i) changes in temperature,
 (ii) changes in substrate concentration,
 (iii) presence of inhibitors, (C)
affect enzyme-mediated reactions.

4 (a) Describe the factors which influence enzyme activity, relating these specifically to the protein nature and catalytic function of enzymes.
(b) Discuss the importance of enzymes in metabolism, illustrating your answer by reference to **three** *intracellular* enzymic reactions. (JMB)

4 Nucleic acids and protein synthesis

4.1 Nucleic acids 54

   4.1.1 Deoxyribonucleic acid 54
   4.1.2 Ribonucleic acid 56
   4.1.3 DNA as the hereditary substance 57
   4.1.4 Replication of DNA 58

4.2 Protein synthesis 60

   4.2.1 Role of the nucleic acids 60
   4.2.2 Gene mutations 64
   4.2.3 Regulating protein synthesis 66

4.3 Genetic engineering 68

   4.3.1 Recombinant DNA technology 68
   4.3.2 Production of insulin by genetic engineering 70
   4.3.3 Other examples of genetic engineering 71

4.4 Genetic fingerprinting 73

Summary 74

Questions 74

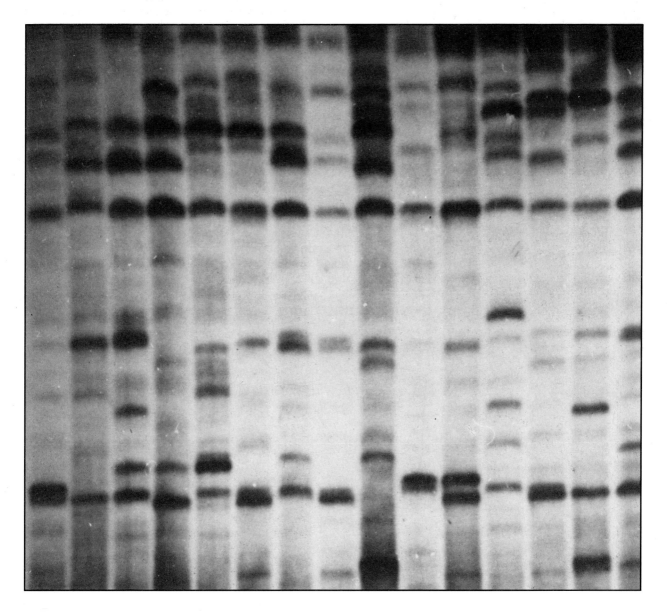

4 Nucleic acids and protein synthesis

In 1869 a German physician named Meischer reported on a chemical analysis he had made of the nuclei of human pus cells collected from bandages. He found the nuclei to be rich in nitrogen, sulphur and phosphorus. Meischer set a pattern for investigating the composition of the nuclei of cells using the methods of the analytical chemist.

During the next eighty years analytical chemists revealed small but significant facts about the composition of nuclei. Crude extracts of nuclei were found to be acidic in reaction, hence the term **nucleic acids** to describe the most important substances present in nuclei. Gradually the building blocks which make up the nucleic acids were identified. But the way in which the building blocks were put together and the roles of the nucleic acids were still poorly understood until the middle of this century. It was then that giant strides in our knowledge and understanding of the nucleic acids were made.

The turning point came in the early 1950s when James Watson and Francis Crick proposed a structure for **deoxyribonucleic acid (DNA)**. The discovery of how exact copies of DNA are made in living cells soon followed and its biological significance was quickly realised. An even greater leap forward was the unravelling of the genetic code, the 'secret of life' as it has been called. We shall see in this chapter that important steps continue to be made in unlocking the secrets of nucleic acids, and which promise to be of enormous benefit to mankind.

4.1 Nucleic acids

Deoxyribonucleic acid (DNA) and **ribonucleic acid (RNA)** are the two main kinds of nucleic acid. DNA is found mainly in the nuclei of cells, with small amounts in mitochondria and chloroplasts. RNA occurs mainly in the cytoplasm, particularly at the ribosomes.

Both DNA and RNA are polymers, and their building blocks are called nucleotides. For this reason the nucleic acids are described as **polynucleotides**.

4.1.1 Deoxyribonucleic acid

A **nucleotide** is made up of three parts, a **pentose sugar**, a **nitrogenous base** and **phosphate**. In DNA the sugar is **deoxyribose** (Fig 4.1). Four different nitrogenous bases occur in DNA (Fig 4.2). **Adenine (A)** and

Fig 4.1 Molecular structure of deoxyribose

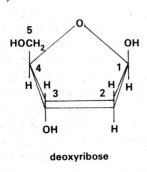

deoxyribose

Fig 4.2 Nitrogenous bases of DNA

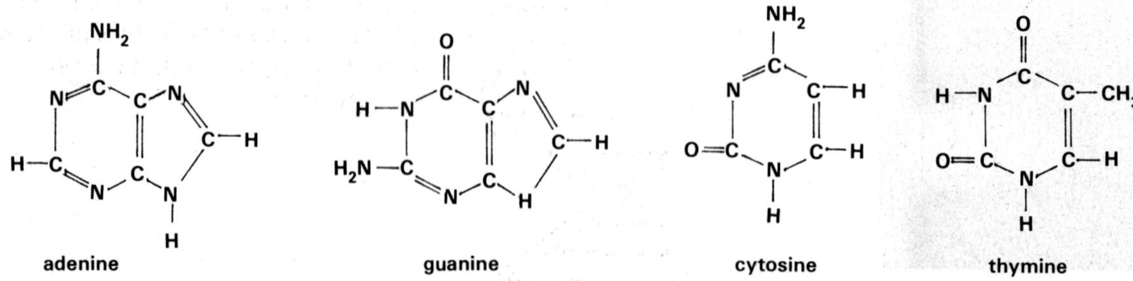

adenine   guanine   cytosine   thymine

guanine (G) belong to a group of compounds called **purines**. **Cytosine (C)** and **thymine (T)** are **pyrimidines**. The sugar, nitrogenous base and phosphate are bonded as shown in Fig 4.3 to form a nucleotide. In a DNA molecule the phosphate groups join nucleotides together by their sugar molecules (Fig 4.4). Hence DNA is a polynucleotide.

Fig 4.3 A nucleotide   **Fig 4.4** Part of a polynucleotide

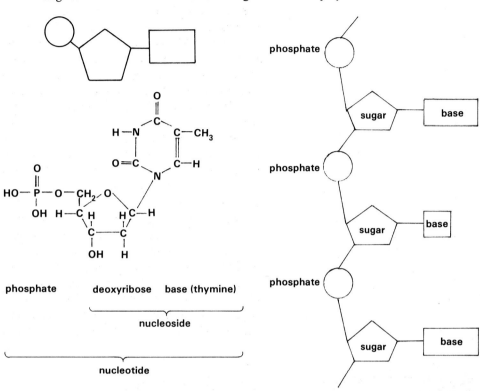

During the early 1950s biochemists established that the number of adenine molecules in DNA from different organisms is the same as the number of thymine molecules. The ratio of cytosine and guanine molecules is also 1:1. However, the ratio of $(A + T):(C + G)$ varies according to the source of the DNA. At this time Rosalind Franklin was working on the structure of DNA using X-ray crystallographic techniques (Chapter 2). Her results were part of the evidence which, in 1953, led James Watson and Francis Crick to propose that a DNA molecule consists of two polynucleotide strands each coiled in a **right-handed helix**. The two strands of the double helix are held together by **hydrogen bonding** between the nitrogenous bases of adjacent nucleotides. For each complete twist of the double helix there are 10 pairs of nucleotides, each twist measuring 3.4 nm in length (Fig 4.5).

Fig 4.5 Watson–Crick model of DNA

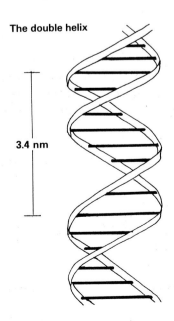

The double helix

3.4 nm

It is important to appreciate that the two polynucleotide strands of a DNA molecule are not identical. They are **complementary**. Where adenine occurs in one strand, thymine is found in the other. Where cytosine appears in a strand, guanine is present in the complementary strand (Fig 4.6(a)). Watson and Crick were aware that such a structure could explain how exact copies of a DNA molecule can be made (section 4.1.4). Altogether there can be several thousand pairs of nucleotides in a single molecule of DNA. The exact number depends on the origin of the DNA. The sequence in which the pairs of bases occurs also differs in DNA from different species (Fig 4.6(b)). In theory the number of permutations of the bases is infinite.

Fig 4.6 (a) Pairing of the bases in the polynucleotides of DNA

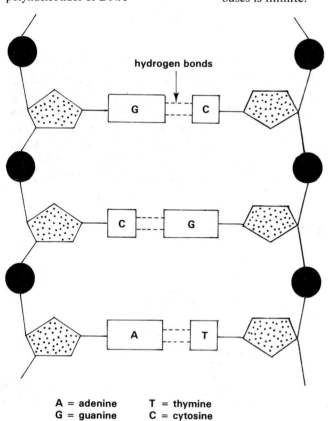

A = adenine   T = thymine
G = guanine   C = cytosine

Fig 4.6 (b) Sequence of base pairs in DNA from two species

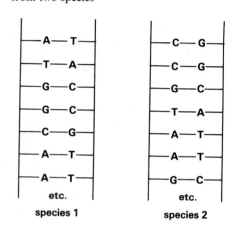

species 1   species 2

4.1.2 Ribonucleic acid

There are three different kinds of ribonucleic acid, **messenger RNA (mRNA), transfer RNA (tRNA)** and **ribosomal RNA (rRNA)**. The nucleotides of which they are made contain the sugar **ribose**, not deoxyribose as in DNA. Another difference is that thymine is not found in RNA. The nitrogenous base **uracil (U)** is present instead (Fig 4.7).

Fig 4.7 Molecular structure of ribose and uracil

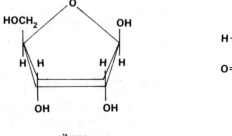

ribose

uracil

1 Messenger RNA

Between 3 and 5% of the RNA in a cell is **mRNA**. The molecules of mRNA are single, helical strands made of up to several thousand nucleotides (Fig 4.8). Messenger RNA is made in the nucleus. From there it passes through the nuclear membrane to the ribosomes where triplets of bases in the mRNA act as **codons** in the synthesis of proteins (section 4.2.1).

Fig 4.8 Messenger RNA. There are as many different kinds of mRNA as there are polypeptides. They differ in the sequence of codons in their molecules

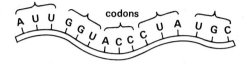

2 Transfer RNA

Transfer RNA makes up between 10 and 15% of a cell's RNA content. The single strand of 75–90 nucleotides which make up a **tRNA** molecule is wound into a double helix which usually has three prominent bulges (Fig 4.9). One of the free ends of every tRNA molecule ends with nucleotides containing the following order of bases **A←C←C←**, with **A** at the very end. There are at least twenty different kinds of tRNA. They differ in the sequence of base triplets making up the **anticodons** by which tRNA binds to the codons of mRNA during the synthesis of proteins (section 4.2).

Fig 4.9 Transfer RNA

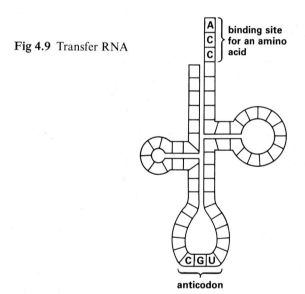

3 Ribosomal RNA

The many thousands of nucleotides which make up a molecule of **rRNA** are wound into a complex structure consisting partly of single and partly of double helices. Ribosomal RNA is made in the nucleus under the control of the nucleoli. It enters the cytoplasm and binds with protein molecules to become ribosomes. Over half the mass of a ribosome consists of rRNA and it makes up more than 80% of the total RNA in a cell. Even so, the precise function of rRNA is still not known.

4.1.3 DNA as the hereditary substance

In 1928 Fred Griffith, an English bacteriologist, noted two distinct strains of the bacterium *Pneumococcus*. Each bacterium of the **S-strain** is enclosed in a capsule, and when inoculated on to an agar-based medium it grows into a smooth, glistening colony. The other, called the **R-strain**, lacks a capsule and its colonies have a rough, dull appearance. Griffith also observed that mice injected with the S-strain soon developed pneumonia and died. The R-strain is non-pathogenic and mice injected with it showed no symptoms of illness. However, when heat-killed S-pneumococcus was mixed with the R-strain and then injected into the mice, the animals died of pneumonia. Moreover Griffith was able to isolate live S-pneumococcus from the mice subjected to this treatment (Fig 4.10). He was, however, unable to explain his findings.

Fig 4.10 Griffith's experiment

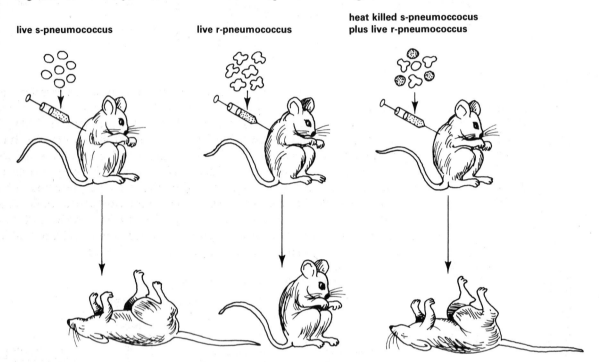

live s-pneumococcus

live r-pneumococcus

heat killed s-pneumoccocus
plus live r-pneumococcus

The reason for the peculiar behaviour of Griffith's *Pneumococcus* was given by Avery in 1943. Avery reported that the non-pathogenic R-strain could be converted to the S-strain by simply adding DNA from the S-strain to the medium in which the R-strain was growing. Furthermore, the newly-acquired characteristic was permanent and was passed to subsequent generations when the bacterium reproduced. Avery gave the name **transformation** to the process whereby the R-strain was converted to the S-strain. Here was a convincing piece of experimental evidence to suggest that DNA is the substance responsible for the characteristics of living organisms.

4.1.4 Replication of DNA

Watson and Crick's proposed double helix provided a strong hint as to how exact copies of DNA are normally produced, but it was a few years later before there was any experimental proof.

The hydrogen bonds between the base pairs break and the two polynucleotide strands unwind. Each of the strands acts as a molecular mould or **template** on which a new polynucleotide strand is made. The new strands are complementary to the original strands (Fig 4.11). Synthesis of DNA is controlled by an enzyme called **DNA ligase**. The process gives rise to two molecules of DNA identical to each other and which are replicas of the original molecule. This mode of DNA **replication** is described as **semi-conservative**. Compare it with conservative replication as in Fig. 4.12.

Fig 4.11 Replication of DNA

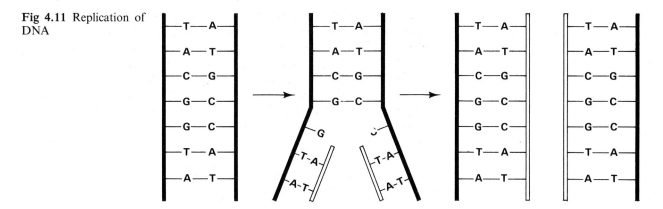

Fig 4.12 Semi-conservative and conservative replication

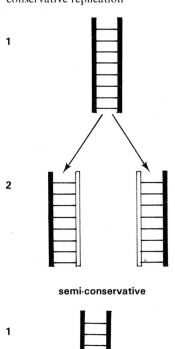

1

2

semi-conservative

1

2

conservative

Refer to Fig 4.12 and make drawings to show the result of replication of the DNA molecules shown in stage 2 in both cases.

Experimental evidence that DNA replication is semi-conservative was obtained in 1957 by Meselson and Stahl. They grew the bacterium *Escherichia coli* for several generations in a medium containing $^{15}NH_4Cl$ as the only source of nitrogen. The DNA of the bacterium was thus **labelled** with ^{15}N, the heavy isotope of nitrogen. The bacteria were then transferred to a medium in which $^{14}NH_4Cl$ was the only source of nitrogen. As *E. coli* multiplied the density of its DNA was measured. After one generation the density of the DNA was intermediate between DNA of the bacterium made with the heavy isotope ^{15}N and DNA made using ^{14}N, the normal isotope of nitrogen (Fig 4.13). The result was possible only if the DNA of *E. coli* had replicated in a semi-conservative way.

Fig 4.13 Meselson's and Stahl's experiment

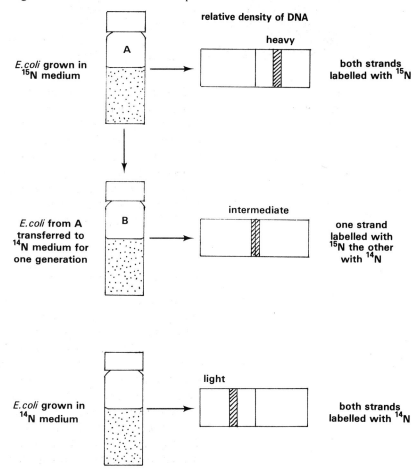

4.2 Protein synthesis

If amino acids, the building blocks of proteins, are labelled with a radio nuclide and are fed to rats, the ribosomes of the rats' liver cells soon become radioactive. Furthermore, peptide links are formed between the amino and carboxyl groups of amino acid molecules incubated with a suspension of ribosomes, RNA and ATP. Evidence of this sort indicates that the ribosomes are the sites of protein synthesis. But what parts do RNA and ATP play in the synthesis of proteins?

Explain how amino acids which are fed to rats soon appear in their livers.

4.2.1 Role of the nucleic acids

Proteins are polymers consisting of one or more polypeptides in which the amino acids are joined together in a specific linear sequence and folded in a particular three-dimensional way (Chapter 2). Many different kinds of protein can be made inside a living cell. The accuracy with which ribosomes repeatedly assemble such complex molecules suggests that some sort of control system is at work inside cells coding the production of proteins. The code, called the **genetic code**, is the sequence of nitrogenous bases in DNA. How is information contained in DNA conveyed to and interpreted by the ribosomes?

In working out the mechanism of protein synthesis, molecular biologists were faced with the problem of explaining how the nucleotides found in DNA could provide a code for the synthesis of many different kinds of protein. Single nucleotides can hardly be the basis of the code as there are just four different nucleotides but at least twenty different amino acids. Neither does it seem logical that pairs of nucleotides e.g. AA are

Table 4.1 The genetic code

UUU	Phe	UCU	Ser	UAU	Tyr	UGU	Cys
UUC	Phe	UCC	Ser	UAC	Tyr	UGC	Cys
UUA	Leu	UCA	Ser	UAA	Term.	UGA	Term.
UUG	Leu	UCG	Ser	UAG	Term.	UGG	Trp
CUU	Leu	CCU	Pro	CAU	His	CGU	Arg
CUC	Leu	CCC	Pro	CAC	His	CGC	Arg
CUA	Leu	CCA	Pro	CAA	Gln	CGA	Arg
CUG	Leu	CCG	Pro	CAG	Gln	CGG	Arg
AUU	Lle	ACU	Thr	AAU	Asn	AGU	Ser
AUC	Lle	ACC	Thr	AAC	Asn	AGC	Ser
AUA	Lle	ACA	Thr	AAA	Lys	AGA	Arg
AUG	Met	ACG	Thr	AAG	Lys	AGG	Arg
GUU	Val	GCU	Ala	GAU	Asp	GGU	Gly
GUC	Val	GCC	Ala	GAC	Asp	GGC	Gly
GUA	Val	GCA	Ala	GAA	Glu	GGA	Gly
GUG	Val	GCG	Ala	GAG	Glu	GGG	Gly

How many amino acids are shown in Table 4.1?

Term. = termination codon

responsible as this would give only $4^2 = 16$ possible combinations of two nucleotides. It was therefore deduced that combinations of at least three nucleotides e.g. AGA are required. The **triplet code** provides $4^3 = 64$ different permutations from the four nucleotides. The term triplet refers to the fact that the basic coding unit (codon) for an amino acid is a sequence of three nucleotides.

In the early 1960s Nirenberg and his colleagues began to experiment with the production of artificial polypeptides using ribosomes extracted from *Escherichia coli*. When incubated with a mixture of amino acids, ATP and synthetically-made mRNA containing only the nitrogenous base uracil, the ribosomes synthesised a polypeptide made solely from the amino acid called phenylalanine. The triplet code for phenylalanine is therefore **UUU**. Nirenberg also discovered that **AAA** is the triplet code for lysine and **CCC** for proline. Since then the triplet codes for all twenty amino acids have been worked out. Methionine and tryptophan each have a single triplet code, the other eighteen amino acids have more than one triplet permutation. Some have just two alternative triplets, others have up to six. For this reason the genetic code is described as **degenerate**. A few of the sixty-four triplets, called **non-sense triplets**, do not code for any amino acid (Table 4.1). They signal the termination of synthesis of a polypeptide.

Protein synthesis takes place in three distinct stages.

1 Transcription

Before a protein is synthesised, mRNA is made inside the nucleus. The double helix of DNA unwinds and single-stranded mRNA molecules are produced, part of one of the DNA strands acting as a template (Fig 4.14). Remember that the mRNA is complementary to its DNA template. In this way the genetic code is **transcribed** as the sequence of bases in mRNA. Synthesis of mRNA is controlled by the enzyme **RNA ligase**. The mRNA now passes through the pores of the nuclear membrane into the cytoplasm. Here one end of the mRNA molecule becomes attached to a ribosome.

Fig 4.14 Transcription

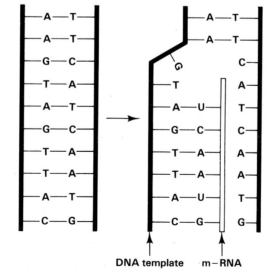

DNA template   m-RNA

2 Activation

During **activation**, energy from ATP (Chapter 5) is used to combine tRNA molecules with amino acid molecules. There are at least twenty different kinds of tRNA, the important difference between them being the sequence of nitrogenous bases in their anticodons (section 4.1.2). Each type of tRNA binds with a specific amino acid. The amino acid molecules join to the free ends of the tRNA molecules where bases A←C←C← are found (Fig 4.15). This rules out the possibility that the anticodon determines the amino acid with which the tRNA combines. The tRNA–amino acid complexes now move to the ribosomes.

Fig 4.15 Activation

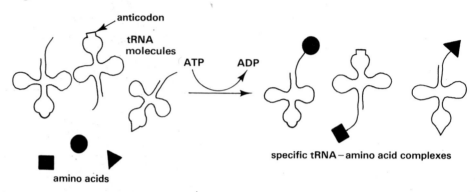

3 Translation

It is now necessary for the transcribed genetic code to be **translated**. A knowledge of protein structure (Chapter 2) is assumed in the following description of **translation**. Starting at one end of an mRNA molecule, a ribosome works its way along, positioning the **anticodon** of each tRNA molecule on to a complementary **codon** of the mRNA strand (Fig 4.16).

Fig 4.16 Translation

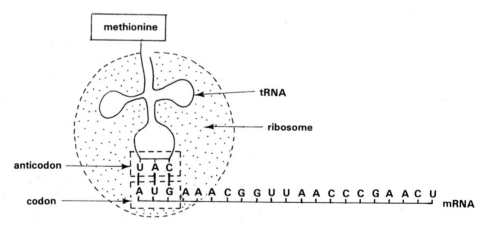

For example, the triplet code for methionine, the first amino acid in a polypeptide chain, is AUG. The tRNA–methionine complex having the anticodon UAC is brought to the end of the mRNA molecule where the codon AUG is located. The ribosome binds the codon and anticodon, then moves on to the next triplet of the mRNA strand. Here **codon–anticodon binding** again takes place and the next amino acid molecule is brought into position. A peptide link is formed with methionine and the polypeptide chain begins to take shape. As soon as an amino acid is linked to its neighbour its tRNA partner is released back into the cytoplasm to pick up another molecule of the same amino acid (Fig 4.17). The ribosome continues to work its way along the mRNA strand until it reaches a triplet for which there is no anticodon. This is the signal that the polypeptide chain is complete. The part of a DNA molecule which codes the synthesis of a polypeptide is usually called a **gene**.

Fig 4.17 Polypeptide synthesis

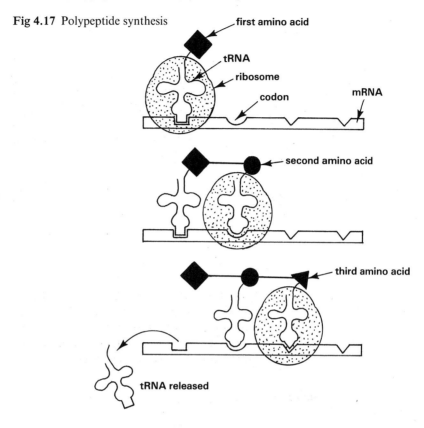

Polysomes, groups of ribosomes connected by a common strand of mRNA, are often seen in cells (Chapter 6). This arrangement may mean that several polypeptides are made at the same time on one mRNA molecule (Fig 4.18). When synthesis is completed, the polypeptides are moved from the ribosomes to the cytoplasm and constructed into proteins for internal use by the cell or for secretion.

Fig 4.18 Polypeptide synthesis by a polysome

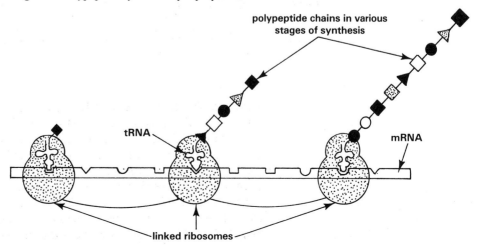

4.2.2 Gene mutations

Sometimes mistakes arise when DNA replicates. There are several ways in which errors can occur. One pair of bases may be replaced by another, for example C:G may be replaced by A:T. A pair of nucleotides may be lost or an extra pair added (Fig 4.19). Whatever the cause, the new DNA is not an

Fig 4.19 Gene mutations

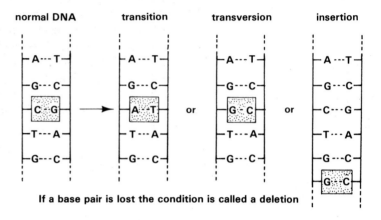

exact copy of the original. Such changes are called **gene mutations**. When a gene mutates the changes in sequence of the bases in DNA causes a complementary change in base sequence in a codon of mRNA. The altered codon may be translated as **non-sense**, causing synthesis of a polypeptide with one or more amino acids missing. Alternatively the codon is translated as **mis-sense** in which case another amino acid is substituted in or added to the polypeptide chain (Fig 4.20). Protein molecules built from such polypeptides are usually defective and cannot carry out their normal functions. **Abnormal haemoglobins** and **inborn errors of metabolism** are examples of the consequences of gene mutations.

Fig 4.20 Some consequences of gene mutations

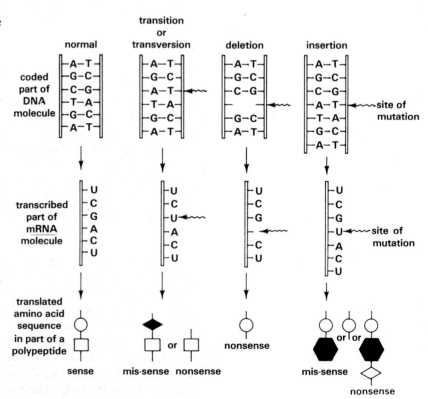

1 Abnormal haemoglobins

People who have **sickle cell disease** produce haemoglobin S (HbS) instead of HbA. In tissues where there is a low oxygen tension, HbS molecules become rigid chains, causing the red blood cells to become sickle-shaped. Such cells can block vital blood vessels. Haemoglobin S also has a lower affinity for oxygen than HbA (Chapter 9). In each of its β-globin chains HbS has the amino acid valine instead of glutamic acid (Fig 4.21). Understanding the genetic code now makes it possible to explain the biochemical difference between HbS and HbA. The codon for glutamic acid is GAG (Table 4.1) which corresponds to the triplet CTC in the DNA template. If the triplet is changed to CAC, the corresponding codon in mRNA becomes GUG which is the codon for valine. The kind of mutation seen here is **transversion**. The effect is that HbS is synthesised instead of HbA.

Fig 4.21 Chromatograms of haemoglobins. The haemoglobins were first broken into peptide fragments by enzyme action, then subjected to two-way chromatography

normal haemoglobin

haemoglobin S

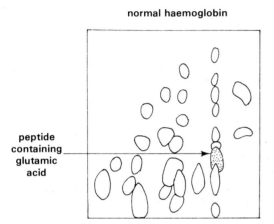

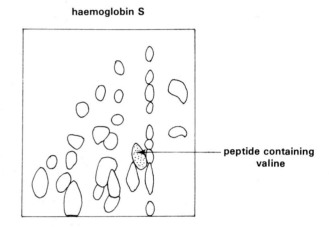

peptide containing glutamic acid

peptide containing valine

2 Inborn errors of metabolism

A number of human diseases can be attributed to the inability to synthesise specific enzymes or to the synthesis of enzymes which have reduced catalytic powers.

A. Phenylketonuria

In the liver of a normal person the essential amino acid phenylalanine is converted to another amino acid tyrosine by the enzyme **phenylalanine hydroxylase**. People with **phenylketonuria (PKU)** cannot bring about the conversion. It is assumed that this is because they cannot synthesise an active form of the enzyme. During foetal life it is not a problem because the mother's liver carries out the reaction, but after birth phenylalanine begins to accumulate in the child's blood. Some phenylalanine is converted to phenylpyruvic acid which is excreted in the urine. However, the remaining phenylalanine prevents the child's brain from absorbing sufficient amounts of other essential amino acids from its blood. As a result the brain and other organs and tissues such as muscles and cartilage fail to grow and develop normally.

Unless PKU is detected soon after birth, the child usually becomes mentally retarded and cannot walk properly. At one time children with PKU were identified by testing their urine for phenylpyruvic acid with iron(III) chloride, $FeCl_3$. However, the acid is not in sufficient quantities for diagnosis until the child is several weeks old. By that time brain damage may have already begun. These days a blood test makes it possible to identify children with PKU at birth. They can then be fed a special diet containing just enough phenylalanine to allow normal growth without it accumulating in the body. The diet can be stopped at about nine years of age when the brain should be fully developed. Females with PKU must return to the diet when they become pregnant. If they do not, the high concentration of phenylalanine in their blood would damage the brain of the developing foetus.

B. Alkaptonuria

The urine of people with **alkaptonuria** turns black when exposed to light. Chemical analysis shows the urine to contain a high concentration of **homogentisic acid**, also called **alkapton**. The acid is produced in the liver from excess amounts of tyrosine. In normal people the breakdown of alkapton to carbon dioxide and water is catalysed by enzymes. In people with alkaptonuria, failure to produce one of the enzymes causes alkapton to accumulate. In early life there are no apparent ill-effects. Later on, alkapton is deposited in cartilage causing the tip of the nose and the pinnae of the ears to turn black. A painful form of arthritis often accompanies these changes for which there is no known cure.

C. Galactosaemia

Children with **galactosaemia** are often apparently normal when they are born, but within a few weeks they begin to vomit much of the milk they drink and fail to thrive. If the condition is not diagnosed early the child may become blind and mentally retarded. Galactosaemia occurs when a child's liver fails to produce the enzyme which catalyses the conversion of **galactose** to glucose. In the small intestine, lactose (milk sugar) is hydrolysed by the enzyme lactase to form galactose (Chapter 15). If this cannot be converted to glucose, it accumulates in the blood and causes the damage described earlier. The symptoms disappear if the condition is diagnosed early and the child is given a lactose-free diet. It is necessary to continue with the diet throughout life.

Such genetic defects may be cured in the not too distant future by **gene therapy** in which the faulty gene is replaced soon after birth or even prenatally with a correct one, thus enabling the body to make the normal enzyme. Diagnosing the precise nature of many genetic defects is possible using DNA probes (section 4.4).

4.2.3 Regulating protein synthesis

It is necessary for cells to regulate the amounts and types of proteins they make because enzymes and metabolites need to be produced only as and when required. The efficiency of metabolism is thus optimised and a steady state is maintained. One device which helps in **cellular homeostasis**, the maintenance of a steady state in living cells, is the temporary inhibition of an enzyme when the concentration of a substrate becomes too high (Chapter 3). Another is to regulate the synthesis of enzymes.

In the late 1950s two French scientists, François Jacob and Jacques Monod, reported that the bacterium *Escherichia coli* was able to produce several enzymes 'on demand'. *E. coli* can synthesise all the metabolites and enzymes it requires from a medium consisting of glucose, mineral salts and water. Jacob and Monod observed that if lactose was substituted for glucose, the *E. coli* makes an additional enzyme called β-D-galactosidase which hydrolyses lactose to glucose and galactose. When returned to the glucose medium the bacterium stops making β-D-galactosidase. The presence of lactose in the medium induces *E. coli* to make an enzyme capable of hydrolysing lactose. This is an example of **enzyme induction**.

E. coli can make all the amino acids it requires from a medium containing NH_4^+ as the only source of nitrogen. However, if the amino acid arginine is added to the medium the bacterium stops making the enzymes needed for the synthesis of arginine. In this case the presence of arginine in the medium inhibits the production of an enzyme, an example of **enzyme repression**.

Jacob and Monod produced a hypothesis to explain enzyme induction and repression (Fig 4.22). They proposed that the sequence of amino acids in an enzyme is determined by a **structural gene** which can be switched on or off by a **regulator gene**. The regulator gene transcribes the production of mRNA which has the codons to make a **repressor protein**. The repressor moves from the ribosomes where it is made to the nucleus. Here it binds with the structural gene, stopping it from transcribing the manufacture of mRNA used in the synthesis of an enzyme. As a consequence, production of the enzyme is repressed. However, if an inducer substance is present it combines with the repressor. The inducer–repressor complex is unable to prevent transcription of the structural gene. Synthesis of the enzyme can then proceed. The formation of the inducer–repressor complex must be reversible because enzyme repression occurs when the inducer is removed. Induction and repression are thus linked. Induction is freedom from repression.

Fig 4.22 The Jacob–Monod hypothesis

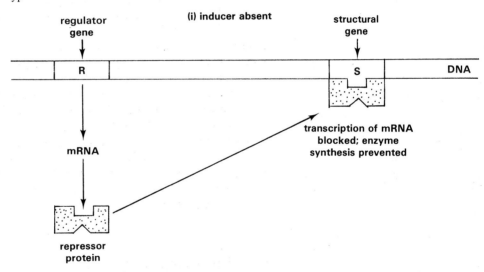

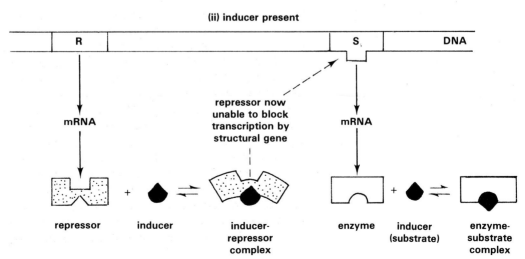

Recently it has been shown that induction and repression of enzymes also occurs in the cells of humans. Liver cells are particularly efficient at regulating the types and amounts of enzymes they produce. The liver can therefore adjust to the fluctuating levels of nutrients it receives from the gut (Chapter 15).

Enzyme induction and repression are also important in the differentiation of tissues in living organisms. How a one-celled fertilised egg grows into a multicellular body made up of a number of different types of tissue is a process about which a lot has yet to be learned. However, induction and repression of enzymes probably enables the various types of tissue to differentiate by allowing different combinations of genes to be expressed at different stages during growth and development. A good example of the differential action of genes is seen in the synthesis of haemoglobin. In the foetal stage of development we produce **foetal haemoglobin (HbF)**. Each molecule contains two α-globin and two γ-globin polypeptide chains. The haemoglobin we produce after birth has two α-globin and two β-globin chains in each molecule. This is haemoglobin A (HbA). Foetal haemoglobin has a higher affinity for oxygen than adult haemoglobin and is thus able to absorb oxygen across the placenta from haemoglobin in the mother's blood (Chapters 8 and 25).

A better understanding of gene regulation may make it possible to switch off defective genes and to switch on genes which are not functioning. In this way it may be possible to prevent production of abnormal haemoglobins and cure inborn errors of metabolism.

4.3 Genetic engineering

Avery in 1943 showed that it was possible to transform the genetic make-up (genome) of the R-strain of *Pneumococcus* by adding to its culture medium DNA from the S-strain (section 4.1.3). Since then sophisticated procedures have been devised whereby the genomes of experimental organisms such as bacteria and yeasts can be altered with great precision. The formation of new genomes by the isolation of nucleic acid molecules from a donor and their introduction into a host organism in which they do not occur naturally but in which they are able to replicate is known as **genetic engineering**. The new combination of nucleic acid so formed is called **recombinant DNA**.

4.3.1 Recombinant DNA technology

In recent years several important discoveries have led to a very rapid development of **recombinant DNA technology**. A stumbling block which hindered past progress was the fact that when foreign DNA is introduced into a host, the DNA is often broken down or restricted from replicating. However a strain of *Escherichia coli* has now been found which accepts non-self DNA from many sources, including humans and allows it to replicate. Another vital step forward has involved the identification of a group of about 300 enzymes called **restriction endonucleases** which are able to break double-stranded DNA from the donor at specific points. The same endonuclease is then used to fragment strands of DNA from the host. Because of the specificity of enzyme action, the base sequence at the ends of the DNA from both sources may be complementary. They are said to have **sticky ends**. Recombinant DNA is formed when the two samples of DNA are incubated with **DNA ligase**, an enzyme which catalyses binding of the pieces of DNA at their sticky ends (Fig 4.23).

Fig 4.23 Making recombinant DNA. Only a small length of DNA from host and donor are shown. Elsewhere the sequence of bases in the DNA from the two sources is different

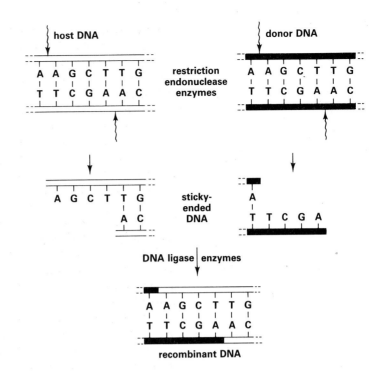

Fig 4.24 Gene cloning

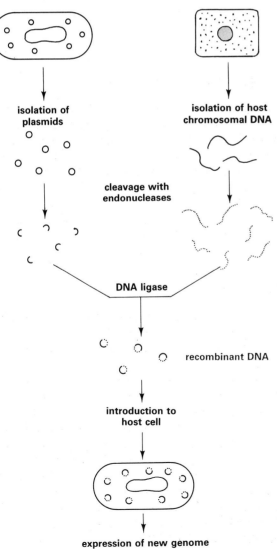

Although it is possible to transform host cells directly in this way, the process is often more successful when the donor DNA is introduced into the host in a **vector**. Bacteriophages (Chapter 28) and plasmids are commonly used as vectors. **Plasmids** are small circular double-stranded DNA molecules found in the cytoplasm of many kinds of bacteria. They can be isolated by lysing bacterial cells, then separating them from other cellular components by density gradient ultra-centrifugation (Chapter 6). Recombinant DNA is then prepared from the plasmid and host chromosomal DNA. When *E. coli* is the host, a culture is first suspended in a cold solution of 0.1 M CaCl$_2$ then warmed at 37°C for five minutes to allow the recombinant DNA to be taken up (Fig 4.24). The cells are then transferred to nutrient broth and incubated when they multiply and the recombinant DNA replicates (**gene cloning**). At the same time the transferred gene enables the host to produce a protein which hitherto it was unable to make. The protein may be an enzyme which the host uses internally to make a new product or it may be secreted into its surroundings. In this way the host acquires new inheritable characteristics. Using such techniques it is theoretically possible to clone the entire genome of any organism. For a complete human genome about 700 000 clones would be required!

4.3.2 Production of insulin by genetic engineering

In 1980 at Guy's Hospital London, seventeen volunteers were injected with insulin from an unusual source. They were the first humans to be treated with insulin produced by *E. coli*. Before this time, insulin for clinical use had come from the pancreas of slaughtered animals. Unsuccessful attempts had been made to culture insulin-secreting cells from the human pancreas as a means of increasing the amount of insulin in production. However it was found that the gene for insulin production could be transferred to *E. coli*. Even more exciting was the fact that when subsequently cultured the bacterium produced and secreted insulin into the culture medium. From there the hormone could be readily extracted and purified. How was it done?

A human cell possesses thousands of genes, hence the chance of finding and isolating the gene for insulin production is rather slim. At any given time only a few genes code the synthesis of proteins such as insulin, each of which uses an unique mRNA molecule for its transcription. It is therefore easier to isolate the mRNA which transcribes insulin production. There are 51 amino acid residues in each molecule of insulin (Fig 2.26). This requires a messenger RNA molecule containing $51 \times 3 = 153$ bases to code its production. Density gradient ultracentrifugation is used to separate this mRNA from molecules of other molecular mass in a homogenate of insulin-producing cells extracted from a human pancreas.

The next step is to use the mRNA as a template for synthesising single-stranded DNA. This is the reverse of the normal transcription process and is catalysed by **reverse transcriptase enzymes**. A complementary strand of DNA is then built on to the single-stranded DNA with the aid of **DNA polymerase enzymes**. The double stranded DNA so formed is called **copy DNA (cDNA)**. It is a copy of the gene for insulin production (Fig 4.25).

Fig 4.25 Production of copy DNA

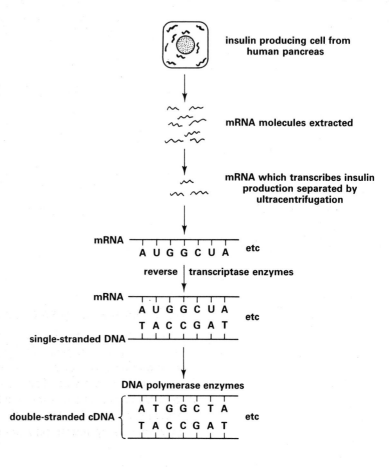

insulin producing cell from human pancreas

mRNA molecules extracted

mRNA which transcribes insulin production separated by ultracentrifugation

mRNA
A U G G C U A   etc

reverse | transcriptase enzymes

mRNA
A U G G C U A   etc
T A C C G A T

single-stranded DNA

DNA polymerase enzymes

double-stranded cDNA {
A T G G C T A   etc
T A C C G A T
}

Plasmids are now removed from *E. coli* cells and the cDNA inserted into them as described previously using restriction endonuclease and DNA ligase enzymes. The plasmids are then added to a culture of *E. coli*, whereupon some of the bacteria absorb the recombinant DNA containing the insulin-producing gene. As these bacteria multiply the gene is cloned. Only a few of the bacteria in a culture are transformed in this way. They can be identified using an immunological technique which relies on the ability of specific antibodies to recognise and to adhere to insulin molecules secreted by the producer bacteria. The insulin molecules are then made to bind to another set of radionuclide-labelled antibodies. When transferred to a photographic films the radionuclide causes the film to develop black spots. In this way the position of the insulin-secreting bacteria can be located (Fig 4.26). They are then transferred to a fresh medium and grown as a pure culture, the insulin-producing efficiency of which is carefully determined.

Fig 4.26 Identification of insulin-secreting cells of *E. coli*

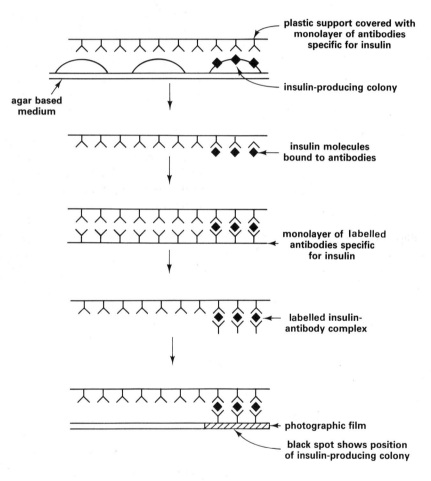

4.3.3 Other examples of genetic engineering

When attacked by viruses, mammalian tissues produce interferon, an antiviral agent (Chapter 10). Until recently it has not been possible to manufacture adequate quantities of interferon for treating humans suffering from viral diseases. The very small amounts available came mainly from leucocytes taken from blood given by donors. However, within the past few years molecular geneticists have succeeded in incorporating the gene for making interferon from human leucocytes into the DNA of the bacterium *Escherichia coli*. When *E. coli* containing the gene is grown in laboratory conditions it secretes interferon into the medium. Mass culture of *E. coli* with recombinant DNA has thus substantially increased the manufacture of an important life-saving drug. Comparable procedures have been used to make anti-haemophilic factor, a variety of antibodies, and human growth hormone. An important advantage of such products is that they do not include the impurities which are present in preparations made by traditional means. For example many haemophiliacs have contracted AIDS on being transfused with blood products derived from an infected donor.

Another field in which genetic engineering promises immense benefits to the human race is in crop improvement. For example, the roots of leguminous plants form a symbiotic relationship with the bacterium *Rhizobium leguminosarum* (Fig 31.30). The bacterium can fix nitrogen gas and some of the organic nitrogenous substances it makes are used by the leguminous host plants. The hosts as a result grow more profusely. Because legumes, such as peas, beans and clover, are extensively grown crop plants, the symbiotic relationship is of great importance to food production. But what of the many other types of crop plant which do not have this facility? The soil in which they are grown has to be treated with expensive artificial fertilisers or with manure (Chapter 14). Clearly it would be advantageous if non-legumes too could fix gaseous nitrogen which is plentiful in the atmosphere. For this reason attempts have been made to transfer the genes for nitrogen fixation into the cells of cereals and other non-leguminous plants.

There is, however, another side to genetic engineering. Concern has been expressed about the danger of producing organisms whose new genomes could pose a threat to the well-being of humans. One of the organisms widely used in such experiments is the gut bacterium *Escherichia coli* (Fig 4.27) which lives commensally in the bowels of humans and other

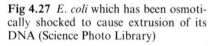

Fig 4.27 *E. coli* which has been osmotically shocked to cause extrusion of its DNA (Science Photo Library)

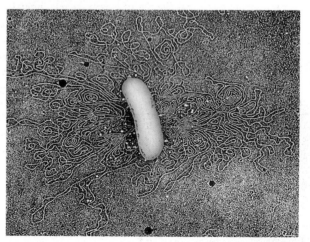

vertebrates. What would be the consequence if a strain of genetically-engineered *E. coli* which was highly pathogenic became resident in a human population? As Dr Erwin Chargaff, an eminent molecular geneticist has put it:

'You cannot recall a new form of life. Once you have constructed a viable *E. coli* cell into which a piece of eukaryotic DNA has been spliced, it will survive you and your children and your children's children. An irreversible attack on the biosphere is something so unheard of, so unthinkable to previous generations, that I could only wish that mine had not been guilty of it.'

The potential hazards of such experiments have led to a call for strict precautions in certain types of genetic engineering. In Britain the Genetic Manipulation Advisory Committee was set up to identify the precautions required to prevent the escape of genetically engineered microbes and to suggest the kinds of organisms to be used. These days most genetic engineers agree that the benefits which may accrue from such experiments greatly outweigh the potential risks which are generally thought to be negligible. Although genetic engineering has been going on for almost a decade, no serious problems have yet been encountered.

4.4 Genetic fingerprinting

Human DNA contains repetitive pieces called **minisatellites** which are distributed in a unique way in every individual. It is comparable to a fingerprint and, like fingerprints, can be used to identify a person with 100% certainty.

The distribution pattern of minisatellites in human DNA is ascertained using a **DNA probe**. The probe is prepared from ^{32}P labelled viral DNA broken into sticky-ended fragments using specific **restriction endonuclease enzymes**. The same enzymes are also used to cut up human DNA extracted from body tissues such as blood, semen or hair. The fragments of human DNA are first separated by electrophoresis in an agar gel and blotted onto a thin nitrocellulose filter. The filter is then washed with a solution containing the probe, when both types of DNA adhere by their sticky ends. The positions of the radioactive complexes are finally determined by placing the filter on an X-ray film. Radiation emitted by the hybrid DNA causes dark bands to appear on the film (Fig 4.28). The banding pattern is different for every human other than identical twins.

Minisatellites are inherited, children having some from their mother and some from their father. The genetic fingerprint of each child is however different. Hence the results of the technique can can provide strong evidence in resolving cases of disputed parenthood. Linked genes on human chromosomes can also be detected using DNA probes.

Genetic fingerprinting also has applications in forensic science. Already several serious crimes including murder have been solved by analysing the DNA of the criminal and matching the results obtained from blood and semen left at the scene of the crime.

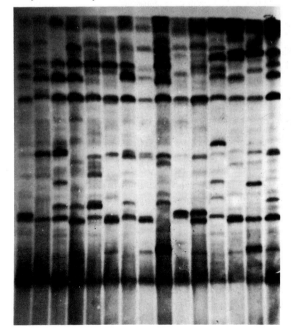

Fig 4.28 Genetic fingerprints of fifteen humans (courtesy R. Dainty). Note that each pattern of banding is unique for each person

SUMMARY

Nucleic acids are the substances which enable living organisms to inherit their characteristic features. There are two main kinds, deoxyribonucleic and ribonucleic acids. Both are polymers made of a pentose sugar, phosphate and nitrogenous bases. DNA is a double stranded polymer in which the bases occur in specific pairs, adenine with thymine and cytosine with guanine. Differences between organisms are substantially due to differences in the sequence of base pairs. Copies of DNA are made by semi-conservative replication, one of the strands serving as a molecular template on which the complementary strand is built.

DNA provides a triplet genetic code for regulating protein synthesis in which there are two key steps, transcription and translation. Transcription is a means of re-writing the base sequence of DNA as a sequence of bases in messenger RNA. Translation of the re-written code into the sequence of amino acids in a polypeptide involves the use of transfer RNA. The sequence of bases in DNA which codes the synthesis of a polypeptide is a gene. Changes to the code, gene mutations, result in the synthesis of unusual proteins e.g. sickle cell haemoglobin.

Gene induction and repression are ways of regulating protein synthesis and they contribute to cellular homeostasis. Genetic engineering is at the forefront of many new and major developments in the field of biotechnology. It promises substantial benefits to mankind especially in drug production and crop improvement.

QUESTIONS

1 Copy the following account of DNA and protein synthesis, filling in the spaces with the most appropriate word or words to complete the account.
The DNA molecule is composed of _____ sugar, phosphoric acid and four types of _____ base. Within this molecule the bases are arranged in pairs held together by _____ bonds. For example, adenine is paired with _____ . Adenine and guanine are examples of a group of bases called _____. The two strands of nucleotides are twisted around one another to form a double _____, and in each turn of the spiral there are _____ base pairs.
The DNA controls protein synthesis by the formation of a template known as _____. Compared with DNA, the sugar component of this template is _____, the base _____ occurs instead of _____ and the molecule consists of a _____ chain of nucleotides. The template is stored temporarily in the _____, before passing out into the cytoplasm through the _____. It becomes associated with organelles called _____, which supply the _____ required for protein synthesis. Transfer RNA molecules, each with an attached _____, are lined up on the surface of the template according to their _____ of _____ bases. The amino acids are joined in a chain by _____ links to form a polypeptide molecule. (L)

2 (a) Biochemical analysis of a sample of DNA showed that 33 % of the nitrogenous bases was guanine. Calculate the percentage of the bases in the sample which would be adenine. Explain how you arrived at your answer.
(b) What name is given to the triplet of three bases which designates individual amino acids?
(c) If the triplet of mRNA bases which designates the amino acid lysine is AAG (where A = adenine and G = guanine), what is the complementary triplet of three bases on the tRNA molecule? Give a key for the letters that you use. (AEB 1984)

3 The diagram shows the information flow in protein synthesis.

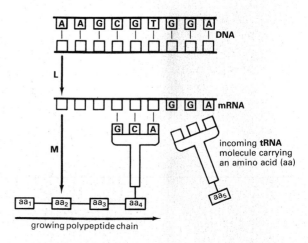

growing polypeptide chain

(a) Write in the boxes in the diagram the missing initial letters of the bases on the (i) DNA strand, (ii) mRNA.
(b) (i) Write in the boxes in the diagram the missing initial letters of the three bases on the incoming tRNA molecule.
(ii) What name is given to this triplet of bases on the tRNA molecule?
(c) Name the process shown by (i) arrow L, (ii) arrow M.
(d) In which organelle does the process shown by arrow M take place? (AEB 1987)

4 (a) Describe the modern ideas on the structure of DNA and the evidence for them.
(b) Explain how DNA is replicated. (WJEC)

5 Describe the structure, functions and interactions of cellular organelles involved in protein synthesis. (C)

5 Energy conversion in living cells

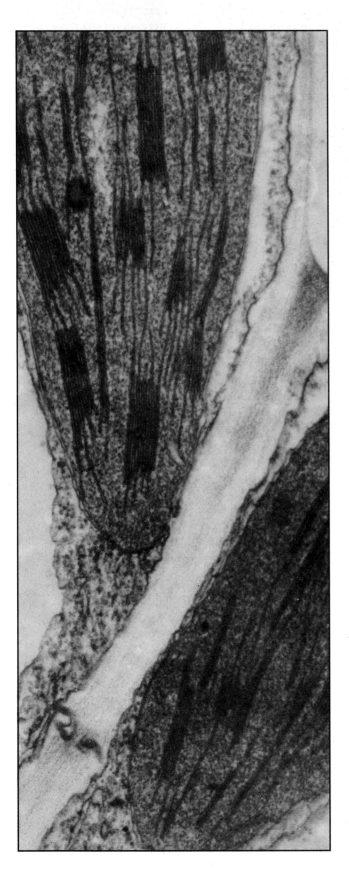

5.1 Energy changes in living cells   76

5.2 Photosynthesis   76

   5.2.1 Light-dependent reactions   76
   5.2.2 Light-independent reactions   82

5.3 Respiration   84

   5.3.1 Adenosine triphosphate   84
   5.3.2 Oxidation of respiratory substrates   85
   5.3.3 Energy yields from respiratory substrates   88
   5.3.4 Measuring the rate of respiration   90
   5.3.5 Respiratory quotient   91
   5.3.6 Basal metabolic rate   92

Summary   93

Questions   93

5 Energy conversion in living cells

Energy exists in many forms. Moving bodies such as atoms and molecules have **kinetic energy**. Light, heat, electricity and mechanical energy are all forms of energy which stem from atomic and molecular movement. The atoms in every molecule are joined to each other by electro- and covalent bonds. Every atom is held together by the electrostatic attraction between its nucleus and the cloud of electrons which orbit it. Both are examples of **potential energy**, a store of energy which is released when atoms and molecules are split.

One form of energy can be converted to another. For example, when we switch on an electric light, the bulb converts electrical energy to light. At the same time some of the energy appears as heat, causing the bulb to become hot. This familiar example is a good illustration of the two **laws of thermodynamics**. The first law states that in converting energy the amount obtained is the same as the amount used. With a light bulb, the amount of light and heat produced is equal to the amount of electricity used. The second law states that when energy conversion occurs some of the energy produced always appears as heat.

5.1 Energy changes in living cells

In biochemical reactions substrates are used and products formed. The amount of energy binding together the atoms of the product molecules is always different from that in the substrate molecules. This is in keeping with the second law of thermodynamics. The energy which binds atoms in molecules is called **free energy**. Hence every biochemical reaction is accompanied by a change in free energy, denoted by the symbol ΔG. The change in free energy is caused by alterations in the position of atoms in the reacting molecules. In energy-using (**endergonic**) reactions the product molecules have more free energy than those of the substrate and ΔG has a positive value. In contrast, the products of energy-releasing (**exergonic**) reactions have less free energy than the substrate molecules and ΔG is negative.

Biochemical rections are reversible and do not often occur in isolation. They are usually part of reaction sequences called **metabolic pathways**. The products of one reaction are the substrates for the next. So long as the free energy change for each reaction is known it is possible to predict the overall direction of the pathway (Fig 5.1).

Fig 5.1 Free energy changes in a metabolic pathway of four reactions

$$A + B \underset{-2000\,\text{kJ}}{\overset{\Delta G}{\rightleftharpoons}} C + D$$
$$D + E \underset{-3000\,\text{kJ}}{\overset{\Delta G}{\rightleftharpoons}} F + G$$
$$G + H \underset{+2000\,\text{kJ}}{\overset{\Delta G}{\rightleftharpoons}} I + J$$

(ΔG values are given per mole of product used in the next step)

The sum of the ΔG values is determined, in this case $-3000\,\text{kJ}\,\text{mol}^{-1}$. Because the value is negative the overall direction the pathway takes is from left to right. At equilibrium the concentration of A and B will be less than I and J. Thus energy using reactions can be made to occur when linked to energy-releasing reactions. In living cells, the link between most energy-using and energy-releasing reactions is **adenosine triphosphate ATP** (section 5.3).

5.2 Photosynthesis

The sun is the main source of energy available to the living world. Solar energy is converted to the potential energy of organic molecules by green plants in a complicated pathway of reactions called **photosynthesis**. One of the products of photosynthesis is the hexose sugar glucose. It has been calculated that 2880 kJ of solar energy is used in synthesising each mole of glucose:

$$6CO_2 + 6H_2O \xrightarrow{\Delta G + 2880\,kJ} C_6H_{12}O_6 + 6O_2$$

Photosynthesis consists of two main phases, the **light-dependent** and **light-independent** (dark) reactions.

5.2.1 Light-dependent reactions

White light is a mixture of light of different wavelengths (Fig 5.2). The energy of light is inversely proportional to its wavelength (measured in nanometres, nm), and so light of short wavelength has more energy than light of long wavelength. Green plants convert the energy of some wavelengths of visible light into the energy which bonds together the atoms of complex organic molecules.

Fig 5.2 The visible spectrum

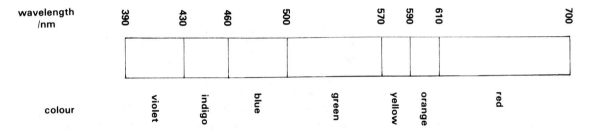

1 Absorption and action spectra

Green plants do not use energy from all of the components of white light for photosynthesis. This was neatly demonstrated by the German plant physiologist H. T. Engelmann in 1882. He placed filaments of the green alga *Cladophora* in a drop of water on a microscope slide. The filaments were illuminated with light of different wavelengths. He then watched the distribution of aerobic bacteria in the water. Engelmann noted that the bacteria clustered near to the filaments when blue light (450 nm) or red light (650 nm) was used (Fig 5.3).

Fig 5.3 Engelmann's experiment

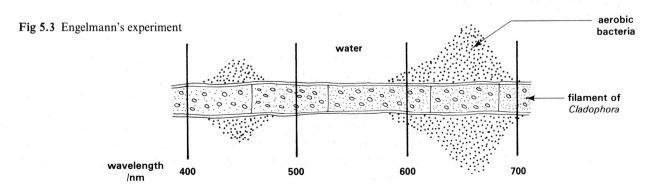

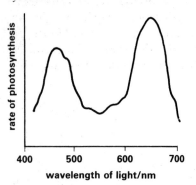

Fig 5.4 An action spectrum for photosynthesis

rate of photosynthesis

400   500   600   700
wavelength of light/nm

Knowing that the alga gave off oxygen as it photosynthesised, Engelmann deduced that blue and red light are the most effective for photosynthesis. It has since been shown that photosynthesis in nearly all green plants occurs most rapidly when they are illuminated by blue and red light.

An **action spectrum** is produced when the rate of photosynthesis is plotted against wavelength of light (Fig 5.4). The reason why light from the blue and red parts of the spectrum is effective is that it is absorbed efficiently by the pigments contained in the chloroplasts of green plants. The photosynthetic pigments most commonly present are the **chlorophylls** and **carotinoids** (Table 5.1). They can be readily extracted by grinding up chopped leaves in an organic solvent such as propanone. The pigments in the extract can then be separated by paper chromatography (Fig 5.5).

Table 5.1 Photosynthetic pigments commonly found in leaves of green plants

Group	Pigment	Colour
Chlorophylls	chlorophyll *a*	green
	chlorophyll *b*	green
Carotinoids	xanthophyll	yellow
	carotene	orange

Fig 5.5 Chromatographic separation of photosynthetic pigments

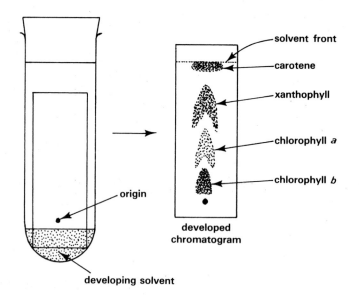

solvent front

carotene

xanthophyll

chlorophyll *a*

chlorophyll *b*

origin

developed chromatogram

developing solvent

Fig 5.6 Structure of a chlorophyll molecule

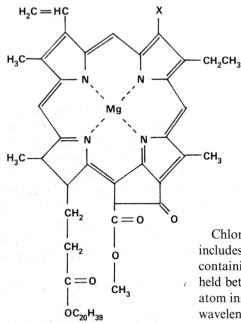

In chlorophyll *a* X = CH_3; in chlorophyll *b* X = CHO

empirical
formula $C_{55}H_{72}O_5N_4Mg$ $C_{55}H_{70}O_6N_4Mg$

Chlorophylls belong to a group of chemicals called **porphyrins**, which includes cytochromes and haemoglobin. They all have four nitrogen-containing pyrrole rings (Fig 5.6). In the chlorophylls a magnesium atom is held between the rings. In cytochromes and haemoglobin there is an iron atom instead. Chlorophylls are green because they reflect green light. The wavelengths they absorb most strongly are in the blue and red parts of the visible spectrum.

Carotene and xanthophyll are long-chain **hydrocarbons** (Fig 5.7). Their colours indicate that they reflect orange and yellow light respectively. If the percentage light absorption is plotted against wavelength of light, an **absorption spectrum** is obtained for each pigment (Fig 5.8). Notice the close correlation between the absorption and action spectra. Chlorophylls absorb both red and blue light, whereas the carotinoids absorb mainly blue light.

Fig 5.7 Structure of a β-carotene molecule

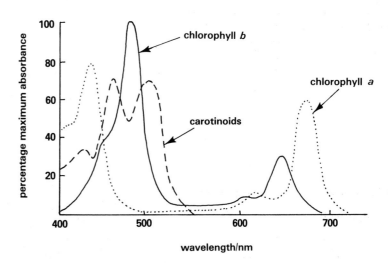

Formula of carotene: $C_{40}H_{56}O_2$

Fig 5.8 Absorption spectra of chloroplast pigments

2 Electron transfer in chloroplasts

What happens when chloroplast pigments absorb light? One clue comes from the behaviour of the pigments extracted from leaves in organic solvents such as ethanol and propanone. When the solution is placed in white light the pigments give off red light. This is called fluorescence. Before illumination, the electrons in the outer shell of magnesium atoms in the chlorophyll molecules are in the ground-state energy level. Light energy temporarily raises the energy level, causing the electrons to become displaced. In their high-energy state the electrons are said to be **excited**. Red light appears during fluorescence because, shortly after being displaced, the electrons fall back into the ground state. The extra energy they held is emitted as red light.

What has just been described is of course an experimental observation. In living tissues the solvent is water and not an organic solvent. In nature, we do not see green plants emitting red light when they photosynthesise. Also, green plants give off oxygen during photosynthesis. These facts strongly suggest that in living plants the energy of the excited electrons in chlorophyll molecules is linked to other reactions that involve the release of oxygen. Where does the oxygen come from, and what happens to the excited electrons?

Some idea of what happens to the excited electrons was established in 1937 by the English biochemist Robin Hill. He observed an illuminated suspension of chloroplasts in water evolving oxygen, and at the same time transferring electrons to iron(III) ions, reducing them to iron(II) ions. This phenomenon, called the **Hill reaction**, suggests that chloroplasts contain acceptors of the excited electrons from chlorophyll molecules. In the 1950s it was discovered that the main acceptor of these excited electrons was **nicotinamide-adenine dinucleotide phosphate (NADP)**.

3 Photolysis

Having given electrons to $NADP^+$, chlorophyll molecules are in an electron-deficient (oxidised) state. How is the pigment reduced to its former, stable condition? One way it could be reduced is to receive electrons from another source. Water is the most likely source of the electrons. Water is a weak electrolyte and a small proportion of its molecules dissociate into hydrogen ions (H^+) and hydroxyl ions (OH^-). When electrons are removed from hydroxyl ions by oxidised chlorophyll, oxygen is given off:

$$2OH^- + \text{oxidised chlorophyll} \rightarrow \tfrac{1}{2}O_2 + H_2O + \text{reduced chlorophyll}$$

Equilibrium is re-established as more water molecules dissociate. Thus one result of light absorption by chlorophyll is a rapid splitting of water molecules into hydrogen ions and oxygen. This is called **photolysis** (photo = light; lysis = to split). Using isotopes it is possible to prove that the oxygen given off in photosynthesis comes from water. When heavy water containing the heavy isotope of oxygen ^{18}O is supplied to photosynthesising plants instead of normal water (containing the isotope ^{16}O), heavy oxygen is given off:

$$CO_2 + 2H_2{}^{18}O \rightarrow {}^{18}O_2 + (CH_2O) + H_2O$$

If a control experiment is also run in which the ^{18}O is supplied in carbon dioxide, while normal water is provided, the evolved oxygen is of the ^{16}O type.

What happens to the hydrogen ions from photolysed water is explained in the next section.

4 Photophosphorylation

The Z-scheme of Hill and Bendall explains what happens to some of the excited electrons displaced from chlorophyll molecules when light is absorbed (Fig 5.9). According to the scheme there are two photosystems PS1 and PS2. The most abundant pigment in PS1 is chlorophyll *a* with lesser amounts of chlorophyll *b* and some carotene. In PS2 chlorophyll *a* is again most abundant but a substantial amount of chlorophyll *b* is also present together with some xanthophyll. The two systems are interconnected by an electron transport chain in which cytochromes are among the electron carriers. As the electrons pass through the chain they provide energy for the generation of ATP from ADP and inorganic phosphate.

Fig 5.9 Z-scheme for non-cyclic photophosphorylation

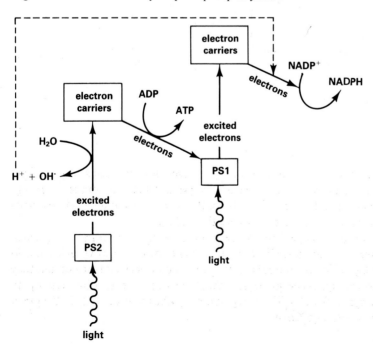

Production of ATP in this way is called **photophosphorylation** as it is made possible as a direct consequence of light absorption. Photolysis is linked with PS2. Electrons leaving PS1 combine with hydrogen ions from photolysed water to reduce $NADP^+$ to NADPH. The main carrier of electrons here is ferredoxin. The complete sequence of events known as **non-cyclic photophosphorylation** is shown in more detail in Fig 5.10.

Fig 5.10 Details of non-cyclic photophosphorylation

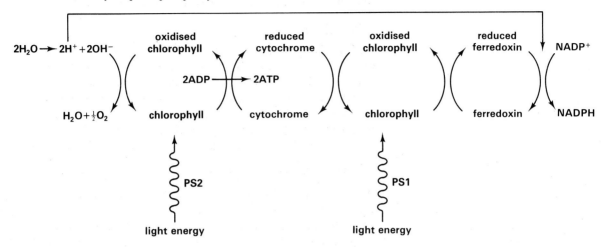

ATP can also be generated by another route called **cyclic photophosphorylation** (Fig 5.11). Here excited electrons from PS1 are recycled back to oxidise chlorophyll in the same system by electron carriers, notably cytochromes. For each molecule of ATP formed by the cyclic route, two are generated in non-cyclic photophosphorylation.

Fig 5.11 Cyclic photophosphorylation

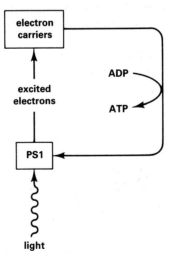

The important products of the light-dependent reactions of photosynthesis are thus ATP, NADPH and oxygen. ATP is a source of energy required for the dark reactions. NADPH is also used in the dark reactions to reduce phosphoglyceric acid (section 5.2.2).

No mention has yet been made of the role of the carotinoid pigments. They are thought to shield the chlorophylls from excessive oxidation in intense light. The carotinoids are not involved in electron transfer, but they can absorb light energy which they transfer as excitation energy to chlorophyll molecules. This is why photosynthesis occurs only in the green parts of variegated leaves.

5.2.2 Light-independent reactions

1 Fixation of carbon dioxide

For a long time, plant scientists had little idea of what happens to carbon dioxide in photosynthesis. In the late 1940s Professor Melvin Calvin of the University of California decided that one way to find out was to allow plants to photosynthesise using carbon dioxide labelled with the radionuclide ^{14}C. Calvin experimented with unicellular green algae such as *Chlorella* and *Scenedesmus* which he grew in a mineral solution held in flat glass containers he called 'lollipops' (Fig 5.12). The isotope was added as sodium hydrogencarbonate which breaks down to form $^{14}CO_2$. After a

Fig 5.13 Autoradiographs of photosynthetic products containing ^{14}C in *Scenedesmus* (courtesy Dr M Tribe)

(a) After 5 seconds

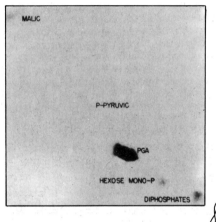

(b) After 15 seconds

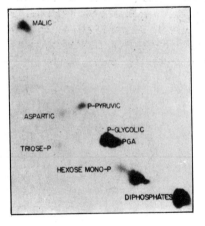

(c) After 60 seconds

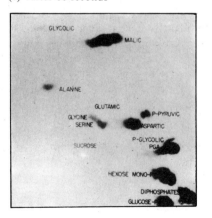

Fig 5.12 Calvin's experiment

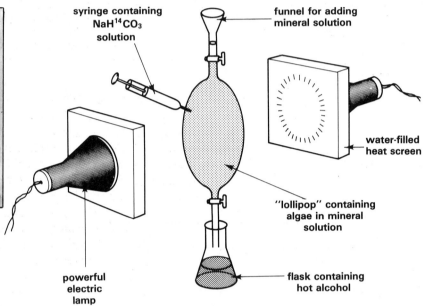

syringe containing NaH^{14}CO$_3$ solution

funnel for adding mineral solution

water-filled heat screen

"lollipop" containing algae in mineral solution

powerful electric lamp

flask containing hot alcohol

fixed period of illumination the algae were rapidly killed by running the cell suspension into a flask of hot ethanol. The cell contents were then analysed by autoradiography to find out which substances were labelled.

First the cell extracts were separated by paper chromatography. A variety of known organic compounds was run on the chromatogram at the same time in order to identify the components of the cell extracts. The chromatogram was then placed on a sheet of X-ray film. Radiations emitted by radionuclides cause black spots called **fogging** to appear on the film (Fig 5.13). When the fog marks on the autoradiogram are checked against the positions of the known compounds on the chromatogram it is possible to identify which of the components of the cell extracts are probably labelled with the isotope ^{14}C.

Calvin argued that after only a short period of illumination the radionuclide should appear in the first products of carbon dioxide fixation. With longer exposures to light, the intermediate and end products of photosynthesis should be labelled. Table 5.2 shows a summary of his findings. These reactions take place in the stroma of chloroplasts (Chapter 6). The first product of carbon dioxide fixation is phosphoglyceric acid (PGA). But what does CO_2 combine with to form PGA? Because each PGA molecule contains three carbon atoms Calvin thought it was logical to search for a CO_2-acceptor molecule containing two carbon atoms. Despite much effort, no such compound was found in green plants. Later a pentose sugar, ribulose bisphosphate (RBP) was shown to be the main CO_2 acceptor in many plant species. Each molecule of RBP combines with one of carbon dioxide to form two molecules of PGA. But how is PGA built up into carbohydrates? Furthermore, how is a constant supply of RBP maintained so that CO_2 fixation can continue indefinitely?

Table 5.2 Components of algal cell extract labelled with radioactive carbon after different periods of photosynthesis in the presence of $^{14}CO_2$

Time of exposure to light after isotope added/s	Main substances containing ^{14}C
5	phosphoglyceric acid (PGA)
15	PGA, hexose phosphates
60	PGA, hexose phosphates, sucrose, amino acids
300	PGA, hexose phosphates, sucrose, starch, amino acids, proteins, lipids

Fig 5.14 (a) The Calvin cycle

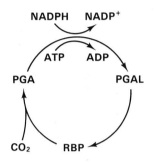

2 Link between light and dark stages

So far there seems to be no connection between the light-dependent reactions, and carbon dioxide fixation which is light-independent. However NADPH and ATP, the products of the light stage, are used in the dark stage. PGA is reduced by NADPH to form the triose sugar phosphoglyceraldehyde (PGAL). The reaction is endergonic and requires ATP for it to proceed. A sixth of the PGAL molecules formed are built into hexose sugars, the remainder are used to resynthesise the carbon dioxide acceptor, RBP. These events are sometimes called the **Calvin cycle** (Fig 5.14(a)). The link between the light and dark stages of photosynthesis is summarised in Fig 5.14(b). Hexoses are condensed into disaccharides and polysaccharides. Amino acids and lipids are formed in other linked pathways.

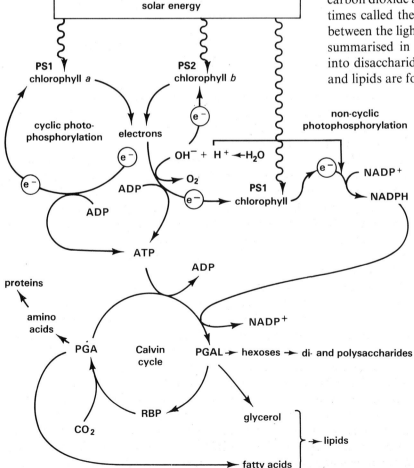

Fig 5.14 (b) Link between the light and dark stages of photosynthesis

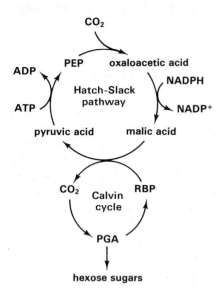

Fig 5.15 The Hatch–Slack (C$_4$) pathway

CO$_2$

PEP → oxaloacetic acid

ADP

ATP

Hatch-Slack pathway

NADPH

NADP$^+$

pyruvic acid

malic acid

CO$_2$

Calvin cycle

RBP

PGA

hexose sugars

3 Other pathways for carbon dioxide fixation

The fixation of carbon dioxide in some green plants occurs in additional ways from that described above. The plants include some important tropical crops such as sugar cane, maize and sorghum. When given ^{14}CO$_2$ the carbon radionuclide soon appears in oxaloacetic acid as well as PGA. Oxaloacetic acid contains four carbon atoms in each of its molecules compared with three in PGA. Hence this pathway is sometimes called the **C4 pathway**. It is alternatively known as the **Hatch-Slack pathway** and its link to the Calvin cycle is shown in Fig 5.15.

The source of energy for photosynthesis in C4 plants is sunlight, and ATP and NADPH are again the products of the light-dependent reactions. Some ATP is used to phosphorylate pyruvic acid, and the phosphoenolpyruvic acid (PEP) so formed is used to fix CO$_2$ alongside RBP. When PEP fixes CO$_2$, oxaloacetic acid is produced in C4 plants. NADPH reduces the oxaloacetic acid so formed to malic acid. The latter then releases CO$_2$ which is fixed by RBP to form PGA. The fate of PGA is the same as that described for C3 plants.

Among the products of CO$_2$ fixation in all green plants is hydroxyethanoic acid. The acid is immediately oxidised in C3 plants in a process called **photorespiration**, and carbon dioxide is given off. Up to 30 % of carbon dioxide fixed in photosynthesis can be recycled in this way. Consequently a considerable amount of the solar energy used to fix carbon dioxide is wasted. C4 plants photorespire less, so their photosynthesis is more efficient in making raw materials for growth. Furthermore, with a rise in temperature the rate of photorespiration increases more rapidly than carbon dioxide fixation. This is mainly why C4 plants are much more productive than C3 plants, in tropical climates.

5.3 Respiration

Respiration is the oxidation of energy-rich substrates. Carbohydrates, sugars especially, are the respiratory substrates most used, although lipids and proteins can also be oxidised. Polysaccharides are first hydrolysed to the hexose sugar glucose, lipids to glycerol and fatty acids, proteins to amino acids. Most forms of life are **aerobes**. They respire using oxygen.

Many aerobic organisms are **facultative anaerobes**. They can respire for short periods of time in the absence of oxygen. A few species of bacteria are **strict anaerobes** and respire only in the absence of oxygen. Whatever the kind of respiration, it is important to understand what energy is provided and how this energy is used in processes which could not occur without respiratory energy. In this context it is useful to know more about adenosine triphosphate.

5.3.1 Adenosine triphosphate

Adenosine triphosphate (ATP) has the structure shown in Fig 5.16. It is readily hydrolysed to adenosine diphosphate (ADP) and inorganic phosphate. The reaction has a ΔG value of -30.66 kJ mol^{-1} in standard conditions (25°C, 1 atm pressure, pH 0, substrate and product concentration 1.0 M). Such a combination of conditions is unlikely to occur in living cells. For example, most of our cells have a temperature of 37°C and pH of 7 or thereabouts. Here the ΔG value for the hydrolysis of ATP may be as high as -50 kJ mol^{-1}. An equivalent amount of free energy change occurs when ADP is hydrolysed to adenosine monophosphate (AMP). However, when the phosphate group of AMP is released by hydrolysis the free energy change is less than half this value. For these reasons the bonds between the phosphate groups of ATP are held by what are called **high energy bonds**, and ATP is often written as:

$$A—P \sim P \sim P$$

where A = adenosine   P = phosphate
        $\sim$ = high energy bond   — = low energy bond

Free energy changes of comparable magnitude occur when ATP is formed from ADP and inorganic phsophate but here the ΔG value is positive. Hence ATP is well suited to act as a link between energy-releasing and many of the energy-consuming reactions which occur in living cells including such processes as active transport (Chapter 6) and the synthesis of nucleic acids and proteins (Chapter 4). Respiration is the means whereby ATP is generated for such purposes. Other links are used in particular instances. For example, phosphocreatine is the main provider for muscle contraction in mammals (Chapter 17).

Fig 5.16 (a) ATP simplified

phosphate groups

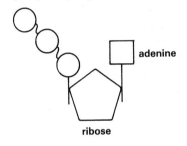

ribose

adenine

Fig 5.16 (b) Structure of a molecule of ATP

3 phosphate groups

adenine

deoxyribose

adenosine

5.3.2 Oxidation of respiratory substrates

In anaerobic conditions, energy-rich substrates such as glucose are oxidised to lactic acid or to ethanol and carbon dioxide. Lactic acid is a product of anaerobic respiration in some bacteria and in animal cells. Ethanol and carbon dioxide are made by yeasts and higher plant cells when oxygen is absent. The overall processes can be summarised as follows:

Lactic fermentation: $C_6H_{12}O_6 \rightarrow 2CH_3CHOHCOOH$
                         glucose   lactic acid

Alcoholic fermentation: $C_6H_{12}O_6 \rightarrow 2CH_3CH_2OH + 2CO_2$
                         glucose   ethanol   carbon
                                                           dioxide

The equations give us no idea of the amount of energy released, nor do they tell us anything about the way in which the substrate is oxidised. What is more, they disguise the fact that the two processes are very similar, each sharing a number of common steps. **Glycolysis** (glyco = sugar; lysis = to split) is the name given to the common steps. Glycolysis also takes place when sugars are respired aerobically.

Fig 5.17 Glycolysis

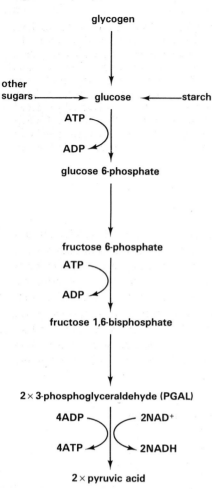

1 Glycolysis

Fig 5.17 shows the more important common steps in the oxidation of glucose, whether in anaerobic or aerobic conditions. Far from producing energy, some of the earlier reactions need an input of energy. The energy comes from the hydrolysis of high-energy bonds of ATP. Without a supply of energy from ATP, the reactions could not proceed. The splitting of each fructose bisphosphate molecule yields two molecules of phosphoglyceraldehyde (PGAL), which is then oxidised to pyruvic acid. Oxidation of PGAL involves the removal of hydrogen and is catalysed by dehydrogenase enzymes (Chapter 3). At the same time the coenzyme nicotinamideadenine dinucleotide NAD^+ becomes reduced to NADH. Oxidation of each PGAL molecule releases enough energy for two ATP molecules to be synthesised from ADP and inorganic phosphate.

The fate of the pyruvic acid depends on the organism in which it was produced and on whether oxygen is available or not. If lactic acid is a product of anaerobic respiration, the NADH formed in glycolysis transfers its hydrogen to pyruvic acid, and lactic acid is formed. This is what happens in mammalian muscle cells when they have an oxygen debt (Chapter 8). Bacteria which sour milk also make lactic acid when oxygen is not available.

If yeasts and higher plant cells are kept without oxygen, pyruvic acid is split into carbon dioxide and ethanal. The ethanal is then reduced to ethanol by hydrogen from NADH (Fig 5.18). Either way, NAD^+ is remade and once again acts as a hydrogen acceptor when more PGAL is oxidised. The enzymes which control glycolysis are found in the cytoplasm.

In aerobic conditions the pyruvic acid is taken instead into the Krebs cycle which occurs in mitochondria (Chapter 6).

Fig 5.18 Fate of pyruvic acid in the absence of oxygen

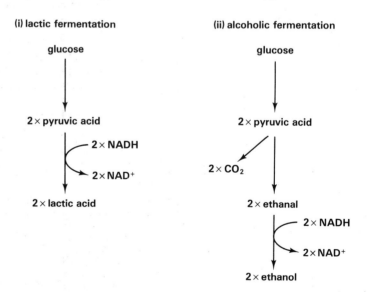

2 The Krebs cycle

At the beginning of this century, Thunberg devised a special tube; with it he discovered that animal tissues contain dehydrogenase enzymes which catalyse the transfer of hydrogen from carboxylic acids (Fig 5.19). In 1935 Szent-Gyorgyi noticed that dehydrogenation of succinic acid in muscle is blocked by the competitive inhibitor malonic acid. Furthermore, the

86

Fig 5.19 A Thunberg tube

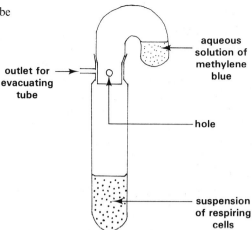

aqueous
solution of
methylene
blue

outlet for
evacuating
tube

hole

suspension
of respiring
cells

The top of the tube is rotated so that the hole
coincides with the outlet and the air in the
tube is sucked out using a vacuum pump.
The top is then turned to seal the contents
from the external air. Finally the tube is
inverted so that the blue dye mixes with the
respiring cells. The dye is gradually reduced
to colourless leuco-methylene blue. If air is
left in the tube gaseous oxygen is reduced
instead and the dye remains in its oxidised
blue state

inhibition stopped respiration in the muscle. Two years later Hans Krebs
showed that respiration in pigeon breast muscle was stimulated by a
specific variety of carboxylic acids. Krebs went on to carry out a series of
brilliant experiments from which he deduced that, in aerobic conditions,
pyruvic acid combines with oxaloacetic acid to form citric acid. The citric
acid is then dehydrogenated via a number of intermediate compounds back
to oxaloacetic acid. One of the intermediates is succinic acid. Carbon
dioxide is released in these reactions. It was some time later before it was
found that pyruvic acid is converted to acetyl coenzyme A before it enters
the **Krebs cycle** (Fig 5.20).

Fig 5.20 The **Krebs** cycle and some
important biochemical pathways to
which it is linked. The acids ionise and
their ions are actually used in the reactions, eg

citric acid $\rightleftharpoons$ citrate
pyruvic acid $\rightleftharpoons$ pyruvate

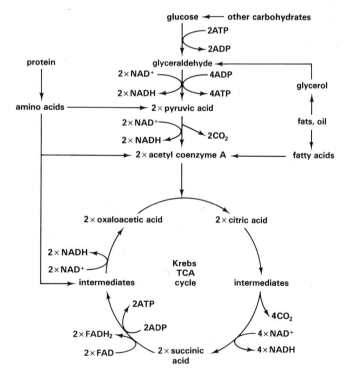

glucose ⟵ other carbohydrates
2ATP
2ADP
glyceraldehyde ⟵
protein
$2 \times NAD^+$ ⟶ 4ADP
glycerol
$2 \times NADH$ ⟵ 4ATP
amino acids ⟶ $2 \times$ pyruvic acid
fats, oil
$2 \times NAD^+$
$2 \times NADH$ ⟵ $2CO_2$
$2 \times$ acetyl coenzyme A ⟵ fatty acids

$2 \times$ oxaloacetic acid $2 \times$ citric acid

$2 \times NADH$
$2 \times NAD^+$

Krebs
TCA
cycle

intermediates   intermediates

2ATP
2ADP $4CO_2$
$2 \times FADH_2$ $4 \times NAD^+$
$2 \times FAD$ ⟶ $2 \times$ succinic $4 \times NADH$
acid

In the late 1940s mitochondria, separated from other cell components by
differential centrifugation (Chapter 6), were found to break down pyruvic
acid to carbon dioxide and water. The dehydrogenase enzymes which
catalyse the oxidations in the Krebs cycle are thus in the mitochondria.
Biochemists are now fairly sure that the enzymes are in the matrix enclosed
by the inner membrane of mitochondria.

In the Krebs cycle oxidation of carboxylic acid molecules is catalysed by dehydrogenase enzymes. Dehydrogenation of the acids is accompanied by reduction of NAD^+ to NADH. The fate of NADH in aerobic conditions is, however, different from NADH formed in anaerobic conditions. When oxygen is available the hydrogen from NADH is passed via a succession of hydrogen acceptors until it finally reduces oxygen to water. The succession is called a **respiratory** or **electron transport chain** (Fig 5.21). After NADH the hydrogen is passed to flavin-adenine dinucleotide (FAD). The hydrogen atoms now split into electrons and protons. Reduced FAD ($FADH_2$) transfers the electrons to the next acceptors called cytochromes.

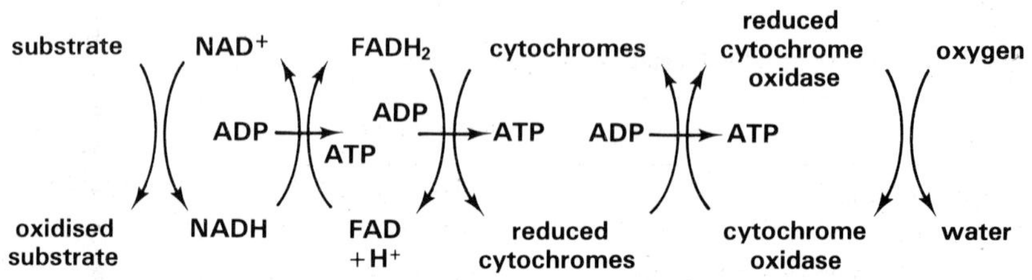

Fig 5.21 A respiratory chain

Reduced cytochromes then pass the electrons to cytochrome oxidase. Finally reduced cytochrome oxidase transfers the electrons back to the protons. The hydrogen atoms so produced, reduce oxygen to form water. Each of the reactions in the chain is an oxidation–reduction reaction. As the hydrogen atoms and later the electrons move along the chain, the reduced acceptors are oxidised and more of them can be taken into the chain.

What is particularly important is that three of the oxidation–reduction reactions in an electron transport chain yield enough energy to make ATP from inorganic phosphate ions and ADP. This type of reaction is called **oxidative phosphorylation**. It takes place in the oxysomes projecting from the cristae of mitochondria (Chapter 6). For each molecule of NADH taken into a respiratory chain, three molecules of ATP are formed. In one of the oxidations in the Krebs cycle, FAD is reduced (Fig 5.20). Only two molecules of ATP are produced for each molecule of $FADH_2$ entering a respiratory chain.

Note that the aerobic oxidation of a molecule of glucose causes the reduction of ten molecules of NAD^+ and two of FAD. The twelve molecules of reduced coenzymes reduce twelve atoms (six molecules) of oxygen to produce twelve molecules of water. The summary equation for aerobic respiration of a mole of glucose is thus:

$$C_6H_{12}O_6 + 6O_2 + 6H_2O \rightarrow 6CO_2 + 12H_2O$$

5.3.3 Energy yields from respiratory substrates

How many ATP moles are formed from each mole of glucose oxidised aerobically? How does this figure compare with the number obtained when glucose is oxidised anaerobically? The breakdown of one mole of glucose in glycolysis uses energy from the terminal high-energy bonds of two moles of ATP. On the other hand, enough energy is released in glycolysis to produce four moles of ATP from ADP and inorganic phosphate (P_i):

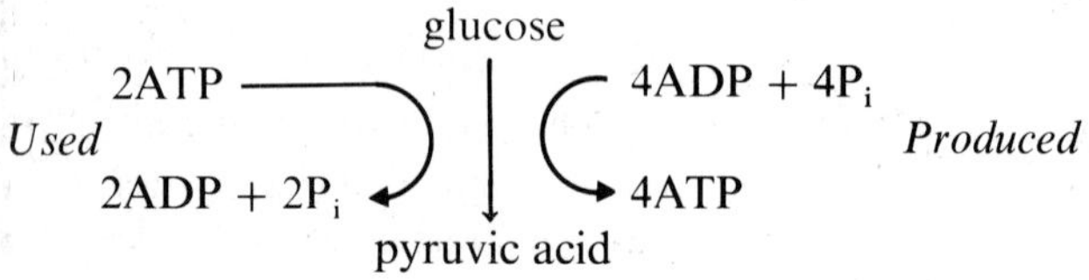

The net release of energy from one mole of glucose broken down to pyruvic acid, therefore, is the energy released when the terminal high-energy bonds of two moles of ATP are later hydrolysed. Thus cells respiring anaerobically release $2 \times 30.66\,kJ$ of energy from every mole of glucose oxidised.

In aerobic conditions, ATP is also made mainly in the respiratory chains into which NADH and $FADH_2$ from the Krebs cycle are fed. Altogether, 38 ATP moles are made from each mole of glucose oxidised aerobically, 8 from glycolysis and 30 from the Krebs cycle (Table 5.3). Cells respiring aerobically thus release $38 \times 30.66\,kJ$ of energy from a mole of glucose. Aerobic respiration is clearly much more efficient than anaerobic respiration in releasing energy from respiratory substrates.

Table 5.3 Origin of ATP formed in aerobic respiration of glucose

Source	No. of ATP molecules
direct synthesis in glycolysis	2 (net gain)
2NADH formed in glycolysis	6 (2×3)
8NADH formed in Krebs cycle	24 (8×3)
2$FADH_2$ formed in Krebs cycle	4 (2×2)
direct synthesis in Krebs cycle	2
Total	38

Table 5.4 Net gain of ATP for respiration of a molecule of tripalmitin

glycerol→$CO_2 + H_2O$	20 ATP
$3 \times$ palmitic acid→$CO_2 + H_2O$	390 ATP
Total	410 ATP

Fig 5.22 A simple calorimeter

When the supply of electricity is switched on the heating element becomes red hot and the sample ignites in the stream of oxygen. The heat given off by the burned sample causes the temperature of the water to rise. The increase in temperature, volume of water and mass of sample are used to work out its calorific value.

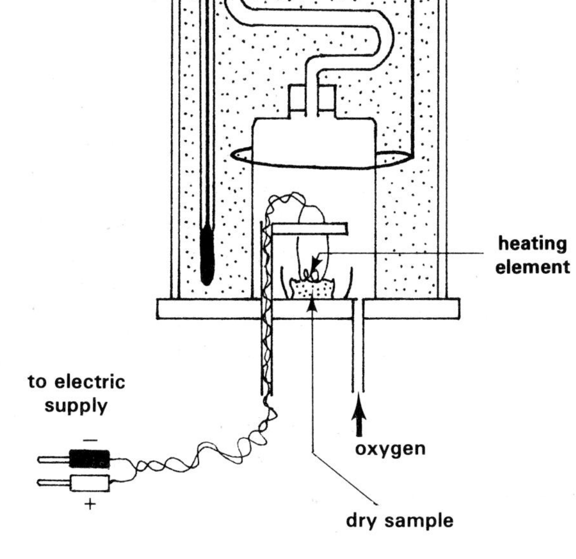

For a trigylceride consisting of glycerol and palmitic acid, the net gain of ATP is shown in Table 5.4. However, a molecule of tripalmitin ($C_{51}H_{95}O_6$) is nearly five times heavier than a glucose molecule ($C_6H_{12}O_6$). Hence on a mass-for-mass basis, lipids yield a little more than twice as much energy as carbohydrates. The energy yield from protein is about the same as for a similar mass of carbohydrate. When burned to carbon dioxide and water in a **calorimeter** (Fig 5.22), a mole of glucose releases $2880\,kJ$ of heat energy. This is the amount of energy potentially available to do work in a living cell. We can now calculate the efficiency of energy change when glucose is respired aerobically:

$$\text{Efficiency of energy change} = \frac{38 \times 30.66}{2880} \times 100 = 44\%.$$

This means that less than half of the potential energy in glucose is used to do work.

You may wonder what has happened to the rest of the energy in the respiratory substrate. The Second Law of Thermodynamics states that whenever one form of energy is changed into another, some of the energy is converted into heat. In aerobic respiration, over half of the energy in the substrate is released as heat. Heat energy helps to maintain the constant body temperature of warm-blooded animals (Chapter 20). However, remember that heat energy cannot be used to do work in living organisms.

Many of the biochemical reactions in glycolysis and the Krebs cycle are reversible. They constitute a metabolic 'hub' which meets the cell's immediate energy needs. Excess input of energy-rich substrates such as glucose leads to their storage as fat. At times of shortage, reserves of energy can be drawn on by converting fat to sugar.

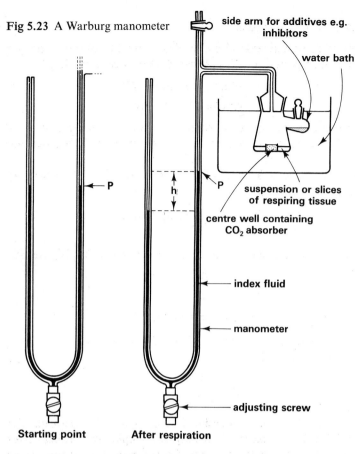

Fig 5.23 A Warburg manometer

side arm for additives e.g. inhibitors

water bath

P

h

P

suspension or slices of respiring tissue

centre well containing CO_2 absorber

index fluid

manometer

adjusting screw

Starting point

After respiration

5.3.4 Measuring the rate of respiration

In the 1930s Otto Warburg devised a constant-volume manometer to measure the rate of oxygen uptake by slices of living tissue. The apparatus is often called a **Warburg manometer** (Fig 5.23). To begin with, the index fluid in both arms is brought to the same height at the reference point (P). As the tissue respires the carbon dioxide it evolves is removed by the potassium hydroxide in the centre well. The volume of air enclosed in the manometer is thus reduced. After a given time, the adjusting screw is used to return the index fluid in the right arm to P. The reduction in pressure of the air in the manometer causes the height of the fluid in the left arm to fall. The change in height (h) is multiplied by the flask constant (K) to determine the volume of oxygen consumed. Oxygen consumption is usually expressed as mm^3 oxygen absorbed $h^{-1} g^{-1}$ of tissue. Note that the volume of air in the manometer is always the same on each occasion a reading is taken. Hence the name constant-volume manometer. A control manometer, called a **thermobarometer** lacking the tissue, is set up at the same time. It is necessary because changes in air pressure and temperature during the experiment would alter the volume of air in the manometer. Such changes have to be accounted for in the calculations.

Simple respirometers (Fig 5.24) are often used in school and college laboratories to measure respiratory rates. A known mass of germinating seeds or small invertebrate animals such as woodlice, is placed in the **respiration chamber**. The organisms are given time to adjust to the chosen temperature and the valves are then closed. Once again the potassium hydroxide absorbs evolved carbon dioxide, so the volume of air in the respiration chamber decreases as oxygen is consumed. Consequently the manometer fluid moves towards the respiration chamber. Knowing the distance moved by the meniscus of the fluid and the bore of the manometer tube, the rate of oxygen uptake can be calculated. Alternatively the manometer fluid can be returned to its original height by pushing the plunger in the syringe. The volume of air required to do this can then be measured directly on the syringe. The **compensation tube** functions as a thermobarometer. It contains the same volume of air as the respiration chamber. Thus changes in air pressure or temperature during the experiment have the same effect on the volume of air in both tubes. In this way the changes cancel out each other.

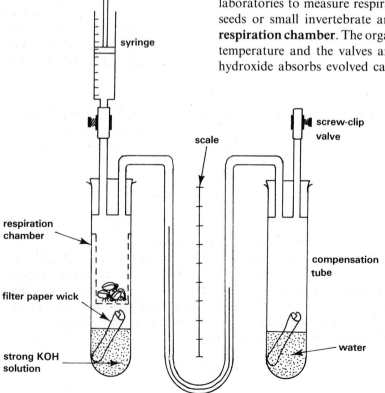

syringe

scale

screw-clip valve

respiration chamber

compensation tube

filter paper wick

water

strong KOH solution

manometer

Fig 5.24 A simple respirometer

5.3.5 Respiratory quotient

The volume of carbon dioxide given off divided by the volume of oxygen taken up in a fixed period is called the **respiratory quotient, RQ**:

$$RQ = \frac{\text{volume } CO_2 \text{ given off}}{\text{volume } O_2 \text{ taken up}}$$

RQ values indicate the type of respiration, aerobic, anaerobic or both. They also tell us which substrates or combination of substrates are oxidised.

In aerobic conditions an RQ of 1.0 indicates that the respiratory substrate is a carbohydrate:

$$\underset{\text{carbohydrate}}{C_6H_{12}O_6} + 6O_2 \rightarrow 6CO_2 + 6H_2O$$

For each volume of oxygen taken up, a similar volume of CO_2 is given off:

$$RQ = \frac{6}{6} = 1.0$$

When a lipid is oxidised aerobically, the RQ is 0.7:

$$\underset{\text{tristearoylglycerol}}{C_{57}H_{110}O_6} + 81.5O_2 \rightarrow 57CO_2 + 55H_2O$$

$$RQ = \frac{57}{81.5} = 0.7$$

For the aerobic oxidation of protein, an RQ of 0.99 is obtained. The aerobic breakdown of a mixture of carbohydrates, lipids and proteins gives an RQ of 0.8–0.9. Animals fed a balanced diet use mainly carbohydrates and lipids in equal amounts as respiratory substrates and have an RQ of about 0.85.

A mixture of aerobic and anaerobic respiration takes place when there is a shortage of oxygen. Here the RQ is usually greater than 1.0. The exact value depends on the respiratory substrates used and on the relative rates of the two types of respiration. For example, when anaerobic and aerobic respiration of glucose occur at similar rates in plant cells, an RQ of 1.33 is obtained:

Anaerobic respiration: $\qquad C_6H_{12}O_6 \qquad\qquad \rightarrow 2CO_2 + 2CH_3CH_2OH$
Aerobic respiration: $\qquad\quad C_6H_{12}O_6 + 6O_2 \rightarrow 6CO_2 + 6H_2O$

$$\overline{\qquad\qquad Total \qquad\qquad 6O_2 \rightarrow 8CO_2 \qquad\qquad}$$

$$RQ = \frac{8}{6} = 1.33$$

Would you expect the same value in animal cells?

Simple respirometers can be used to determine the RQ of living organisms. One is set up in the way described earlier to measure oxygen consumption, say $a\,mm^3\,h^{-1}$. The other lacks the KOH solution and thus measures the difference in volumes of oxygen consumed and carbon dioxide evolved, say $b\,mm^3\,h^{-1}$. The differences in readings make it possible to calculate the RQ as follows:

$$RQ = \frac{a - b}{a}$$

What might the RQ be if aerobic respiration takes place more rapidly, or more slowly than anaerobic respiration of glucose?

The RQ of starving animals is between 0.9 and 1.0. Can you explain why this is so.

Use the formula to calculate RQ values from the following readings:

(i) $a = 10\,mm^3\,h^{-1}$; $b = 0\,mm^3\,h^{-1}$
(ii) $a = 10\,mm^3\,h^{-1}$; $b = +3\,mm^3\,h^{-1}$
(iii) $a = 10\,mm^3\,h^{-1}$; $b = -3\,mm^3\,h^{-1}$

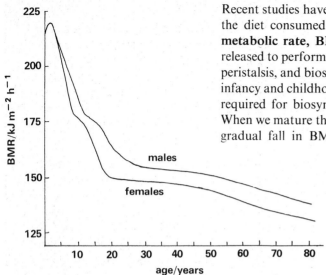

Fig 5.25 The effect of age on BMR

5.3.6 Basal metabolic rate

Recent studies have shown that about two thirds of the energy content of the diet consumed by a resting person is used to maintain the **basal metabolic rate, BMR.** This is the basic rate at which energy must be released to perform vital functions such as beating of the heart, breathing, peristalsis, and biosynthesis of proteins and other important molecules. In infancy and childhood the BMR is relatively high as much of the energy is required for biosynthesis of cellular components necessary for growth. When we mature the BMR levels off until middle age. In old age there is a gradual fall in BMR as metabolism beings to slow down (Fig 5.25).

Throughout life, men usually have a higher BMR than women. This is because men generally have less fat per unit body mass and surface area. Calculated on the basis of unit lean body mass, the BMR is similar in both sexes.

Table 5.5 Daily energy expenditure for average men and women (from Taylor 1978)

Occupation	Sleep 8 h (basal)/kJ	Work 8 h/kJ	Leisure 8 h/kJ	Total 24 h/kJ
Men				
sedentary	2000	3500	5500	11 000
moderately active	2000	5000	5500	12 500
very active	2000	7500	5500	15 000
Women				
home or office	1750	3500	3750	9 000

Table 5.6 Recommended daily energy intakes for people of different ages (from Taylor 1978)

Age	Energy intake/ kJ kg^{-1} body mass
0–3 months	500
6–9 months	460
1–2 years	430
3–4 years	410
4–5 years	400
adult man	180

The expenditure of energy in an active person includes the amount used in body movement. Hence the total energy requirement of any individual depends on a variety of factors including age, sex, body mass and the extent of physical activity engaged in (Tables 5.5 and 5.6). The energy requirements of women increase in pregnancy to support foetal growth and, after birth, to meet the demands of lactation. This is why the World Health Organisation recommends an additional energy intake of 1500 kJ per day for pregnant women, especially during the second half of the gestation period.

If the intake of energy-producing foods is more than immediate requirements, the energy is stored, mainly as fat. Consequently body mass increases. Some people can respire the excess intake in brown fat (Chapter 20), so do not put on weight. The present UK Government recommendations are that the energy intake for women should be between 7030 and 10 510 and for men between 10 080 and 14 070 kJ per day depending on how active they are. Research carried out on women in recent years at the Dunn Nutrition Unit, Cambridge has shown that physical activities take up less energy than was previously thought. The study indicates that the Government figures are between 1260 and 1680 kJ a day too high. It is for this reason that slimmers' diets based on Government recommendations may not result in a decrease in body mass. Finding an appropriate intake of energy is a useful contribution to maintaining health. Many cardiovascular and respiratory ailments of modern societies are thought to be caused by excess body mass.

SUMMARY

Energy occurs in many forms which are interconvertible. Whenever one form of energy is converted to another some energy appears as heat. Changes in free energy occur when biochemical reactions occur because of alterations in positions of atoms in the reacting molecules. Exergonic reactions show a reduction in free energy and occur spontaneously. Endergonic reactions show an increase in free energy and require an input of energy in order to proceed. They can be made to proceed if linked with exergonic reactions. In living organisms the main link is ATP which is generated in the exergonic reactions of respiration and used in endergonic processes such as protein and nucleic acid synthesis, muscle contraction and active transport across membranes.

The sun is ultimately the main source of energy for most forms of life. Solar energy is used in the light-dependent reactions of photosynthesis to yield oxygen, hydrogen ions and electrons. The hydrogen ions and some of the electrons are used to produce NADPH and ATP in the two photosystems of non-cyclic photophosphorylation. More ATP is generated in cyclic photophosphorylation. In the light-independent reactions the products of the light dependent reactions NADPH and ATP are used to reduce carbon dioxide to form a range of organic molecules, notably carbohydrates. In this way solar energy is converted to the free energy which binds the atoms of the organic products of photosynthesis.

Respiration is the oxidation of organic molecules, principally carbohydrates. The free energy changes of some of the reactions which make up the metabolic pathways of respiration result in the synthesis of ATP.

In glycolysis which does not require the presence of oxygen gas, hexose sugars are split to form pyruvic acid which is further metabolised to ethanol and carbon dioxide in alcoholic fermentation, or lactic acid in lactic fermentation. The changes in free energy of such forms of anaerobic respiration generate only four molecules of ATP, whilst the early steps of glycolysis use two ATP molecules for each hexose molecule respired.

If oxygen gas is available the pyruvic acid derived from glycolysis is taken instead into the Krebs cycle where it is gradually decarboxylated and dehydrogenated. The latter reactions are linked to respiratory (electrontransport) chains in which ATP is synthesised and oxygen gas reduced to water. Forty molecules of ATP are generated for each molecule of hexose sugar oxidised in aerobic respiration.

QUESTIONS

1 In 1882 Engelman placed the green alga *Cladophora* in a suspension of aerobic bacteria. After different parts of the algal filament had been exposed to light of different wavelengths, he observed the distribution of the bacteria. His results are represented in the following diagram.

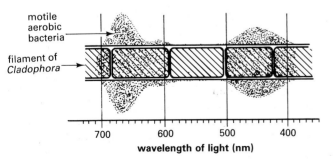

(a) Name the factor in the bacterial environment that determines their distribution.
(b) (i) Describe the relationship between wavelength of light and photosynthesis in the alga, as suggested by the distribution of the bacteria.

(ii) State the **most** effective wavelength of light for photosynthesis in the alga.
(c) The cellular components of a cabbage leaf were separated into fractions in ice-cold isotonic buffer solution. The fraction containing chloroplasts was suspended in isotonic buffer solution containing dilute methylene blue, divided into four equal parts and treated as follows:

| | | Colour of solution | |
Part	Conditions	After 5 min	After 45 min
1	Darkness at 5 °C	blue-green	blue-green
2	Darkness at 25 °C	blue-green	blue-green
3	Light at 5 °C	blue-green	pale-green
4	Light at 25 °C	pale-green	pale-green

(i) Outline a procedure for separating the chloroplasts from other cell organelles.
(ii) State why it was necessary to suspend the cells in ice-cold buffer solution during the separation.
(iii) Explain the purpose of adding methylene blue.

(d) Discuss the effects on the activity of the chloroplast suspension of (i) temperature, (ii) light.
(e) Suggest **two** ways in which the design of the investigation could be improved. (WJEC)

2 (a) The graph shows the absorption spectra of chlorophyll a, chlorophyll b and xanthophyll.

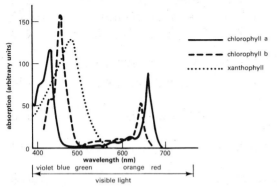

(i) What is the effect of visible light on chlorophyll a?
(ii) How do chlorophyll b and xanthophyll contribute to photosynthesis?
(b) (i) What is meant by *photophosphorylation*?
(ii) How does cyclic photophosphorylation differ from non-cyclic photophosphorylation?
(c) Which **two** products of the light-dependent stage are used in the light-independent stage (dark stage) of photosynthesis?
(d) The graph shows the effect of light intensity on the rate of photosynthesis at different temperatures and different concentrations of carbon dioxide.

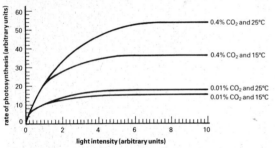

With the help of information in the graph, explain how temperature, carbon dioxide and light intensity interact to control the rate of photosynthesis. (AEB 1984)

3 The diagram below shows an outline of cellular respiration.

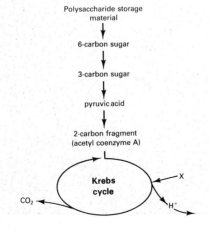

(a) Name a polysaccharide commonly stored in (i) green plants and (ii) mammals.
(b) (i) Name the process by which the 6-carbon sugar is converted to pyruvic acid.
(ii) Where in the cell does this process occur?
(iii) Why is ATP used in this process?
(c) Name the compound formed from pyruvic acid in muscle cells under conditions of oxygen debt.
(d) Name the type of enzyme involved at stage **X**.
(e) What happens finally to the hydrogen ions released from the Krebs cycle?
(f) Make a labelled drawing of a mitochondrion, and on your drawing indicate where ATP synthesis occurs. (L)

4 The apparatus shown below was used to measure the uptake of oxygen by respiring seeds.

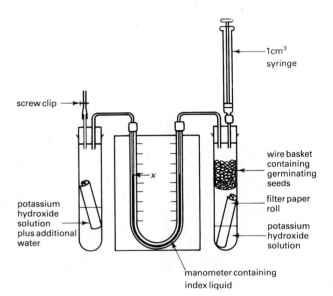

(a) (i) Explain the function of the roll of paper in each tube.
(ii) Explain why additional water was placed in the left hand tube.
(iii) What would you expect to happen to the level of liquid at **x** as respiration proceeds?
(b) List the steps you would take when setting up the apparatus before taking readings.
(c) Outline how you would use the same apparatus to determine the volume of carbon dioxide produced by the seeds.
(d) The following results were obtained from two experiments using different kinds of germinating seeds.

	Volume of oxygen consumed (cm^3)	Volume of carbon dioxide produced (cm^3)
Germinating seeds A	9.6	10.4
Germinating seeds B	9.1	6.3

Calculate the respiratory quotient (R.Q.) and suggest the type of substrate being respired for (i) germinating seeds A; (ii) germinating seeds B.
(e) Suggest how you might use the apparatus to measure the rate of anaerobic respiration of plant material.
(WJEC)

6 Cell structure

6.1 Cell structure as seen with
the light microscope 97

6.2 Cell ultrastructure as revealed
by the electron microscope 99

6.2.1 The cell membrane 102
6.2.2 Endoplasmic reticulum 105
6.2.3 The Golgi body 106
6.2.4 Lysosomes 107
6.2.5 Mitochondria 108
6.2.6 Peroxisomes 109
6.2.7 Nucleus 109

6.2.8 Centrioles, cilia and flagella 110
6.2.9 Chloroplasts 111
6.2.10 The cytoskeleton 112
6.2.11 The plant cell wall 113

6.3 Comparison of light and
electron microscopes 114

6.4 Cell fractionation 116

Summary 117

Questions 118

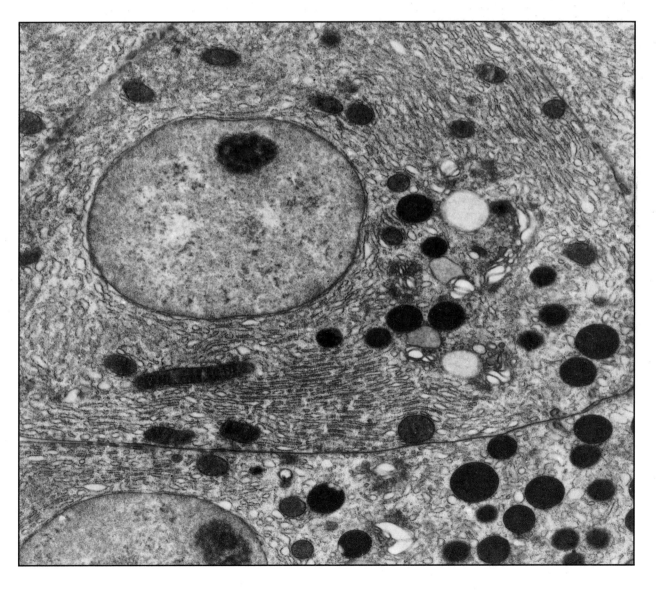

6 Cell structure

What is known about cell structure has largely depended on the development of microsopes and microscopical techniques. A **simple microscope** was invented by Galileo in 1610, but there are no reports that he used it to examine living organisms. In 1676 a Dutch draper, van Leeuwenhoek, whose hobby was the grinding of lenses, used one of his simple microscopes (Fig 6.1) to examine rainwater in which grains of pepper had been soaked. He observed a variety of unicellular organisms which he called 'animalcules'. It is now known that among the organisms he saw were bacteria. A decade earlier Robert Hooke in England had made a **compound microscope** with which he observed, among other things, thin slices of cork tissue. He saw that the cork was porous, rather like a honeycomb, consisting of a great many small compartments which he called **cells**.

Fig 6.1 Van Leeuwenhoek's simple microscope

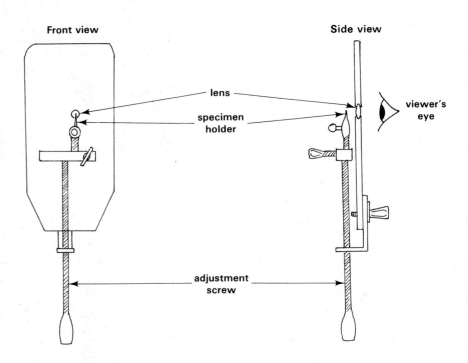

During the next 150 years, with further improvement of the compound microscope, a great deal was learned of the structure of cells and tissues from many plants and animals. In the 1840s the **cell theory** of Schleiden and Schwann became generally accepted by biologists. It stated that cells are the basic structural units of all living organisms. For some time after the theory was first proposed there was a tendency for biologists to emphasise the importance of individual cells; their structure and activity were thought to mirror that of the whole organism. Recently there has been more interest in the ways in which different types of cell interact in the functioning and development of multicellular organisms.

6.1 Cell structure as seen with the light microscope

The compound microscope is used by most biologists to examine cells and tissues (Fig 6.2). With this instrument it is possible to observe living material. Good images showing much detail can be obtained especially if the microscope is of the **phase-contrast** type (Fig 6.3). A phase-contrast microscope exaggerates small differences in the refractive index of cell components to create an image in which the components are clearly distinguished by the human eye.

Fig 6.2 A modern compound microscope (courtesy Vickers Instruments)

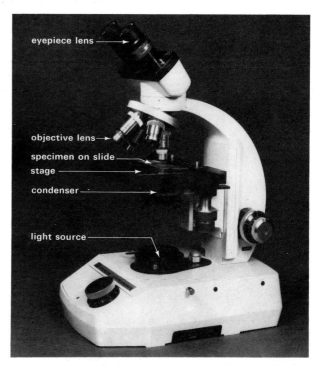

eyepiece lens

objective lens

specimen on slide

stage

condenser

light source

Fig 6.3 An epithelial cell from the mouth lining as seen using (a) a conventional light microscope (b) a phase contrast microscope, × 900

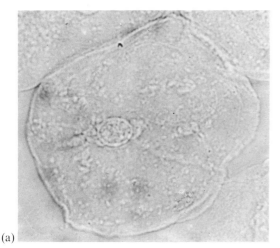

(a)

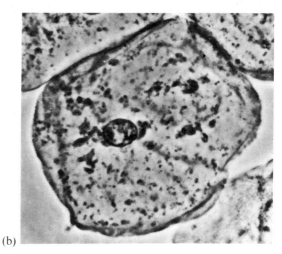

(b)

Nevertheless, most of what was known about the structure of cells up to the 1940s was obtained from observations on dead tissue which had been treated with preservatives. In the technique, which had been developed for over a hundred years, the tissue is first immersed in a fixative such as methanal in order to prevent deterioration and to keep the structure as life-like as possible. Following **fixation**, the tissue is **dehydrated** with ethanol and then **cleared** with an organic solvent such as dimethylbenzene which is miscible with paraffin wax. The next stage is to **embed** the tissue in molten

97

wax which, on hardening, supports the tissue while thin sections, 2–10 μm thick, are cut using a **microtome**. After attaching the sections to microscope slides, the wax is removed before the tissue is **stained**.

Staining enables cell components to be differentiated when the material is examined microscopically. Any of the above steps can lead to distortion of the specimen, so it is necessary to guard against artificial structures, called **artefacts**, which appear in the specimen during treatment but are not present in living cells. Using such methods, coupled with observations on living cells, it is possible to build up a fairly detailed picture of cell structure (Fig 6.4).

Fig 6.4 Structure of animal and plant cells revealed by a compound microscope

(a) (i) Section of a liver cell, × 1200

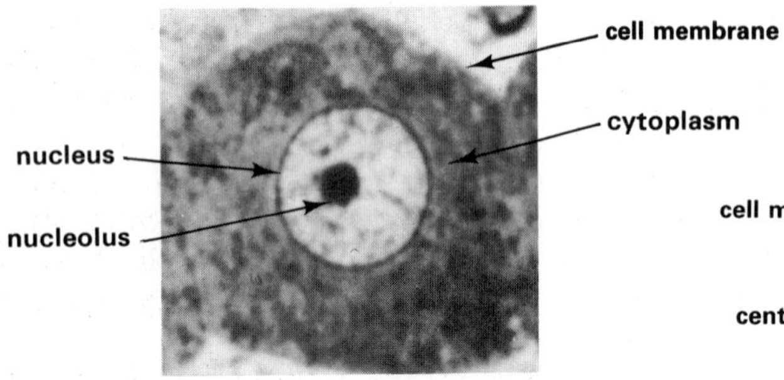

cell membrane

cytoplasm

nucleus

nucleolus

(a) (ii) Cell from gastric gland

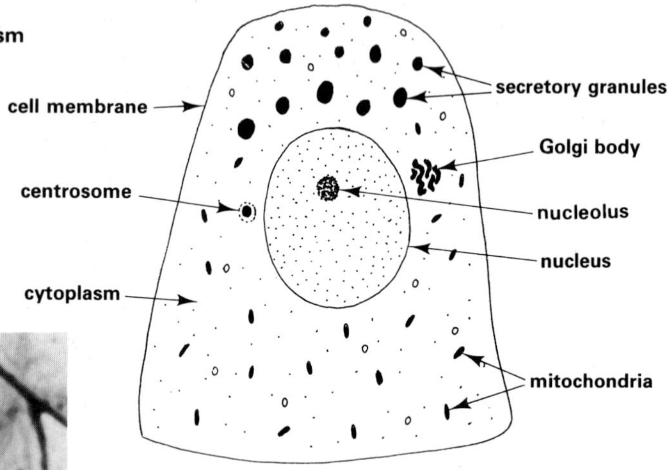

cell membrane

centrosome

cytoplasm

secretory granules

Golgi body

nucleolus

nucleus

mitochondria

(b) (i) Section of a leaf mesophyll cell, × 1200

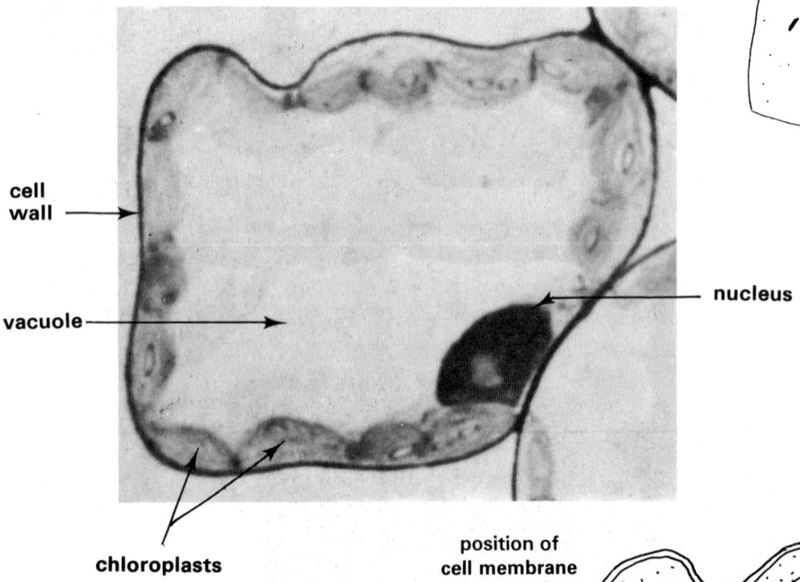

cell wall

vacuole

chloroplasts

nucleus

(b) (ii) A leaf mesophyll cell

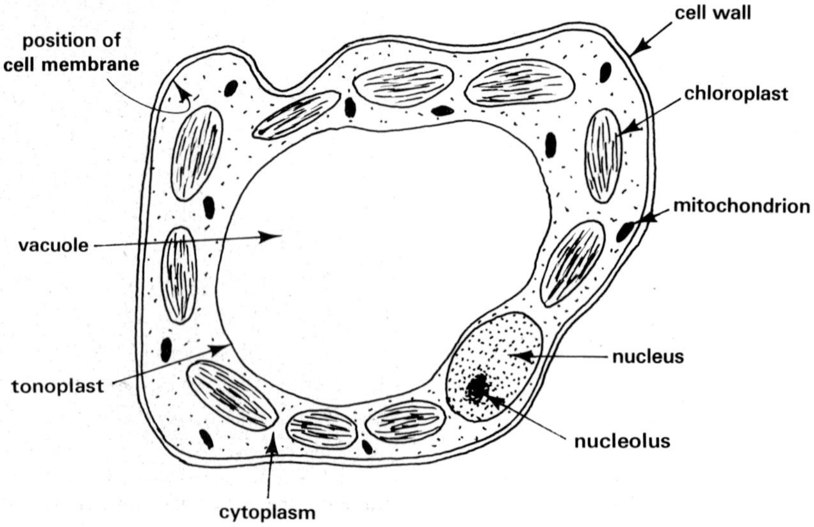

position of cell membrane

vacuole

tonoplast

cytoplasm

cell wall

chloroplast

mitochondrion

nucleus

nucleolus

The living material of cells is called **protoplasm** and is enclosed in a **cell membrane**. In plant cells a **cell wall**, mainly of cellulose, surrounds the cell membrane. Adjacent plant cells are held together by a thin layer composed mainly of calcium pectate and known as the **middle lamella**. Other distinctive features of some plant cells are pigment-containing bodies called plastids, the most common of which are the green **chloroplasts**, and also large sap-filled **vacuoles**.

Both plant and animal cells contain a **nucleus** at some stage in their development. The nucleus is surrounded by a **nuclear membrane** and contains granular chromatin in which one or more dense areas known as **nucleoli** are suspended. During nuclear division the nucleoli disappear and the chromatin appears as thread-like **chromosomes** (Chapter 7).

The protoplasm outside the nucleus is called **cytoplasm**. At the end of the nineteenth century, Camillo Golgi stained brain cells with silver salts, and observed a cytoplasmic structure which looked like a tiny net. It was subsequently called the **Golgi body**.

Mitochondria are also structures of the cytoplasm common to most cells, and are just visible as tiny granules with a compound microscope. **Storage materials** are often seen in the cytoplasm. In plant cells **starch** is the main storage substance, while in animal cells granules of **glycogen** are commonly found.

Although all living cells have many common features, there is no such thing as a typical or generalised cell. Attention has already been drawn to the main differences between plant and animal cells. It is important to realise that multicellular plants and animals consist of a variety of cell types. Reference is made in succeeding chapters to the ways in which cell structure is related to function.

6.2 Cell ultrastructure as revealed by the electron microscope

Since the 1950s biologists using the electron microscope (Fig 6.5) have made tremendous strides in our knowledge of the detailed structure of cells.

Fig 6.5 A modern electron microscope (courtesy Kratos Ltd)

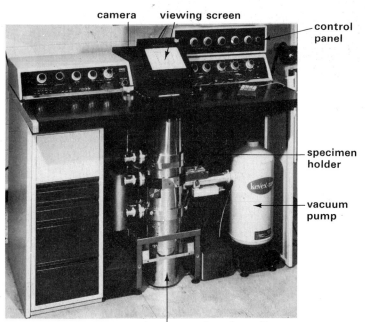

camera   viewing screen   control panel   specimen holder   vacuum pump

housing for condenser objective and projector lenses

The preparation of specimens for sectioning prior to examination with an electron microscope is in some ways similar to the method used for light microscopy, but there are important differences.

To create an image, the electron microscope directs a beam of electrons at and through the material. The image results from the way the material scatters the electrons. Atmospheric atoms and molecules would interfere by also scattering the electron beam, so the material has to be held in a vacuum. Therefore it has always to be prepared in a way which resembles the technique used by light microscopists for preserved specimens. For fixation, osmium oxide is often used. It binds to lipids and proteins, making them electron-dense. Such cell components scatter electrons strongly and appear as dark areas in the image. Otherwise, glutaraldehyde is used which fixes the material without rendering it electron-dense.

The sample is next **dehydrated** with ethanol as described for light microscopy, then **cleared** ready for **embedding**. Paraffin wax breaks if cut into very thin sections, so it is unsuitable as an embedding substance for electron microscopy. Instead, clear plastic epoxy resins such as Araldite are used. The material is cleared in liquid resin which is hardened by gentle heat.

The sample, now embedded in a tough, clear supporting substance, can be cut into very thin slices. Sections of cells and tissues must be extremely thin (0.01–0.5 μm thick), to allow some electrons to pass through them. Sectioning is carried out with an **ultra-microtome**, with the blade usually made by breaking a thick piece of glass to give a hard cutting edge.

The sections are then transferred to tiny circular copper grids on which they may be **stained**. Solutions of lead salts are taken up by lipid components in the specimen, whilst uranyl salts react with proteins and nucleic acids. The effect is to make the lipids, proteins and nucleic acids electron-dense so that they contrast with other cell components in the final image. Such staining is not necessary if osmium oxide is used as a fixative. A special holder is used to place the sections, still supported on their grids, in the electron microscope which is evacuated of air before the electron beam is switched on (Fig 6.6).

When electron microscope techniques were first used, both plant and animal cells were shown to have a detailed structure previously unsuspected. Not only were some components discovered for the first time;

Fig 6.6 Stages in the preparation of thin sections for electronmicroscopy: *left* cutting sections with an ultramicrotome, *right* mounting sections on to a copper grid

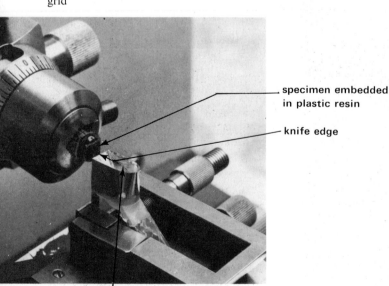

specimen embedded in plastic resin

knife edge

water bath for sections to float in

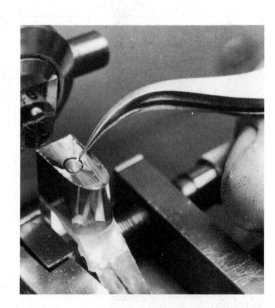

Fig 6.7 (a) Electronmicrograph of a section through a pancreatic cell, × 10 000

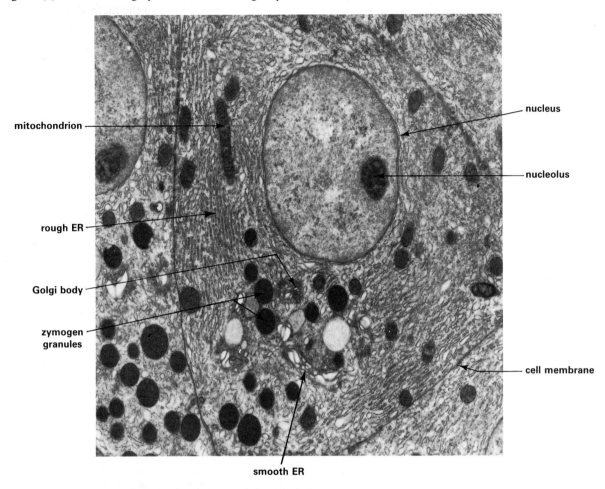

mitochondrion

nucleus

nucleolus

rough ER

Golgi body

zymogen
granules

cell membrane

smooth ER

Fig 6.7 (b) Electronmicrograph of a section through parts of two adjacent cells from a tobacco leaf, × 15 000

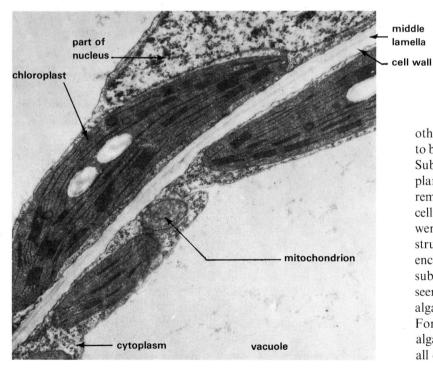

part of
nucleus

middle
lamella

cell wall

chloroplast

mitochondrion

cytoplasm   vacuole

others already well known were found to be very complex in structure (Fig 6.7). Subcellular components common to plant and animal cells were seen to be remarkably similar in appearance. The cells of bacteria and blue-green algae were found to be relatively simple in structure. They lack a membrane-enclosed nucleus and have few of the subcellular structures called organelles seen in fungi, protozoa, most kinds of algae, multicellular plants and animals. For this reason bacteria and blue-green algae are termed **prokaryotic** whereas all other organisms are **eucaryotic**.

101

6.2.1 The cell membrane

The outer boundary of the protoplast, the **cell membrane**, is invisible with the light microscope. Even so, its presence can be inferred because protoplasm leaks out of animal cells when the cell surface is punctured. Overton in 1895 suggested that the membrane was made of fatty substances. Other workers later deduced that two layers of lipid were present in the cell membrane. In 1935 Danielli and Davson proposed a model for membrane structure in which a **lipid bilayer** was coated on either side with **protein** (Fig 6.8). Mutual attraction between the hydrocarbon chains of the lipids, and electrostatic forces between the protein and the 'heads' of the lipid molecules, were thought to maintain the stability of the membrane.

Fig 6.8 The Danielli–Davson model of the cell membrane

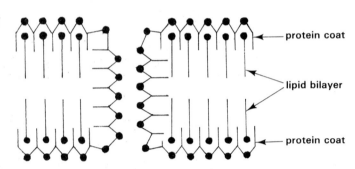

protein coat

lipid bilayer

protein coat

Using evidence from electronmicrographs Robertson, in 1960, proposed a **unit membrane hypothesis** based on the observation that all membranes of cells have a comparable appearance when viewed with the electron microscope. The two outer layers of protein are each about 2 nm thick and appear densely granular. They enclose a clear central area about 3.5 nm wide consisting of lipid (Fig 6.9). The term **unit membrane** is now used to denote all cell membranes which are similar to this in structure.

Fig 6.9 Electronmicrograph of a section through microvilli from intestinal epithelium, × 180 000. A unit membrane surrounds each microvillus

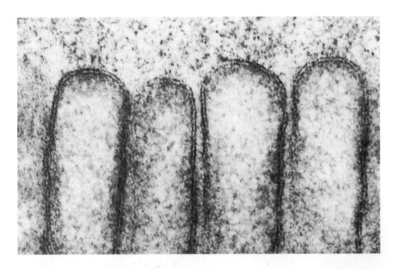

Much has since been learned about the composition and probable organisation of cell membranes. The lipids are mainly **phospholipid** molecules which are polar at the phosphate group end (Chapter 2). There is considerable variation in the fatty acid content of membrane lipids from the cells of different species. The saturated fatty acids of some lipids attract molecules of **cholesterol**. The amount of protein relative to the quantity of lipid also varies from one cell type to another.

Proteins in the outer parts of the membrane may even differ from those of the inner part. Many are **carrier proteins** which transport substances across the membrane. Others are enzymes which catalyse biochemical reactions at the cell surface. The current view of membrane structure which is generally held is the **fluid-mosaic model** proposed in 1972 by Singer and Nicholson. It suggests that the protein component of the membrane is patchy rather like a mosaic. Constant movement of the phospholipid molecules gives the membrane fluidity. The cell membrane is therefore a dynamic rather than a static structure. Some proteins on the exterior of cell membranes are **antigenic**. Lymphocytes have antibodies attached to the cell membrane. Antigens and **antibodies** provide a means whereby our cells can distinguish self from non-self (Chapter 10). The way in which the components of cell membranes are now thought to be arranged is shown in Fig 6.10.

Fig 6.10 The fluid-mosaic model of the cell membrane

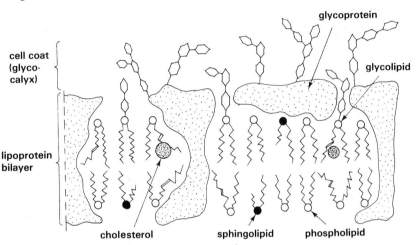

The exterior of the cell membrane of animal cells is called the **cell coat**. It is made of **mucopolysaccharides, glycolipids, glycoproteins** and **hyaluronic acid**. The cell coat is sticky, enabling animal cells to adhere to each other.

The cell membrane is much more than just a protoplasmic boundary. It provides a means of controlling the passage of materials both into and out of the cell. Some materials are taken in by **phagocytosis** and **pinocytosis**. Phagocytosis is a mechanism that enables large suspended particles to be taken wholesale into cells. Pinocytosis is the intake of droplets of liquid by the formation of tiny pockets in the membrane. The cell membrane regulates the passage of water and dissolved substances into or out of the cell. Water passes through the membrane by **osmosis** (Chapter 1). Water-soluble substances cross the membrane by **diffusion**, by **facilitated transport** or by **active transport**. It is now generally agreed that many water-soluble solutes are transported through the membrane by carrier proteins. Lipid-soluble compounds can pass more quickly through membranes by dissolving in the phospholipid layer.

Diffusion is the random movement of ions, atoms or molecules. It results in such particles moving from places where they are highly concentrated to where their concentration is less. For example, oxygen diffuses through the cell membrane of the epithelium lining our alveoli and into blood circulating in nearby capillaries. Carbon dioxide diffuses in the opposite direction (Chapter 8). Carrier proteins may bind with and facilitate the diffusion of some kinds of particles across the plasma membrane (Fig 6.11). The absorption of most nutrients from the gut occurs by facilitated transport (Chapter 15). Active transport requires the use of ATP to move particles across the plasma membrane against a concentration gradient. Reabsorption of sodium ions from the renal filtrate, and the sodium pump in nerve impulse transmission, are examples of active transport (Chapters 11 and 17 respectively).

Fig 6.11 (a) Relative permeability of cell membrane

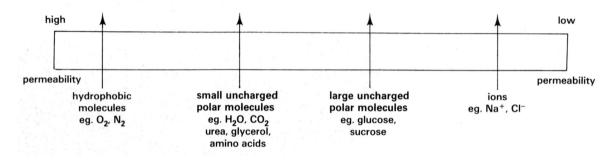

(b) Transmembrane transport mechanisms

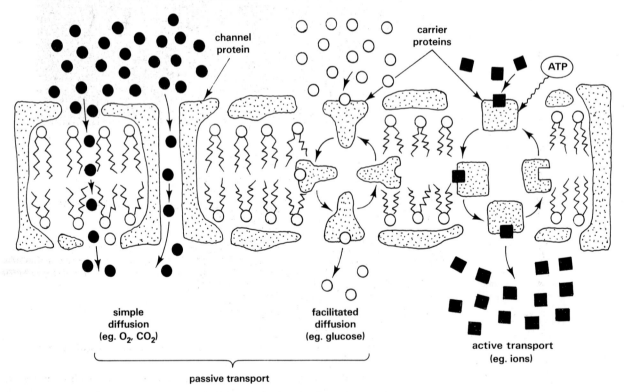

(b)

Passive mechanisms move substances along a concentration gradient using the kinetic energy of the transported atoms, molecules or ions.
Active transport uses energy from ATP to move substances against a concentration gradient.

6.2.2 Endoplasmic reticulum

Biologists once regarded the cytoplasm as a homogeneous jelly. This view was radically changed when sections of cells were examined with electron microscopes. Extending throughout the cytoplasm is a three-dimensional network of sac-like and tubular cavities called **cisternae** bounded by a unit membrane. These structures are collectively called the **endoplasmic reticulum (ER)** (Fig 6.12). In places the membranes are covered on the cytoplasmic side with ribosomes, the **rough ER**. Ribosomes are small bodies about 15 nm in diameter. They consist of protein and ribosomal RNA. Elsewhere the ribosomes are lacking, the **smooth ER**. The total area of the ER membranes in a cell of volume $5000 \, \mu m^3$ can be as much as $40\,000 \, \mu m^2$.

Fig 6.12 (a) Electronmicrograph of a section through rough ER, $\times 100\,000$

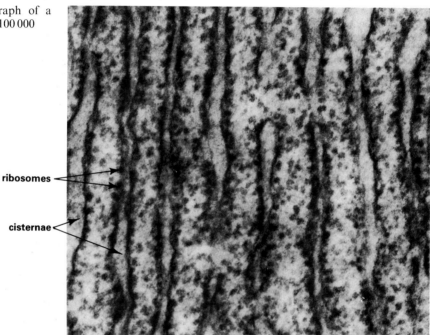

Fig 6.12 (b) Diagram showing three-dimensional structure of ER

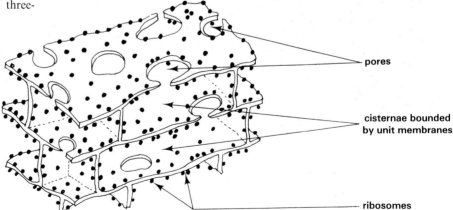

If labelled amino acids are introduced into live cells, radioactivity first appears in the ribosomes. Within a few minutes it is found in the membrane-enclosed sacs of the rough ER. The reason for this is that proteins are produced at the ribosomes, threaded through the membrane, and are stored temporarily in the sacs of the rough ER before they are used inside the cell or are secreted to the exterior. It is thus not surprising that the rough ER is very prominent in **enzyme-secreting** cells such as those of the pancreas. The enzymes enter vesicles formed by the Golgi body before they are moved through the cell membrane by reverse pinocytosis.

The smooth ER is prominent in **steroid-secreting** cells such as the interstitial cells of the testes and the adrenal cortex, and also in cells concerned with lipid metabolism such as the epithelial cells of the intestine. The smooth ER also gives rise to the Golgi body (section 6.2.3). Both types of ER are continuous with the nuclear membrane.

The main functions of the ER are to provide a relatively large surface area for synthesis, and to permit the rapid transport of such molecules within the cell, and from the inside to the outside of the cell.

6.2.3 The Golgi body

The **Golgi body** was discovered in brain cells. Since then it has been seen in cells from almost every group of living organisms. In transverse section the Golgi body often appears as closely packed, parallel curved pockets (Fig 6.13). The pockets are bounded by unit membranes and are called

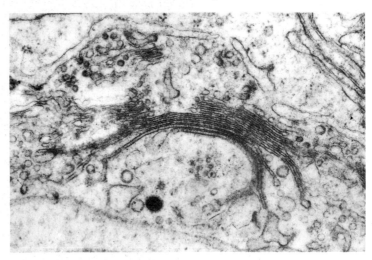

Fig 6.13 Electronmicrograph of a section through a Golgi body, × 24 000

cisternae. From the edges of the cisternae tiny **vesicles** arise. Some of the vesicles become lysosomes (section 6.2.4), some fuse with and enlarge the cell membrane, others carry secretions to the cell membrane for release to the exterior. Recently it has been shown that the cisternae are net-like (Fig 6.14). Like the endoplasmic reticulum, Golgi bodies are well developed in cells whose secretions include glycoproteins (Chapter 2). In the Golgi body carbohydrate is added to protein coming from the ER, and the glycoprotein product is secreted at the cell surface. **Mucus** is a typical glycoprotein secreted by goblet cells which abound in the respiratory and gastro-intestinal tracts of mammals.

Fig 6.14 Diagram showing three-dimensional structure of a Golgi body

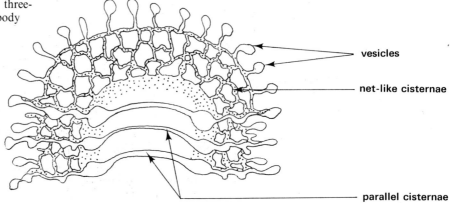

vesicles

net-like cisternae

parallel cisternae

6.2.4 Lysosomes

The term **lysosome** was given by Christian de Duve in 1955 to tiny organelles containing hydrolytic enzymes lying near the nucleus of liver cells (Fig 6.15). Electronmicrographs show **primary lysosomes** as small vesicles bounded by a double unit membrane arising from the edge of the Golgi body. Larger **secondary lysosomes** are formed by fusion of primary lysosomes with small vacuoles. The vacuoles arise by infolding of the cell membrane. Primary lysosomes may also fuse with **autophagosomes** which are internally-formed membranous pockets enclosing worn-out organelles such as mitochondria and ribosomes (Fig 6.16).

Fig 6.15 Electronmicrograph of a section through lysosomes, × 28 000

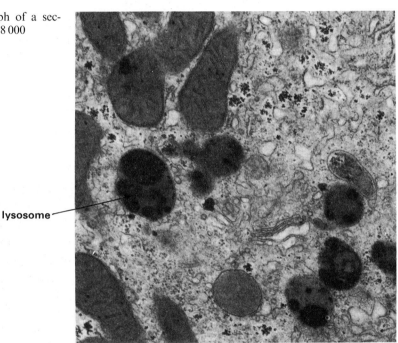

lysosome

Fig 6.16 Summary of the functions of lysosomes

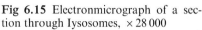

pinocytosis

phagocytosis

secretion

excretion

autophagosome

secondary lysosomes

worn out mitochondrion

primary lysosomes

vesicle

Golgi body

The main function of lysosomes is the digestion of particles made in the cell or taken into the cell from outside. A large variety of enzymes has been demonstrated in lysosomes from different sources. Many are lipases, carbohydrases and peptidases which hydrolyse lipids, carbohydrates and proteins respectively. Lysosomal enzymes cause the destruction of foreign particles such as bacteria engulfed by phagocytes, and the breakdown of ageing organelles in all cells.

In old and diseased cells, enzymes released internally by lysosomes bring about self-destruction, **autolysis**, of the protoplast. Erosion of cartilage in rheumatoid arthritis is an example of such activity.

6.2.5 Mitochondria

The presence of small elongated bodies in the cytoplasm of plant and animal cells was first reported in the 1850s. The name **mitochondria**, meaning thread-granules, was later given to these structures which are just visible with a light microscope. It was not until a hundred years later that thin sections of mitochondria examined under the electron microscope showed that they are complex. They have a smooth outer unit membrane and a much folded inner unit membrane of relatively large surface area (Fig 6.17). Many stalked spherical bodies called **oxysomes** are attached to the folds which are known as **cristae**. The inner membrane encloses a space called the **matrix** containing enzymes and DNA. The DNA codes the synthesis of proteins in mitochondrial membranes. In this way mitochondria replicate when a cell divides.

Fig 6.17 (a) Electronmicrograph of a section through a mitochrondrion, × 81 000

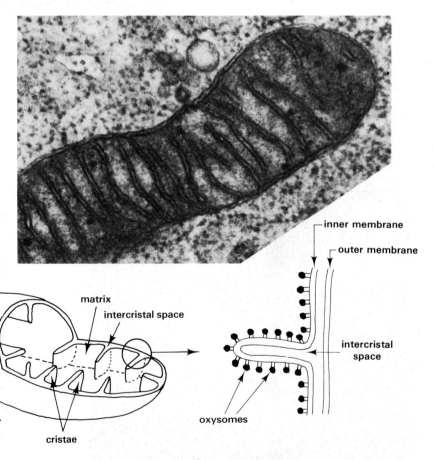

Fig 6.17 (b) Diagram showing three-dimensional structure of a mitochondrion

matrix

intercristal space

inner membrane

outer membrane

intercristal space

oxysomes

cristae

Measure in mm the width of the mitochondrion shown in Fig 6.17(a). Multiply by 1000 to convert to μm. Now divide by the magnification to find the width in μm.

The shape of mitochondria is variable, from rod-shaped to spherical, spiral and even cup-shaped. They also vary in size depending on their source. The mitochondria from liver cells measure $1-2\,\mu m$ long and $0.3-0.7\,\mu m$ wide, while those from pancreatic cells, although of the same width, are up to $10\,\mu m$ long. Cells which are metabolically very active contain large numbers of mitochondria. The tubule cells of kidney nephrons, muscle fibres and axon terminals, for example, are packed with mitochondria.

Mitochondria provide sites isolated from the cytoplasm on which the enzyme-catalysed reactions of **aerobic respiration** occur. For instance, mainly at the oxysomes, **adenosine triphosphate (ATP)** is produced from adenosine diphosphate (ADP) and inorganic phosphate ions. Conversion of ADP to ATP is an energy-consuming reaction. Energy for the conversion comes from the oxidation of energy-rich substrates such as pyruvic acid derived from glycolysis of sugars in the cytoplasm. ATP provides the cell with energy for energy-consuming processes such as active transport and muscle contraction. The biochemistry of respiration is described in detail in Chapter 5.

6.2.6 Peroxisomes

Peroxisomes are also called **microbodies**. They are tiny vesicles about the size of lysosomes but bounded with a single unit membrane. The vesicles contain the enzyme **catalase** which catalyses the breakdown of hydrogen peroxide, a by-product of aerobic respiration. Hydrogen peroxide is toxic if allowed to accumulate, so is best disposed of as quickly as it forms.

$$2H_2O_2 \rightarrow 2H_2O + O_2$$

In some instances the hydrogen peroxide is used to oxidise other molecules. For example, about half of the ethanol we consume is oxidised to ethanal in the peroxisomes of our liver cells. This is one of several ways in which our liver detoxifies unwanted molecules (Chapter 15).

Fig 6.18 Electronmicrograph of a section through a nucleus, $\times 14\,000$

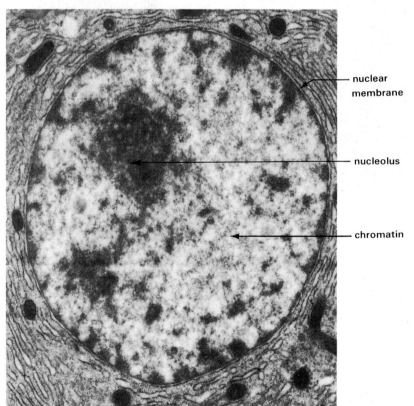

nuclear membrane

nucleolus

chromatin

6.2.7 Nucleus

The **nucleus** is usually the largest of the cell's organelles and can thus be seen with a light microscope. Van Leeuwenhoek is accredited with the discovery of the nucleus in the red cells of salmon blood towards the end of the seventeenth century. A distinct nucleus is present at some stage in the cells of all forms of life apart from bacteria, blue-green algae and viruses. Although usually more or less spherical, the nucleus can be more complex in shape. For example, the nucleus of a neutrophil white blood cell is lobed (Chapter 10).

In electronmicrographs the nucleus is seen to be bounded by a double-layered **nuclear membrane** (Fig 6.18). Each layer is a unit membrane, the outer one often covered with ribosomes and continuous with the endoplasmic reticulum. Between the two layers is a **perinuclear space** about 20 nm wide.

A prominent feature of the nuclear membrane is the presence of numerous **pores**. They may occupy up to 15 % of the membrane's surface area, each pore being approximately 50 nm in diameter (Fig 6.19). The

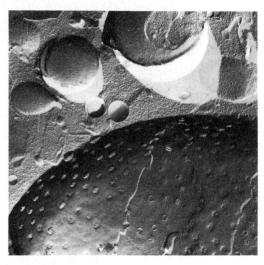

Fig 6.19 Electronmicrograph of a freeze-etched nuclear membrane showing nuclear pores, × 45 000

pores provide routes for the passage of large molecules, such as messenger RNA, from the nucleus to the cytoplasm and vice versa. Inside the nuclear membrane are two main ingredients, **nucleic acids** and protein. Both RNA and DNA are present, the nucleus being the main store of the cell's DNA. Nuclear DNA is bonded to a number of proteins collectively called **histone**. When the nucleus is not dividing the nucleic acid–protein complex appears as tiny granules of **chromatin**. Among the chromatin granules are one or more densely granular bodies called **nucleoli** composed mainly of DNA.

During nuclear division, the nuclear membrane and nucleoli disappear and the chromatin becomes visible as thread-like bodies called **chromosomes** (Chapter 7). At telophase the nucleoli reappear. It is thought that nucleoli are the sites where ribosomal RNA is made.

The way in which the nucleus transmits hereditary materials is described more fully in Chapter 7. In Chapter 4 you can read how the nucleic acids provide the genetic code for protein synthesis at the ribosomes.

6.2.8 Centrioles, cilia and flagella

Centrioles are characteristic of animal and fungal cells but are not found in plants. A pair of **centrioles**, each with its long axis at 90° to the other, usually lies near the nucleus. A centriole is a cylinder made of nine tubular filaments about 0.2 μm in length. At very high resolution, each filament can be seen to consist of three fused hollow fibrils (Fig 6.20(a)).

During nuclear division the centrioles divide and a pair of them moves to each pole of the cell. They produce a system of **microtubules** called **spindle fibres** radiating towards the equator of the cell. Chromosomes become attached to the spindle equator before migrating to the poles of the cell, seemingly connected to the microtubules.

In some cells, centrioles divide to produce **basal bodies** from which flagella and cilia develop. **Cilia** contain longitudinal fibrils, two in the

Fig 6.20 (a) Electronmicrograph of a section through a pair of centrioles, × 65 000. The left centriole has been cut transversely, the right longitudinally

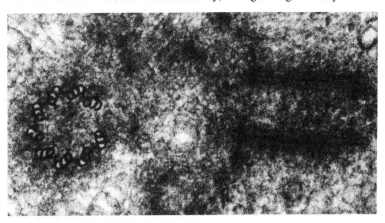

Fig 6.20 (b) Electronmicrograph of transverse section through cilia, × 5000 (courtesy Biophoto Associates)

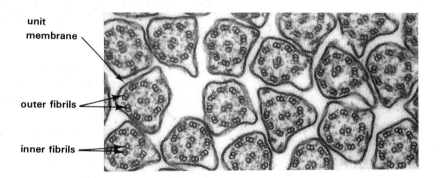

centre surrounded by nine outer pairs, all enclosed in a unit membrane (Fig 6.20(b)). Alternate contraction and relaxation of the fibrils causes rhythmical bending of the cilia. Ciliary movement wafts liquids and particles in suspension over the surface of the cell. **Flagella** are whip-like organelles, relatively long compared with cilia, but with the same ultrastructure. They are used in locomotion by human sperm (Chapter 25).

6.2.9 Chloroplasts

In photosynthetic plants the light-absorbing pigments are housed in complex organelles called **chloroplasts**. The chloroplasts of flowering plants are shaped like biconvex lenses 4–10 μm in diameter and 2–3 μm thick. They are found chiefly in the mesophyll cells of leaves. Being so large, chloroplasts can be seen with a light microscope and they have been studied from the mid-seventeenth century.

Thin sections of chloroplasts viewed under an electron microscope show them to be bounded by a double unit membrane. The outer membrane is smooth while the inner is extended inwards as a system of layers called **lamellae** in which the photosynthetic pigments are located. In places the lamellae appear as flat discs known as **grana** piled on top of each other. These are connected by intergrana lamellae. The entire system of internal membranes is suspended in an aqueous matrix called the **stroma** which contains protein and DNA. Following a period of illumination, photosynthetic end-products such as starch grains and lipid globules appear in the stroma (Fig 6.21).

Fig 6.21 (a) Electronmicrograph of a section through chloroplasts, × 30 000

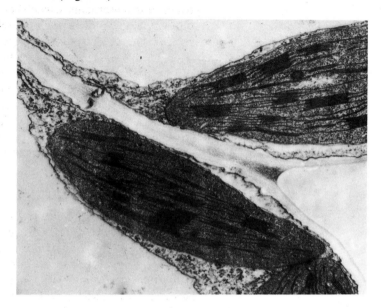

Fig 6.21 (b) Drawing of a thin section through a chloroplast

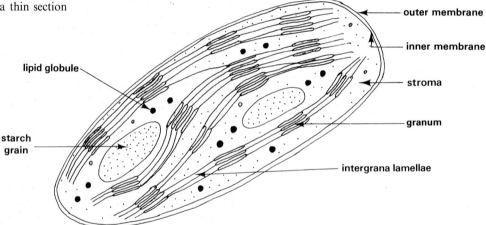

Chlorophyll and other photosynthetic pigments in the membranes are concentrated in the grana which are just visible with a light microscope as darker green spots inside chloroplasts. Chloroplasts provide sites on which the biochemical and photochemical reactions of **photosynthesis** can proceed independent of those going on in the rest of the cytoplasm. Details of the reactions are given in Chapter 5. In the grana especially, solar energy is used to produce **reduced nicotinamide-adenine dinucleotide phosphate (NADPH)** and **ATP** in the light-dependent reactions of photosynthesis. The stroma contains the enzymes necessary for the light-independent reactions in which NADPH, carbon dioxide and ATP are used to synthesise energy-rich organic compounds such as sugars and starch. Chloroplasts are thus the organelles in which energy from sunlight is converted to chemical bond energy.

6.2.10 The cytoskeleton

Many kinds of cells are capable of changing their shape and can alter the position of their organelles. These activities are brought about by a network of cytoplasmic threads, the two most important of which are microfilaments and microtubules. They make up the **cytoskeleton** which is linked to other structures such as the cell membrane by accessory proteins.

Muscle cells contain a permanent cytoskeleton of **actin** and **myosin** protein **microfilaments** which slide between each other to effect muscle contraction (Chapter 17). The hollow **microtubules** of cilia are made of a protein called **tubulin**, the overlapping molcules of which slide over each other in ciliary movement. In less specialised cells the cytoskeleton is transient and assembled only when required. An example is the ring of actin microfilaments which appear just inside the cell membrane at the equator of the cell after nuclear division. Their contraction brings about cytoplasmic cleavage (Chapter 7). The spindle microtubules which form during nuclear division are assembled from a pool of tubulin, a process organised by the centriole (section 6.2.8).

6.2.11 The plant cell wall

A characteristic of plant cells is that they have a **cell wall**. Structurally it is like fibreglass, consisting of fibres enmeshed in an amorphous matrix. The fibres are of **cellulose**, an unbranched polymer of β-D-glucose (Chapter 2). Each fibre is made of several hundred microfibrils in which about 2000 cellulose molecules are held together by hydrogen bonds (Fig 6.22). The matrix consists of **pectic acid** and its salts calcium and magnesium pectate, and **hemicelluloses** which are polymers of various pentose and hexose sugars. Pectic substances also make up most of the middle lamella which binds adjacent plant cells to one another.

Fig 6.22 Diagrammatic representation of the structure of a cellulose fibre

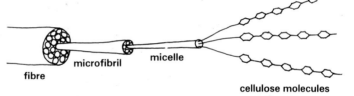

fibre   microfibril   micelle   cellulose molecules

In young cells the wall is thin and called the **primary wall**. The cellulose fibres are orientated at random (Fig 6.23). As a cell grows, more cellulose fibres are laid down on the inside of the primary wall, forming a thicker **secondary wall**. The cellulose fibres of the secondary wall are closely packed and laid down in an orderly way. The wall of fibrous cells may become impregnated with an alcohol polymer called **lignin** which gives great strength. In cork tissues, **suberin** impregnates the cell wall, while in the outer walls of epidermal cells of leaves and young stems **cutin** appears instead. Both suberin and cutin are waxy materials which provide an effective waterproof covering to the aerial surface of plants.

Fig 6.23 Electronmicrographs of (a) the primary wall, and (b) the secondary wall of a plant cell

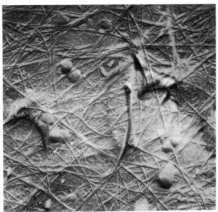

(a)

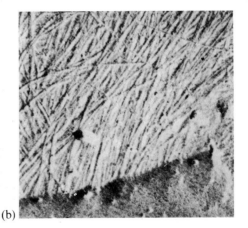

(b)

6.3 Comparison of light and electron microscopes

Why is it that so much more detail can be made out when specimens are examined with an electron microscope as opposed to a light microscope? Before we can answer this question it is necessary to compare the principles on which light and electron microscopes work.

Fig 6.24 shows the paths of radiation in the two instruments. In both microscopes a tungsten filament lamp is used as a **source of radiation**, but whereas the light microscope uses **visible light** to create the final image, the electron microscope uses a beam of **electrons**. The radiation is focused on to the specimen by a **condenser**, which in the light microscope consists of thick glass lenses mounted beneath the stage. In electron microscopes the condenser is a vertical magnetic field produced by a large cylindrical electromagnet which straightens and intensifies the electron beam. The light rays or electrons, as the case may be, now pass through the specimen, after which the radiation is focused by an **objective lens**. An **eyepiece lens** in the light microscope further enlarges the image. The human eye cannot see electrons, so in the electron microscope the final image is focused by a **projector lens** on to a viewing screen coated with a fluorescent compound such as zinc sulphide. When irradiated with electrons, the fluorescent substance emits light visible to the human eye. A comparable process is used to create a picture on a television screen. With both microscopes it is possible to photograph the final image to produce **photomicrographs** and **electronmicrographs**.

Fig 6.24 Comparison of the components and pathways of radiation:

(a) of an electron microscope

(b) of a conventional light microscope

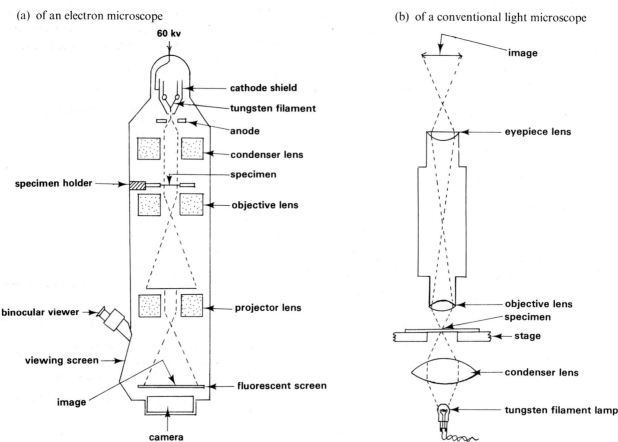

Perhaps the most obvious difference in the final images is the extent to which the specimens are magnified. With a high quality compound microscope fitted with an oil-immersion objective, a magnification of about 1500 times is possible. An electron microscope can magnify up to 500 000

times. However, it is the **resolving power** of the electron microscope rather than its magnifying power which enables it to produce images containing so much more detail. Resolving power or **resolution** is the ability to make out bodies which lie close to one another as separate entities.

The power of resolution (R) of a microscope depends mainly on the wavelength of the radiation used and on the **numerical aperture** (NA) of the objective lens.

$$R = \frac{0.5\,\lambda}{n \,.\, \sin \theta}$$

where λ = wavelength of radiation

n = refractive index of medium between specimen and objective lens

$\theta = \frac{1}{2}$ angle of aperture (Fig 6.25(a))

The expression $n \,.\, \sin \theta$ is the NA of the objective. The smaller the value of R, the better is the resolving power of the microscope. Using a compound microscope fitted with an oil-immersion objective, $R = 210\,\text{nm}$ (Fig 6.25(b)). In other words this instrument is theoretically capable of resolving particles lying as close as 210 nm to one another. In practice it is difficult to resolve as well as this when viewing unstained preparations with a conventional compound microscope.

The **phase-contrast microscope** provides better powers of contrast but does not improve resolution (Fig 6.3). In this instrument a circular **backstop** is fitted into the substage to create a hollow beam of light which is then focused onto the specimen. Light rays not refracted by the specimen are then put out of phase of the refracted rays by an objective lens which has a circular groove cut in it (Fig 6.26). The effect is that structures can be seen

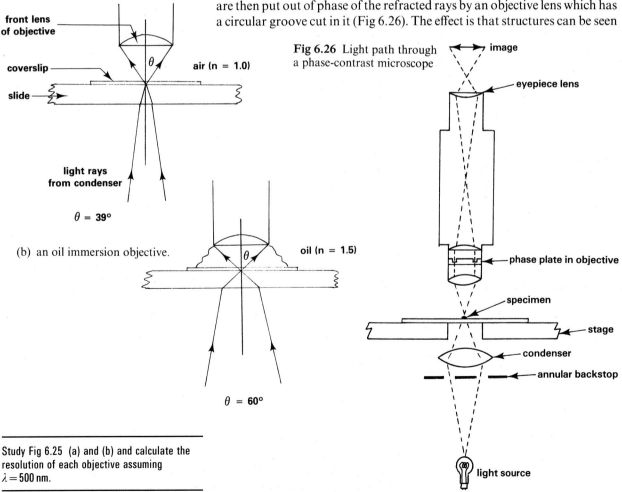

Fig 6.25 Angle of aperture:

(a) a high power dry objective

front lens of objective

coverslip

slide

air (n = 1.0)

light rays from condenser

$\theta = 39°$

(b) an oil immersion objective.

oil (n = 1.5)

$\theta = 60°$

Fig 6.26 Light path through a phase-contrast microscope

image

eyepiece lens

phase plate in objective

specimen

stage

condenser

annular backstop

light source

Study Fig 6.25 (a) and (b) and calculate the resolution of each objective assuming $\lambda = 500$ nm.

which are difficult to make out with ordinary bright-field illumination. It does not however show any further detail of structure.

With an electron microscope λ is 0.05 nm, which is 10 000 times shorter than the average wavelength of white light. Other things being equal, this should produce a corresponding increase in resolution, but there are technical difficulties preventing this from being realised in practice. Present-day electron microscopes have resolving powers of about 0.5 nm, about 400 times greater than the light microscope. Compare these figures with the resolving power of the human eye which is about 1.0×10^8 nm. The major disadvantage of the electron microscope is that it cannot be used to examine live specimens. Why is this so?

6.4 Cell fractionation

The wealth of information which now exists on cell structure has clearly depended on technological improvements in microscopy. The electron microscope is the most sophisticated of instruments available to date for probing the fine structure of cells. Useful as it is to know the detailed structure of cells, electronmicrographs tell us little about the functions of the various cellular components. Cell biologists therefore make use of a variety of techniques in investigating the relationship between structure and function.

How is so much known about the functions of the various components of cells? One technique which has been of enormous help involves separating or **fractionating** the organelles. The activities of the organelles can then be studied without interference from all of the other reactions which take place in whole cells.

Live tissue is first chopped up in a cold **isotonic buffer** solution. The isotonic solution prevents distortion of the organelles. The chopped tissue is then ground up in an **homogeniser**. A domestic blender can be used for this purpose but it usually breaks most of the organelles. Sophisticated work employs a motor-driven ground glass pestle which fits into a tube. This type of homogeniser develops shearing forces just sufficient to rupture the cells. Cells can also be ruptured using **ultrasonic waves**. The homogenate is then transferred to a **centrifuge** in which the mixture is spun at specific speeds at which organelles are known to sediment separately. The main factors governing sedimentation are the magnitude of the centrifugal force, which depends on the spinning speed, and the size and density of the suspended organelles relative to the medium in which they are suspended. Exact times and speeds of centrifugation vary from one tissue to another and are determined by trial and error.

Centrifugal forces of up to 1000 times the force of gravity $(1000 \times g)$ can be attained with simple bench centrifuges used in school laboratories. Large organelles such as nuclei and chloroplasts can be sedimented by spinning at 500–600 g for 5–10 minutes. If the supernatant liquid is spun at 10 000–20 000 g for 15–20 minutes, mitochondria and lysosomes are sedimented. Fragmented endoplasmic reticulum with attached ribosomes, collectively called microsomes, can be sedimented from the supernatant liquid by spinning it for 60 minutes in an **ultracentrifuge** (Fig 6.27) in which forces of 100 000 g

Compare the advantages and limitation of light microscopes and electron microscopes.

Fig 6.27 (a) Diagram of the interior of an ultracentrifuge

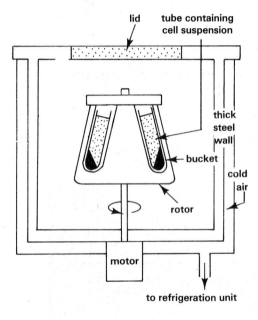

lid   tube containing cell suspension

thick steel wall

bucket

cold air

rotor

motor

to refrigeration unit

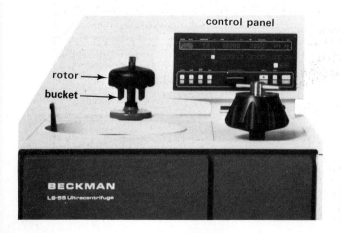

control panel

rotor

bucket

BECKMAN
L8-55 Ultracentrifuge

Fig 6.27 (b) Upper part of an ultracentrifuge (courtesy Beckman – RIIC). The homogenised cell suspension is placed in the buckets for centrifugation

and above are developed. Fig 6.28 summarises the various steps in fractioning cell organelles by **differential centrifugation**. Separating organelles in this way can lead to their physical and chemical damage. They may not then function as they usually do in the intact cell where their activities are closely co-ordinated. Nevertheless, by suspending the separated components in a medium which closely resembles the intracellular environment, some of the functions of organelles can be investigated.

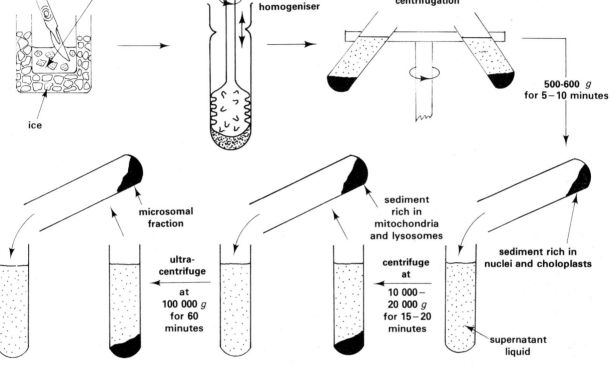

Fig 6.28 Major stages in the fractionation of organelles by differential centrifugation

tissue chopped in cold isotonic buffer solution

ice

pestle homogeniser

centrifugation

500-600 g for 5–10 minutes

microsomal fraction

ultra-centrifuge at 100 000 g for 60 minutes

sediment rich in mitochondria and lysosomes

centrifuge at 10 000– 20 000 g for 15–20 minutes

sediment rich in nuclei and choloplasts

supernatant liquid

SUMMARY

The basic structural and functional unit of living organisms is the cell. Much of what is known about cell structure has come from observations with microscopes. The resolving power of the electron microscope reveals that all cells have a complex ultrastructure. Within cells, different activities go on in specialised organelles. In this way a vast array of biochemical reactions can proceed without interference. Many of the organelles are common to most cells but are developed to different extents according to the cell's functions. Others are unique to particular kinds of cells.

The living matter of all cells is bounded by a cell membrane. The current view of the structure of cell membranes is centred on the fluid-mosaic model. By a variety of mechanisms including osmosis, diffusion, active and facilitated transport the cell membrane regulates the movement of substances into and out of the cell. Proteins on the exterior of the cell membrane are part of a cell recognition system. Membrane-bound enzymes catalyse reactions at the cell surface.

Among the organelles are:

  (i) The endoplasmic reticulum (ER) which is of two varieties—rough and smooth. The role of the ER is to act as an internal transport system for proteins and steroids made within its cisternae
 (ii) Golgi body which is concerned with the synthesis of glycoproteins such as mucus
(iii) Lysosomes which contain hydrolytic enzymes; they serve to digest solid particles
 (iv) Mitochondria and chloroplasts which are energy converters
  (v) Peroxisomes in which catalase acts on hydrogen peroxide, often simultaneously detoxifying unwanted substances such as ethanol ▶

QUESTIONS

1 The diagram below shows a model of the plasma membrane.

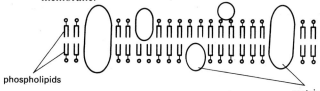

phospholipids

protein

(a) Describe the structure of a phospholipid.
(b) Phospholipids are polar molecules. Give *two* reasons why this is important in the formation of plasma membranes.
(c) Give *two* functions of the protein molecules in the membranes.
(d) What is meant by each of the following terms in relation to the uptake of substances through the plasma membrane?
 (i) facilitated diffusion,
 (ii) active transport.   (L)

2 The following passage refers to pinocytosis. Read the passage carefully and then answer the questions that follow.

 Pinocytosis is the engulfing of a purely liquid sample of the environment by a cell. The cell membrane produces folds that surround the material and the folds fuse together enclosing the material in a vacuole in the cell. By this method cells may take in useful solute macromolecules much too large to cross the cell membrane in the usual way.
 Pinocytosis does not occur continuously but is stimulated by a variety of agents that may be present in the external medium. For example, proteins and amino acids stimulate cells to commence pinocytosis, as do a number of inorganic and organic cations. Anionic substances and negatively charged macromolecules, such as nucleic acids, are without effect. In addition, proteins only induce pinocytosis when they are at a pH such that they bear a net positive charge. It appears that the stimulation of pinocytosis requires binding of the positive ion to the cell surface prior to the initiation of cell membrane activity.
 In a sense, pinocytosis can be regarded as an active, but not selective, transport mechanism. The event does require metabolic energy and, at least in principle, can lead to movement of a molecule against its concentration gradient. Obviously, the pinocytotic vacuole does not select

which of the molecules in the medium it will include, so that the uptake reflects, very strictly, the percentage composition of the medium. In any case, pinocytosis is inhibited by agents that interfere with cellular-energy production, such as carbon monoxide and cyanide. These inhibit mitochondrial oxidation and, therefore, ATP synthesis.

(a) (i) Distinguish between the terms *phagocytosis* and *pinocytosis*.
(ii) Give **two** examples of phagocytosis.
(b) (i) List **four** agents which stimulate pinocytosis.
 (ii) What characteristic do these agents all have in common?
(iii) Suggest a reason why this characteristic may stimulate pinocytosis.
(c) List **three** mechanisms by which small molecules, such as glucose, enter cells.
(d) Comment on the significance of the statement 'pinocytosis can be regarded as an active, but not selective, transport mechanism'.
(e) Name the reverse process to pinocytosis and give an example.   (O&C)

3 Give an illustrated account of the structure and function of the following cell organelles:
(a) rough endoplasmic reticulum;
(b) cilia;
(c) chloroplasts.   (WJEC)

4 (a) (i) How have our ideas on the structure of plasma membranes been influenced by the introduction of the electron microscope?
(ii) Describe the *fluid mosaic model* of a plasma membrane.
(b) How do substances move across cell membranes?
(O)

5 (a) Make a large labelled drawing of a mitochondrion to show its structure.
(b) Outline the processes which take place within a mitochondrion.
(c) Give *three* types of cell in a mammal where you would expect to find large numbers of mitochondria and explain the significance of this in each example you have given.
(L)

7 Cell division

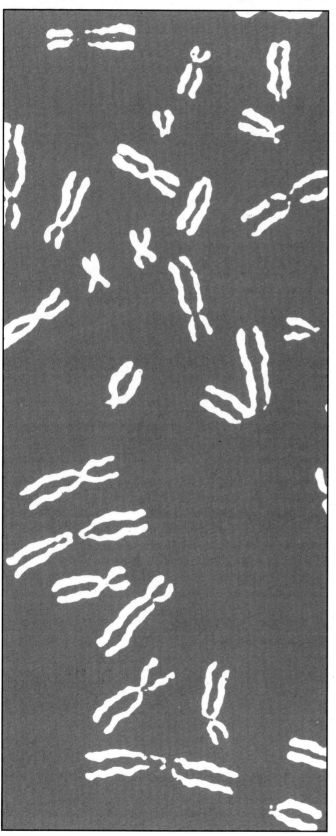

7.1 Mitosis 120

    7.1.1 Mitotic nuclear division 121
    7.1.2 Cytoplasmic cleavage 123
    7.1.3 The cell cycle 124
    7.1.4 The significance of mitosis 125

7.2 Meiosis 126

    7.2.1 The first meiotic division 126
    7.2.2 The second meiotic division 128
    7.2.3 The significance of meiosis 130

Summary 131

Questions 132

7 Cell division

Some types of cells divide constantly throughout life. Typical examples are our bone marrow cells and the epithelial cells which line our gut and cover our skin. Others such as neurones, muscle cells and red blood corpuscles do not divide further when they have matured. Cells also differ greatly in the rate at which they divide. The mean generation time for various kinds of human cells range from as little as 8 hours to over 100 days.

Cell division occurs in two main steps—nuclear division followed by the cleavage of the cytoplasm (**cytokinesis**).

Fig 7.1 Human chromosomes, × 1000

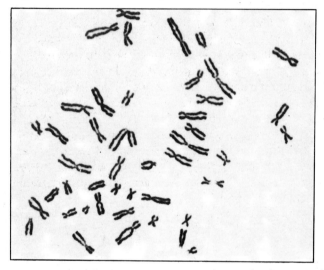

There are two types of nuclear division, **mitosis** and **meiosis**. Mitosis takes place when new cells are added to multicellular organisms as they grow and when tissues are repaired or replaced. Meiosis occurs in the production of gametes by organisms which reproduce sexually.

During both types of division the DNA and histone proteins of the cell nucleus can be seen as threads called **chromosomes** (Fig 7.1). Chromosomes can be stained and are visible using a conventional compound microscope (chromo = coloured; soma = body). The number of chromosomes in the nucleus is fixed for each species of living organism. What is more, the chromosomes can be arranged in **homologous pairs**. In humans for example, the body cells contain 46 chromosomes, 23 homologous pairs.

Mitosis normally ensures that the cells produced contain exactly the same number of chromosomes as the cells from which they were formed. The cells of the bone marrow of humans, for example, constantly give rise to cells which have 46 chromosomes. On the other hand, cells produced by meiosis have half the chromosome number. For example, human sperm and eggs each have 23 chromosomes. However, there is much more to it than this. Let us take a close look at the two types of cell division to see in what other ways they differ.

7.1 Mitosis

Before starting to divide, a cell is at the **interphase** stage. The nucleus appears as a granular body. Inside the nuclear membrane are one or more dense nucleoli (Fig 7.2). The absence of any visible signs of activity

Fig 7.2 A cell at interphase

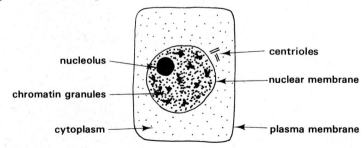

nucleolus

chromatin granules

cytoplasm

centrioles

nuclear membrane

plasma membrane

disguises the fact that intense metabolism is taking place. It is during interphase that replication of DNA and synthesis of nuclear proteins occurs, new ribosomes are made, and mitochondria and centrioles divide. The proteins which later make up the microtubules of the spindle are also made at interphase, although the spindle is not yet constructed. The energy for the various forms of activity comes from respiration. It is therefore not surprising that the respiratory rate of cells at interphase is very rapid.

During both mitosis and meiosis the nucleus divides first, followed by cleavage of the cytoplasm.

7.1.1 Mitotic nuclear division

For convenience of description it is usual to separate mitotic division of the nucleus into four main stages. It is possible to see the various stages by microscopic examination of stained preparations of bone marrow cells or of growing embryos or other suitable material (Fig 7.3). In this way what appears to be a series of static events can be observed. However, mitosis is an active process, the different stages merging into each other. Time-lapse photomicrography provides a means of seeing the dynamic nature of mitosis. Dividing cells are placed in a nutrient solution on a microscope slide and photographed through the microscope every 30 seconds or so, over a period of several hours or even days. The exposed film is developed and run through a ciné projector. It then becomes apparent that during division cells are particularly active.

Fig 7.3 Mitotic division in a fish embryo, × 1000

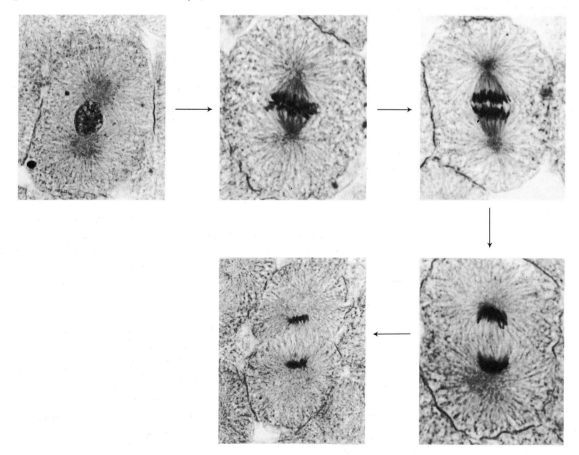

1 Prophase

At the start of **prophase**, chromosomes become visible inside the nuclear membrane. At first they are long, thin, entangled threads. As time goes on the threads become shorter and thicker. The chromosomes disentangle and can be seen as separate structures. As the chromosomes become visible the nucleoli gradually disappear. The centrioles which duplicated at interphase begin to migrate to opposite ends (poles) of the cell. As they move apart the centrioles lay down microtubules which extend from one pole of the cell to the other. The microtubules are the **spindle**, a fibrous structure which is widest at the centre (equator) of the cell. A mass of microtubules called an **aster** may also radiate from the centrioles at each of the poles (Fig 7.4(a)).

Fig 7.4 (a) Stages of mitotic nuclear division (only four chromosomes shown for clarity)

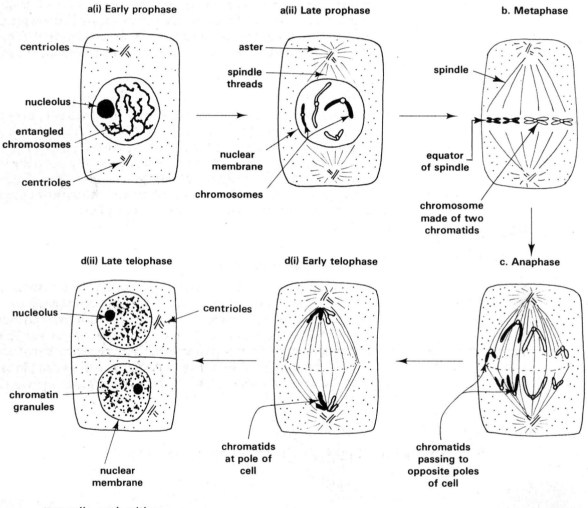

2 Metaphase

The nuclear membrane breaks down. How and why this occurs is still a matter for debate. Mitochondria often gather near the nuclear membrane at this stage. They may provide energy for some of the spindle microtubules to pull the nuclear membrane apart.

Each chromosome can now be seen to consist of two threads called **chromatids** joined at a **centromere** (Fig 7.4(b)). Unlike the rest of the chromosome, the centromere is not easily stained. Its position differs from one chromosome to another. Independently of each other the chromosomes become attached by their centromere to the equator of the spindle.

Fig 7.4 (b) A chromosome

3 Anaphase

The centromere of each chromosome splits and the chromatids move to opposite poles of the cell. Separation of the chromatids appears to be caused by a shortening of the spindle microtubules to which the centromeres are attached. Consequently the chromatids are dragged, centromere first, away from the equator of the spindle. During their passage the chromatids slide over other spindle microtubules which extent from pole to pole.

4 Telophase

The two groups of chromatids come together at opposite poles. Each group becomes surrounded by a newly formed nuclear membrane. It is not clear whether the new membranes are put together from fragments of the nuclear membrane destroyed in prophase or are made anew. Whatever their origin, they are assembled from pieces of membrane. Inside the nuclear membranes the chromatids become uncoiled, nucleoli reform and the nucleus takes on the granular appearance it had at interphase.

7.1.2 Cytoplasmic cleavage

Soon after nuclear division the cytoplasm is separated into two more or less equal parts, each part enclosing one of the newly formed nuclei.

During **cytoplasmic cleavage**, a ring of **microfilaments** appears around the middle of the animal cell just inside the plasma membrane. A shallow **furrow** develops in the membrane, possibly caused by contraction of the filaments. Further shortening of the filaments ultimately pinches the cytoplasm into two more or less equal parts, each part surrounding a nucleus (Fig 7.5(a)).

Fig 7.5 (a) Cytoplasmic cleavage in a dividing bone marrow cell, × 1800

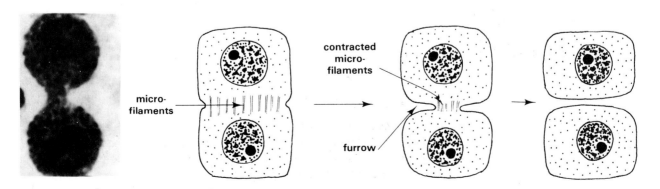

123

In the cells of higher plants a series of flat vacuoles surrounded by unit membrane, probably made by the Golgi body, appear at the middle of the cell. The vacuoles extend across an area of dense cytoplasm called the **phragmoplast** which contains numerous microtubules, the remains of the spindle. The membranes around the vacuoles fuse to form the **cell plate**, a double unit membrane which grows outwards and joins with the plasma membrane of the dividing cell. Cell wall materials are laid down between the two membranes of the cell plate and cleavage of the cytoplasm is thus achieved. By this time the phragmoplast has disappeared (Fig 7.5(b)).

Fig 7.5 (b) Cytoplasmic cleavage in a plant cell

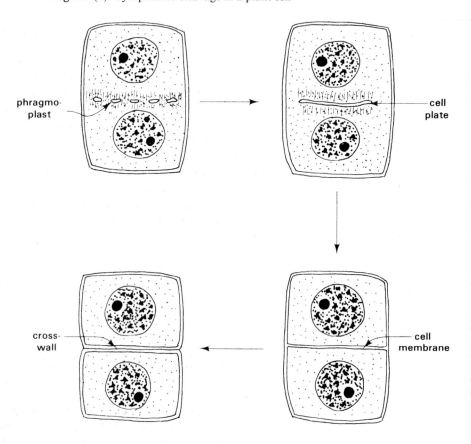

7.1.3 The cell cycle

When incubated at their optimum temperature and supplied with ample nutrients and space, some kinds of human cell grow and divide by mitosis every 24 hours. Interphase occupies by far the longest part of the **cell cycle** (Fig 7.6) and can be divided into the following stages:

i. The G_1 **stage**, in which synthesis of RNA and proteins occurs. It takes about 8 hours to complete.
ii. The **S stage**, in which replication of DNA is completed and duplication of histone proteins occurs. It lasts about 6 hours.
iii. The G_2 **stage**, in which synthesis and replication of organelles such as mitochondria occurs. About 4 hours are normally required for this stage to be completed.

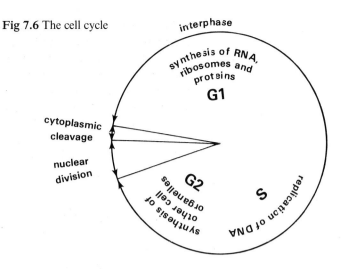

Fig 7.6 The cell cycle

Mitotic nuclear division and cytoplasmic cleavage occupy the remainder of the cycle. In the body of an adult human, few kinds of cells cycle at this rate. Skin cells divide on average once a week, whereas brain cells never multiply. A better understanding of the factors which regulate the cell cycle would be extremely valuable. Restoring growth to vital, damaged organs such as the brain and heart and checking growth of tumours may then be possible. Success in this direction to date has centred mainly an **anti-mitotic drugs** such as vinblastine and vincristine. They bind to tubulin molecules, thus preventing the formation of the microtubules which make up the spindle. In this way they stop the otherwise unchecked cell division, a typical feature of cancers.

7.1.4 The significance of mitosis

The essential feature of mitosis is that it provides a means of distributing the hereditary material DNA equally between two cells. This does not mean that the DNA content is halved at each mitotic division. The hereditary material is normally reproduced in each cell exactly as it was in the parent cell.

One way of finding out when DNA replicates is to measure the amount of DNA in a cell at different times during mitosis. Fig 7.7 shows the results of such an investigation. The amount of DNA exactly doubles during interphase and is halved at anaphase. In this way each of the new cells receives the same amount of DNA as in the parent cell. It does not, however, prove that an exact copy of DNA is distributed to the new cells.

Fig 7.7 Changes in the amount of DNA in a nucleus during mitosis

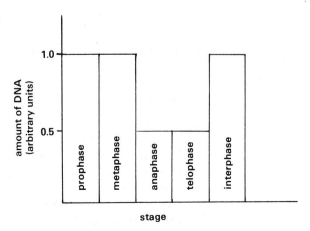

Another approach is to use radionuclides to label the DNA. Cells are grown in a nutrient solution containing thymine labelled with tritium (^{3}H), a radionuclide of hydrogen. As the cells divide, the labelled thymine is incorporated into their DNA, and after a while nearly all the DNA is radioactive. The cells are then placed in a non-radioactive medium. At each subsequent division some of the cells are removed and placed in photographic emulsion. Radiations given off by the labelled DNA cause the emulsion to develop, giving labelled chromosomes a dark colour when viewed under a microscope. The results of such an experiment are shown in Fig 7.8. They prove conclusively that replication of DNA does occur during the cell cycle.

Fig 7.8 Chromosomes of a Chinese hamster after labelling with ^{3}H thymidine (based on autoradiographs)

(a) Metaphase immediately after labelling

(b) Metaphase of next division

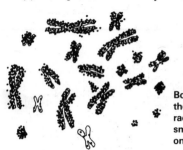

Both chromatids of most of the chromosomes are very radioactive. A few of the smaller chromosomes are only lightly labelled.

Only one chromatid of most of the chromosomes is very radioactive.

The significance of what has just been described is that mitosis normally gives rise to cells with the same combination of genes, the **genome**. Successful genomes can be perpetuated generation after generation in organisms which multiply asexually. In this way breeders can maintain pure strains of many useful plants which can be propagated vegetatively. Some of these are important crop plants such as the potato.

Even so, variations do arise in organisms which do not reproduce sexually. The variation is due in some instances to gene mutations (Chapter 4). In others it is caused by chromosome mutations (Chapter 26).

7.2 Meiosis

Meiosis occurs in the formation of gametes in organisms which reproduce sexually. In meiosis, nuclear division is followed by cytoplasmic cleavage, but in contrast with mitosis there are two nuclear divisions not one. Thus four nuclei are formed from a cell which undergoes meiosis, not two as in mitosis. However, there are other differences which are just as important. The interphase stage of meiosis is the same as in mitosis.

7.2.1 The first meiotic division

Once more, for convenience of description, division of the nucleus is separated into a number of stages. The Roman numeral I is placed after each of the stages in the first meiotic division of the nucleus to distinguish them from the stages of the second meiotic division.

1 Prophase I

The events of **prophase I** are much more complex than those which occur in prophase of mitosis. At the start of prophase I the chromosomes appear as long, thin entangled threads inside the nuclear membrane (Fig 7.9(a)). They then come together in pairs called **bivalents**. Each bivalent is a **homologous pair** of chromosomes. The homologous chromosomes are

Fig 7.9 (a) Stages of the first meiotic division (only four chromosomes shown for clarity)

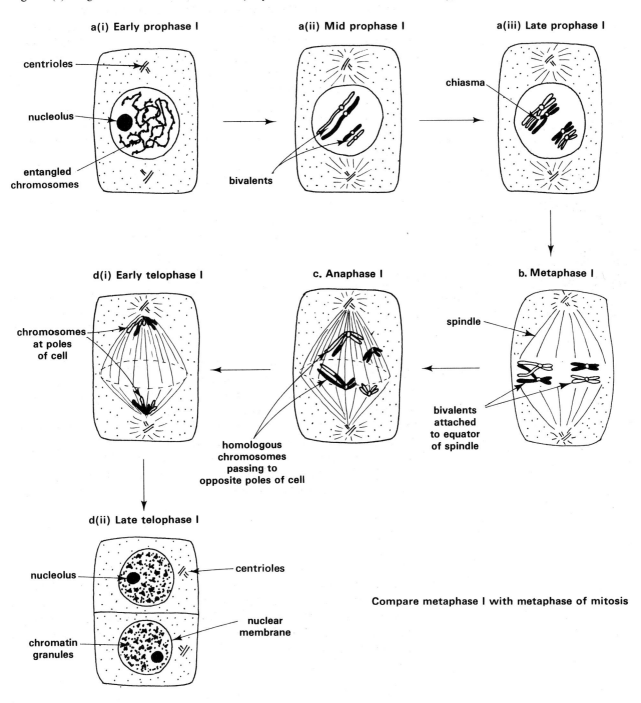

a(i) Early prophase I

centrioles

nucleolus

entangled
chromosomes

a(ii) Mid prophase I

bivalents

a(iii) Late prophase I

chiasma

d(i) Early telophase I

chromosomes
at poles
of cell

c. Anaphase I

homologous
chromosomes
passing to
opposite poles of cell

b. Metaphase I

spindle

bivalents
attached
to equator
of spindle

d(ii) Late telophase I

nucleolus

chromatin
granules

centrioles

nuclear
membrane

Compare metaphase I with metaphase of mitosis

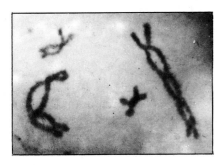

positioned so that their centromeres are adjacent. Gradually the chromosomes shorten and thicken. At this stage the two chromatids making up each chromosome can be clearly seen. Cross-links called **chiasmata** frequently develop between chromosomes and their homologous partners. Bivalents of long chromosomes often display several chiasmata (Fig 7.9(b)).

Fig 7.9 (b) Chiasmata in grasshopper chromosomes (courtesy Philip Harris Biological Ltd)

127

Pairing of homologous chromosomes and the formation of chiasmata do not occur during prophase of mitosis. However, two other events which take place in mitotic prophase also occur in prophase I. The nucleoli disappear and a spindle is laid down in the cytoplasm by the centrioles which divided at interphase.

2 Metaphase I

The nuclear membrane breaks down and the bivalents become attached by their centromeres to the microtubules at the equator of the spindle. One chromosome of each bivalent is directed towards one of the poles of the cell, its homologous partner towards the other pole. It is important to realise that each bivalent is orientated at random with respect to each of the other bivalents.

3 Anaphase I

Shortening of the spindle microtubules drags the homologous chromosomes of each bivalent apart, pulling them to opposite poles of the cell.

4 Telophase I

The two groups of chromosomes come together at opposite poles. Each group becomes surrounded by a new nuclear membrane. The chromosomes uncoil, the nucleoli reappear and the nuclei take on a granular appearance. Cleavage of the cytoplasm may occur as in mitosis.

The two nuclei have half the number of chromosomes as the nucleus from which they were derived. For this reason the first division of meiosis is sometimes called **reduction division**. A short interphase may follow, but there is no replication of DNA as happens during interphase of mitosis. Often there is no cleavage of the cytoplasm and the two nuclei proceed directly to the second division of meiosis.

7.2.2 The second meiotic division

The Roman numeral II is used after each of the stages to distinguish them from the stages of the first meiotic division.

1 Prophase II

Chromosomes appear in both of the nuclei formed in the first division of meiosis. The centrioles move to opposite poles of the cells, laying down the microtubules of the spindle. There is no pairing of chromosomes and chiasmata do not develop as in prophase I (Fig 7.10).

2 Metaphase II

The nuclear membranes disappear and the chromosomes become attached by their centromeres to the microtubules at the equators of the spindles. The two chromatids of each chromosome are now easily seen. What is not so obvious is that the chromosomes are orientated at random with respect to one another.

3 Anaphase II

The centromeres of the chromosomes break in two and the chromatids are pulled, centromere first, towards opposite poles of the cell.

4 Telophase II

The chromatids come together at opposite poles of the cells. Here they become surrounded by nuclear membranes, uncoil, and the nucleoli appear. The spindle disappears and cleavage of the cytoplasm follows.

What are the similarities and differences between the events of prophase I of meiosis and prophase of mitosis?

In what important way does anaphase II of meiosis differ from anaphase of mitosis?

Altogether four cells, each with half the number of chromosomes, are produced from each cell which divides by meiosis. The events of the second meiotic division are similar to those of mitosis, except that there are half the number of chromosomes.

Fig 7.10 Stages of second meiotic division

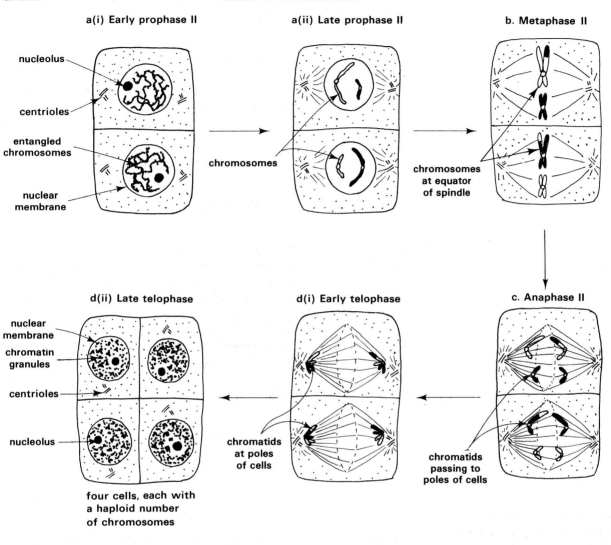

a(i) **Early prophase II**

nucleolus

centrioles

entangled chromosomes

nuclear membrane

a(ii) **Late prophase II**

chromosomes

b. **Metaphase II**

chromosomes at equator of spindle

d(ii) **Late telophase**

nuclear membrane

chromatin granules

centrioles

nucleolus

four cells, each with a haploid number of chromosomes

d(i) **Early telophase**

chromatids at poles of cells

c. **Anaphase II**

chromatids passing to poles of cells

Examine the photographs in Fig 7.11. Which stage of meiosis is shown in each picture?

Fig 7.11 Meiotic division in grasshopper testis (courtesy Philip Harris Biological Ltd)

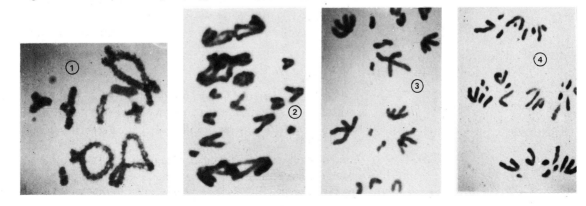

7.2.3 The significance of meiosis

Whilst mitosis normally produces cells with an exact replica of the genetic material found in the parent cell, meiosis gives rise to cells with half the amount of genetic material. Our body cells have a **diploid** number of chromosomes, gametes have a **haploid** number. At fertilisation the diploid number is restored. Meiosis thus ensures that the chromosome number is normally kept constant in each generation (Fig 7.12).

Fig 7.12 Mammalian life cycle

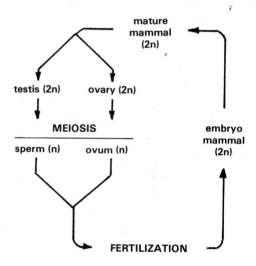

Another significant feature of meiosis is that it produces gametes with varied combinations of genes. There are two important events in meiosis which create new genomes.

1 Crossing over

Crossing over takes place when bivalents appear in prophase I. Chiasmata are formed and homologous chromosomes exchange genes (Fig 7.13). The homologous chromosomes later separate and end up in different gametes. As a result of crossing over, linked genes are parted and gametes with new genomes are thus produced (Chapter 26).

Fig 7.13 Crossing over. Because of crossing over, genes carried on the same chromosone (linked genes) are separated.

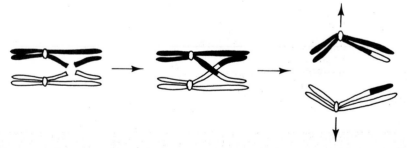

2 Random orientation of chromosomes

The separation of a pair of homologous chromosomes at anaphase I is independent of the separation of other pairs. As the chromosomes are orientated at random, the alleles on one pair of homologous chromosomes separate independently of the alleles on other (Fig 7.14). Because of **random assortment** a vast permutation of genes is possible in the gametes. Crossing over and random assortment ensure that progeny resulting from sexual reproduction are genetically different from their parents. This is why no two persons, unless they are identical twins, have the same genome. Clearly meiosis is important in producing **genetic variation**. Variation can also arise from gene and chromosome mutations (Chapters 4 and 26 respectively).

Fig. 7.14 Random orientation of chromosomes. The genes on one pair of homologous chromosomes segregate independently of those on the other pair. For clarity crossing over has not been shown

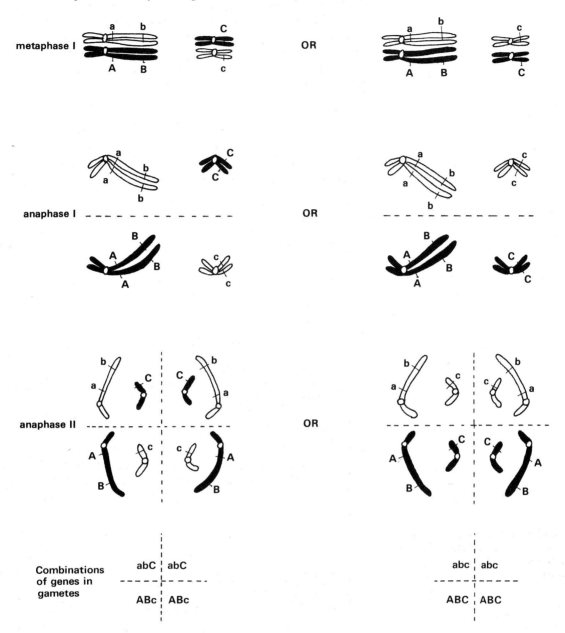

SUMMARY

Cell division occurs in two main steps – nuclear division followed by cytokinesis. Mitosis is the form of nuclear division which occurs when somatic cells divide. The sequential phases of mitosis are prophase, metaphase, anaphase and telophase. They result in the production of cells which have the same number of chromosomes as the parent cell. Growth and cell replacement are two activities which are dependent on this form of nuclear division. Between mitotic nuclear divisions a cell doubles its mass and duplicates its contents during interphase. Studies of the cell cycle reveal that many somatic cells spend a substantial proportion of their time at the interphase stage.

Meiosis is the form of nuclear division which produces gametes. It comprises two successive divisions of the nucleus. In the first meiotic division, the pairing of homologous chromosomes, their random ▶

131

orientation and subsequent crossing over create nuclei with half the original chromosome number which are genetically varied.

The second meiotic division is comparable to mitosis except that the chromosome number is half that of a somatic cell. Hence four genetically different nuclei are formed from each nucleus undergoing meiosis. This form of nuclear division is an integral part of the life-cycle of all organisms which reproduce sexually. It ensures that the chromosome number is kept constant from generation to generation.

QUESTIONS

1 Copy Table 1 and indicate by a tick or ticks (√) in the columns on the right which of the following statements is or are true of **mitosis,** of the **first division of meiosis** and of the **second division of meiosis.**   (AEB 1985)

2 Each of the diagrams labelled **A** to **K** represents a different stage of meiosis in a nucleus with two pairs of chromosomes.

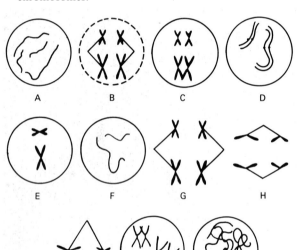

(a) Arrange the stages in the correct sequence for meiosis by listing the letters given to the diagrams, in order, beginning with interphase.
(b) (i) At which stage does crossing-over occur?
(ii) State precisely what crossing-over involves.
(iii) What is the result of a cross-over?   (C)

3 Copy and complete Table 2 to compare mitosis and meiosis:
(a) Fill in each box using a tick if the feature is correct and a cross if it is incorrect.
(b) State *three* ways in which genetic variation may result from the occurrence of meiosis.
(c) Explain how reproduction in a fern may result in variation among the offspring.   (L)

4 (a) State precisely where meiosis occurs in a flowering plant.
(b) With the aid of annotated diagrams, describe the process of meiosis.
(c) What is the significance of meiosis?   (C)

Table 1

Statement	Mitosis	First division of meiosis	Second division of meiosis
(a) DNA replication occurs between the previous division and this division			
(b) Homologous chromosomes lie together in pairs at the equator of the spindle			
(c) The centromere splits and the products move to opposite poles			
(d) Two chromatids involved in a crossover are separated			

Table 2

Feature	Mitosis	Meiosis
(i) Involves two successive nuclear divisions		
(ii) Chromosomes replicate before they become apparent in a stained cell		
(iii) Does not occur in a haploid cell		
(iv) Involves the formation of chiasmata		
(v) Leads to random assortment of chromosomes		
(vi) Involves the separation of sister chromatids		
(vii) Occurs during gamete formation in a mammal		
(viii) Protein synthesis occurs during the process		
(ix) Splitting of centromere is followed by anaphase		
(x) Daughter nuclei have identical genetic composition		
(xi) Occurs during vegetative growth		
(xii) A mutation may occur during the process		

8 Gas exchange in mammals

8.1 Gas exchange and evaporation 134

8.2 Breathing 136

    8.2.1 The breathing mechanism 136
    8.2.2 Control of breathing 139
    8.2.3 Oxygen debt 142

8.3 Oxygen uptake and carriage 142

    8.3.1 Oxygenation of the blood 142
    8.3.2 Gas exchange in the tissues 144
    8.3.3 Adaptations to high altitude 147

8.3.4 Pregnancy and fetal
      haemoglobin 149
8.3.5 Myoglobin 150

8.4 Respiratory physiology of
    diving mammals 151

    8.4.1 Aquatic mammals 151
    8.4.2 Diving and man 152

Summary 153

Questions 153

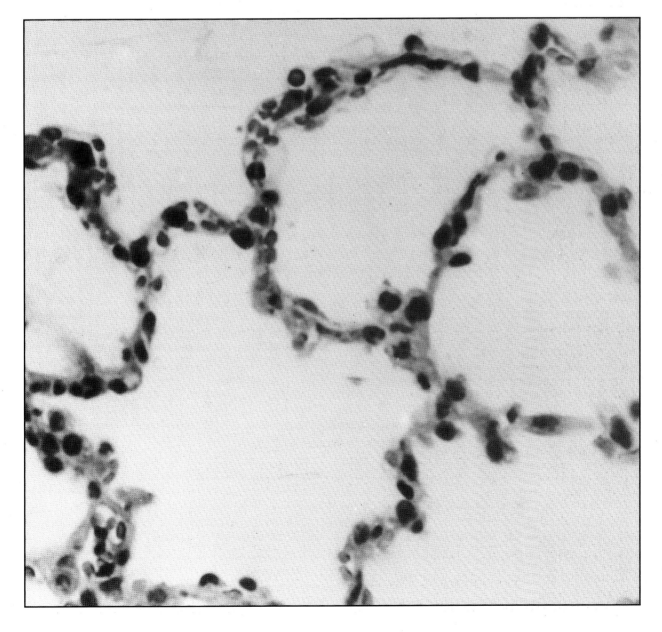

8 Gas exchange in mammals

Mammals respire aerobically so they need a constant supply of oxygen. The role of oxygen in cellular respiration is described in Chapter 5. Oxygen is consumed by mammals at a fairly rapid rate. This happens because respiration provides heat which helps to maintain a stable and relatively high body temperature (Chapter 18).

The exchange of respiratory gases between any animal's body fluids and the surroundings takes place at the respiratory surface. Mammalian respiratory surfaces have several important features which are summarised in Table 8.1.

What properties must a respiratory surface have in order to allow adequate respiratory gas exchange?

Table 8.1 Main properties of a respiratory surface

Property	Function
Large surface area	facilitates high rate of exchange
Moist surface	oxygen must dissolve before entering blood
Thin	short diffusion path (diffusion is relatively slow)
Permeable	respiratory gases must pass through it
Well ventilated	efficient delivery of oxygen to (and carbon dioxide from) the surface
Good blood supply	facilitates oxygen delivery to (and carbon dioxide from) the tissues

8.1 Gas exchange and evaporation

Table 8.2 Loss of water vapour from several mammals

Species	Evaporative water loss as a percentage of total water loss
kangaroo rat	68
rat	65
camel	51
human	42
donkey	36

Mammals obtain their oxygen from the atmosphere. Oxygen makes up about 21 % of the atmosphere's volume, the rest being nearly all nitrogen. Before oxygen gas can diffuse through the outer surface of an organism and into its tissues, oxygen must first dissolve in water. Consequently the surface through which oxygen is absorbed must be moist. The moisture on any surface exposed to the air will usually evaporate. Some $500 \, cm^3$ water are lost daily from the lungs of a man out of a total water loss of about $2000 \, cm^3$. The rest is lost through the skin and in the urine and faeces. Water loss by **evaporation** may thus be a major physiological problem during gas exchange (Table 8.2).

No biological system is perfect and physiological adaptations are usually compromises between advantage and disadvantage. To obtain enough oxygen to meet their metabolic needs, air-breathing animals such as mammals bring the air into direct contact with a relatively large area of moist lung surface. In doing so, water evaporates from the lungs resulting in substantial water loss. The rate of evaporation from a moist surface is affected by environmental factors such as temperature, atmospheric humidity and air movement. The ways in which such factors affect the loss of water vapour from terrestrial plants are described in Chapter 12. Climate similarly affects the rate of water loss from terrestrial animals.

Because mammals can regulate their breathing rate there is some control over water loss from their lungs. There is no such control in many terrestrial invertebrates such as earthworms and lower vertebrates such as amphibians which rely on a moist skin for gas exchange. Evaporative water loss is one of the main factors responsible for restricting such animals to moist habitats. Many mammals however live in deserts where evaporation is very rapid. Desert-living mammals possess adaptations which enable them to succeed where others would dry up and die within hours.

Fig 8.1 Kangaroo rat (Bruce Coleman: John Cancalosi)

How can kangaroo rats live in the desert without drinking any water at all?

One of these mammals is the kangaroo rat, *Dipodomys*, an inhabitant of the deserts of the south-west United States of America (Fig 8.1). The kangaroo rat has such fine regulation of its water content that although it lives in the desert it does not drink water. Since it has to match its water loss with a limited water intake the control of evaporation from its lungs is critical. Kangaroo rats have an efficient heat-exchange mechanism in their nasal passages which reduces the rate of water varpour loss during breathing. When breathing in, moisture on the inner surfaces of its nasal cavities evaporates, lowering the temperature of its nasal membranes. When breathing out, the moisture in the exhaled air condenses on the cooled nasal linings. Thus oxygen is taken in from the atmosphere with less of the accompanying water loss than occurs in most mammals (Fig 8.2).

Fig 8.2 Moisture retention in the kangaroo rat during breathing

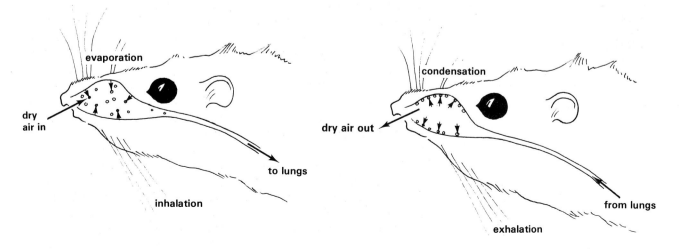

8.2 Breathing

8.2.1 The breathing mechanism

Fig 8.3 The position of the lungs and air passages in the human thorax

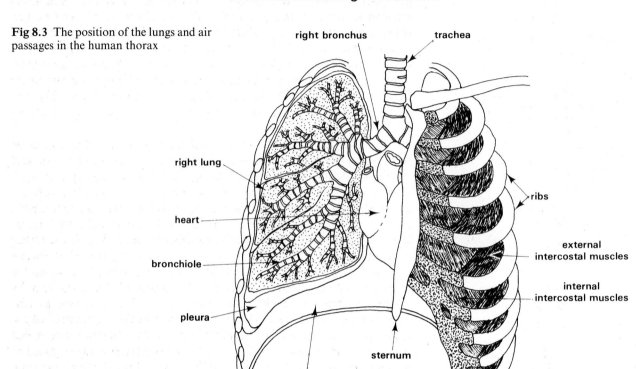

Mammalian **lungs** are compact organs. They are situated in the **thorax** which is separated from the abdominal cavity by a muscular **diaphragm** (Fig 8.3). The lungs consist of millions of microscopic air sacs called **alveoli**. Each alveolus is surrounded by a network of fine capillary blood vessels. The walls of the alveoli and the blood capillaries are made of **squamous epithelium**. Squamous epithelium is sometimes called **pavement epithelium** because its cells are flat like pavement slabs. Transport of respiratory gases across squamous epithelium is relatively fast because the distance across flat cells is relatively short. (Fig 8.4). Cells in the alveolar

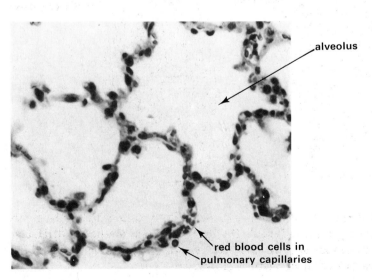

Fig 8.4 (a) Photomicrograph of a thin section of cat lung, × 200

Fig 8.4 (b) Squamous epithelium

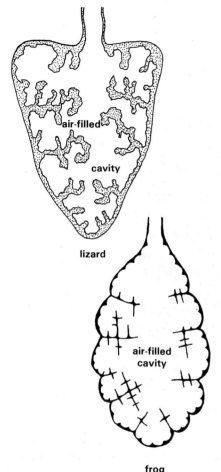

lizard

frog

walls secrete a phospholipid called surfactant (surface active agent). It lowers the surface tension of the water layer lining the alveoli and thus prevents the alveoli from collapsing. It is across the delicate membranes separating blood from the alveolar air that diffusion of gases occurs. Oxygen is taken in from the alveolar air, carbon dioxide is released into it.

Mammalian lungs have a considerable surface area for diffusion. For each gram of body mass there are about $7\,cm^2$ of lung surface in humans, $13\,cm^2$ in seals and $100\,cm^2$ in bats. This is a distinct improvement on the lungs of lower vertebrates which are little more than simple sacs (Fig 8.5). In these animals the area of lung surface per unit mass of body is much less than in mammals.

Breathing is a complex muscle-controlled activity in which the volume of the thorax is rhythmically increased and decreased. Air is sucked into and pushed out of the lungs when a mammal breathes. Breathing is brought about by movements of the ribs and diaphragm. Two sets of muscles, the **external** and **internal intercostal muscles** which work antagonistically, move the ribs. When the external intercostals contract, the internal intercostals relax and the ribs are forced obliquely out and up. Because they adhere to the inner wall of the thorax the lungs are stretched. At the same time the diaphragm contracts and pulls the lungs downwards towards the abdomen. These movements increase the volume of the lungs, thereby lowering the pressure of the air in them. As a result air is drawn into the lungs through the **trachea**, **bronchi** and **bronchioles**. Relaxation of the diaphragm and external intercostals and the elastic recoil of the stretched lungs return the thorax to its previous state. The pressure now applied to the lungs forces air from the lungs to the outside through the air passages (Fig 8.6). Forced expiration is brought about by contraction of the internal intercostals. The trachea, bronchi and bronchioles are lined with pseudo-stratified ciliated columnar epithelium (Fig 8.6(b)). The cells appear to be in

Fig 8.6 (a) The process of breathing

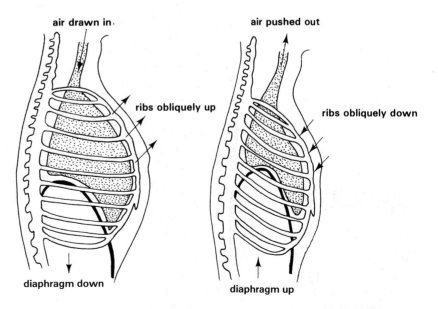

air drawn in

ribs obliquely up

diaphragm down

air pushed out

ribs obliquely down

diaphragm up

Fig 8.6 (b) Pseudostratified ciliated columnar epithelium

several layers (stratified) but each cell is in contact with the basement membrane (pseudostratified). The back and forth movement of the cilia pushes mucus towards the pharynx. The mucus traps inhaled dust and any other particles such as micro-organisms.

During each ventilation only about 10 % of the lungs' total air capacity is changed in most resting mammals. The volume of air exchanged with the atmosphere at each breathing cycle is called the **tidal volume**. A greater volume of air can be exchanged by forced breathing. The maximum volume of air breathed out after breathing in to the fullest extent is called the **vital capacity**. About 70 % of the air breathed in enters the alveoli. The rest occupies **dead space** mainly in the bronchi and bronchioles. The total capacity of the lungs of an adult human is about $6 \, dm^3$. Consequently during each ventilation of the lungs, only a small proportion of the alveolar air is replaced.

Fig 8.7 A spirometer being used to measure lung volumes (courtesy Philip Harris Biological Ltd.)

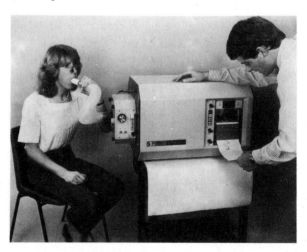

Tidal volume, vital capacity and other lung capacities can be measured using a **spirometer** (Fig 8.7). A spirometer is an air-filled chamber which rises and falls when a subject breathes into an attached mouthpiece. Movement of the chamber is recorded as a trace on a moving paper (Fig 8.8). Spirometers are used in hospitals to measure lung capacities of patients suffering from respiratory diseases.

Fig 8.8 A spirometer tracing typical of a normal human adult. The inspiratory reserve volume is the maximum volume of air which can be inhaled during forced breathing. Similarly the expiratory reserve volume is the maximum volume of air exhaled. The residual volume is the minimum volume of air always present in the lungs.

What volume of air do we normally breathe in and out?

How much can we breathe in and out if we force our breathing?

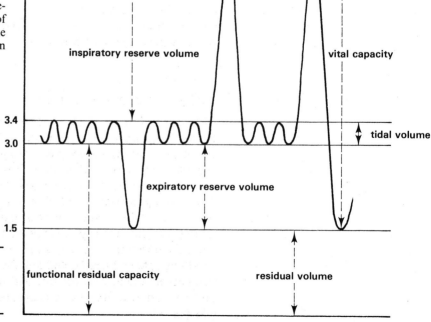

The tidal ventilation of lungs is very different from the continuous flow ventilation of the gills of many aquatic animals. In fish for example, oxygenated water is continuously moved over the surface of the gills and carbon dioxide is taken away in the same movement (Fig 8.9). Continuous

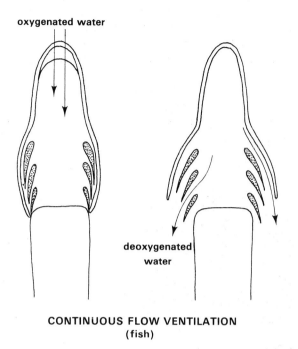

Fig 8.9 A comparison of continuous flow and tidal ventilation

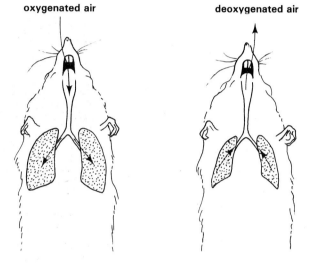

flow ventilation is more efficient than tidal ventilation in supplying oxygen from the environment to the respiratory surfaces. The oxygen available to air-breathers is $21\,cm^3$ in every $100\,cm^3$ of the atmosphere. The oxygen available to aquatic animals is only about $0.5–0.9\,cm^3$ in every $100\,cm^3$ of water (depending on temperature).

8.2.2 Control of breathing

The oxygen demand of an organism depends on the metabolic rate of its tissues. During periods of increased activity a mammal ventilates its lungs rapidly, so meeting the increased oxygen demand. Faster breathing also eliminates the extra carbon dioxide produced by increased respiration. Conversely the breathing rate slows down when the demand for oxygen drops during periods of rest. For example, a young woman resting on a bicycle may take in about $300\,cm^3$ oxygen per minute. When pedalling rapidly, oxygen consumption may rise to about $1500\,cm^3$ per minute. The fivefold increase in oxygen uptake is brought about by nearly trebling the breathing rate and nearly doubling the tidal volume. She breathes faster and deeper. Changes in the rate and depth of breathing in response to exercise and rest are also accompanied by changes in the heart rate (Chapter 9).

The muscles which bring about breathing movements are activated by nerve pathways which originate in a region of the hindbrain called the **respiratory centre**. The centre comprises three distinct areas called the **medullary rhythmicity area**, the **apneustic area** and the **pneumotaxic area** (Fig 8.10). The medullary rhythmicity area controls the basic rhythm of breathing. Nerve impulses from the rhythmicity area bring about the rhythmic cycle of inspiration and expiration. It is thought that the rhythmicity area contains two nerve circuits. One circuit is responsible for causing inspiration and inhibiting expiration, the other acts in the opposite way.

Fig 8.10 Main areas of the respiratory centre

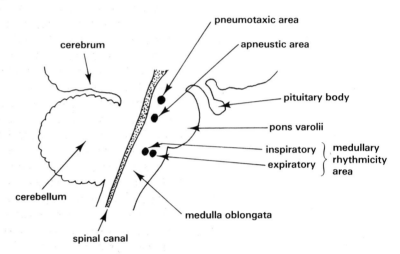

During quiet breathing, inspiration lasts about 2 seconds and expiration about 3 seconds. This rhythm produces about 12 ventilations a minute. In certain circumstances the quiet breathing rhythm can be modified. We may breathe faster and deeper, for example during physical exercise or emotional excitement. We may just decide to breathe at a different rate or even to hold our breath. Stimulation of the rhythmicity area by nerve impulses from the apneustic area can alter the depth of breathing. The tidal volume is increased. Stimulation of the rhythmicity area by impulses from the pneumotaxic area can alter the breathing rate. Suppose we breathe twice as deeply and three times as rapidly as during quiet breathing, the total volume of gases exchanged between the atmosphere and the lungs would increase by six times. The **minute volume** or **pulmonary ventilation, $\dot{V}$** (the dot above the V indicates a measurement made over a period of time) is the total volume of gases breathed in one minute.

Pulmonary ventilation = tidal volume × breathing rate

Typical adult values during quiet breathing might be:

$$\dot{V} = 400 \, cm^3 \times 15 \text{ ventilations per minute}$$
$$= 6000 \, cm^3 \text{ per minute}$$
$$= 6 \, dm^3 \text{ per minute}$$

During exercise these values may rise to:

$$\dot{V} = 800 \, cm^3 \times 25 \text{ ventilations per minute}$$
$$= 20\,000 \, cm^3 \text{ per minute}$$
$$= 20 \, dm^3 \text{ per minute}$$

If we subtract the dead space from these volumes we obtain the **alveolar ventilation, a$\dot{V}$**. This is the volume of gases that enters the alveoli each minute.

What would be the values for alveolar ventilation during quiet breathing and exercise, using the figures for breathing rate and tidal volume given on the right? Assume the dead space is 150 cm³.

The respiratory centre responds to a variety of sensory signals. These signals indicate changes in the body that demand an alteration of breathing. The most important of these is the concentration of hydrogen ions in the blood which may rise because of an increase in tissue respiration in the muscles during exercise. A rise in blood carbon dioxide, called **hypercapnia**, causes an increase in the concentration of hydrogen ions in the blood (section 8.3.2). The respiratory centre is sensitive to a rise in hydrogen ion concentration in the blood flowing through it. It responds by causing an increase in the rate and depth of breathing. The response is appropriate as the extra carbon dioxide produced during exercise must be excreted by the lungs. Also more oxygen is required to sustain the increase in tissue respiration. When the deeper, faster breathing has eliminated the extra carbon dioxide from the blood, the respiratory centre reverts to its resting rhythm and quiet breathing resumes.

Another factor that can affect breathing is the oxygen tension in the blood. In the walls of the aortic arch and carotid bodies are **chemoreceptors** which are sensitive to changes in blood oxygen tension. However, they respond only to a relatively large drop in oxygen tension. This is less important than the respiratory centre's direct response to a rise in blood hydrogen ion concentration.

Baroreceptors, also in the aortic arch and carotid bodies, are sensitive to changes in blood pressure (section 9.1.1 and Fig 8.11). Sensory nerve impulses are transmitted to the respiratory centre from the baroreceptors. A sudden rise in blood pressure may thus result in a decrease in breathing rate, and a drop in blood pressure may result in an increased breathing rate (Fig 8.12, see next page).

Fig 8.11 Position of the baroreceptors (stippled) in the cat

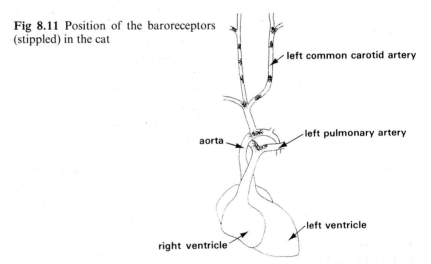

left common carotid artery

aorta

left pulmonary artery

left ventricle

right ventricle

A rise in body temperature, as occurs during severe exercise or fever, causes an increase in the breathing rate. Conversely, a fall in body temperature slows the rate of breathing. A sudden drop in temperature caused by plunging into cold water, for example, may cause breathing to stop altogether, **apnoea**.

Stretch receptors are present in the walls of the bronchi and bronchioles. When stimulated by overstretching during excessive inspiration, sensory impulses are transmitted from the receptors along afferent branches of the vagus nerve to the respiratory centre. There the inspiratory and apneustic areas are inhibited and expiration occurs. This mechanism is called the **Hering–Breuer reflex**. It may be a protective device that prevents excessive inflation of the lungs. It is not thought to be important in the normal regulation of breathing.

Breathing and gas exchange are adjusted to meet the body's immediate needs. As with other homeostatic mechanisms the regulation of breathing involves **negative feedback** controls. The rate at which oxygen is taken in is regulated by factors which reflect the need for oxygen (Fig 8.12).

Fig 8.12 (a) The main components of an automatic control system

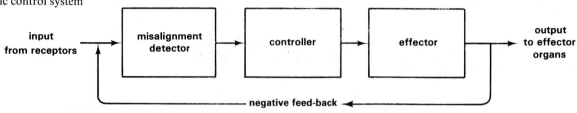

Fig 8.12 (b) Factors affecting the respiratory centre

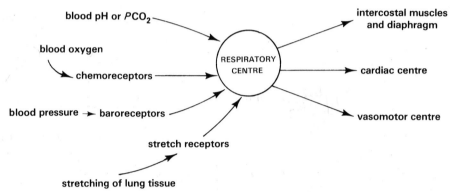

8.2.3 Oxygen debt

For short periods at a time the mammalian body can engage in severe exercise which requires more oxygen than can be provided by ventilating the lungs, whatever the rate of breathing. During intense exercise the reserves of free energy such as ATP and phosphocreatine are depleted. Anaerobic respiration occurs in the cells and lactic acid is produced (Chapter 5). During short periods of intense exercise an **oxygen debt** is built up. The debt is repaid by continued rapid and deep breathing when the period of exercise ends. The extra oxygen then absorbed by the blood corresponds to the oxygen debt and is used to oxidise the lactic acid produced in anaerobic respiration.

Table 8.3 Oxygen-carrying capacity of the blood of several mammals

Species	cm^3 oxygen per $100\,cm^3$ blood
cat	15.0
rabbit	15.6
sheep	15.9
horse	16.7
kangaroo rat	17.5
rat	18.6
sea lion	19.8
human	20.0
porpoise	20.7
fox	21.7
llama	23.4
seal	29.3

8.3 Oxygen uptake and carriage

One of the main functions of the lungs is to allow atmospheric oxygen to come into close proximity with the blood which can then absorb oxygen and transport it to the body organs and tissues.

8.3.1 Oxygenation of the blood

The **red blood cells (erythrocytes)** make up about 45 % of the total blood volume in man. Red cells contain a respiratory pigment called **haemoglobin**. The main component of blood is **plasma**, a mixture of organic and inorganic materials dissolved in water. Oxygen is not very soluble in water. At 35 °C, $100\,cm^3$ of water contains about $0.5\,cm^3$ oxygen. The blood of some mammals, such as the seal, can carry as much as $29\,cm^3$ oxygen per $100\,cm^3$ of blood. The ability of the blood to carry so much oxygen is due to haemoglobin (Table 8.3).

Haemoglobin is a complex molecule made of four similar **sub-units**. Each sub-unit consists of a molecule of the protein **globin** attached to a central porphyrin group called **haem** which contains an iron(II) atom (Fig 8.13).

Fig 8.13 Haemoglobin structure. Each molecule consists of four sub-units. Each sub-unit contains an iron-centred haem group attached to the protein globin. The total relative molar mass is about 68 000

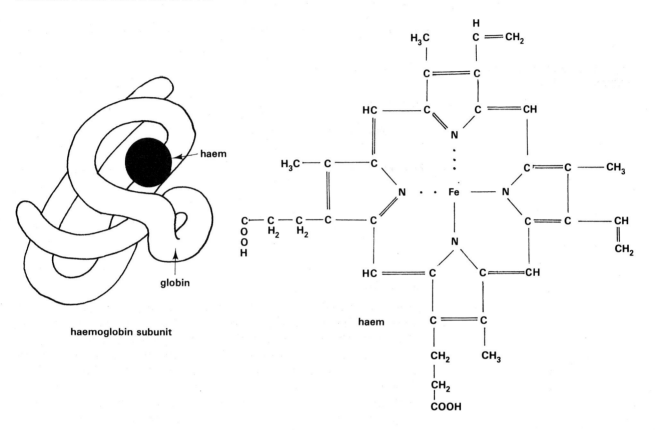

haemoglobin subunit

In the capillaries of the lungs oxygen combines loosely with haemoglobin to form **oxyhaemoglobin**. The iron of the haem groups is not oxidised and remains as iron(II) (Fig 8.14). Certain chemicals can oxidise the iron(II) to iron(III) producing derivatives of haemoglobin which cannot carry oxygen.

Fig 8.14 Oxygenation of haemoglobin

$$\text{haemoglobin} + O_2 \underset{\text{deoxygenation}}{\overset{\text{oxygenation}}{\rightleftharpoons}} \text{oxyhaemoglobin}$$

$$Hb + O_2 \rightleftharpoons HbO_2$$

$$HbO_2 + O_2 \rightleftharpoons HbO_4$$

$$HbO_4 + O_2 \rightleftharpoons HbO_6$$

$$HbO_6 + O_2 \rightleftharpoons HbO_8$$

Carbon monoxide poisoning, for example, is caused by the formation of carboxyhaemaglobin containing iron(III). Traces of carboxyhaemaglobin occur in the blood of non-smokers whereas up to 15 % is present in the blood of smokers. Traffic exhaust fumes contain a lot of carbon monoxide. Oxygenation of haemoglobin takes place in the blood capillaries surrounding the alveoli. Here the haemoglobin becomes saturated with oxygen. The relationship between oxygen tension and haemoglobin saturation is usually shown as an **oxygen dissociation curve**(Fig 8.15).

Fig 8.15 Oxygen dissociation curve of human haemoglobin

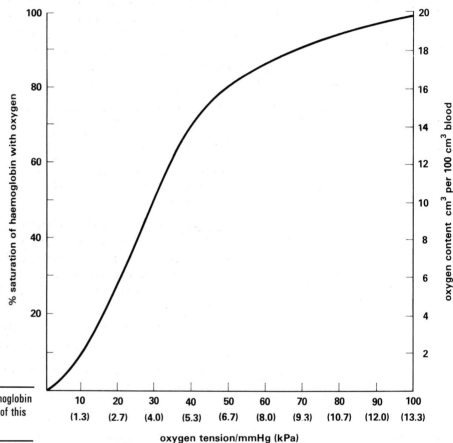

The oxygen dissociation curve for haemoglobin is very steep. What is the significance of this in the lungs and tissues?

In air with a pressure of about 100 kPa (1 atm) the partial pressure exerted by the 21 % oxygen present is about 21 kPa. In the lungs, air breathed in is diluted with air in the dead space. Consequently the partial pressure of oxygen in the alveoli is lowered to about 13.3 kPa. Even so, blood circulating in the lung capillaries normally becomes fully saturated with oxygen from the alveoli. The oxygen tension in the blood becomes 13.3 kPa.

8.3.2 Gas exchange in the tissues

Carbon dioxide is produced in body organs as a result of tissue respiration. The carbon dioxide diffuses through the walls of blood capillaries into the blood. In the red cells an enzyme called **carbonic anhydrase** greatly accelerates the chemical combination of carbon dioxide with water to form carbonic acid:

$$\underset{\substack{\text{carbon dioxide}}}{CO_2} + \underset{\substack{\text{water}}}{H_2O} \overset{\text{carbonic anhydrase}}{\rightleftharpoons} \underset{\substack{\text{carbonic acid}}}{H_2CO_3}$$

The carbonic acid partially dissociates into hydrogen (H^+) and hydrogencarbonate ($HCO_3{}^-$) ions:

$$\underset{\substack{\text{carbonic acid}}}{H_2CO_3} \rightleftharpoons \underset{\substack{\text{hydrogen ion}}}{H^+} + \underset{\substack{\text{hydrogencarbonate} \\ \text{ion}}}{HCO_3{}^-}$$

The $HCO_3{}^-$ ions readily diffuse out of the red cells into the plasma. It is as $HCO_3{}^-$ ions that most of the carbon dioxide is carried to the lungs to be excreted (Table 8.4). In the lungs the reactions described above are reversed and the carbon dioxide gas so formed diffuses into the alveoli and is breathed out.

Table 8.4 Transport of carbon dioxide in the blood

	Percentage of total carbon dioxide in blood
CO_2 in solution	7
carbon dioxide haemoglobin	7
hydrogencarbonate	86

In response to the rapid loss of $HCO_3{}^-$ ions from the red cells, an equal number of chloride ions (Cl^-) diffuse in the opposite direction. The **chloride shift** prevents electrochemical imbalances, particularly pH changes, which could adversely affect the red cells.

The H^+ ions which stay in the red cells cause the release of oxygen from oxyhaemoglobin:

$$\underset{}{H^+} + \underset{\substack{\text{oxyhaemoglobin}}}{HbO_2} \rightleftharpoons \underset{\substack{\text{haemoglobinic acid}}}{HHb} + \underset{\substack{\text{oxygen}}}{O_2}$$

Gas exchange in the tissues takes place very rapidly. The reactions involved are summarised in Fig 8.16.

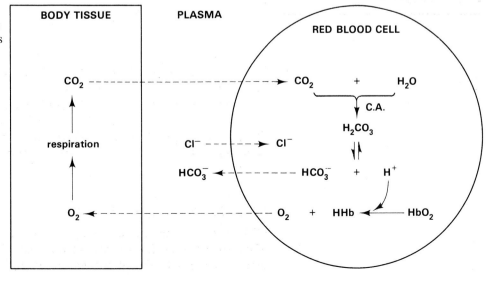

Fig 8.16 Summary of red cell chemistry related to the carriage of respiratory gases

The main factors controlling the release of oxygen to the tissues are the drop in oxygen tension and the rise in carbon dioxide tension produced by tissue respiration. As the carbon dioxide tension of the tissues and capillary blood rises the affinity of oxyhaemoglobin for oxygen is lowered. The phenomenon is called the **Bohr effect**. The oxygen dissociation curve moves to the right (Fig 8.17).

Fig 8.17 The effect of carbon dioxide tension on the oxygen dissociation curve of haemoglobin. The carbon dioxide tension in human arteries is about 40 mmHg (5.3 kPa). Tissue respiration increases this to about 46 mmHg (6.1 kPa). The oxygen tension of arterial blood is about 100 mmHg (13.3 kPa). Tissue respiration lowers this to about 40 mmHg (5.3 kPa)

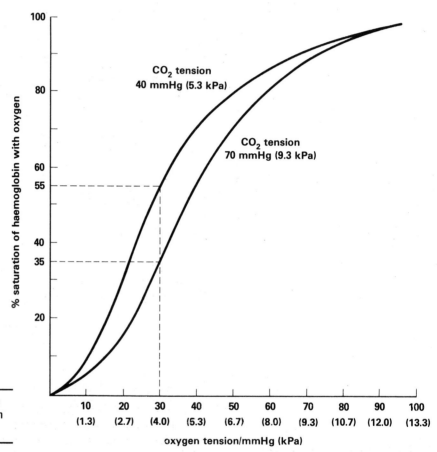

How do carbon dioxide, pH and temperature affect the delivery by haemoglobin of oxygen to the tissues?

Thus increased carbon dioxide production by respiring tissues causes the blood to release more oxygen. For example, assume that in a tissue somewhere in the body the oxygen tension is 4.0 kPa. If the carbon dioxide tension increases from 5.3 kPa to 9.3 kPa then, from the oxygen dissociation curves, it can be seen that the saturation of haemoglobin is lowered from 55 to 35 %. As a direct result of such an increase in carbon dioxide production some 20 % of the haemoglobin releases its oxygen. The more carbon dioxide a tissue produces the greater is this effect. Consequently the tissues and organs such as muscles and the liver which have a relatively high respiration rate cause a rapid release of oxygen from the blood supplying them. Conversely those tissues with a low oxygen demand cause oxygen to be released less rapidly from the blood.

The carriage and exchange of respiratory gases also contributes to the regulation of the pH of blood and the body tissues (Chapter 1).

A rise in blood temperature lowers the affinity of haemoglobin for oxygen, thus causing unloading from the pigment (Fig 8.18). Increased tissue respiration which occurs in skeletal muscles during exercise generates heat. The consequent temperature rise causes the release of extra oxygen from the blood.

Fig 8.18 Effect of increased temperature on the position of the haemoglobin–oxygen dissociation curve

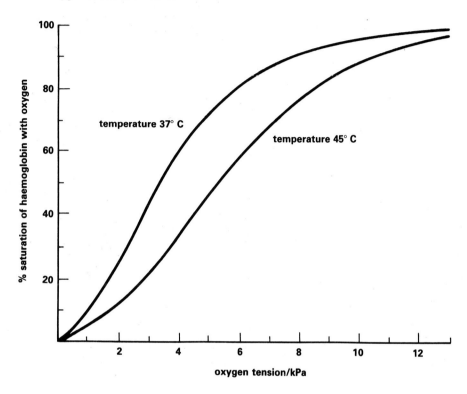

8.3.3 Adaptations to high altitude

As altitude is gained atmospheric pressure drops. The drop in atmospheric pressure is significant for organisms living at high altitudes because the volume of oxygen in the atmosphere at high altitudes is less than at sea level (Table 8.5).

Table 8.6 Numbers of red cells in the blood of several mammals at sea level and at high altitude

Species	Altitude	Average red cell numbers $\times 10^{12}$ per dm^3
human	sea level	5.00
	5333 m	7.37 (residents)
		5.95 (transients)
sheep	sea level	10.50
	4673 m	12.05
rabbit	sea level	4.55
	5303 m	7.00
llama	sea level	11.40
	2800 m	12.30

Table 8.5 Approximate partial pressures of oxygen and nitrogen in the atmosphere at sea level and at 4848 metres

	Percentage	Partial pressure at sea level/kPa	Partial pressure at 4848 metres/kPa
oxygen	21	21	11
nitrogen	79	79	42
barometric pressure (approx)		100	53

The amount of haemoglobin and the number of red cells in the blood of mammals increase when they live at high altitudes (Table 8.6). This is a direct response to the lower partial pressure of oxygen in the air. Red cell production occurs in the red bone marrow and is controlled by a hormone called **erythropoietin** which is made in the kidneys. Secretion of erythropoietin is stimulated by low tensions of oxygen (Fig 8.19). However, when a mammal moves from low to high altitude time is needed for an increase of haemoglobin and red cells. The problem was highlighted at the 1968 Olympic Games in Mexico City, 2460 metres above sea level. Athletes who were natives of high altitude countries or who had spent several months acclimatising to high altitude prior to the Games performed quite a lot better than those who had not.

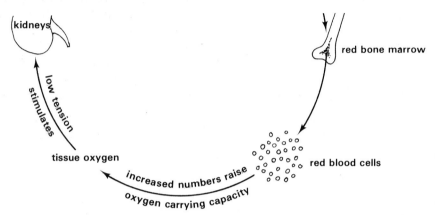

Fig 8.19 Hormone control of red cell production (erythropoiesis). Erythropoiesis is a continuous process and red cells have a limited life in the circulation, about 4 months in humans

Another adaptation which helps to overcome low oxygen tensions is seen in mammals living at high altitude. These animals possess haemoglobin which loads more readily with oxygen in the lungs. Haemoglobin of this sort has a dissociation curve to the left of normal haemoglobin (Fig 8.20).

Fig 8.20 (a) Oxygen dissociation curves of llama and horse haemoglobins

Fig 8.20 (b) Llama (Biofotos)

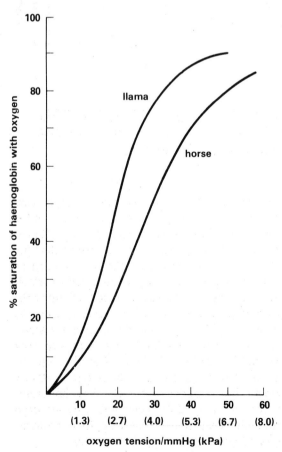

8.3.4 Pregnancy and fetal haemoglobin

During pregnancy several physiological changes take place inside the mother's body concerned with the provision of oxygen for the developing embryo. The changes occur partly as a result of the increased output of certain hormones during the gestation period (Chapter 25). For instance, the developing fetus in the uterus progressively displaces the mother's abdominal organs. The displaced organs push against the mother's diaphragm, as well as elsewhere, thus lowering the ventilating capacity of her lungs. Nevertheless, her rate of oxygen consumption increases by about 20 % as a result of widening of the thorax and an average 40 % increase in breathing rate.

Fig 8.21 Oxygen dissociation curves of the adult and fetal haemoglobins of sheep

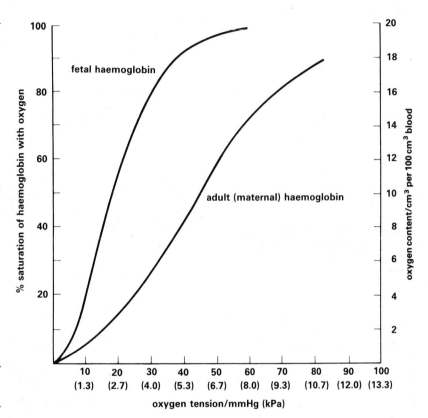

Why is fetal haemoglobin replaced by adult haemoglobin?

Transport of the extra oxygen, partly to the fetus, is helped by an increase of about 1 dm³ in the mother's total blood volume. The extra blood is circulated more rapidly than normal by increased cardiac output (Chapter 9) which rises from an average of 4.9 dm³ per minute to about 6 dm³ per minute. As it develops in the uterus, the embryo is attached to its mother by a **placenta** (Chapter 25). Inside the placenta the maternal blood vessels lie very close to the fetal blood vessels. A two-way exchange of nutrients, wastes and respiratory gases occurs across the membranes separating the two circulations. Because of its rapid growth and development a mammalian fetus has a very high oxygen demand. Unloading of oxygen from the mother's blood to the fetal circulation is brought about because the fetal red cells contain a variant of haemoglobin which has a greater affinity for oxygen than the mother's haemoglobin (Fig 8.21). This important functional difference between fetal and adult haemoglobin results from very small differences in the amino acid sequences of their globin proteins. **Fetal haemoglobin** is replaced by adult haemoglobin at birth.

8.3.5 Myoglobin

The muscles of mammals contain a red pigment called **myoglobin** which is structurally similar to one of the four sub-units of haemoglobin (Fig 2.32). Myoglobin has a higher affinity for oxygen than haemoglobin has and unloads its oxygen only when the blood is almost fully deoxygenated. This property is reflected in the oxygen dissociation curve for myoglobin which is further to the left than the curve for haemoglobin (Fig 8.22).

Fig 8.22 Oxygen dissociation curve of myoglobin. That of haemoglobin is shown for comparison

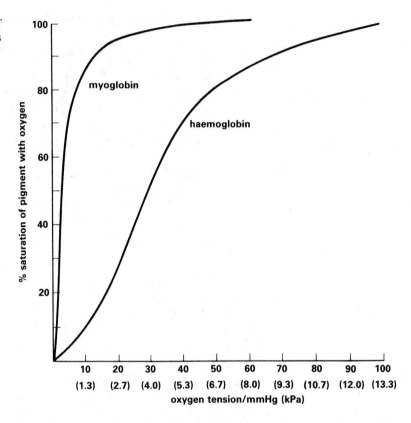

Not all changes in the haemoglobin molecule are advantageous. A serious disease of man called **sickle cell disease** results from the production of a haemoglobin variant differing from normal haemoglobin by only a single amino acid out of several hundred in the whole molecule (Chapter 4). At low oxygen tensions, haemoglobin S crystallises in the red cells distorting them into a sickle shape (Fig 8.23). Many sickle cells are destroyed in the circulation causing an **anaemia**, which considerably lowers the oxygen-carrying capacity of the blood. Some sickle cells block the capillaries in vital organs and the disease may thus be fatal.

The incidence of haemoglobin S in certain parts of Africa and Asia is very high compared with other areas. People who inherit the gene for haemoglobin S from only one parent display the heterozygous condition called **sickle cell trait**. Sufferers of sickle cell trait are resistant to some forms of malaria and are

Fig 8.23 Photomicrograph of a smear of human blood showing sickle cells ($\times 750$)

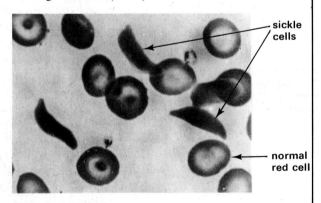

therefore more likely to survive to adulthood. The resistance explains the high frequency of haemoglobin S among natives of regions where malaria is prevalent (Chapter 26).

Oxymyoglobin acts as an oxygen reserve in skeletal muscles which have a high oxygen demand during exercise. In this way the muscles can be provided with oxygen even though the blood flowing through them has all oxygen removed from it. Mammalian muscle displays several other adaptations which enable them to sustain intense activity for short periods of time (Chapter 16).

8.4 Respiratory physiology of diving mammals

Many marine mammals can remain submerged in water for long periods, often at great depths (Table 8.7). Diving involves several adaptations of the respiratory system and tissue physiology. For example, the tidal volume is about 80 % of the total lung capacity in the porpoise compared with only 10 % in man. This means that any oxygen debt built up in a porpoise during prolonged submergence in water can be repaid quickly on surfacing. There are, however, other adaptations which are just as important.

Table 8.7 Diving times of several mammals

Species	Average duration of dive/min
human	2.5
seal	15
finback whale	30
sperm whale	60–90
bottlenose whale	120+

8.4.1 Aquatic mammals

P. F. Scholander was one of the first to investigate the physiology of diving mammals, mainly seals. The blood of the seal has a very high oxygen-carrying capacity, $29.3 \, cm^3$ oxygen per $100 \, cm^3$ blood (Table 8.3). Even though there is a lot of oxygen in the seal's blood when it submerges, much of its tissue respiration during long dives is anaerobic. This is because breathing may stop for up to fifteen minutes when the seal is under water. Lactic acid and carbon dioxide are produced as a result of anaerobic respiration (Chapter 5). The lactic acid can increase to seven times its normal concentration in the blood of seals without ill-effect. The respiratory centre of seals must be far less sensitive than that of non-diving mammals to high concentrations of carbon dioxide in the blood.

Another feature of seals when diving is a remarkable drop in heart rate from about 150 beats per minute to about 10 beats per minute. Blood is also redirected in the vascular system so that the supply to the brain is maintained while the amount of blood pumped to other parts of the body is lowered. In this way the oxygen goes to organs which cannot do without it. Seals also expel air from their lungs as they dive, lowering the volume of gases which are compressed on submerging. The danger of bubbles of nitrogen forming in the blood and tissues when the seal surfaces is thus minimised.

The mechanisms described above ensure the efficient use of oxygen during submergence and allow anaerobic respiration without damage to the tissues. Less is known of such mechanisms in whales. However, the muscles of whales contain high concentrations of myoglobin.

How can seals live under the sea without breathing for up to fifteen minutes?

8.4.2 Diving and man

There has been much interest recently in the physiological problems of human diving. One reason for concern was the high rates of accidents and deaths among divers engaged in the first phase of oil exploration in the North Sea off the coast of Britain.

The main problems relate to the effects of pressure from the surrounding water as divers descend (Fig 8.24). The pressure increases the volumes of gases entering the blood from the alveoli. At depths below about 60 m the increased oxygen content of the blood is such that the tissues receive more oxygen than they normally need. The increased supply of oxygen leads to abnormal metabolism and damage to brain cells

Fig 8.24 Relationship between pressure and depth beneath the sea

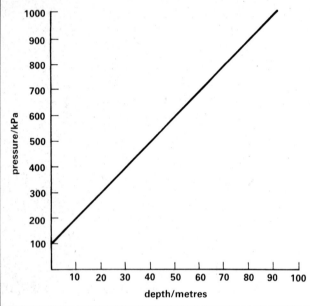

can occur. Consequently the diver can quickly lose control of his actions. Oxygen poisoning may be overcome by adjusting the oxygen content of a diver's breathing mixture as he descends.

High levels of dissolved nitrogen in the body have an anaesthetic effect on the central nervous system. At depths below 60 m **nitrogen narcosis** can make a diver unconscious. Prior to this, a state similar to drunkenness can give a diver a false sense of security in which he cannot make rational decisions. In this state, he may endanger his life. Substitution of helium for nitrogen in the breathing mixture eliminates the danger of nitrogen narcosis. One amusing, if inconvenient, side effect of the use of helium for such purposes is that it affects the voice which becomes very high pitched and squeaky.

Rapid **decompression**, on sudden surfacing, can cause a further problem, that of gas bubble formation in the tissues and body fluids. At a depth of 60 m the nitrogen in a diver's body is at a pressure of about 517 kPa, more than five times atmospheric pressure at sea level. At the surface however, nitrogen is at a pressure of 81 kPa. Sudden decompression causes rapid conversion of dissolved nitrogen in the blood into bubbles of nitrogen. The bubbles cause severe pain which in turn gives rise to terrible contortions of the body. For this reason, the condition is called the **bends**. The effects on the central nervous system can be severe and permanent. The bends or **decompression sickness** is prevented by controlling decompression at a rate slow enough for the extra nitrogen in the body to be gradually eliminated in the normal way by the lungs.

SUMMARY

 Mammals exchange respiratory gases with the atmosphere. In order to obtain oxygen, the respiratory surfaces in the lungs are moist. The moisture evaporates during breathing. Water loss is a physiological price paid for obtaining oxygen from the air.

Breathing is brought about by stretching the lungs inside the rib cage. Intercostal muscles pull the ribs outwards and the diaphragm pushes down into the abdomen. Elastic recoil of the stretched lung tissue forces air out of the lungs before the breathing cycle begins again.

The physiological demand for oxygen changes with stress. Breathing is modified so as to increase or decrease gas exchange in different circumstances. The respiratory centre in the hind brain transmits impulses to the muscles responsible for breathing. Sympathetic stimulation increases breathing rate and tidal volume. Parasympathetic stimulation slows the rate and lowers the volume. The respiratory centre responds to a variety of stimuli which indicate the need to modify breathing. They include blood hydrogen ion concentration, blood carbon dioxide, blood oxygen, blood pressure and the degree of stretching in lung tissue.

Once in the blood, oxygen is transported to the tissues in combination with haemoglobin in the red blood cells. Respiration in the tissues creates conditions which cause release of oxygen from haemoglobin. The conditions include a fall of blood oxygen tension, a rise of blood carbon dioxide tension, a fall of blood pH and a rise of blood temperature.

At high altitudes there is less oxygen in the atmosphere than at sea level. Mammals adapt to altitude by increasing the numbers of red cells and the quantity of haemoglobin in their blood. Mammals which live permanently at high altitude possess a variant of haemoglobin which combines very readily with oxygen.

The haemoglobin in foetal blood combines with oxygen more readily than does the mother's haemoglobin. It enables oxygen to cross the placenta from mother to foetus.

Myoglobin in the muscles takes oxygen from the blood. Oxymyoglobin releases oxygen when the oxygen tension is very low. It represents a store of oxygen which becomes available when the muscles are stressed in severe exercise.

Diving mammals possess adaptations which enable long submerged periods without breathing. The adaptations include a large tidal volume, a high oxygen carrying capacity of the blood, a low sensitivity of the respiratory centre to blood carbon dioxide, a drop in heart rate and a redirection of blood to the brain at the expense of other parts of the body.

QUESTIONS

1 The figure shows the oxygen dissociation curves for myoglobin (**A**) and for haemoglobin at three different partial pressures of carbon dioxide (**B** at 15 mm of mercury, **C** at 40 mm and **D** at 70 mm).
(a) By reference to curve **B**, indicate why haemoglobin efficiently carries and releases oxygen.
(b) Explain the significance of the fact that
(i) with an increase in partial pressure of carbon dioxide the oxygen dissociation curves for haemoglobin shift to the right,
(ii) the oxygen dissociation curve for myoglobin (**A**) is displaced well to the left of that for haemoglobin.
(c) In what tissues does myoglobin commonly occur?
(C)

2 (a) What are the properties of respiratory surfaces?
(b) Describe the mechanisms by which the respiratory surfaces are ventilated in (i) a fish and (ii) a mammal.
(c) Compare the properties of air and water as respiratory media. (L)

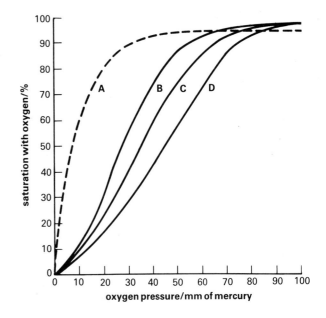

3 The oxygen dissociation curves below represent the relationship between the partial pressure of oxygen and the percentage oxygen saturation of two respiratory pigments. Curve **A** shows the response of myoglobin in muscle and curves **B**, **C** and **D** the response of haemoglobin in the blood at three different partial pressures of carbon dioxide.

Key

A = myoglobin in muscle

B = haemoglobin at CO_2 partial pressures 2666 Pa (20 mmHg)

C = haemoglobin at CO_2 partial pressures 5332 Pa (40 mmHg)

D = haemoglobin at CO_2 partial pressures 10664 Pa (80 mmHg)

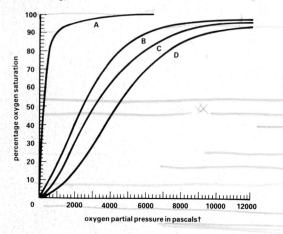

oxygen partial pressure in pascals†

(a) For curve **B**, briefly describe the effect that increasing partial pressure of oxygen has on the saturation of the respiratory pigment.

(b) Over which range of partial pressures of oxygen does the most rapid reaction with haemoglobin occur for (i) curve **B** and (ii) curve **D**?

(c) Where in the body of a mammal is the partial pressure of carbon dioxide likely to be high, as in curve **D**?

(d) If the blood when fully saturated with oxygen is able to carry 100 cm³ (ml) of oxygen per dm³ (litre), calculate the volume of oxygen released per dm³ (litre) from the blood when blood that is 90 % saturated flows into a tissue where the partial pressure of oxygen is 4000 Pa (30 mmHg) and that of carbon dioxide is 5332 Pa (40 mmHg).

(e) Suggest reasons for the differences in curve **A** and curve **D**.

(f) From curves **B**, **C** and **D** it is clear that increasing partial pressure of carbon dioxide affects the ability of the respiratory pigment to combine with oxygen. Give the name for this phenomenon.

(g) Curve **A** is similar to a curve obtained when investigating the oxygen carrying capacity of the respiratory pigment in an aquatic worm which burrows in mud. Explain how this curve indicates the worm's adaptation to its environment. (L)

4 In fish, oxygen is transported in the blood in the form of oxyhaemoglobin.

The table below shows the percentage saturation of blood with oxygen of a teleost (bony) fish after equilibrating with oxygen of different partial pressures. The experiment was carried out at two different partial pressures of carbon dioxide.

Partial pressure of oxygen in Pa†	Percentage saturation of blood with oxygen	
	Partial pressure of carbon dioxide at 500 Pa	Partial pressure of carbon dioxide at 2600 Pa
500	30	5
1000	70	13
2000	90	24
3000	96	33
4000	98	41
5000	99	48
7000	100	60
9000	100	69
11 000	100	76
13 000	100	81

†*A pascal (Pa) is a unit of pressure. A pressure of 100 000 pascals is approximately equal to atmospheric pressure (760 mmHg).*

(a) Present the data in a suitable graphical form.

(b) Calculate the difference in percentage saturation of blood with oxygen at the two different partial pressures of carbon dioxide at an oxygen partial pressure of 5500 Pa.

(c) With reference to the graph, describe the effects of different partial pressures of carbon dioxide on the percentage saturation of blood with oxygen.

(d) Explain how the properties of the haemoglobin molecule are affected by changes in the oxygen and carbon dioxide partial pressures.

(e) Explain how changes in oxygen content of the blood at different partial pressures of carbon dioxide are important in the release of oxygen to the tissues of the fish.

(f) What information do experiments of this type give about the environmental conditions in which fish would maintain a high level of growth as required in commercial fish farming? (L)

9 Circulation of blood and lymph

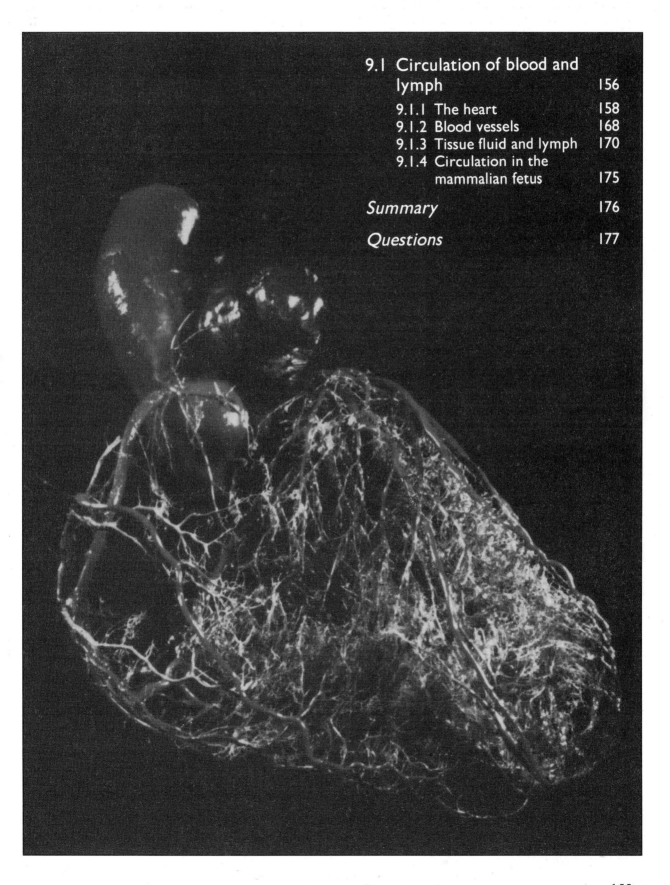

9.1	**Circulation of blood and lymph**	**156**
	9.1.1 The heart	158
	9.1.2 Blood vessels	168
	9.1.3 Tissue fluid and lymph	170
	9.1.4 Circulation in the mammalian fetus	175
	Summary	176
	Questions	177

9 Circulation of blood and lymph

Oxygen in the lungs and many nutrients in the gut enter the blood mainly by diffusion through the alveolar and intestinal surfaces respectively. Movement of dissolved metabolites around the body by diffusion alone, however, would be a very slow process. An inactive earthworm for example could only obtain about 10 % of its oxygen requirement if it had to rely on diffusion to distribute the gas throughout its body. Consequently all but the smallest of animals need a rapid means of internal transport. In mammals and other vertebrates transport is carried out mainly by the **blood**.

Blood distributes nutrients and respiratory gases to the tissues, transports hormones to target organs and carries metabolic wastes from the tissues to excretory organs. Blood also plays an important part in the defence against disease and in the repair of injured body tissues and in the distribution of heat.

9.1 Circulation of blood and lymph

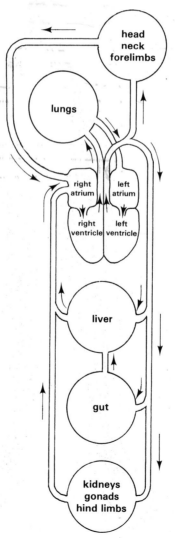

Fig 9.1 Diagram of the blood circulation of mammals

The volume of blood in mammals is substantial (Table 9.1). To act as a transport system large volumes of fluid require a powerful pump, the **heart**. From the heart blood is pumped through an extensive system of **arteries** and **arterioles** to all the organs of the body. It then passes through microscopic **blood capillaries** where metabolites and waste products are exchanged with the tissue fluid and cells of the organs. Blood drains from the organs in **venules** and **veins** which carry it back to the heart. Figs 9.1 and 9.2 illustrate the extent and complexity of the mammalian blood vascular system.

Table 9.1 Blood volume as a percentage of body mass in several mammals

Species	Blood volume as per cent body mass
goat	6.1
rabbit	6.5
human	7.0
cow	7.5
guinea pig	7.5
pig	8.0
dog	9.4

Fig 9.2 The main arteries and veins in humans. The pulmonary and gastrointestinal vessels are omitted for clarity.

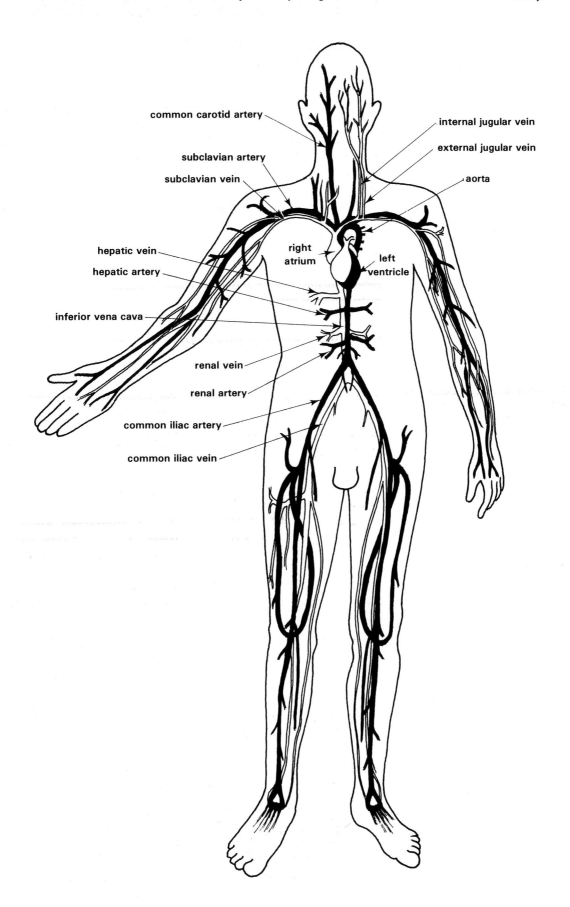

common carotid artery

subclavian artery

subclavian vein

hepatic vein

hepatic artery

inferior vena cava

renal vein

renal artery

common iliac artery

common iliac vein

internal jugular vein

external jugular vein

aorta

right atrium

left ventricle

9.1.1 The heart

The mammalian heart is a remarkable organ. Functionally it is divided into right and left halves. The right half collects blood from the general body circulation and pumps it to the lungs where it is oxygenated (Chapter 8). The left half of the heart receives oxygenated blood from the lungs and pumps it into the general (systemic) circulation (Fig 9.3).

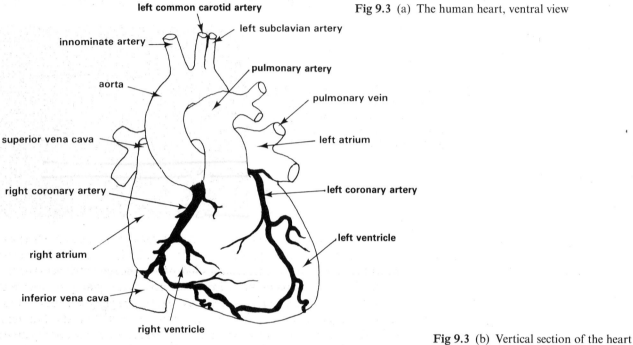

Fig 9.3 (a) The human heart, ventral view

Fig 9.3 (b) Vertical section of the heart

1 The cardiac cycle

The **cardiac output** is the volume of blood pumped from each ventricle in a minute. When the body is at rest the cardiac output of an adult human is about 4.9 dm³ per minute. During severe exercise the output can rise to 30 dm³ per minute. The pumping of blood requires considerable and sustained expenditure of energy. The mechanical work in pumping blood is performed by the **cardiac muscle** in the walls of the heart's four chambers. It makes up the **myocardium**. Cardiac muscle is **myogenic**. Even if isolated from all its nerve supply it continues to contract and relax rhythmically.

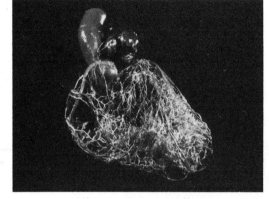

Fig 9.3 (c) Resin cast of the blood vessels of the human heart (courtesy Professor McMinn, Department of Anatomy, The Royal College of Surgeons of England)

The **cardiac cycle** begins with atrial contraction. The electrochemical basis of muscle contraction is described in Chapter 17. The stimulus for the heart's rhythmic beat comes from a small part of the right atrium called the **sinoatrial node (pacemaker)**. From the pacemaker waves of electrical activity similar to nerve impulses spread out very rapidly over both atria. Each wave contracts the atrial muscle, forcing blood in the **atria** through a pair of **atrio-ventricular valves** into the **ventricles**. **Semi-lunar valves** prevent a backflow of blood into the veins supplying blood to the heart when the atria contract (Fig 9.5(a)). The non-conductive septum between the atria and ventricles prevents the cardiac impulse in the atrial muscles from spreading directly into the ventricles. However, a second node, the **atrio-ventricular (AV) node**, also in the wall of the right atrium, picks up the atrial impulse. The AV node transmits the cardiac impulse along the conductive **bundle of His** and its branches in the interventricular septum. When the impulse reaches the apex of the heart it spreads rapidly up the ventricular walls in a network of conductive **Purkinje fibres** (Fig 9.4). The arrival of the cardiac impulse in the ventricles stimulates contraction there.

Fig 9.4 Conductive pathway of the cardiac impulse (modified from Julian)

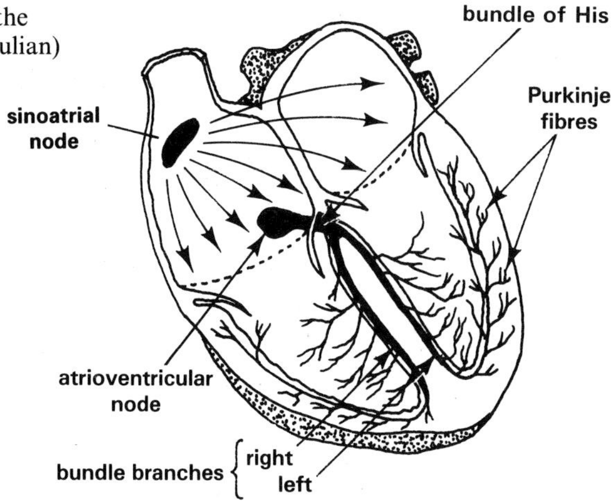

bundle of His

Purkinje fibres

sinoatrial node

atrioventricular node

bundle branches { right left

Blood is forced upwards from the ventricles into the pulmonary arteries leading to the lungs and the aorta leading to the rest of the body (Fig 9.5(b)). The atrio-ventricular valves prevent blood flowing back into the atria when the ventricles contract. When the ventricles relax, pulmonary and aortic semi-lunar valves prevent blood flowing back into the ventricles. The valves of the heart are non-muscular structures. They respond passively to blood pressure changes brought about by contraction and relaxation of the heart's chambers (Fig 9.6).

Fig 9.5 (a) The cardiac cycle begins with the cardiac impulse passing through the atria. Atrial constriction forces blood into the ventricles. The aorta and pulmonary arteries are omitted for clarity

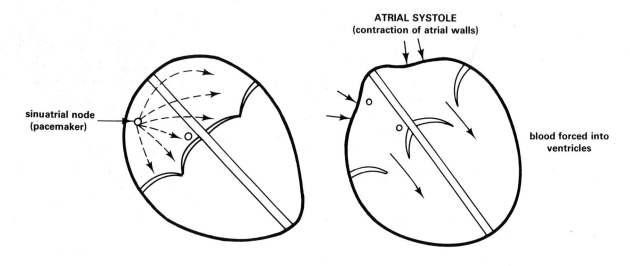

ATRIAL SYSTOLE
(contraction of atrial walls)

sinuatrial node
(pacemaker)

blood forced into
ventricles

Fig 9.5 (b) The cardiac impulse reaches the ventricles along the conductive bundle of His. Ventricular contraction forces blood into the aorta and pulmonary arteries

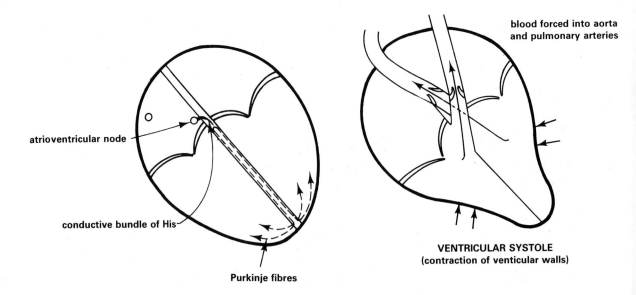

blood forced into aorta
and pulmonary arteries

atrioventricular node

conductive bundle of His

Purkinje fibres

VENTRICULAR SYSTOLE
(contraction of venticular walls)

160

Fig 9.6 The passive action of the heart's valves

(a) The chordae tendineae prevent the mitral valve's leaflets from inverting when the left ventricle contracts. The action is helped by contraction of the papillary muscles which pull on the chordae. When the left atrium contracts and the ventricle relaxes, the mitral's leaflets open passively

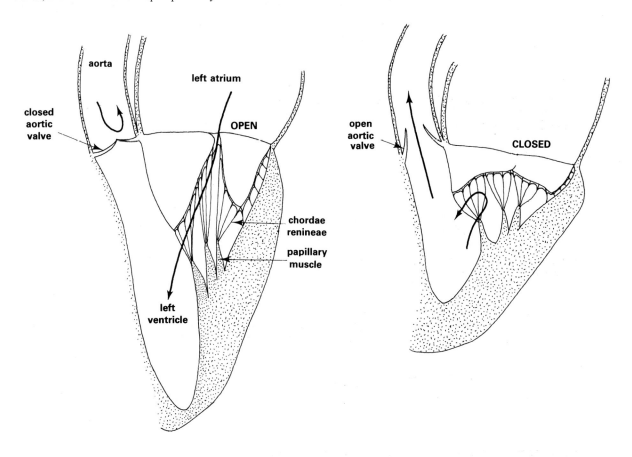

(b) Thickened ridges on the aortic valve's leaflets prevent them from inverting when the left ventricle relaxes. When the ventricle contracts blood pushes the aortic valve open.

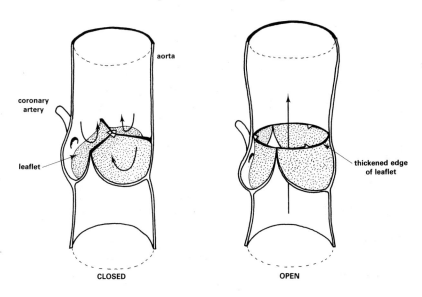

161

The electrical events which occur during each cardiac cycle can be displayed as an **electrocardiogram (ECG)**. Taking an ECG involves placing metal electrodes at specific sites on the skin. The simplest system uses four electrodes, one on the right wrist, one on the left wrist and one on the left ankle. The fourth electrode is connected from the right ankle to earth so that any electromagnetic disturbances in the room which are picked up by the body do not interfere with the recording (Fig 9.7(a)).

The cardiac impulse spreads through the heart starting at the pacemaker. Small voltages also spread through the body fluids which, because they contain electrolytes, are electrically conductive. Voltage changes therefore occur at the body surface and correspond to the intensity and direction of the cardiac impulse in the heart. The electrical changes are measured as voltage differences between pairs of electrodes called **leads**. The differences are usually recorded on moving paper (Fig 9.7(b)). A typical ECG consists of **waves** and **complexes** which correspond to particular cardiac events. The P-wave is generated by the cardiac impulse passing over the atria. The QRS complex is generated by the passage of the impulse down the bundle of His into the Purkinje network and through the walls of the ventricles. The T-wave corresponds to the electrical recovery of the ventricles. Deviations from the normal ECG are readily seen. Certain heart abnormalities result in typical abnormal wave forms (Fig 9.7(c)).

Fig 9.7 (a) Patient with electrodes in position for the recording of his ECG

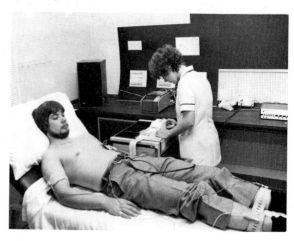

Fig 9.7 (b) A typical normal ECG pattern

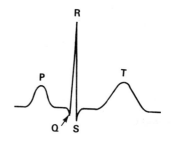

Fig 9.7 (c) Some abnormal ECG patterns

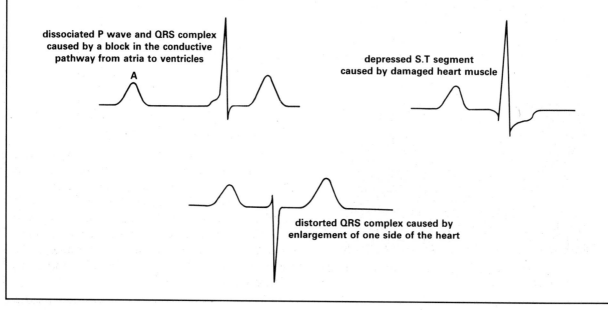

dissociated P wave and QRS complex caused by a block in the conductive pathway from atria to ventricles

depressed S.T segment caused by damaged heart muscle

distorted QRS complex caused by enlargement of one side of the heart

2 Pressure changes in the heart

Contraction of the atria, **atrial systole**, increases the pressure of blood in the atria. The rise in pressure forces the atrio-ventricular valves open and blood flows into the ventricles. Movement of blood from atria to ventricles is not entirely due to atrial systole. In a human heart about 70 % of the atrial blood passes into the ventricles as they relax, **ventricular diastole**, just before the atria contract. When the ventricles contract, **ventricular systole**, the pressure of the blood in the ventricles increases much more than atrial blood pressure. This is because the ventricles, especially the left one, have thicker and more muscular walls. During ventricular contraction blood pressure in humans rises from zero to about 16 kPa (120 mm Hg) in the left ventricle and about 3.3 kPa (25 mm Hg) in the right ventricle. Blood enters the ventricles and stops for a moment before it is pumped into the arteries. The pumping action of the ventricles is so strong that it can be felt as the **pulse** in arteries far away from the heart. Although blood enters the arteries in spurts, its flow is continuous. How is this possible?

As the ventricles contract and the semi-lunar valves of the aorta and pulmonary arteries open, blood in the ventricles and arteries is continuous. The blood pressure in the arteries is now the same as in the ventricles. Blood under pressure thus flows through the arteries causing the arterial walls to bulge. When the ventricles relax the main source of blood pressure is removed and the elastic arterial walls **recoil** keeping the blood in the arteries under pressure. In this way a continuous flow of blood is maintained. When the ventricular pressure falls to zero between contractions, aortic pressure drops to about 10.7 kPa (80 mm Hg) and pulmonary artery pressure to about 1.1 kPA (8 mm Hg). The term **blood pressure** as it is commonly used, refers to the pressure of blood in the aorta and the main arteries.

3 Heart sounds

As the blood flows through the heart, **heart sounds** are produced as the valves open and close. The sounds can be heard with the aid of a **stethoscope** and are often described by the words **lub** and **dup**. Simultaneous closure of the atrio-ventricular valves when the ventricles contract makes the first sound, lub. The second sound, dup, 0.3 seconds later is caused by the simultaneous closure of the aortic and pulmonary valves. Extra sounds called **murmurs** can be heard if the blood rushes into the ventricles faster than normal, possibly because one of the valves is defective.

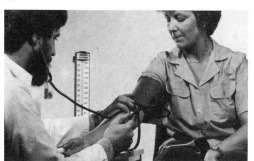

Fig 9.8 A sphygmomanometer being used to take the blood pressure of a patient

Blood pressure in the brachial artery of the arm is often measured using a **sphygmomanometer**:

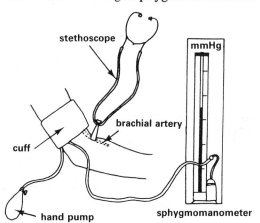

When the body is at rest the human heart beats about 70 times a minute and each cardiac cycle lasts for about 0.8 seconds. During the cycle the atria contract for about 0.1 seconds, the ventricles for 0.3 seconds. When we consider that the heart may beat more than 200 times per minute during severe exercise and that it beats without stopping throughout life it is evident that the heart is a remarkably durable and efficient pump.

4 Control of cardiac activity

Contraction of the cardiac muscle is **myogenic** because it generates impulses spontaneously and can therefore contract without any external stimulus being applied. An entire heart severed from its nerve supply and removed from the body may continue to beat rhythmically.

Nevertheless the heart is connected to sympathetic and parasympathetic nerves of the autonomic system (Figs 9.9 and 18.13). The nerves arise in the **medulla oblongata** of the brain. The sympathetic pathway begins in the **cardio-acceleratory centre**, the parasympathetic pathway in the **cardio-inhibitory centre**. Activation of the cardiac pacemaker by the **accelerator nerve** (sympathetic) increases the rate at which cardiac impulses are generated and hence the rate of heart beat. Branches of the accelerator nerve also stimulate the ventricular walls and increase the power of contraction, hence the stroke volume and cardiac output also increase. Activation of the pacemaker by the **vagus nerve** (parasympathetic) decreases the heart rate.

Fig 9.9 Neuronal pathways controlling the heart (after Green)

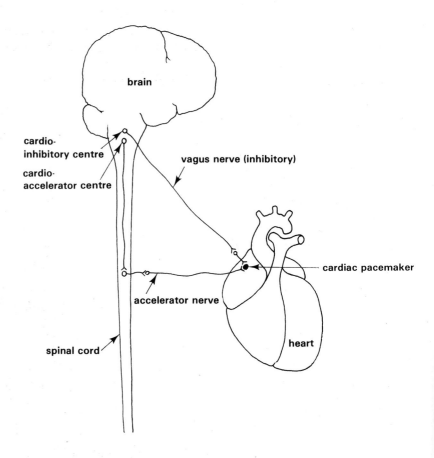

Cardiac output increases when we engage in physical exercise. How is the heart controlled and what causes cardiac output to fall after the exercise stops?

Although contraction of cardiac muscle is myogenic, the nerves supplying the heart allow cardiac activity to be modified in different circumstances. For example, during exercise cardiac output increases, thus supplying extra oxygen to the skeletal muscles and removing the extra carbon dioxide produced by them as a result of the exercise. Conversely cardiac output drops during periods of rest. The maximum amount by which the cardiac output can increase is called the **cardiac reserve**. It is usually about 400 % in humans but may be about 600 % in an athlete.

There are several mechanisms which control the heart's activity.

Starling's law of the heart

Starling's law states that:

> **the power of cardiac contraction is directly related to the length of the cardiac muscle fibres.**

If more blood enters the heart from the veins, the cardiac muscle fibres in the myocardium are stretched more and this induces the fibres to contract with greater force.

5   Baroreceptors

Small receptors sensitive to stretching are found in the walls of the aortic arch, the carotid sinuses, the venae cavae and the right atrium (Fig 9.10).

Fig 9.10 Position of the baroreceptors (stippled) in the cat

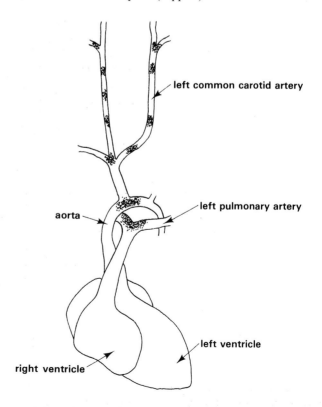

left common carotid artery

left pulmonary artery

aorta

left ventricle

right ventricle

They are called **baroreceptors**. If blood pressure increases in any of these vessels, the baroreceptors are activated and they transmit sensory impulses to the cardiac centres. Cardiac activity is decreased appropriately.

6 Adrenaline

In times of stress adrenaline is secreted into the blood by the medullae of the adrenal glands (Chapter 21). Adrenaline increases the excitability of the pacemaker and speeds up the heart rate. A similar substance called **noradrenaline** is released onto the pacemaker by the sympathetic (cardiac) nerve endings on the arrival of impulses. The inhibitory parasympathetic (vagus) nerve releases acetylcholine onto the pacemaker, resulting in a decrease in heart rate (Chapter 18).

7 Other factors

A drop in oxygen tension or a rise in carbon dioxide tension of the blood, directly affect the cardio-acceleratory centre and result in an increased heart rate. The same effect results from a drop in blood pH.

A high concentration of potassium ions in the blood interferes with the nerve impulses that control the pacemaker, causing the heart to slow down. Raised levels of blood sodium ions also cause the heart to slow down because excess sodium ions interfere with the action of calcium ions in muscle contraction. (Chapter 17).

Emotional excitement and a rise in body temperature cause the heart rate to speed up. Conversely, feelings of grief and depression may slow the heart rate down.

8 Vasomotor control

How are the actions of the lungs and the heart and blood vessels related?

A further effect of stimulation of the baroreceptors is the transmission of nerve impulses to the **vasomotor centre**. Like the cardiac centre the vasomotor centre is found in the medulla oblongata of the brain. Sympathetic nerves originating in the centre lead to arterioles. By secreting noradrenaline from their endings the nerves bring about contraction of the smooth muscle in the walls of the arterioles. As the muscle is arranged in a circular fashion its contraction results in narrowing of the arterioles called **vasoconstriction**. Conversely, decreased sympathetic activity from the vasomotor centre results in **vasodilation** (Fig 9.11).

Fig 9.11 Photomicrographs of arterioles seen in transverse section

(a) dilated, ×400

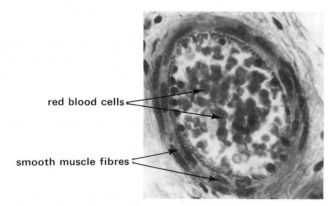

red blood cells

smooth muscle fibres

(b) constricted, ×270

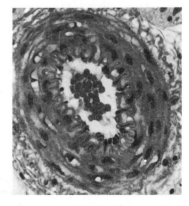

166

High blood pressure stimulates the baroreceptors which suppress sympathetic output from the vasomotor centre. The resulting vasodilation lowers the blood pressure, an appropriate response to high blood pressure. Low blood pressure has the opposite effect (Fig 9.12).

Fig 9.12 The main factors controlling mammalian blood pressure

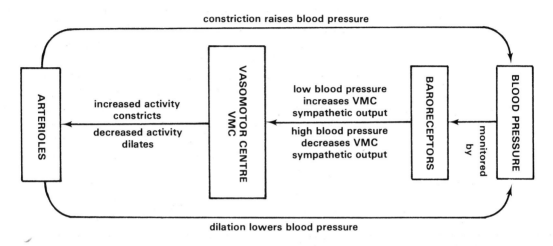

As with the cardiac and respiratory centres, the vasomotor centre is stimulated by changes in the tensions of oxygen and carbon dioxide in the blood. Generally all three centres respond to the same physiological stimuli (Fig 9.13). Similar responses can be produced by secretions from the adrenal glands (Chapter 21).

Fig 9.13 The control of breathing, heart rate and blood pressure in mammals

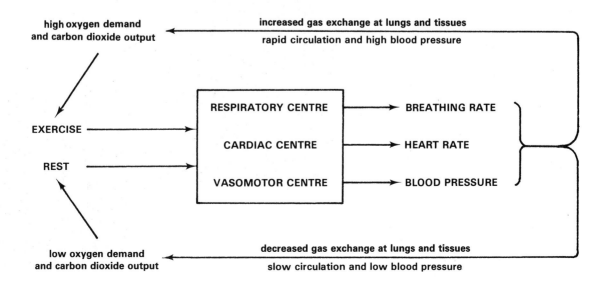

9.1.2 Blood vessels

Blood flows from the ventricles into **arteries**, then **arterioles**, **blood capillaries** and **venules**, and is finally returned to the heart in **veins**. Blood vessels have properties that help the circulation and allow the blood to perform many of its functions. The walls of the larger blood vessels conform to a basic structural pattern (Fig 9.14). There are three main layers. On the inside the **tunica intima** (t. interna) consists of a single layer of flat cells, the **endothelium**, supported by a basement membrane and **connective tissue** containing **collagen**. The middle layer is the **tunica media** which contains smooth muscle and fibres of collagen and **elastin**. The tunica media varies in thickness in different vessels. It is completely absent in blood capillaries but is the thickest layer in the elastic arteries such as the aorta. Collagen provides strength and prevents excessive stretching of the blood vessels. In contrast, elastin can be stretched, thus allowing vasodilation. The outermost layer is called the **tunica adventitia** (t. externa). It also contains collagen fibres and is relatively tough. In the larger vessels with relatively thick walls, there are small blood vessels in the tunica adventitia. They are called the **vasa vasorum**, meaning the vessel's vessels. They deliver nutrients and oxygen to the tissues of the large blood vessels. This arrangement is comparable to the way in which the coronary arteries supply blood to the wall of the heart.

Fig 9.14 Structure of the wall of (a) elastic and (b) muscular arteries

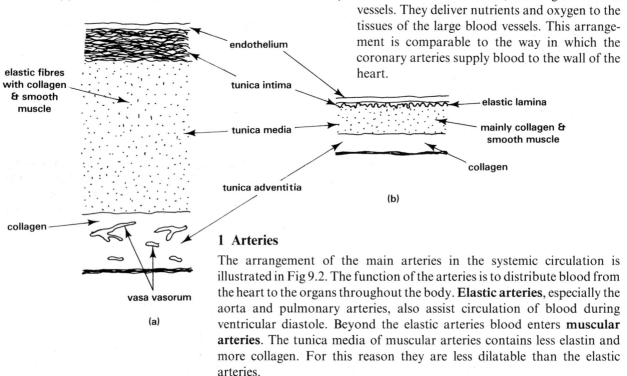

(a)

(b)

1 Arteries

The arrangement of the main arteries in the systemic circulation is illustrated in Fig 9.2. The function of the arteries is to distribute blood from the heart to the organs throughout the body. **Elastic arteries**, especially the aorta and pulmonary arteries, also assist circulation of blood during ventricular diastole. Beyond the elastic arteries blood enters **muscular arteries**. The tunica media of muscular arteries contains less elastin and more collagen. For this reason they are less dilatable than the elastic arteries.

Sometimes arteries become diseased. Deposits called **atheromatous plaques** may form in the tunica intima. They cause a narrowing, **stenosis**, of the affected vessel. Blood flow is impaired (Fig 9.15). The deposits may also give rise to intravascular thrombi (section 10.1.3). High concentrations of cholesterol and saturated fats in the blood are thought to cause such deposits.

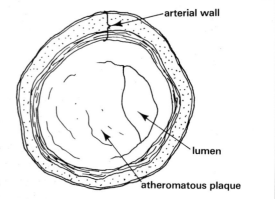

Fig 9.15 Stenosis of an artery due to atheromatous plaques

2 Arterioles

Arterioles are found inside the tissues. They are the terminal branches of arteries. Smooth muscle fibres are relatively abundant in arteriole walls. By contracting or relaxing, the fibres can alter the diameter of the **lumen** through which the blood flows, and thus the blood pressure (Fig 9.11).

3 Blood capillaries

Capillaries are the smallest blood vessels, having a diameter as small as 3 to 4 μm. This is about half the diameter of a red blood cell (Fig 9.16). Some blood capillaries are up to ten times larger than this and are called **sinusoids**, such as those in the liver (Fig 15.29(a)). The capillary wall consists of a single layer of endothelium and presents little barrier to the movement of many metabolites across the wall. The main function of blood capillaries is to allow an orderly interchange of metabolites between blood and tissue fluid (section 9.1.3). The capillary system in any organ is much branched and is often called the **capillary bed**. No cell in the organ is more than a few cells distant from the nearest blood capillary. Once in the tissue fluid, metabolites move to and from the cells mainly by diffusion.

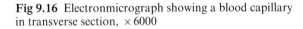

Fig 9.16 Electronmicrograph showing a blood capillary in transverse section, × 6000

red blood cell

nucleus of
capillary wall cell

4 Venules and veins

The positions of the main veins in the systemic circulation are illustrated by Fig 9.2. Venules and veins collect blood of low pressure and return it to the heart. Blood in the veins flows in steady streams rather than spurts as in arteries. The blood in the centre of a vein flows fastest, whereas that just inside the vein wall flows very slowly.

Fig 9.17 Roles of semi-lunar valves and the contraction of skeletal muscle in regulating blood flow through veins

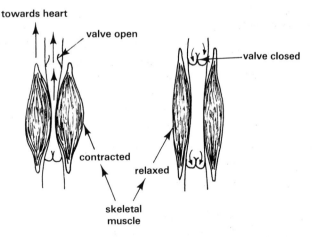

Several factors cause venous blood to flow towards the heart. One is the negative pressure created in the thorax every time air is drawn into the lungs (Chapter 8). When breathing out the thoracic pressure is raised. Blood in the veins is also prevented from flowing away from the heart by semi-lunar valves (Fig 9.17). Many large veins, especially in the limbs, lie between skeletal muscles. Contraction of the muscles squeezes the veins and forces blood through them towards the heart.

The important properties of the different blood vessels are summarised in Fig 9.18.

Fig 9.18 Main properties of the different kinds of vessels in the blood vascular system

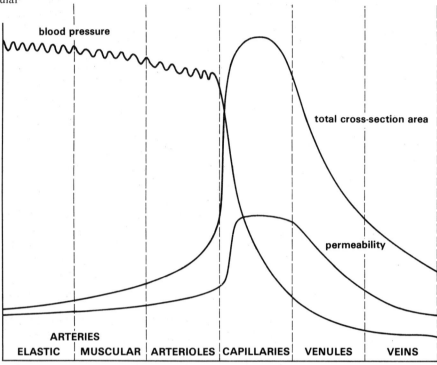

direction of blood flow through circulation

9.1.3 Tissue fluid and lymph

Blood distributes oxygen and metabolites to the body's tissues and removes their products including wastes. However, tissue cells are not in direct contact with the blood. They are bathed in **tissue fluid** which is an intermediary between the blood and the cells.

1 Interchange between blood and tissue fluid

The passage of metabolites from the blood to the tissues is not haphazard. What controls the process?

Metabolites pass between the blood and the tissue fluid through the walls of the blood capillaries. Capillary walls are freely permeable to substances of relatively small molecular size. The walls of other blood vessels are relatively impermeable to the blood constituents (Fig 9.18). Substances of relative molar mass greater than about 65 000 and most kinds of blood cell normally remain inside the capillaries. The molecules of most blood proteins are too large to pass into the tissue fluid.

Interchanges between blood and tissue fluid are brought about by the interaction of the **pressure potential (PP)** and the **solute potential (SP)** of the blood and tissue fluid. PP is the hydrostatic pressure exerted on a fluid by its surroundings, e.g. blood pressure created by elastic recoil of blood vessels. The main contributors to SP are ions and molecules of large relative molar mass, especially proteins which attract water. The actions of these forces are summarised in Fig 9.19. Because of the pumping action of the heart and recoil of the elastic arteries, blood leaving arterioles has a relatively high potential pressure at about 3.29 kPa. Plasma proteins, especially albumen, cause the solute potential of the blood to be about −3.68 kPa. In the tissue fluid the pressure potential is estimated at about

-0.83 kPa, slightly less than atmospheric pressure. The solute potential in the tissues is also low at about -0.66 kPa. This is because only few proteins can pass through capillary walls from the blood into the tissue fluid.

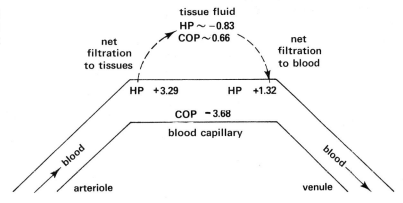

Fig 9.19 Summary of the forces affecting the interchange of materials between blood and tissue fluid (figures from Guyton)

The resultant of the pressures is called **filtration pressure (FP)**:

$$FP = \begin{bmatrix} SP + PP \\ of\ blood \end{bmatrix} - \begin{bmatrix} SP + PP \\ of\ tissue\ fluid \end{bmatrix}$$
$$= [-3.68 + 3.29] - [-0.66 + (-0.83)]\ kPa$$
$$= [-0.37] - [-1.49] = 1.12\ kPa$$

The positive filtration pressure at the arteriolar end of the capillary network causes water, oxygen and other metabolites to be forced from the blood into the tissue fluid.

As the blood flows through to the venous end of the capillary network its pressure potential drops to about 1.32 kPa. All other pressures remain fairly constant and so the filtration pressure is now given by:

$$FP = [-3.68 + 1.32] - [0.66 + (-0.83)]$$
$$= [-2.36] - [-1.49]$$
$$= -0.87\ kPa$$

Because the filtration pressure at the venous end of the capillary network is negative, water, excess metabolites, tissue products and wastes are forced from the tissue fluid into the blood. However, there is a greater loss from the blood to the tissue fluid at the arteriolar end of the capillaries than returns to the blood in the venous capillaries. For this reason the tissues gradually accumulate fluid at the blood's expense.

The worked example for filtration pressure using the terms 'hydro-static pressure' and 'colloid osmotic pressure' as used in the previous edition was:

$$FP = \begin{bmatrix} HP\ of & - & HP\ of \\ blood & & tissue\ fluid \end{bmatrix} - \begin{bmatrix} COP\ of & - & COP\ of \\ blood & & tissue\ fluid \end{bmatrix}$$
$$= [25 - (-6.3)] - (28 - 5)$$
$$= 31.3 - 23 = 8.3\ mmHg$$

$$FP = [10 - (-6.3)] - (28 - 5)$$
$$= 16.3 - 23$$
$$= -6.7\ mmHg$$

2 Lymph

The excess fluid absorbed from the blood in the tissues enters **lymphatic capillaries** (Fig 9.20). They drain the excess fluid, now called **lymph**, into a system of vein-like vessels known as the **lymphatic system** (Fig 9.21). Lymph is produced at a rate of about $1.5\,cm^3$ each minute. The main lymphatic vessels contain semi-lunar valves which ensure lymph flows in one direction. Lymph is pushed through the lymphatic system when the surrounding tissues and muscles squeeze on the lymphatic vessels.

Fig 9.20 (a) Diagrammatic relationship between the blood vascular and lymphatic systems

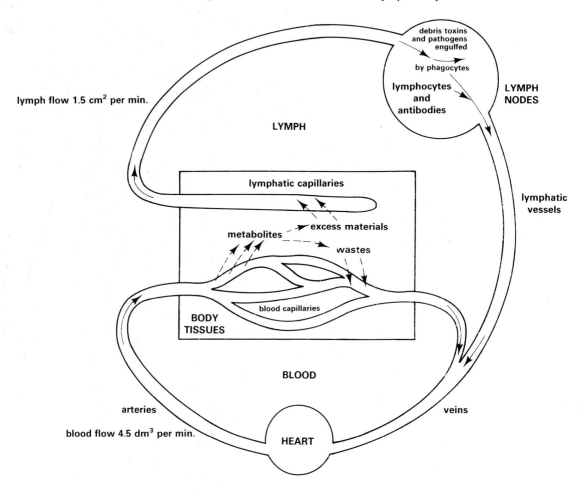

Fig 9.20 (b) Photomicrograph of a thin section of reticular connective tissue, $\times 100$

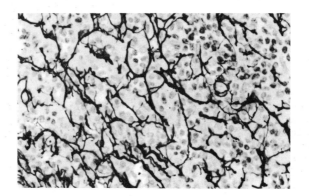

172

Fig 9.21 Main vessels of the lymphatic system. Those of the left limbs are omitted for clarity

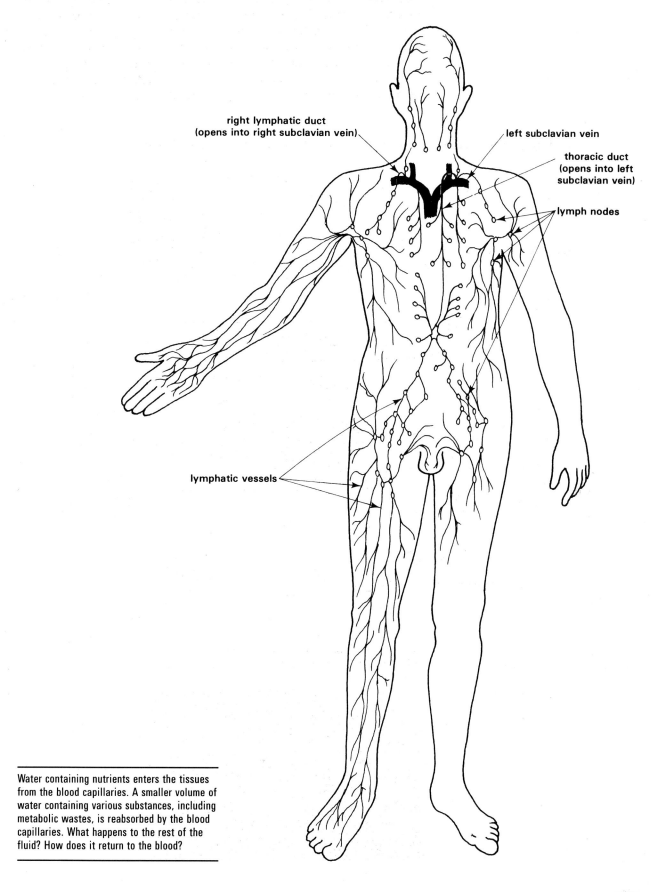

right lymphatic duct
(opens into right subclavian vein)

left subclavian vein

thoracic duct
(opens into left
subclavian vein)

lymph nodes

lymphatic vessels

Water containing nutrients enters the tissues from the blood capillaries. A smaller volume of water containing various substances, including metabolic wastes, is reabsorbed by the blood capillaries. What happens to the rest of the fluid? How does it return to the blood?

Lymph eventually returns to the blood. This is because the thoracic lymph ducts open into the subclavian veins. Thus a constant blood volume is maintained and the tissues are not saturated with excess fluid.

If lymph return is blocked, such as by parasitic worms in **elephantiasis**, then tissue fluid accumulates. This is called **oedema** (Fig 9.22). Oedema can also result from imbalances in the pressure potential and solute potential of the blood and tissue fluids, or from changes in the permeability of blood capillary walls. Severe protein deficiency is sometimes seen in the diets of children in some third world countries. The deficiency leads to **kwashiorkor**. Among the many harmful effects is severe oedema caused by a low solute potential in the blood and excess retention of fluid in the tissues.

Fig 9.22 Elephantiasis

On its return journey to the blood, lymph flows through one or more **lymph nodes**. They remove particulate debris, microbes and toxins from the lymph and release **antibodies** and **lymphocytes** into the lymph (Chapter 10). Lymph nodes contain **reticular connective tissue** which has many fibres of the protein **reticulin** (Fig 9.20(b)). It is delicately branched and supports many cells. Reticular connective tissue is also found in organs such as the liver and endocrine glands.

9.1.4 Circulation in the mammalian fetus

The mammalian fetus obtains its oxygen and nutrients from the **placenta** in the mother's uterus (Chapter 25). The placenta also removes wastes from the fetal blood. Maternal and fetal blood vessels come into close proximity to one another in the placenta (Fig 9.23). The two-way exchange occurs mainly by diffusion. Thus the functions of the fetal lungs, kidneys and gut are performed by the placenta.

Fig 9.23 Human fetus and placenta inside the uterus

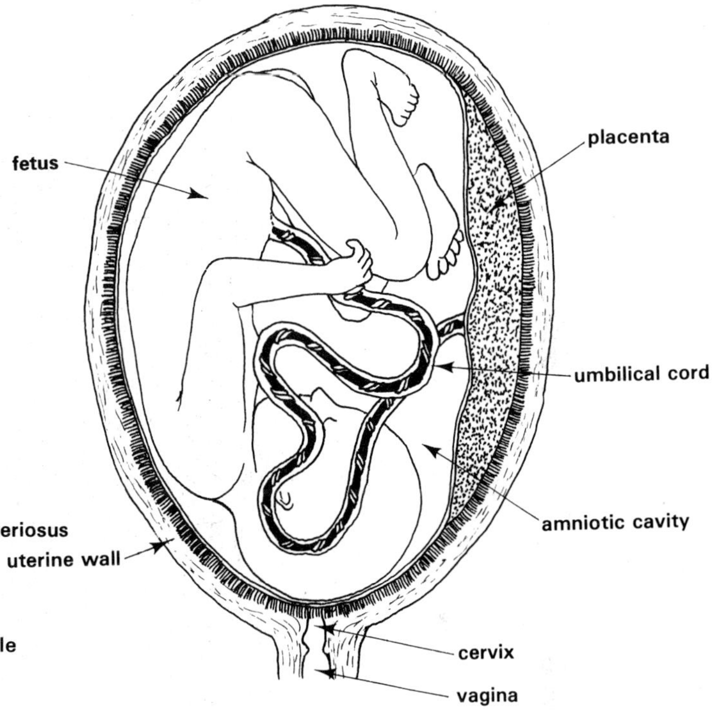

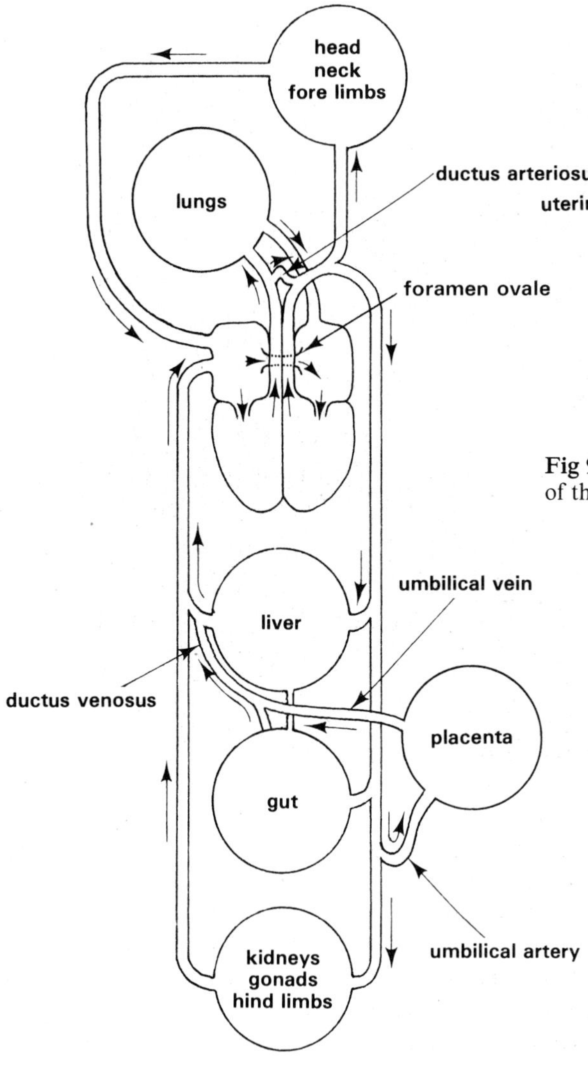

Fig 9.24 Diagram of the blood circulation of the mammalian fetus

The pattern of circulation in the fetal body is different from that of an adult. Fig 9.24 illustrates the circulatory system of a human fetus. Compare it with Fig 9.1. An obvious difference between the fetal and adult circulation is the presence of **umbilical blood vessels** in the fetus which carry fetal blood to and from the placenta. The umbilical vessels are found in the umbilical cord. The other main differences are concerned with the transfer of the functions of the fetal lungs to the placenta. The pulmonary circulation in the fetus is almost completely by-passed due to a **ductus arteriosus** between the pulmonary arteries and the aorta, and a hole called the **foramen ovale** between the right and left atria. Such an arrangement involves mixing of oxygenated blood from the placenta and deoxygenated blood from the fetal tissues in the posterior vena cava and in the heart. The relative loss of efficiency in the transport of oxygen is compensated for by the presence of **fetal haemoglobin** which combines more readily with oxygen than does adult haemoglobin (Chapter 8). Further, the heart of the fetus does not pump blood to its unaerated lungs so energy is saved. The **ductus venosus** directs blood from the placenta and the fetal gut to the posterior vena cava, thus by-passing the fetal liver. Since most of the functions of the fetal liver are performed by the mother's liver, a supply of blood to the liver of the fetus is less important. The fetal liver is, however, the only source of fetal blood cells.

Shortly after birth the ductus venosus, ductus arteriosus and foramen ovale normally close. Failure to do so is the cause of severe circulatory problems some of which can be corrected by surgery.

SUMMARY

Blood is a mass transport system. It is pumped by the heart which is a double pump. The left side collects oxygenated blood from the lungs into an upper chamber, the left atrium. The blood passes through the mitral valve into the left ventricle which pumps it through the aortic valve to all parts of the body via the aorta. The myocardium receives blood through the coronary arteries. The right atrium collects deoxygenated blood from the body through the venae cavae and from the myocardium through the coronary sinus. The blood passes through the tricuspid valve into the right ventricle which pumps it through the pulmonary valve to the lungs via the pulmonary arteries.

The heart valves are fibrous flaps which respond passively to blood pressure changes. They prevent blood flowing backwards in the heart. Closure of the valves creates the 1st and 2nd heart sounds.

The cardiac impulse begins in the sino-atrial node (pacemaker) in the right atrium. Atrial depolarisation activates the atrio-ventricular node also in the right atrium. The impulse conducts along the bundle of His and right and left bundle branches to the apex; then into the ventricle walls through the Purkinje network. The conduction of the cardiac impulse in this way brings about orderly contraction (systole) and relaxation (diastole) of the heart's chambers.

The cardiac centres in the hind brain transmit impulses to the heart. Sympathetic stimulation increases heart rate and the power of contraction. Parasympathetic (vagus) stimulation slows the heart and it pumps with less force. The cardiac centres respond to a variety of stimuli which indicate the need to modify cardiac activity. They include blood pressure, adrenaline, blood oxygen and carbon dioxide tensions, blood pH, potassium and sodium ions and body temperature.

Arteries leaving the heart have elastic walls. After stretching when the ventricles pump blood into them, they recoil and push the blood onwards between heart beats. Arterioles can dilate (vasodilation) or constrict (vasoconstriction) thus changing the resistance to blood flow and hence blood pressure. The changes are caused by nerve impulses from the vasomotor centre in the hind brain. Blood flow in venules and veins is caused by several factors including; negative pressure in the thorax due to breathing, muscles squeezing on the vessels, semilunar valves in veins preventing back flow.

Exchange of metabolites between blood and tissues occurs across capillary walls. Pressure potential (caused by the heart) and solute potential (caused mainly by the plasma proteins) combine to effect the exchange. There is a net loss of substances from blood to the tissue fluids. The excess is taken into the lymphatic capillaries and returned to the blood as lymph via the lymphatic system. ▶

> Circulation in the fetus differs from that in the adult because the fetal lungs and liver are largely by-passed and metabolites are exchanged with the mother at the placenta. Shortly after birth the ductus venosus, ductus arteriosus and foramen ovale close. Proper circulation is established to the liver and lungs. Oxygenated blood in the left side of the heart is separated from deoxygenated blood in the right side. The umbilical supply to the placenta ceases.

QUESTIONS

1 The diagram presents a schematic arrangement of some mammalian organs and their associated blood vessels.

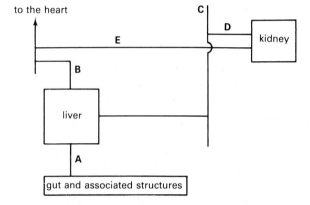

For each of questions (a) to (d), select the letter (**A–E**) which identifies the blood vessel which carries blood that is most likely to have or to transport

(a) the greatest concentration of insulin,

(b) the lowest percentage mass of haemoglobin saturated with oxygen,

(c) the greatest total volume per unit time,

(d) the highest average temperature. (JMB)

2 The table gives data for a well-trained endurance athlete and an untrained individual working at a standard rate.

	Endurance athlete	Untrained individual
Oxygen uptake (dm³ minute⁻¹)	3.023	3.024
Volume of blood pumped per beat (dm³)	0.156	0.120
Heart rate (beats minute⁻¹)		180
Cardiac output (dm³ minute⁻¹)	19.5	

(a) Complete the table by calculating the missing values.

(b) What would be the cardiac output of each individual with a rate of 200 heartbeats per minute?

(i) endurance athlete,

(ii) untrained individual.

(c) Suggest **one** way in which the muscles of the endurance athlete obtain more oxygen from the blood than those of the untrained individual. (AEB 1986)

3 (a) Draw a large, labelled diagram of a mammalian heart to show the detailed internal structure.

(b) Describe the mechanism by which the heart beat is (i) initiated and (ii) controlled.

(c) Discuss the advantages to a mammal of a double circulation. (C)

4 (a) Describe the mechanism of the heart beat in a mammal and the ways in which it is controlled.

(b) What part do the blood vessels play in the circulation of the blood?

(c) Explain the relationship between blood and tissue fluid. (L)

5 (a) Describe the structure and action of the mammalian heart.

(b) How is the heartbeat initiated and controlled? (L)

6 The diagram below shows the volume changes in cm³ in the left ventricle of a human heart in a single cardiac cycle.

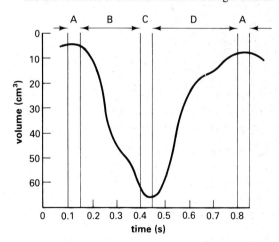

(a) Give the letter **A**, **B**, **C** or **D** to indictate during which phase:

(i) the bicuspid (mitral) valve is open;

(ii) the aortic valve is open;

(iii) the bicuspid (mitral) valve is open;

(iv) the aortic valve closes.

(b) State the function of:

(i) chordae tendineae;

(ii) the sino-atrial (auricular) node;

(iii) the Purkinje fibres.

(c) What is the heart rate in beats per minute of the cardiac cycle shown in the diagram? (O)

7 The diagrams show parts of the blood-circulatory systems of the human fetus and adult.

(a) (i) Describe four circulation differences visible in the diagrams between the fetus and adult.

(ii) Explain the functional significance of each of these differences.

(b) The fetus most obtain all its oxygen from the mother via the placenta. How is the efficient transfer of oxygen from the maternal to the fetal blood achieved in the placenta?

(c) Describe and account for the changes that occur in the fetal circulation at and soon after birth.

(AEB 1986)

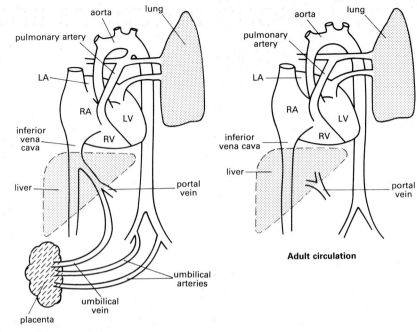

Fetal circulation

Adult circulation

10 Blood functions and immune system

10.1 Blood cells   180

    10.1.1 Red blood cells   180
    10.1.2 White blood cells   181
    10.1.3 Platelets and coagulation   182
    10.1.4 Plasma proteins   184

10.2 Immune system   185

    10.2.1 Acquired immunity   186
    10.2.2 Blood groups   194
    10.2.3 Tissue typing   197

Summary   197

Questions   198

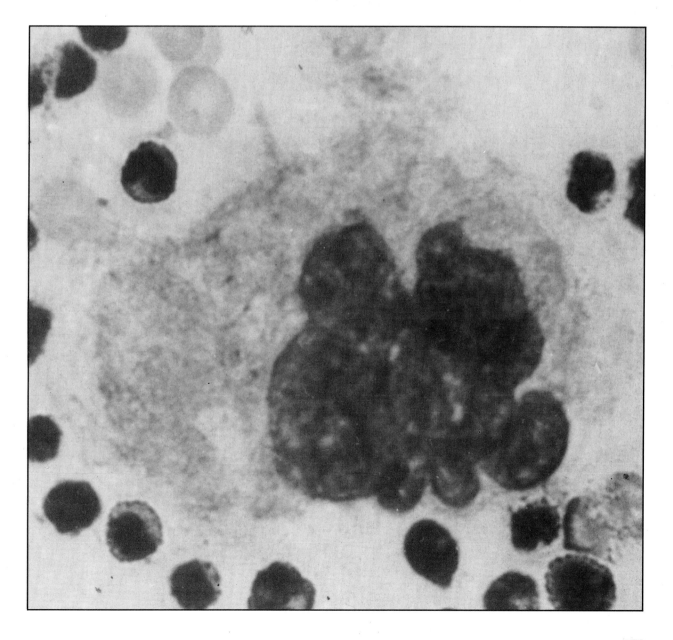

10 Blood functions and immune system

10.1 Blood cells

Fig 10.1 Photomicrograph of a smear of human blood showing (a) a neutrophil, (b) a lymphocyte, (c) a monocyte, and (d) an eosinophil, all × 1500

Under the microscope **blood cells** are seen to be of several types. They make up about 45 % of the volume of mammalian blood (Table 10.1 and Fig 10.1).

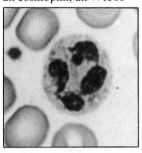
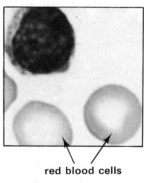
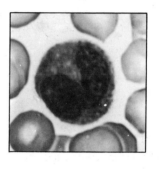

red blood cells

(a)   (b)   (c)   (d)

Table 10.1 Summary of the numbers and functions of human blood cells

Cell type	Average numbers × 10^9 dm^{-3} adult human blood	Functions
red cells	3900–6500	oxygen carriage
neutrophils	2.50–7.50	phagocytosis
eosinophils	0.04–0.44	reduce inflammation; dissolve clots
basophils	0–0.10	unknown
monocytes	0.20–0.80	phagocytosis
lymphocytes	1.50–3.50	immune response
platelets	150–400	coagulation

10.1.1 Red blood cells

By far the most numerous of the blood cells are the **red cells, erythrocytes**. Their colour comes from the respiratory pigment **haemoglobin** which they contain. Human red cells each contain about 30 pg of haemoglobin. Red cells occupy between 40 and 60 % of the total blood volume in different mammals and their main function is to carry oxygen. Mammalian red cells are biconcave discs and have a relatively great surface area in relation to their volume. They are thus ideally shaped for the uptake of oxygen when in the pulmonary capillaries (Fig 10.2). The function of haemoglobin in this respect is described in Chapter 8.

Fig 10.2 Scanning electronmicrograph of some human red blood cells, × 1700

The absence of a nucleus in red cells provides extra space for haemoglobin. The anucleate condition does have its disadvantages, however. In particular, red cells have only a relatively short lifetime. Red cells in humans survive for about four months and have to be replaced continuously to keep a constant number in circulation. It is estimated that human red cells are made at a rate of about 9000 million per hour. Red cell production, **erythropoiesis**, occurs in the liver in the foetus but in adult life is restricted to the red marrow of certain bones. They include the cranium, vertebrae, ribs, sternum, parts of the pelvic girdle and some limb bones. Old red cells are engulfed by phagocytic cells in the liver, spleen and red bone marrow. The haemoglobin is broken down and the iron it contains is retained for further haemoglobin synthesis. The rest of the pigment is excreted from the liver in bile (Chapter 15).

Impairment of the oxygen-carrying capacity of the blood is called **anaemia**. Anaemia can be caused by physical blood loss, **haemorrhage**, or functional blood loss, **haemolysis**. In haemolysis, red cells are destroyed in the blood vessels and the haemoglobin is broken down in the blood plasma. Decreased blood production is another cause of anaemia. Anaemias in this category result from inadequate production of red cells and haemoglobin in the bone marrow. **Impaired erythropoiesis** is often caused by deficiency of a vital nutrient in the diet. Iron deficiency is the commonest cause of human anaemia in the world today. Vitamin B_{12} is necessary for erythropoiesis. Absorption of vitamin B_{12} from the gut depends on the presence of a substance called **intrinsic factor (IF)** which is secreted in gastric juice (Chapter 15). Absence of IF leads to poor absorption of vitamin B_{12} and is the cause of **pernicious anaemia**. Production of abnormal haemoglobin is the cause of **sickle cell anaemia** (Chapters 6 and 8).

10.1.2 White blood cells

Far less numerous, but no less important than the red blood cells are the white blood cells collectively called **leucocytes**. They include **neutrophils,, lymphocytes, monocytes, eosinophils** and **basophils**. They are involved in defending the body against disease. The lymphocytes are produced in the lymph nodes. Other white cells are made in the bone marrow and some in connective tissues elsewhere in the body. White cells have a limited life time in the circulation and old white cells are disposed of by phagocytes in the liver and spleen.

Neutrophils and monocytes exhibit **amoeboid movement**. By extending pseudopodia they can encircle and engulf bacteria and other particles, a process called **phagocytosis** (Fig 10.3). The phagocytic behaviour of living neutrophils can be observed microscopically and recorded by time-lapse photomicrography. Studies of this sort show that neutrophils can locate

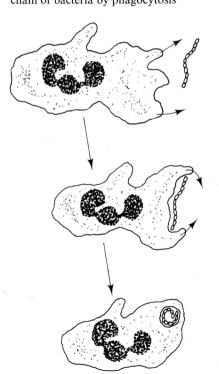

Fig 10.3 (a) A neutrophil engulfing a chain of bacteria by phagocytosis

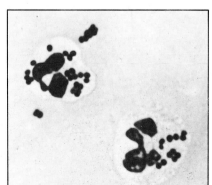

Fig 10.3 (b) Photomicrograph showing human neutrophils containing phagocytised bacteria, $\times 1400$

bacteria at a distance and actively seek them out for phagocytosis. This response is called **chemotaxis** and seems to be triggered by chemicals released by the bacteria. Engulfed bacteria are taken into large food vacuoles where secretions of the white cells' lysosomes kill the bacteria after several minutes. When the bacteria are dead they are discharged from the trailing ends of leucocytes by reverse phagocytosis. Phagocytosis of some strains of virulent bacteria requires the presence of antibodies (section 10.2).

The numbers of phagocytes in the blood often increase in response to bacterial infection. Phagocytes can move in and out of blood capillaries by squeezing between the cells of capillary walls. They appear rapidly at sites of localised infection. Monocytes found in the tissues are sometimes called **macrophages**. As well as mobile phagocytes there are many fixed phagocytic cells in the mammalian body. Patches of fixed phagocytes are found in the lymph nodes, gut wall, alveoli, liver, spleen and red bone marrow. Fixed phagocytes form the **reticulo-endothelial system** which removes particulate debris including old blood cells from the blood and lymph.

Phagocytosis is thus a widespread and important means of neutralising infective agents and removing particulate debris generally from the body fluids.

Basophils are thought to be **mast cells** which have entered the blood. Mast cells are found in connective tissues. The numbers of eosinophils may rise during an **allergic response** (section 10.2.1).

Lymphocytes originate in the bone marrow but are produced in large numbers in the nodes of the lymphatic system (Fig 9.21). Lymphocytes produce and may carry **antibodies** and are part of the body's immune response to infection (section 10.2).

10.1.3 Platelets and coagulation

As well as red and white cells, mammalian blood contains numerous anucleate cellular fragments called **platelets (thrombocytes)**. Human blood contains between 150 and 400 thousand platelets per mm³. They originate in the red bone marrow from large cells called **megakaryocytes** from which the platelets are budded off as fragments of cytoplasm (Fig 10.4).

Fig 10.4 Photomicrographs showing (a) a megakaryocyte in a bone marrow smear from a guinea-pig, × 750, (b) a smear of human blood with thrombocytes (platelets), × 1500

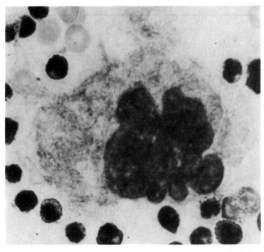

(a)

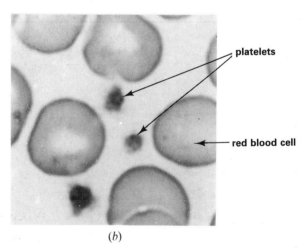

(b)

Platelets are involved in the **coagulation** of blood, a response to injury which prevents excessive blood loss. Coagulation is a complex sequence of chemical reactions which results in the deposition of insoluble fibrous proteins to form a **clot**. It is the clot which blocks damaged blood vessels and may stop bleeding. Platelets stick to the surfaces of damaged vessels. Healthy blood vessels secrete into the blood a substance called **prostacyclin**. Prostacyclin prevents thrombocytes from sticking together and to the walls of blood vessels. In the absence of prostacyclin, however, agglutination of platelets occurs. They then burst releasing platelet factors into the plasma. The **platelet factors** react with other **blood factors** in plasma to produce **thromboplastin**. Thromboplastin can also be produced when certain **tissue factors** react with the blood factors, a reaction which occurs for example when blood vessels are injured. Thromboplastin, which originates from within the blood, is called **intrinsic thromboplastin**. **Extrinsic thromboplastin** is found in the tissues.

The next stage of coagulation is the conversion of **prothrombin**, a plasma protein, into an active enzyme called **thrombin**. Thromboplastin activates the conversion which requires the presence of **calcium(II) ions**, a normal constituent of plasma. The final stage is the conversion by thrombin of the soluble plasma protein **fibrinogen**, into insoluble **fibrin**. The entire clotting mechanism is summarised in Fig 10.5. Fibrin is a fibrous protein which precipitates as strands around the platelets and damaged edges of the vessel. The meshwork of fibrin traps blood cells which increase the size of the clot, eventually sealing off the damaged vessel (Fig 10.6). Burst platelets also release a substance called **serotonin** (5-hydroxytryptamine). Serotonin causes vasoconstriction which may be enough to stop blood loss from damaged arterioles. The stoppage of blood flow is called **haemostasis**.

Important though coagulation is it is equally important that coagulation is restricted to sites of injury. A complex set of factors exists in mammalian blood to ensure that this normally happens. Fibrin inhibits excessive conversion of fibrinogen by **negative feedback**:

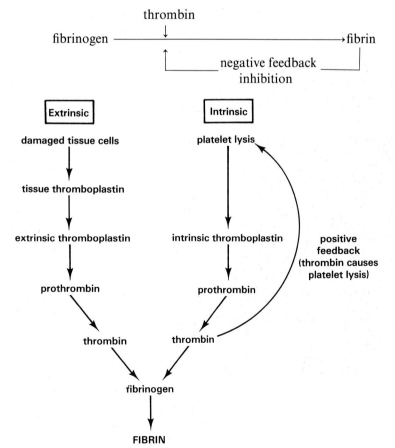

Fig 10.5 Summary of the coagulation mechanism

Enzymes capable of dissolving fibrin are carried in blood. The granules of eosinophils contain an inactive precursor called **profibrinolysin**. At sites of fibrin deposition profibrinolysin released by eosinophils is converted to **fibrinolysin** which helps to dissolve a clot after it has done its job.

The inner surfaces of blood vessels can become damaged, sometimes because of deposition of fatty substances particularly **cholesterol**. Deposits of this kind prevent secretion of prostacyclin and can trigger off coagulation, resulting in the formation of a clot inside the vessel. The clot is called an **intravascular thrombus**. Such a condition, commonly called **thrombosis**, can be very dangerous. For example, a thrombus in an artery feeding a vital organ such as the heart, can cause a sudden and fatal stoppage of blood to the organ. Coronary thrombosis accounts for a large number of deaths in industrialised countries. High-fat diets, sedentary jobs and tobacco smoking are among the main causes. An intravascular thrombus can occur in other vessels in the body such as a leg vein. If dislodged by the blood flow the thrombus is carried in the blood stream as an **embolus** which may eventually come to rest in the arteries of a vital organ. An embolus could for example be taken to the pulmonary arteries causing a **pulmonary embolism** which can cause death. The treatment of thrombosis and embolism involves administration of various **anticoagulants** such as warfarin which prevents coagulation by blocking the production of prothrombin in the liver. Other anticoagulants are used to dissolve the clot.

Evidently, normal coagulation depends on the delicate balance of many factors.

Fig 10.6 Scanning electronmicrograph showing some of the components of a blood clot, × 1000

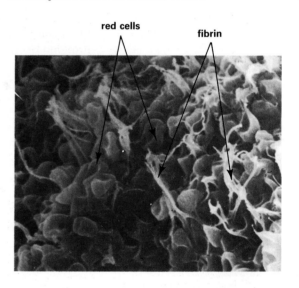

red cells   fibrin

10.1.4 Plasma proteins

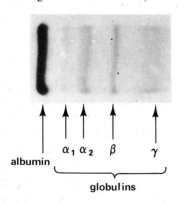

Fig 10.7 Main fractions of plasma proteins separated by electrophoresis, normal pattern (courtesy Mr R. Dainty, Department of Biochemistry, University of Nottingham Medical School).

albumin α_1 α_2 β γ

globulins

Blood plasma contains a large quantity of proteins normally amounting to about 70 to 90 g in each dm^3. Nearly all the plasma proteins are made in the liver. They circulate in the blood only for a limited period after which they are broken down, also in the liver (Chapter 15). The most abundant plasma protein is **albumen**, accounting for about 36 to 52 g per dm^3. The rest consists of **globulins**, accounting for about 24 to 37 g per dm^3. When subject to electrophoresis (Chapter 2) the plasma proteins separate into several fractions (Fig 10.7). Each of the plasma proteins has a specific function. Prothrombin and fibrinogen are involved in coagulation (section 10.1.3). The gamma-globulin fraction contains antibodies called **immunoglobulins** (section 10.2). Most plasma proteins also act as carriers of metabolites. For example, transferrin carries iron (Table 10.2).

Collectively all the plasma proteins exert a **solute potential**. This is important in controlling the interchange of water and dissolved solutes between the blood and the tissue fluid (section 9.1.3) and the ultrafiltration of blood in the kidneys (Chapter 11). Another collective property of the plasma proteins is their contribution to the viscosity and density of the blood which are important in determining the pattern of blood flow in the vessels.

Table 10.2 Some of the main metabolites transported in the blood by plasma proteins

Protein	Metabolites
albumen	hormones, e.g. thyroxine and steroid hormones drugs, e.g. aspirin, penicillin vitamins A and C acetylcholine, bilirubin, calcium, copper, zinc
α_1-lipoprotein	phospholipids, cholesterol, hormones, vitamins A and E
transcortin	cortisol
B_{12}-binding protein	vitamin B_{12}
thyroxin-binding protein	thyroxine
α_2-macroglobulin	insulin
β-lipoprotein	phospholipids, cholesterol, hormones, vitamins A and E, free fatty acids
transferrin	iron
haemopexin	haem

10.2 Immune system

Immunity is the capacity to recognise intrusion of material foreign to the body and to mobilise cells and cell products to remove that particular foreign material with greatest speed and effectiveness. Much of the body's resistance to pathogenic (disease-causing) microbes is brought about by several general body functions. Mechanisms include phagocytosis, the acidity of gastric juice, the hydrolytic action of the enzyme muramidase in tears and resistance of the epidermis of skin to penetration.

Acquired immunity results from the actions of **antibodies**. Antibodies are proteins produced in the lymph nodes in response to the presence of **antigens** to provide **active immunity**. A fetus or the new-born infant may also obtain antibodies from its mother, across the placenta or in breast milk respectively. Antibodies obtained in this way provide **passive immunity**. They are not produced by the individual's own immune system. Passive immunity is short-lived because the transferred antibodies are soon destroyed in the liver. For this reason new-born babies are usually resistant to a variety of diseases for only about three months. At this age a programme of vaccination is normally begun to give active immunity against serious microbial diseases. Active immunity is usually long-lasting. The antibody content of the blood rises after initial exposure to an antigen, but usually falls quite quickly afterwards. Following subsequent exposure to the same antigen the antibody level rises further and stays high for a long time (Fig 10.8).

Newborn babies are resistant to a variety of diseases. How is this when they have not yet been exposed to the organisms that cause the diseases? How do the babies lose their resistance after about three months? How can the resistance be restored by the use of vaccines?

Fig 10.8 Changes in the level (titre) of antibody in the blood following two injections of antigen

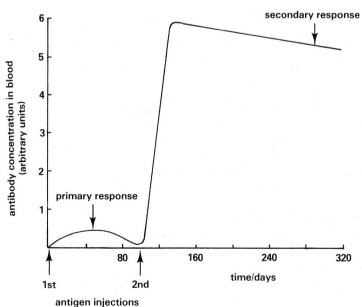

10.2.1 Acquired immunity

Any substance which triggers off an active immune response is called an antigen. Antigens are mainly proteins or even protein fragments. **Non-self** antigens form a part of the outer surface of bacteria, fungi and viruses, or may be secreted as toxins by pathogenic micro-organisms. **Self** antigens are found in the membranes of our body cells. We do not normally produce antibodies in response to self antigens. Blood transfusion and organ transplantation are ways in which our antigens could be introduced into someone else. To the recipient, our antigens are non-self. An immune response normally follows exposure to non-self antigens. Any particular antibody reacts with only one kind of antigen. Exposure to a different antigen induces the production of a different antibody. Like enzymes, antibodies are specific.

Clearly the ability to recognise and distinguish between self and non-self antigens is vital to the proper functioning of our immune system. **Lymphocytes** are stimulated by antigens to produce an immune response. They originate in the bone marrow from **stem cells** similar to those which give rise to the other blood cells. However, unlike the other blood cells, lymphocytes migrate to the lymph nodes where they mature and are produced in large numbers throughout life. The advantage of the lymph nodes being the sites of sensitivity to non-self antigens is that the lymphatic system drains lymph from all the body's tissues. Antigens anywhere in the body should therefore be detected.

There are two kinds of lymphocyte, **T-cells** and **B-cells**. T-cells are so called because they are processed in the **thymus** gland before entering the lymph nodes. They are thymus-dependent lymphocytes. Removal of the thymus gland before birth prevents T-cell formation. The thymus gland is prominent early in life, reaching its maximum size during puberty (Fig 10.9). By the time physical maturity is established the thymus is usually atrophied. T-cells are produced throughout life in the lymph nodes from the lymphocytes that were processed in the thymus.

Fig 10.9 Position of the thymus gland in a child

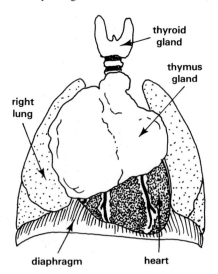

B-cells are thought to be processed in the **bone marrow**. T-cells account for about 80 % of lymphocytes in the blood, the rest are B-cells. T- and B-cells are involved in two distinct immune mechanisms called the cellular and humoral responses respectively.

1 Cellular immune response

The **cellular** immune or **cell-mediated response** is usually brought about by the presence in our bodies of cells with non-self antigens on their surfaces. Such cells may be part of a transplanted organ. They may be cancer cells or self cells whose antigens have been changed to non-self after a viral infection.

Non-self antigens entering the lymph nodes **sensitise** T-cells. Large numbers of these cells are then quickly produced by mitosis, many of them entering the bloodstream. Such a population of identical cells is called a **clone**.

Several types of T-cells can be recognised in a clone. **Killer** T-cells attach to the invading cells onto which they secrete a number of **cellulotoxic** substances. The substances may be enzymes from the killer cells' lysosomes (Chapter 6). Other substances released by killer T-cells attract **macrophages** and activate phagocytosis. Macrophages are thought to be monocytes which have invaded the tissues. Other substances secreted by T-cells are called **lymphokines** such as **interferon**. It prevents viral replication.

Helper T-cells help the plasma cells produced by B-lymphocytes to secrete antibodies.

Suppressor T-cells suppress the activity of killer T-cells and B-cells. Interaction between suppressor T-cells and helper T-cells regulates the immune response.

Memory T-cells retain the ability to recognise the non-self antigen in the future. Hence, subsequent exposure to the same antigen normally brings about a rapid cellular response. Immunity of this kind is thus conferred for a long time, often for life (Fig 10.10).

Fig 10.10 Summary of the cellular response

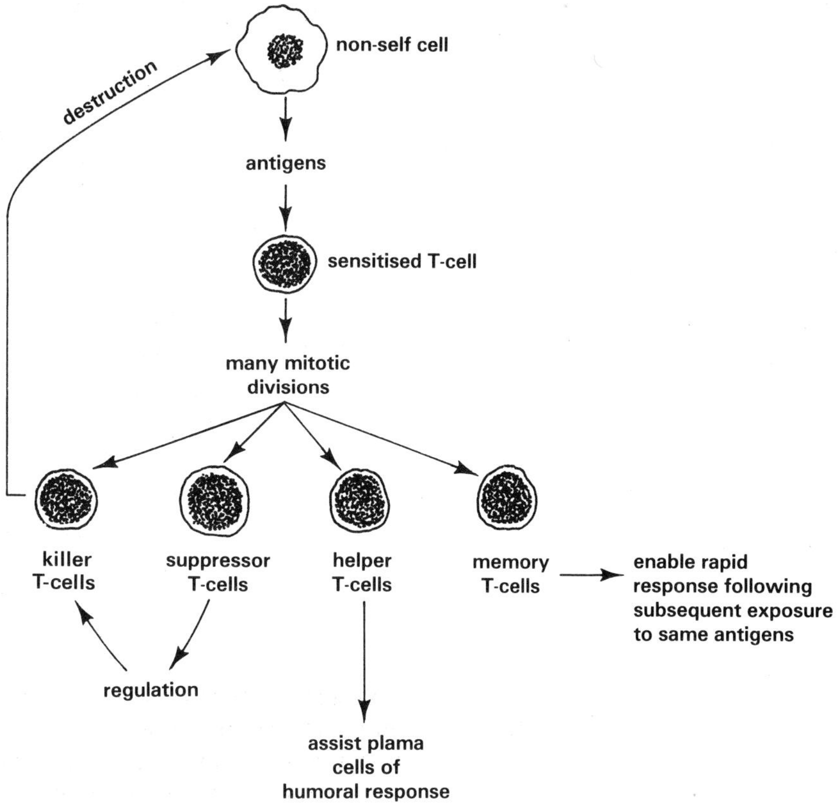

187

2 Humoral immune response

The **humoral response** usually occurs when pathogenic micro-organisms get into our bodies. On entering the lymph nodes they activate B-cells to divide many times by mitosis and swell to become large **plasma cells**

Fig 10.11 Electronmicrograph of a plasma cell, × 15 000

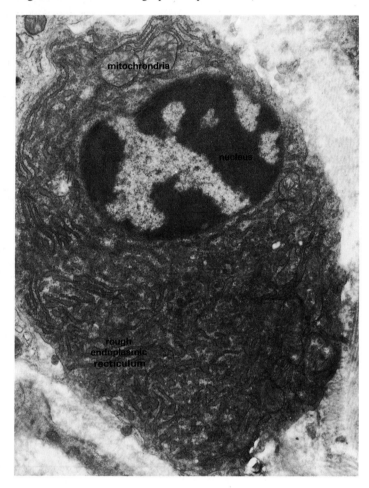

mitochrondria

nucleus

rough endoplasmic recticulum

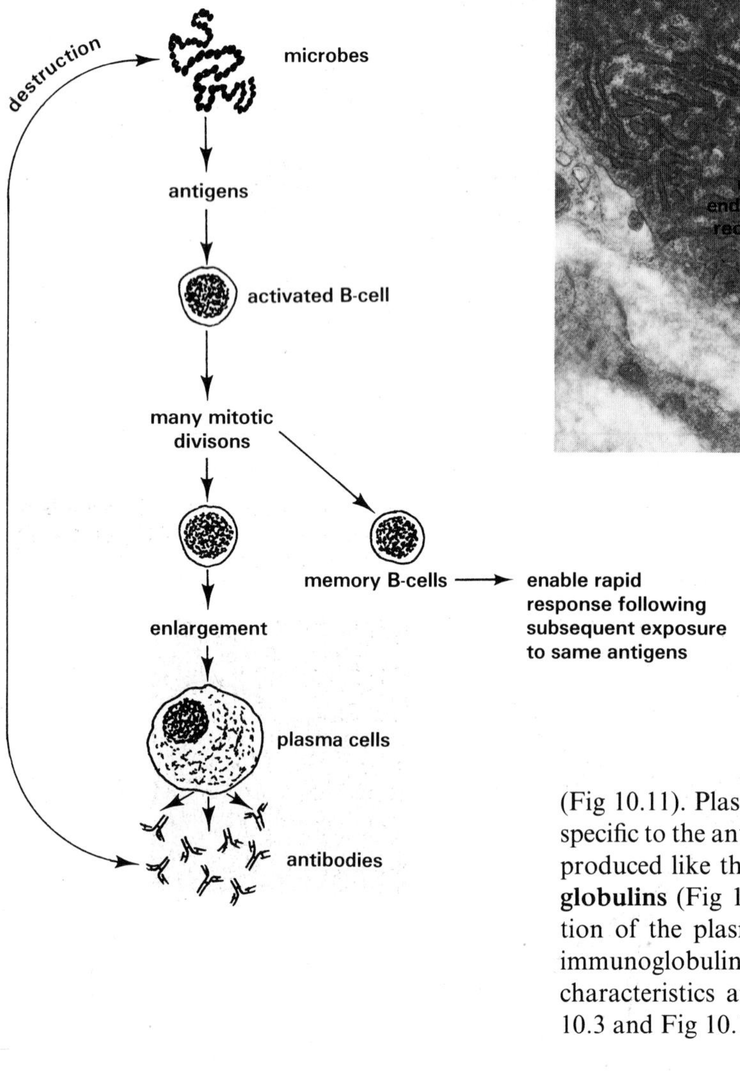

destruction

microbes

↓

antigens

↓

activated B-cell

↓

many mitotic divisons

↓

memory B-cells ⟶ enable rapid response following subsequent exposure to same antigens

enlargement

↓

plasma cells

↓

antibodies

Fig 10.12 Summary of the humoral immune response

(Fig 10.11). Plasma cells produce large quantities of antibody specific to the antigen which triggered the response. Antibodies produced like this enter the blood and circulate as **immuno-globulins** (Fig 10.12). They form the **gamma-globulin** fraction of the plasma proteins. There are several categories of immunoglobulins called IgG, IgM, IgA, IgE and IgD. Their characteristics and main functions are summarised in Table 10.3 and Fig 10.13.

Table 10.3 Main categories of immunoglobulins

IgG	Represents about 85 % of the total immunoglobulin fraction of plasma proteins. Relative molar mass is about 160 000. IgG includes many different antibodies, e.g. anti-D.
IgM	Consist of a cluster of five IgG-like antibodies joined at the centre. IgM includes anti-A and anti-B antibodies.
IgA	Much less abundant than IgG in blood but is found in tears, secretions of the naso-pharynx and in breast milk. May provide passive immunity for the new-born as well as protecting the eyes and respiratory tract from infection.
IgE	Includes the reagin-type antibodies involved in allergic responses. IgE antibodies may cross the placenta to provide passive immunity for the fetus.
IgD	A category of immunoglobulin of uncertain function.

Fig 10.13 Immunoglobulin structures

(a) Structure of IgG. The antigen-binding sites are variable, thus allowing different antibodies to attach to different antigens. The antigen-binding site of any one antibody is fixed thus explaining the specificity of antigen-antibody reaction

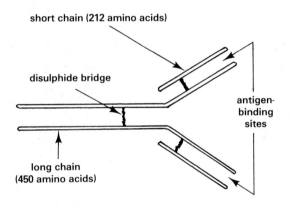

(b) Structure of IgM

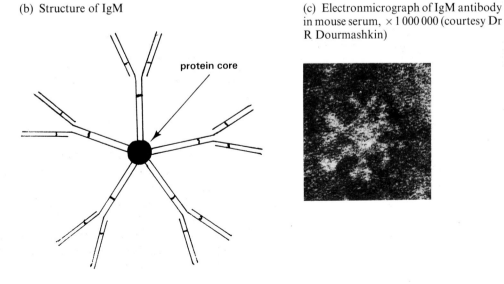

(c) Electronmicrograph of IgM antibody in mouse serum, × 1 000 000 (courtesy Dr R Dourmashkin)

In recent years a technique has been developed to fuse lymphocytes with tumour cells. The resulting **hybridomas** can be grown indefinitely. They secrete antibody specific to the antigen that activated the lymphocytes used in the hybridomas. Such **monoclonal antibody** can be purified and used to provide immunity to certain diseases.

Fig. 10.14 Electronmicrograph of the surface of a red blood cell membrane showing the holes produced by the combined action of antibody and complement, ×400 000 (courtesy Dr R Dourmashkin)

Fig 10.15 Antigen–antibody lattice formation

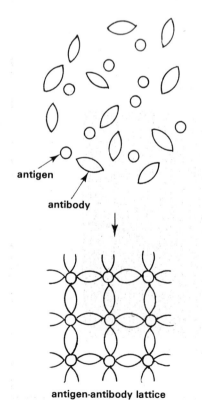

Blood plasma contains at least 15 different proteins collectively called **complement**. They may be activated by the binding of immunoglobulin molecules to surface antigens such as may be found on bacteria. When activated, the complement proteins interact to help the humoral immune response. One reaction causes destruction of the bacterial cell wall, leading to its lysis and death (Fig 10.14). Complement may also attach to macrophages and neutrophils, promoting phagocytosis. Some of the activated B-cells do not develop into plasma cells. They remain in the lymph nodes as **memory B-cells**. Like memory T-cells they can recognise the original non-self antigens in the future. They bring about a rapid humoral immune response following subsequent infection involving the same antigen.

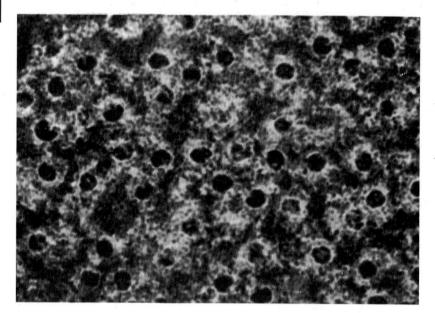

3 Antigen–antibody reactions

Antibodies react with antigens in a variety of ways. The reactions, which usually destroy the antigens, include the following.

i. Neutralisation results in the **inactivation** of toxins produced by pathogenic bacteria.

ii. Precipitation Each IgG antibody molecule can react with two molecules of antigen. Hence, many cross-linkages can be made to form a large **lattice** precipitate (Fig 10.15). Antibodies which cause precipitation to occur are called **precipitins**. Lattices are readily phagocytosed by macrophages.

iii. Agglutination **Agglutinin** antibodies cause cells coated with non-self antigens to clump together. The clumping is called **agglutination** and the antigens are called **agglutinogens**. Agglutinated bacteria are susceptible to phagocytosis.

iv. Lysis Antibodies attached to antigens on cell surfaces may cause the cells to rupture. This is called **lysis** and leads to cell death. Complement proteins help bring about lysis (Fig 10.14).

Agglutination reactions also form the basis of blood grouping procedures. Modern pregnancy tests are based on **agglutination inhibition**. Human chorionic gonadotrophic hormone (HCG) appears in the urine of pregnant women. In the test, anti-HCG antibody is added to the urine. Small latex particles coated with HCG are added next. Any HCG in the urine binds with the antibody. Consequently the HCG attached to the latex has no antibody with which to produce agglutination of the particles. Agglutination occurs however if the urine lacks HCG (Fig 10.16).

Fig 10.16 Sequence of events in an agglutination inhibition test: (a) antibody added to test sample; (b) latex particles coated with antigen are added; (c) agglutination of latex particles – negative, agglutination inhibited – positive; (d) results of a pregnancy test

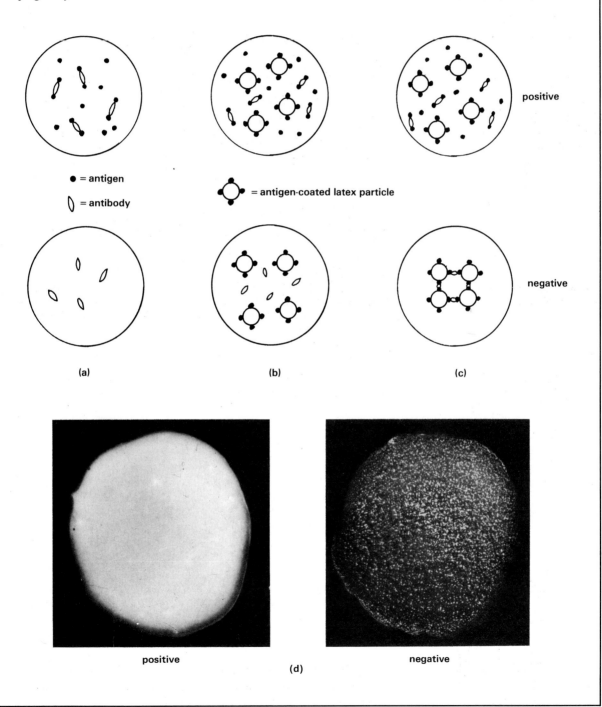

191

4 Autoimmune responses

Self antigens do not normally initiate an immune response. Sometimes, however, immune responses result in the destruction of self cells and tissues. They are called **autoimmune responses**. They may be caused by changes in self antigens brought about by the actions of viruses.

There are many diseases which have recently been explained on this basis. For example, diabetes mellitus results from the body's inability to produce enough insulin (Chapter 21). This may follow destruction by antibodies of β-cells in the islets of Langerhans. Another autoimmune disease of the endocrine system is myxoedema, caused by antibody activity in the thyroid gland.

The most frequent cause of kidney failure in Britain is glomerulonephritis, an autoimmune disease. Another autoimmune response is antibody neutralisation of receptor sites on muscle membranes (Chapter 17). This leads to poor neuromuscular transmission and a muscle wasting disease called myasthenia gravis.

5 Allergy

About 30 % of the population of Britain has an **allergy**. Allergies include hay fever, asthma, childhood eczema and food allergies. An allergy is an immune response to an antigen called an **allergen**, to which most people show no reaction. Allergens are found on pollen grains, fungal spores, house dust, feathers, fur and in a variety of foods. Allergens on pollen for example, become attached to the mucus membranes in the breathing passages. The presence of an allergen in a person who suffers from an allergy stimulates the production of antibodies called **reagins**. Reagins belong to the IgE category of immunoglobulins (Table 10.3). They circulate in the blood and become attached to **mast cells** throughout the body, particularly in the skin and mucus surfaces of the mouth, nose and breathing passages. The reagins can remain for years in these tissues which are said to be **hypersensitive**. Later, whenever there is exposure to the allergen, an allergen–reagin reaction takes place. The reaction triggers a vigorous response involving the rupture of the mast cells and the release of **histamine**. Histamine causes inflammation of the affected tissues, constriction of the bronchi leading to breathing difficulties, and excessive secretion of mucus. Eosinophils increase in number during allergic responses. They are thought to have an anti-inflammatory effect by absorbing histamine from the tissues.

Some people gradually become accustomed to an allergen if it is presented to them in gradually increased doses. They become **desensitised**. Children often grow out of an allergy for this reason. Sensitivity to different allergens may be determined by performing a **skin prick test**. Small amounts of allergens are scratched into the skin. A weal appears around the scratch in response to a substance to which the person is allergic. The patient may then be dosed with the allergen to produce desensitisation.

There is considerable interest in the possible link between certain food allergens and hyperactivity in children. Some highly-strung children with undesirable behaviour have calmed down after removal of the allergens from their diet.

6 Vaccines

Among the main weapons in our armoury against disease are **vaccines**. Vaccines are made from micro-organisms having the antigens which stimulate the body's immune system. The organisms are treated so that they do not give rise to disease when administered to a patient. The antibodies produced as a result of vaccination give immunity against disease following subsequent exposure to disease-causing micro-organisms. Unfortunately it is not yet possible to make vaccines against all diseases caused by micro-organisms.

In 1797 Edward Jenner, a Gloucestershire country doctor, was the first to use a vaccine successfully in preventing a human disease. He noticed that dairymaids who had milked cows suffering from cowpox were far less susceptible to the much more virulent and often fatal smallpox, although they often showed mild symptoms of the disease, such as hand sores. Jenner removed some of the liquid from a sore on the hand of a dairymaid and scratched it into the skin of a young boy, James Phipps. Later Jenner inoculated Phipps with material from the sore of someone suffering from smallpox. The boy was found to be immune to smallpox. Jenner had used an antigen to produce immunity against a dangerous pathogen. It is only relatively recently that mass vaccination programmes have eliminated smallpox all over the world. The reaction of the body to cowpox vaccine is an example of **active immunisation**. The body actively makes its own antibodies against the antigen.

Passive immunity is important in a fetus which absorbs antibodies from its mother across the placenta. The mother's milk also contains antibodies which provide passive immunity in the gut of new-born infants.

In Britain, children are normally offered a vaccination programme which gives them protection against childhood diseases that were once widespread killers (Table 10.4).

Table 10.4 Vaccination schedule recommended in Britain by the Department of Health and Social Security

Age	Vaccine	Notes
During the 1st year of life.	diphtheria, tetanus, pertussis and poliomyelitis	1st dose at 3 months of age; 2nd dose 6 8 weeks later; 3rd dose after a further interval of 4–6 months.
During the 2nd year of life.	measles	
At school entry or entry to nursery school.	diphtheria, tetanus, poliomyelitis	Allow an interval of at least 3 years after completing the basic course.
Between 11 and 13 years of age.	tuberculosis (BCG)	Leave an interval of at least 3 weeks between BCG and rubella vaccination.
Between 10 and 13 years of age (girls only).	rubella	
Between 15 and 19 years of age or on leaving school.	poliomyelitis, tetanus	

The first vaccines were used by Edward Jenner in the late eighteenth century. He deliberately infected a young boy with smallpox, one of the main killer diseases of his day. Jenner's pioneering work has now led to the eradication of smallpox from the human population. Was Jenner right to put a young life at risk?

Vaccines are produced in several ways including:

1 The micro-organisms may be killed in such a way that their antigens are not affected and can bring about the required immune response. Vaccines of this type are used to provide immunity against influenza, typhoid fever and cholera.

2 Some diseases are caused by **toxins** produced by the infecting micro-organisms. Toxins may be treated to make harmless **toxoids** which can initiate the immune response. Vaccines of this kind include diphtheria and tetanus toxoids.

3 Pathogens may be **attenuated** (weakened) by injecting them into some other animal such as cattle or horses. Living attenuated organisms contained in vaccines multiply within the body but do not cause disease. They include oral polio vaccine, smallpox and measles vaccines and BCG vaccine against tuberculosis. *BCG* is derived from *Bacillus Calmette Guérin*, the strain of bacterium used in the vaccine's manufacture.

7 AIDS

In recent years a disease of the immune system has spread throughout the world and given rise to much concern. The disease is called AIDS which stands for Acquired Immune Deficiency Syndrome. AIDS is caused by an RNA retrovirus, that is a virus which injects its own RNA into the nucleus of the host cell (section 25.1.1). The virus is called HIV which stands for Human Immunodeficiency Virus. The host cells are Helper T-lymphocytes. These cells normally help in both the humoral immune response (antibody production) and the cellular immune response. People who have been infected with HIV usually produce HIV antibody. However, the virus can exist in the presence of the antibody without losing its ability to produce Aids or to infect others.

HIV has been found in many different body fluids of infected people. Transmission of the virus, though, occurs via blood or semen. Routes of transmission include anal intercourse between homosexual men, vaginal intercourse between an infected bisexual man and a woman, antenatal infection of a foetus by an infected mother, transfusion of unscreened blood products, and the sharing of infected needles by drug addicts.

The majority of people who are infected show no symptoms and so there are no reliable figures for people carrying the virus. Official figures in the UK suggest that up to July 1986 over 4000 people had been infected with HIV. However, it is estimated that by November 1986 over 30 000 people may have been infected. By the end of November 1986, there had been 599 confirmed cases of Aids in the UK since 1983; 296 of these had died.

Infection with HIV does not automatically mean that Aids will develop. The incubation period is between 15 months and 5 years. People with the disease suffer from opportunistic diseases and may develop a rare vascular tumour called Kaposi's sarcoma.

10.2.2 Blood groups

Table 10.5 Red cell antigens of the main blood group systems in humans

System	Antigens
ABO	A_1 A_2 A_3 A_x B and others
Rhesus	D C c C^W C^X E D^U and others
MNSs	M N S s
P	P_1 P^k
Lutheran	Lu^a Lu^b
Kell	K k Kp^a Kp^b Js^a Js^b
Lewis	Le^a Le^b
Duffy	Fy^a Fy^b
Diego	Di^a Di^b
Yt	Yt^a Yt^b
I	I i
Xg	Xg^a
Kidd	Kj

People can be classified into one of several **blood groups** depending on the presence or absence of certain antigens (agglutinogens) on the red blood cells. Antibodies (agglutinins) may also be present in the plasma. A large number of blood group systems has been described (Table 10.5). The **ABO** and **Rhesus** systems are particularly important. Transfusion from a person of one blood group into someone of another group may be fatal.

1 ABO system

Our red cells have one, both or neither of two agglutinogens called **A** and **B**. Correspondingly, our plasma contains one, both or neither of two agglutinins called **anti-A** or α and **anti-B** or β. The agglutinins are immunoglobulins of the IgM category (Table 10.3 and Fig 10.13). They appear soon after birth and are present throughout life. They are called **isoantibodies** as distinct from **immune antibodies** which appear after an immune response. Production of isoantibodies may decline and disappear in old people.

Cells with A agglutinogen belong to **group A**, those with B belong to **group B** and those with both agglutinogens belong to **group AB**. Cells with neither agglutinogen belong to **group O**. Anti-A and anti-B agglutinins are distributed in such a way that they do not normally come into contact with their specific agglutinogen (Table 10.6). Landsteiner first described groups A, B and O in 1900 and the AB group in 1901. His discoveries led to a dramatic rise in the success rate of blood transfusions.

Table 10.6 The main antigens and antibodies of the human ABO blood group system

Blood group:	O	A	B	AB
Red cell antigen	—	A	B	A + B
Plasma antibody	anti-A + anti-B	anti-B	anti-A	—
British population/%	46.7	41.7	8.6	3

Techniques were devised to establish the ABO groups of donors and recipients for transfusion. Only blood of the same group within the ABO system as the recipient is normally transfused. **Incompatible** transfusions may bring A (or B) agglutinogen into contact with anti-A (or anti-B) agglutinin. If this happens, the transfused cells are made to agglutinate by the antibody and they haemolyse. The **transfusion reaction** is normally fatal.

People of group O have red cells which possess neither A nor B and have been described as **universal donors**. Similarly, people of group AB have plasma containing neither anti-A nor anti-B. They have been called **universal recipients**. These terms have fallen into disuse nowadays, since the term *universal* means *in all possible circumstances* and takes no account of blood group systems other than ABO.

After Landsteiner's discoveries it was noticed that the cells from some group A people react more strongly than those of others when mixed with anti-A. This led to the discovery of sub-groups of A and AB called A_2 and A_2B (Table 10.7). They were first described in 1911 by Von Dungern and Hirszfeld.

Table 10.7 Sub-groups of the ABO blood group system

Group	Sub-group	Approximate percentage of UK population
A	A_1	34
	A_2	8
AB	A_1B	2.6
	A_2B	0.4

2 Rhesus system

The rhesus blood group system was first described by Landsteiner and Levine in 1940. They injected some red cells from rhesus monkeys into rabbits. They then exposed human red cells to blood serum extracted from the rabbits. The cells from about 85% of the people whose blood was tested agglutinated. Their cells must therefore have had the same agglutinogen as that on the monkey cells. This was called the **rhesus factor** or agglutinogen **D**. It is now known that there are many variants of D (Table 10.5). Cells that possess D are **rhesus positive**, **Rh +**, those without are **rhesus negative**, **Rh −**. There is no corresponding isoantibody. However, transfusion of Rh + cells into a Rh − individual may stimulate a humoral immune response in the recipient. The **anti-D** agglutinin produced causes agglutination and haemolysis of the transfused cells, a transfusion reaction. By accounting for the rhesus as well as the ABO groups of patients requiring blood, adverse reactions following transfusions have become very rare.

Complications arise sometimes when a Rh − woman bears a Rh + fetus. During normal pregnancy there is no mixing of the fetal and maternal blood cells, although the two circulations run close together in the placenta (Chapter 25). However, when the child is born, the severe contractions of the uterine wall may squeeze significant numbers of fetal red cells into the mother's blood. Some days later the mother's immune system may produce anti-D antibody. Anti-D belongs to the IgG category of immunoglobulins (Table 10.3 and Fig 10.13). It is capable of crossing the placental membranes and entering the blood of any fetus the woman may bear in the future. If the fetus is Rh + then the anti-D causes a transfusion reaction in the fetal circulation. The reaction is called **haemolytic disease of the newborn, HDNB**. It can be treated successfully by giving the fetus several transfusions while it is developing inside the uterus.

An alternative is to avoid the problem of HDNB by injecting a Rh − mother with anti-D immunoglobulin immediately she has given birith to her Rh + child. Any of the fetal cells which may have entered her blood are thus destroyed before they trigger the immune response.

HDNB arises in only about 10 % of cases like that described above. It is mainly because D-carrying fetal cells may be destroyed by the mother's

Haemolytic disease of the newborn, HDNB, can occur in the second or subsequent Rhesus positive fetus of a Rhesus negative woman. Why does HDNB only occur in about 10 % of such cases?

own ABO isoantibodies. If the child belongs to group A, for example, and the mother is group O, then her anti-A haemolyses the fetal cells, regardless of rhesus group (Table 10.6).

HDNB caused by ABO incompatibility between mother and fetus is rare. The ABO isoantibody molecules (IgM) are far too large to cross the placental membranes.

3 Blood grouping procedures

Techniques for establishing blood groups involve exposing batches of red cells to a variety of antisera containing known specific blood group agglutinins. Agglutination of the cells indicates the agglutinogens they possess and hence the groups to which they belong.

What pattern of agglutination would you expect in an ABO slide test for a group AB patient? What about patients of the other ABO groups?

i. ABO grouping A simple technique can be performed on a glass slide. It is rapid and useful in emergencies. Four clean slides are arranged as in Fig 10.17. The first slide is used to test the patient's red cells, the other slides are used as controls. Red cells are suspended in isosmotic saline (0.9 % aqueous) to wash any antibodies off them and to provide an approximately 10 % (by volume) suspension. The patient's red cells and the control cells are mixed in appropriate known antisera and left undisturbed for a few minutes. The slides are then examined for agglutination either with the naked eye or with the aid of a microscope. The controls are checked first. In the **positive controls**, A_2 cells and B cells are mixed with anti-A and anti-B agglutinins respectively and should agglutinate. In the **negative controls**, A_2 cells and B cells are mixed with anti-B and anti-A respectively and should not agglutinate. Fig 10.18 illustrates the results which might be expected for a group A patient.

Fig 10.17 ABO grouping, slide method

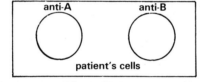

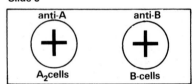

Fig 10.18 ABO grouping, slide method results for a group A patient

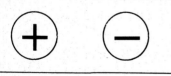

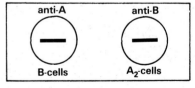

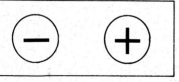

ii. Rhesus grouping Unlike the isoantibodies of the ABO system, anti-D is an immune antibody. Anti-D is an IgG immunoglobulin and is usually incapable of causing the agglutination of D-cells (Rh +) on its own. For this reason it is called an **incomplete** antibody. The cells become coated (sensitised) by anti-D. Agglutination may occur when another protein is added, usually albumen.

In a simple test, a drop of 20 % bovine albumen is added to a drop of antiserum containing anti-D. To this is added a drop of the patient's cells in saline (30–40 % suspension). After 3 to 5 minutes incubation at room temperature, agglutination indicates rhesus positive cells.

10.2.3 Tissue typing

Just as the antigens on red blood cells can be used to categorise people within the major blood groups, antigens on other body cells may be used to determine their **tissue type**. They are called **histocompatibility antigens**. Transplantation of incompatible antigens leads to an immune response and **rejection** of the transplant. The histocompatibility antigens are determined by several hundred genes at loci called **HLA** on chromosome 6. HLA stands for **human lymphocyte antigens**.

HLA-A and HLA-B are the loci of antigens which produce the strongest transplantation reactions and are thus most often used in tissue typing. Typing can be performed by exposing some of the patient's lymphocytes to a range of specific HLA antisera and complement. The presence on the lymphocytes of an HLA antigen specific to the antibody in the antiserum leads to lysis of the lymphocytes. Lysis is detected by use of a blue dye which is absorbed only by lysed cells. The different HLA-A and HLA-B antigens important in transplantation have been numbered. They include HLA-A1, 2, 3, 9, 10, 11, 28 and 29 and HLA-B5, 7, 8, 12, 13, 14, 18 and 27. There are two antigen specifications for each locus. Consequently, an individual who is heterozygous for each locus has, for HLA specificities, two HLA-A and two HLA-B from each parent. There are other HLA loci, such as HLA-C. They determine relatively weak antigens and are less important in transplantation.

There is very great variation in HLA types between different people. Even closely related persons are rarely identical in this respect. The closer the HLA match between donor and recipient in transplantation, the greater is the likelihood of success. Despite efforts including international cooperation to match donors to recipients across the world, immune rejection is still the main hazard in transplantation. In addition to tissue typing, **immunosuppression** is used to help prevent rejection. After receiving the transplant the patient may be treated with a variety of drugs that suppress the immune response. Unfortunately, the drugs are usually not specific in their actions. The patient's natural defences to many pathogenic microbes may be lowered. Contraction of otherwise trivial diseases may thus threaten life.

Nonetheless, many people who were near to death have been able to live many years of active life with a transplant. In 1985, a man from South Wales who was given the heart of a 16-year-old donor a year earlier, entered and completed the Boston marathon in America.

The classification of people into different tissue types depends on the same principles as classifying them into different blood groups. Why is tissue-typing more difficult than blood grouping? What problems are involved in organ transplantation that do not occur in blood transfusion?

SUMMARY

Red blood cells carry oxygen from the lungs to the tissues. Platelets are involved in coagulation. White blood cells are concerned with body defences. Neutrophils and monocytes are chemotactic and phagocytic; they locate, move towards and engulf particles such as bacteria. Eosinophils absorb toxins such as histamine. Basophils may be mast cells which release histamine during allergic reactions. Lymphocytes are concerned with immune responses.

Coagulation prevents excess blood loss following injury to the vessels. Platelets agglutinate at sites of injury and release a number of chemical substances. 5-hydroxytryptamine causes vasoconstriction. Reaction with blood (intrinsic) or tissue (extrinsic) factors results in the formation of thromboplastin. Plasma prothrombin is converted to thrombin in the presence of thromboplastin and calcium ions. Thrombin converts plasma fibrinogen to fibrin. Insoluble fibrin strands deposit at sites of injury forming a clot with trapped blood cells. Blood flow stops. This is haemostasis. Injury to vessels may not involve haemorrhage; vessel walls may thicken (fatty deposits); intravascular blood clots may form called thrombi; these cause thrombosis. Dislodged thrombi called emboli eventually block small arteries or arterioles causing embolism.

▶

► Plasma proteins perform several collective functions; transport of metabolites, solute potential pressure, blood viscosity and density.

Cellular immune responses result in production of sensitised T-lymphocytes following exposure to non-self antigens. Killer T-cells destroy the source of antigens by cellulotoxic action. Memory T-cells enable rapid response to same antigens in the future.

Humoral immune responses result in production of antibodies (immunoglobulins) from plasma cells (derived from B-lymphocytes) following exposure to non-self antigens. Memory B-cells allow rapid response to the same antigens in the future. Antigen–antibody reactions destroy the source of antigens. There are several mechanisms including agglutination, precipitation, neutralisation and lysis; phagocytosis of debris. Complement proteins help antibody activity. Immune system does not usually respond to self-antigens.

Allergy is a hypersensitive immune response to antigens normally tolerated by most people.

Vaccines are harmless preparations of antigens which trigger an immune response but do not cause disease.

Blood group systems are based on red blood cell antigens. ABO system:
Group A; red cell Ag. A, plasma Ab. anti–B
Group B; red cell Ag. B, plasma Ab. anti–A
Group AB; red cell Ags. A and B, no plasma antibodies
Group O; no red cells Ags., plasma Abs. anti–A and anti–B

Rhesus system: Rh positive; red cell Ag. D
Rh negative; no red cell Ag.

Compatibility must be established before transfusion between donor and recipient, otherwise antibodies may destroy transfused cells.

Tissue types are similar to blood groups; based on tissue cell antigens. Transplant rejection caused by cellular immune response to non-self antigens on transplant cells.

QUESTIONS

1 The three tubes X, Y and Z shown right contain mammalian blood that is fresh (**X**), has been centrifuged (**Y**), and had been allowed to clot (**Z**).

(a) Identify the components **A**, **B**, **C** and **D**.
(b) How does the composition of **D** differ from that of **A**?
(c) Name *four* types of cell or cell derivatives found in mammalian blood and indicate *one* function of each.
(d) What happens to the relative numbers of cells of different types in the blood of a person with a bacterial infection?
(e) Explain what happens to the blood of people who move from low altitudes to live at high altitudes. (L)

2 Describe the parts played by the *blood* of a mammal in
(a) the transport of respiratory gases,
(b) temperature control,
(c) protection. (L)

3 (a) Describe the essential features of the immune system in mammals.
(b) Give an account of the ABO blood group system in humans and explain why certain ABO group donations cause agglutination in the recipient, while others do not.
(c) Besides blood, other tissues can be transplanted from one mammal to another. Discuss the problems associated with such procedures and the steps taken to minimise transparent failure. (JMB)

4 The diagrams below represent some of the constituents of mammalian blood.

  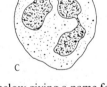

A   B   C

(a) Copy and complete the table below giving a name for each of the constituents **A**, **B** and **C**, a place in the body where they are produced, and *one* function of each.

Constituent	Name of constituent	Site of production	Function
A			
B			
C			

(b) Describe the sequence of events which occurs during the clotting of blood. (L)

11 The renal system

11.1 Anatomy of the renal system 200

11.2 The functions of nephrons 203

   11.2.1 Ultrafiltration 204
   11.2.2 Direct secretion 205
   11.2.3 Selective reabsorption 205
   11.2.4 Role of medulla in controlling water retention 207
   11.2.5 Regulation of nephron function 210
   11.2.6 Extreme cases of water economy 211

11.3 The role of the kidneys in acid–base balance 212

11.4 Replacement of kidney function 214

   11.4.1 Kidney machines 214
   11.4.2 Kidney transplantation 216

Summary 217

Questions 217

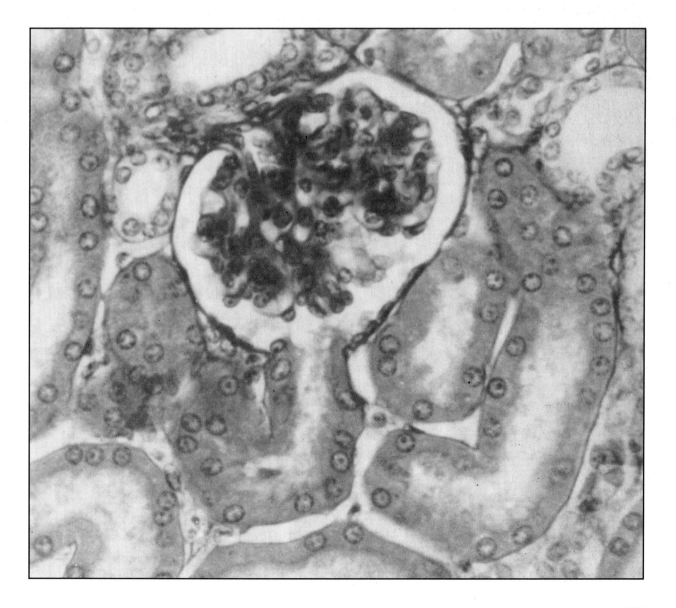

11 The renal system

The complex and numerous activities of the mammalian body require a stable environment in which to take place. Even slight chemical or physical changes can upset the smooth functioning of the body. Many of the physiological activities which regulate the constitution of the body's internal environment do so to within relatively narrow limits. Maintaining a stable internal environment is called **homeostasis**.

The very functions and activities which need a stable environment change it continuously. Metabolism involves the production of wastes. If wastes accumulate in the body they can become toxic. Consequently the body has to eliminate or **excrete** its wastes. **Excretion** is

> **the elimination of any substances which are present in the body's tissues in concentrations exceeding normal levels, whether metabolic wastes or not.**

Among the organs concerned with excretion and homeostasis are the kidneys which form part of the **renal system**.

11.1 Anatomy of the renal system

The **kidneys** are paired organs found in the abdominal cavity. They are usually embedded in fat and held firmly in position by the peritoneum, a thin layer of tissue lining the abdominal cavity (Fig 11.1).

Fig 11.1 The renal system

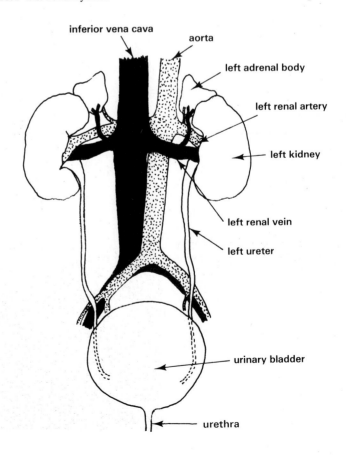

inferior vena cava

aorta

left adrenal body

left renal artery

left kidney

left renal vein

left ureter

urinary bladder

urethra

Each kidney receives blood through a **renal artery** and is drained of blood by a **renal vein**. In kidneys the blood circulates in a network of arterioles, capillaries and venules which surrounds numerous microscopic urinary tubules called **nephrons**. The nephrons remove excess and unwanted materials from the blood. The excretory products collect as **urine** which passes from the kidneys through the **ureters**. Urine is stored temporarily in the **urinary bladder** before it is eliminated from the body. Emptying of the bladder is called **micturition** and is controlled by the autonomic nervous system (Chapter 18). The exit from the bladder into the **urethra** is closed by contraction of rings of muscle called the **bladder sphincters**. As the bladder fills, cells in its wall sensitive to stretching trigger off a reflex action which results in relaxation of the bladder sphincter. Simultaneous contraction of the smooth muscle in the bladder wall forces the urine out through the urethra. Micturition can be controlled by voluntary nervous activity which is learned by humans in early life (Fig 11.2). The urinary bladder is lined with **transitional epithelium**. It consists of several layers of cells of similar shape and size. When the bladder is full of urine the epithelium stretches and becomes a single layer of cells. After urine has passed, the bladder wall relaxes and the epithelial lining is then of several layers (Fig 11.2(b)).

Fig 11.2 (a) Autonomic supply to the urinary bladder and urethra

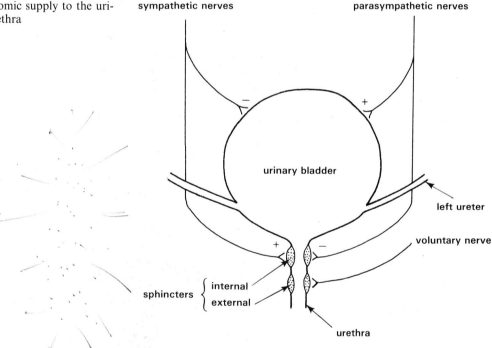

Fig 11.2 (b) Photomicrograph of a thin section of transitional epithelium in urinary bladder, × 75

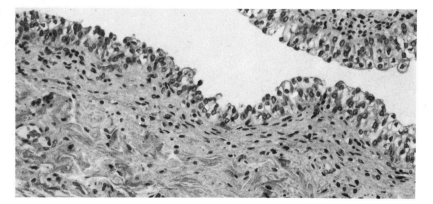

Each human kidney is a compact organ measuring about 7–10 cm long and 2.5–4 cm wide in an adult. Inside, the tissues are in distinct regions. There is an outer dark **cortex** and an inner lighter **medulla** (Fig 11.3 (a) and (b)). The internal appearance is due to the arrangement of the blood vessels and nephrons which make up most of the organ.

Fig 11.3 (a) Diagrammatic vertical section of a human kidney. Note the arrangement of the medulla into a series of pyramids

Fig 11.3 (b) Photomicrograph of a vertical section of a rat kidney, × 5

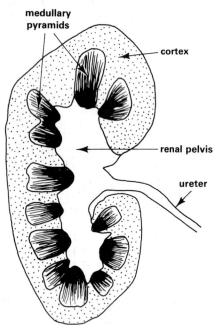

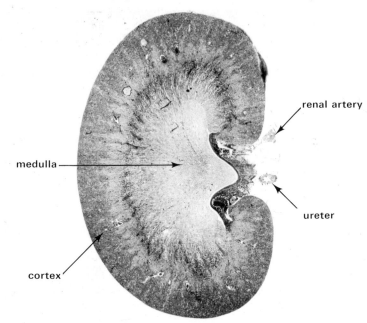

There are about a million nephrons in each human kidney. Nephrons are very small thin-walled tubules between 2 and 4 cm long (Fig 11.4 (a)). At one end is a cup-shaped **renal capsule (Bowman's capsule)** which encloses a small group of capillaries called a **glomerulus**. A renal capsule and the glomerulus together are called a **Malpighian body**. The capsule leads into a coiled structure called the **proximal convoluted tubule** which opens into the **loop of Henle**. The loop of Henle consists of a descending limb and an ascending limb and leads into a second coiled tube, the **distal**

Fig 11.4 (a) Photomicrograph of a thin section of monkey kidney in the cortical region, × 300

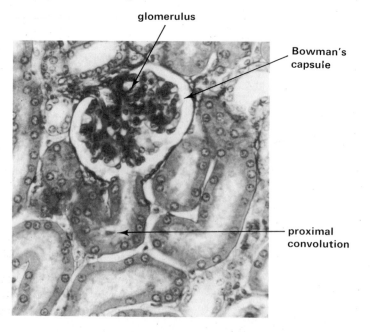

convoluted tubule. The distal convoluted tubules of several nephrons join a common **collecting duct** and many collecting ducts lead through the medulla to the **renal pelvis**.

There are two types of nephrons depending on the length and nature of their loops of Henle. **Cortical nephrons** have 'short-reach' loops which project to the boundary between the outer and inner zones of the medulla.

Juxtamedullary nephrons have 'long-reach' loops which extend deeper into the medulla, usually to the tips (papillae) of the pyramids (Fig 11.4(b)). In human kidneys about 85% of the nephrons are cortical and 15% are juxtamedullary. The proportions are different in the kidneys of other mammals. In some desert-living rodents. For example, nearly all the nephrons have very long loops of Henle (Fig 11.14).

Fig 11.4 (b) Cortical and juxtamedullary nephrons and associated blood vessels

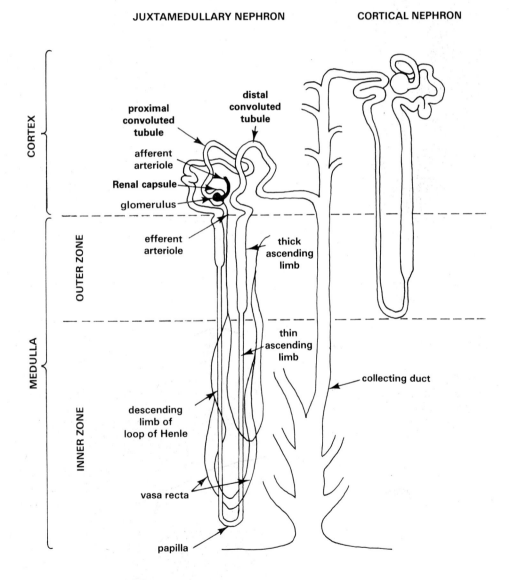

11.2 The functions of nephrons

Much of what is known of the functions of kidneys come from analysis of the fluid inside nephrons. Delicate micropipettes are inserted at various points into nephrons of experimental animals. The fluid is drawn off

Fig 11.5 Removal of a sample of filtrate for analysis

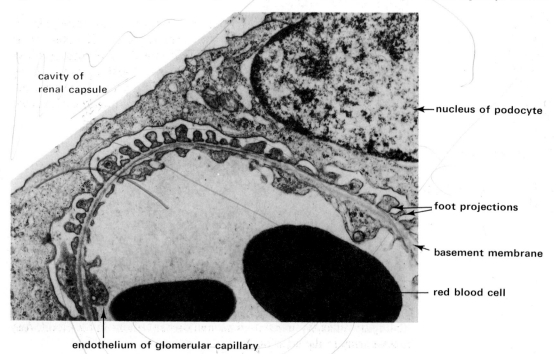

micropipette

glass rod

carefully and analysed to determine what changes have occurred during its passage through a nephron (Fig 11.5).

11.2.1 Ultrafiltration

The first activity of a nephron is the **ultrafiltration** of blood brought to the renal capsule by arterioles. The arterioles branch into the capillaries of the glomeruli, tightly nestled in the capsules. In humans about 20 % of the blood plasma which enters the kidneys is filtered. Electron microscope studies of the capsules show that the only effective barrier between the blood in the glomeruli and the cavity of the capsules is a thin porous **basement membrane** (Fig 11.6). The basement membrane of the capsule is

Fig 11.6 (a) Diagram of the barrier between the glomerular blood and the filtrate in the renal Bowman's capsule

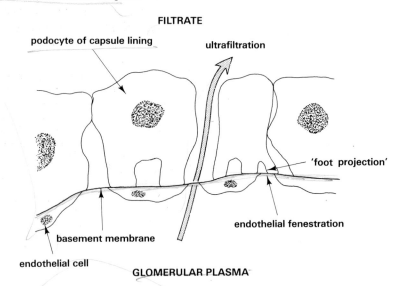

FILTRATE

podocyte of capsule lining

ultrafiltration

'foot projection'

endothelial fenestration

basement membrane

endothelial cell

GLOMERULAR PLASMA

Fig 11.6 (b) Electronmicrograph showing a section through a podocyte and a glomerular capillary, × 12 000

cavity of renal capsule

nucleus of podocyte

foot projections

basement membrane

red blood cell

endothelium of glomerular capillary

permeable to some blood constituents but not to others. The pressure potential of blood in the glomerular capillaries is relatively high, at about 5.92 kPa (45 mmHg). This is partly because the diameter of the afferent arterioles is greater than that of the efferent arterioles. Blood pressure is normally maintained by the pumping action of the heart. If the blood pressure falls too low, for example when much blood is lost in an accident, temporary use of a kidney machine may be required until the patient's blood volume and pressure are restored (section 11.4.1).

Because the blood in the glomerulus has a high pressure potential it is filtered through the basement membrane. Blood cells and substances of high relative molar mass, such as most of the plasma proteins, are too large to pass through the pores of the basement membrane. The chemical composition of the glomerular **filtrate** is thus virtually the same as the plasma minus its proteins. The composition of urine, however, is very different. Hence the filtrate is modified considerably while passing along the nephrons. The volume of urine excreted is also much less than the volume of filtrate produced in a given time (Table 11.1).

Table 11.1 Some blood constituents and the quantities filtered and reabsorbed by the human kidneys in a day

Constituents	Amount in filtrate/g	Amount in urine/g	Percentage reabsorption
Sodium ions	600	6	99
potassium ions	35	2	94
calcium ions	5	0.2	96
glucose	200	0	100
urea	60	35	42
water	$180 \, dm^3$	$1.5 \, dm^3$	99

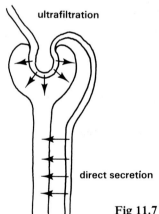

ultrafiltration

direct secretion

Fig 11.7 Means by which substances enter the nephrons from blood

11.2.2 Direct secretion

About 80 % of the blood plasma which enters the human kidneys is not filtered from the glomeruli into the renal capsules. However, some substances in the blood may be discharged into the nephrons by direct secretion, mostly into the proximal convoluted tubules, as well as by ultrafiltration (Fig 11.7). Substances excreted in this way include uric acid. **Direct secretion** enables greater quantities of such wastes to be eliminated than by ultrafiltration alone.

11.2.3 Selective reabsorption

Changes in the composition of the filtrate begin in the proximal convoluted tubules. Here the epithelial cells of the nephron wall reabsorb a large proportion of the filtrate, passing it back into the blood flowing in the surrounding vessels. **Selective reabsorption** of individual substances is at a rate just sufficient to maintain normal concentrations in the blood. Any excesses stay in the nephron.

Since the transfer of materials from the filtrate into the blood is against a concentration gradient, selective reabsorption is active (Fig 11.8). The energy for **active uptake** is provided by respiration in the nephron's cells which contain many mitochondria. The efficiency of reabsorption is helped by the presence of numerous **microvilli** which greatly enlarge the surface area through which the materials pass (Fig 11.9).

Fig 11.8 Active solute reabsorption from the proximal convoluted tubules against a concentration gradient as the levels of reabsorbed materials exceed those remaining in the filtrate

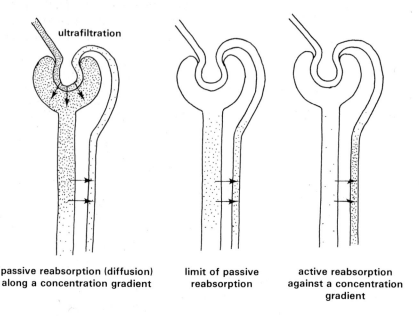

ultrafiltration

passive reabsorption (diffusion) along a concentration gradient

limit of passive reabsorption

active reabsorption against a concentration gradient

Fig 11.9 Electronmicrograph of a cell lining the proximal convoluted tubule of a nephron, × 5800

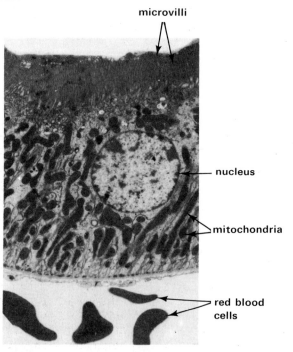

microvilli

nucleus

mitochondria

red blood cells

Our kidneys filter about 120 cm³ of fluid each minute. However, only about 1 cm³ of urine is produced each minute. What happens to the other 119 cm³ of fluid?

In humans about 120 cm³ of water pass into the nephrons every minute. Of this, about 10 cm³ per minute are reabsorbed passively from the proximal convoluted tubules. The water potential gradient between the remainder of the filtrate in the proximal convolution and the blood in surrounding capillaries promotes **osmosis**.

About 19 cm^3 of every 20 cm^3 of water left in the nephrons are reabsorbed every minute from the distal convoluted tubules and collecting ducts. The extent to which water is reabsorbed from these parts of the nephrons depends on the body's state of hydration. If the body's water content is below normal, the walls of the distal convoluted tubules and collecting ducts become very permeable to water. If the body contains sufficient water, however, the walls of the distal convoluted tubules are less permeable to water. Less water is then reabsorbed into the blood. Excess water remains in the urine to be excreted (Fig 11.10).

Fig 11.10 Some stages in the establishment of the medullary gradient by the loop of Henle of a cortical nephron. (b) and (d) show the effects of water reabsorption from the descending limb and solute reabsorption from the ascending limb. (a) and (c) show the effect of fluid flow within the loop. Figures are in units of mosmol per kg H$_2$O. (Modified from *Best and Taylor's Physiological Basis of Medical Practice*, eleventh edition, 1985, Williams and Wilkins)

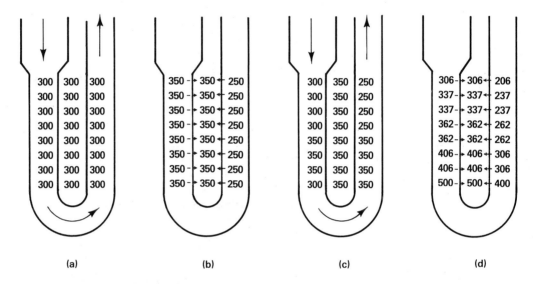

(a)   (b)   (c)   (d)

11.2.4 Role of the medulla in controlling water retention

The loops of Henle are important in creating a water potential gradient between the renal fitrate and the medullary tissue fluid, that is, the **peritubular fluid**. The gradient enables water to be reabsorbed from the nephrons beyond the proximal convolutions.

In all nephrons, the descending limb of the loop is permeable to water but relatively impermeable to solutes. In cortical nephrons, establishing a water potential gradient requires an active transport mechanism along the whole length of the ascending limb. Sodium chloride is pumped out of the limb into the surrounding peritubular fluid. The increased solute concentration in the peritubular fluid lowers its water potential causing water to leave the descending limb by osmosis. In this way the solute concentration of the filtrate in the descending limb is increased. The fluid moves along the loop and more sodium chloride is actively removed from it in the ascending limb. More water is then removed passively from the descending limb, and so on (Fig 11.10). Consequently a solute concentration gradient is established in the peritubular fluid of between about 300 mosmol per kg H$_2$O at the top, to about 600 mosmol per kg H$_2$O around the hairpin of the loop.

The creation of the solute concentration gradient is possible because of the flow of fluid in opposite directions in the adjacent limbs of the loops of Henle. For this reason it is called a **counter-current multiplier**.

In juxtamedullary nephrons, the descending limb of the loop of Henle is longer. Also there is a thin segment of the ascending limb which is not found in cortical nephrons (Fig 11.4(b)). The thin segment cannot transport sodium chloride actively into the peritubular fluid, although solutes may be reabsorbed passively by diffusion. As in cortical nephrons, sodium chloride is pumped out of the renal filtrate in the thick segment of the ascending limb. Water then enters the peritubular fluid by osmosis from the descending limb to surrounding peritubular fluid. This dilution of the peritubular fluid creates a concentration gradient for urea which diffuses into the peritubular fluid from the medullary collecting ducts. Hence the water potential gradient in the medulla is established by active sodium chloride reabsorption in the outer zone, and also by passive urea reabsorption in the inner zone (Fig 11.11).

Fig 11.11 The role of urea in establishing the water potential gradient in the medulla of the kidney. Although some urea enters the thin ascending limb of the loop of Henle there is a net loss of solute (NaCl) by diffusion. Figures are in units of mosmol per kg H_2O. (Modified from *Best and Taylor's Physiological Basis of Medical Practice*, eleventh edition, 1985, Williams and Wilkins)

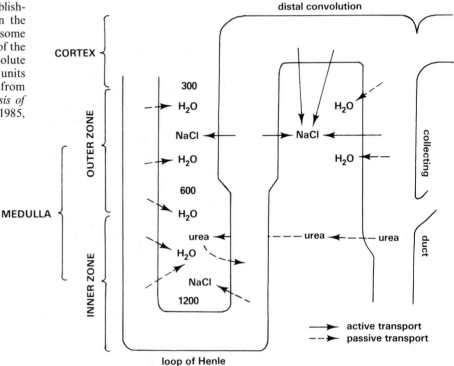

The solute concentration gradient surrounding juxtamedullary nephrons is from about 300 mosmol per kg H_2O at the top to about 1200 mosmol per kg H_2O around the hairpin of the loop. Clearly the greater the concentration gradient between the fluid in the descending limb and the peritubular fluid (horizontally) the greater the reabsorption of water. Long loops of Henle establish a greater concentration gradient in the medulla (vertically) than short ones.

The movement of water from the descending limb into the peritubular fluid could result in a water potential of zero. However, the gradient is maintained by another counter-current mechanism in the medullary blood vessels. The **vasa recta** vessels bring blood into the medulla from the efferent arterioles of the juxtamedullary nephrons. Blood flows relatively slowly in the vasa recta which carry about 10 % of the total kidney blood flow. Like the loops of Henle, the vasa recta turn and then pass back into the cortex (Fig 11.4(b)). Blood entering the vasa recta from the cortex has the same solute concentration as the peritubular fluid in the cortex. The walls of the vasa recta are very permeable to water and solutes of small relative molar mass. As the blood flows deeper into the medulla it becomes surrounded by tissue fluid with an increasing concentration of solute. Hence a lower water potential. The resulting water potential gradient causes water to leave the blood by osmosis. Solutes also enter the blood from the peritubular fluid. The blood becomes more concentrated. This exchange continues as the blood approaches the hairpin of the vasa recta. The solute concentration of the blood is always slightly different from that of the peritubular fluid because of the time taken to reach equilibrium and the fact that the blood is continuously flowing.

Fig 11.12 Water and solutes exchange between the blood in the vasa recta and the peritubular fluid, thus maintaining the medullary gradient (from *Best and Taylor's Physiological Basis of Medical Practice*, eleventh edition, 1985, Williams and Wilkins)

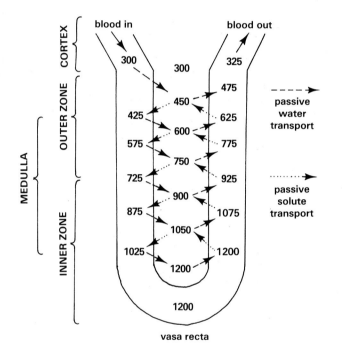

The blood turns the hairpin and begins to flow back towards the cortex. As blood ascends the medulla it absorbs water by osmosis from the peritubular fluid; solutes leave the vasa recta by diffusion (Fig 11.12). This passive **counter-current exchange** preserves the water potential gradient in the medulla. There is a slightly greater absorption of water by the ascending vasa recta than is lost to the peritubular fluid from the descending vasa recta. This is because the plasma pro-

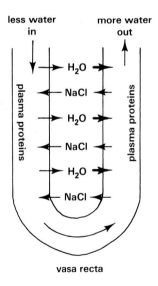

Fig 11.13 The effect of plasma proteins on the exchange of water between the peritubular fluid and the blood in the vasa recta.

teins in the vasa recta attract water. The proteins cannot leave the blood. They counteract the osmotic loss from the descending vasa recta but add to the osmotic gain in the ascending vasa recta (Fig 11.13). Consequently there is a net loss of water from the kidney's medulla whilst most of the solutes responsible for the water potential gradient in the medulla remain. Although some solutes are removed in the blood the gradient is preserved. The mechanism allows water which is absorbed from the descending limb of the loop of Henle and the collecting duct to be retained in the body.

Sodium chloride is actively reabsorbed from the fluid in the distal convolution and collecting duct. In the presence of **antidiuretic hormone, ADH,** the collecting duct becomes permeable to water and urea. Water is reabsorbed by osmosis, urea by diffusion. The water concentration of urine is thus further reduced. In the absence of vasopressin less water and urea are reabsorbed and more are excreted.

The urea reabsorbed into the peritubular fluid stimulates water reabsorption by osmosis from the descending limbs of the loops of Henle in juxtamedullary nephrons. The water concentration of the fluid in the loops is thus further reduced. When the filtrate enters the solute-permeable thin segments of the ascending limbs, sodium chloride is passively reabsorbed. Some of the urea in the peritubular fluid may also enter the ascending thin segments by diffusion to be recycled to the collecting ducts (Fig 11.11).

The ability to concentrate urine and hence conserve body water is important to mammals living in arid conditions. It is therefore of interest that the nephrons of many desert mammals have unusually long loops of Henle (Fig 11.14).

Fig. 11.14 Relative proportions of the loops of Henle in the kidneys of three mammals living in different conditions

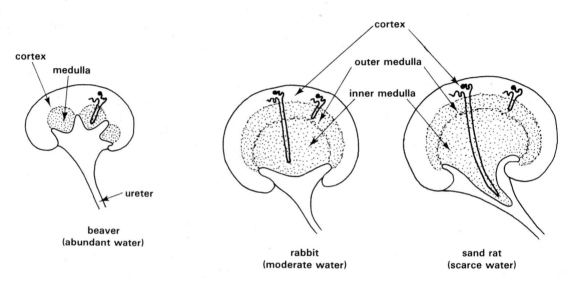

beaver
(abundant water)

rabbit
(moderate water)

sand rat
(scarce water)

11.2.5 Regulation of nephron function

Active reabsorption of some of the solutes in the renal filtrate is regulated by hormones (Chapter 21). Reabsorption of water from the proximal convoluted tubules is by osmosis and normally accounts for the reabsorption of over 80 % of the water in the filtrate. Water reabsorption from the distal parts of the nephrons is also by osmosis The extent of water reabsorption depends on the body's state of hydration and the permeability of the walls of the distal convoluted tubules and collecting ducts.

Beneath the **hypothalamus** in the brain and projecting downward from it is an endocrine gland called the **pituitary body** (Chapter 21). The hypothalamus contains the **osmoregulatory centre** which is sensitive to the concentration of sodium chloride and hence the water potential of blood flowing through it. If the water potential is low, the sodium chloride concentration is high. The pituitary body, stimulated by the hypothalamus, releases **antidiuretic hormone (ADH).** ADH increases the permeability to water of the walls of the distal convoluted tubules and collecting ducts, encouraging reabsorption of water into the blood. Consequently a small volume of concentrated urine is eliminated, a condition called **antidiuresis**. Conversely, if the body's water content is high, ADH output diminishes. The rate of water reabsorption from the distal ends of the nephrons then slows down. Urine flow increases and the urine becomes diluted, a condition called **diuresis** (Fig 11.15).

Fig 11.15 Mechanism controlling water reabsorption from the distal parts of the nephrons

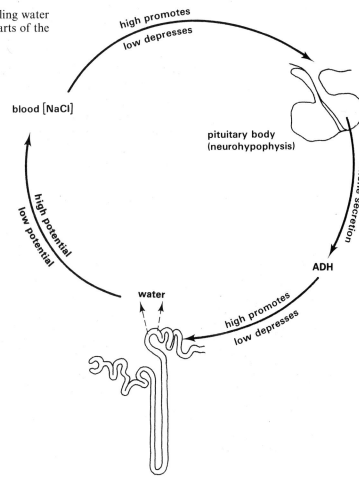

Reduced ADH output can result in a daily urine production of up to ten times the average 1.5 dm³. The condition is called **diabetes insipidus** and is quite distinct from diabetes mellitus (Chapter 16).

How does the body detect and monitor its water content? How is the information translated into permeability changes in the walls of the distal convolutions and collecting ducts?

The role of the hormone renin in the regulation of the body's water content is described in Chapter 21.

11.2.6 Extreme cases of water economy

The camel is an animal which seems ideally fitted to life in the desert. Indeed, the camel can survive for many days without water. The exact length of time depends mainly on temperature and the availability of food. Camels have been reported to travel more than 900 km between watering points without drinking, taking three weeks to complete the trip.

For a long time it was assumed that the camel's hump, or even its stomach acted as a kind of water store. Investigations of the stomach ruled out such a function and the hump is filled with fatty tissue. Certainly the fat can be respired with the production of water. However, in breathing, to obtain the oxygen needed for respiration, even more water is lost from the lungs by evaporation (Chapter 8). Schmidt-Nielsen and others have investigated the water relations of desert animals including camels. Their findings indicate that a camel's nephrons are particularly efficient in reabsorbing water from the glomerular filtrate. After allowing for differences in body size, the daily urinary water loss of a camel is only about 25 % that of a human.

More significant, however, is the camel's very great **tolerance of dehydration**. A camel can survive water loss of up to one third of its total body mass with no apparent ill-effects. This remarkable tolerance of dehydration cannot be matched by any other mammal. It is estimated that a water loss of 15 to 20 % of total body mass would be fatal to a human. Recent work has revealed the presence of a special protein in the membranes of the red blood cells of camels. The protein strengthens the red cells and protects them against collapse when the surrounding plasma becomes concentrated because of water loss. Most mammals sweat and pant both of which help maintain a constant body temperature. By sweating and panting very little camels preserve water, but in doing so camels have to tolerate wide internal temperature fluctuations. A camel's temperature can vary from about 34.5 °C to 40.5 °C, a 6 °C variation which would be very harmful to other mammals (Chapter 20).

Another mammal well fitted for survival in desert conditions is the kangaroo rat, *Dipodomys* (Chapter 8). This rodent has such a well-balanced water economy that it does not normally drink at all. Its total water requirement is provided by the small amount of moisture in the rather dry vegetation it eats and more important, the moisture produced by respiration of its food.

$$C_6H_{12}O_6 + 6O_2 \rightarrow 6CO_2 + 6H_2O$$

Survival on **oxidation water** depends on reducing water loss to an absolute minimum. *Dipodomys* has highly efficient kidneys which produce very concentrated urine. The kangaroo rat also keeps to a minimum the amount of water vapour it breathes out (Chapter 8).

In most terrestrial animals, drinking sea water raises the osmotic potential of the blood plasma, resulting in dehydration of the body tissues by osmosis. Schmidt-Nielsen predicted that kangaroo rats concentrate their urine so much that they should be able to drink sea water without ill effect. To test the prediction he fed experimental kangaroo rats with soya beans, the high protein content of which resulted in a high concentration of blood urea. This led to excessive water loss during urination since urea is toxic and is eliminated by the kidneys. Kangaroo rats which were given sea water to drink after this treatment recovered just as well as a control group provided with fresh water. The experiment demonstrates the ability of kangaroo rats to eliminate excess salts, such as sodium chloride, in forming urine with a low water concentration.

11.3 The role of the kidneys in acid–base balance

Buffers such as the carbonic acid–hydrogencarbonate acid–base pair are limited in the extent to which they can cope with excess hydrogen ions in the body. The final regulation of the pH of body fluids is carried out by the lungs and kidneys. Gas exchange in the lungs disposes of carbon dioxide (Chapter 8). The build-up of carbonic acid in the tissues with accompanying high concentrations of hydrogen ions is thus prevented.

Relatively small but significant quantities of acids enter the body daily in food. These and other acids are disposed of by the kidneys in such a way as to keep basic ions like sodium (Na^+) in the body. Acid excretion in the kidneys takes place in the distal convoluted tubules of nephrons. Here, inside the tubule cells the enzyme **carbonic anhydrase** acts on carbonic acid. Hydrogen (H^+) and hydrogencarbonate (HCO_3^-) ions are formed. The carbonic acid comes from the carbon dioxide produced by the tubule cells as they respire (Fig 11.16). The hydrogen ions are secreted into the

Fig 11.16 Production of hydrogencarbonate and hydrogen ions from respiration

respiration $----\rightarrow$ $CO_2 + H_2O$

$$H_2CO_3$$

partial dissociation

$$H^+ + HCO_3^-$$

tubule where they meet and react with disodium hydrogenphosphate in the glomerular filtrate. The hydrogen ions replace some of the sodium ions, producing sodium dihydrogenphosphate. Sodium ions are then actively reabsorbed into the blood where they combine with the hydrogencarbonate ions from the tubule cells (Fig 11.17).

If the acid content of the blood is very high, the resulting low pH of the glomerular filtrate stimulates production of ammonia by the tubule cells. The ammonia completely replaces sodium ions from disodium hydrogenphosphate in the filtrate, resulting in increased acid excretion (Fig 11.18).

> The pH of human blood lies normally between 7.35 and 7.45. A drop in pH below 7.30, a condition called **acidosis**, or a rise above 7.50 called **alkalosis**, results in metabolic malfunction. So sensitive are the body's physiological processes that failure to regulate the pH of the blood between about 7.0 and 8.0 is usually fatal.

Fig 11.17

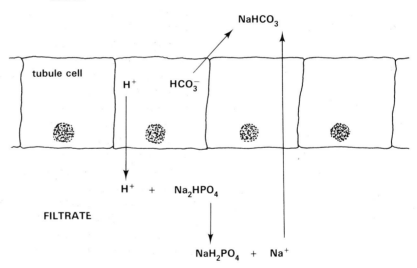

Fig 11.18

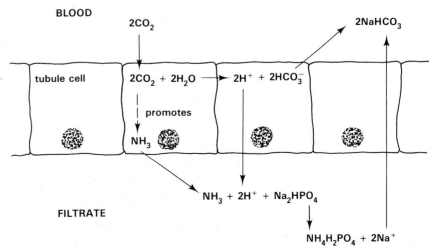

213

11.4 Replacement of kidney function

Kidney failure can be treated in either of two main ways: artificial **dialysis**, that is by **kidney machine**, or kidney **transplantation**.

11.4.1. Kidney machines

A kidney machine is a mechanical device through which a patient's blood passes. The blood leaves the body usually from an artery in the forearm and returns to a nearby vein (Fig 11.19). Inside the machine the blood flows

Fig 11.19 (a) Patient connected to a kidney machine

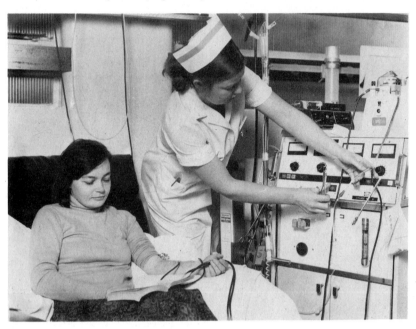

Fig 11.19 (b) Diagrammatic arrangement of the apparatus causing dialysis of a patient's blood in a kidney machine

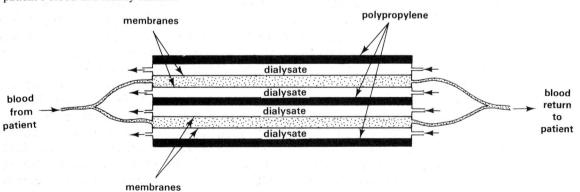

over or between membranes which separate it from an aqueous **dialysing fluid** containing dissolved sugars and salts in concentrations normally found in blood. Soluble constituents in the blood in excess of normal concentrations diffuse across the membrane into the dialysing fluid. In this way wastes like urea which accumulate in the body are extracted. Blood cells and proteins remain in the blood. The process is called renal dialysis (Fig. 11.20).

Fig 11.20 Mechanism of dialysis. Normal (for the blood) concentrations of dialysable substances in the dialysing fluid promote the diffusion of excesses from the blood across the membrane. Dialysis continues until the concentrations of dialysable substances on either side of the membrane are equal

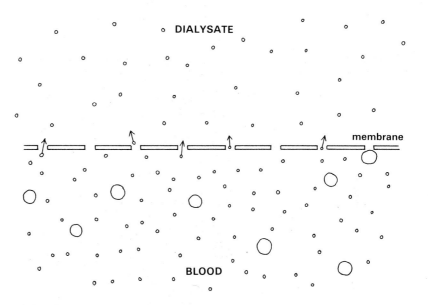

By using a clip to partially obstruct the tube carrying blood back to the patient's vein the blood pressure in the machine is increased. This makes it possible to remove water from the blood by ultrafiltration. Water is removed from the blood in a comparable way in a real kidney.

The pH of the blood falls from 7.4 to about 7.3 between dialysis treatments. This is because the body's store of hydrogencarbonate ions is used to buffer the blood and in so doing is lowered to about 75% of its normal value. The kidney machine replaces hydrogencarbonate used up in this way. Earlier machines had sodium hydrogencarbonate in the dialysing fluid but sodium ethanoate is used now. The ethanoate diffuses in to the blood and is metabolised in the body to hydrogencarbonate.

A patient usually spends several hours twice a week connected to a kidney machine, during which time the dialysing fluid drains the blood of excess and toxic constituents. However, man-made machines are not perfect. The activities of kidneys are regulated in response to changes in the body's internal environment. As yet kidney machines are not sufficiently sophisticated to be self-regulating. Furthermore, the machines are not designed to perform functions other than excretion. Kidneys functioning abnormally usually stop secreting the hormone **erythropoietin** which is necessary for normal red blood cell production (Chapter 10). Because of this, patients requiring dialysis are often anaemic. Nevertheless, despite their shortcomings kidney machines have prolonged the lives of a large number of people.

Kidney machines do not replace all the kidney functions. Why are patients with failed kidneys often anaemic?

11.4.2 Kidney transplantation

Transplantation involves the transfer of a healthy kidney from one person called the **donor** into the body of the patient whose kidneys have failed, the **recipient**. The problems associated with kidney machines are eliminated, since the transplanted kidney takes over all the functions of the failed kidneys. The surgical procedures of kidney transplantation are relatively straightforward (Fig 11.21). However, as with the transplantation of any body organs, a major problem which has to be overcome is **immune rejection** (Chapter 10). The recipient's immune system recognises the transplant as non-self and reacts by the cellular immune response. T-lymphocytes invade the tissues of the transplanted organ and destroy it. The rejection problem can be largely avoided by matching as closely as possible the tissue type of the donor to that of the recipient. If relatively few of the transplant's antigens are non-self to the recipient, the immune response is less likely. In addition to tissue type matching, the recipient may be treated with immunosuppressant drugs. These inhibit the immune response. However, immunosuppression lowers the body's general defences, and antibodies which would normally protect the recipient against infections are not produced. The patient is thus prone to infections. Treatment usually includes a combination of tissue type matching and limited immunosuppression (Chapter 10).

Fig 11.21 A transplanted kidney connected to the recipient's blood and urinary systems

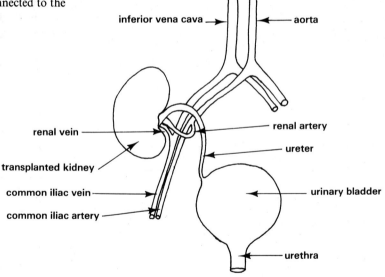

The main problem for kidney transplantation today is the shortage of suitable donor kidneys. There is an international organisation which keeps on file the tissue type of patients who require new kidneys. When kidneys become available, say from road traffic accident victims, their tissue type is determined and the international agency is informed immediately. Kidneys may then be flown from one country to another in order to provide a transplant for a suitable patient who may be thousands of miles away.

SUMMARY

Excretion is the elimination from the body of any substance which is present in excess of its normal levels. Blood is supplied to paired kidneys by renal arteries and drained by renal veins. Urine leaves the kidneys in paired ureters; it accumulates in the urinary bladder prior to periodic elimination from the body (micturition).

Nephrons are the functional units of the kidneys. Ultrafiltration of metabolites from blood in glomeruli takes place in the renal capsules; it is facilitated by high blood pressure and a thin capsular membrane. Some metabolites enter the nephrons by direct secretion.

Selective reabsorption (into the blood) occurs from the proximal convoluted tubules solutes actively, water by osmosis. The absorptive surface area is increased by microvilli in the proximal tubule cells.

The counter-current mechanism in the parallel loops of Henle creates a medullary water potential gradient; this allows osmotic reabsorption of water from the distal convoluted tubules and collecting ducts.

Mammals living in arid conditions possess relatively long loops of Henle; these maximise water retention. Some mammals, such as the camel, tolerate relatively great water loss.

Water retention (or excretion) is controlled by antidiuretic hormone (vasopressin) from the posterior pituitary; it is secreted when the body's hydration is low (blood Na high); it increases permeability to water of the walls of the distal convoluted tubules and collecting ducts. Water responds to the medullary gradient and is absorbed; conversely when the body is over-hydrated.

Kidneys help to regulate blood pH; nephron cells produce hydrogen carbonate which enters the blood replacing that filtered into the urine.

Excretory function may be replaced by kidney machines. Excess metabolites are removed from the patient's blood by dialysis across thin membranes. Alternatively, transplantation of kidneys from organ donors replaces total kidney function. The main problems are immune rejection of the transplant and the availability of suitable donor kidneys.

QUESTIONS

1 The table below gives a comparison between the quantities of various substances in the glomerular filtrate and urine in Man.

Constituents	Quantities passed into filtrate of glomerulus day^{-1}	Quantities passed in urine day^{-1}
Na$^+$	600 g	6 g
Glucose	200 g	nil
Urea	60 g	35 g
Water	180 dm^3	1.5 dm^3

(a) With reference to the figures given in the table, suggest **two** major functions of the kidney.

(b) Given that:

(i) protein is present in the blood plasma, explain why there is no protein in the glomerular filtrate,

(ii) glucose is present in the glomerular filtrate but is absent in the urine, explain what happens to the glucose in the glomerular filtrate.

(c) In relative terms, only a small quantity of sodium ions are excreted in the urine. Describe the role of these ions in the functioning of the kidney.

(d) In what part of the nephron does the major decrease in volume of the glomerular filtrate occur? (C)

2 The diagram shows part of a nephron of a mammalian kidney.

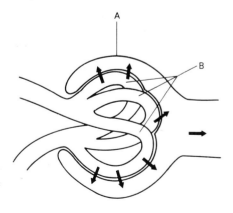

(a) Name the structures labelled **A** and **B**.

(b) Name the process, indicated by the arrows, taking place between **B** and **A**.

(c) State **two** conditions which are necessary for this process to occur. (C)

3 Many of the metabolic reactions that occur in organisms produce hydrogen ions which could change the pH of body fluids.
(a) Name **two** substances or groups of substances which act as buffers in mammals.
(b) Describe how the kidney helps to maintain the pH of the blood at a constant level.   (AEB 1987)

4 The diagrams show nephrons from the kidneys of three different kinds of organism.

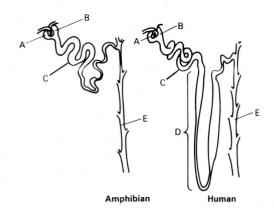

Amphibian   Human

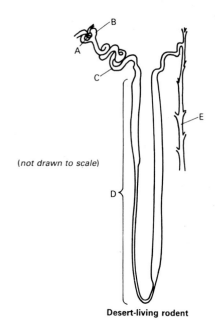

(not drawn to scale)

Desert-living rodent

(a) Identify parts **A** to **E**.

(b) (i) Account for the absence of **D** in the amphibian, for the presence of **D** in the human and for the relative length of **D** in the desert-living rodent.
(ii) Explain the roles played by parts **D** and **E** in regulating the volume of urine produced in humans.
(c) Explain how glucose is reabsorbed in the nephron of humans.
(d) In the nephrons of sea-living bony fish the parts **A** and **B** are small. Osmoregulation is mainly accomplished by secretion of salt and nitrogenous waste through the gills into the current of water that flows over them. What are the special problems of osmoregulation which are overcome by fish living in sea water?   (AEB 1986)

5 The table below shows the concentration (in g/100 cm^3 fluid) of some of the constituents in the blood plasma, the kidney glomerular filtrate and the urine of man.

Constituents	Plasma	Glomerular filtrate	Urine
Protein	7.00	0.00	0.00
Urea	0.03	0.03	2.00
Glucose	0.10	0.11	0.00
Sodium	0.32	0.33	0.60

(a) Explain how urea is formed.
(b) Suggest a reason why no protein is present in the glomerular filtrate.
(c) Name *two* proteins in the blood plasma and indicate their function.
(d) How does the composition of plasma differ from that of serum?
(e) Explain why the concentration of sodium ions in the urine is not the same as in the glomerular filtrate.
(f) How could you test for the presence of glucose in the urine?
(g) If the pressure in the renal artery decreases, state the effect this would have on the volume of urine produced and give *one* reason for your answer.   (L)

6 (a) (i) Why is it necessary for a mammal to dispose of waste materials?
(ii) Give examples of the processes and products involved in the disposal of these waste products.
(b) With the aid of annotated diagrams, describe the micro-anatomy of a mammalian nephron (uriniferous tubule) and its blood supply.
(c) Explain carefully how the nephron carries out its various functions.   (O)

7 (a) What is homeostasis?
(b) Describe the parts played by each of the following in the homeostatic mechanisms of a mammal: (i) kidney, (ii) liver.   (L)

12 Uptake and transport of water in flowering plants

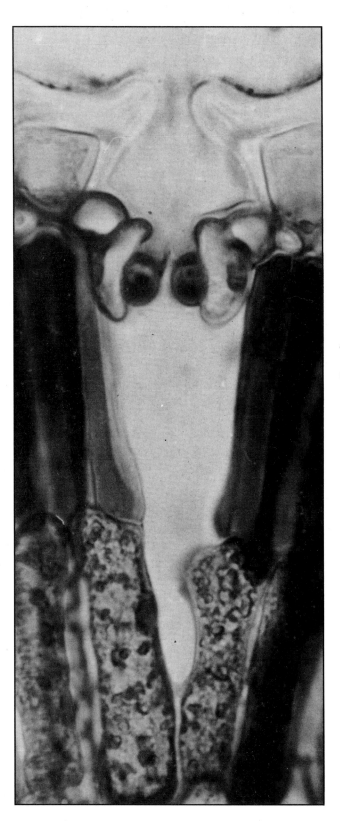

12.1	Transpiration		220
12.2	Factors affecting the transpiration rate		222
	12.2.1	Climate	222
	12.2.2	Leaf structure	225
	12.2.3	Xerophytic plants	226
12.3	The effects of transpiration		228
	12.3.1	Water movement through the shoot system	228
	12.3.2	Water absorption by roots	230
12.4	Root pressure		233
12.5	Tracheids, vessels and capillarity		234
Summary			237
Questions			237

12 Uptake and transport of water in flowering plants

Water is essential to all forms of life. In plants, as in animals, water performs important functions as a solvent, as a thermal buffer and as a raw ingredient in hydrolytic reactions catalysed by enzymes (Chapter 3). There are some uses to which water is put in green plants alone. Water provides hydrogen to reduce the products of carbon dioxide fixation in photosynthesis (Chapter 5). In non-woody land plants the turgidity of water-containing cells helps to hold the shoot system erect in the atmosphere. Not surprisingly, lack of sufficient water even for short intervals is one of the major factors which limits the growth of crops in many parts of the world.

Most of the water required by a land plant is absorbed by its roots from the soil. The rate of water uptake is largely determined by the rate at which water is lost to the atmosphere from the shoot system.

12.1 Transpiration

A major problem confronting all land-dwelling organisms is how to avoid desiccation. The chief advantages of a terrestrial as opposed to an aquatic existence for plants are direct access to sunlight and increased availability of oxygen for respiration. However, in making the most of these advantages land plants continually lose water vapour to the atmosphere. Indeed, as much as 99 % of the water absorbed by a terrestrial plant can be lost by evaporation from the shoot system. **Transpiration** occurs most rapidly when the stomata are open and the internal tissues of stems and leaves are in contact with the atmosphere. Well over 90 % of the total water loss from a leafy shoot is due to **stomatal transpiration**. Water vapour may also pass directly through the epidermis of the shoot system, but this route is normally sealed by the waxy cuticle. **Cuticular transpiration** usually amounts to less than 5 % of the total water vapour loss. Woody stems of deciduous plants lose small amounts of water vapour through lenticels by **lenticular transpiration** before leaf fall. Afterwards all of their water loss is through the lenticels.

network of minor veins

main vein

Fig 12.1 Venation of a leaf of elder, × 1

Because there are normally more stomata on the leaves than elsewhere on the shoot system it is evident that most of the water vapour is lost from the leaves. In many dicotyledonous species the leaves are broad, flattened organs with a vast network of minor veins (Fig. 12.1). Even the smallest vein contains transporting cells which supply the mesophyll tissue with water and dissolved minerals and remove soluble photosynthetic products.

A conspicuous feature of the spongy mesophyll is the presence of numerous intercommunicating air pockets (Fig 12.2). Water constantly evaporates into the air pockets from the walls of the mesophyll cells.

Fig 12.2 (a) Transverse section through part of the blade of a leaf of a dicotyledonous plant, × 100

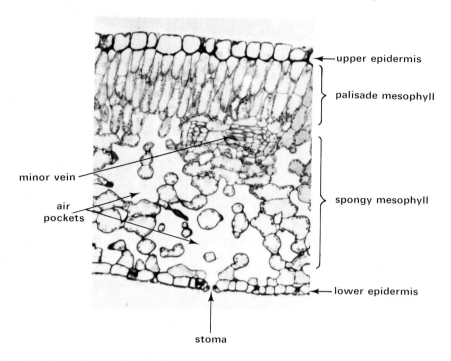

Fig 12.2 (b) Air pockets in the spongy mesophyll

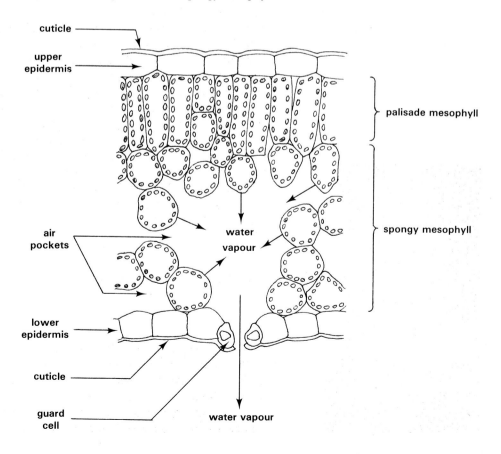

Since the mesophyll air pockets are enclosed spaces, the air in them is usually saturated or nearly so with water vapour. The air inside leaves therefore has a relatively high water potential (Chapter 1). The water potential of the atmosphere surrounding the shoot system fluctuates according to prevailing climatic conditions. In hot, damp greenhouses and in tropical jungles the atmosphere is extremely humid with a fairly high water potential. However, in most terrestrial habitats the atmosphere is generally much less humid and therefore has a relatively low water potential. Consequently when the stomata are open an area of high water potential is brought into contact with an area of low water potential and water vapour rapidly **diffuses** into the surrounding air down a **water potential gradient**. Clearly any factor which alters the size of the gradient will influence the rate of transpiration.

12.2 Factors affecting the transpiration rate

Climate is the product of a large number of interacting factors, notably the intensity and duration of sunlight, wind movement and precipitation, especially rainfall. The water potential of the atmosphere is determined by such interactions. As we have just seen, this affects the size of the water potential gradient between leaves and the surrounding atmosphere. However, the situation is complex because climatic factors may not affect leaves and the atmosphere in the same way or to the same extent. Although climate profoundly influences transpiration rates, other factors such as leaf structure, shape and physiology also determine the rate at which leaves lose water vapour to the atmosphere.

12.2.1 Climate

1 Solar radiation

Energy as heat and as light comes from the sun. About 10 % of **infra-red radiation**, the main source of heat from the sun, falling on leaves is absorbed and is therefore potentially available to raise the temperature of leaves. However, most of this energy provides latent heat for the evaporation of water from the leaf cells. A drastic increase in leaf temperature is thus averted. Nevertheless, leaf temperatures of between 5–8 °C above atmospheric temperature are common even in Britain. The effect of such temperature differences is described more fully in the next section.

Light intensity affects transpiration by controlling the degree and to some extent the pattern of stomatal opening (Chapter 13). The stomata of most plants open fully in daylight hours, bringing the saturated air enclosed in a leaf into direct contact with the surrounding atmosphere. This is why there is usually a close correlation between the rate of transpiration and light intensity (Fig 12.3). Exceptions are succulent xerophytes such as cacti whose natural habitat is the desert. They close their stomata in light, thereby minimising transpiration in the hotter part of the day (section 12.2.3).

Fig 12.3 Correlation between rate of transpiration of oats and light intensity (after Sutcliffe 1968)

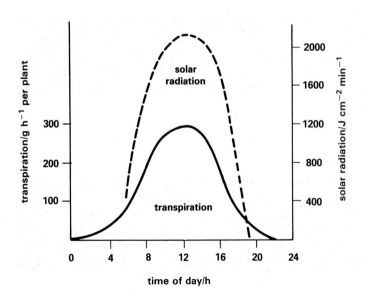

2 Temperature

A rise in temperature provides additional **kinetic energy** for the movement of water molecules. The effect is to accelerate the rate of evaporation of water from the walls of mesophyll cells and, if the stomata are open, to speed up the rate of diffusion of water vapour into the surrounding atmosphere.

The partial pressure of water vapour in the air increases with temperature as more water evaporates into air. Table 12.1 compares the water vapour pressure at two temperatures of a fully saturated atmosphere, as exists inside a leaf, with one which is 60 % saturated, as may occur outside a leaf. The water content of air is usually expressed as % relative humidity (RH) and is measured as water vapour pressure.

Table 12.1 Comparison between water vapour pressure of air at 100 and 60%RH at two temperatures

Temp/°C	Water vapour pressure/kPa		Water vapour pressure gradient/kPa
	100 %RH 60		
15	1.70	1.02	0.68
20	2.33	1.40	0.93

Use the data in Table 12.1 to calculate the percentage increase in water vapour pressure gradient for the 5 °C rise in temperature.

The air inside a leaf normally remains saturated with water vapour when its temperature is raised because of increased evaporation of water from the walls of the mesophyll cells. Thus, as the leaf temperature rises so does the water vapour pressure of the air in the leaf. It is not always likely that a similar situation exists in the surrounding atmosphere. The water vapour pressure here will increase only if there is a nearby source of liquid water from which evaporation can take place. Transpiration from plants and evaporation from the soil and from lakes and oceans are the main sources of water vapour in the atmosphere. If the upper layers of the soil are dry following a long period without rainfall, a rise in temperature has little or no effect on the water vapour pressure of the atmosphere in a terrestrial environment. In hot, dry weather, therefore, the water vapour pressure gradient between leaf and atmosphere is great and transpiration is very rapid.

The effect of infra-red radiation in raising leaf temperatures by several degrees above ambient accelerates the rate of transpiration. However, when a leaf is hotter than the air around it, convection currents develop bringing cooler air in contact with the leaf. To some extent the cooling effect counteracts the expected increase in the transpiration rate by lowering the leaf temperature.

3 Air movement

Transpiration in still air results in the accumulation of a **boundary layer**, several millimetres thick, of saturated air at the surfaces of leaves. The boundary offers resistance to the diffusion of water vapour through stomata and cuticle, thereby reducing the transpiration rate. Movement of the surrounding air reduces the thickness of the boundary layer, thus maintaining a high water potential gradient near to the leaf surface (Fig 12.4)

Fig 12.4 Effects of air movement on thickness of boundary layer

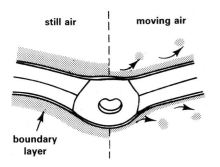

still air

moving air

boundary layer

However, air movement generally brings cool air into contact with leaves. As we have already seen, this may lower the temperature of leaves. The outcome of these two contrary phenomena determines to what extent the rate of transpiration is affected. Some species close their stomata when exposed to high wind speeds. Nevertheless, continuous exposure to wind of high velocity generally causes wilting of leaves. This is because the rapid loss of water vapour by transpiration is not balanced by absorption of water from the soil.

4 Atmospheric humidity

The water vapour content of air is usually expressed as **percentage relative humidity (%RH)**:

$$\%RH = \frac{\text{water vapour pressure of air at t°C}}{\text{water vapour pressure of saturated air at t°C}} \times 100$$

We have noted that the air inside a leaf is more or less saturated with water vapour. During rapid transpiration the water content of the air in the intercellular spaces adjacent to stomata falls to about 96% RH. This lowers the water potential by as much as -5580 kPa. Elsewhere in the leaf the air remains saturated at 100% RH. Conversely the humidity of the atmosphere surrounding a leaf is changeable. Values exceeding 70 %RH are rare outdoors in Britain while figures as low as 30 %RH can occur following a long spell of hot, dry weather. In either case the water potential gradient between leaf and atmosphere is very substantial and when the stomata are open water vapour rapidly diffuses from the leaf. As the %RH of the atmosphere falls the water potential gradient increases. There is thus a direct relationship between atmospheric humidity and the rate of transpiration (Table 12.2 and Fig 12.5).

Table 12.2 Water vapour pressure of air at four different humidities (readings taken at 20 °C)

%RH	Water vapour pressure/kPa	Water vapour pressure gradient/kPa
100	2.33	2.33 − 2.33 = 0.00
70	1.63	2.33 − 1.63 = 0.70
60	1.40	2.33 − 1.40 = 0.93
50	1.16	2.33 − 1.16 = 1.17

Fig 12.5 Effect of atmospheric humidity on the rate of transpiration

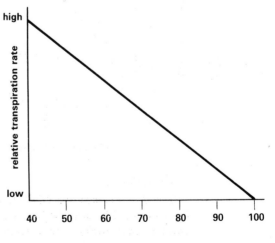

12.2.2 Leaf structure

Brown and Escombe in 1900 investigated the rates of diffusion of gases through membranes perforated by pores varying in size and number per unit area. Their findings of the movement of carbon dioxide and its implications in the absorption of the gas by leaves for photosynthesis are described in Chapter 13. Part of their work involved a study of the diffusion of water vapour. They observed that, providing the pores were not too close to one another, the rate of diffusion of water vapour through a perforated membrane is as much as 50 % of the total water loss from a body of water of surface area similar to the membrane. This occurred despite the fact that the pores occupied less than 3 % of the area of the membrane. Stomata usually account for about 5 % of the total area of a leaf. It may therefore be deduced that the epidermis is only a limited barrier to water loss when the stomata are open. The rate of diffusion of water vapour through stomata results, however, from the interaction of a number of factors.

1 Pore size

Fig 12.6 Diffusion of water vapour through a pore

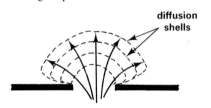

As molecules of a gas such as water vapour escape through a pore into still air they spread out into concentric, hemispherical zones called **diffusion shells** (Fig 12.6). The water potential of each shell diminishes the further it is from the pore. Water vapour molecules continually diffusing from the outer shell into the atmosphere are replaced by molecules diffusing from an inner shell. A continuous water potential gradient thus exists across the diffusion shells.

The rate of diffusion **per unit area of pore** is greater through small pores than through large pores (Chapter 13). Furthermore the diffusion rate is proportional to the **diameter** of the pore (Table 13.1). In still air a reduction in pore size is less effective in lowering the transpiration rate than in moving air (Fig 12.7). Why do you think this so?

Fig 12.7 Effect of pore size in controlling transpiration rate in *Zebrina* (after Bange, 1953)

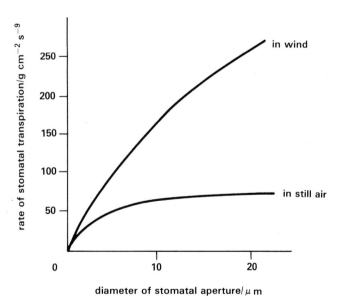

2 Pore density

Where stomata are spaced wide enough apart there is no interference between the diffusion shells of water vapour above each pore. However, where stomata are close to one another, as occurs in most land plants, there may be considerable **overlap** between the diffusion shells above adjacent

Fig 12.8 Diffusion of water vapour through adjacent pores

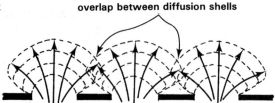

overlap between diffusion shells

stomata (Fig 12.8). The effect is to form one large area from which water vapour diffuses into the atmosphere. Because the rate of diffusion per unit area from a large area is less than that from a number of smaller areas of equivalent area the transpiration rate is reduced.

Other ways in which leaf structure affect transpiration are mentioned in the next section.

12.2.3 Xerophytic plants

There is a considerable variety of structure, stomatal distribution and physiology among higher plants. The differences often play a key part in determining the rate of transpiration. Species which inhabit areas where water is scarce are of particular interest in this respect. They are generally called **xerophytes** as opposed to most land plants known as **mesophytes** which inhabit areas where ample water is available and **hydrophytes** which live totally or partially submerged in water.

Fig 12.9 (a) Transverse section through a stoma of *Hakea*, × 375

Fig 12.9 (b) Transverse section through part of a leaf of *Nerium*, × 100

pit in which water vapour collects

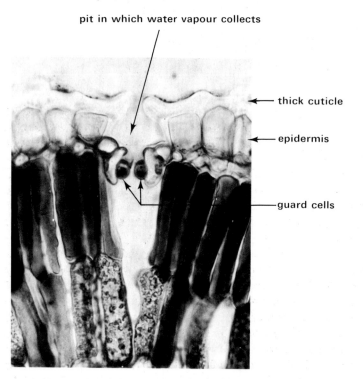

thick cuticle

epidermis

guard cells

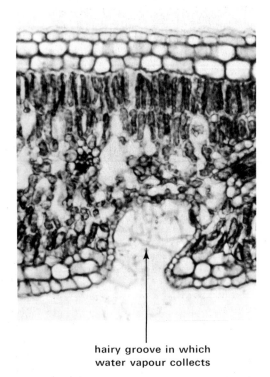

hairy groove in which water vapour collects

Frequently the guard cells of xerophytes are sunk in pits or in grooves well below the surface of the leaf or stem (Fig 12.9). Water vapour escaping through stomata accumulates immediately above the guard cells to form a **thick boundary layer** which reduces the transpiration rate even in moving air. Other modifications, such as **hairiness** of the shoot system and a relatively **thick cuticle**, also assist in minimising the rate at which water vapour is lost. In very dry weather the leaves of some xerophytes such as marram grass can even be **rolled** into a cylinder with the stomata on the inside. Water vapour is trapped inside the cylinder, so building up the humidity of the air into which transpiration takes place. Ultimately the water potential inside the cylinder becomes sufficiently high to stop transpiration (Fig 12.10).

Fig 12.10 Transverse section of a rolled leaf of marram grass, × 40. When the hinge cells lose water vapour excessively they become flaccid and the leaf rolls up as shown

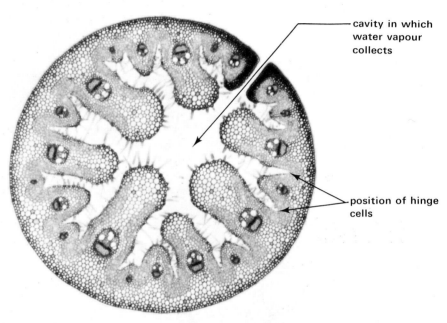

cavity in which water vapour collects

position of hinge cells

Fig 12.11 Desert cactus (Bruce Coleman: Jen & Des Bartlett)

Table 12.3 Transpiration ratios for three categories of plants (from Evans 1978)

Category	Ratio
CAM	30/116
C4	50/116
C3	150/250

What other criteria would need to be considered in deciding the suitability of the three categories of plants for crop production?

Cacti (Fig 12.11) are often thought of as plants ideally suited to conditions of drought. They are **succulent** plants and store water in a special tissue inside their bodies. The cylindrical shape of many cacti means that for each unit volume of tissue there is a small unit of surface area through which water vapour can be lost. Not all xerophytes are succulent. Some such as the creosote bush are highly lignified and can **tolerate** extremes of desiccation without wilting.

Perhaps even more significant is the rate at which some xerophytes fix carbon dioxide compared with mesophytes. Cacti and many other succulent xerophytes fix CO_2 in a pathway called **crassulacean acid metabolism (CAM)**. They can make photosynthetic end-products more efficiently than C3 plants from the same volume of carbon dioxide. This means that their stomata need remain open for only short periods of time each day to acquire sufficient carbon dioxide. They take in carbon dioxide at night when the air is cooler and conditions are less likely to encourage transpiration. Such adaptations considerably lessen the volume of water vapour transpired. The **transpiration ratio** expresses the mass of water transpired per gramme of dry matter made in photosynthesis (Table 12.3). The efficiency with which crops use water will be an extremely important criterion in the future, when agriculture will have to compete even more intensively with industry and the growing human population for water supplies. To date, the pineapple is the only CAM species which is cultivated on a large scale.

12.3 The effects of transpiration

One of the effects of transpiration has already been mentioned: the cooling effect which prevents drastic rises in leaf temperature. There is also little doubt that transpiration is the main factor contributing to water movement through plants.

12.3.1 Water movement through the shoot system

A **potometer** (Fig 12.12(a)) can be used to investigate the rate of absorption of water by a plant shoot. As the shoot transpires, the water vapour it has lost is replaced by liquid water drawn in from the potometer. What is more, the rate of absorption is affected by the same factors which influence the transpiration rate. If the leafy shoot is replaced with a water-filled porous pot (Fig 12.12(b)) comparable results are obtained. As water evaporates from the tiny holes of the porous pot it is replaced by liquid water entering the pores by **capillarity**. The capillary movement of water is due partly to forces of **adhesion** between water molecules and the walls of the pores and of **cohesion**, the forces of attraction holding the water molecules to each other. It is therefore logical to deduce that the forces responsible for water movement in the pot are also at work in the shoot system of a plant. Water evaporating from the porous walls of leaf mesophyll cells is replaced by capillarity. But where does the water come from and what effect does the refilling of the pores in the mesophyll cell walls have on the rest of the plant?

Fig 12.12 (a) A potometer

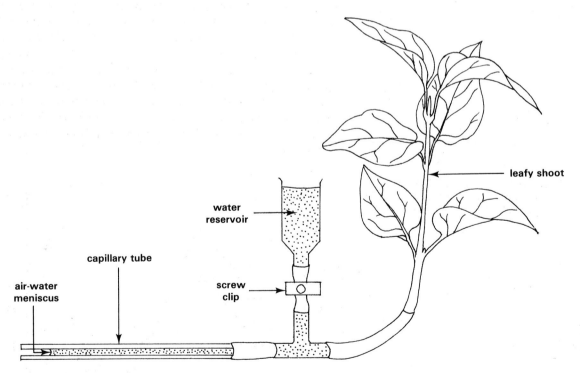

leafy shoot

water reservoir

capillary tube

air-water meniscus

screw clip

With the screw clip closed, the rate at which the meniscus moves along the capillary tube indicates the rate at which the leafy shoot is absorbing water. When all the water has been absorbed from the capillary tube it can be refilled by opening the screw clip.

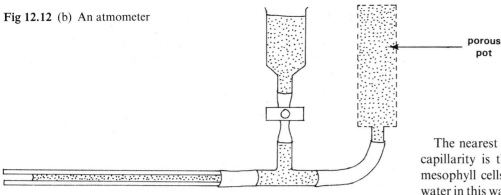

Fig 12.12 (b) An atmometer

porous pot

Readings are taken as with the potometer

The nearest reservoir of water for capillarity is the protoplasts of the mesophyll cells. The effect of losing water in this way is to lower the water potential of mesophyll cells. If a source of water of higher potential is available nearby the protoplasts of the mesophyll cells could absorb water through their plasma membranes by **osmosis**. Alternatively water could be drawn by **capillarity** from the source directly into the mesophyll cell walls where evaporation is taking place. Whichever route is taken, the water comes from a minor vein containing water-conducting xylem elements. One effect of transpiration therefore is to create **leaf suction** which removes water from the veins of leaves. In turn the leaf veins become replenished by absorbing water from the xylem elements of the stem. Again the movement of water from stem to leaf is along a water potential gradient.

The cut shoot in the potometer has direct access to water in the apparatus and can draw freely on it. Does a similar situation exist in a whole plant? Some idea can be obtained by comparing the rates of transpiration and water absorption of an intact plant over a period of time (Fig 12.13). Absorption lags behind transpiration, indicating that the roots are a barrier to a continuous flow of water triggered by transpiration. The effect of leaf suction and root resistance working in opposite directions places the water columns in the vessels and tracheids under strain or **tension**. When this happens the pressure potential of the xylem elements has a negative value. The columns are sufficiently elastic to take the strain so they do not normally break and the flow of water, **the transpiration stream**, continues. What prevents breakage of the water columns? The answer to this question is the **cohesive forces** which exist between water molecules (Chapter 1).

Fig 12.13 Comparison of the rates of transpiration and water absorption by sunflower plants (after Kramer, 1937)

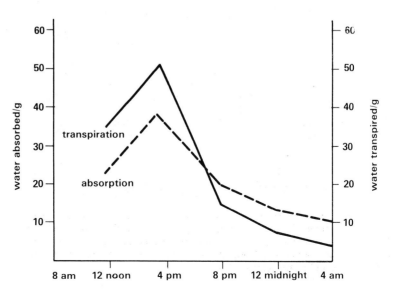

The account of water transport given on pp. 228–9 is a summary of the **transpiration–cohesion–tension theory** described in 1894 by Dixon and Joly. What is of particular interest is that the proposed mechanism can account for the movement of water to the tops of extremely tall trees such as the redwood, *Sequoia*, which frequently grows to a height of well over 100 m (Fig 12.14). At sea level a suction force of about − 100 kPa can raise a column of water to a height of only 10 m. This is quite inadequate to move water to the top of a tall tree. Water potentials as low as − 3000 kPa have been demonstrated in the mesophyll tissue of the leaves of tall trees. This means that leaves can generate a suction capable of pulling water to a height of 300 m. The tension required to stretch to breaking point columns of sap from the xylem of trees has been shown to be about 30 000 kPa. This is well in excess of the tension normally generated in plants.

Fig 12.14 A giant California redwood Sequoia tree. (Photo Paul Conklin, permission Colorific)

What we have seen so far is that transpiration can lead to an upward movement of water through the shoot system. But what about water transport in roots?

12.3.2 Water absorption by roots

Water is absorbed by the outermost layer of the root which is in direct contact with the soil. Between 4 and 10 mm behind the root tip the **epidermal cells** have **root hairs** which increase the surface area of the root several fold (Fig 12.15). Because of the unceasing apical growth of the root tip, a root hair zone has a functional life of only a few weeks. It is continually replaced by a new zone in the freshly extended part of a root. In older parts of the root the epidermis is replaced by a suberised **exodermis**.

Fig 12.15 (a) Transverse section through the root hair zone of a broad bean root, ×40

Fig 12.15 (b) Close-up view of some root hairs, ×100

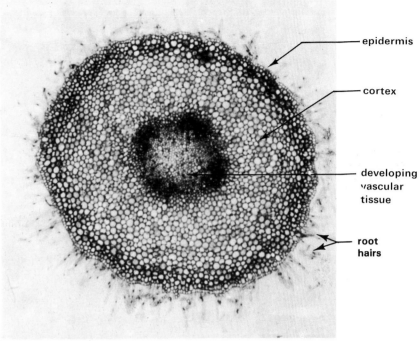

epidermis

cortex

developing vascular tissue

root hairs

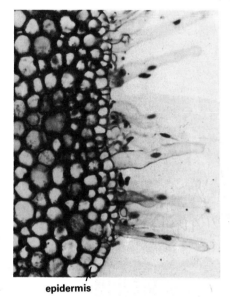

epidermis

If secondary thickening occurs a **periderm** becomes the outermost layer (Chapter 24). It is commonly thought that water uptake by a root is confined to the root hair zone. Yet it has been shown that water absorption can occur rapidly through the surface of older parts of the root even where an exodermis or a periderm is present (Fig 12.16).

Fig 12.16 Water uptake by different parts of broad bean roots (after Brouwer, 1965)

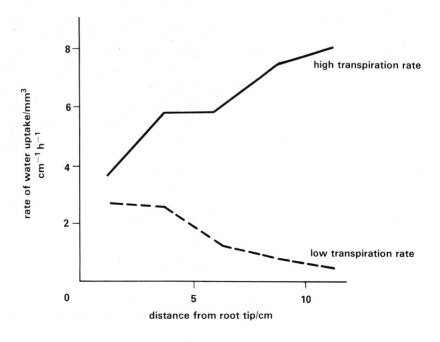

high transpiration rate

low transpiration rate

rate of water uptake/mm^3 cm^{-1} h^{-1}

distance from root tip/cm

Fig 12.17 (a) A young endodermal cell

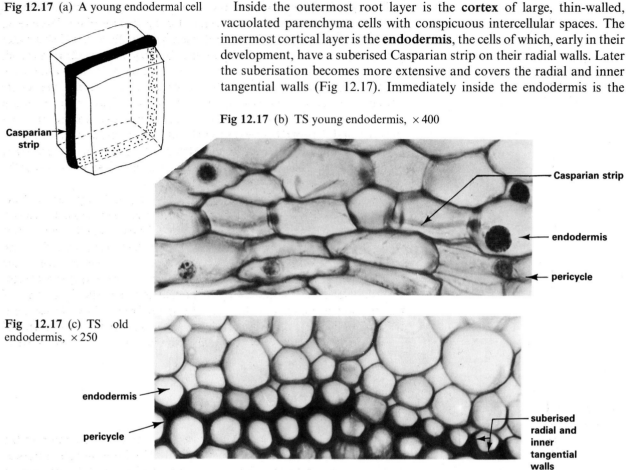

Casparian strip

Inside the outermost root layer is the **cortex** of large, thin-walled, vacuolated parenchyma cells with conspicuous intercellular spaces. The innermost cortical layer is the **endodermis**, the cells of which, early in their development, have a suberised Casparian strip on their radial walls. Later the suberisation becomes more extensive and covers the radial and inner tangential walls (Fig 12.17). Immediately inside the endodermis is the

Fig 12.17 (b) TS young endodermis, × 400

Casparian strip

endodermis

pericycle

Fig 12.17 (c) TS old endodermis, × 250

endodermis

pericycle

suberised radial and inner tangential walls

pericycle, a region one to several cells wide of small, thin-walled parenchyma cells. The pericycle is the site of origin of lateral roots which may add greatly to the absorptive surface area of the root system. At the core of the root is the **vascular tissue**. In the root hair zone of the root of a dicotyledonous plant there are between two and five groups of **protoxylem** elements. At a later stage the central **metaxylem** elements differentiate and the primary xylem looks star-shaped in cross section (Fig 12.18).

Fig 12.18 Transverse section through a root of a dicotyledonous plant, × 200

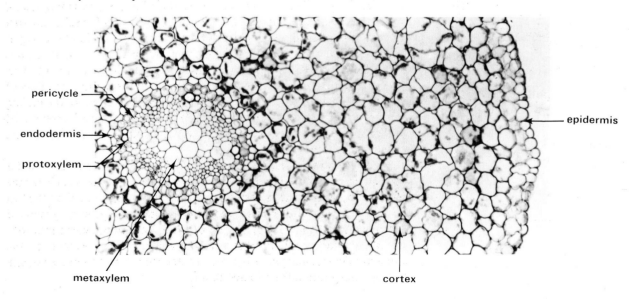

pericycle

endodermis

protoxylem

epidermis

metaxylem

cortex

The transpiration stream drags columns of water upwards through the network of xylem elements which extend throughout the plant. The stream lowers the water potential in the root xylem sufficiently to establish a water potential gradient down which water is ultimately absorbed from the soil. There are forces which lower the water potential of the soil solution, thereby opposing absorption. The opposing forces include the solute potential of the soil solution, which is usually significant only in certain habitats such as salt marshes. Capillary and inbibitional forces also hold the soil solution in and between mineral particles and the humus fraction of soil. Nevertheless the transpiration stream can generate the necessary **water potential gradient** for water to pass from the soil solution across the root cortex and into the root xylem elements.

The classical view is that water moves through the living tissues of the root cortex from vacuole to vacuole by **osmosis**. However, as in the leaves, it is possible that cell to cell transport by **capillarity** along the cell walls is also important. The experts do not agree on which of the routes is the more important. The wall route cannot be taken at the endodermis where the waxy Casparian strip is impervious to water. At the endodermis water probably flows by osmosis through the cell membrane. The significance of this pathway in controlling the entry of dissolved mineral ions is discussed in Chapter 14. Eventually the absorbed water reaches the xylem elements of the root from where it moves upwards into the veins of the shoot system.

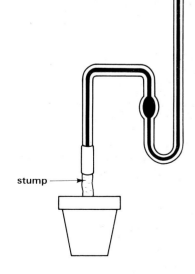

Fig 12.19 A root pressure manometer

mercury column
pushed up by
sap exuding
from stump

stump

12.4 Root pressure

Few plant physiologists would challenge the role of transpiration in the movement of water through terrestrial plants. Yet there are other forces at work in plants which can take part in the flow of water. If, for example, a glass tube is attached to the stump of a freshly decapitated plant such as a vine, sap is pushed up the tube to a height of several metres (Fig 12.19).

Clearly the flow of sap from the stump cannot be linked with transpiration. The force responsible is called **root pressure**, a form of pressure potential which raises the water potential of the root sap. In some species, notably low-growing plants such as grasses, root pressure is sufficiently strong at times to cause exudation of liquid water from special porous structures called **hydathodes** located at the edges of the leaves (Fig 12.20). This phenomenon is called **guttation**. It is therefore possible that the flow of water through plants may sometimes be the combined result of the pulling effect of transpiration and the pushing effect of root pressure. However, the available evidence suggests that the forces generated by transpiration are usually more important.

Root pressure is most evident in conditions favouring rapid water absorption but limiting transpiration. For example, it is responsible for the rising of sap in the trunks of deciduous trees during spring before the leaf buds have opened. But even in these circumstances the pressure generated is rarely more than 200 kPa, a figure of minor importance compared with the transpiration pull. Furthermore the volume of sap exuded by the stumps of decapitated plants is generally little more than 5% of the volume lost by transpiration from intact plants.

The development of root pressure seems to depend on the active secretion of mineral ions into the root xylem elements from surrounding parenchymatous **transfer cells** (Fig 12.21). The effect of mineral secretion is to lower the water potential of the xylem sap leading to the absorption of water from the soil down a water potential gradient. Respiration by the transfer cells provides the energy required for movement of the ions against a concentration gradient. It is not surprising therefore that root pressure disappears when plant roots are treated with respiratory inhibitors such as potassium cyanide.

Fig 12.20 A hydathode in TS from the edge of a cabbage leaf

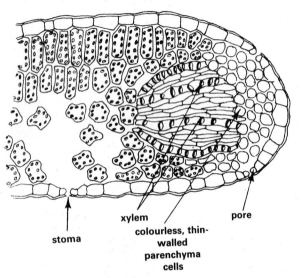

stoma

xylem

colourless, thin-walled parenchyma cells

pore

Fig 12.21 Electronmicrograph of a section through a transfer cell from the pericycle of a root of clover (courtesy Dr. G. Briarty)

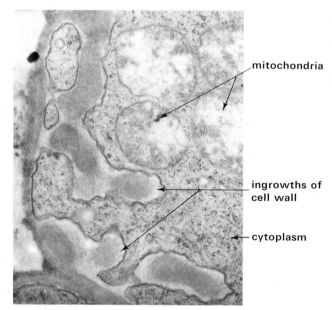

mitochondria

ingrowths of cell wall

cytoplasm

The cell wall ingrowths increase the surface area of the cell membrane for the transfer of solutes. The mitochondria provide the energy for active secretion of solutes

12.5 Tracheids, vessels and capillarity

There is little doubt that xylem is the main water-carrying tissue of higher terrestrial plants. Proof of this comes from experiments in which the xylem elements of leafy shoots are blocked with wax and then given access to water. The leaves soon wilt compared with untreated shoots or where tissues other than xylem are blocked. As wax does not penetrate the lignified walls of xylem elements it is clear that most of the water passes through the cavities of the xylem cells and not up the walls in a wick-like manner. These facts have recently been confirmed using isotopes of water.

The chief water-transporting cells found in flowering plants are **tracheids** and **vessels**. Tracheids were the water-carrying cells of the earliest land plants where they fulfilled the dual role of support and water conduction. They are still the main water-carrying cells in pteridophytes and gymnosperms. In conifers such as the pine, *Pinus*, mature tracheids are very long, narrow, angular cells measuring up to 5 mm long and 30 μm in diameter (Fig 12.22(a)). Stiffening of the walls with lignin and interlocking of the tapered end walls are significant features in the role of tracheids as supporting cells. Water movement in the cavities of tracheids is helped by the lack of protoplasts. The presence of large, bordered pits allows the passage of aqueous solutions between adjacent tracheids.

Fig 12.22 (a) Tracheids and vessels

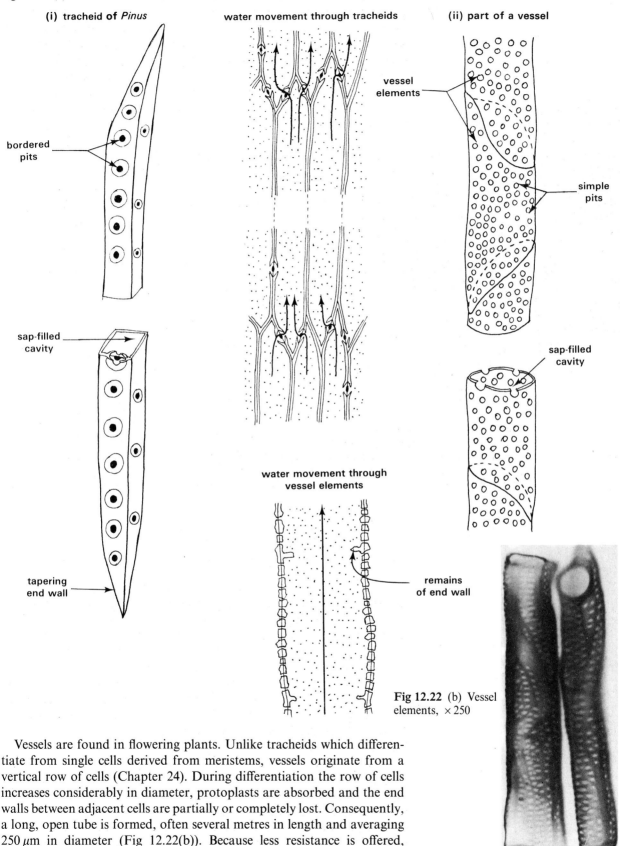

(i) tracheid of *Pinus*

water movement through tracheids

(ii) part of a vessel

bordered pits

vessel elements

simple pits

sap-filled cavity

tapering end wall

sap-filled cavity

water movement through vessel elements

remains of end wall

Fig 12.22 (b) Vessel elements, × 250

Vessels are found in flowering plants. Unlike tracheids which differentiate from single cells derived from meristems, vessels originate from a vertical row of cells (Chapter 24). During differentiation the row of cells increases considerably in diameter, protoplasts are absorbed and the end walls between adjacent cells are partially or completely lost. Consequently, a long, open tube is formed, often several metres in length and averaging 250 μm in diameter (Fig 12.22(b)). Because less resistance is offered, aqueous solutions can move three to six times more quickly through vessels compared with tracheids. As in tracheids, the walls of vessels are thickened with cellulose and hardened with lignin. The extent of wall thickening and lignification varies according to the stage of development of a plant organ when differentiation takes place. Protoxylem vessels which differentiate close to the apices of roots and stems have bands or spirals of strengthening material. This enables the unthickened parts to be enlarged as the organs grow in length. Metaxylem vessels differentiate further back from root and shoot apices where growth in length has ceased. The walls of metaxylem vessels are more extensively thickened and cannot be enlarged.

Water rises in the bore of a fine tube by **capillarity** and plant physiologists have wondered whether capillarity is significant in the ascent of sap through the narrow cavities of xylem elements. Experiments have shown that water can rise by capillarity to a height of 3 m in glass tubes with a bore comparable to the lumen of a narrow tracheid. However, in wider tubes with bores similar to the lumen of a vessel the capillary rise is only a few centimetres. Even allowing for the fact that the many crevices in the walls of the transporting elements add to the capillary force it is generally considered that capillarity does not play an important role in the upward movement of sap through the xylem elements of land plants.

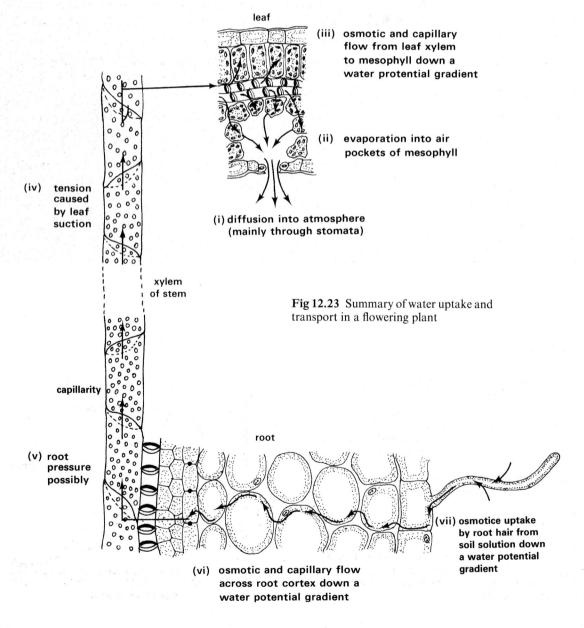

Fig 12.23 Summary of water uptake and transport in a flowering plant

We have seen that the absorption of water and its transport through the body of a flowering plant is the consequence of several phenomena which are summarised in Fig 12.23. It is important to remember that water moves from one part of a plant to another and out into the atmosphere down **water potential gradients**. Heat from the sun is responsible for the largest water potential gradient between the shoot system and the atmosphere. Whereas emphasis in this chapter has been placed on the movement of water it should be mentioned that substances dissolved in the water are also moved simultaneously. The transport of mineral ions, the main category of dissolved substances carried in xylem elements, is described in Chapter 14.

SUMMARY

Terrestrial plants lose water vapour from their leaves and stems by transpiration. Over 90% of the water loss normally occurs through the stomata. Transpiration is the diffusion of water vapour down a water potential gradient through the external tissues of the shoot system.

Climate affects the rate of transpiration which can be measured in the laboratory using a potometer. Temperature, air movement, atmospheric humidity and the duration and intensity of sunlight are the most important factors in this context.

When stomata are open, the epidermis of the leaves of mesophytes offers little barrier to transpiration. In contrast, xerophytes have various structural and physiological features which limit transpiration thus enabling them to resist desiccation in arid environments.

Transpiration stimulates evaporation of water from cells and thus helps prevent overheating of leaves in hot weather. By substantially lowering the water potential of leaf cells transpiration also creates the required water potential gradient for water and dissolved minerals to be drawn through the xylem elements of the plant in the transpiration stream. Vessels are structurally suited to permit a more rapid movement of water than are tracheids.

Roots absorb water by osmosis in response to a lowering of the water potential in their vascular tissue generated by the transpiration stream. Root hairs have a large surface area through which much of the absorbed water is taken in.

QUESTIONS

1 The graph shows the rate of transpiration and rate of water uptake of a sunflower plant at different times of day.

(a) (i) Describe the relation between the two curves. What does this relation suggest?
(ii) Define 'root pressure' and 'transpiration pull' and explain their roles in water uptake and water movement in plants.
(b) Explain how wind, air temperature, light intensity and humidity influence the rate of transpiration. (AEB 1985)

2 (a) What is meant by the term *water potential*?
(b) Describe the path taken by water through an angiosperm plant, from the soil to the atmosphere.
(c) Explain the mechanisms involved in this movement, in terms of water potential. (C)

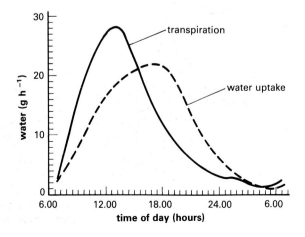

3 (a) There are wide differences in the rates at which the leaves of various plant species lose water. Such differences may be attributable to structural features of the leaf. The following drawings represent sections of a leaf *Fagus* sp. which loses water relatively quickly and another leaf *Hakea* sp. which loses water slowly.

(i) Explain fully how the stomata of *Hakea* sp. allow far less water to pass out of the leaf than do the stomata of *Fagus* sp.

(ii) Suggest how **one** other feature, visible in the diagrams, might act to cut down the rate of transpiration.

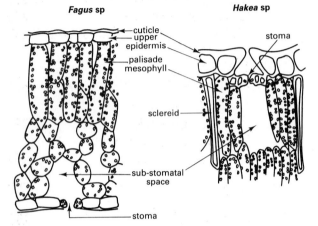

Fagus sp   **Hakea sp**

(b) The sclereids are thick walled, lignified elements. Suggest an explanation for their presence in *Hakea* sp. and their absence in *Fagus* sp.

(c) Which of these two leaves is likely to have the smaller surface area/volume ratio? Give a reason for your choice.

(d) The following figure shows the variation in the rate of water movement in *Fagus* sp.

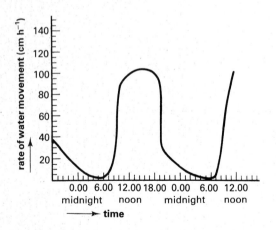

(i) Describe the variation in the rate of water movement over a 24 hour period.

(ii) Give an explanation for this variation.

(iii) Suggest *two* environmental factors, in addition to any given in your answer in (ii) which will influence the rate of transpiration. (WJEC)

4 A dicotyledonous plant shoot was set up as shown in the apparatus below. Movement of a bubble introduced into the capillary tube was recorded over a fixed time period for each of five sets of conditions.

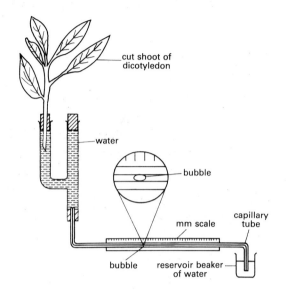

The experiment was carried out under the following environmental conditions, numbered 1–5.
1. Still air, leaves untreated
2. Moving air (using a fan), leaves untreated
3. Still air, leaves vaselined on upper sides only
4. Still air, leaves vaselined on lower sides only
5. Still air, leaves vaselined on upper and lower sides

Temperature and light intensity were held constant throughout the experiment.

(a) State precisely what physiological process the apparatus is designed to measure.

(b) The results are identified by the key letters **V–Z** but they are not necessarily in the order in which the experiment was conducted.

Copy and complete the table below to indicate in each case which of the conditions 1–5, produced the results **V–Z**.

Result	Distance, in mm, moved by bubble after 120 seconds	Conditions
V	3	
W	10	
X	150	
Y	45	
Z	60	

(c) Briefly explain the results of the experiment.

(d) Explain why it is important to keep light intensity constant during the experiment. (L)

13 Photosynthesis and translocation of photosynthetic products

13.1 The leaf as a photosynthetic organ 240

    13.1.1 Penetration of light 241
    13.1.2 Supply of water 241
    13.1.3 Uptake of carbon dioxide 242
    13.1.4 Opening and closing of stomata 244

13.2 Factors affecting the rate of photosynthesis 246

    13.2.1 Temperature 247
    13.2.2 Light intensity 248
    13.2.3 Concentration of carbon dioxide 249

13.3 Productivity of plants and plant communities 249

    13.3.1 Leaf area index 250
    13.3.2 Unit leaf rate 250
    13.3.3 Net primary productivity 250

13.4 Translocation of photosynthetic products 251

    13.4.1 The pathway of translocation 251
    13.4.2 The mechanisms of translocation 256

Summary 259

Questions 259

13 Photosynthesis and translocation of photosynthetic products

Animals and green plants need a supply of raw materials and a source of energy to synthesise the many complex organic substances from which protoplasm is made. Animals get their energy from organic substances in their diets. Green plants obtain their energy from sunlight. The process whereby green plants use solar energy to produce organic molecules is called **photosynthesis**. Carbon dioxide and water are the raw ingredients for the process, which can be summarised simply in the following equation:

$$\text{energy} + CO_2 + H_2O \rightarrow (CH_2O) + O_2$$

from sunlight   carbon dioxide   water   complex organic molecules   oxygen

Photosynthesis is of the greatest importance to all living organisms. The organic substances made in photosynthesis can later be used by animals and other non-photosynthetic organisms. Thus in all ecosystems green plants are the **producers** on which the consumer organisms depend for energy and raw ingredients. On a global scale carbon dioxide is used in photosynthesis at a rate which more or less balances its output from respiration and the burning of fossil fuels such as coal and oil. What is more, the release into the environment of oxygen made in photosynthesis compensates for the uptake of oxygen by living organisms for respiration. Photosynthesis therefore helps to maintain an equilibrium or **steady state** in the environment.

Details of the way in which sunlight is converted into chemical bond energy in green plants are given in Chapter 5. In this chapter we shall examine how the structure of terrestrial plants enables them to obtain the raw materials and sunlight for photosynthesis. We shall also look at the ways in which various factors affect the rate of photosynthesis and what happens to photosynthetic products.

13.1 The leaf as a photosynthetic organ

The main organs of photosynthesis in terrestrial plants are the **leaves**. It is in the leaves that **chloroplasts** are present in abundance. If they are to function efficiently chloroplasts must have an adequate supply of **water** and **carbon dioxide**. They must also receive **light** of suitable wavelength and intensity. The structure of the leaves of mesophytes (Chapter 12) enables the photosynthetic requirements to reach the chloroplasts efficiently.

13.1.1 Penetration of light

The leaves of mesophytic plants are broad, thin, flattened structures. They thus have a relatively large surface area through which light can pass. The thinness of leaves means that light reaching the leaf surface has to pass only a short distance before reaching the mesophyll tissue where the chloroplasts are mainly located. Penetration of light to the mesophyll tissue is helped by the **transparency** of the leaf epidermis (Fig 13.1(a)).

Inside chloroplasts the photosynthetic pigments are spread out in thin layers (Chapter 6). This arrangement presents a relatively extensive surface area for the absorption of sunlight. Chloroplasts can also change their positions to make the best use of available light (Fig 13.1(b)).

Fig 13.1 (a) Passage of light through the outer surface of a leaf

Fig 13.1 (b) Movement of chloroplasts in response to light intensity

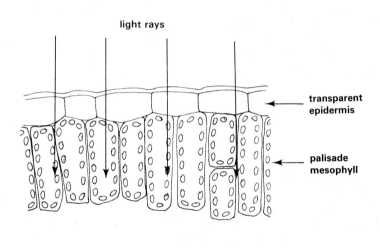

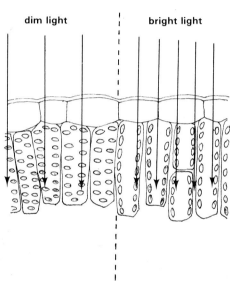

Fig 13.2 Leaf mosaic of sycamore

The leaves of many plants grow so that the leaf blade is usually at 90° to the sun's rays. Shading by leaves on the same shoot is often avoided because the leaves are normally arranged in a **mosaic** (Fig 13.2). These adaptations help to ensure that the maximum amount of available sunlight is received by each unit area of leaf surface. Leaves which are shaded, **shade leaves**, have a larger surface area than those exposed to full sunlight, **sun leaves**. Shade leaves are also thinner and have more chloroplasts. Because of this shade leaves make efficient use of the dim light they receive.

13.1.2 Supply of water

Water is carried into a leaf through a main vein in the midrib. An extensive network of minor veins arising from the main vein penetrates the mesophyll tissue (Fig 12.1). In an oak leaf for example, each mm^2 of leaf area contains a total length of about 10 mm of vein. Under normal circumstances the **xylem elements** of the veins maintain a continuous flow of water to the leaf tissues (Chapter 12). Another important function of the leaf veins is to carry away the organic products of photosynthesis to other parts of the plant. Phloem elements in the veins are responsible for this task (section 13.4).

13.1.3 Uptake of carbon dioxide

The waxy cuticle covering the epidermis of leaves is permeable to carbon dioxide in some plants. In other species carbon dioxide reaches the mesophyll tissue mainly through the **stomata**. There are three main patterns of stomatal distribution in leaves. In many mesophytic plants

Fig 13.3 (a) (i) Surface view of stoma of privet

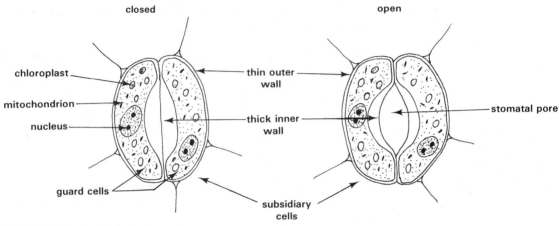

Fig 13.3 (a) (ii) TS stoma of privet

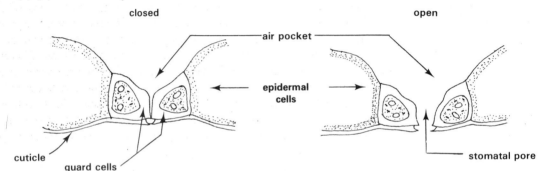

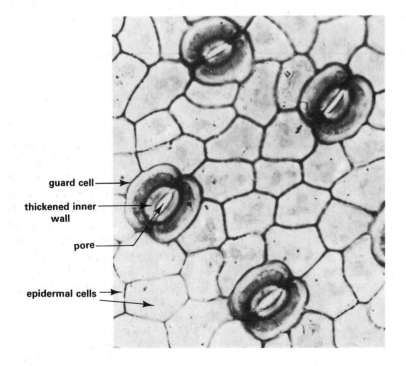

stomata are confined to the lower epidermis. The leaves of some meso-phytes have a few stomata on the upper epidermis too. Equal numbers of stomata occur on both surfaces in many monocotyledonous species. Each stoma consists of a pair of **guard cells** between which a **pore** is formed when the stoma is open. The walls of the guard cells are unevenly thickened. The inner walls where the two cells face each other is much thicker than the outer walls (Fig 13.3). The guard cells are nucleate and several small chloroplasts are often present in the dense cytoplasm.

Fig 13.3 (b) Surface view of open stomata on the leaf epidermis of a dicoty-ledonous plant, × 300

It might be expected that the tiny stomatal pores should offer considerable resistance to the passage of carbon dioxide from the surrounding air into the mesophyll tissue. However, experiments carried out by Brown and Escombe at the end of the last century demonstrate that gases can pass quickly through small pores (Table 13.1). The data shows that for a given area of pore, carbon dioxide flows more rapidly through small pores than through large pores. Stomatal pores are on average 10–30 μm long and 5–10 μm wide when fully open. The rate of gas flow per unit area through stomatal pores is much more rapid than that shown for the smallest pores used in the experiments.

Table 13.1 Diffusion of carbon dioxide through pores of different diameter (after Brown and Escombe, 1900)

Diameter of pore (mm)	cm^3 of CO_2 diffusing h^{-1}	cm^3 diffusing $h^{-1} cm^{-2}$ of pore
22.70	0.24	5.9×10^{-4}
12.06	0.10	8.8×10^{-4}
6.03	0.06	21.0×10^{-4}
3.23	0.04	49.0×10^{-4}
2.00	0.02	64.0×10^{-4}

Plot the data given in Table 13.1 as a graph. Use the graph to calculate the rate at which carbon dioxide passes through pores 0.2 mm in diameter. Pores of this size are at least ten times larger than stomatal pores.

Further experimental investigations by Brown and Escombe showed that the rate of diffusion of gases through porous membranes is also affected by the closeness of the pores (Table 13.2). It may therefore be anticipated that the uptake of carbon dioxide by a leaf to some extent depends on the number of stomata per unit area of leaf. Stomatal densities of between 50 and 200 per mm^2 of leaf area are commonly found on the leaves of mesophytes. The combined area of open pores is usually between 0.5 and 1.5 % of the total leaf area. Experiments and observations such as these suggest that atmospheric carbon dioxide enters the air spaces of leaves when the stomata are open almost as though there is no epidermis present at all.

Table 13.2 Diffusion of carbon dioxide through membranes with different densities of pores (after Brown and Escombe, 1900)

No. of pores cm^{-2} of membrane	% area perforated	Diffusion as % of value without membrane
100.00	11.3	87.6
25.00	2.8	63.7
11.11	1.25	44.0

The mesophyll tissue is permeated by air-filled spaces which amount to 30–40 % of the total leaf volume in many species. The internal **air spaces** are very important because they allow carbon dioxide taken in from the atmosphere to diffuse rapidly to photosynthesising cells. If the carbon dioxide had to diffuse through water instead it would move 10 000 times less quickly. However, once inside the mesophyll cells carbon dioxide has to move through the aqueous cell solution to reach the chloroplasts. As the chloroplasts are situated just inside the cell membrane the distance that carbon dioxide passes through water is minimal. Uptake of carbon dioxide by the mesophyll cells is enhanced by the enzyme **carbonic anhydrase** which catalyses the formation of carbonic acid:

$$CO_2 + H_2O \underset{}{\overset{\text{carbonic anhydrase}}{\rightleftharpoons}} H_2CO_3$$

carbon dioxide   water   carbonic acid

In the air spaces of an actively photosynthesising leaf the concentration of carbon dioxide is lowered to 0.01 % compared with 0.03 % in the external air. The result is that atmospheric carbon dioxide diffuses down a **concentration gradient** through stomatal pores into the air spaces of the mesophyll tissue.

13.1.4 Opening and closing of stomata

The stomata of most terrestrial plants open in light and close in darkness. Water loss due to transpiration is thus confined to periods when the plant is obtaining carbon dioxide from the surrounding air. Conservation of water is essential in a terrestrial environment because of the dehydrating effect of the atmosphere. Some xerophytic plants open their stomata only at night when the drying power of the air is less intense (Chapter 12).

Changes in turgidity of the guard cells bring about the opening and closing of stomatal pores. When the guard cells are turgid the pores are open. The guard cells of closed stomata are flaccid. Stomata close in wilted plants because the guard cells lose turgidity following excessive water loss by evaporation. The reasons for changes in the turgidity of guard cells of non-wilted plants, however, are still a matter for debate. The factors which are thought to be mainly responsible include photosynthesis by the guard cells, carbon dioxide concentration of the air inside leaves and the concentration of mineral ions in the guard cells.

1 Photosynthesis by the guard cells

The earliest theories state that the formation of sugars by photosynthesis in the guard cells is the key factor causing stomata to open. The water-soluble sugars lower the water potential of the guard cells causing water to enter by osmosis from surrounding epidermal cells down a water potential gradiant. When exposed to light therefore the guard cells gradually increase in turgidity and the stomata open. However, not all guard cells contain chloroplasts. Think of those on the non-green parts of variegated leaves. Even guard cells with chloroplasts are unlikely to make sugars quickly enough to bring about the required decrease in water potential necessary for opening stomata.

2 Carbon dioxide concentration inside leaves

Guard cells are sensitive to changes in the concentration of carbon dioxide in the air around them. High concentrations of carbon dioxide cause stomata to close, low concentrations cause them to open. What is more, there are fluctuations in the concentration of carbon dioxide in and around leaves each day. In darkness, when photosynthesis stops, respiratory carbon dioxide accumulates. During the daytime, however, carbon dioxide is used for photosynthesis. The concentration of carbon dioxide inside the leaf is then much lower than at night. But how can changes in carbon dioxide concentration bring about changes in turgidity of the guard cells? After a period of darkness guard cells often contain starch grains which are changed to sugar on exposure to light. Could it be that the conversion of starch to sugar is affected by carbon dioxide concentration, and if so how?

Leaf mesophyll cells contain the enzyme **carbonic anhydrase** which catalyses the reaction between carbon dioxide and water to form carbonic acid. The acid dissociates weakly into hydrogencarbonate (HCO_3^-) and hydrogen (H^+) ions:

$$H_2CO_3 \rightleftharpoons HCO_3^- + H^+$$

carbonic acid   hydrogencarbonate ion hydrogen ion

A high concentrations of hydrogen ions therefore accumulate in leaf cells, guard cells included, in darkness. When light is available the carbonic acid is used for photosynthesis thereby lowering the hydrogen ion concentration $[H^+]$ in leaf cells during the daytime. But how are $[H^+]$ changes in leaf cells linked with changes in turgidity of the guard cells?

The starch-sugar conversion in guard cells is catalysed by a group of enzymes. Like most enzyme-catalysed reactions, the conversion is affected by pH. In acid conditions, high $[H^+]$, the enzymes convert glucose phosphate to starch. However at a lower $[H^+]$ the direction of the reaction is reversed and sugar accumulates in the guard cells:

$$\text{starch} \underset{\text{high } [H^+]}{\overset{\text{low } [H^+]}{\rightleftharpoons}} \text{glucose phosphate}$$

The daytime accumulation of sugar lowers the water potential of the guard cells. The guard cells take in water down a water potential gradient from nearby epidermal cells and become turgid. In darkness the sugar is converted to starch which, being insoluble in water, causes the water potential of the guard cells to be raised. Water now passes from the guard cells into surrounding epidermal cells down a water potential gradient. The resulting flaccidity of the guard cells closes the stomatal pores. The combined effects of light and carbon dioxide concentration on stomatal movements are shown in Fig 13.4.

Fig 13.4 Effect of light and CO_2 concentration on stomatal movement

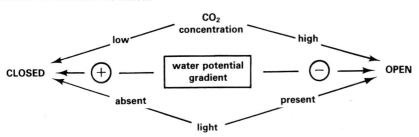

Plausible as this explanation is, it does not account for recent observations showing that there are fluctuations in the concentrations of certain mineral ions in guard cells during the opening and closing of stomata.

3 Mineral ion concentration of guard cells

Increases in mineral ion content of the guard cells, especially of potassium ions (K^+), accompany stomatal opening (Fig 13.5). ATP is produced rapidly in guard cells exposed to light. It is thought that the ATP may be used to work an **ion pump** which draws in K^+ ions from surrounding epidermal cells. The accumulation of K^+ ions could create the necessary water potential gradient which would lead to the increase in turgidity of the guard cells and stomatal opening.

Fig 13.5 Relative amounts of potassium (continuous line) and phosphorus (dotted line) in guard cells of open and closed stomata (after Humble and Raschke, 1971)

However, there are simultaneous increases in concentration of other substances in the guard cells of open stomata. One of the substances is malic acid, a product of the breakdown of starch. Some plant physiologists think that an increase of hydrogen ions (H^+) in the guard cells, due to accumulation of malic acid is prevented by exchanging the H^+ ions for K^+ ions from surrounding epidermal cells. The water-soluble malate ions help to lower the water potential of the guard cells leading to stomatal opening.

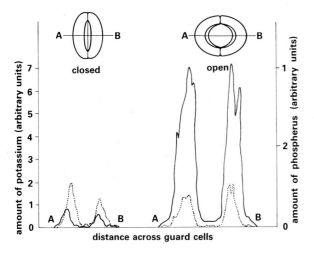

Although stomata have been studied for a very long time it is evident that there is still much to be learned about the mechanisms which control stomatal opening and closing. It may even be that different mechanisms operate in different species or that a combination of mechanisms is involved.

13.2 Factors affecting the rate of photosynthesis

The rate at which oxygen is given off by green plants can be used to measure the rate of photosynthesis. Aquatic plants such as *Elodea* are particularly suitable for this purpose. Bubbles of oxygen are given off from the cut end of the stem of *Elodea* when it is immersed in illuminated pond water or a dilute solution of sodium hydrogencarbonate (Fig 13.6). Using this method in the 1930s the British plant physiologist F. F. Blackmann investigated the effect of light intensity on the rate of photosynthesis.

Fig 13.6 Measuring the rate of photosynthesis of the pondweed *Elodea*. The intensity of light is varied by increasing or decreasing the distance (D) between the lamp and the plant. Light intensity (I) is inversely proportional to the square of the distance $\left(I \propto \dfrac{1}{D^2} \right)$. The results of the experiment are shown in Fig 13.7

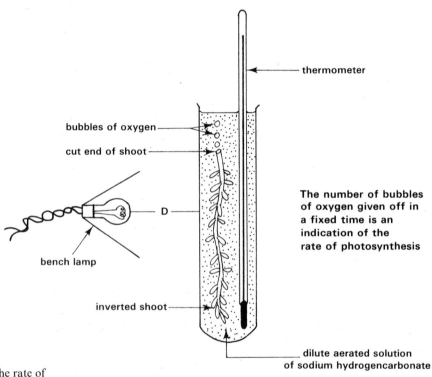

thermometer

bubbles of oxygen

cut end of shoot

D

The number of bubbles of oxygen given off in a fixed time is an indication of the rate of photosynthesis

bench lamp

inverted shoot

dilute aerated solution of sodium hydrogencarbonate

Fig 13.7 Effect of light intensity on the rate of photosynthesis

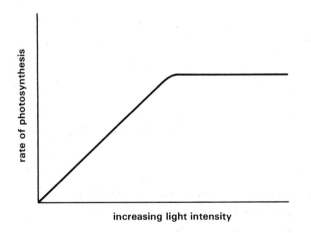

rate of photosynthesis

increasing light intensity

At low intensities of light Blackmann observed a straight line relationship between the rate of photosynthesis and light intensity. However, once a critical light intensity is reached the rate of photosynthesis remains constant. The results surprised Blackmann who reasoned that if photosynthesis is driven by solar energy then it should go faster if more intense light is provided. He concluded that factors other than the availability of light limit the rate of photosynthesis at high light intensities.

Knowing that carbon dioxide is an essential raw ingredient for photosynthesis and that the process is probably catalysed at least in part by enzymes, Blackmann went on to investigate the effects of carbon dioxide concentration and temperature on the rate of photosynthesis. The results of these investigations are shown in Figs 13.8 and 13.9. Clearly photosynthesis is controlled by a combination of factors and it is apparent that the photosynthetic rate is limited by whichever factor is nearest its minimum value. This is a good example of the **law of limiting factors**.

Fig 13.8 Effect of carbon dioxide concentration on the rate of photosynthesis

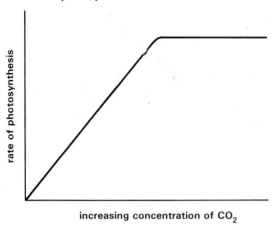

Fig 13.9 Effect of temperature on the rate of photosynthesis

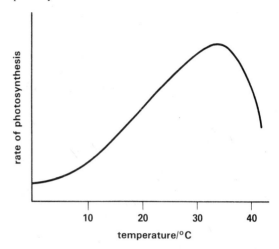

Blackmann did not attempt to find out whether the availability of water affects the rate of photosynthesis. In non-wilted plants there is always sufficient water for photosynthesis. The closure of stomata of wilted leaves may however limit the rate of photosynthesis by preventing the diffusion of carbon dioxide to the mesophyll tissue. In normal circumstances temperature, light intensity and carbon dioxide concentration are the factors most likely to limit the rate of photosynthesis.

13.2.1 Temperature

Photosynthesis consists of light-dependent and light-independent stages (Chapter 5). It is the enzyme catalysed reactions of the **light-independent stage** which are affected by temperature. The temperature coefficient, Q_{10} (Chapter 3), for the light-independent stage is between 2 and 3. Fig 13.10 shows the effect of temperature on photosynthesis at high and low light intensities. Notice that increasing the temperature has no effect on the rate of photosynthesis at low light intensity. Why is this so? For C4 plants the optimum temperature for photosynthesis is between 35–40 °C and for C3 plants it is 20–25 °C.

Fig 13.10 Effect of temperature on the rate of photosynthesis at two intensities of light

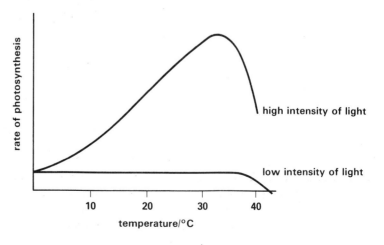

13.2.2 Light intensity

Light reaching the shaded leaves of a plant is less intense than light received by leaves exposed to full sunlight. Shade leaves compensate for this by structural adaptations which enable them to make the most of dim light (section 13.1.1). In some habitats such as dense beech forests and coniferous woodland the shade of the trees prevents the growth of herbaceous plants which normally appear as a field layer in open woodland where light can penetrate to the ground (Fig 13.11).

Fig 13.11 Effect of shading on the field layer of a beech wood. The deep shade cast by beech trees prevents the development of a field layer, so the ground is bare (Bruce Coleman: Eric Crichton)

In darkness photosynthesis cannot take place, although other metabolic processes such as respiration go on just as they do in daylight. In the absence of light, green plants give off carbon dioxide made in respiration. When light is available respiratory carbon dioxide provides a proportion of a plant's photosynthetic needs. The rest of the carbon dioxide is absorbed from the surrounding air. For a plant to grow it has to synthesise organic materials more rapidly than it oxidises them in respiration. When photosynthesis and respiration occur at rates such that there is no gain or loss of organic matter a plant is at its **compensation point**. The time taken for a plant which has been in darkness to reach the compensation point is called the **compensation period**. The most important factor governing the length of the compensation period in natural conditions is light intensity. Shade leaves have shorter compensation-periods than sun leaves (Fig 13.12). This is a physiological adaptation which enables shade leaves to make efficient use of light of low intensity.

Fig 13.12 Effect of light intensity on CO_2 exchange by sun and shade leaves

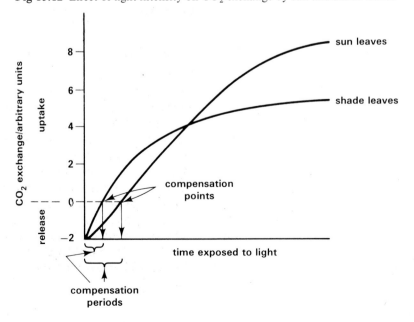

13.2.3 Concentration of carbon dioxide

On a warm sunny day the concentration of carbon dioxide in the air is probably the factor which limits photosynthesis more than any other. Enriching air with carbon dioxide has a significant effect on crop plants grown in greenhouses (Table 13.3). It is now common practice for commercial growers of salad plants to raise the concentration of carbon dioxide of the air in greenhouses during daylight. The gas is either pumped in directly or is released by the burning of fuels such as paraffin, propane and natural gas which are used to heat greenhouses. Concentrations of up to 0.2% carbon dioxide produce increases in crop yield which are economically worthwhile.

Table 13.3 Yield of lettuces and tomatoes grown in normal air and in air enriched with carbon dioxide

Crop	Without added CO_2	With added CO_2	Yield
lettuces	0.9	1.1	Fresh mass/kg per 10 heads
tomatoes	4.4	6.4	Fresh mass/kg per plant

Of course, increasing the concentration of atmospheric carbon dioxide for crops grown in the open air is impracticable. On average the amount of carbon dioxide in the atmosphere outdoors is 0.03%. It is controlled mainly by the rate at which living organisms give off carbon dioxide from respiration compared with the rate at which it is used by green plants for photosynthesis. An old-fashioned way of improving crop production in greenhouses was to add large quantities of manure to the soil. Carbon dioxide, released by soil micro-organisms which decomposed the manure, accumulated in the air and stimulated photosynthesis resulting in bigger crops.

The Calvin cycle for carbon dioxide fixation (Chapter 5) functions at relatively high CO_2 concentrations. In contrast, the Hatch–Slack pathway enables carbon dioxide to be fixed in C4 species even when the air in the intercellular spaces of their leaves contains very little of the gas. In the mesophyll cells of such plants the Hatch–Slack pathway fixes carbon dioxide, and the malic acid so formed is then transported to sheath cells surrounding minor veins. There it releases carbon dioxide which is fixed in the Calvin cycle. This metabolic adaptation is one of the reasons why C4 plants such as maize and sugar cane grow more quickly than C3 species when exposed to strong sunlight.

13.3 Productivity of plants and plant communities

So far we have concentrated on the leaves of terrestrial plants as photosynthetic organs. It is just as important to consider photosynthesis in whole plants. After all, the entire plant is potential food for consumer organisms. Plants normally grow with others of the same species or of different species in **plant communities**. A field of wheat is an example of a man-made plant community. Natural forest and woodland are plant communities not made by man. The efficiency with which whole plants and plant communities produce dry matter determines how much food is available for the higher trophic levels in an ecosystem.

It is only recently that biologists have begun to appreciate which factors cause some plants and plant communities to produce dry matter more efficiently than others. Studies on crop plants show that two factors of fundamental importance to crop yield are **leaf area index** and **unit leaf rate**.

13.3.1 Leaf area index

Plants with a large surface area of leaves and other parts which can photosynthesise may be expected to produce more dry matter than plants having shoot systems with a small surface area. The area of leaves available for photosynthesis can be expressed as the **leaf area index (LAI)**:

$$LAI = \frac{\text{total leaf area of plant}}{\text{area of ground covered by plant}}$$

It determines the amount of light intercepted by the shoot system of a plant. During the early stages of growth, crop plants have small LAI values because each plant has only a few small leaves and is surrounded by a patch of bare ground. As growth proceeds and the shoot system enlarges, the LAI increases. Maize grown in the U.S.A. achieves a maximum LAI of 4. Some crop plants have larger values than this. Sugar cane for example has a maximum LAI of about 7.

The shape of the shoot system is particularly important in determining the leaf area index of a plant. Plants which can be grown close to each other and which have leaves held vertically have higher LAI values than those with horizontally-held or drooping leaves. With this in mind plant breeders have produced new varieties of wheat and rice with erect leaves. The new varieties grow so that there is little mutual shading of their leaves and they yield extremely good crops of grain.

13.3.2 Unit leaf rate

Whatever the LAI value, increases in organic matter occur efficiently only if most of the photosynthetic products are converted to plant tissue or storage materials. If most of the products of photosynthesis are respired dry matter accumulates slowly.

The term **unit leaf rate (ULR)** expresses the efficiency of dry matter accumulation by green plants. ULR is the mean rate of increase in dry mass of a whole plant per unit area of leaf. The ULR of a plant can be calculated from measurements of the leaf area and dry mass of a representative sample of plants at different stages of growth. Most species of crop plants grown in temperate areas have mean ULR values of between 4 and $6 \, g \, m^{-2} \, day^{-1}$. In ideal growing conditions, when there are no factors limiting the rate of photosynthesis, the ULR can reach $15 \, g \, m^{-2} \, day^{-1}$. Even so, some species have higher ULR values than others. Because they do not photorespire and have very short compensation periods C4 plants such as sugar cane and maize have a much greater unit leaf rate than most C3 plants.

13.3.3 Net primary productivity

Synthesis of dry matter by green plants is called **primary production**. The total amount of dry matter produced per unit area of ground per year is called **gross primary productivity**. Some of the dry matter is used by green plants in respiration. What is left is called **net primary productivity (NPP)** and it is this which is available for consumer organisms including man.

$$NPP = LAI \times ULR$$

Clearly plants which quickly achieve high LAI values and which sustain an efficient ULR over a long growing period are highly productive. In attempting to raise the amount of food for human consumption plant breeders now have such factors foremost in mind.

13.4 Translocation of photosynthetic products

Leaves can be thought of as a **source** of photosynthetic products. Although leaves can store organic compounds such as starch, the products of photosynthesis are generally exported fairly quickly to tissues which actively use them or where they can be stored. Such places are called **sinks**. The meristems of root and shoot apices and the vascular and cork cambium (Chapter 24) are among the more important sinks for photosynthetic products. It is there that the organic materials made by leaves are used to produce new cells and provide the energy for growth. Most green plants have some means of storing food which can be used at a later stage. The cotyledons and endosperm of seeds are the only sites of food storage in many species (Chapter 23). In others, an enlarged part of the vegetative plant body acts as a storage organ (Fig 13.13).

The long distance movement of metabolites from sources to sinks is called **translocation**. The term translocation is often used to describe the movement of all water-soluble substances in the vascular tissue of plants. Here we shall confine our account mainly to the movement of organic substances made in photosynthesis.

Fig 13.13 Storage organs of flowering plants

bulb of onion

tap root of carrot

stem tuber of potato

13.4.1 The pathway of translocation

There is a lot of experimental evidence which suggests that the **phloem** is the tissue mainly concerned with the translocation of organic substances. An elegant technique applied to translocation studies makes use of the fact that phloem elements are penetrated by the mouthparts of **aphids** such as greenfly and blackfly when they are feeding on plant juices (Fig 13.14). If a

Fig 13.14 (a) Aphids feeding on a rose shoot

Fig 13.14 (b) Ventral view of aphid × 50

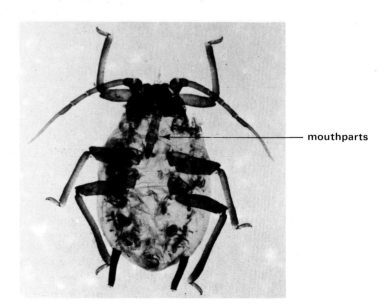

mouthparts

feeding aphid is anaesthetised and the mouthparts carefully cut across, the stylets left in the plant act as a tiny sampling tube at the end of which droplets of phloem sap soon appear (Fig 13.15). Analyses show that the phloem sap, unlike xylem sap, is rich in sugars, mainly sucrose, amino acids and potassium and phosphate ions (Table 13.4). Furthermore, the composition of the sap varies according to the photosynthetic activity of the plant. Sucrose for example is at its highest concentration a few hours after sunrise.

Fig 13.15 (a) Aphid feeding on phloem sap (modified after Dixon, 1973)

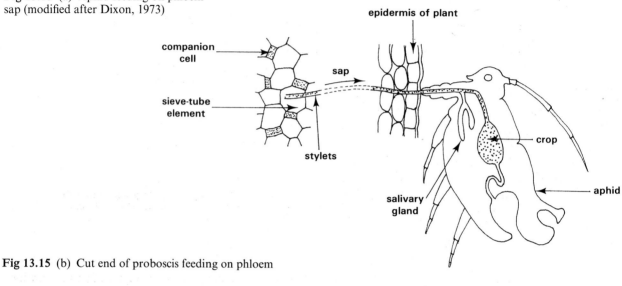

Fig 13.15 (b) Cut end of proboscis feeding on phloem

Table 13.4 Composition of phloem sap of the castor oil plant *Ricinus communis* (after Hall and Baker 1972)

Component	Concentration/mg cm^{-3}
sucrose	80–106
protein	1.45–2.20
amino acids	5.2 (as glutamic acid)
carboxylic acids	2.0–3.2 (as malic acid)
phosphate ions	0.35–0.55
sulphate ions	0.024–0.048
chloride ions	0.355–0.675
hydrogencarbonate ions	0.010
potassium ions	2.3–4.4
sodium ions	0.046–0.276
calcium ions	0.020–0.092
magnesium ions	0.109–0.122
ammonia	0.029
ATP	0.24–0.36
auxin	10.5×10^{-6}
gibberellin	2.3×10^{-6}
cytokinin	10.8×10^{-6}

Fig 13.16 Investigating the spread of radioactive carbon in a plant

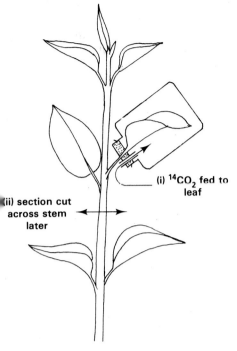

(i) $^{14}CO_2$ fed to leaf

(ii) section cut across stem later

Another technique which has been used to great advantage in recent years involves supplying carbon dioxide containing the **radionuclide** ^{14}C to illuminated plant leaves. The isotope is fixed in the organic products of photosynthesis which are then translocated to other parts of the plant. The spread of radioactivity can be followed and is found almost entirely in the phloem (Fig 13.16).

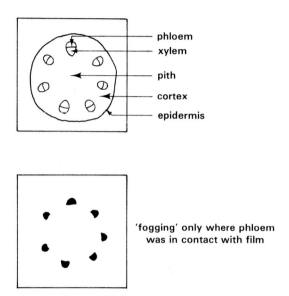

phloem
xylem
pith
cortex
epidermis

'fogging' only where phloem was in contact with film

Explain the difference in the results of the analysis of phloem in ringed shoots shown in Fig 13.17.

How did plant physiologists manage to learn anything about translocation before such sophisticated techniques were developed? They relied mainly on **ringing experiments** in which cylinders of bark were removed from woody stems and the contents of the phloem above and below the cylinder were later analysed (Fig 13.17). Although crude, the ringing method provided some evidence that phloem is the main tissue in which organic compounds are translocated.

Fig 13.17 A ringing experiment

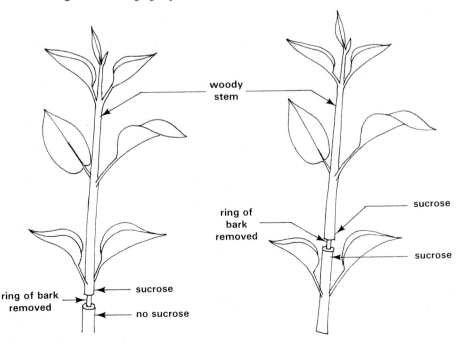

woody stem

ring of bark removed

sucrose

sucrose

ring of bark removed

sucrose

no sucrose

In flowering plants there are two conspicuous types of phloem cell concerned with translocation, sieve-tubes and companion cells. They lie alongside one another, derived from the same meristematic cell by longitudinal division (Chapter 24).

1 Sieve tubes

Fig 13.18 (a) LS of phloem as seen with light microscope

Sieve tubes are composed of a series of **sieve-tube elements** joined end to end (Fig 13.18). Each element is 10–50 μm in diameter and 150–1000 μm long. When first produced from a meristematic cell by mitosis, sieve-tube elements have a large nucleus and the usual cytoplasmic organelles.

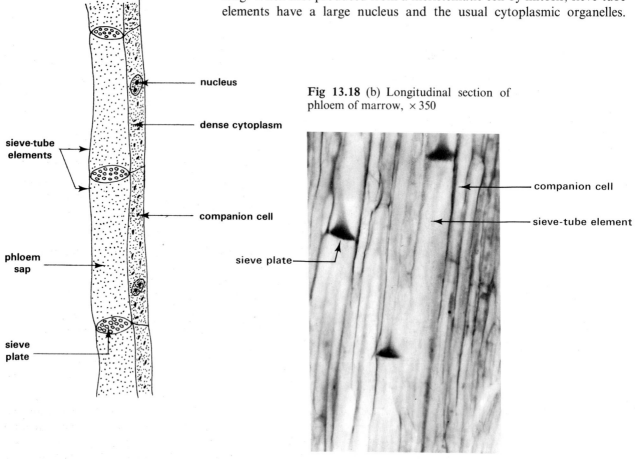

nucleus

dense cytoplasm

sieve-tube elements

companion cell

phloem sap

sieve plate

Fig 13.18 (b) Longitudinal section of phloem of marrow, × 350

companion cell

sieve-tube element

sieve plate

Fig 13.18 (c) Transverse section of phloem of marrow, × 800

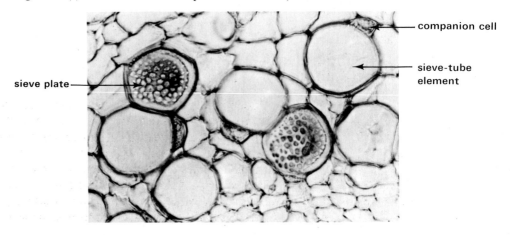

sieve plate

companion cell

sieve-tube element

However, at maturity, when translocating organic materials rapidly in solution, the nucleus, ribosomes, Golgi body and tonoplast degenerate and the side walls are thickened mainly with cellulose. Small mitochrondria remain, as does part of the endoplasmic reticulum which exists as parallel layers near the side walls. Plastids containing small starch grains are also present and the lumen is filled with slimy sap containing fibrils of a material called **phloem protein** (Fig 13.19).

Fig 13.19 Ultrastructure of phloem (based on electronmicrographs)

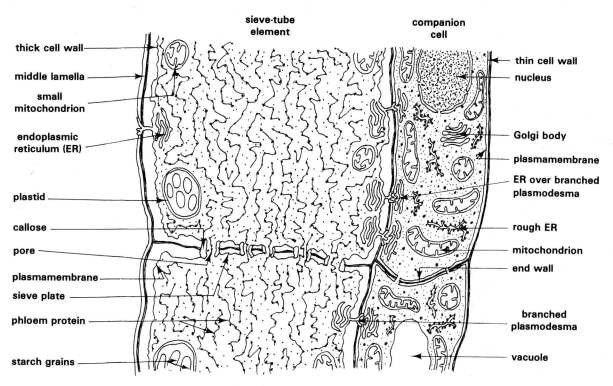

During differentiation the end walls between adjacent sieve-tube elements are modified to form **sieve plates** perforated by pores. The pores are lined with deposits, which vary in thickness, of a glucose polymer called **callose**. The mean diameter of the pores is 2–6 μm but where thick deposits of callose occur the pores appeard to be blocked.

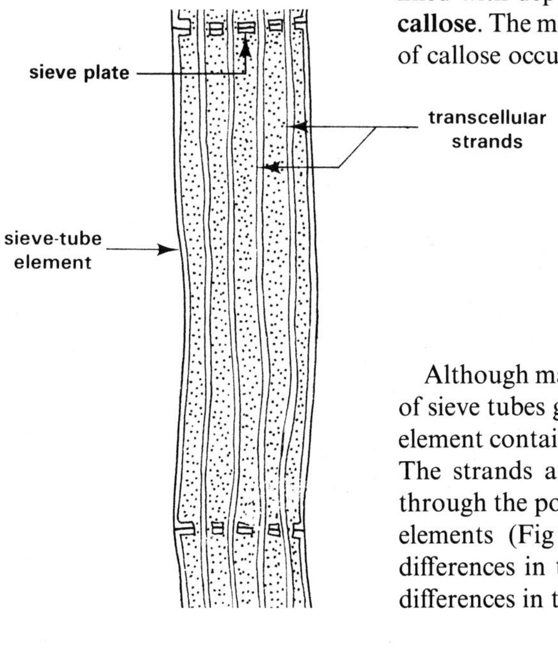

Fig 13.20 Longitudinal section of part of sieve tube showing transcellular strands (based on photomicrographs)

Although many biologists accept the interpretation of the ultrastructure of sieve tubes given above, there are others who state that each sieve-tube element contains a number of parallel **transcellular strands** 1–7 μm wide. The strands are thought to be hollow, membranous tubes which pass through the pores of sieve plates and extend through a series of sieve-tube elements (Fig 13.20). Such a divergence of opinion may be due to differences in the species used for investigation. It may also result from differences in the techniques used to examine sieve tubes.

255

Fig 13.21 Electronmicrograph of a section of part of a transfer cell from lentil, × 20 000.

Note the ingrowths of the cell wall which increase the surface area of the plasma membrane for the transfer of solutes. What functions may the mitochondria have?

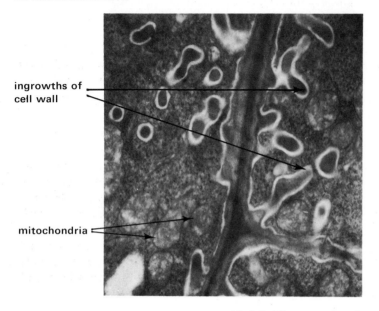

ingrowths of
cell wall

mitochondria

2 Companion cells

A **companion cell** has a thin cellulose wall enclosing a protoplast which contains a prominent nucleus, numerous mitochondria, a well-developed endoplasmic reticulum with attached ribosomes, a Golgi body, small vacuoles and colourless plastids. The protoplasts of companion cells and adjacent sieve-tube elements are connected by many plasmodesmata (Fig 13.19) suggesting that transfer of substances between the two types of cell can occur. At the ends of minor leaf veins the companion cells are much bigger compared with the sieve-tube elements than elsewhere in the plant. These companion cells also have ingrowths of their cell walls which greatly increase the surface area of the plasma membrane for the uptake of substances from surrounding mesophyll cells. The name **transfer cell** is given to specialised companion cells of this sort (Fig 13.21).

13.4.2 The mechanism of translocation

A common method used to determine the rate of translocation in phloem is to measure the rate at which growing organs increase in dry mass. If the area of cross-section of phloem supplying the organ is also measured it is possible to calculate the rate at which organic matter is transferred per unit area of phloem in a known period of time (Table 13.5). Another technique is to follow the spread of radioactivity in a plant after a labelled compound such as $^{14}CO_2$ has been supplied to a photosynthesising leaf. This procedure has been particularly rewarding when used in conjunction with aphids (Fig 13.22). The results of such experiments clearly demonstrate that translocation is a rapid process, much too rapid to be explained by diffusion. Sugars such as sucrose are translocated at a rate of $25-200 \, \text{cm h}^{-1}$ in sieve tubes compared with a meagre $0.2 \, \text{mm day}^{-1}$ possible by diffusion.

Fig 13.22 Use of aphids and radioactive carbon to measure the rate of translocation.

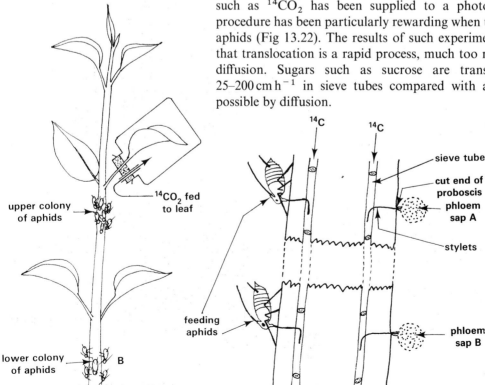

upper colony
of aphids

$^{14}CO_2$ fed
to leaf

lower colony
of aphids

B

feeding
aphids

^{14}C

^{14}C

sieve tube

cut end of
proboscis

phloem
sap A

stylets

phloem
sap B

Phloem sap at **A** and **B** is analysed for radioactivity at frequent intervals. The time taken for the radioactivity to spread from **A** to **B** is a measure of the rate of translocation

Table 13.5 Rates of translocation measured by transfer of dry mass

Species	Organ	Mass transfer/ g dry mass cm^{-2} phloem h^{-1}
potato	tuber	2.1–4.5
marrow	fruit	3.3–4.8
grasses	leaves	4.4–14.9

Having dismissed diffusion as a possible mechanism of translocation it is not easy to provide a really convincing alternative. There are many theories but relatively few facts to support them. One of the most favoured explanations of translocation is the mass-flow hypothesis.

1 Mass-flow hypothesis

Fig 13.23 shows the principle of the **mass-flow hypothesis**. The sap in the sieve-tube elements next to leaf mesophyll cells, the **source**, is rich in organic solutes and thus has a low water potential. Water is absorbed into the sieve-tube elements by osmosis from surrounding tissues. A high pressure potential is thus created near the source and the solution inside the sieve-tube elements is pushed towards a **sink** along a pressure potential gradient. In this way organic solutes can be carried downwards to the roots of the plant or upwards to developing flowers, fruits and young leaves, all of which are sinks (Fig 13.24).

Fig 13.23 Model to demonstrate the principle of the mass-flow hypothesis.

Fig 13.24 Mass-flow of organic solutes in a plant

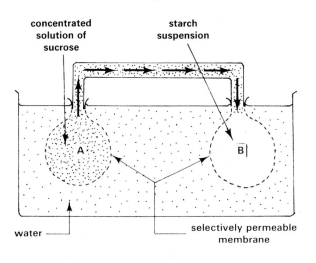

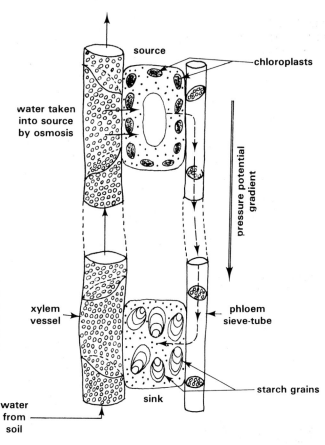

The sucrose solution, having a relatively low water potential, draws water into **A** by osmosis. The resultant rise in pressure potential in **A** causes the sucrose solution to enter the connecting tube and pass into **B**. Starch, being osmotically inactive, does not draw water into **B** from the container. The model could be made into a continuous-flow system if sugar could be continually added to **A** and removed from **B**

Table 13.6 Rates of oxygen uptake by various plant tissues (after Coult 1971)

Tissue	mm³ O_2 consumed $g^{-1}h^{-1}$
vascular tissue	800
whole leaves	400
petioles	200
carrot taproot	30–40

The rapid and continued exudation of phloem sap from the severed mouthparts of feeding aphids shows that the contents of the sieve-tube elements have a high pressure potential. Furthermore, suitable concentration gradients of translocated substances exist between sources and sinks. Even so, the hypothesis does not account for all that is known of the physiology of phloem. There are doubts too that the structure of sieve tubes in all plants is suitable for mass-flow. The main points of criticism of mass-flow are as follows:

i. Phloem tissue has a relatively high rate of oxygen consumption (Table 13.6) and translocation is slowed down or stopped altogether if respiratory poisons such as potassium cyanide enter the phloem. Observations of this sort suggest that metabolic energy is used in translocation. The companion cells produce energy because they contain numerous mitochondria. Yet the mass-flow hypothesis is based on a passive, physical phenomenon and does not suggest a role for companion cells.

ii. It has recently been shown that plant hormones such as indoleacetic acid (IAA) assist the loading of sugars into sieve-tubes and their unloading into sinks. The mass-flow hypothesis makes no mention of hormones in translocation.

iii. The sieve plates may offer a resistance which is greater than could be overcome by the pressure potential of the phloem sap. Higher pressure potentials than have been recorded would be necessary to squeeze sap through partially blocked pores in sieve plates. However, in view of the controversy centred around sieve plate structure this criticism may not be valid for all plants.

Fig 13.25 Alternative mechanisms of translocation

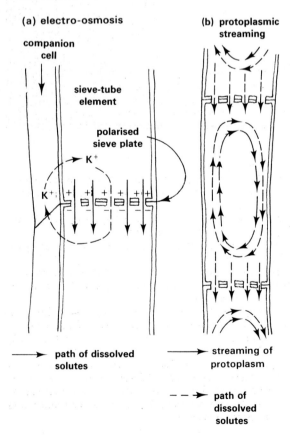

(a) electro-osmosis

companion cell

sieve-tube element

polarised sieve plate

K^+

K^+

→ path of dissolved solutes

(b) protoplasmic streaming

→ streaming of protoplasm

--→ path of dissolved solutes

2 Alternative hypotheses

There has been no shortage of alternative ideas about mechanisms of phloem transport. One suggestion is that respiratory energy produced by companion cells maintains an electrical potential difference across sieve plates. The potential difference is achieved by the active removal of potassium ions (K^+) from one side of the plate by the companion cells and their secretion on the other side. The movement of K^+ ions through the pores of the sieve plate rapidly draws molecules of water and dissolved solutes through the pores. This phenomenon is called **electro-osmosis**. Experimental evidence to support the theory is sparse. Even so it is known that potassium ions stimulate the loading of phloem with sugars. Furthermore, potassium uptake is promoted by IAA.

Circulation of protoplasm has been observed in sieve-tubes and some physiologists have thought that this may be involved in translocation. It is unlikely though that **protoplasmic streaming** on its own would account for the speed at which substances are normally translocated.

Finally, supporters of the existence of transcellular strands have proposed that solutions may be pumped through the strands by **peristaltic waves** of contraction passing along the strands!

The various ideas embodied in the alternative hypotheses are summarised in Fig 13.25. There is obviously much debate surrounding the structure and physiology of phloem. At

Fig 13.25 (contd.)

(c) peristalsis in transcellular strands

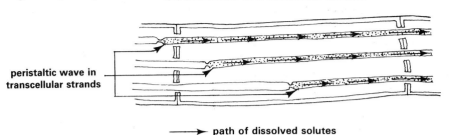

peristaltic wave in
transcellular strands

⟶ **path of dissolved solutes**

present, plant physiologists do not have a complete understanding of what appears to be a sophisticated and subtle mechanism of translocation.

SUMMARY

Green plants are heterotrophes, making a range of organic molecules by photosynthesis. They are the producers on which all other organisms in any ecosystem depend as a source of energy. Water and carbon dioxide are the raw materials required for photosynthesis which uses energy from sunlight. An important by-product of photosynthesis is gaseous oxygen which the majority of living organisms use in respiration.

The thin, broad leaves of mesophytes are structured such as to absorb efficiently solar energy and take in carbon dioxide from the air. They are often positioned in a mosaic to minimise overlap. Leaves which grow permanently in shade have more chloroplasts than sun leaves. They are also thinner and have a larger surface area thus enhancing light absorption.

Carbon dioxide for photosynthesis diffuses from the air into terrestrial plants mainly through leaf stomata. The rate of stomatal diffusion of carbon dioxide is rapid and enhanced by the enzyme carbonic anhydrase in the mesophyll cells. In mesophytes stomata usually open during daylight hours. Biologists do not agree as to the precise causes of stomatal opening and closing. One explanation is based on changes in carbon dioxide content in the air; another on fluctuations in the concentration of potassium ions in the guard cells of stomata.

The rate of photosynthesis is affected by temperature, light intensity and carbon dioxide concentration in the environment. Leaf area index and unit leaf rate are factors which have a fundamental effect on the rate at which the products of photosynthesis accumulate in plants.

Photosynthetic products are moved by translocation in phloem sieve tubes from leaves to other parts of the plant. The mass-flow hypothesis is favoured by most plant physiologists as offering the best explanation to date of the mechanism of translocation.

QUESTIONS

1 The table shows the rates at which carbon dioxide is taken up ($+$) and released ($-$) from the stem of a herbaceous plant and from a single leaf of the same species at different light intensities.

Light intensity (arbitrary units)	Uptake ($+$) and release ($-$) of carbon dioxide (mg $50\,cm^{-2}\,h^{-1}$)	
	Stem	Leaf
0.0	-0.5	-0.5
1.0	-0.2	$+0.6$
2.5	$+0.3$	$+2.8$
4.0	$+0.8$	$+4.6$
5.0	$+1.0$	$+5.3$
7.0	$+1.6$	$+6.0$
11.0	$+2.5$	$+6.3$

(a) Graph the data on a single set of axes.
(b) (i) Explain the term *compensation point*.
 (ii) What is the compensation point of this stem?

(iii) Calculate the rate at which carbon dioxide is used in photosynthesis by $50\,cm^2$ of the leaf at a light intensity of 1 arbitrary unit.
(c) Explain the shape of the graph line for the leaf above a light intensity of 5 arbitrary units.
(d) Explain, in terms of anatomical and physiological factors, why
 (i) the leaf takes up carbon dioxide faster than the stem;
 (ii) the leaf and the stem release carbon dioxide at the same rate in darkness.
(e) Suggest **three** practical difficulties you would meet in conducting an experiment to obtain data of the kind given in the table. (AEB 1985)

2 A plant is made up of tissues that are either net producers or net consumers of carbohydrates. Mature leaves in the light are almost always net producers and are referred to as *sources*. Most other plant tissues are net consumers and are referred to as *sinks*.

259

In an experiment, the rate of photosynthesis of the uppermost leaf of wheat was measured in plants treated as follows:

Plant (1) plants with developing grain removed and all leaves in the light,
Plant (2) plants with developing grain removed and all other leaves, except the uppermost one, kept in the dark,
Plant (3) control (intact) plants.

The results are given in the figure.

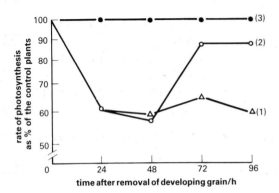

(a) What is the effect of the removal of developing grain on the rate of photosynthesis of the uppermost leaf of wheat when all leaves have been in the light (**Plant (1)**)?
(b) Suggest a reason for this effect.
(c) What is the effect of the removal of developing grain on the rate of photosynthesis of the uppermost leaf of wheat when all the other leaves have been kept darkened (**Plant (2)**)?
(d) Suggest a reason for this effect. (C)

3 The graph below shows how net primary productivity of the marine alga *Hormosira banksii* varies with depth and sea temperature.

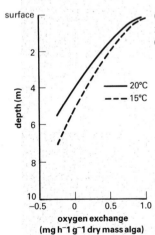

(a) What is meant by the term net primary productivity?
(b) (i) Explain why oxygen exchange can be taken as an indicator of net primary productivity.
(ii) On the graph, oxygen exchange is expressed as $mg\,h^{-1}\,g^{-1}$ dry mass of alga. Describe how you would find the dry mass of a sample of the fresh material in the laboratory.

(iii) Using the graph, state the relationship between net primary productivity and depth. Suggest an explanation for this relationship.
(c) (i) Show clearly on each curve the position of the compensation point. Give the reason for your choice of position.
(ii) Account for the difference in the positions of the compensation points at each temperature.
(iii) Would *H. banksii* be able to survive at greater depths in warmer or colder seas? Explain. (JMB)

4 The apparatus below was used to investigate the translocation of carbohydrate in the pea plant *Pisum sativum*. A chosen leaf was allowed to photosynthesise in the presence of radioactive carbon dioxide for 30 minutes. The plastic bag was then removed from the plant which was left for 24 hours before harvesting. The radioactivity of different parts of the plants was counted.

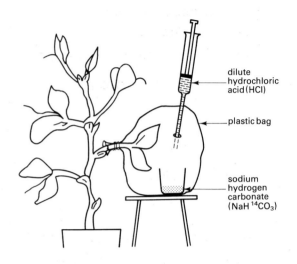

(a) (i) Write an equation to show how the radioactive carbon dioxide was produced.
(ii) State **one** assumption which is made when using radioactive isotopes in this kind of investigation.
(b) (i) Indicate a precaution that should be taken when removing the plastic bag.
(ii) Suggest why the plant was left for 24 hours after the removal of the plastic bag.
(c) This treatment was applied to two fruiting plants. In one, a leaf from the upper part of the stem was fed with radioactive carbon dioxide whereas in the other, the treated leaf was on the lower part of the stem. The amounts of radioactivity found in various parts of the plants are shown below.

Radioactivity (counts min^{-1})

Parts of plants	Plant A (Leaf on upper part of stem treated)	Plant B (Leaf on lower part of stem treated)
Shoot apex	1123	759
^{14}C treated leaf	11325	11372
Untreated leaf	234	168
Stem	816	1160
Pod	9055	4937
Roots	842	2700

(i) Name the **two** parts of plant **A** to which most carbohydrate was translocated.
(ii) Suggest an explanation for his finding,
(iii) What are the main differences between the pattern of radioactive carbohydrate translocation in plants **A** and **B**?
(iv) Give an explanation for these differences.
(d) Describe briefly an experimental method to test the hypothesis that phloem is the pathway for the translocation of ^{14}C labelled carbohydrate. (WJEC)

14 Mineral nutrition of flowering plants

14.1 Mineral requirements of
flowering plants   262

    14.1.1 Water culture experiments   263

14.2 Functions of minerals in
flowering plants   264

    14.2.1 Macronutrients   264
    14.2.2 Micronutrients   267

14.3 Toxic minerals   267

14.4 Soil as a source of minerals   267

14.5 Mineral absorption by plants   268

    14.5.1 Roots as organs of mineral
        absorption   268
    14.5.2 Mechanisms of mineral
        absorption   269

14.6 Transport of minerals in
plants   272

    14.6.1 Passage across the root
        tissues   272
    14.6.2 Movement in the shoot
        system   272
    14.6.3 Circulation of minerals in
        plants   273

14.7 The nutrient-film technique   274

Summary   275

Questions   275

Glasshouse Crops Research Institute

14 Mineral nutrition of flowering plants

The great improvement in crop yields (Table 14.1) which has come from a combination of mechanisation on the farm, the application of fertilisers and pesticides to the land and the breeding of high-yielding varieties of crop plants has been called the **green revolution**. Fertilisers improve the fertility of soils, that is the ability of soils to sustain vigorous, healthy plant growth. **Minerals**, inorganic chemical compounds, are used extensively as **fertilisers** in agriculture and horticulture. Mixtures of minerals called compound fertilisers are applied to the soil before and at various times during the growth of a crop (Fig 14.1). Few arable farmers in developed countries now grow crops without treating their land with fertilisers, especially those containing nitrogen, phosphorus and potassium.

Table 14.1 Mean yields of some important crops grown in the UK over a 20-year period

| Year | Yield (tonnes Ha^{-1}) | | | |
	Wheat	Barley	Oats	Potatoes
1963	4.16	3.61	3.89	22.08
1973	4.36	3.97	3.84	30.40
1983	6.31	4.63	4.38	29.55

Fig 14.1 (a) A bag of fertiliser

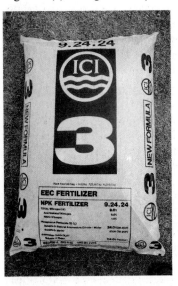

Fig 14.1 (b) Artificial fertiliser being applied to the land

It is therefore evident that minerals perform a crucial role in the growth of flowering plants. But what minerals are required and what part do they play in the functioning of plants?

14.1 Mineral requirements of flowering plants

Seventeenth-century records indicate that scientists were then aware that plant growth depends on a supply of minerals. Yet it was two centuries later before any real progress was made in identifying precisely which minerals are required by higher plants.

14.1.1 Water culture experiments

Table 14.2 Knop's culture solution (1865)

Mineral	g dm⁻³ water
KNO_3	0.2
$Ca(NO_3)_2$	0.8
KH_2PO_4	0.2
$MgSO_4 \cdot 7H_2O$	0.2
$FePO_4$	0.1

In 1860 the German plant physiologist Julius von Sachs described a system for growing flowering plants without soil (Fig 14.2). A few years later Knop published the formula of a liquid medium which was suitable for growing a wide variety of plants (Table 14.2). The main advantage of using water culture techniques to investigate a plant's mineral requirements is that the chemical composition of the liquid medium can be exactly fixed. Such techniques make it possible to find out how plants respond to any combination of minerals.

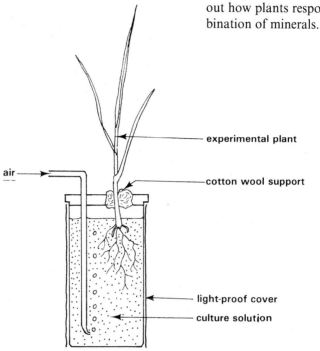

Fig 14.2 Equipment for water culture investigation using grass plants.

air —

experimental plant

cotton wool support

light-proof cover

culture solution

For what reasons is:
(a) the solution aerated?
(b) the light-proof cover fitted to the culture vessel?

Sachs, Knop and others discovered that given a supply of carbon dioxide, water and light, flowering plants require seven chemical elements in relatively large amounts if they are to show healthy and vigorous growth. The seven **macronutrients** as they have since been called are **nitrogen (N), phosphorus (P), potassium (K), calcium (Ca), magnesium (Mg), sulphur (S)** and **iron (Fe)**. It was later found that relatively small amounts of a number of other elements are also needed for normal plant growth. These are called **micronutrients** or trace elements. **Manganese (Mn), zinc (Zn), copper (Cu), boron (B), molybdenum (Mo)** and **chlorine (Cl)** are the most important micronutrients.

The reason why Sachs and his contemporaries failed to discover the importance of micronutrients was because the chemicals they used, although the purest available at the time, probably contained small amounts of trace elements as impurities. Liquid media used in present-day water culture experiments are prepared from pure chemicals dissolved in de-ionised water (Table 14.3). Glassware is cleaned with hot 50 % hydrochloric acid and then thoroughly washed with de-ionised water. Otherwise the vessels may be a source of minerals which the experimental plants can absorb.

Table 14.3 A modern culture solution (modified from Hewitt 1974)

Mineral	g dm⁻³ water		g dm⁻³ water
Macronutrients:			
KNO_3	0.404	K	0.156
		N	0.057
$Ca(NO_3)_2$	0.656	Ca	0.160
		N	0.113
$MgSO_4 \cdot 7H_2O$	0.368	Mg	0.036
		S	0.048
$NaH_2PO_4 \cdot H_2O$	0·208	P	0.041
Fe citrate $\cdot 5H_2O$	0·03350	Fe	0.005
Micronutrients:			
$MnSO_4 \cdot 4H_2O$	0.00223	Mn	0.00055
$ZnSO_4 \cdot 7H_2O$	0.00029	Zn	0.00006
$CuSO_4 \cdot 5H_2O$	0·00025	Cu	0.00006
H_3BO_3	0.00310	B	0.00054
$Na_2MoO_4 \cdot 2H_2O$	0.00012	Mo	0.00005
$NaCl$	0.00580	Cl	0.00350

The effects of mineral deficiencies on plants grown in water culture can be assessed in a variety of ways. One of the simplest is to compare the fresh and dry mass of whole plants or the organs of plants kept in a medium deficient in one or more elements with similar plants grown in a complete medium (Table 14.4). Differences in growth pattern and leaf colour can also be noted.

Table 14.4 Effect of mineral deficiency on growth of lettuce (after Strafford 1963)

Elements in solution	Average fresh mass/g	
	Shoots	Roots
Macronutrients only	71.4	14.5
Macronutrients plus Zn, Cu, Mn, B, Cl	105.7	22.0

Good subjects for water culture work are seeds with small food reserves because they contain only small amounts of stored minerals. Many grasses produce seeds of this kind. Large seeds and cuttings are less suitable as they often have a substantial reserve of one or more minerals.

14.2 Functions of minerals in flowering plants

Valuable as the results of water culture experiments are, they rarely give a complete picture of the roles of minerals in the functioning of a plant. The results often need to be supplemented with data from metabolic studies in which the rates of processes such as respiration, photosynthesis and protein synthesis are measured. Analysis of the chemical composition of experimental and control plants may also be required.

14.2.1 Macronutrients

With the exception of iron, **macronutrients** are normally required at concentrations of 0.04–$0.2\,g\,dm^{-3}$ in mineral culture solutions to sustain vigorous, healthy plant growth (Table 14.3). About $0.005\,g\,dm^{-3}$ of iron gives satisfactory results. For this reason iron is sometimes described as a micronutrient.

1 Nitrogen

Nitrogen is absorbed by plant roots as nitrate ions (NO_3^-) or as ammonium (NH_4^+) ions. Insufficient NO_3^- or NH_4^+ in the soil is usually one of the factors which most limits the growth of crops on cultivated land (Fig 14.3). Some important crop plants such as clover, beans and peas have symbiotic nitrogen-fixing bacteria living in nodules in their roots (Fig 31.30). The bacteria can use nitrogen gas to form organic nitrogenous compounds, some of which are used by the higher plant partner.

In flowering plants nitrogen is found in amino acids, proteins, nucleic acids and chlorophyll. The functions of these important compounds are described in Chapters 3, 4, 5 and 6. Nitrate ions absorbed by roots are first reduced to ammonia which combines with carboxylic acids to form amino acids. Ammonium ions are used directly for amino acid synthesis.

Typical symptoms of nitrogen deficiency are reduced growth of all organs and chlorosis, a yellowing of the leaves due to inadequate production of chlorophyll. Chlorosis appears first in older leaves.

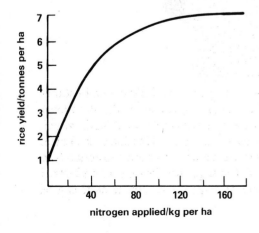

Fig 14.3 Effect of nitrogen concentration on the growth of rice (after Bland & Bland 1983)

2 Phosphorus

Phosphorus is absorbed from the soil as dihydrogenphosphate ions ($H_2PO_4^-$). Nucleic acids, phospholipids and ATP are among the phosphorus-containing compounds found in plants. The roles of these substances are described in Chapters 4 and 5.

Table 14.5 shows the beneficial effect of applications of phosphate fertiliser on crop growth.

Table 14.5 Yields of potatoes and barley from plots of land subject to different rates of applications of P fertiliser

Rate of application of P fertiliser ($kg\,Ha^{-1}$)	0	14	28	56
Yield (tonnes Ha^{-1}):				
Potatoes	13	29	32	32
Barley	2.03	2.66	2.98	3.49

Plot a graph of the data shown in Table 14.4. Summarise the differences in response of the two crops to the fertiliser treatment.

Fig 14.4 Spring barley showing effect of phosphorus deficiency (Crown copyright)

Control plants   P-deficient plants

Severe lack of phosphorus obviously affects processes which use energy provided by ATP. Poor growth, especially of roots (Fig 14.4), and a reduction in uptake of all minerals are the usual symptoms of phosphorus deficiency.

3 Potassium

Potassium is absorbed from the soil as potassium ions (K^+). Whereas nitrogen and phosphorus appear in complex organic compounds in flowering plants potassium does not. Nevertheless potassium is found in green plants in large amounts. An important function of potassium is to activate enzymes (Chapter 3). Over forty enzymes depend on potassium for optimum activity. Potassium is also probably essential for translocation in phloem sieve-tubes (Chapter 13).

A feature of plants grown in a potassium-deficient environment is the mottled appearance of the older leaves. The leaves may also display chlorosis (Fig 14.5).

Fig 14.5 Apple leaves showing effect of potassium deficiency (Crown copyright)

The powerful impact of nitrogen, phosphorus and potassium in stimulating the growth of crops can be seen in the results of the Broadbalk Experiment (Table 14.6, p. 266). Broadbalk is a large field at Rothamsted, Hertfordshire belonging to the adjacent Experimental Station of the Lawes Agricultural Trust. Various fertiliser regimes have been applied annually to different plots of the field since 1844. Consequently the long term effects of the use of N, P and K on the yield of crops grown on the plots can be assessed.

Calculate the percentage increase or decrease of yield for each of the fertiliser regimes compared with the untreated plots. Draw a bar chart of your results. Comment on the design of the experiment. What factors other than yield may be of importance?

Table 14.6 Mean yields of potatoes and wheat, Broadbalk 1980–82

Annual application (kg Ha^{-1})			Yield (tonnes Ha^{-1})	
N	P	K	Potatoes	Wheat
0	0	0	8.47	1.69
96	0	0	8.30	3.68
0	77	107	16.63	2.04
96	77	107	38.57	6.60

4 Calcium

Calcium is taken up by plant roots as calcium ions (Ca^{2+}). The main function of calcium in flowering plants is to form salt linkages between pectic acid molecules in the middle lamella which binds adjacent cells to each other (Fig 14.6). Some enzymes are also stabilised by calcium.

Fig 14.6 The role of calcium in the middle lamella

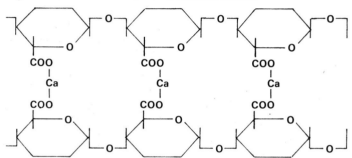

Typical deficiency symptoms for calcium are chlorosis of young leaves, poor root growth and die-back of shoots due to death of apical buds.

5 Magnesium

Magnesium is absorbed from the soil as magnesium ions (Mg^{2+}). In green plants magnesium forms part of the chlorophyll molecule where it donates electrons in the light-dependent reactions of photosynthesis (Chapter 5). Magnesium also activates certain enzymes. Pronounced chlorosis beginning between the veins of older leaves is the main symptom of magnesium deficiency.

6 Sulphur

Sulphur is taken up by plant roots as sulphate ions (SO_4^{2-}). It is needed for the formation of amino acids such as methionine and cysteine which contain thiol (—SH) groups. Sulphur-containing amino acids are present in most proteins. For this reason the symptoms of sulphur deficiency are very similar to those of nitrogen deficiency.

7 Iron

Iron is absorbed mainly as iron(II) ions (Fe^{2+}). Iron is present in cytochromes which act as electron carriers in photosynthesis and respiration (Chapter 5) and in the reduction of nitrate ions. Iron is also required for the synthesis of chlorophyll and to activate certain enzymes.

Characteristic signs of iron deficiency are chlorosis, especially in young leaves, and inhibition of photosynthesis and respiration.

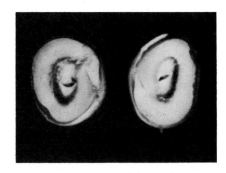

Fig 14.7 Cotyledons of pea seeds showing effect of manganese deficiency (Crown copyright). The symptoms shown are sometimes called marsh spot. They are typical of peas grown in reclaimed marshland soil which contains little manganese

A large number of chemical elements given to plants in very small quantities stimulate plant growth. However, not all of these elements are essential for plant metabolism. The **micronutrients** essential for plant growth are manganese, zinc, copper, boron, molybdenum and chlorine. Water culture experiments show that they are required at much lower concentrations than the macronutrients (Table 14.3).

The main functions of the micronutrients are summarised in Table 14.7. Chlorosis can be a symptom of insufficient Zn, Cu, Mo or Cl. Even so, the deficiency symptoms for most of the micronutrients are so characteristic that certain plants may be used as indicator species in diagnosing soils which are deficient in trace elements (Fig 14.7).

Table 14.7 Roles of micronutrients in plant metabolism

Element	Roles
Mn	Activates carboxylase enzymes. May act as an electron donor for excited chlorophyll *b*.
Zn	Present in the enzyme carbonic anhydrase and also in various dehydrogenase enzymes.
Cu	Present in a number of oxidase enzymes. Also found in plastocyanin which acts as an electron carrier for chlorophyll *a*.
B	Not fully known. Essential for meristem activity and for growth of pollen tubes.
Mo	Required for nitrogen fixation. Activates the enzyme nitrate reductase.
Cl	Essential for oxygen evolution in photosynthesis.

14.3 Toxic minerals

In recent years large-scale reclamation of derelict industrial sites has taken place in Britain and in other parts of the world. One of the difficulties in work of this sort is to get vegetation to grow in such places where the soil, mainly derived from spoil heaps, is sometimes contaminated with heavy metals such as copper, lead, zinc and nickel. In spoil heaps the concentrations of these elements are toxic to most green plants.

Old spoil heaps from heavy metal workings are colonised by a select group of plants called **metallophytes** because of their tolerance to high concentrations of heavy metals. It is said that prospectors often located surface deposits of ores rich in heavy metals by the type of vegetation growing on the sites. One approach to the problem of getting vegetation to grow on spoil heaps has been to plant metallophytes. Varieties of grasses such as *Festuca ovina* and *Agrostis canina* tolerant of high concentrations of heavy metals have been used successfully in projects of this kind.

14.4 Soil as a source of minerals

Soil is much more complex than solutions used in water culture experiments. A fertile soil consists of a **mineral skeleton** derived from weathered rock, **organic matter** mainly as dead and decaying plant material, **water, air** and **living organisms** notably bacteria, fungi and soil-dwelling animals. The interaction of the various components mainly determines the availability of minerals to the roots of higher plants.

In natural soils nearly all the nitrogen comes from the breakdown of organic matter by **decomposer organisms** such as bacteria and fungi. Most of the sulphur, phosphorus, calcium, magnesium and potassium are also of organic origin. The rest, together with iron and trace elements, are obtained from weathering of the mineral skeleton.

Cations such as Ca^{2+}, Mg^{2+} and K^+ are attracted to and held by the negative charges of clay and humus particles. However, the attraction is not permanent and the elements can be released by **ion exchange**. Anions such as NO_3^-, SO_4^{2-} and Cl^- are not usually held in this way and are found dissolved in soil water. Phosphorus is normally found in soil as insoluble phosphates of calcium, iron and aluminium.

In an undisturbed terrestrial ecosystem the soil is in a state of equilibrium. Minerals absorbed by plants are returned to the soil when the plants die and are decomposed. On the contrary, substantial quantities of mineral elements are removed annually from cultivated soils. Most of the loss is due to the high mineral content of the crop at harvesting (Table 14.8). Thus to maintain the fertility of land used for growing crops, it is necessary to replenish the soil with minerals. This is done by applying **manures, composts** or **mineral fertilisers**.

Table 14.8 **Mineral content of leaves of sugar beet (after Wallace 1961)**

Element	Ca	Mg	K	P	Fe	Mn	B
% dry mass	2.64	0.55	4.21	0.35	0.0125	0.0046	0.0029

Among the organic materials that gardeners use as sources of plant nutrients are **hoof and horn** and **dried blood**. However these are in short supply and make up less than 0.1 % of the UK fertiliser market. Farmers make use of **manure** from their livestock. This contributes about half of the nitrogen applied annually to the land, the remainder coming from mineral fertilisers.

Some people will not eat foodstuffs grown on land which has been treated with pesticides and mineral fertilisers. They rely on **organic farming** to produce crops reared in soil kept fertile by manures. It is doubtful however that we could grow enough food at a reasonable price to meet the requirements of the UK population by this method.

14.5 Mineral absorption by plants

Roots are the main organs concerned with mineral uptake by terrestrial plants although minerals can be absorbed through intact leaves. Foliar sprays containing fertilisers are now quite widely used to feed crops grown on a large scale.

13.5.1 Roots as organs of mineral absorption

The absorption of water by roots is described in Chapter 12. Minerals are taken into roots in aqueous solution yet there is no direct link between the absorption of minerals and water. Water uptake is controlled mainly by the rate at which a plant transpires. Mineral uptake can be rapid even when transpiration has stopped. Indeed, submerged aquatic plants absorb minerals even though they do not transpire. Most of a terrestrial plant's mineral requirements are absorbed by its roots from the topsoil. It is here that roots grow profusely and minerals are most abundant.

Roots which penetrate deeper into the soil absorb mainly water. By growing constantly in length, roots continually occupy new areas of soil from which minerals can be absorbed. All parts of the root system do not absorb minerals with equal efficiency. Mineral uptake takes place most rapidly in the zone of elongation just behind the apices of roots where root hairs are present (Fig 14.8). The root hairs increase the area of contact between the soil and the root surface.

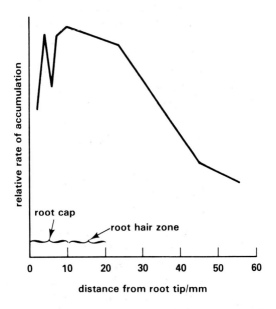

Fig 14.8 Mineral accumulation by different parts of barley roots (after Kramer, 1969)

The roots of many terrestrial plants form a mutualistic association with the hyphae of certain species of soil-dwelling fungi. The association is called a **mycorrhiza**. Mycorrhizal roots absorb minerals more efficiently than ordinary roots (Table 14.9). What do you think are the reasons for this?

Table 14.9 Uptake of N, P and K by *Pinus strobus* (after Hatch 1937)

| Type of root | Nutrients absorbed/% dry mass | | |
	N	P	K
mycorrhizal	1.24	0.196	0.744
non-mycorrhizal	0.85	0.074	0.425

14.5.2 Mechanisms of mineral absorption

Minerals are usually absorbed against concentration gradients. Not all minerals are absorbed at the same rate and uptake is most rapid by tissues respiring aerobically. Let us examine two of the mechanisms which could account for mineral absorption.

1 Diffusion

Diffusion is the random movement of ions or molecules down a concentration gradient. Most of the available evidence indicates that the concentration of mineral ions inside plants is higher than in their surroundings (Fig 14.9). Hence minerals entering plants usually move against concentration gradients – quite the reverse of diffusion. It is also difficult to explain why respiring cells absorb minerals more rapidly than non-respiring cells if diffusion is the mechanism of mineral absorption.

Fig 14.9 A comparison of the concentrations of various ions in the sap of the alga *Nitella* and the pond water in which it lives (after Hoagland, 1944)

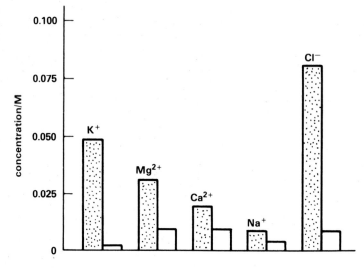

2 Active uptake

When roots are transferred from distilled water to a dilute solution of a mineral salt the respiratory rate of the roots increases. Furthermore the increase in respiration is proportional to the rate of ion absorption (Fig 14.10). It is thought that respiratory energy is used to absorb ions against a concentration gradient.

Fig 14.10 Relationship between the rate of respiration and bromide uptake by carrot root discs (after Steward, Berry and Broyer, 1936)

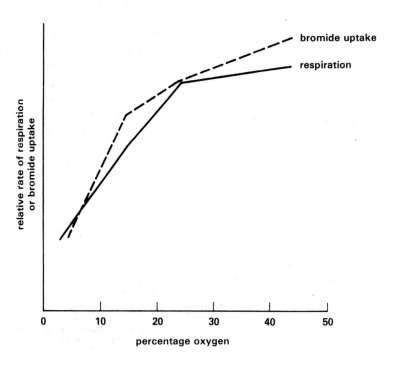

(a) active transport (against a concentration gradient)

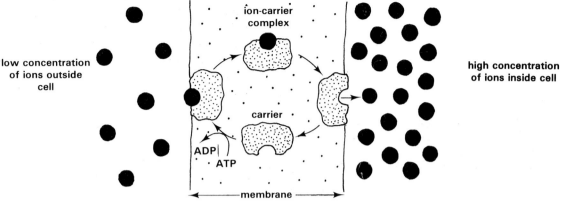

Energy from ATP is used to form the ion-carrier complex.

(b) facilitated transport (down a concentration gradient)

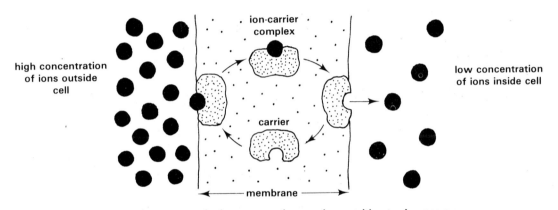

The carrier conveys the ions across the membrane without using energy.

Carrier molecules are probably used to transport mineral ions into cells (Fig 14.11). Evidence that carriers are involved has come from experiments in which the rates of ion uptake are measured in the presence of increasing concentrations of mineral ions (Fig 14.12). The results are similar to the relationship between the rate of an enzyme-catalysed reaction and substrate concentration (Chapter 3). The similarity in the relationships suggests that ions combine with carriers in much the same way as enzyme-substrate complexes are formed. In Chapter 3 it is explained how inhibitors compete for active sites on enzyme molecules. A comparable phenomenon occurs in mineral uptake when the presence of one mineral sometimes inhibits the uptake of another. Why do you think this occurs?

Fig 14.12 Effect of concentration of ions on ion uptake

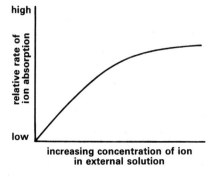

271

14.6 Transport of minerals in plants

Because they absorb minerals, roots can be thought of as a **source** of minerals for the rest of the plant. Actively growing tissues which have a demand for minerals can be thought of as **sinks**. Let us see how minerals move from a source to sinks.

14.6.1 Passage across the root tissues

After absorption by the epidermal cells of roots, minerals eventually enter the root xylem elements. Plant physiologists are still uncertain as to how the minerals cross the intervening tissues. The path of least resistance across the root cortex is probably through the cytoplasm, movement from cell to cell taking place along plasmodesmata. Alternatively mineral transport may take place along the walls of the cortical cells. At the endodermis the wall route is blocked by the suberised **Casparian strip**. It is likely that minerals pass through the selectively permeable cell membrane of the endodermal cells. Entry of minerals into the pericycle cells is controlled in this way. Movement of minerals from the pericycle to xylem elements is thought to occur by **active transport**. The pumping of ions into the vascular tissue uses respiratory energy and is probably responsible for root pressure (Chapter 12).

Not all of the minerals absorbed by roots end up in the plant's vascular tissue. Some minerals are retained by the root cells for metabolic purposes. Metallophytes often store large deposits of heavy metals in the vacuoles and cell walls of the root cortex.

14.6.2 Movement into the shoot system

The ascent of solutes in the xylem elements of the stem can be demonstrated by placing the stalk of a leafy transpiring shoot in a solution of a dye such as eosin. Microscopic examination of thin sections of the stem cut a few hours later reveals the presence of dye in the vessels and tracheids. Some of the dye is found in the cell walls but the results of recent experiments with radionuclides indicate that the cavities of tracheids and vessels are the most important route for mineral transport. The ascent of minerals is a **passive** process dependent on transpiration. A correlation between the rate of transpiration and the upward movement of solutes has been demonstrated with indicators such as dyes and radionuclides. Although many minerals ascend the xylem as inorganic ions, some are carried as organic compounds. While most of the sulphur is transported as sulphate ions, a small proportion is moved as sulphur-containing amino acids such as cysteine and methionine. Nitrate ions are generally converted in the roots to amino acids.

On reaching a sink in the **transpiration stream**, minerals are moved into the metabolising tissue. **Transfer cells** which have a large surface area to volume ratio are found in the minor veins of many species of flowering plants and are thought to transport minerals from the xylem into metabolising tissue. Because transfer cells have large numbers of mitochondria it is probable that **active transport** is involved. Some lateral transport also takes place between xylem elements and adjacent tissues such as the vascular cambium and phloem.

14.6.3 Circulation of minerals in plants

What happens to individual mineral nutrients after they have been taken into a sink varies considerably. Calcium and iron accumulate in young leaves and display little movement into new organs formed later. The current needs of a plant for Ca and Fe are met by absorption from the soil. Thus deficiency symptoms of these elements often appear in the young leaves at shoot apices.

Most minerals are relatively mobile and after entering initial sinks they then pass into developing sinks such as young leaves, flowers and fruits. For this reason deficiency symptoms of N, K, Mg, P and S are often seen in mature leaves. It has been shown that phosphate ions are re-exported in phloem sieve-tubes and it is likely that other mobile elements follow the same pathway. The internal recycling of minerals enables plants to withdraw nutrients from old leaves before they fall off. The yellowing of older leaves for example is accompanied by the breakdown of nitrogenous compounds including chlorophyll and withdrawal of the end products into other parts of the plant.

The main processes and pathways thought to be involved in mineral uptake and transport are summarised in Fig 14.13.

Fig 14.13 Summary of mineral uptake and transport in a flowering plant

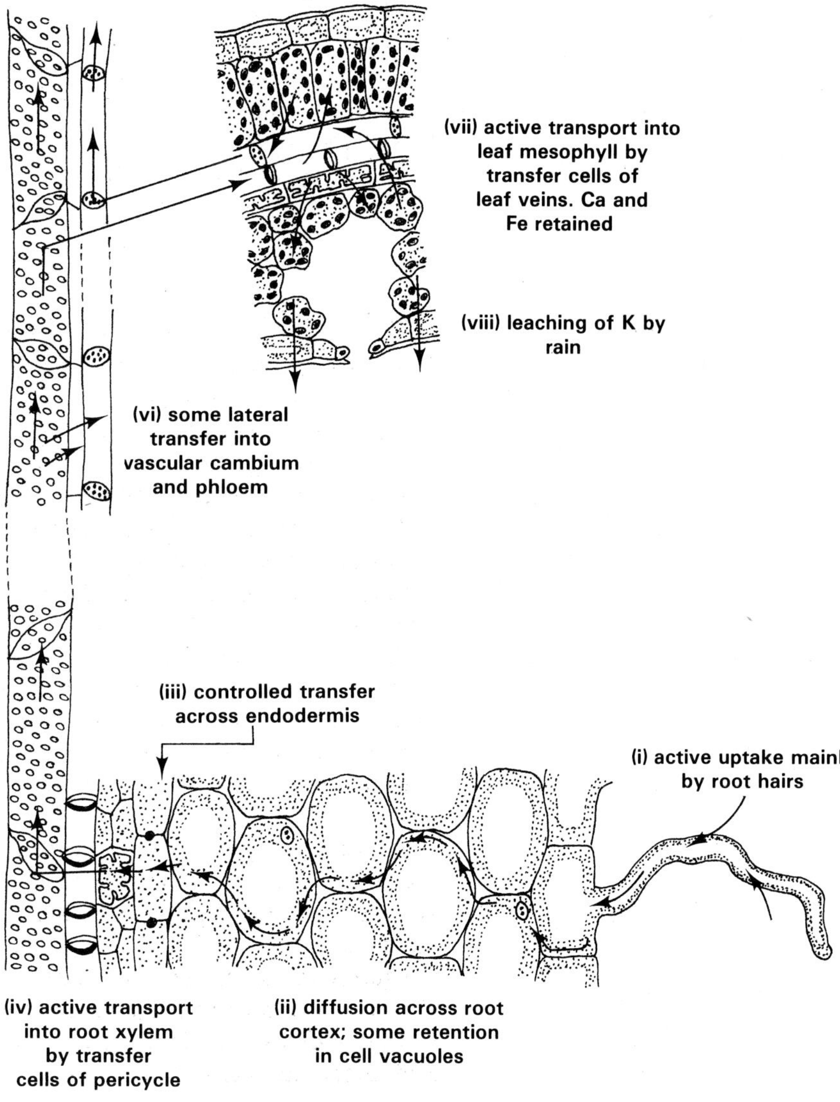

(ix) export of N, P, K, S and Mg to younger organs via phloem sieve-tubes

(vii) active transport into leaf mesophyll by transfer cells of leaf veins. Ca and Fe retained

(viii) leaching of K by rain

(v) passive ascent up xylem of stem in the transpiration stream

(vi) some lateral transfer into vascular cambium and phloem

(iii) controlled transfer across endodermis

(i) active uptake mainly by root hairs

(iv) active transport into root xylem by transfer cells of pericycle

(ii) diffusion across root cortex; some retention in cell vacuoles

14.7 The nutrient-film technique

This chapter ends as it began with water culture techniques. In recent years a system has been developed at the Glasshouse Crops Research Institute in Britain for the large-scale cultivation of crops without soil. The system is called the **nutrient-film technique**. The plants are grown in plastic troughs through which circulates a film of water containing balanced amounts of macro- and micronutrients (Fig 14.14). So successful is the system that commercial production of tomatoes, cucumbers and strawberries has now begun using the nutrient-film technique. One of the important advantages of this form of farming is that it dramatically increases crop production.

Fig 14.14 (a) and (b) The nutrient film technique (courtesy Glasshouse Crops Research Institute)

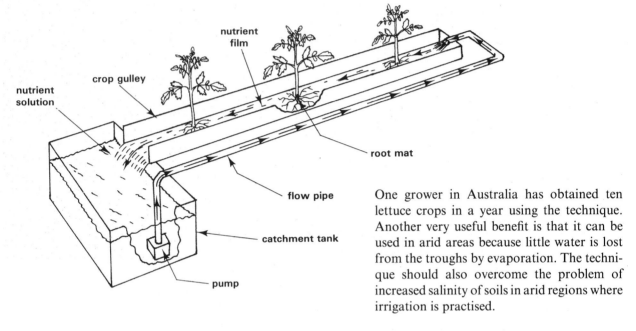

One grower in Australia has obtained ten lettuce crops in a year using the technique. Another very useful benefit is that it can be used in arid areas because little water is lost from the troughs by evaporation. The technique should also overcome the problem of increased salinity of soils in arid regions where irrigation is practised.

SUMMARY

Water culture experiments have enabled physiologists to identify the importance of mineral nutrients in plant metabolism and growth. Seven elements, called macronutrients, are required by green plants in substantial quantities. Smaller amounts of many other elements called micronutrients are also needed.

Long term agricultural trials such as those conducted at Rothamsted Experimental Station have revealed the significance of nitrogen, phosphorus and potassium in particular in boosting the yield of crop plants. It is for this reason that mineral fertilisers applied in bulk to the land contain these three elements.

Manure is another source of minerals used on a large scale to maintain soil fertility. Compost, hoof and horn and dried blood are organic materials which gardeners use more widely to provide mineral nutrients for their crops. Small scale organic farming has become popular in recent years as a means of growing crops without applying fertilisers and pesticides to the land.

Roots are the main organs of mineral absorption, though intake through the leaf surface can also occur. The rate of mineral uptake is most rapid in the root hair zone which has a large surface area for absorption. Mycorrhizae, associations between roots and fungi, are also efficient at absorbing minerals from the soil. The mechanism of mineral absorption is an active process using energy derived from respiration. Active transport is also the means whereby minerals move across the pericycle of roots on their way to the xylem in which they are passively circulated in the transpiration stream. From there, active transport is again used to transfer minerals to metabolising tissues.

Crop plants such as tomatoes and cucumbers can be grown commercially in a shallow running solution containing a balanced range of minerals. The nutrient film technique, as it is called, offers a means of successfully raising crops in parts of the world where water is scarce.

QUESTIONS

1 An experiment was carried out to investigate the movement of phosphate in bean plants.

Twelve intact plants wre placed for 24 hours with their roots immersed in a nutrient solution containing phosphate labelled with radioactive phosphorus (^{32}P). The plants were then transferred to a non-radioactive nutrient solution. The leaves of six plants were covered with aluminium foil to exclude the light.

The radioactivity in the leaves of all twelve plants was measured over a period of seven days. The radioactivity present in the leaves of all plants was measured at intervals in terms of counts per minute (c.p.m.). The results are shown in the graph below, the values recorded being the mean value for the six plants in each group.

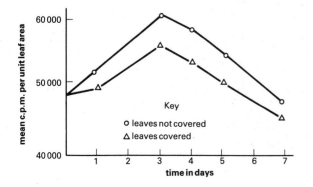

(a) Explain why several plants were used for each treatment.
(b) Explain briefly how the ^{32}P passed from the nutrient solution into the leaves.
(c) Suggest why the amount of ^{32}P continued to increase in the leaves even though the plants had been transferred to the non-radioactive solution.
(d) How would the amount of ^{32}P accumulated in the leaves have been affected if the plants had been exposed to moving air? Explain your answer.
(e) Suggest a reason for the difference in amount of radioactivity measured in the two sets of leaves.
(f) (i) In which tissue would the radioactive phosphorus have been transported out of the leaf?
(ii) Outline an experiment which could be carried out to support your answer.
(g) What would be the effect of the darkening on the amount of starch and soluble sugars in the covered leaves?
(h) The mass flow theory for translocation in plants proposes that movements of organic compounds occur as a result of pressure potential gradients set up by gradients in soluble sugars. Other compounds would move with the carbohydrate stream.

Do the results of this experiment provide evidence in support of the mass flow theory? Give a reason for your answer. (L)

2 Gardening magazines urge us to treat our lawns with fertilizer. How would you test the hypothesis that the growth of grass in a lawn (as measured by dry mass)

increases with increasing concentration of fertilizer applied? (Nuffield/JMB)

3 A series of experiments was carried out to investigate the effects of ions on the growth of bean seedlings. The apparatus used in the investigation is shown below.

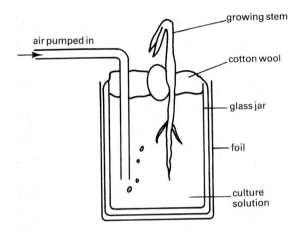

Some of the seedlings were grown in complete culture solution with all essential ions present, whereas other seedlings were grown in solutions deficient in either nitrate or phosphate ions.

The length of the stem was measured at 5-day intervals and the results are given in the table below. Average values of ten seedlings were taken.

Time in days	Length of stem in mm		
	Complete culture solution	Deficient in nitrate ions	Deficient in phosphate ions
5	10	5	5
10	20	7	10
15	25	10	17
20	50	17	25
25	78	18	35
30	103	18	50
35	125	18	62
40	130	18	65
45	140	17	66
50	143	17	66

(a) Present the data in a suitable graphical form.
(b) What is the difference in stem length after 37 days between seedlings grown in solutions deficient in phosphate and those deficient in nitrate?
(c) Comment on the following points in the experimental method.
 (i) Foil was placed around each culture jar.
 (ii) Air was pumped into the solutions.
 (iii) Average values of 10 seedlings were taken.
(d) Comment on the growth changes shown by the seedlings in the complete culture solution.
(e) Suggest reasons why the growth curve of seedlings grown in the complete culture solution differs from those grown in solutions deficient in (i) nitrate ions and (ii) phosphate ions. (L)

4 An experiment was carried out on the effect of different concentrations of nutrient solution on the growth of grass. Seeds were germinated on filter paper and healthy seedlings were transferred to pots of washed sand. All the seedlings were maintained for nine days on 1/64th of the recommended concentration of a nutrient solution. Thereafter different pots were given the same volume of different concentrations of the original nutrient solution. Pots were watered daily and nutrient solution was added twice weekly. The nutrient solution contained potassium, nitrogen, sulphur, calcium, magnesium, phosphorus, sodium, iron, chlorine, copper and trace elements. The table shows the height (in mm) of the tallest grass at different nutrient concentrations.

Nutrient concen- tration	Height of tallest grass (mm)					
	Day 0	Day 20	Day 40	Day 60	Day 80	Day 100
×16	12	20	38	45	50	45
×4	12	23	42	70	90	100
×1	12	25	45	78	90	95
×1/4	12	16	18	20	20	20
×1/16	12	13	15	16	17	18

After Austin and Austin, J. Ecol., 1980

(a) Plot these data on a single set of axes.
(b) Give **three** environmental conditions which it would be important to control.
(c) Why were the seedlings maintained for nine days before the start of the experiment?
(d) Why was washed sand rather than loam used in the pots?
(e) (i) What are trace elements?
 (ii) Give **one** example of a trace element.
 (iii) Give **one** role each of nitrogen, magnesium and calcium in these seedlings.
(f) Give an osmotic explanation of why the tallest plants in the strongest solution must have died between 80 and 100 days after the start of the experiment. (AEB 1985)

15 The mammalian alimentary system

15.1	Nutritional requirements	278
	15.1.1 Organic nutrients	278
	15.1.2 Vitamins	279
	15.1.3 Inorganic nutrients	280
	15.1.4 Diet	281

15.2	Dentition	282
	15.2.1 Structure and development of teeth	282
	15.2.2 Variety of mammalian teeth	283

15.3	Digestion	287
	15.3.1 Mouth	288
	15.3.2 Stomach	289
	15.3.3 Rumination	291

	15.3.4 Small intestine	291
	15.3.5 Intestinal micro-organisms	294
	15.3.6 The caecum	295

15.4	Absorption	295
	15.4.1 Fate of lipids	297
	15.4.2 Fate of protein	297
	15.4.3 Fate of carbohydrate	297
	15.4.4 Colon	298

15.5	The liver	298
	15.5.1 Liver structure	298
	15.5.2 Liver functions	300

| *Summary* | | 302 |

| *Questions* | | 304 |

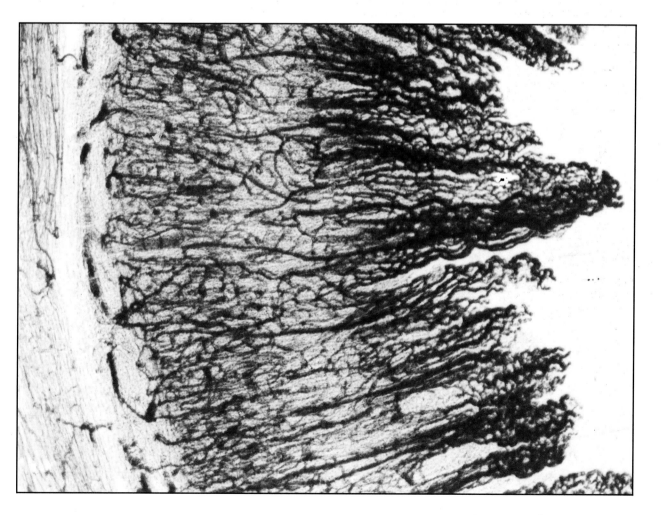

15 The mammalian alimentary system

A constant supply of organic and inorganic nutrients is required to sustain the many and complex activities which take place in the mammalian body.

Organic nutrients such as proteins and lipids are the raw materials for the synthesis of cellular components in growth and repair. The energy needed to drive synthetic processes comes mainly from the oxidation of sugars in cellular respiration. Mammals also require energy for movement and to help maintain a constant body temperature (Chapter 20). Vitamins are organic nutrients which participate in a wide range of body functions.

Inorganic nutrients (minerals) are needed to create and maintain electrical potentials across the membranes of nerve and muscle cells (Chapter 17). The coagulation of blood (Chapter 10), activation of enzymes, growth of teeth and bones and maintenance of the solute potential of the body fluids, are among the many other essential activities which depends on a supply of minerals.

Mammals obtain nutrients from their diet which consists of food and water. Food is a mixture of organic and inorganic nutrients. Water contains dissolved minerals.

15.1 Nutritional requirements

The nutritional requirements of mammals can be divided into three categories, **organic nutrients, vitamins** and **inorganic nutrients**.

15.1.1 Organic nutrients

The chemical nature of the main organic constituents of the diet is described in Chapter 2.

1 Carbohydrates

Starch, sugars and **cellulose** are the main carbohydrates we eat. About 50 % of the energy from the average human diet in Britain comes from starch and sugars. However, a larger proportion of energy may be derived from dietary carbohydrates in people who lead very active lives. Reserves of carbohydrate are stored as **glycogen** in the liver and skeletal muscles. Cellulose is an important constituent of **dietary fibre** (section 15.4.4).

2 Proteins

Proteins are necessary in the diet as a source of **amino acids**. Amino acids may also be respired, but only about 5 % of the human body's energy requirement normally comes from amino acids. Amino acids are used mainly for the synthesis of our proteins. A constant supply of amino acids is needed for the production of enzymes and the proteins in cell membranes. Growth and repair especially depend on the presence of proteins in the diet. Proteins are also made in secretory cells for export to other parts of the body. Plasma proteins in the blood, for example, are made in the liver. Antibodies are proteins made in the lymph nodes, and some hormones are short-chain polypeptides. Transamination in the liver enables **non-essential amino acids** to be made from others (section 15.5.2). **Essential amino acids** cannot be made in this way and are vital constituents of the diet (Table 15.1). The average adult man requires about 440 µg of protein

Table 15.1 Essential amino acids

lysine	phenylalanine	leucine
threonine	methionine	isoleucine
tryptophan	histidine	valine

Why do men require more protein in their diet than women? Why is more protein required by women during pregnancy? Why do children require a lot of protein in their diet?

kg^{-1} body mass each day. Women need about 10 % less as their bodies contain less tissue in which protein synthesis occurs. More protein is required during pregnancy for fetal development. Children have a relatively higher protein requirement than adults because they are constantly forming new tissues throughout the body as they grow and develop.

3 Lipids

Fats and **oils** are the main lipids we eat. They can be used as respiratory substrates and account for about 45 % of the energy in the average diet in Britain. Some lipids are sources of **essential fatty acids** such as arachidonic acid. Many fatty acids are **non-essential** because they can be synthesised in our bodies. The synthesis of cell membranes requires lipids as well as proteins. The fatty membranes of the myelin sheath around certain nerve cells are important in impulse transmission (Chapter 17). Surplus fats are stored beneath the skin and around internal body organs such as the kidneys. As well as being a foodstore, fat acts as a heat insulator which helps to keep a constant body temperature (Chapter 20).

15.1.2 Vitamins

Vitamins are a range of substances necessary for good health. They affect a variety of body functions and are required only in trace quantities. Some vitamins are soluble in fats and are absorbed from the gut dissolved in lipids. **Fat-soluble vitamins** can be stored in the liver and therefore need not be consumed regularly. Other vitamins are water-soluble and are absorbed into the blood dissolved in water. **Water-soluble vitamins** are readily excreted in urine and must, therefore, be consumed regularly.

1 Fat-soluble vitamins

i. Vitamin A (retinol) is made from the plant pigment β-carotene and is needed for synthesis of the visual pigments. Deficiency of vitamin A can lead to night blindness (Chapter 19). Vitamin A is also essential for maintaining epithelial tissues in good condition. Infection of the epithelium of the respiratory tract often results from serious lack of vitamin A in the body.

ii. Vitamin D is a group of several sterols including **ergocalciferol** (vitamin D_2). Ergocalciferol is produced by the action of ultraviolet radiation from the sun on ergosterol in the skin. Vitamin D_2 stimulates the absorption of calcium from the gut. Deficiency of vitamin D_2 can lead to the release of calcium from the bones under the influence of parathormone (Chapter 21). Malformation of the skeleton due to vitamin D_2 deficiency is called rickets.

iii. Vitamin E The function of vitamin E in the human body is not fully understood. However, a deficiency of it may lead to degeneration of muscle tissue, destruction of liver cells and anaemia.

iv. Vitamin K is necessary for the production of prothrombin in the liver. Deficiency of vitamin K thus leads to impaired coagulation of the blood (Chapter 10).

2 Water-soluble vitamins

i. Vitamin B is a complex of several vitamins, most of which are components of co-enzymes (Chapter 3). Vitamin B_1 (**thiamin**) is required to make the co-enzyme of carboxylase enzymes. Deficiency of vitamin B_1 leads to beri-beri which is characterised by paralysis, oedema and heart failure.

Vitamin B_2 (**riboflavine**) is part of the co-enzyme FAD used as a hydrogen acceptor in Krebs cycle (Chapter 5). Vitamin B_6 is required for protein metabolism including transamination (section 15.5.2). Vitamin B_{12} (**cyanocobalamin**) is necessary for the normal maturation of red blood cells and the health of the nervous system and mucosae. Deficiency of vitamin B_{12} causes anaemia and degeneration of the nerve cord. **Nicotinamide** and **nicotinic acid** are other B-group vitamins. They are required to make the hydrogen acceptor co-enzymes NAD and NADP. Folic and pantothenic acids also belong to this group. **Folic acid** is essential for normal red blood cell maturation. **Pantothenic acid** is a component of acetyl co-enzyme A (Chapter 5).

ii. Vitamin C (ascorbic acid) is thought to act as an electron carrier in respiration. It also stimulates synthesis of collagen fibres. Deficiency of vitamin C causes scurvy, symptoms of which include breakdown of connective tissues and blood vessels. Scurvy was once common among sailors who were deprived of fresh fruit containing vitamin C while on long voyages. Bleeding of the gums and loosening of the teeth are symptoms of scurvy. Internal bleeding also occurs in many tissues. Where it occurs near the surface, bruised patches appear in the skin, a sympton often seen in elderly people whose consumption of fresh fruit and vegetables may be inadequate. Recent evidence suggests that a high daily intake of vitamin C in the diet may have many beneficial effects, including reduction of blood cholesterol.

15.1.3 Inorganic nutrients

Minerals are inorganic nutrients which participate in a wide variety of body functions. Some minerals called **macronutrients** are needed in relatively large quantities. Others called **trace elements** are needed in small amounts.

1 Macronutrients

Sodium ions are more abundant than any other cation in the body fluids. Sodium ions play an important role in the electrochemical activities of nerves and muscles (Chapter 17). Sodium and chloride ions are the main determinants of the solute potential of tissue fluid and blood plasma. **Potassium** ions are the most abundant of cations in cells. Potassium ions are involved, like sodium, in nerve and muscle impulse transmission. **Calcium** and **phosphate** ions are the main mineral constituents of bone and teeth. Calcium ions are also required for coagulation of blood and for muscle contraction.

2 Trace elements

Trace elements must be provided in the diet to maintain health. Large quantities of trace elements can be poisonous. **Cobalt** is a constituent of vitamin B_{12} (cyanocobalamin) and is vital for the production of haemoglobin and red blood cells. **Iodine** is a constituent of the hormone thyroxine and is absorbed from the blood by the thyroid gland. Lack of iodine in the diet leads to decreased thyroxine production and the enlargement of the thyroid gland called a goitre (Chapter 21). **Copper** is vital for the activation of a variety of enzymes and for the production of haemoglobin. **Iron** is a constituent of haemoglobin and myoglobin (Chapter 8) and cytochromes (Chapter 5). Reserves of iron are stored in the liver (section 15.5.2). **Zinc** is an activator of several enzymes, including carbonic anhydrase, and is required for synthesis of the hormone insulin (Chapter 21).

What is meant by the term 'a balanced diet'?

15.1.4 Diet

The food we eat constitutes our **diet**. Different foods contain different nutrients. To obtain a balanced diet, that is, all the nutrients in the amounts required to maintain health, we must eat a variety of different foods.

Among the most important foods are **cereals** such as wheat, maize and rice. They contain many vitamins, much starch and protein, but are deficient in lysine, an essential amino acid. Cereals are also poor sources of minerals (Table 15.2).

Table 15.2 Nutritional make-up (per 100 g) of some of the main constituents of the human diet (data from Taylor, after *Manual of Nutrition*, HMSO 1976)

Item of diet	Energy /kJ	Protein /g	Fat/g	Carbohy- drate/g	Minerals (Ca + Fe)/mg	Vitamins A/µg	D/µg	B₁/mg	B₂/mg	Nicotinic acid/mg	C/mg
apples	197	0.3	0	12.0	4.3	5	0	0.04	0.02	0.1	5
bananas	326	1.1	0	19.2	7.4	33	0	0.04	0.07	0.8	10
beef	940	18.1	17.1	0	8.9	0	0	0.06	0.19	8.1	0
bread, white	1068	8.0	1.7	54.3	101.7	0	0	0.18	0.03	2.6	0
bread, wholemeal	1025	9.6	3.1	46.7	31.0	0	0	0.24	0.09	1.9	0
butter	3006	0.5	81.0	0	15.2	995	1.25	0	0	0.1	0
cabbage	66	1.7	0	2.3	38.4	50	0	0.03	0.03	0.5	23
carrots	98	0.7	0	5.4	48.6	2000	0	0.06	0.05	0.7	6
cheese, cheddar	1708	25.4	34.5	0	810.6	420	0.35	0.04	0.50	5.2	0
cod	321	17.4	0.7	0	16.3	0	0	0.08	0.07	4.8	0
cream, double	1848	1.8	48.0	2.6	65.0	420	0.28	0.02	0.08	0.4	0
eggs	612	12.3	10.9	0	56.1	140	1.50	0.09	0.47	3.7	0
liver	1020	24.9	13.7	5.6	22.8	6000	0.75	0.27	4.30	20.7	20
margarine	3019	0.2	81.5	0	4.3	900	8.00	0	0	0.1	0
milk	274	3.3	3.8	4.8	120.1	40	0.03	0.04	0.15	0.9	1
oranges	150	0.8	0	8.5	41.3	8	0	0.10	0.03	0.3	50
parsnips	210	1.7	0	11.3	55.6	0	0	0.10	0.09	1.3	15
peas	208	5.0	0	7.7	14.2	50	0	0.25	0.11	2.3	15
potatoes, boiled	339	1.4	0	19.7	4.5	0	0	0.08	0.03	1.2	10
rice	1531	6.2	1.0	86.8	4.4	0	0	0.08	0.03	1.5	0
sugar, white	1680	0	0	100.0	1.0	0	0	0	0	0	0
tomatoes	52	0.8	0	2.4	13.4	117	0	0.06	0.04	0.7	21

Nuts and **pulses** (leguminous seeds) such as beans, lentils and peas are valuable sources of protein. Soya beans and groundnuts also contain much oil.

Vegetables include cabbage, carrots, cassavas, cauliflowers, potatoes, swedes, turnips and yams. Most root vegetables contain much carbohydrate and fibre but are poor in protein, fat and vitamins. Leafy vegetables contain many vitamins and minerals and are a good source of fibre too. Over-cooking, however, often destroys and removes water-soluble vitamins (section 15.1.2).

Much of the carbohydrate in the diet of the Western world is provided by refined **sugar** which is little more than pure carbohydrate. Honey, brown sugar, treacles and molasses are less refined and contain many minerals.

Most **fruits** contain much carbohydrate and water; many also provide vitamin C.

Fats and **oils** are concentrated sources of energy. They may also contain fat-soluble vitamins (section 15.1.2). Generally fats of animal origin such as butter, lard and suet are more saturated than those from plants (Chapter 2). The high incidence of coronary disease in many parts of the Western world has been linked to excessive dietary intake of saturated fats. In much of Africa and southern Italy where plant oils such as olive oil, corn oil and sunflower oil are consumed instead, coronary disease is rare.

Meat and **fish** are valuable sources of high quality protein. This means they provide all the essential amino acids we require. Meat and liver contain much iron as well as other minerals and many vitamins. **Eggs** are rich in protein, fat and vitamins. The human diet often contains **milk** from cows, goats and sheep. Milk contains carbohydrate (lactose), fat, protein, vitamins and minerals, especially calcium and phosphate. It provides all the nutritional requirements for a young baby.

Fig 15.1 Development of teeth. The dental lamina develops from epithelium (a) and (b) to produce an enamel organ (c). The organ is responsible for the production of the tooth (d). Permanent tooth buds develop below the milk teeth (e)

15.2 Dentition

Ingested food first enters the mouth where it is chewed. **Chewing** makes it easier to swallow food and helps the action of digestive enzymes secreted by the various glands which open into the gut. The cutting and grinding of food which takes place when it is chewed is the function of the **teeth**.

15.2.1 Structure and development of teeth

Tooth buds in the developing jaw give rise to the teeth. The buds each contain an **enamel organ** in which **ameloblast cells** produce **enamel** made of calcium salts. Enamel is the hardest material in the body and coats the protruding surfaces called the **crowns** of the teeth. Inside the cup-shaped enamel organ, **odontoblast cells** of the dental papilla produce **dentine** (Fig 15.1). During tooth development, sockets of bone grow around the **roots** of the teeth (Fig 15.2).

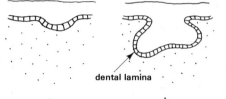

dental lamina

a   b

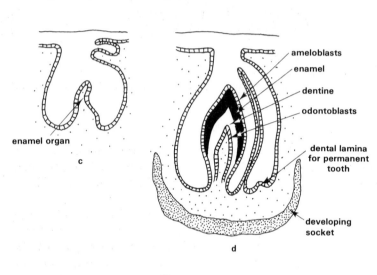

enamel organ

c

ameloblasts
enamel
dentine
odontoblasts
dental lamina for permanent tooth
developing socket

d

Fig 15.2 Diagrammatic vertical section through a molar. The root is held in the bony socket by cementum joined to a fibrous connective layer called the periodontal membrane. Elasticity of the fibres in this membrane allows limited movement of the root in the socket, thus acting as a 'shock absorber' during chewing

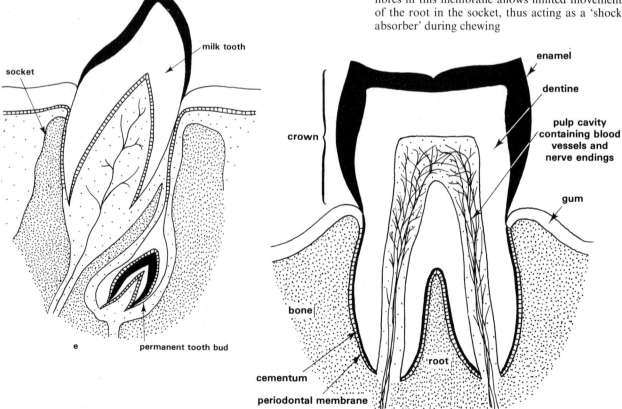

milk tooth

socket

permanent tooth bud

e

enamel
dentine
pulp cavity containing blood vessels and nerve endings
crown
gum
bone
cementum
periodontal membrane
root

During our lives we have two sets of teeth. The first are called **milk teeth**. They are later replaced by the **permanent teeth**. Tooth replacement begins usually at about five years of age and takes several years to complete. The final molars are called the **wisdom teeth** and usually emerge in early adult life. Permanent teeth develop from separate rows of tooth buds which arise under the buds from which the milk teeth grow (Figs 15.1 and 15.3).

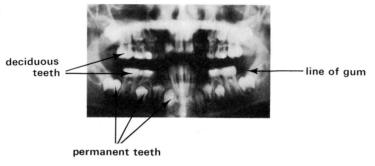

Fig 15.3 X-ray photograph showing the dentition of an 8-year-old child. Note the presence of milk (deciduous) teeth as well as permanent teeth (courtesy Mr Gould, Dental Department, General Hospital, Nottingham)

15.2.2 Variety of mammalian teeth

Mammals are **heterodont**. Their dentition includes several kinds of teeth. Moving from the front of the jaw to the rear there are **incisors, canines, premolars** and **molars**. The various kinds of teeth perform different functions during feeding.

Carnivores eat meat which is often captured alive. The teeth pierce and grip the prey to prevent it escaping and rip and cut it when chewing. In the dog these functions are performed mainly by the canine and **carnassial** teeth respectively. The latter are the fourth upper premolar and the first lower molar teeth (Fig 15.4). Attachment of the lower jaw to deep grooves on the sides of the skull of the dog restricts the jaw to up and down movement, ideal for cutting flesh (Fig. 15.5). The molar and premolar teeth act like scissors, the upper teeth biting outside the lower ones.

Fig 15.4 Dentition of the dog. The dental formula is $\frac{3}{3}\frac{1}{1}\frac{4}{4}\frac{2}{3}$. The formula refers to the number of incisors, canines, premolars and molars on each side of the upper and lower jaws

Fig 15.5 Vertical biting motion of the lower jaw in carnivores

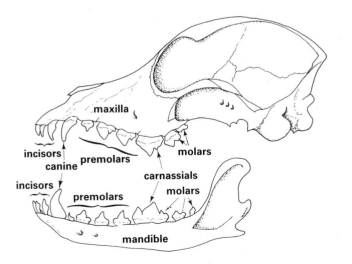

Herbivores eat vegetation. The canine teeth are absent and a toothless space, the **diastema**, separates the incisors from the premolars (Fig 15.6). In sheep the upper incisors are also missing. When grazing, the lower set of incisors bite against a horny pad in the upper jaw. The main chewing action in herbivores involves the premolars and molars. Because herbivores chew almost constantly the premolars and molars become very worn. The wearing is uneven, as the enamel on the outer surfaces of the teeth is harder than the dentine inside. Consequently the grinding surfaces of the premolars and molars develop a ridged pattern (Fig 15.7).

Fig 15.6 Dentition of the sheep. The dental formula is $\frac{0}{3}\frac{0}{1}\frac{3}{3}\frac{3}{3}$. Compare with Fig 15.4

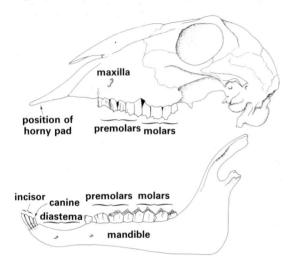

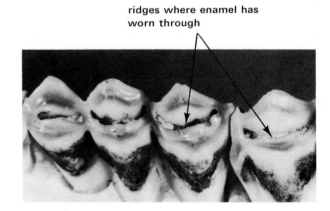

Fig 15.7 The grinding surfaces of a sheep's molars

ridges where enamel has worn through

Attachment of the lower jaw to the skull allows a circular jaw motion in the horizontal plane. The ridges of the upper and lower premolars and molars slide over one another with the food in between (Fig 15.8). In this way vegetation is ground to very fine particles before swallowing. Humans are called **omnivores** because we usually eat meat, fruits and vegetables. We have a relatively unspecialised dentition (Fig 15.9).

Fig 15.8 Circular grinding motion of the lower jaw in herbivores

Fig 15.9 Dentition of a human. The dental formula is $\frac{2}{2}\frac{1}{1}\frac{2}{2}\frac{3}{3}$. Note the relatively unspecialised teeth compared with the dog and sheep (Figs 15.4 and 15.6)

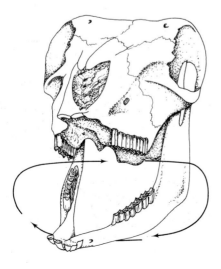

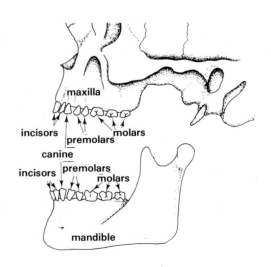

284

The fossil record shows that primitive mammals had a **dental formula** of:

$$\dfrac{3 \quad 1 \quad 4 \quad 3}{3 \quad 1 \quad 4 \quad 3}$$

That is, three incisors, one canine, four premolars and three molars on each side of both upper and lower jaws, a total of 44 teeth. Present-day mammals have either this full dentition or, more usually, a reduced number of teeth. Some species of whales and a few other mammals have more teeth.

Modern mammals are thought to have evolved from early insect-eating species. Mammals belonging to the order Insectivora which includes shrews, hedgehogs and moles usually have a full dentition (Fig 15.10).

Fig 15.10 Dentition of the hedgehog

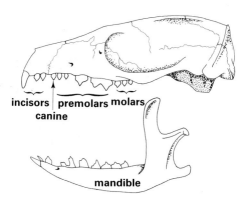

Rodents, such as rats, mice, squirrels, voles and beavers have incisors with **open roots** which allow the incisors to grow continually. Constant growth of the incisors is necessary to replace the tooth substance worn away by the gnawing action of the jaws. Enamel is present only on the outer surface of the incisors and a sharp sloping cutting edge is created as the teeth are worn. Rabbits and hares, which belong to the order Lagomorpha, have the same pattern of incisor growth. However, lagomorphs have a small second pair of incisors just behind the first pair (Fig 15.11).

Fig 15.11 (a) Dentition of the rat. The dental formula is $\frac{1}{1}\,\frac{0}{0}\,\frac{0}{0}\,\frac{3}{3}$

Fig 15.11 (b) Dentition of the rabbit. The dental formula is $\frac{2}{1}\,\frac{0}{0}\,\frac{3}{2}\,\frac{3}{3}$

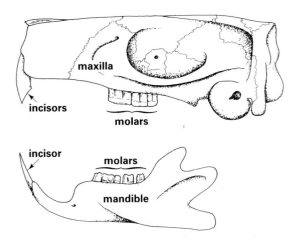

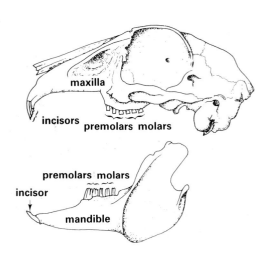

Toothed whales and dolphins are mostly flesh-eaters. They have very long jaws with rows of many peg-like teeth which are used to grip prey. Whalebone whales on the other hand feed on plankton. Only the fetus whalebone whale has teeth. In the adult, rows of large keratinous plates hanging transversely from the roof of the mouth trap plankton from sea water drawn into the mouth. The muscular tongue then forces sea water out of the mouth and the food is swallowed (Fig 15.12).

Fig 15.12 (a) Dentition of a toothed whale

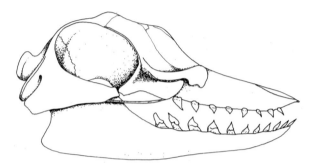

Fig 15.12 (b) Skull of a right whale, showing the baleen plates used for sieving food from the sea

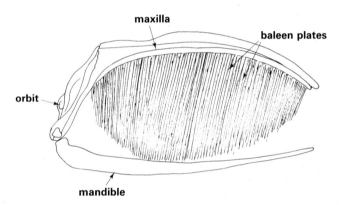

Elephants have huge molar teeth with very large grinding surfaces. The large grinding area enables elephants to cope with the quantities of food necessary to support their massive bodies. The tusks of elephants are a single pair of continually growing upper incisors. They are composed of solid dentine except for a cap of enamel at their tips (Fig 15.13).

Fig 15.13 Dentition of the elephant

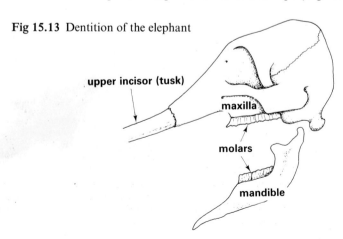

Fig 15.14 (a) Diagram of the digestive system

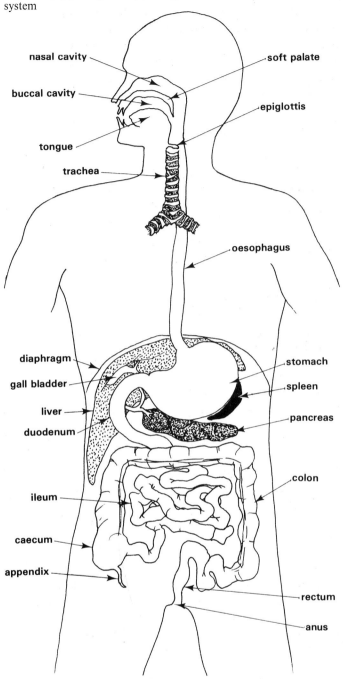

- nasal cavity
- buccal cavity
- tongue
- trachea
- soft palate
- epiglottis
- oesophagus
- diaphragm
- gall bladder
- liver
- duodenum
- ileum
- caecum
- appendix
- stomach
- spleen
- pancreas
- colon
- rectum
- anus

15.3 Digestion

Many of the organic nutrients in food, such as proteins and polysaccharides, are substances of high relative molar mass. They are insoluble in water and cannot pass through the membranes of body cells. It is necessary for such substances to be broken down into units small enough to be absorbed into the body fluids in which they can be transported to all parts of the body. The breaking down of food is called **digestion**. It takes place in the gut where food is acted on by hydrolytic enzymes (Chapter 3).

The digestive system is illustrated in Fig 15.14. The enzymes are secreted by a variety of **exocrine glands** located in the digestive system. The glands develop from epithelia and secrete their products through a system of ducts (Fig 21.2). They are classified according to the shape of the secretory structure and the complexity of their ducts (Fig 15.14(b)).

Fig 15.14 (b) The main kinds of exocrine glands, many of which are found in the digestive system

simple tubular
eg. gastric pit

simple coiled tubular
eg. sudorific gland

simple alveolar
eg. mucus gland

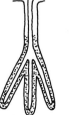

simple branched tubular
eg. fundic gland in stomach

simple branched alveolar
eg. meibomian gland in eyelids

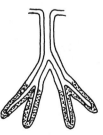

compound tubular
eg. Brunner's gland

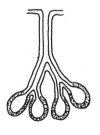

compound alveolar
eg. lactating mammary gland

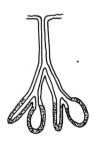

compound tubular-alveolar
eg. sub-maxillary gland

287

15.3.1 Mouth

The smell and sight of food, as well as the mechanical stimulation of food in the mouth, triggers a reflex action which results in the secretion of **saliva**. The reflex ensures a flow of saliva when food is in the mouth. Secretion of saliva slows down when we are not eating. Saliva enters the mouth from three pairs of salivary glands (Fig 15.15).

Fig 15.15 Position of the salivary glands

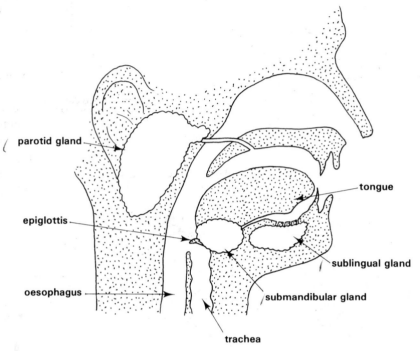

An enzyme called **salivary amylase** is usually found in saliva. Salivary amylase hydrolyses glycosidic linkages in **starch**, breaking it down to the disaccharide sugar **maltose** (Chapter 2). Salivary amylase probably contributes little to digestion. This is because we hold food in the mouth for only a short time. Salivary amylase works best in neutral conditions and the pH of saliva is about 7. After swallowing, food passes down the oesophagus into the stomach where the pH is much lower. In acid conditions salivary amylase becomes inactive. Saliva **lubricates** the pharynx and oesophagus, making it easier for food to be swallowed.

The food is moulded into a ball (bolus) by the tongue, then pushed into the pharynx. Contraction of the pharyngeal wall pushes the bolus into the oesophagus. Rhythmical contractions of the oesophagus, called **peristalsis**, move the bolus towards the stomach. Behind the bolus the circular muscle layer contracts, while the longitudinal muscle layer relaxes. The effect is to constrict the gut behind the bolus, thus pushing it along. Peristalsis is a property of all parts of the gut and helps to move food along the alimentary canal. During swallowing the larynx is raised and the **glottis** is covered by the **epiglottis**, so preventing food from entering the trachea (Fig 15.16).

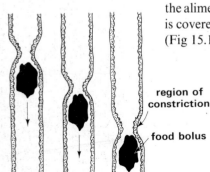

region of constriction

food bolus

Fig 15.16 Peristalsis

Fig 15.17 (a) Main regions of the stomach

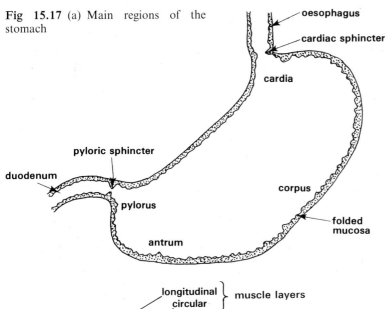

oesophagus
cardiac sphincter
cardia
pyloric sphincter
duodenum
pylorus
antrum
corpus
folded mucosa

15.3.2 Stomach

The first region of the gut where any significant digestion of food takes place is the **stomach**. It is a muscular bag with a volume of about two litres. It has a delicate inner folded membrane called the **gastric mucosa**. **Gastric juice** is secreted by numerous microscopic **gastric pits** embedded in the mucosa (Figs 15.17 and 15.18). The juice is very acid with pH of about 2. The acidity is caused by **hydrochloric acid** secreted by **oxyntic cells**. It is thought the main function of gastric acid is to kill microbes which enter the body in our food. **Peptic cells** secrete **pepsinogen** which, on contact with the acid, is converted to the peptidase enzyme **pepsin**. Pepsin hydrolyses peptide linkages in proteins to produce polypeptides.

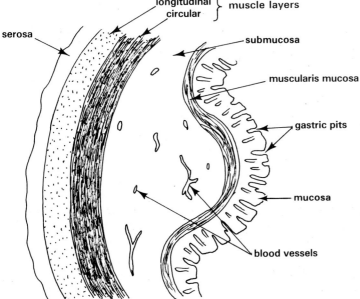

longitudinal
circular
} muscle layers
serosa
submucosa
muscularis mucosa
gastric pits
mucosa
blood vessels

Fig 15.17 (b) Three layers of muscles in the wall bring about stomach movements which mix the food with gastric juice during digestion. See Fig 15.20

Fig 15.18 (b) Photomicrograph showing the gastric glands in a thin section of cat stomach, × 400

Fig 15.18 (a) Diagram of a gastric gland

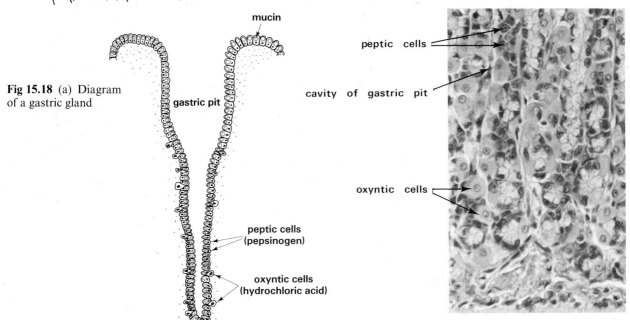

mucin
gastric pit
peptic cells (pepsinogen)
oxyntic cells (hydrochloric acid)
peptic cells
cavity of gastric pit
oxyntic cells

The corrosive and digestive properties of hydrochloric acid and pepsin place the delicate gastric mucosa at risk. Protection is given by a slimy glycoprotein called **mucin**. Mucin is secreted by **goblet cells** in the mucosa and forms a layer of **mucus** over the stomach lining. Mucus protects against mechanical as well as chemical injury. If the protection is not effective the mucosa and stomach wall are attacked by the gastric juice, causing an **ulcer** to form.

Gastric secretion is partly controlled by autonomic reflexes. Secretion of gastric juice begins when food is in the mouth. The presence of food in the stomach triggers the stomach lining to produce a hormone called **gastrin** which enters the blood. Gastrin stimulates continued secretion of gastric juice after the faster nervous mechanism has started it off (Fig 15.19).

Fig 15.19 Relative importance of nervous (vagus) and hormonal (gastrin) influence on gastric juice secretion

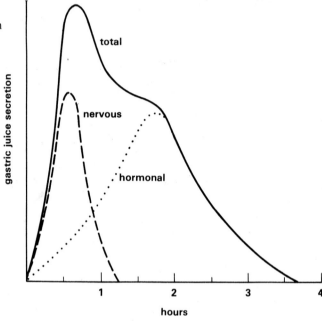

How is gastric juice secreted before food even enters the stomach?

During the three or four hours food is held in the human stomach, rhythmic muscular contractions of the stomach wall churn the food, mixing it with gastric juice. Consequently more protein molecules in the food are brought into contact with gastric enzymes in a given time, so promoting digestion. The food is gradually converted into a creamy, acid suspension called **chyme** (Fig 15.20).

Fig 15.20 Churning action of the stomach during digestion (from X-ray photographs)

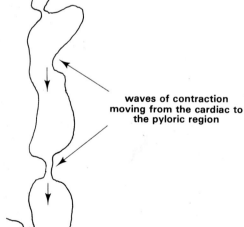

closed sphincters

waves of contraction moving from the cardiac to the pyloric region

15.3.3 Rumination

In **ruminant mammals** such as cattle and sheep, the stomach is large and modified for the digestion of vegetation. Cellulose in plant cell walls is the main carbohydrate constituent of the food eaten by ruminants. The ruminant stomach consists of four chambers, the **rumen, reticulum, omasum** and the **abomasum** (Fig 15.21).

Fig 15.21 Ruminant stomach. Continuous arrows indicate the course of food in the rumen and reticulum after swallowing. Following a return to the mouth for cudding, food enters the true stomach by a course shown by a broken arrow

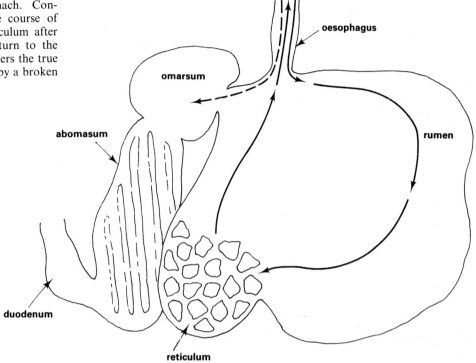

Food is passed first into the largest chamber, the rumen, where huge populations of **anaerobic cellulolytic bacteria** begin the fermentation of cellulose. Fermentation continues in the reticulum into which the contents of the rumen are later directed. The partially digested food is then forced back into the mouth for further grinding called **chewing the cud**. Thorough grinding of plant material is essential if the cellulolytic bacteria are to be effective. On reswallowing, the semi-liquid cud is directed by closure of a groove on the side of the reticulum into a small chamber, the omasum. Inside the omasum, water is pressed out of the cud and is absorbed. The solid material passes into the abomasum. This is the true stomach and secretes an acid peptic juice.

The bacterial fermentation which occurs in the rumen produces sugars, methane gas and carboxylic acids such as ethanoic acid. Bacteria entering the abomasum with the cud are digested and add to the material available for absorption.

15.3.4 Small intestine

During gastric digestion, food is kept in the stomach by the constriction of two **sphincters**. These are circular muscles which seal off the stomach where it is joined to the oesophagus and small intestine (Fig 15.17(a)). Relaxation of the **pyloric** sphincter allows the propulsion of chyme, a little at a time, into the small intestine by peristalsis of the involuntary muscles in the gastric wall.

From the stomach, chyme passes first into the **duodenum** which is about 30 cm long. Beyond the duodenum is the **jejunum**, and then the **ileum** making up the rest of the small intestine (Fig 15.14). The total length of the small intestine is about 6 m.

Digestion occurs mainly in the small intestine where enzymes complete the chemical breakdown of food into a state suitable for absorption. The enzymes come from two sources, the pancreas and the glands in the wall of the small intestine. Food is also exposed to a non-enzymic fluid called bile made in the liver (Fig 15.22).

Fig 15.22 Anatomical relationship between the liver, gall bladder, pancreas, stomach and small intestine (not drawn to scale)

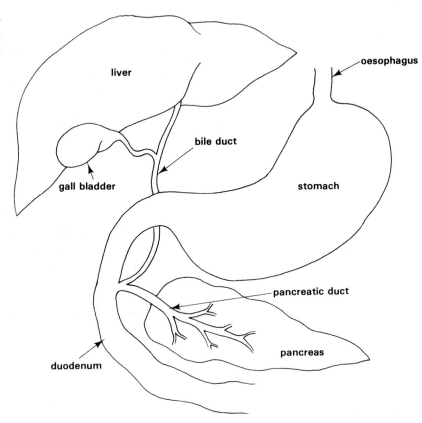

1 Bile

Bile is a greenish fluid containing **bile pigments** which are the excretory products of the breakdown of the haem part of haemoglobin. Though containing no enzymes, bile helps digestion in several important ways. First, it contains sodium hydrogencarbonate which gives it a pH of between 7 and 8. This is the optimum pH for the action of pancreatic and intestinal enzymes. Bile is therefore important in **neutralising** the acid chyme from the stomach. The second digestive function of bile is due to the **bile salts**, sodium and potassium glycocholate and taurocholate. They **emulsify** lipids, causing them to break down into numerous small droplets, about 0.5–1.0 μm in diameter. Emulsification provides a relatively large surface area of lipid for the action of lipase enzymes and hence speeds up digestion of fats and oils.

2 Pancreatic juice

The **pancreas** is situated just beneath the stomach and is connected to the small intestine by a pancreatic duct through which **pancreatic juice** is discharged. The bile duct joins the pancreatic duct. The endocrine component of the pancreas, the islets of Langerhans (Chapter 21), plays no part in digestion.

Bile contains no digestive enzymes, yet is important to the process of digestion. What digestive role does bile play?

Pancreatic juice contains four enzymes. **Pancreatic amylase** hydrolyses glycosidic linkages in starch, converting it to maltose in the same way as salivary amylase. Because food remains in the duodenum for some time, there is more opportunity for hydrolysis of starch in the small intestine than in the mouth. **Pancreatic lipase**, probably a group of several lipase enzymes, hydrolyses lipids to fatty acids and glycerol (Fig 15.23).

Fig 15.23 Summary of the actions of pancreatic amylase and lipase

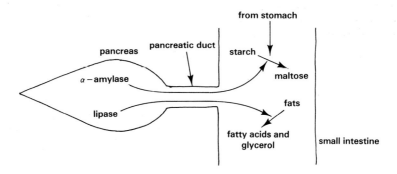

Two inactive enzyme precursors are also found in pancreatic juice. They are **trypsinogen** and **chymotrypsinogen**. Trypsinogen is converted to the active enzyme **trypsin** by an activator called **enterokinase** made in the intestinal glands. Trypsin activates the conversion of chymotrypsinogen to **chymotrypsin**. Trypsin and chymotrypsin are enzymes having an effect on proteins similar to that of gastric pepsin. They bring about a partial breakdown of proteins to polypeptides (Fig 15.24).

Fig 15.24 Summary of the actions of trypsin and chymotrypsin

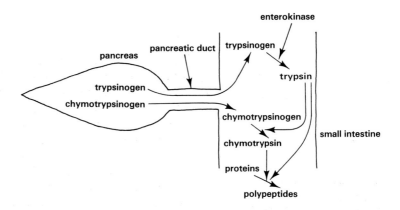

3 Intestinal fluid

Embedded in the intestinal walls is a great number of microscopic pits called the **crypts of Lieberkühn**. In the duodenum, coiled **Brunner's glands** in the sub-mucosa secrete alkaline mucus into the crypts. Enzymes from the crypts complete the process of digestion started by the mouth, stomach and pancreas. Intestinal juice, called **succus entericus**, contains a large number of enzymes which act on carbohydrates, polypeptides and lipids. The enzymes and their activities are summarised in Table 15.3. Some of the final stages in the digestion of carbohydrates, fats and polypeptides occur in the cells lining the intestinal mucosa (section 15.4).

Table 15.3 Summary of the action of enzymes secreted by the small intestine

Enzymes	Substrates	Products
amylase	starch	maltose
maltase	maltose	glucose
sucrase (invertase)	sucrose	glucose and fructose
lactase	lactose	glucose and galactose
peptidases	polypeptides	amino acids
lipases	fats	fatty acids and glycerol

The mechanisms controlling the release of bile, pancreatic juice and succus entericus are complex. They are triggered by the presence of chyme in the small intestine. Chyme stimulates the release of at least three hormones from the intestinal mucosa into the blood. **Pancreozymin** (cholecystokinin-pancreozymin, CCK-PZ) triggers the release of bile from its temporary store, the gall bladder, and the release of pancreatic enzymes. **Secretin** stimulates the flow of an alkaline fluid from the pancreas. **Gastric secretin** from the stomach wall has an effect similar to secretin. Finally **enterocrinin** stimulates the intestinal glands to secrete their digestive juice (Fig 15.25).

Fig 15.25 Summary of the actions of gastrointestinal hormones involved in the release of bile and the digestive juices of the stomach, pancreas and small intestine

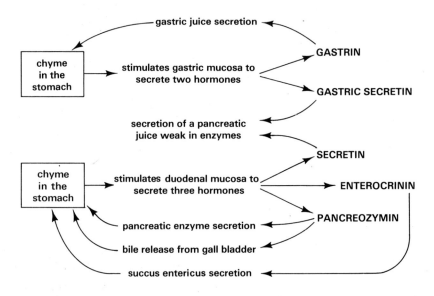

15.3.5 Intestinal micro-organisms

There are countless micro-organisms on our bodies and in our digestive system. They perform a variety of functions that are poorly understood. Nonetheless they are known to be vital to our normal development.

There are normally relatively few micro-organisms in the stomach, duodenum, jejunum and upper ileum. Many of those present have probably been swallowed. Gastric juice is an effective barrier to the entry of unwanted micro-organisms into the gut. Micro-organisms indigenous to the digestive system are frequently present in the lower ileum and are always found in the caecum and colon. They usually closely adhere to the mucosal surface where they metabolise bile pigments and acids and produce gases such as carbon dioxide and hydrogen sulphide. They also degrade enzymes secreted in the various digestive juices. Some intestinal bacteria can synthesise vitamins such as K, B_{12} thiamin and riboflavin (section 15.1.2).

Occasionally the intestinal micro-organisms may cause disease. They can remove excessive quantities of nutrients from the gut and cause mild toxaemias. There is some evidence of a possible link between the activities of some kinds of intestinal micro-organisms and cancer of the colon.

15.3.6 The caecum

In non-ruminant herbivores such as rabbits and horses fermentation of cellulose occurs in the caecum. The caecum is a bag-like structure and is the first part of the large intestine (Fig 15.26). Inside the caecum **cellulolytic bacteria** perform the same function as those in the rumen of cattle and sheep (section 15.3.3). In many herbivores digestion and fermentation render the food soluble enough to be absorbed on its first passage through the gut. In others such as rabbits, the food must pass twice through the gut before it is sufficiently broken down for absorption to take place. For this reason rabbits eat the faeces formed from food which has passed once through the gut.

Fig 15.26 The relative proportions of the caecum in (a) rabbit and (b) human. The drawings are not to the same scale (after Clegg and Clegg)

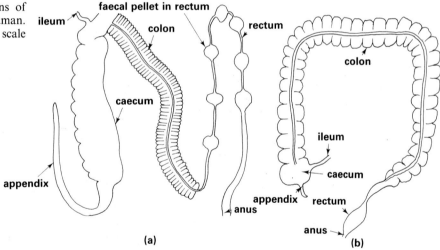

15.4 Absorption

Peristalsis of the small intestine brings chyme into contact with fresh secretions of intestinal fluid. However, movement by peristalsis is relatively slow, about $1-2\,cm\,s^{-1}$. More effective mixing of food with intestinal fluid is brought about by **segmentation** of the gut. This happens when the circular muscle layer contracts locally, pinching the gut into short segments. Consequently food in the small intestine is squeezed rapidly back and forth. In this way more enzyme molecules come into contact with the food in a given time. As a result, chyme is turned into a watery emulsion called **chyle**.

The absorption of nearly all the digestive products takes place through the extensive mucosa mainly in the duodenum and the first part of the jejunum. The internal surface area of the small intestine is greatly enlarged

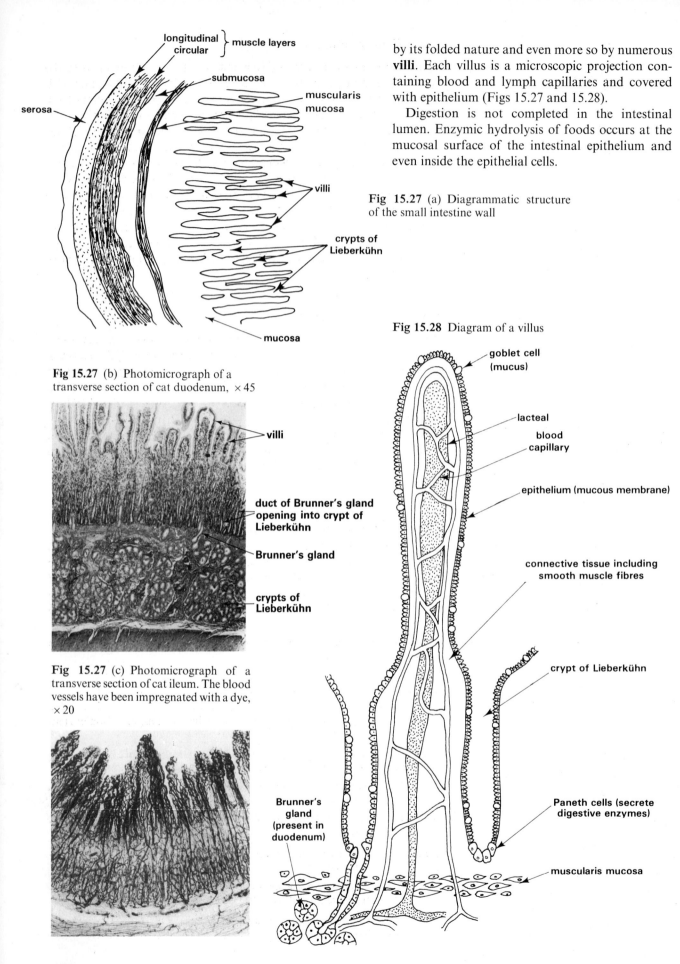

longitudinal ⎫
circular ⎬ muscle layers

submucosa

serosa

muscularis
mucosa

villi

crypts of
Lieberkühn

mucosa

by its folded nature and even more so by numerous **villi**. Each villus is a microscopic projection containing blood and lymph capillaries and covered with epithelium (Figs 15.27 and 15.28).

Digestion is not completed in the intestinal lumen. Enzymic hydrolysis of foods occurs at the mucosal surface of the intestinal epithelium and even inside the epithelial cells.

Fig 15.27 (a) Diagrammatic structure of the small intestine wall

Fig 15.27 (b) Photomicrograph of a transverse section of cat duodenum, × 45

villi

duct of Brunner's gland opening into crypt of Lieberkühn

Brunner's gland

crypts of Lieberkühn

Fig 15.27 (c) Photomicrograph of a transverse section of cat ileum. The blood vessels have been impregnated with a dye, × 20

Fig 15.28 Diagram of a villus

goblet cell (mucus)

lacteal

blood capillary

epithelium (mucous membrane)

connective tissue including smooth muscle fibres

crypt of Lieberkühn

Brunner's gland (present in duodenum)

Paneth cells (secrete digestive enzymes)

muscularis mucosa

15.4.1 Fate of lipids

Emulsification increases the surface area of fats and oils exposed to the action of pancreatic and intestinal lipases. Bile salts, in combination with lecithin or polar lipids such as monoglycerides, act as powerful emulsifiers. Fatty acids derived from dietary lipids are insoluble in water. They are **solubilised** through the formation of small, stable **micelles** of about 4–5 nm diameter. Each micelle consists of a shell containing monoglycerides, free fatty acids, cholesterol and fat-soluble vitamins. In this form the products of fat digestion are transported to the surfaces of the intestinal epithelial cells into which they are absorbed. Absorption is thought to be by diffusion. Di- and triglycerides are formed inside the epithlial cells and incorporated into envelopes of protein, cholesterol and phospholipid called **chylomicrons** which can be more than 100 nm across. Chylomicrons enter the intestinal **lacteals** which are lymphatic capillaries (Fig 15.28).

15.4.2 Fate of protein

Free amino acids and short-chain polypeptides remain in the small intestine after the action of gastric and pancreatic proteolytic enzymes. Complete hydrolysis of polypeptides involves intestinal enzymes which are active in the plasma membrane of the epithelium and in the epithelial cytoplasm. Polypeptides are hydrolysed to free amino acids which are transferred into the blood vessels of the villi. Amino acid transfer is enhanced by the presence of sodium ions (Na^+). Some long-chain polypeptides and even traces of intact dietary protein may be absorbed into the blood from the intestinal lumen.

15.4.3 Fate of carbohydrate

Carbohydrate can only be absorbed into the blood in appreciable quantities in the form of monosaccharides. Most carbohydrate digestion occurs in the jejunum, mainly at the epithelial surface. Pancreatic amylase has been found on the jejunal surface. The transfer of the products of carbohydrate digestion into the blood involves a special mechanism that is not fully understood. It requires metabolic energy and may be enhanced by sodium ions.

Some of the factors affecting the absorption of dietary constituents are summarised in Table 15.4.

Table 15.4 Factors affecting the absorption of nutrients

Glucose	Active absorption may be linked with the transport of sodium ions across the membranes of mucosal cells.
Amino acids, some peptides	Active absorption may be affected by absorption of sugars and may also be linked with sodium ion transport through the mucosa.
Vitamin B_{12}	Absorption dependent upon presence of intrinsic factor (IF) in gastric juice.
Vitamin K	Much produced by intestinal micro-organisms. Absorption promoted by bile.
Calcium ions	Active absorption promoted by vitamin D_2 and probably also by parathormone.
Iron ions	Absorbed in the iron(II) state. Rate of absorption depends on degree of saturation of transferrin with iron in the blood.
Water	Passive absorption by osmosis depends upon solute absorption.

15.4.4 Colon

It is the large intestine, especially the colon, that the absorption of water takes place. The digestive secretions of the gut add several litres of fluid to the contents of the intestine. If most of the water were not reabsorbed, the loss would seriously reduce the body's water content. Faecal water accounts, on average for about 4–8 % of the total water loss in humans. Sodium, chloride and hydrogencarbonate ions are also absorbed from the contents of the colon.

In recent years a lot of interest has been expressed in the **fibre** content of our diet. Fibre is a complex of many substances of vegetable origin, mainly polysaccharide carbohydrates such as cellulose, hemicelluloses, pectins, gums and mucilages. Non-carbohydrate components of fibre include lignin. Most Western diets are relatively low in fibre compared with those of rural populations in Africa. In the West there is a higher risk of intestinal diseases such as constipation, diverticular disease, haemorrhoids (piles) and cancer of the colon. Why this should be so is not understood, but it has been shown that fibre decreases intestinal transit time and increases the faecal mass. Fibre absorbs water and swells, filling and stretching the gut. A stretched gut stimulates peristalsis, hence waste food is held for a shorter time in the large intestine. Also, fibre combines with fatty acids and lowers their rate of absorption. A full gut triggers the satiety ('full-up') reflex too, thus suppressing the desire to eat.

Unabsorbed matter including the excretory products of bile is passed to the exterior through the **anus** as **faeces**.

15.5 The liver

Several chapters in this book refer to **homeostasis**, the maintenance of a steady internal environment. Regulating the composition of blood and tissue fluid is an important aspect of homeostasis. The composition of blood leaving the gut in the hepatic portal vein is largely determined by the diet and thus the substances absorbed from the small intestine. However, the blood which enters the general circulation is often very different in composition. The hepatic portal vein carries blood to the liver. One of the functions of the liver therefore is to regulate the composition of blood. The liver is the largest organ in the body and performs many functions, not all of which are concerned with homeostasis.

15.5.1 Liver structure

The liver consists of a large number of **lobules** (Fig 15.29). Each lobule contains many vertical plates of liver cells,

Fig 15.29 (a) Diagram of a liver lobule

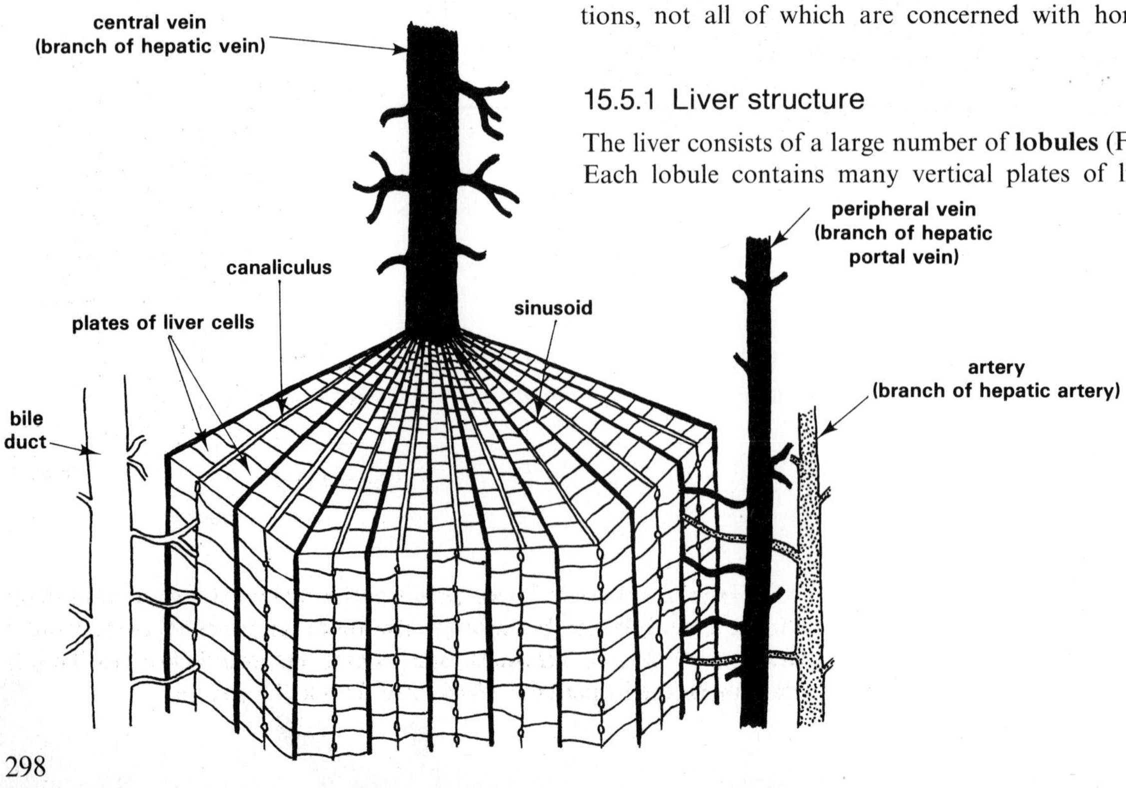

central vein (branch of hepatic vein)

canaliculus

plates of liver cells

bile duct

sinusoid

peripheral vein (branch of hepatic portal vein)

artery (branch of hepatic artery)

arranged radially around a central blood vessel which is a branch of the **hepatic vein**. The hepatic vein drains blood from the lobules and releases it into the general circulation. The blood supply to each lobule is from two sources, the **hepatic artery** and the **hepatic portal vein**. Branches of these vessels are found between the liver lobules. The artery delivers oxygenated blood from the general circulation and the portal vein delivers food-laden blood from the gut. Blood from the hepatic artery and hepatic portal vein flows between the plates of liver cells inwards to the central vein in channels called **sinusoids** (Fig 15.29(a) and (c)).

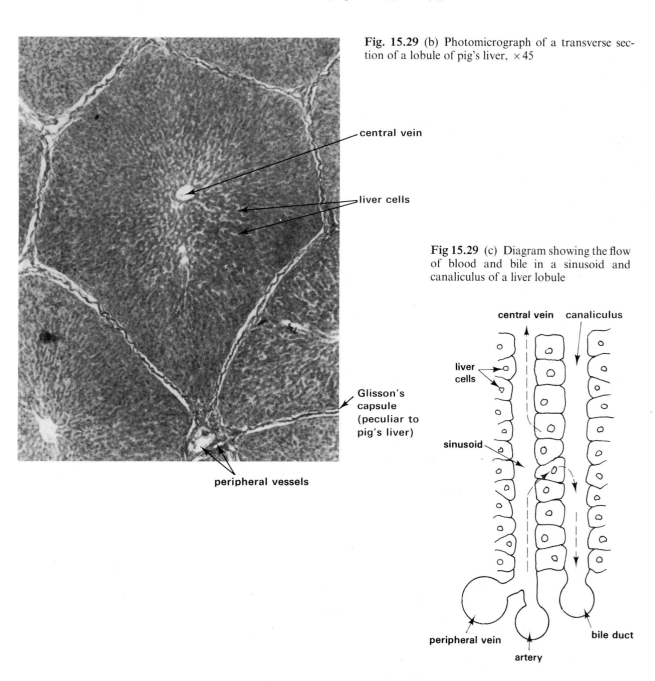

Fig. 15.29 (b) Photomicrograph of a transverse section of a lobule of pig's liver, ×45

Fig 15.29 (c) Diagram showing the flow of blood and bile in a sinusoid and canaliculus of a liver lobule

Between the plates of liver cells are other channels called **canaliculi** which receive **bile**. The bile moves outwards to the periphery of the lobules where it collects into **bile ducts**. Bile is stored temporarily in a sac-like **gall bladder** before its periodic release into the small intestine.

15.5.2 Liver functions

The functions of the liver are numerous and vital to life. For convenience they are described under three main headings.

1 Metabolism of absorbed food

i. Carbohydrate metabolism   Soluble sugars, mainly glucose, are carried from the gut to the liver in the hepatic portal vein. Following a carbohydrate-rich meal the concentration of glucose in blood going to the liver is relatively high. When absorption is completed the glucose level drops. However, the concentration of glucose in the general blood circulation remains fairly stable. This is because excess glucose absorbed after a meal is converted in the liver cells to a storage polysaccharide called **glycogen**. The conversion is controlled by the hormone **insulin** from the islets of Langerhans in the pancreas (Chapter 21).

Glucose carried in the general blood circulation is used for tissue respiration throughout the body. It is replaced mainly from the liver's store of glycogen. The liver thus regulates the concentration of glucose in the general circulation even though the rate of glucose absorption from the gut varies.

ii. Protein metabolism   Unlike carbohydrate, protein is not stored in the body. Some amino acids from the gut pass through the liver and enter the **amino acid pool** in the general circulation. Amino acids are absorbed by the body cells and used for protein synthesis. The liver cells are important sites of **protein synthesis**, producing the plasma proteins. A stable **protein pool** is established in the blood. Synthesis is balanced by an equal breakdown of plasma proteins which have a limited lifetime in the circulation.

Plasma proteins at the end of their lifetime, together with excess amino acids absorbed from the gut, are deaminated by liver cells. **Deamination** is the removal of amino groups and their conversion to ammonia. The remains of the acid molecules enter the Krebs cycle and are used to produce respiratory energy. Ammonia is very toxic and enters a sequence of reactions called the **ornithine cycle** which converts ammonia to the less toxic **urea** (Fig 15.30). Urea is released into the general circulation and is excreted by the kidneys (Chapter 11).

Fig 15.30 Ornithine cycle

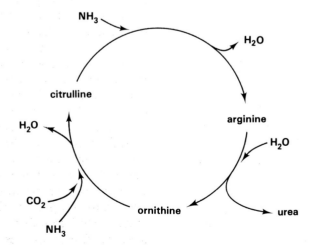

Urea is the main nitrogenous excretory product in mammals. Insects, reptiles and birds excrete their nitrogen mainly as uric acid. Uric acid is less toxic than urea and almost insoluble in water. Their urine is semi-solid, containing little water. The excretion of uric acid can be seen as an adaptation to life on land where water loss in urine can be a severe physiological problem. Some animals, such as non-chordate animals living in fresh water, excrete nitrogen in the form of ammonium compounds. These are very toxic compared with urea and uric acid and require great dilution during their excretion. Relatively great water loss accompanies the excretion of ammonium compounds.

About half of the twenty amino acids used for protein synthesis are non-essential: they can be produced from other amino acids in the liver by **transamination**. Liver cells contain a number of transaminase enzymes which transfer amino groups from amino acids to carboxylic acids, thus producing new amino acids. For example, glutamic-oxaloacetic transaminase (GOT) catalyses the following reaction:

$$\text{glutamic} + \text{oxaloacetic} \underset{}{\overset{\text{GOT}}{\rightleftharpoons}} \text{aspartic} + \alpha\text{-ketoglutaric}$$

$$\begin{array}{cccc} \text{acid} & \text{acid} & \text{acid} & \text{acid} \\ \text{(amino acid)} & \text{(carboxylic acid)} & \text{(amino acid)} & \text{(carboxylic acid)} \end{array}$$

Such enzymes enable the liver cells to make a wide range of amino acids, even though the variety of amino acids in the diet is limited.

iii. Lipid metabolism Fats and oils absorbed from the gut can be converted to glycogen which is stored in the liver. Fat-soluble vitamins such as vitamins A and D are also stored in liver cells. **Cholesterol**, a sterol carried in the blood, is used in a variety of syntheses, particularly cell membrane production. The liver makes cholesterol which is important when the dietary intake is inadequate. Excess cholesterol absorbed from the gut is eliminated in bile. Gross excesses of cholesterol, however, may be precipitated as gallstones in the bile duct and gall bladder.

2 Iron metabolism

Some of the liver cells lining the sinusoids are part of the body's **reticulo-endothelial system**. The system removes particles of debris from the blood. The **Küpffer cells** are phagocytic liver cells which remove old red cells from the blood. After old red cells have been engulfed, the haemoglobin is split into two, an iron–globin complex and an iron-less haem group. The haem is then converted to **bilirubin**. Bilirubin combines with glucuronic acid to form the bile pigment **bilirubin diglucuronide** which is excreted in bile. About 7–8 g haemoglobin are removed from the circulation in this way each day.

Any physical obstruction, such as **gallstones** in the bile ducts, or an increased breakdown of red cells, results in an increase in the concentration of bilirubin in blood. The skin takes on a yellow colour and the condition is called **jaundice**. The iron–globin complex is metabolised further in the liver. The globin is broken down into amino acids which are used for synthesis of proteins such as those of the blood plasma. The iron is retained for the production of fresh haemoglobin. Haemoglobin synthesis and red cell production (erythropoiesis) occur in the liver of the fetus. In the adult these processes are restricted to the red bone marrow (Chapter 10). Iron released from broken-down haemoglobin becomes attached to a plasma protein called **transferrin** and is transported in the blood to the bone marrow. Excess iron is stored in the liver cells as **ferritin** and **haemosiderin**.

The liver has been described as a filter. How does the liver help to regulate the concentrations of carbohydrates, fats and proteins in the systemic blood?

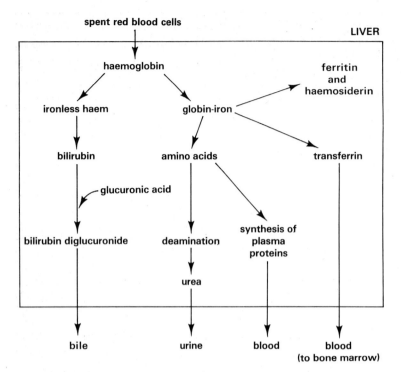

spent red blood cells

LIVER

haemoglobin

ferritin and haemosiderin

ironless haem → bilirubin

globin-iron → amino acids, transferrin

glucuronic acid

bilirubin diglucuronide

deamination

synthesis of plasma proteins

urea

bile   urine   blood   blood (to bone marrow)

Fig. 15.31 Summary of haemoglobin breakdown in the liver and the fate of the products

The total iron content of the adult body is normally between 3 and 5 g. Of this, about 1.5–5 g is in haemoglobin and about 1–1.5 g is stored in the liver. Pregnancy and lactation can lower a woman's total iron by as much as 20%. This is why iron-containing tablets are given to pregnant women. Fig. 15.31 summarises some of the factors involved in iron metabolism.

3 Detoxification

Many chemical substances which pass through the liver in the blood are modified by the liver cells. The substances include a variety of hormones. For example, much of the insulin from the pancreas is broken down by enzymes in the liver. Many sex hormones are also inactivated in the liver and excreted in the bile or released into the blood and excreted by the kidneys. Some chemicals are destroyed by liver cells, others are combined with various substances to render them less toxic. For example, benzene-carboxylic acid, a commonly used food preservative, is joined to the amino acid glycine to form *N*-benzoylglycine which is excreted in the urine. Certain chemicals such as tetrachloromethane, trichloromethane and ethanol damage the liver cells.

Within limits, then, the liver acts as a filter, removing toxic substances from the blood, making them less harmful and preparing them for excretion. The liver performs a variety of other functions which are described elsewhere in the book. Notable among them is the contribution the liver makes to heat production in the body (Chapter 20).

SUMMARY

Nutritional requirements: organic, vitamins and inorganic nutrients.

Organic

Carbohydrates: main energy source in respiration.
Proteins: source of amino acids, enzymes, structures, e.g. cell membranes, plasma proteins, antibodies.
Amino acids: essential ones must be provided in the diet; non-essential ones can be made from the essential ones in the liver.
Lipids: fats and oils; energy source in respiration, cell membranes, myelin, steroids (hormones), heat insulation.
Essential and non-essential fatty acids (as for amino acids).

Vitamins: fat-soluble, A, D, E and K; water-soluble, B and C.

Vit. A (retinol): visual pigments, epithelia.

Vit. D (ergocalciferol, D2): produced in the skin from ergosterol in UV light; calcium absorption in gut; deficiency results in rickets.

Vit. E: deficiency leads to degeneration of muscle and liver; causes anaemia.

Vit. K: prothrombin production in liver; deficiency leads to impaired coagulation.

Vit. B complex: many coenzymes.

B1 (thiamin): deficiency leads to beri-beri; paralysis, oedema and heart failure.

B2 (riboflavin): B6 required in protein metabolism.

B12 (cyanocobalamin): red cell maturation; deficiency leads to anaemia.

Vit. C (ascorbic acid): involved in respiration and collagen production; deficiency leads to scurvy.

Inorganic

Macronutrients: Sodium and potassium (impulse transmission and solute potential of body fluids).

Calcium and phosphate (bone and teeth, blood coagulation, muscle contraction).

Trace elements: Large quantities are poisonous.

Cobalt (in vit. B12), iodine (thyroid hormones), copper and zinc (enzyme activation), iron (haemoglobin, myoglobin, cytochromes).

Digestion

Teeth develop from buds in the gums; they have hard enamel coats and soft dentine; the pulp cavity contains blood vessels and nerve endings. In some mammals, the first set of teeth (milk) are replaced by permanent teeth. Mammals are heterodont; different types of teeth perform different functions, e.g. incisors (gripping), canines (piercing), premolars and molars (biting, ripping and chewing). Chewing surfaces often wear away in herbivores, producing abrasive ridges; teeth grow continuously.

Hydrolytic enzymes from digestive glands of gut render food in an absorbable form.

Mouth – saliva: α-amylase (starch→maltose).

Stomach – gastric juice: pepsin (proteins→polypeptides); hydrochloric acid (kills micro-organisms in food); intrinsic factor (absorption of vit. B12).

Rumen – stomach modified for digestion of vegetation; cellulose is hydrolysed by bacteria and partially digested food chewed as cud.

Pancreas – pancreatic juice: α-amylase (starch→maltose); lipase (fats→fatty acids and glycerol); trypsin and chymotrypsin (proteins→polypeptides).

Bile from liver – non digestive: $NaHCO_3$ (neutralises gastric acid); bile salts (emulsify fats); bile pigments (excretion of iron-less haem from haemoglobin).

Intestinal juice: α-amylase (starch→maltose); maltase (maltose→glucose); sucrase (sucrose→glucose and fructose); lactase (lactose/glucose and galactose); peptidases (polypeptides→amino acids); lipases (fats→fatty acids and glycerol).

The control of digestive juice secretion involves nerve reflex (vagus activation following tactile stimulation of mouth and stomach) and hormones: gastrin (gastric juice), gastric secretin and secretin (pancreatic juice weak in enzymes), enterocrinin (intestinal juice), pancreozymin (bile release and pancreatic enzymes).

Micro-organisms are mostly in the lower ileum, caecum and colon: they degenerate enzymes, metabolise bile pigments and synthesise some vitamins. Cellulolytic bacteria aid cellulose digestion in the stomach (rumen) or caecum (e.g. rabbits).

Most absorption occurs through the mucosa of the duodenum and jejunum. A large absorptive surface area is provided by villi. Enzymic hydrolysis of food occurs on the mucosal surface and inside the epithelial cells.

Lipids: solubilised micelles are absorbed into epithelial cells where chylomicrons are formed and absorbed into lacteals.

Proteins: complete hydrolysis occurs in epithelial plasma membranes and cytoplasm; absorption into blood is enhanced by sodium; some direct absorption of entire protein molecules.

Carbohydrates: absorption mainly of monosaccharides; complete digestion occurs on epithelial surface. ▶

► Colon: water, sodium and hydrogen carbonate absorption.

Liver: divided into lobules; central hepatic vein and peripheral hapatic portal vein, artery and bile duct.

Liver functions: metabolism of absorbed food:
Carbohydrates: excess glucose stored as glycogen (controlled by insulin).
Protein: plasma protein synthesis; transamination; deamination of excess amino acids and urea formation.
Lipids: fats and oils can be converted to glycogen; fat-soluble vitamins stored in liver cells; cholesterol synthesis; excess cholesterol excreted in bile.
Iron metabolism: spent red blood cells engulfed by Kupffer cells; iron-less haem converted to bile pigments (excess accumulates in jaundice); iron stored as ferritin and haemosiderin.
Detoxification: many poisons and hormones are modified by liver cells.

QUESTIONS

1 Describe the following processes in a mammal:
(a) digestion of starch,
(b) absorption of the products of starch digestion,
(c) control of the level of glucose in the blood. (L)

2 (a) Make a labelled low power plan of a transverse section of the ileum to show the main layers.
(b) Describe the functioning of the main layers of the ileum after a meal.
(c) Why is lymphoid tissue associated with the ileum?(L)

3 The diagram below represents the stomach of a sheep.

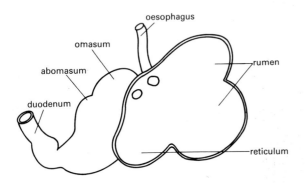

(a) Describe the treatment of food which occurs in the rumen.

(b) Explain what is meant by chewing the cud and indicate its significance.

(c) The abomasum can be described as the true stomach. Explain what this means.

(d) Indicate the ways in which the alimentary tract of a dog differs in structure from that of a sheep. Your answer should include reference to teeth. (L)

4 (a) What is the functional importance of the double blood supply to the mammalian liver?

(b) Complete the table below to identify those processes occurring in the liver which fit the descriptions stated.

Description	Process
1. Essential for the digestion of lipids	
2. Role in clotting blood	
3. Uses carbon dioxide in the detoxification of ammonia	
4. Storage as insoluble material within liver cells	
5. Helps in the maintenance of endothermy (homeothermy)	
6. Occurs only during the embryo stage	

(c) How does the liver assist in the mobilisation of energy reserves for use during periods of starvation?

(d) Suggest how the removal and destruction of worn out erythrocytes (red blood cells) by the liver cells may be important for efficient functioning of the kidneys. (L)

5 (a) Explain the concept of essential and non-essential amino acids.
(b) Describe how the composition of the following can be investigated.
(i) A mixture of proteins, and
(ii) a single protein.
(c) Describe the digestion of proteins in a mammal and explain how the mammal avoids digesting itself. (JMB)

16 Skeletal and muscle systems

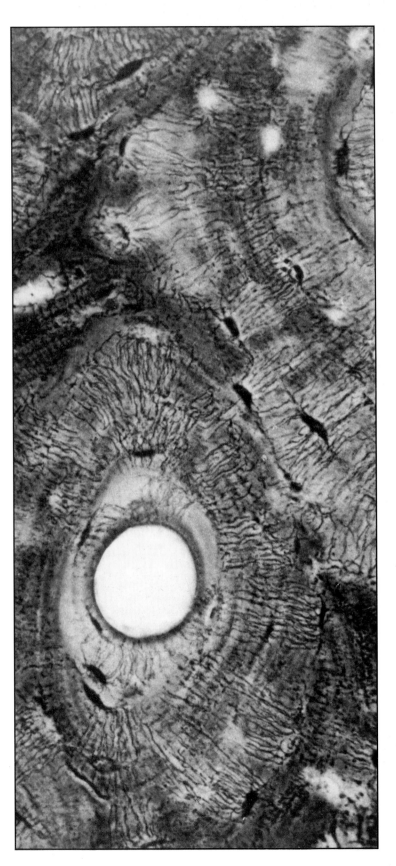

16.1 The skeleton   306

    16.1.1 The skull   307
    16.1.2 Vertebrae   308
    16.1.3 Rib cage   310
    16.1.4 Girdles   310
    16.1.5 Limbs   311

16.2 Cartilage and bone   313

    16.2.1 Cartilage   313
    16.2.2 Bone   313

16.3 Joints and movement 315

    16.3.1 Joints   315
    16.3.2 Movement   316

Summary   318

Questions   318

16 Skeletal and muscle systems

Mammals move their bodies by coordinated contraction and relaxation of muscles attached to the bones of the **skeleton**. Bones articulate with each other at **joints**. Movable joints allow considerable manoeuvrability while retaining the necessary strength to support the body's mass during movement. An important feature of the human skeleton is that it has evolved to allow **bipedal** movement, that is, walking upright on the hind limbs.

Fig 16.1 The skeleton, ventral view

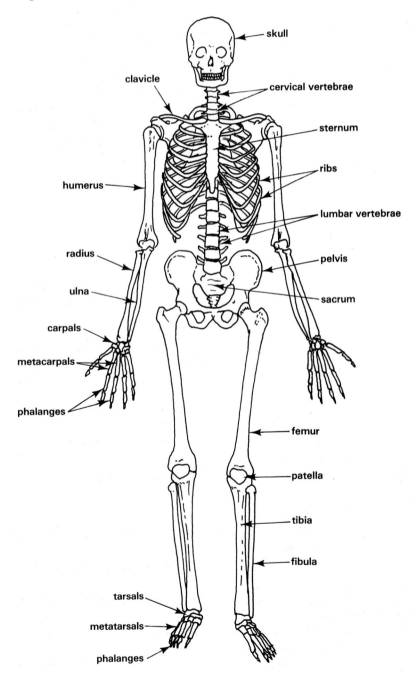

skull
clavicle
cervical vertebrae
sternum
ribs
humerus
lumbar vertebrae
radius
pelvis
ulna
sacrum
carpals
metacarpals
phalanges
femur
patella
tibia
fibula
tarsals
metatarsals
phalanges

16.1 The skeleton

The skeleton consists of bone and cartilage (section 16.2). They provide support and protection for the body's organ systems. Many of our blood cells are produced in the marrow of our bones (Chapter 10). Some bones perform specialised functions. For example, the middle ear ossicles conduct sound waves to the cochlea (Chapter 19). The skeleton (Fig 16.1) may be considered in two main parts. The **axial skeleton** consists of the skull, vertebrae and rib cage. The girdles and limbs constitute the **appendicular skeleton**.

16.1.1 The skull

The **skull** consists of 22 bones (Fig 16.2 and Table 16.1). Enclosing the brain, eyes and ears is the **cranium**. The human has a cranial capacity of about 1500 cm³ compared with 400–500 cm³ for apes such as chimpanzees and gorillas. The cranial bones include the large, upright **frontal** bones which give humans the characteristic high forehead. At the back of the cranium the **occipital** bones lack the ridges for attachment of powerful

Fig 16.2 Skull (a) frontal view (b) side view

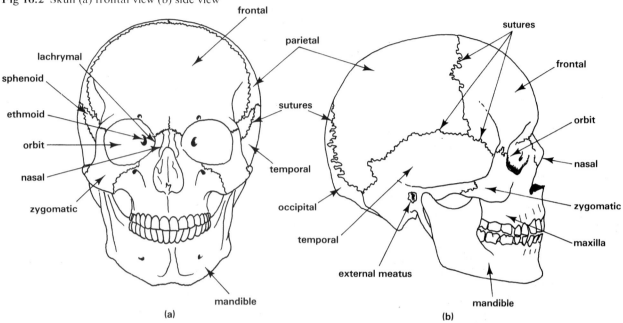

Table 16.1 Main functions of the skull bones

Name	Main function(s)
frontal bone	Forehead, anterior cranium, roofs of orbits, anterior floor of cranium. Contains cavities (frontal sinuses) which act as sound chambers, giving resonance to the voice.
parietal bones	Sides and roof of cranium.
temporal bones	Lower sides and part of floor of cranium. Enclose the inner ears. External auditory meati lead to middle ears. On mastoid processes behind external meati are points of attachment for neck muscles.
occipital bone	Posterior and part of floor of cranium. Foreamen magnum in base through which passes medulla oblongata.
sphenoid bone	Part of the floor of cranium. Part of the floors and sides of orbits. Houses the pituitary body. Part of the walls of the nasal cavities.
ethmoid bone	Part of the floor of cranium between the orbits. Main supporting structure of the nasal cavities.
nasal bones	Part of the bridge of the nose and upper part of face.
maxillae	Upper jaw bone; part of floors of orbits, part of roof of mouth, most of the hard palate, part of walls and floor of nasal cavities. Upper teeth set into maxillae.
malar (zygomatic) bones	Cheekbones; part of walls and floors of orbits.
mandible	Lower jaw; the only movable skull bone. Condylar processes articulate with temporal bones. Lower teeth set into mandible. Movements enable feeding, chewing and speech.
palate bones	Posterior part of hard palate. Part of floor and walls of nasal cavities. Separate nasal and oral cavities.
lachrymal bones	Part of medial wall of orbits.
inferior turbinated bones	Part of wall of nasal cavities. Like parts of the ethmoid bone, they allow circulation and filtration of inhaled air before it enters the respiratory passages.
vomer	Part of nasal septum; together with cartilage it divides nose into left and right.
hyoid bone	Found in neck between manidble and larynx. Supports tongue.

Fig 16.3 Skull of a new born baby showing fontanelles

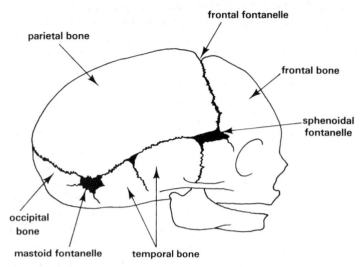

neck and jaw muscles that occur in apes. The **foramen magnum** is on the underside of the skull, thus allowing considerable reduction of musculature at the nape of the neck. Inserted on the mastoid swellings at the base of the skull are the sternomastoid muscles which allow free rotation of the head. Many of the skull bones are held together by immovable joints called **sutures**.

The skull bones develop from connective tissue and are not fully developed at birth (section 16.2.2). Between the cranial bones of a new-born baby there are membrane-filled spaces called **fontanelles** (Fig 16.3). They gradually close as the cranial bones grow and usually disappear by about 12 months of age.

The shortened **facial** bones give humans a more or less vertical face, apart from the nose, which extends the nasal channel. Inhaled air can be filtered and warmed before entering the respiratory system. The forward pointing **orbits** position the eyes so that their fields of vision overlap, thus allowing binocular vision (Chapter 19). The **jaws** are relatively short compared with other primates and the teeth arranged in semicircular rows. In apes the dental arcade is U-shaped. The lower jaw in humans is not as shortened as the upper and is strengthened on the outside to make the **chin**. In contrast, apes have a strengthening **simian shelf** of bone on the inside of the **mandible**. Their mandibles are larger and more robust than in humans, especially at the posterior ramus.

16.1.2 Vertebrae

The **vertebrae** make up the slightly S-shaped **spinal column** or backbone. In fetal life the thoracic curve of the backbone develops. After birth the cervical curve appears when a baby learns to hold up its head. When a child begins to walk the lumbar curve develops. There are 33 vertebrae in the five main regions of the spinal column: **cervical, thoracic, lumbar, sacral** and **coccygeal** (Fig 16.1). The vertebrae are modified to perform different functions in each region. However, each vertebra conforms to a common structural pattern (Figs 16.4 and 16.5).

The main load-bearing portion of a vertebra is a solid disc of bone called the **centrum**. The centra are relatively broad compared with those of some other vertebrate animals, thus helping to support an upright body. The centra of adjacent vertebrae are separated by **intervertebral discs** of cartilage. The discs act as shock absorbers and allow the spinal column to flex and arch. Sometimes the intervertebral discs become distorted, possibly due to excessive pressure from the spinal column, causing a disc to burst. The deformed disc may push against nearby nerve tissue and cause pain. The condition is often called a **slipped disc**.

Fig 16.4 Basic structure of vertebrae

(a) transverse view   (b) side view

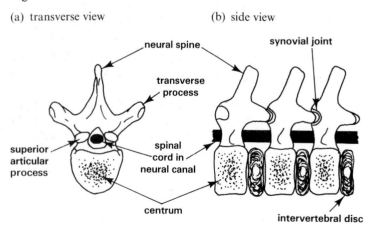

308

Projecting from the centrum dorsally is a **vertebral arch** which encloses the **neural canal**. The canal houses the spinal cord (Chapter 18).

Several bony projections arise from the vertebral arch. On each side there is a **transverse process**. These and a dorsal **neural spine** are points of attachment for muscles. Two **superior** (anterior) and two **inferior** (posterior) **articular processes** articulate with the vertebra above and below (Fig 16.4(b)).

The seven cervical vertebrae possess a relatively small centrum and a relatively large neural canal. The very short, forked neural spine provides sites for the insertion of posterior neck muscles which, in humans, are much less developed than in apes. There is a small canal in each transverse process which encloses the vertebral artery and vein and nerve fibres. The first cervical vertebra is called the **atlas** (Fig 16.5(a)). It articulates with the occipital bone of the skull and the head is thus balanced on the neck when the body is upright. The second cervical vertebra is called the **axis** (Fig 16.5(b)). Its centrum has an anterior projection called the **odontoid process** which fits into a space in the atlas. The atlas has no centrum. In its place is a space which accommodates the odontoid process. The arrangement acts as a pivot allowing the head to rotate sideways (Figs 16.5(c) and 16.17(d)). The odontoid process is separated from the spinal cord in the neural canal by a **transverse ligament**. However, sudden jolting of the neck may push the odontoid process into the spinal cord or even the medulla oblongata of the hind brain (Fig 18.15). The result is usually sudden death. The use of head restrainers on car seats has reduced the number of deaths caused by this type of injury in road traffic accidents. The twelve thoracic vertebrae have relatively long, backward-pointing neural spines (Fig 16.5(d)). They also possess relatively long transverse processes with **facets** for the attachment of the ribs. Facets are also found on the centra.

The five lumbar vertebrae are relatively bulky (Fig 16.5(e)). They have relatively short but sturdy neural spines and transverse processes to which are attached the powerful muscles of the lower back.

The five sacral vertebrae are fused into a single triangular structure called the **sacrum** (Fig 16.5(f)). The fusion of its constituent bones provides the sacrum with strength and rigidity. It acts as a firm point of attachment for the pelvic girdle (section 16.1.4).

The **coccyx** consists of four fused coccygeal vertebrae attached to the posterior apex of the sacrum (Fig 16.5(f)). The coccyx performs no known function in the human skeleton and is a vestigial tail.

How is the basic plan of vertebrae modified in different regions of the vertebral column to perform different skeletal functions?

Fig 16.5 Structures of vertebrae, not drawn to scale.

(a) atlas

(b) axis

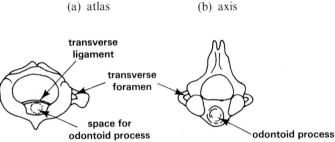

(c) atlas–axis articulation, side view

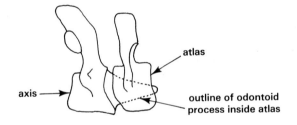

(d) thoracic vertebra

(e) lumbar vertebra

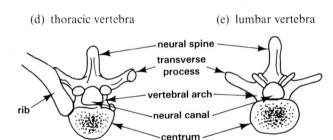

(f) sacrum

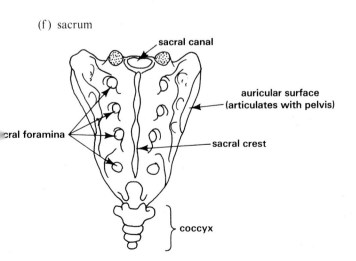

16.1.3 Rib cage

The rib cage of the **thorax** encloses the heart and lungs. Breathing is partly brought about by rhythmical movements of the **ribs**, resulting from the actions of **intercostal muscles** (Fig 8.3). The rib cage also provides support for the pectoral girdle and hence the arms.

There are twelve pairs of ribs, each articulating with a thoracic vertebra (Fig 16.5(d)). The first ten pairs of ribs are attached to the **sternum** (breastbone) by strips of cartilage called the **costal cartilages** (Fig 16.6). The first seven pairs of ribs are called **true ribs** because their costal cartilages attach them directly to the sternum. The costal cartilages of the next three pairs of ribs attach to those of the seventh rib pair. For this reason the eighth, ninth and tenth rib pairs are called **false ribs**. The two most posterior pairs of ribs are not attached to the sternum at all. They are called **floating ribs**.

Fig 16.6 Rib cage, ventral view. Ribs are numbered 1 to 12

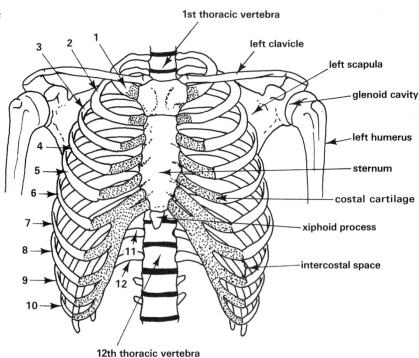

1st thoracic vertebra
left clavicle
left scapula
glenoid cavity
left humerus
sternum
costal cartilage
xiphoid process
intercostal space

12th thoracic vertebra

The sternum consists of three parts. Most anteriorly is the **manubrium**. The clavicles (collar bones) articulate with the manubrium and the scapula (section 16.1.4). The costal cartilages of the first rib pair also attach to the manubrium. The other costal cartilages are attached to the **body** of the sternum. The small **xiphoid process** projects from the posterior edge of the sternal body. Bone marrow samples are often taken from the sternum. The technique is called **sternal puncture** and involves pushing a broad needle into the sternal marrow.

16.1.4 Girdles

The arms articulate with the **pectoral girdle**, the legs with the **pelvic girdle**.

The pectoral girdle consists of two ventral **clavicles** (collar bones) and two dorsal **scapulae** (shoulder blades). The only bony attachment to the rest of the skeleton is by the articulation of the clavicles with the sternum (Fig 16.6). The scapulae are attached to the dorsal rib cage by a complex of muscles. The freedom of movement thus allowed to the arms reflects the swinging habit of our tree-living ancestors.

The pelvic girdle is more firmly attached to the spinal column. This enables the mass of the body to be transmitted to the ground through the pelvic girdles and the legs. The pelvic girdle consists of two **pelvic bones** which attach to the sacrum dorsally (Fig 16.5(f)) and to one another ventrally at the **symphysis pubis**. The pelvic bones are each formed in the embryo by the fusion of three bones called the **ilium, ischium** and **pubis** (Fig 16.7). The pelvic girdle, sacrum and coccyx comprise the **pelvis**.

Fig 16.7 Pelvic girdle, ventral view

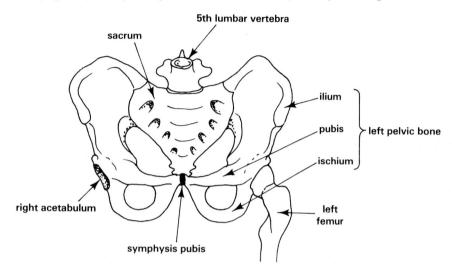

The broad ilium provides a large surface for the attachment of muscles which enable us to adopt a bipedal gait. Especially important are the relatively large gluteal muscles. They give us buttocks, another typical human feature. Also attached to the ilium are the abdominal muscles which carry much of the weight of the viscera of the abdomen and the iliacus muscle, a flexor of the hip.

16.1.5 Limbs

Our arms and legs are **pentadactyl limbs** (Fig 16.8). The upper part of each limb contains a single long bone, the **humerus** in the arm and the **femur** in the leg. The head of the humerus articulates with a depression in the scapula called the **glenoid cavity** (Figs 16.6 and 16.9). The distal end of the humerus articulates with the **ulna** and **radius** at the **elbow**. In the leg the femur (thigh bone) articulates with the pelvic girdle. The head of the femur is ball-shaped and fits into a cavity called the **acetabulum** in the pelvic bone (Figs 16.7 and 16.10). The distal end of each femur articulates with the tibia at the **knee**. The femur is straight, but its articular surface at the knee is at an angle. Hence, although the heads of the femurs are far apart, the knees can be held together. Our lower legs and feet are thus immediately under our centre of gravity when we are standing.

The two bones of the forearms are the ulna and radius. The ulna is the longer of the two and includes the **olecranon process** which makes the elbow. The possession of two freely-articulating bones in the forearm allows a twisting action at the wrist such as may occur when using a screwdriver. The human forearm is much shorter than that of apes. Its muscles are also less strong. For this reason we find it difficult to pull our bodies upwards when swinging from a bar. In contrast, apes can do so with ease.

The two bones of the lower leg are the **tibia** and **fibula**. The tibia (shin bone) is the longer of the two and is the main load-bearing bone of the lower leg. Its upper end articulates with the femur at the knee and its lower

Fig 16.8 Pentadactyl limb, general plan

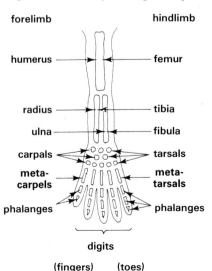

311

end with the bones of the ankle. The **patella** (kneecap) is a small bone that develops in the tendon of the **quadriceps extensor** muscle. This large muscle connects the pelvic girdle and femur to the patella and tibia. It serves to keep the knee extended and checks the forward momentum of the body when walking or running. Contraction of the calf muscles, especially the soleus which extends from the heel to the tibia, creates the propulsive thrust for bipedal movement.

Each wrist contains eight small bones called **carpals**, joined together by ligaments. The five bones in the palms of the hands are called **metacarpals**. The fingers each contain three bones called **phalanges**. The second finger is the largest and the **index** finger is about as long as or longer than the third. The **pollex** (thumb) contains two phalanges (Fig 16.9). The joint between the first carpal and metacarpal enables the thumb to be fully opposed to the fingers. No other primate can do so. This arrangement enables us to hold and manipulate objects with great dexterity. It has enabled humans to manufacture and use elaborate tools and instruments and to write and draw with great accuracy. Such activities distinguish human life from that of all other animals.

Each ankle contains seven small bones called **tarsals**. The largest of these, called the **os calcis** (calcaneum) constitutes the **heel**. The main body of the foot contains five **metatarsals** from which project the toes at the extremities. Like the fingers, each toe contains three phalanges, except the **hallux** (big toe) which, like the thumb, has only two (Fig 16.10). Man is unique in having a big toe longer than his other toes. The articular surface at the ankle joint is at right angles to the tibia. Consequently when we stand, our weight is transferred from the tibia partly forwards to the metatarsals and partly backwards to the os calcis. The foot is arched and acts as an efficient weight-bearing organ. When we walk, our toes remain in contact with the ground, thus maintaining the forward momentum.

Fig 16.9 Skeleton of the arm, dorsal view

Fig 16.10 Skeleton of the leg, ventral view

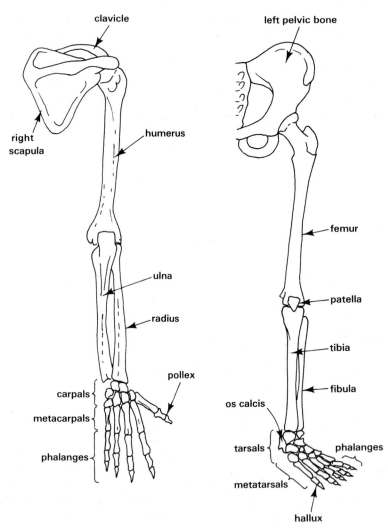

The bones in our limbs are arranged in a pentadactyl pattern which is common to all classes of vertebrate animals. What adaptations of our limbs enable us to perform delicate manipulations with our hands and to walk upright?

16.2 Cartilage and bone

The skeleton is mainly composed of **bone**. In the embryo, however, the first skeletal material to develop is **cartilage**. Embryonic cartilage ossifies during development to become bone. Some bones, called **membrane bones**, arise from embryonic connective tissue and do not begin as cartilage.

16.2.1 Cartilage

Hyaline cartilage consists of a matrix of **chondrin**, containing **collagen**. Chondrin is secreted by cells called **chondroblasts** which lie in the matrix in small clusters (Fig 16.11). Cartilage is bounded by a fibrous layer called the **perichondrium** in which there are blood vessels. It is here that new cartilage is laid down. In adults some hyaline cartilage remains in the skeleton. One of its main functions is to protect bones from abrasion where they articulate at joints (Figs 16.12 and 16.16). Hyaline cartilage also joins the ribs to the sternum (Fig 16.6) and allows limited flexing of the ribs during breathing. Rings of hyaline cartilage in the trachea and bronchi support the air passages and keep them open. **Elastic cartilage** contains yellow elastic fibres arranged in all directions in the matrix. The fibres allow greater flexibility compared with hyaline cartilage. Elastic cartilage is found in the epiglottis and in the pinna of the ear. **Fibro-cartilage** has many white fibres of collagen in the matrix. They resist stretching and give this type of cartilage considerable strength. Fibro-cartilage is found in the intervertebral discs of the spinal column (Fig 16.4(b)).

Fig 16.11 Photomicrograph of hyaline cartilage, × 100

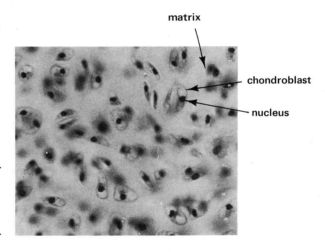

Many parts of our skeleton contain cartilage. What special properties does cartilage have which enables those parts of the skeleton to perform their functions?

16.2.2 Bone

Most bones have a shaft called the **diaphysis**. It is usually hollow and contains the bone **marrow** which produces most of the various kinds of blood cells (Chapter 10). Bone is covered by a tough fibrous membrane called the **periosteum**. At the ends, bones are usually expanded to form the **epiphyses**. These are sites for the attachment of tendons and articulation with other bones (Fig 16.12).

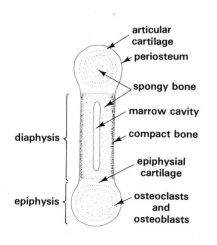

Fig 16.12 Diagram of the main parts of a bone

Fig 16.13 Photomicrograph of compact bone, ×100

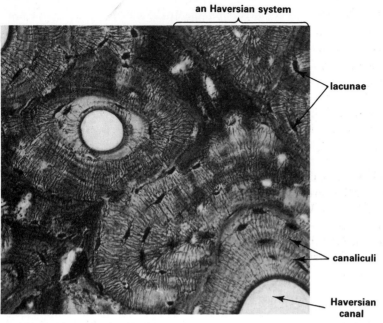

an Haversian system

lacunae

canaliculi

Haversian canal

The substance of bones may be **spongy** or **compact** and is arranged to provide maximum strength with minimum mass. The matrix of the bone is a mesh of collagen fibres impregnated with concentric **lamellae** of calcium salts, especially phosphate. In the matrix is a system of **Haversian canals** arranged parallel to the longitudinal axis of the bone (Fig 16.13). They contain nerves, blood and lymphatic vessels and connect with the periosteum and the marrow. Bone cells called **osteocytes** are found in concentric rings in the lamellae. The cells lie in cavities called **lacunae** which are linked by a series of fine **canaliculi**. The osteocytes secrete the bone matrix. The production of bone from embryonic cartilage is called **endochondral ossification**. The limb bones are formed in this way. Chondroblasts multiply and become arranged in longitudinal columns. Then, starting from the centre, the chondroblasts deposit calcium salts in the matrix. The cartilage thus becomes ossified.

Many of the chondroblasts degenerate and are replaced by large, amoeboid cells called **osteoclasts**. They cause erosion of much of the calcified cartilage which then becomes invaded by blood vessels. Cells called **osteoblasts** now secrete a series of bony columns called **trabeculae** in the centre of the bone shaft. Ossification gradually spreads outwards towards the periosteum. At the same time osteoblasts just beneath the periosteum produce dense bone. Some of the osteoblasts are enclosed in lacunae and become osteocytes (Fig 16.14).

Fig 16.14 Diagram of the stages of ossification

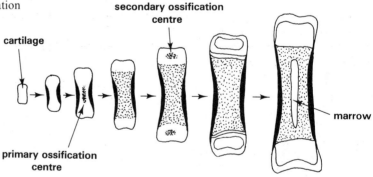

cartilage

secondary ossification centre

primary ossification centre

marrow

The production of bone from embryonic connective tissue is called **intramembranous ossification**. The bones of the skull are formed in this way (section 16.1.1). Fine bundles of white fibres appear in the connective tissue matrix and calcium salts are deposited around them. Osteoblasts then erode much of the calcified matrix. Osteoblasts now form bony trabeculae, the periosteum appears and new bone is laid down beneath it. Osteocytes are active throughout life. Their activities help the healing of bones which have been fractured.

Fig 16.15 Diagram of a joint

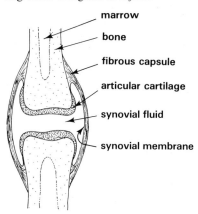

Fig 16.16 Weeping lubrication (from McNeil-Alexander, after McCutchen)

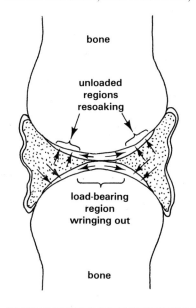

How does the structure of joints allow a variety of different skeletal movements?

16.3 Joints and movement

16.3.1 Joints

The bones of the skeleton articulate at **joints**. The actions of muscles bring about the movement of bones at a **synovial joint** (Fig 16.15). The articulating surfaces of the bone are covered with pads of articular cartilage. The joint is enclosed in a fibrous **synovial capsule** which, with the aid of ligaments, keeps the bones together. Lining the capsule is a **synovial membrane**. It secretes **synovial fluid** into the capsule's cavity. The synovial fluid is a lubricant allowing movement of the bones at the joint.

The exact mechanism of synovial joint action is not fully understood. Lewis and McCutchen have suggested the theory of **weeping lubrication**. The articular cartilage acts like a sponge, absorbing synovial fluid from the synovial cavity. When a load is applied to the joint the articular cartilages of the bones may be pushed together (Fig 16.16). Synovial fluid is squeezed out of the articular cartilage at the point of contact. Since the pores in the cartilage are very small, the fluid is wrung out only very slowly. Consequently the synovial fluid and not the cartilage bears most of the load applied to the joint. There are several different kinds of synovial joint depending on the ways in which the bones move against each other.

i. Gliding movements occur, for example, between the ribs and thoracic vertebrae (Fig 16.5(d) and 16.17(a)). Back and forth or side to side movements occur between the articulating bones.

ii. Hinged joints allow movements which alter the angle in one plane between two articulating bones. An example is the elbow (Figs 16.9 and 16.17(b)).

iii. Ball and socket joints involve the ball-shaped end of one bone moving within a cup-shaped cavity of another bone. Movement including rotation is possible in all directions. An example is the hip joint between the femur and pelvic girdle (Figs 16.7, 16.10 and 16.17(c)).

iv. Pivot joints allow rotation. An example is the articulation of the axis and atlas in the neck. It allows rotation of the head (Figs 16.5(c) and 16.17(d)). Another exmple is the articulation of the elbow between the radius and ulna. It allows the lower arm to be twisted (section 16.1.5).

Fig 16.17 Main types of synovial joint

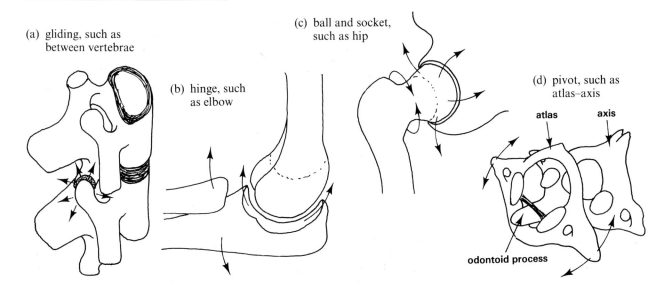

(a) gliding, such as between vertebrae

(b) hinge, such as elbow

(c) ball and socket, such as hip

(d) pivot, such as atlas–axis

Where free movement does not occur, the bones are separated by connective tissue as in the sutures of the skull (Fig 16.2), or by pads of fibro-cartilage as in the joints between vertebrae (Fig 16.4(b)).

The bones of movable joints are held together by **ligaments**. They are made of white fibrous tissue containing non-stretchable collagen fibres (Fig 16.18).

Fig 16.18 (a) Elbow joint showing ligaments

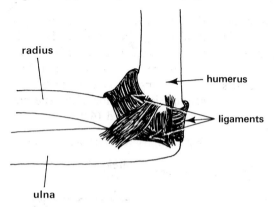

Fig 16.18 (b) White fibrous connective tissue, × 100

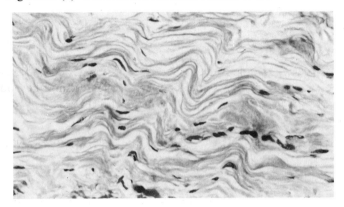

16.3.2 Movement

The movement of bones is brought about by the actions of muscles. Muscles are attached at each end to bones by collagenous fibres called **tendons**. Contraction of the muscle moves the bones in a manner dictated by the joint where they articulate. Most skeletal movement is controlled by **antagonistic** muscles. They are opposing sets of muscles which act in the opposite way at any one time. For example, the upper and lower arm articulate at the elbow. To raise the lower arm, **flexion** requires contraction of the **biceps** muscle. The biceps connects the scapula to the radius. At the same time the **triceps** muscle relaxes (Fig 16.19(a)). The triceps connects the scapula and humerus to the ulna. When the biceps relaxes and the triceps contracts, **extension** of the lower arm occurs (Fig 16.19(b)).

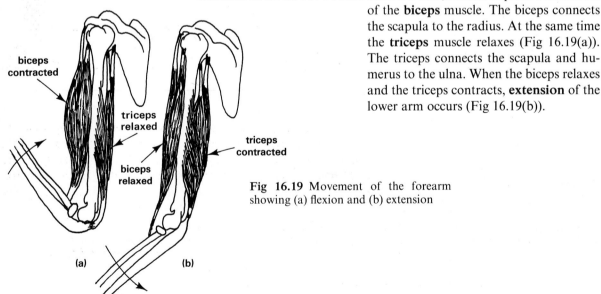

Fig 16.19 Movement of the forearm showing (a) flexion and (b) extension

Similar antagonistic muscle activity occurs at other joints. The coordination of such activity often involves **reflexes** (section 18.2). Reflex pathways to antagonistic muscles include inhibitory and excitatory synapses so that contraction is inhibited in one muscle while it is stimulated in the other (Fig 18.11).

The movements of bones can be described in terms of **levers**. Each joint acts as a **fulcrum** or **pivot**. The force used to move a lever is called the **effort**. It is provided by muscle contraction. The force to be overcome by the effort is called **resistance**. It is provided by the action of gravity on the mass of the body and any additional masses being lifted by the skeleton. Friction at the joint also adds minimally to resistance. Levers are classified according to the relative positions of the effort, resistance and fulcrum.

i. First class levers have the fulcrum in the middle. The effort and resistance are applied at opposite ends. An example is the movement of the skull on the spinal column (Fig 16.20).

ii. Second class levers possess the fulcrum at one end, the resistance in the middle. The effort is applied at the other end. An example is seen when raising the body on the toes (Fig 16.21).

iii. Third class levers are the commonest type in the body. The fulcrum is at one end of the lever and the resistance at the other end. The effort is applied to the middle. An example is the flexion of the forearm (Figs 16.19(a) and 16.22).

Fig 16.20 First class lever

Fig 16.21 Second class lever

Fig 16.22 Third class lever

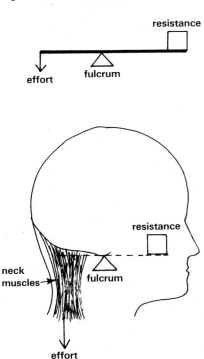

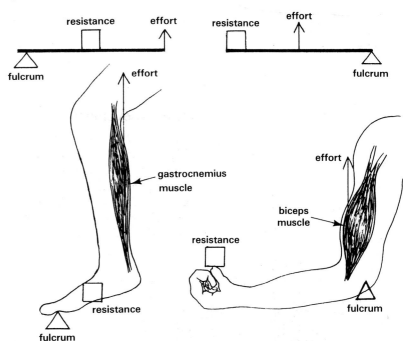

Fig 16.23 Principle of moments. A 0·5 kg load is applied to the lever 20 cm from the fulcrum, F

The advantage of using levers to move large masses has long been realised by engineers. Generally the further away from the fulcrum the effort is applied, the less is the force required to raise the mass; that is, the greater the **leverage**. This phenomenon can be illustrated by the **principle of moments** (Fig 16.23).

(a) A lifting force, x, is applied a further 20 cm away from the load at the other end of the lever from the fulcrum. The lifting force just balances the load when

$$x \times 40 = 0.5 \times 20$$

i.e. $$x = \frac{0.5 \times 10}{40} = 0.25 \text{ kg}$$

(b) If a lifting force, y, is applied 15 cm from the load then it just balances the load when

$$y \times 35 = 0.5 \times 20$$

i.e. $$y = \frac{0.5 \times 20}{35} = 0.28 \text{ kg}$$

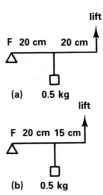

SUMMARY

The axial skeleton consists of the skull, vertebrae and rib cage. The appendicular skeleton consists of the girdles and limbs.

Skull: cranium encloses the brain. Spinal cord leaves the cranium by the foramen magnum. Maxilla (upper jaw) and mandible (lower jaw) support the teeth.

Vertebrae make up the spinal column: cervical, thoracic, lumbar, sacral and coccygeal. Each vertebra is perforated by a neural canal which accommodates the spinal cord. Centra are separated by the intervertebral discs; these allow flexibility and provide support; there are articulating surfaces and neural spines and transverse processes for the attachment of muscles. The first two vertebrae allow the head to swivel; the odontoid process of the axis projects into a space in the atlas. The sacral vertebrae are fused; they provide strong attachment for the pelvic girdle.

The rib cage encloses the thorax. Ribs articulate with the thoracic vertebrae. Intercostal muscles move the ribs during breathing. There is muscular attachment for the pectoral girdle. The sternum attaches to most ribs ventrally by cartilage; flexibility.

Girdles transmit the body mass to the limbs.
Pectoral: clavicles and scapulae; articulate with the fore-limbs.
Pelvic: ilium, ischium and pubis; articulate with the hind-limbs. The pelvic girdle is robust and fused to the vertebral column; transmits the locomotive force from the hind-limbs to the spine.

Limbs: pentadactyl structure; modified in different vertebrates to perform different functions.

Hyaline cartilage: chondrin containing collagen and chondroblasts; bounded by perichondrium.
Elastic cartilage contains yellow elastic fibres; flexibility.
Fibro-cartilage contains white collagen fibres; strength.

Bone: the shaft (diaphysis) containing marrow is surrounded by periosteum. There are epiphyses at the ends to which tendons are attached at the joints. Bone consists of collagen fibres impregnated with concentric lamellae of calcium salts. Haversian canals contain nerves, blood and lymphatic vessels. Lamellae contain osteocytes.

Endochondral ossification is the production of bone from embryonic cartilage. Intramembranous ossification is the production of bone from embryonic connective tissue.

Bones articulate at joints. Synovial joints consist of a capsule which keeps the bones together (aided by ligaments); contains synovial fluid which lubricates movement.
Joint types: gliding, e.g. between ribs and thoracic vertebrae; hinged, e.g. elbow; ball and socket, e.g. hip; pivot, e.g. axis and atlas.

Bones articulating at joints act as levers. Levers minimise the effort required to move loads; principle of moments. Classes of levers: first, e.g. skull moving on the vertebral column; second, e.g. raising the body on the toes; third (commonest), e.g. flexion of the fore-arm.

QUESTIONS

1 The diagram below represents part of a transverse section through hard (compact) bone tissue.

(a) Name the parts labelled **A–F**.
(b) State *one* function in living bone of each of the parts labelled **B**, **C** and **E**.
(c) (i) State *two* structures which are joined by a ligament and *two* structures which are joined by a tendon.
(ii) Compare the roles played by ligaments and tendons in skeletal movement. (L)

2 (a) Give an illustrated account of the structure of
(i) xylem
(ii) compact bone tissue.
(b) How is the structure of these tissues related to the functions they perform? (L)

3 (a) Make a large drawing of a lateral view of a lumbar vertebra. Label four features on your drawing. Annotate these to explain their role in the function of the lumbar vertebra in the vertebral column.
(b) State three differences between the femur and the radio-ulna. Relate these differences to the functions of the specimens in a vertebrate. (L)

17 Nervous control and coordination in mammals
1. Neurones and muscles

17.1 Electrochemical basis of nervous activity 321

  17.1.1 The resting state 322
  17.1.2 Depolarisation 323
  17.1.3 Impulse conduction 324
  17.1.4 Repolarisation 325
  17.1.5 Rate of impulse conduction 326
  17.1.6 Synapses and neuroeffector junctions 329

17.2 Muscle 333

  17.2.1 Skeletal muscle 333
  17.2.2 Cardiac and visceral muscle 335
  17.2.3 Muscle contraction 336

Summary 339

Questions 340

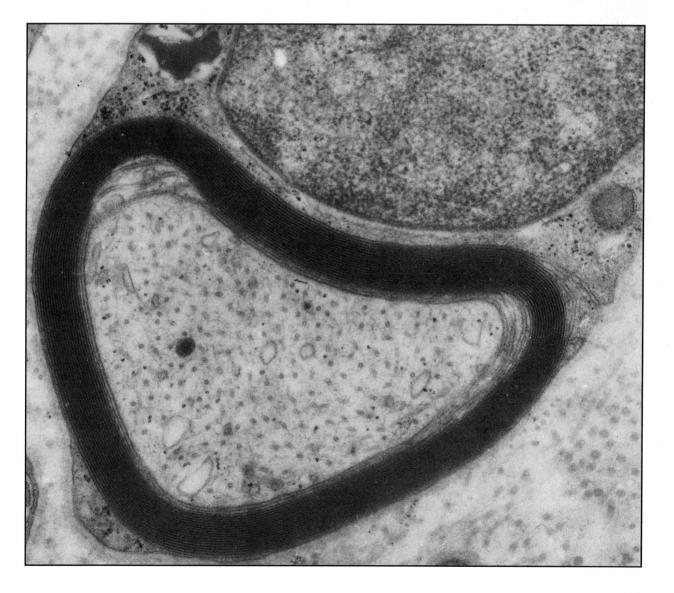

17 Nervous control and coordination in mammals
1. Neurones and muscles

Changes constantly take place inside and outside the body. Mammals respond to the changes and their internal environment remains fairly constant. The ability to respond in this way is called **homeostasis**. An important aspect of homeostasis is the **coordination** of responses. Co-ordination is the ability to make an appropriate response(s) to a stimulus (or stimuli), the body acting as a self-regulating mechanism.

A coordinating system contains several components. There are means

Fig 17.1 Basic components of a feed-back control mechanism

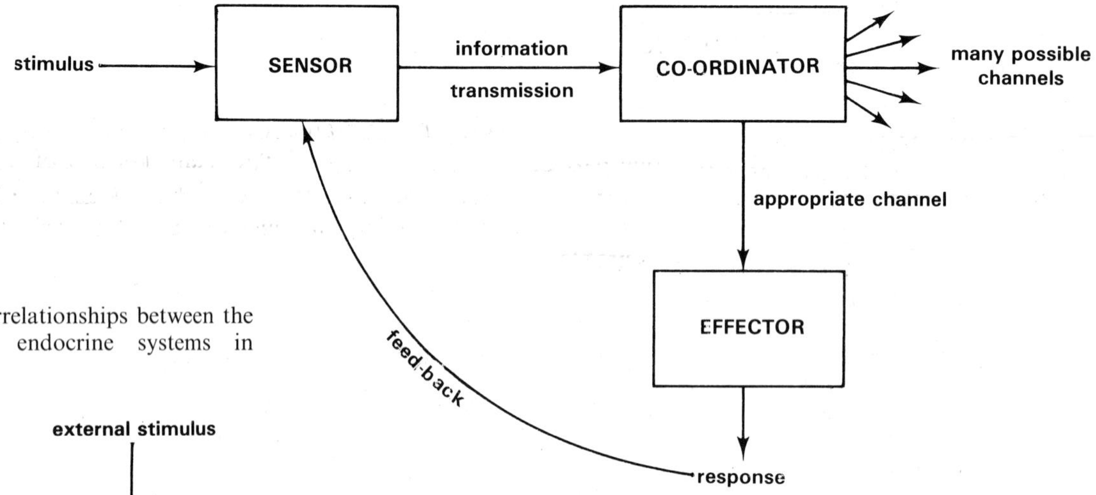

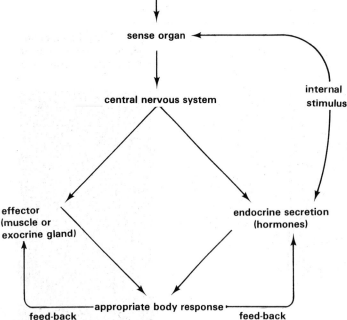

Fig 17.2 Interrelationships between the nervous and endocrine systems in mammals

of **detecting stimuli**, means of **transmitting information** about the stimuli and means of **responding to the stimuli**. The system also has a means of directing information along the most appropriate of many possible channels. Coordination also involves **feedback** which ensures that the degree of response is related to the intensity and direction of stimuli (Fig 17.1). Note the similarity between Fig 17.1 and Fig 8.12 which summarises the mechanism controlling ventilation of the lungs. Other comparable examples are described elsewhere in the book such as control of the heart (Chapter 9) and body temperature (Chapter 20).

Mammals have two main co-ordinating systems, the **nervous system** and the **endocrine system**. Each system has different means of detecting and transmitting information. However, the two systems often work together (Fig 17.2). The endocrine system is the subject of Chapter 21.

17.1 Electrochemical basis of nervous activity

Fig 17.3 (a) Main parts of a neurone

The mammalian **nervous system** contains an organised collection of numerous nerve cells called **neurones** (Fig 17.3). Neurones consist of a bulbous cell body called the **soma** which contains a nucleus. Branched hair-like structures called **dendrites** which receive information from other neurones project from the soma. Some neurones have a long membrane-enclosed cytoplasmic thread called an **axon**. At their ends the axons divide into many branches which nearly touch the dendrites of adjacent neurones. Intricate pathways

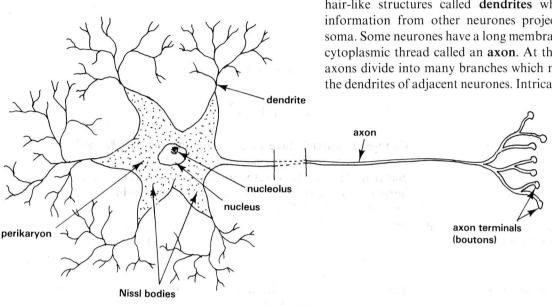

thus exist along which information can be transmitted very rapidly from one part of the body to another. The main function of neurones is to transmit nerve impulses over relatively long distances. The ability to transmit impulses is found only in neurones and muscle cells. These cells are **excitable**.

Fig 17.3 (b) Main categories of neurones: (i) multipolar (commonest), (ii) bipolar and (iii) pseudo-unipolar

(i)

effector neurone

(ii)   (iii)

receptor sensory neurones

Fig 17.3 (c) Pyramid cell from the cerebral cortex

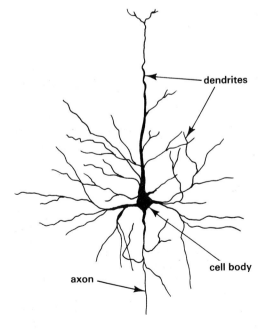

dendrites

cell body

axon

Fig. 17.3 (d) Photomicrograph of a thin section of cerebral cortex of a rat, × 75

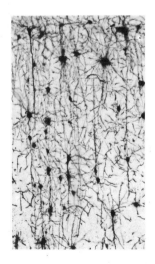

17.1.1 The resting state

All living cells maintain an uneven distribution of different inorganic ions across their membranes. The extracellular concentration of these ions is different from that in the cytoplasm (Table 17.1). This state of affairs is created by the action of the cell membrane which pumps some ions out of the cell and pumps others into the cytoplasm. Metabolic energy provided by ATP is required for the pumping process. Important among the membrane pumps is the **sodium–potassium exchange pump**. It discharges sodium ions (Na^+) to the outside and potassium ions (K^+) to the inside of the cell. The ions are exchanged in equal numbers on a one-for-one basis. The pump creates concentration gradients down which diffusion occurs (Chapter 1). Sodium ions slowly diffuse back into the cytoplasm, and potassium ions slowly leak out of the cell. However, the cell membrane is about fifty times more permeable to potassium ions than it is to sodium ions. Consequently the potassium ions diffuse out of the cell more rapidly than the sodium ions diffuse in. As both ions have a positive charge the outer surface of the cell membrane is positive relative to the inside. The membrane is **polarised** because an electrical potential difference exists across it (Fig 17.4).

When the cell is not transmitting an impulse, the transmembrane potential is called the **resting potential**. The size of the resting potential varies from cell to cell, but for most excitable cells it is about $-70\,mV$. The minus sign indicates that the inside of the cell is negative relative to the outside. It is possible to measure the transmembrane potential directly by connecting an intracellular electrode and an extracellular electrode to an oscilloscope which acts as a sensitive voltmeter. The intracellular electrode is usually made from a very narrow glass tube tapered to a microscopic tip. It is filled with an electrolyte such as potassium chloride solution. The tip of the electrode is then inserted through the membrane, with the aid of a microscope and a device called a micromanipulator (Fig 17.5).

Table 17.1 Concentrations of various ions in cells and tissue fluid

Ion	Cytoplasm/ m mol dm^{-3}	Tissue fluid/ m mol dm^{-3}
sodium	16	140
potassium	100	4.4
chloride	4	103
hydrogencarbonate	8	27

Fig 17.4 Establishment of transmembrane potential by the combined actions of the sodium–potassium exchange pump and diffusion

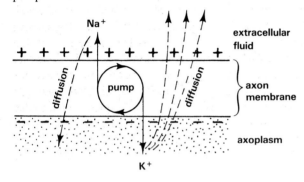

Fig 17.5 Measuring the transmembrane potential of an axon

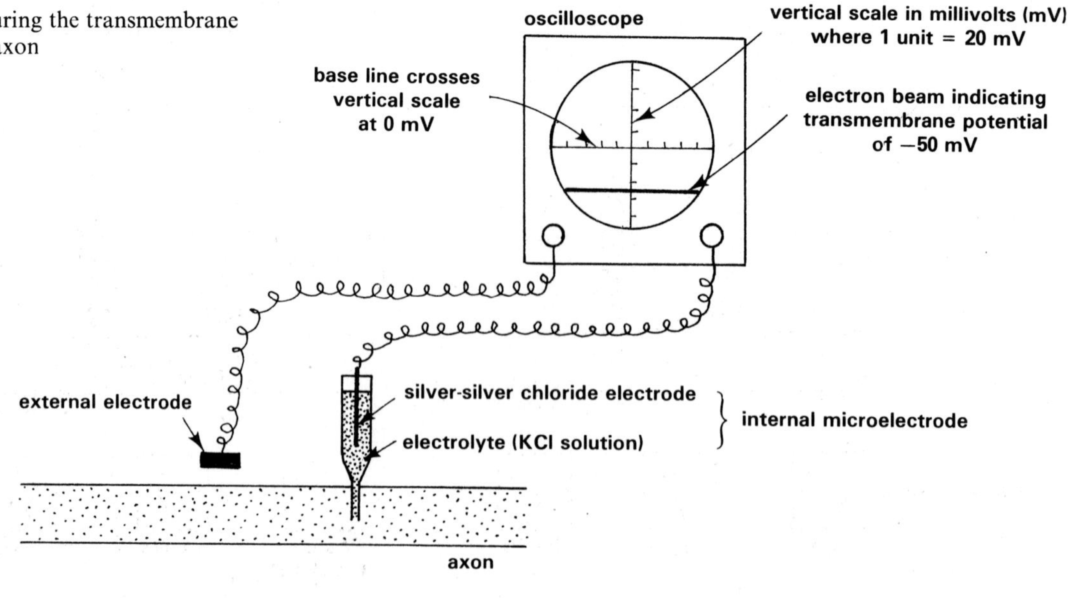

17.1.2 Depolarisation

Cell membranes can be stimulated in various ways, including the application of an electrical discharge through the membrane or the addition of chemical substances to the membrane surface. The primary effect of such stimulation is to cause, at the site of stimulus, a sudden and very brief change in the permeability of the membrane to sodium ions. The membrane becomes more permeable to sodium than to potassium ions. Some of the extracellular sodium ions enter the cytoplasm by diffusion more easily than before. The ions move down a concentration gradient and are also attracted into the cell by the negative charge on the inside of the membrane.

Fig 17.6 Depolarisation of a neurone membrane

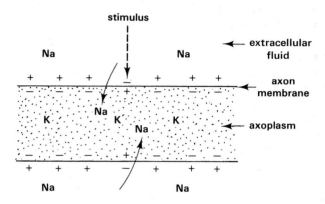

The entry of sodium ions into the cell disturbs the resting potential. A point may eventually be reached when enough sodium ions have entered the cytoplasm to balance the positive charge remaining outside. The entry of more sodium ions causes the inside of the cell to become positive relative to the outside. The membrane is now **depolarised** (Fig 17.6). When enough sodium ions enter the cell to change the transmembrane potential to a certain **threshold** level, an **action potential** arises which generates an **impulse**. At threshold, pores called **sodium gates** open in the membrane and allow a sudden flood of sodium ions into the cell. When the action potential reaches its peak the sodium gates close and no more sodium ions enter. If the depolarisation is not great enough to reach threshold, then an action potential and hence an impulse are not produced. This is called the **all-or-nothing law**. If a stimulus stronger than that necessary to produce an impulse is applied, the frequency of impulse production generally increases (Fig 17.7). In this way the nervous system discriminates between strong and weak stimuli. For most excitable cells the threshold is about $-60\,\text{mV}$. This is 10 mV less than the resting potential of $-70\,\text{mV}$.

Fig 17.7 All-or-nothing law. Raising the stimulus strength from zero does not produce an impulse until a stimulus strength results in the same action potential

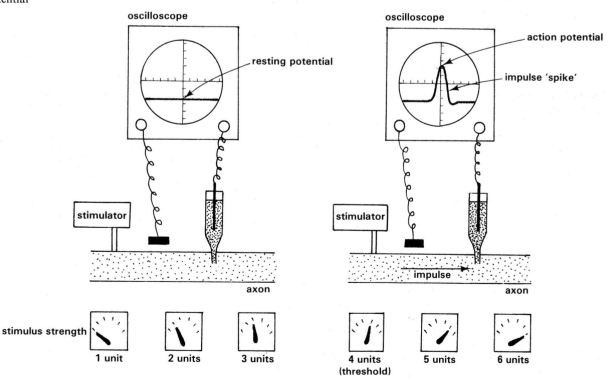

Fig 17.8 Changes in transmembrane potential during impulse transmission

Fig 17.8 illustrates the voltage changes that occur across a neurone's membrane as an impulse passes by.

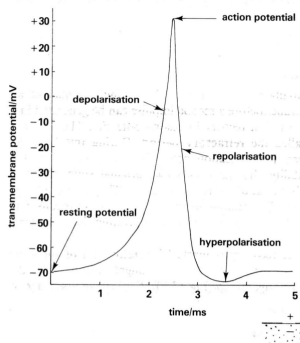

17.1.3 Impulse conduction

Entry of sodium ions into an excitable cell at the point of stimulation sets up a **current** of ions in the extracellular fluid and the cytoplasm (Fig 14.7). The electric current crosses the membrane a short distance in both directions from the point at which the stimulus was applied. Electric current is also a stimulus, so the cell is further stimulated where the current crosses the membrane. The membrane's permeability to sodium ions changes, depolarisation occurs and another action potential is set up. This is followed in turn by more current flow and stimulation of the membrane even further away from the original point of stimulation. This sequence of electrochemical events constitutes the **impulse**.

Fig 17.9 Impulse transmission along an axon

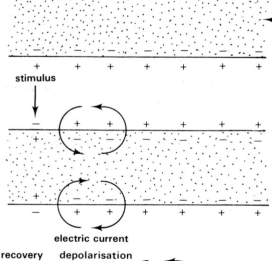

(a)   The resting state

(b)   Depolarisation of the membrane causing the flow of electric current

(c)   Electric current stimulates the membrane nearby causing further depolarisation

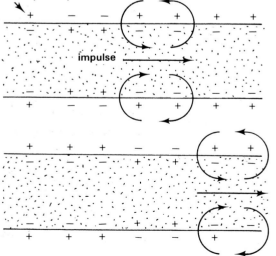

(d)   The process repeats itself along the axon

As neurones are usually stimulated at one end (either the cell body or dendrite), impulses travel in one direction along them. In contrast, muscle fibres are normally stimulated in the middle and so impulses travel in both directions.

17.1.4 Repolarisation

An impulse is essentially a disturbance of the resting potential of an excitable cell's membrane. Before a second impulse can be generated in a cell the membrane must first recover its resting potential. The recovery period in cells is called the **refractory period**. During this time the membrane repolarises. Repolarisation occurs in two stages. First the membrane's permeability changes back to its original state, allowing potassium ions to diffuse out of the cell more easily. The negative charge outside the depolarised cell membrane also attracts the potassium ions. The movement outwards of potassium ions restores the resting electric potential (Fig 17.8).

The second stage of recovery is marked by restoration of the ionic balance across the membrane. It is brought about by the sodium–potassium exchange pump (Fig 17.10). During the first stage of

Why can a second impulse not be generated immediately after a first impulse?

Fig 17.10 Membrane recovery

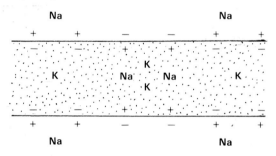

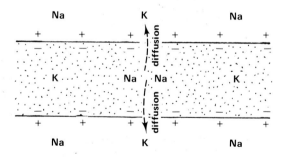

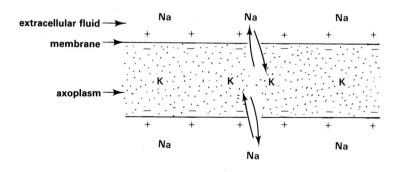

recovery it is impossible for a second impulse to be generated because the membrane is not yet repolarised. The time required for repolarisation is called the **absolute refractory period**. However, during the second stage it is possible to generate an impulse before membrane recovery is completed. This can occur if a stimulus is applied which is strong enough to produce the threshold transmembrane potential (Fig 17.11). The second stage of recovery is called the **relative refractory period**.

Refraction between successive impulses limits the frequency with which impulses can be conducted. Most neurones can conduct up to about 100 impulses per second.

Fig 17.11 Absolute and relative refraction. Impulse generation is possible during the relative refractory period providing the stimulus intensity is great enough (after Vander, Sherman and Luciano)

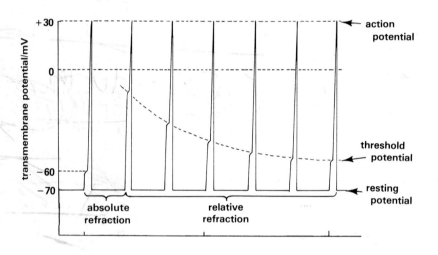

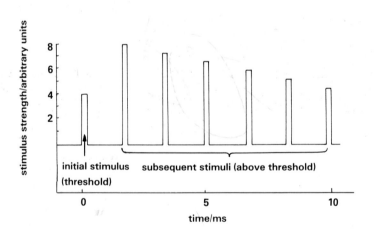

17.1.5 Rate of impulse conduction

Peripheral neurones are surrounded by **Schwann cells** which may respond to the proximity of neurons by growing around their axons. The axon is then wrapped in several membranous layers called the **myelin sheath**. The myelin sheath limits contact between the axon membrane and tissue fluid to

What factors affect the speed of impulse conduction in nerves?

only a few sites called **nodes of Ranvier** (Fig 17.12). Because less of the axon membrane is used in impulse transmission the rate of impulse conduction in myelinated axons is greater than in unmyelinated axons (Fig 17.13).

Fig 17.12 (a) Development of myelin sheath from Schwann cell seen in transverse section

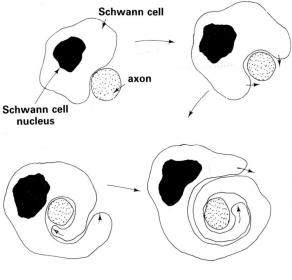

Fig 17.12 (b) Electronmicrograph of a transverse section of a myelinated axon, × 25 000

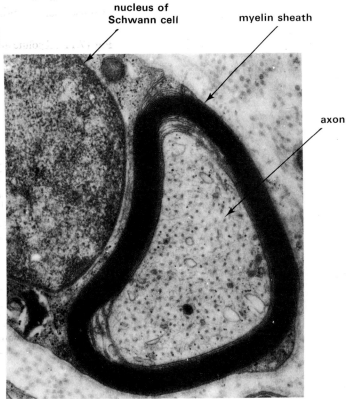

Fig 17.12 (c) Photomicrograph of some human myelinated neurones showing the nodes of Ranvier, × 400

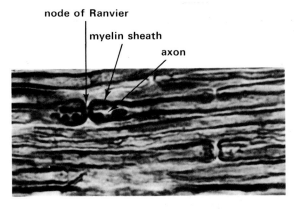

Fig 17.13 Electrochemical events at the nodes of Ranvier during the passage of an impulse along a myelinated axon. Compare with Fig 17.8

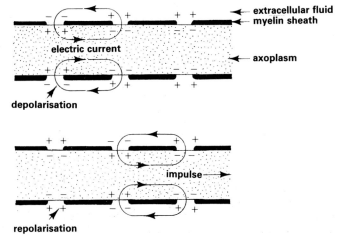

Fig 17.14 Diagram of non-myelinated neurones as seen in transverse section (from various electronmicrographs). Compare with Fig 17.12(b)

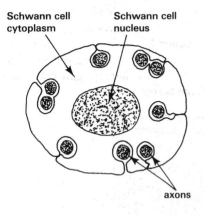

Schwann cell cytoplasm

Schwann cell nucleus

axons

Some neurones do not become myelinated. Axons of small diameter, such as those of the autonomic system, simply lie in shallow grooves in the Schwann cell cytoplasm. Many **non-myelinated** axons may be found in the same Schwann cell (Fig 17.14).

In a limb nerve there are very many neurones. Some conduct impulses down the limb away from the central nervous system (effector neurones) whilst others conduct impulses up the limb towards the central nervous system (receptor neurones). In transverse section it can be seen that the axons vary in diameter (Fig 17.15).

Fig 17.15 Photomicrograph of a transverse section of human median nerve. Note the variety of axon diameters. × 850

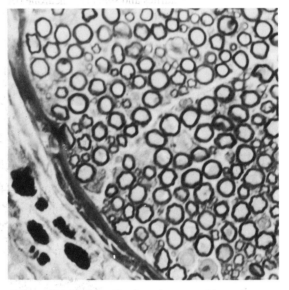

Another factor affecting the rate of impulse transmission is the diameter of the axon. The relationship for large myelinated mammalian nerve fibres is approximately:

$$v \propto 7 \times d$$

where v is the velocity of impulse conduction in metres per second and d is the fibre diameter in μm.

Thus, other factors excepted, a doubling of fibre diameter increases impulse velocity by about 14 times. Table 17.2 shows the conduction velocities of neurones from different animals. It was Helmholz in 1850 who first measured the velocity of a nerve impulse. He obtained a value of about 25 m per second in a frog's nerve. Prior to his experiments impulses were thought to travel at about the speed of light!

Table 17.2 Speeds of impulse conduction in the nerves of various animals (from Bendal 1969)

Tissue	Temperature /°C	Fibre diameter /μm	Impulse velocity /m s^{-1}
crab nerve	20	30	5
squid giant axon	20	500	25
cat nerve (unmyelinated)	38	0.3–1.5	0.7–2.3
cat nerve (myelinated)	38	2–20	10–100
prawn nerve (myelinated)	20	35	20
frog nerve (myelinated)	24	3–16	6–32

17.1.6 Synapses and neuroeffector junctions

How do impulses pass from one neurone to another?

Nerve pathways consist of at least two neurones joined end to end. The junctions of neurones are called **synapses** (Fig 17.16). They are similar in structure and function to the junctions between the last neurone in a pathway and an effector to which impulses are transmitted. Each axon branches and terminates in many bulb-like structures called **boutons**. Electron microscopy reveals many mitochondria and small membranous vesicles inside each bouton. Between the bouton and the cell body or dendrites of a postsynaptic neurone, there is a narrow gap called the **synaptic cleft**, usually about 10 to 20 nm across.

Fig 17.16 (a) Many axons terminate as synapses on the dendrites and soma of a neurone (from photomicrographs)

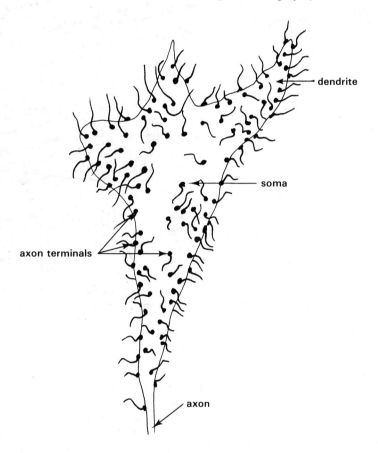

Fig 17.16 (b) Axosomatic and axodendritic synapses (from electronmicrographs)

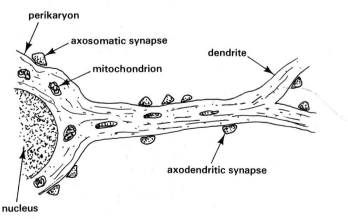

Fig 17.16 (c) Diagram of structures visible in a synapse viewed with the aid of electronmicroscopy. Compare with Fig 17.16 (d)

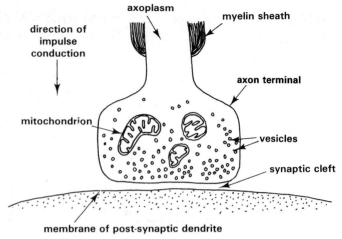

Fig 17.16 (d) Electronmicrograph of a synapse, × 24 000

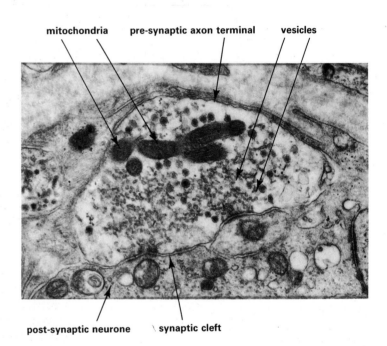

Fig 17.16 (e) Neuromuscular junction as seen with the aid of an electron microscope

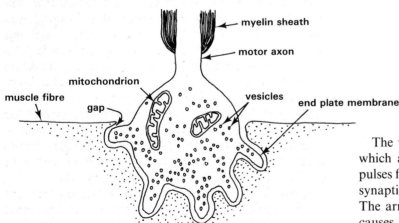

The vesicles contain **transmitter substances** which are involved in the transmission of impulses from the presynaptic neurones, across the synaptic clefts and to the postsynaptic neurones. The arrival of an impulse at an axon terminal causes some of the vesicles to fuse with the presynaptic membrane. The contents of the vesicles are released onto the postsynaptic membrane. Here there are specific **receptor sites** to which the transmitter substance attaches, causing depolarisation of the postsynaptic membrane. An **excitatory**

Fig 17.17 Synaptic transmission

postsynaptic potential (EPSP) is generated and an impulse travels along the postsynaptic neurone (Fig 17.17).

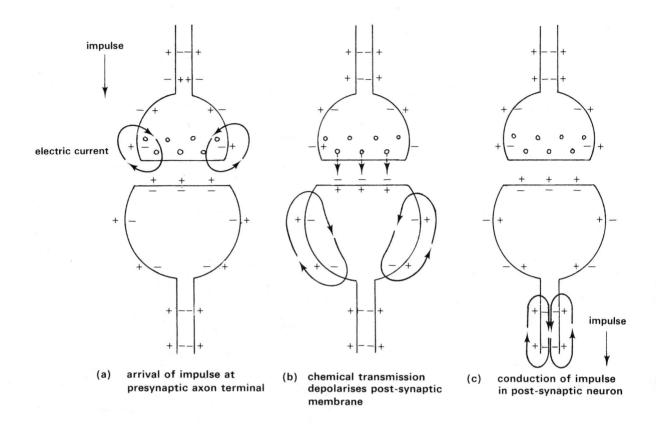

(a) **arrival of impulse at presynaptic axon terminal**

(b) **chemical transmission depolarises post-synaptic membrane**

(c) **conduction of impulse in post-synaptic neuron**

Fig 17.18 Cycle of release, breakdown, reabsorption and resynthesis of synaptic transmitter substance

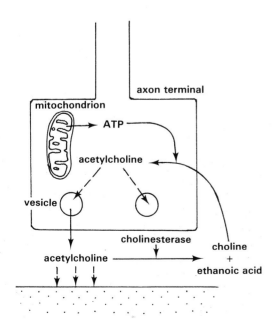

Many potent transmitter substances have been isolated from nerve endings. Most widespread in the nervous system is **acetylcholine**. Another called **noradrenaline** is found in sympathetic nerve endings. Following their release into the synaptic clefts, transmitter substances are quickly degraded by enzymes. Acetylcholine for example is broken down by **acetylcholinesterase** into choline and ethanoic acid. The products cannot cause depolarisation. It is for this reason that the effects of the transmitter substances are only temporary. It means that the arrival of an impulse at a synapse does not result in the generation of many impulses in the postsynaptic neurone. This is important since the frequency of impulses conducting along a pathway reflects the intensity of the original stimulus at the beginning of the pathway. An uncontrolled increase in impulse frequency at synapses could result in the perception of a stronger stimulus than was present in the first place. We would not wish to be deafened by a quiet noise or blinded by dim light.

Active transmitter substance is resynthesised from the inactive products reabsorbed by the neurones and packaged in new vesicles. The energy necessary for resynthesis is provided by the relatively large numbers of mitochondria in the boutons (Fig 17.18). Some synapses in the central nervous system are inhibitory. On the arrival of an impulse, **inhibitory synapses** release transmitter substances that prevent the generation of impulses in the postsynaptic neurone . They do this by **hyperpolarising** the postsynaptic membrane rather than depolarising it. The permeability of the postsynaptic membrane to potassium ions increases. Consequently more of these ions diffuse out of the cytoplasm and an **inhibitory postsynaptic potential (IPSP)** arises (Fig 17.19). Inhibitory synapses are important in nerve pathways and control antagonistic muscles.

Fig 17.19 Action of an inhibitory synapse. Inhibitory transmitter substances include γ-aminobutyric acid and glycine

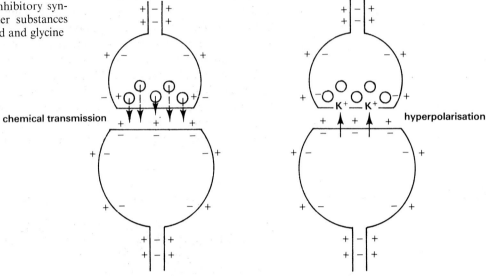

At the end of each nerve pathway **neuroeffector junctions** are found between the last neurone and an effector, such as a muscle. These junctions are similar to synapses in structure and mode of functioning (Fig 17.16(e) and 17.17).

Effector neurones terminate as **motor end plates** on muscle fibres (Fig 17.20). Electron microscopy reveals synapse-like structures including mitochondria, vesicles and a narrow cleft between the axon terminal and the muscle fibre. Stimulation of the muscle by a nerve is achieved by chemical transmitter substances released from vesicles. Acetylcholine is the main neuromuscular transmitter substance. Noradrenaline is the neuromuscular transmitter substance at sympathetic nerve endings.

Fig 17.20 Photomicrograph showing the junctions between effector neurone endings and striped muscle fibres, × 450

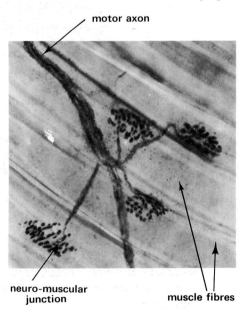

17.2 Muscle

There are three kinds of muscle in our bodies, **skeletal, visceral** and **cardiac**.

17.2.1 Skeletal muscle

Skeletal muscle may be controlled consciously to move the bones of the skeleton. For this reason it is called **voluntary** muscle. The arrangement of the protein filaments that bring about contraction gives skeletal muscle a banded appearance when viewed with the aid of a microscope (Figs 17.22 and 17.26). For this reason it is also called **striped** or **striated** muscle.

Skeletal muscle consists of **fibres** which are arranged in large groups called **fasciculi**. Each fasciculus is surrounded by a fibrous membrane called the **perimysium**. The entire muscle is surrounded by a membrane called the **epimysium**. Tendons are composed of dense collagenous connective tissue. They attach the tapered ends of muscles to the bones (Fig 17.21).

Fig 17.21 Structure of skeletal muscle

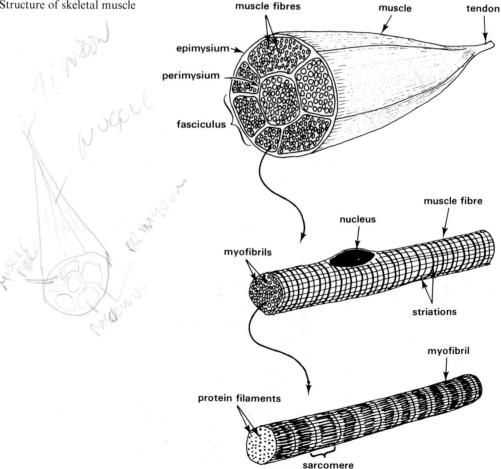

Skeletal muscle fibres are between 10 and 100 μm in diameter and may be up to 300 000 μm long. The fibres have an outer membrane called the **sarcolemma** which contains large numbers of **myofibrils**, each about 2 μm in diameter. Nuclei are scattered under the sarcolemma along the length of each fibre.

The myofibrils are divided into compartments called **sarcomeres** by internal membranous partitions called **Z-lines** (Figs 17.22 and 17.26). There are several bands across the myofibrils and hence the fibres. The **I-bands** are light in appearance and straddle each Z-line. They are regions where only thin filaments of the protein actin are found inside the sarcomeres. Between the I-bands and in the middle of each sarcomere are the **A-bands**. They are dark in appearance and contain the thick filaments of the protein myosin as well as overlapping thin filaments. In the middle of each A-band is a lighter **H-band** where there are thick filaments only (Figs 17.23(a) and 17.26).

Fig 17.22 Photomicrograph of some skeletal muscle fibres teased apart, × 600

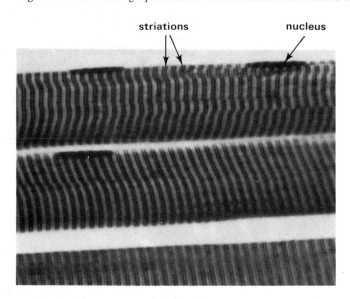

Fig 17.23 Diagram of the microscope apperance of sarcomeres when (a) relaxed and (b) contracted

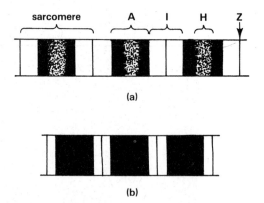

When the muscle fibres contract the thin filaments slide in between the thick filaments and may meet in the middle of each sarcomere. When this happens the H-bands become narrower and may disappear. Similarly, the I-bands become narrower as the Z-lines are drawn closer to each other. Since the filaments do not shorten during contraction the A-bands remain the same length (Fig 17.23(b)).

Skeletal muscle normally only contracts when stimulated by nerve impulses. It is **neurogenic**.

17.2.2 Cardiac and visceral muscle

The walls of the heart are mainly composed of cardiac muscle. Cardiac muscle is striated like skeletal muscle (Fig 17.24). However, the fibres are branched and divided longitudinally by **intercalated discs** into cells each containing a central nucleus. The discs are composed of several layers of membranes which strengthen the fibres and help to conduct the cardiac impulse (Chapter 9). Branching of the fibres enables contraction to spread rapidly and smoothly through the heart wall.

Fig 17.24 Photomicrograph of some cardiac muscle fibres, × 450

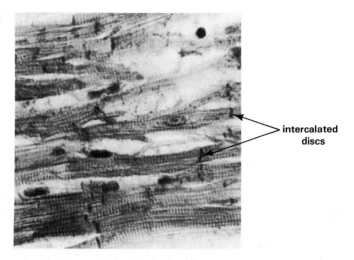

intercalated discs

Cardiac muscle is **involuntary** since its contraction cannot normally be controlled consciously.

Visceral muscle is found in the walls of many tubular structures in the body such as blood vessels, the gut and the urinogenital system. Like cardiac muscle, it is involuntary. Visceral muscle cells are not arranged into fibres. The cells are nucleate, and usually tapered with fine longitudinal striations in the cytoplasm (Fig 17.25).

Fig 17.25 Photomicrograph of some visceral muscle cells, × 1000

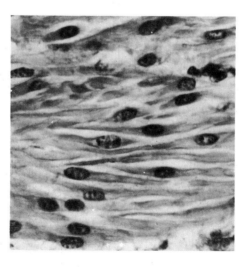

Contraction in visceral muscle involves sliding of actin and myosin filaments. However, they are not so highly organised as in skeletal and cardiac muscle.

Cardiac and visceral muscle can contract and relax rhythmically on their own, even without nervous stimulation. They are **myogenic**. The nerves connected to these muscles control activity rather than initiate it.

17.2.3 Muscle contraction

In striped and cardiac muscle the fibres contain many **myofibrils**. These are divided internally by membranous partitions called Z-lines into a large number of **sarcomeres** (Fig 17.26). Projecting into the **sarcoplasm** of each sarcomere is an array of **thin filaments** consisting of the protein **actin**. Thin filaments are between 5 and 8 nm in diameter. Suspended between them and in the centre of the sarcomeres is an array of **thick filaments** consisting of the protein **myosin**. Thick filaments are between 12 and 18 nm in diameter. Both kinds of filaments run parallel to the longitudinal axis of the myofibril (Fig 17.26(b)).

Why is striped muscle striped?

Fig 17.26 Sarcomeres

(a) Electronmicrograph, ×65 000

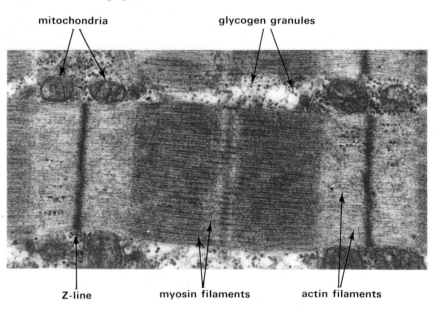

(b) Protein filaments

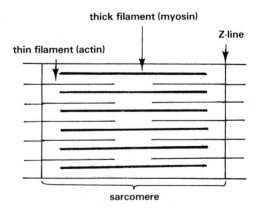

Each thin filament consists of very many molecules of actin arranged in long threads that spiral around each other in pairs (Fig 17.27(a)). Along each thread are many sites where the myosin molecules can become attached to the actin. Another protein called **tropomyosin** is arranged along the actin threads and may cover the myosin-binding sites.

Fig 17.27 Muscle proteins:

(a) thin filaments of actin

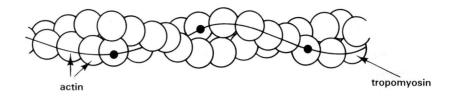

actin   tropomyosin

(a)

(b) thick filaments of myosin

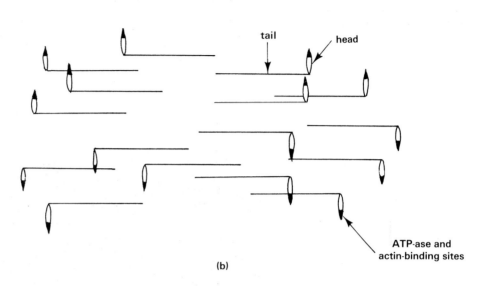

tail   head

ATP-ase and
actin-binding sites

(b)

Fig 17.28 Sliding filaments during muscle contraction

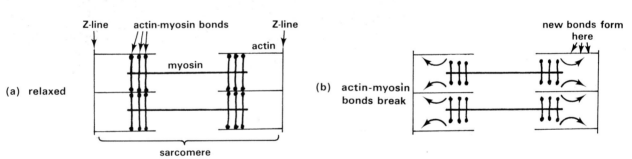

Z-line   actin-myosin bonds   Z-line

actin

myosin

(a) relaxed

sarcomere

(b) actin-myosin
bonds break

new bonds form
here

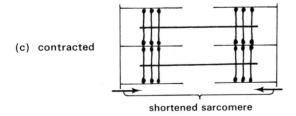

(c) contracted

shortened sarcomere

A thick filament contains about 200 molecules of myosin. Each molecule has a bulbous head and a filamentous tail. They are arranged so that the heads project from the sides and at the ends of the thick filaments (Fig 17.27(b)). On the tips of the myosin heads there are sites which bind them to the thin filaments. About half the myosin heads are attached to thin filaments at any one time. Also present on the myosin heads are molecules of the enzyme **ATP-ase**. They catalyse the removal of terminal phosphate from ATP, producing ADP (Chapter 5). The energy released is used to bring about muscle contraction. When muscle contracts the myosin heads detach from the thin filaments and rotate around in a spiral fashion. Other myosin heads nearer to the Z-lines attach to the thin filaments, drawing them towards the centre of the sarcomeres (Fig 17.28).

337

This **sliding filament** model explains why the Z-lines of each sarcomere are drawn towards each other during muscle contraction. The sarcomeres and hence the myofibrils and fibres shorten. At the level of structure revealed by the light microscope the I-bands and the H-zones become narrower or even disappear when contraction occurs. The A-bands, representing the thick filaments remain unchanged (Fig 17.23).

In the relaxed state, the myosin-binding sites on the thin filaments are covered up by the tropomyosin threads (Fig 17.27(a)). When the muscle is stimulated by a nerve impulse, muscle impulses are generated in the myofibril membranes. The muscle impulses depolarise the myofibril membranes. Sodium and **calcium ions** enter the sacroplasm. Calcium ions are normally excluded from the sarcoplasm by a membrane pump. The calcium ions displace the tropomyosin threads, thus exposing the myosin-binding sites on the thin filaments. Contraction can now occur. This is equivalent to uncovering a door lock so that a key can be inserted, turned and the door opened.

Excessive fluctuations in the calcium ion content of body fluids lead to neuromuscular disturbances (Chapter 21). Muscle is adapted in a number of ways to provide the energy required for contraction. Its red colour is due to the presence of **myoglobin** (Chapter 8). The pigment may become oxygenated to oxymyoglobin by collecting oxygen from oxyhaemoglobin in the muscle's blood supply. Oxymyoglobin is induced to dissociate, releasing its oxygen to the muscle when the oxygen tension falls so low that the blood cannot supply enough oxygen to the muscle (Fig 8.15). Oxymyoglobin is therefore an emergency store of oxygen. Some muscles contain no myoglobin and appear white in colour. Fish contain a lot of white muscle. Another example is breast meat in poultry in contrast to the leg meat which contains myoglobin and is red.

The main respiratory substrate is glucose. Muscle contains the polysaccharide **glycogen** (Fig 17.26(a)) which can be hydrolysed, thus providing glucose for respiration in addition to that supplied in the blood.

Muscle also contains **creatine** which can be phosphorylated by the transfer of phosphate from ATP. When ATP is used to provide energy for muscle contraction the ADP so formed is immediately rephosphorylated by phosphocreatine. In this way a supply of ATP is maintained (Fig 17.29). Phosphocreatine thus acts as an energy store in muscle fibres in addition to ATP (Chapter 5).

Since all the components for the contraction mechanism are present inside the myofibrils, why are muscles not contracted all the time? How does the arrival of impulses at the neuromuscular junctions trigger off contraction? How does contraction occur simultaneously throughout the entire muscle?

Fig 17.29 The role of creatine in the provision of energy for muscle contraction

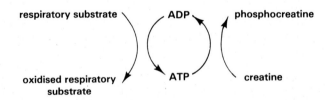

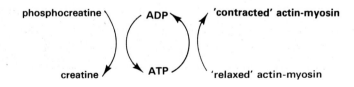

SUMMARY

Homeostasis is the ability to respond to changes in the body's internal environment in such a way as to restrict the changes to within tolerable limits.

Co-ordination is the ability to make an appropriate response, both qualitatively and quantitatively, to a particular stimulus.

Excitable cells are neurones and muscle fibres. They can conduct impulses. Neurones consist of soma, dendrites (receivers) and axon (transmitter).

Impulses depend on a disturbance of the electrochemical properties of the excitable cell membrane. In the resting state, sodium and potassium ions are exchanged in equal numbers (sodium pumped out of and potassium pumped into the cell). This establishes concentration gradients. Sodium ions diffuse back into the cell, potassium ions diffuse out of the cell. The membrane is more permeable to potassium than to sodium and so more potassium moves out of the cell than sodium moves in. This establishes a transmembrane electrical potential difference of about 70 mV.

If the membrane is stimulated, it becomes more permeable to sodium ions. They move into the cell and the membrane depolarises. If depolarisation reaches a threshold value, a sudden greater influx of sodium ions creates an action potential and an impulse is transmitted. The threshold must be reached; all or nothing law.

Movement of ions across the membrane sets up a current which crosses the membrane near the point of original stimulation. The process begins again; permeability change, depolarisation, action potential, more current and so on.

A second impulse cannot pass until the membrane recovers its resting state following the first impulse. The time for recovery is called the refractory period.
First stage (absolute refractory period): repolarisation due to outflow of potassium ions. Membrane charge restored.
Second stage (relative refractory period): sodium–potassium exchange pump restores ion balance. A second impulse is possible if stimulus strength is increased.

Impulse conduction velocity is increased by the myelin sheath. Schwann cells wrap many layers of membrane around the axon. They insulate the axon membrane. Ion and current exchanges only occur at nodes of Ranvier. Impulse velocity is much less in unmyelinated axons. Fibre diameter and temperature also affect impulse velocity.

Axons send impulses to other neurones by synapses. Boutons at axon terminals contain mitochondria and vesicles. Transmitter substances (e.g. acetylcholine) are released into the synaptic cleft on arrival of an impulse. Transmitters may cause depolarisation (excitatory synapses) or hyperpolarisation (inhibitory synapses) of the postsynaptic membrane. Synaptic transmitters are degraded by enzymes (e.g. acetylcholinesterase) and their effect is only temporary. Products of degradation are absorbed and converted back to active transmitters in neurones. Neuromuscular transmission occurs at motor end plates; similar to synapses.

Skeletal muscle consists of fibres which contain myofibrils. Myofibrils consist of sarcomeres which contain thick (myosin) and thin (actin) protein filaments. The interdigitating arrangement of these displays a banded or striated appearance.

The arrival of nerve impulses at motor end plates activates muscle membranes by a transmitter substance (acetylcholine). Muscle impulses cause depolarisation of fibre membranes. Calcium ions enter the fibres. In the presence of calcium ions, ATP-ase enzyme (part of myosin) releases energy from ATP. Thick and thin filaments slide over each other. Sarcomeres and myofibrils shorten.

Skeletal muscle must be activated by nerve impulses; it is neurogenic. Cardiac and visceral muscle are myogenic; they can generate impulses and contract spontaneously. Neurones regulate their activity.

Energy reserves in muscle are increased by stores of oxygen as oxymyoglobin, glucose as glycogen and high energy phosphate as phosphocreatine.

QUESTIONS

1 The diagram shows a section through parts of adjacent nerve cells.

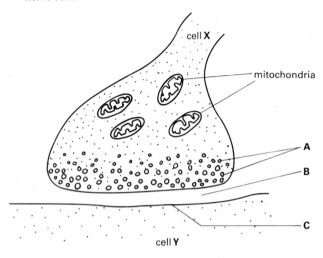

(a) Name parts **A**, **B** and **C**.
(b) Using information from the diagram, explain how a nerve impulse passes from cell **X** to cell **Y**. (AEB 1986)

2 (a) Make a fully-labelled diagram to show the principal muscles and bones involved in movement of the fore-limb of a **named** mammal.
(b)
(i) The most widely accepted theory of muscle contraction is termed '*the sliding filament theory*', Describe, with special reference to the micro-anatomy of striated muscle, how this theory helps to explain muscle contraction.
(ii) In order to initiate contraction, a nerve impulse must arrive at the muscle. Explain how potassium and sodium ions are important in this process. (O)

3 (a) Give an illustrated account of the structure of (i) striated (skeletal) muscle, (ii) cardiac muscle.
(b) Describe how the contraction of striated muscle is initiated. (L)

4 (a) (i) What are the characteristic features of muscular tissues?
(ii) Compare the structure of skeletal (striated) muscle with that of smooth (plain) muscle.
(b) How can the special features of the contraction of cardiac muscle be related to its structure and physiology?
(c) Explain why strenuous exercise may cause fatigue in skeletal muscle. (L)

5 The diagram below illustrates a type of nerve cell found in a mammal.

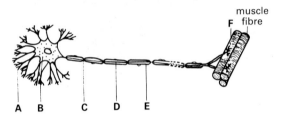

(a) Name the parts labelled **A**–**E**.
(b) What functional type of neurone is shown in the diagram?
(c) State *one* function of each of the parts labelled **A**, **C** and **E**.
(d) Explain in detail how a nerve impulse is transmitted from part **F** to the muscle fibre. (L)

6 With reference to a mammal:
(a) (i) distinguish between the appearance of striated, cardiac and smooth muscle as seen using a light microscope;
(ii) briefly state the functions of each of the three muscle types.
(b) (i) What further details of the structure of striated muscle can be resolved using an electron microscope?
(ii) Explain the contraction of striated muscle in the light of present knowledge. (O)

18 Nervous control and coordination in mammals 2. Nervous integration

18.1	Neuronal pathways	343
	18.1.1 Diverging pathways	343
	18.1.2 Converging pathways	344
	18.1.3 Reverberating pathways	345
18.2	Nerve cord and reflexes	345
18.3	Autonomic system	348

18.4	Brain	350
	18.4.1 Hindbrain	351
	18.4.2 Midbrain	352
	18.4.3 Forebrain	353
	Summary	355
	Questions	356

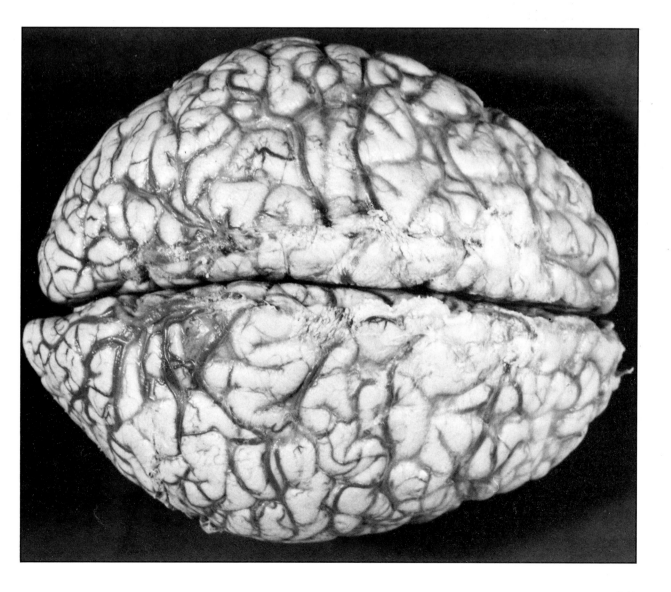

18 Nervous control and coordination in mammals
2. Nervous integration

The nervous system contains countless millions of neurones which are joined by synapses into a large number of circuits or **pathways**. The precise ways in which neuronal pathways are constructed can affect the nature and destination of the information transmitted. Any one neurone may receive impulses from many other neurones that synapse with it. Synapses are formed on either the dendrites or the soma of the postsynaptic neurone (Fig 17.16). Similarly the axon of any one neurone may branch many times and terminate at synapses on many other neurones. The pathways along which impulses are transmitted may thus be complex, and impulses may be sent to any part of the body. In addition, some pathways may be closed at times by the action of inhibitory synapses.

The nervous system consists of several distinct divisions. Each division performs its own particular function and contributes to overall nervous integration. The **central nervous system (CNS)** consists of the brain and spinal cord. The rest of the nervous system is called the **peripheral system**. It includes **somatic nerves** and the **autonomic nervous system**. The autonomic system may be subdivided further into the **sympathetic** and **parasympathetic** divisions (Figs 18.1 and 18.13).

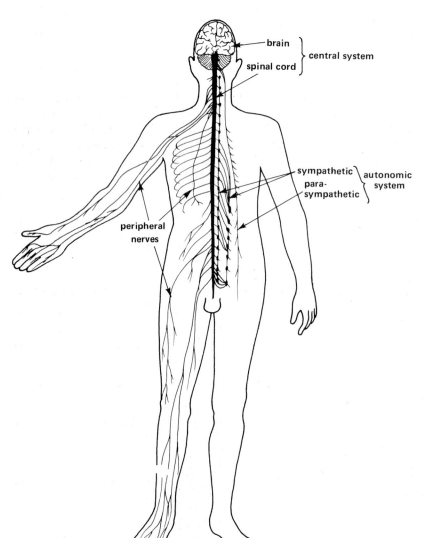

brain
spinal cord
} central system

sympathetic
para-
sympathetic
} autonomic
system

peripheral
nerves

Fig 18.1 Diagram of the main divisions of the nervous system

18.1 Neuronal pathways

Several kinds of neuronal pathways can be described.

18.1.1 Diverging pathways

In a **diverging pathway** the route along which impulses are transmitted divides, enabling information from a single source to be transmitted to a number of destinations (Fig 18.2). There are many instances of nervous coordination in which a variety of responses are appropriate reactions to a single stimulus. An example is the body's responses to changes in blood and skin temperature (Chapter 20).

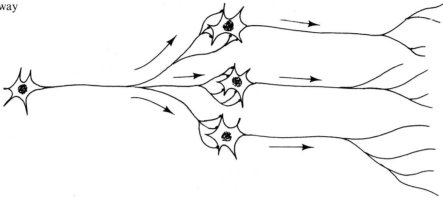

Fig 18.2 Diverging neuronal pathway

At the branches in diverging pathways at least two different zones of activity can be seen (Fig 18.3). Look at the pathways from neurone 1 to neurone B and from neurone 2 to neurone E. There are relatively many synapses joining these neurones *in line* at regions called **discharge zones**. Impulses are readily transmitted through discharge zones. Now look at the pathways from neurone 1 to neurones A and C and from neurone 2 to neurones D and F. There are relatively fewer synapses joining these

Fig 18.3 Discharge zones (dz) and facilitated zones (fz) in diverging neuronal pathways

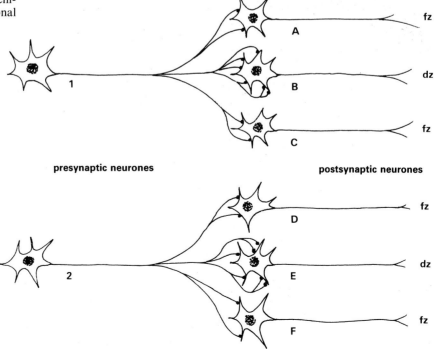

presynaptic neurones

postsynaptic neurones

neurones in regions called **facilitated zones**. Synaptic transmission here may be too weak to generate action potentials in the postsynaptic neurones. However, the postsynaptic membranes may be rendered more susceptible to depolarisation when activated by subsequent stimuli. Hence, impulses are transmitted through facilitated zones only if the presynaptic stimulation is sufficiently frequent.

18.1.2 Converging pathways

In a **converging pathway**, impulses from two or more sources are channelled to the same destination in the body (Fig 18.4). This enables an organ to respond to more than one stimulus. An example is the effect on the heart's activity of changes in the pressure and the gas content of the blood (Chapter 9).

Fig 18.4 Converging neuronal pathway

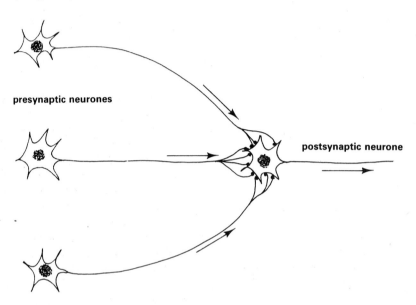

presynaptic neurones

postsynaptic neurone

The more synapses there are in a neuronal pathway, the longer it takes for impulses to travel the entire length of the pathway. This is because the secretion of transmitter chemicals at the synapse is slow relative to impulse conduction along an axon. Examine Fig 18.5. The neurone on the left diverges to several pathways which then converge on the neurone on the right. Because of the different numbers of synapses in the pathways, many impulses arrive at the neurone on the right at different times. The frequency of output is thus greater than the frequency of input.

Fig 18.5 If the input to this circuit is a single impulse, what do you think would be the output?

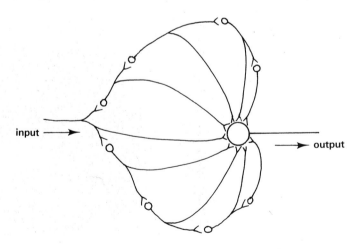

input

output

18.1.3 Reverberating pathways

In a **reverberating pathway** impulses are routed back into the same circuit time and time again (Fig 18.6). Re-routing continues until the neurones or their synapses fatigue. Reverberating pathways are thought to be responsible for the actions of the medullary rhythmicity area in controlling the basic rhythm of breathing (Fig 8.10).

Fig 18.6 Reverberating neuronal pathway. Once activated by an input, the pathway continues to transmit impulses until it fatigues

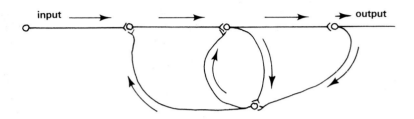

18.2 Nerve cord and reflexes

The great majority of body functions controlled by the nervous system involve **reflexes**. Reflex actions are very rapid, automatic responses to stimuli. Reflex pathways involve the central nervous system but do not necessarily involve conscious awareness. Thus we do not have to think about our limb movements while we are walking, or our rib and diaphragm movements while breathing or our heart's actions in pumping blood. These activities are controlled rapidly and effectively for us by reflex actions. We can thus carry out a conversation with someone, even while we are walking, breathing and maintaining our circulation.

Many reflex pathways involve the spinal cord, a posterior extension of the brain which runs the length of the back. It is enclosed and protected by the vertebrae. The cord in humans is about 130 mm across in the cervical (neck) region and tapers to about 70 mm across in the sacral (pelvic girdle) region (Fig 18.7). Between the vertebrae pairs of spinal nerves arise, one on each side of the spinal cord. Spinal nerves contain both **receptor neurones** and **effector neurones**. The sensory neurones transmit impulses to the spinal cord from receptors. The motor neurones transmit impulses from the spinal cord to effectors.

What are reflexes? What is their significance in our nervous systems?

Fig 18.7 Photomicrograph of the spinal cord of a cat, lumbar region, × 25

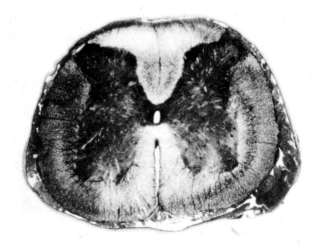

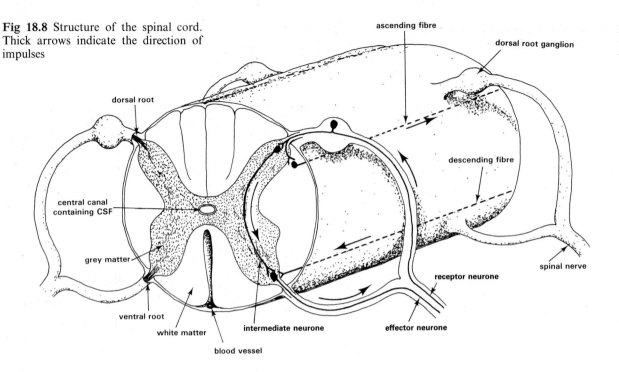

Fig 18.8 Structure of the spinal cord. Thick arrows indicate the direction of impulses

Labels (Fig 18.8): ascending fibre, dorsal root ganglion, dorsal root, descending fibre, central canal containing CSF, grey matter, spinal nerve, ventral root, receptor neurone, white matter, intermediate neurone, effector neurone, blood vessel

Each spinal nerve joins the spinal cord at two points. Receptor neurones are found in the **dorsal root**, effector neurones in the **ventral root** (Fig 18.8). Within the spinal cord the nervous tissue is distributed in two regions. The central **grey matter** contains transverse fibres, the outer **white matter** contains tracts of longitudinal fibres that transmit impulses up and down the cord. The simplest reflex pathways involve only two neurones, joined by a single synapse. An example is the **stretch reflex** (Fig 18.9). Striped muscle contains small receptors called **spindles** that generate sensory impulses when stretched. Impulses are transmitted along receptor neurones and enter the spinal cord through the dorsal root. The receptor neurones terminate in the grey matter where they synapse with motor neurons. The latter transmit impulses out of the nerve cord, through the ventral root and back into the same spinal nerve that carried the sensory impulses. The effector neurones terminate at neuromuscular junctions on the stretched muscle. Contraction of the muscle counteracts the stretching. A **knee-jerk** reflex occurs when the patellar ligament connecting the knee-cap (patella) to the tibia in the lower leg is struck with a special hammer. The patella is pulled and stretches the quadriceps femoris muscle in the upper leg. The reflex results in contraction of the quadriceps muscle and the lower leg jerks outwards.

Fig 18.9 Stretch reflex

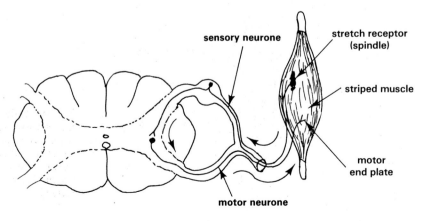

Labels (Fig 18.9): sensory neurone, stretch receptor (spindle), striped muscle, motor end plate, motor neurone

Fig 18.10 Crossed extensor reflex (after Tortora and Anagnostakos)

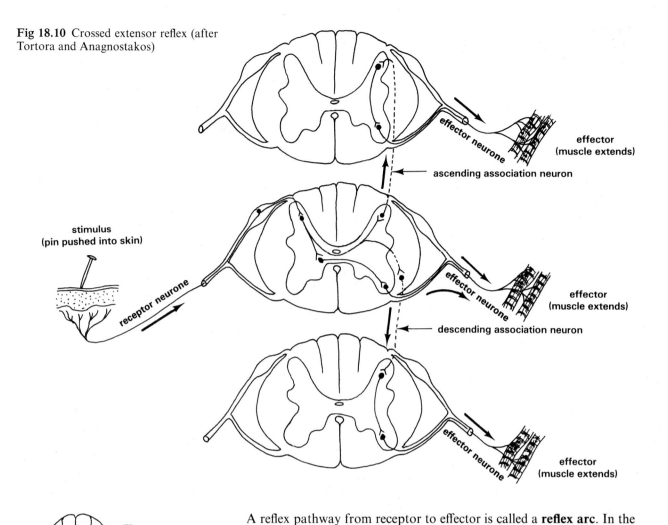

effector neurone

effector (muscle extends)

ascending association neuron

stimulus (pin pushed into skin)

receptor neurone

effector neurone

effector (muscle extends)

descending association neuron

effector neurone

effector (muscle extends)

spinal cord

excited pathway

inhibited pathway

biceps contracts

triceps relaxes

forearm raised

A reflex pathway from receptor to effector is called a **reflex arc**. In the knee-jerk reflex the receptor and effector neurones join the spinal cord on the same side. It is called an **ipsilateral** reflex arc. Some pathways are more complex and include longitudinal tracts in the white matter of the spinal cord. An example is the **crossed extensor reflex** (Fig 18.10). The receptor and effector neurones are on opposite sides of the spinal cord in this **contralateral** reflex arc.

Antagonistic activity of muscles regulates movement and posture. Contraction of one muscle is usually accompanied by relaxation of another. During such activity the simultaneous contraction of both sets of muscles is prevented by the actions of inhibitory synapses (Fig 18.11 and section 17.1.6).

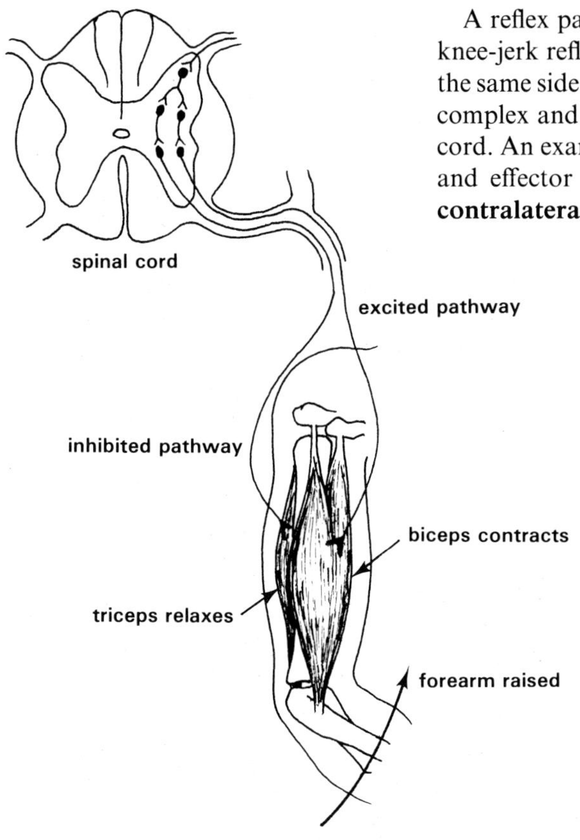

Fig 18.11 Reciprocal inhibition. During body movements involving the actions of antagonistic muscles, one muscle is excited while the other is inhibited. Here the forearm is flexed by contraction of the biceps while the triceps relaxes. Neuronal pathways involving excitatory and inhibitory synapses are thought to control this type of activity

We are born with certain neuronal pathways ready to function immediately. A new-born baby automatically sucks at objects placed in its mouth. This **innate reflex** is vital for the baby to feed. If a baby's body is supported and its feet allowed to touch a solid surface, it will usually move its legs in a characteristic walking action. A baby's hands grip tightly onto objects placed in them. Very young infants acquire additional reflexes that override the innate reflexes at certain stages of development. Observation of innate and acquired reflexes can help in determining the neurological development of babies (Fig 18.12).

Fig 18.12 Some infantile reflexes (after Williams and Wendell-Smith)

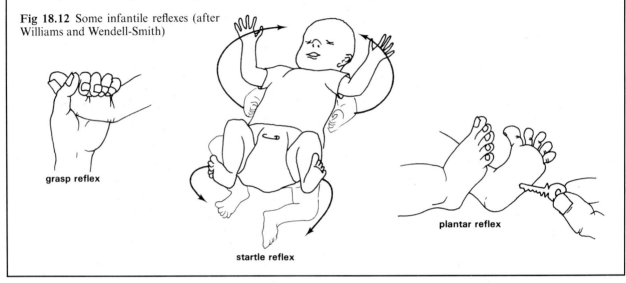

grasp reflex

startle reflex

plantar reflex

It is an interesting exercise to look for Pavlov's conditioning in advertisements. How many of them describe the true nature of the product and how many associate it with something attractive but quite unrelated to the product?

In 1910, Pavlov demonstrated **conditioned reflexes**. He noticed that hungry dogs salivated when presented with food. He also observed that hungry dogs salivated when exposed to a secondary stimulus such as ringing bells at the same time as the primary stimulus, food. Pavlov discovered that, in time, the dogs learned to associate the secondary stimulus with food. They salivated when bells were rung even if food was absent. Knowledge of such **conditioning** is effectively used by the advertising industry.

18.3 Autonomic system

The autonomic nervous system consists of two sets of nerves, **sympathetic** and **parasympathetic** (Fig 18.13). The two sets generally have the opposite effects (Table 18.1). Many organs have double innervation, being supplied by both sympathetic and parasympathetic nerves. The balance between the activity of the two sets of nerves results in coordinated regulation of the organs, usually by reflex action. As with peripheral reflexes, the central nervous system is involved in autonomic activity. Many of the centres coordinating autonomic functions are situated in the hind-brain. They include the cardiac and vasomotor centres (Chapter 9), the respiratory centre (Chapter 8) and the thermoregulatory centres (Chapter 20). These receptors detect body changes that require adjustment of the cardiovascular and gas exchange systems and the skin. Appropriate impulses are conveyed to these systems along autonomic nerves. Unlike somatic reflexes, autonomic reflexes are less easily controlled by conscious activity of the brain. Although it is possible to train an individual to reduce the heart rate, for instance, autonomic function relies almost entirely on reflexes. The ability to control breathing consciously is due to the fact that the muscles involved are supplied by somatic and well as autonomic nerves.

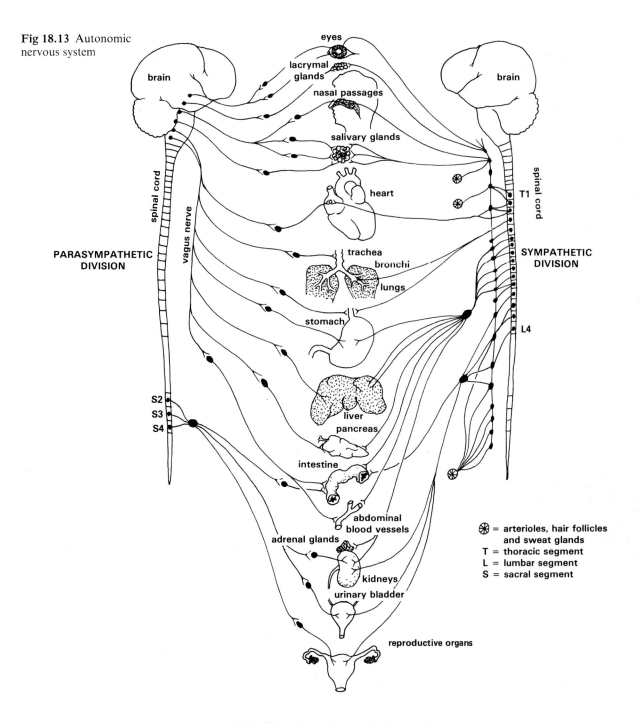

Fig 18.13 Autonomic nervous system

eyes
lacrymal glands
nasal passages
salivary glands
heart
trachea
bronchi
lungs
stomach
liver
pancreas
intestine
abdominal blood vessels
adrenal glands
kidneys
urinary bladder
reproductive organs

brain

brain

spinal cord

spinal cord

vagus nerve

PARASYMPATHETIC DIVISION

SYMPATHETIC DIVISION

T1

L4

S2
S3
S4

⊛ = arterioles, hair follicles and sweat glands
T = thoracic segment
L = lumbar segment
S = sacral segment

Table 18.1 Some of the main effects of the autonomic nervous system

Target	Sympathetic effect	Parasympathetic effect
iris of eye	pupil dilation	pupil constriction
bronchi	—	constriction
heart	increased cardiac output	decreased cardiac output
gut sphincters	increased tone	relaxation
urinary bladder	relaxation of urinary bladder wall	contraction of urinary bladder wall
sweat glands	stimulates secretion	—
salivary glands	decreases secretion	increases secretion
stomach	decreases secretion	increases secretion
pancreas	decreases secretion	increases secretion
genitalia	—	vasodilation in erectile tissue

The opposite effects of sympathetic and parasympathetic nerves result from the different transmitter chemicals secreted at their neuroeffector junctions. Acetylcholine is the transmitter secreted by parasympathetic nerve endings. Parasympathetic nerves are said to be **cholinergic**. Sympathetic nerves secrete noradrenaline and are said to be **noradrenergic** (Fig 18.14).

Fig 18.14 Cholinergic and adrenergic nerves

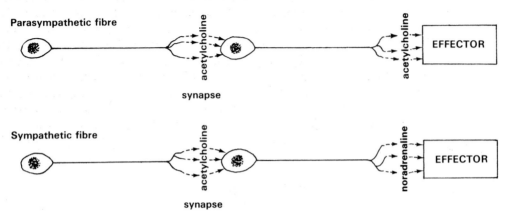

Adrenaline as well as noradrenaline is secreted into the blood from the medullae of the adrenal glands which are modified portions of the sympathetic system. In conditions of stress, adrenalin secretion increases. The significance of this response is discussed in Chapter 16.

18.4 Brain

The adult human brain contains more than a thousand million neurones. It has been estimated that in the cerebral cortex alone there are about $10^{2\,783\,000}$ synapses! Considering these numbers of neurones and synapses in an organ weighing only about 1.3 kg, it is evident how complex the human brain is. Inside the **cranium** of the skull the brain is surrounded by several protective membranes and fluids. The brain is divisible into three main sections: hindbrain, midbrain and forebrain (Fig 18.15).

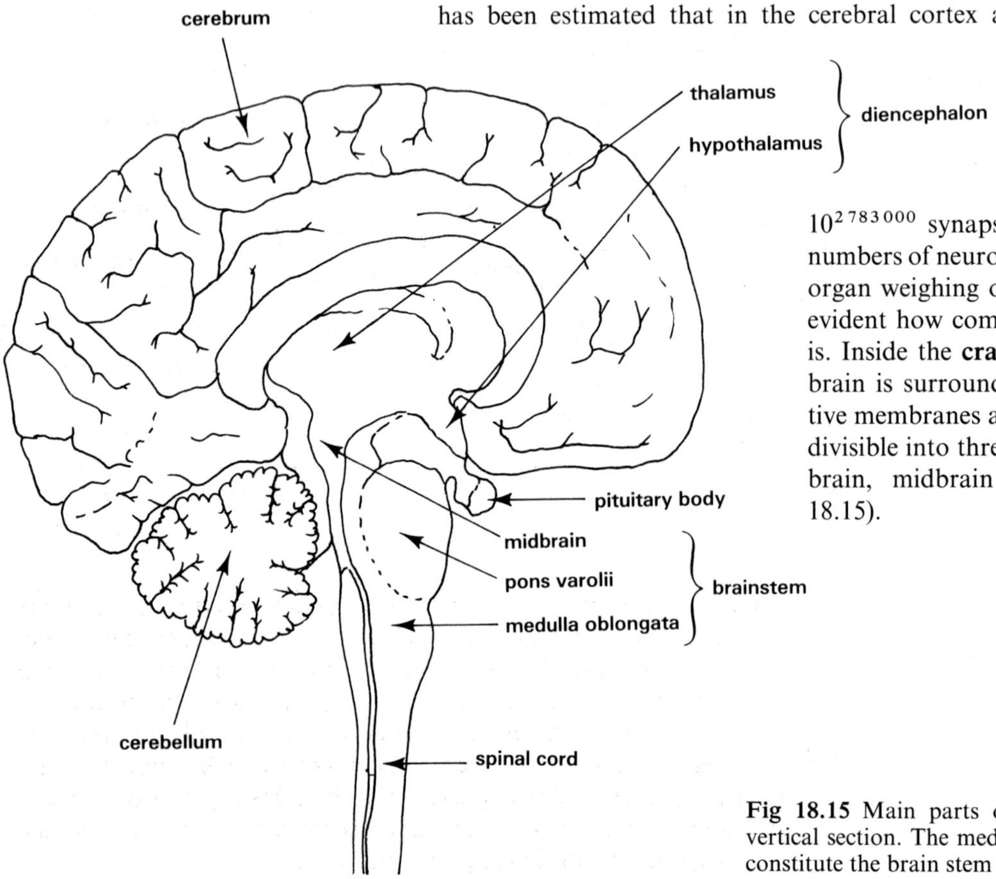

cerebrum

thalamus

hypothalamus

diencephalon

pituitary body

midbrain

pons varolii

medulla oblongata

brainstem

cerebellum

spinal cord

Fig 18.15 Main parts of the brain as seen in vertical section. The medulla, pons and midbrain constitute the brain stem

18.4.1 Hindbrain

There are three main parts to the hindbrain, the **medulla oblongata**, the **pons varolii** and the **cerebellum**. The medulla contains many tracts of longitudinal nerve fibres connecting the spinal cord with the rest of the brain. Many of the fibres cross over from left to right and right to left. The medulla also contains a number of **reflex centres**. The **vital centres** include the cardiac, vasomotor and respiratory centres. (Chapters 10 and 12). The **non-vital centres** are concerned with the reflex control of swallowing, vomiting, hiccupping, coughing and sneezing. **Cranial nerves** VIII, IX, X, XI and XII originate in the medulla (Fig 18.16 and Table 18.2). The pons is on the ventral surface of the medulla. It acts as a bridge between various parts of the central nervous system. There are longitudinal fibres connecting the spinal cord and medulla with the rest of the brain. There are also transverse fibres between the pons and the cerebellum. Cranial nerves V, VI, VII and branches of VIII also originate in the pons (Fig 18.16 and Table 18.2).

Fig 18.16 Cranial nerves (brain viewed from underneath)

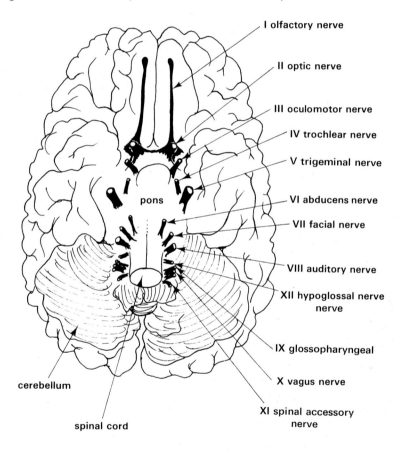

The cerebellum consists of two hemispheres attached to the rest of the hindbrain by three bundles of nerve fibres. The surface of the cerebellum has many ridges called **folia**. It is a motor region concerned with subconscious skeletal movements involved in posture and balance. Sensory information is supplied to the cerebellum from the vestibular apparatus of the inner ear (Chapter 19). The cerebellum is essential for smooth precise movement and delicate manipulations such as playing a piano, typing speech and writing. Without cerebellar control, such movements lack precision and are erratic in range and direction.

Table 18.2 Summary of the actions of the cranial nerves

Nerve	Function	Origin
I Olfactory	Sensory–smell	Olfactory mucosa
II Optic	Sensory–vision	Retina
III Oculomotor	Motor– eyelids & eyeball muscles, ciliary & iris muscles	Midbrain
	Sensory	Eyeball muscles
IV Trochlear	Motor–eyeball muscles	Midbrain
	Sensory	Eyeball muscles
V Trigeminal	Motor–chewing	Pons varolii
	Sensory	Eyelids, eyeballs, lachrymal glands, nasal cavity, forehead, scalp, palate, pharynx, teeth, lips, tongue, cheek
VI Abducens	Motor–eyeball muscles	Pons varolii
	Sensory	Eyeball muscles
VII Facial	Motor–facial, scalp & neck muscles, lachrymal & salivary glands	Pons varolii
	Sensory–taste	Taste buds on tongue
VIII Vestibulocochlear	Sensory–balance	Vestibular apparatus
	Sensory–hearing	Cochlea
IX Glossopharyngeal	Motor–swallowing	Medulla oblongata
	Sensory–taste	Taste buds on tongue
X Vagus	Motor–pharynx, larynx, respiratory tract, lungs, heart, gastrointestinal tract & gall bladder	Medulla oblongata
	Sensory	Organs supplied by motor fibres
XI Accessory	Motor–pharynx, larynx & palate	Medulla oblongata & cervical portion of spinal cord
	Sensory	Muscles supplied by motor fibres
XII Hypoglossal	Motor–tongue	Medulla oblongata
	Sensory	Tongue muscles

18.4.2 Midbrain

As you sit reading this book you are bombarded by a variety of environmental stimuli. The more successfully you concentrate on your reading, the less aware you are of outside distractions. When studying, unwanted stimuli may include the noise of traffic, people talking, a radio or television. You also constantly receive signals from your skin in response to the touch of your clothes or the chair you are sitting on. In other words, of the great number of environmental stimuli you receive at any one time, you may be consciously aware of only a few, whether you are studying or not. This **filtering** prevents overloading your conscious awareness. Most people experience difficulty in concentrating on more than one thing at a time. Confused behaviour results from overloading the brain with too many stimuli. Filtering sensory signals is a function of the **reticular formation** that lies mainly in the midbrain. The reticular formation activates the forebrain with appropriate signals.

Activation of the cerebral cortex by the reticular formation is essential for **wakefulness**. When activation stops, a state of **sleep** follows. The significance of sleep is not fully understood. Nevertheless, sleep is vital for normal activity of the brain. Lack of sleep and excessive nervous fatigue can lead to serious mental disturbances. Sleep deprivation has been used to lower the mental resistance of prisoners prior to interrogation.

On the dorsal surface of the midbrain are four rounded bodies called the **corpora quadrigemina**. They control movements of the eyeballs, head and trunk. Cranial nerves III and IV originate in the midbrain (Fig 18.16 and Table 18.2).

Fig 18.17 Brain viewed from above. Note the extensive folds and grooves

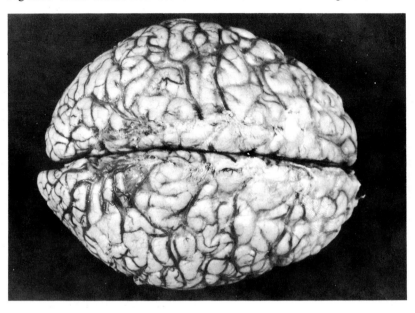

18.4.3 Forebrain

The forebrain consists of two main parts, the **diencephalon** and the **cerebrum**. The diencephalon contains the **thalamus** and **hypothalamus** (Fig 18.15).

The thalamus directs sensory impulses from the lower parts of the brain and the spinal cord to appropriate parts of the cerebrum. It is organised into several masses called **nuclei** which act as the relays for particular sensory pathways (Table 18.3). The thalamus also relays motor impulses to the spinal cord from the cerebrum. Limited sensory awareness of pain, temperature, touch and pressure is provided by the thalamus.

Just beneath the thalamus is the hypothalamus. It contains reflex centres linked to the autonomic system. They include the **thermoregulation centres** (Chapter 20). The **feeding centre** is stimulated when the stomach is empty and gives us the sensation of hunger. The **satiety centre** is stimulated when the stomach is full and inhibits further feeding. The **thirst centre** makes us feel thirsty when stimulated by angiotensin which appears in the blood when the body is short of water (Chapter 21).

Many of the **pituitary hormones** are thought to originate in the hypothalamus. The pituitary body projects beneath the hypothalamus (Chapter 21). The hypothalamus is also concerned with feelings of rage and aggression. One of the hypothalamic centres acts with the reticular formation to regulate wakefulness and sleep.

By far the largest and most highly developed part of the brain is the cerebrum. Most cerebral activity occurs in the outer **cortex** of grey matter about 2 to 4 mm thick. The great mass of the cerebrum is white matter. Because of relatively greater growth of the cortex, the cerebral surface is highly folded. Between the folds, called **gyri**, are very many shallow grooves called **sulci** and deeper **fissures**. Consequently much of the grey matter is hidden beneath the surface (Fig 18.17). The cerebrum is divided into two **hemispheres** by a prominent longitudinal fissure. The two

Table 18.3 Main sensory functions of the thalamic nuclei

Nucleus	Sensory pathways relayed
Median geniculate	hearing
Lateral geniculate	vision
Ventral posterior	general sensations and taste

Fig 18.18 Lobes of the cerebral hemispheres

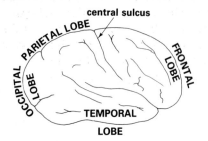

central sulcus

PARIETAL LOBE
OCCIPITAL LOBE
FRONTAL LOBE
TEMPORAL LOBE

hemispheres are connected by a bundle of transverse fibres called the **corpus callosum**. Each cerebral hemisphere is divided into four **lobes** (Fig 18.18). They are the **frontal** at the front, the **parietal** towards the top of the head, the **temporal** on the side and the **occipital** at the rear. The **basal ganglia** are paired masses of neurones in the cerebral hemispheres. They contain receptor and effector fibres connecting the cerebral cortex with the **brainstem** and spinal cord. The brainstem is the midbrain, pons and medulla. The basal ganglia control subconscious skeletal movements such as arm swinging while walking.

The **limbic system** consists of parts of the cerebral hemispheres and diencephalon. It controls the emotional aspects of behaviour such as pleasure, anxiety and pain. The limbic system is also a major centre for **memory** and overall control of **behaviour**.

The cerebral cortex can be divided into three main kinds of functional areas. They are **sensory, motor** and **association** areas (Fig 18.19). The sensory areas are the sites of reception, correlation and interpretation of sensory information, that is, **perception**. The motor areas are the regions

Fig 18.19 Some of the main functional areas of the cerebral cortex

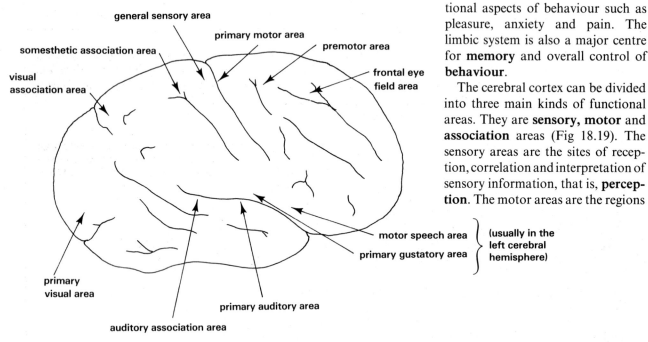

general sensory area

primary motor area

premotor area

somesthetic association area

frontal eye field area

visual association area

visual association area

motor speech area

primary gustatory area

(usually in the left cerebral hemisphere)

primary visual area

primary auditory area

auditory association area

Table 18.4 Summary of the functions of the main areas of the cerebral cortex

Area	Function
General sensory (somesthetic)	Receives sensations from skin, muscle and visceral receptors from all over the body.
Somesthetic association	Receives signals from general sensory area and thalamus – interpretation and integration. Memory of past sensory experiences.
Primary visual	Vision – shape and colour.
Visual association	Receives signals from primary visual area. Recognition of objects by reference to memory of past visual experiences.
Primary auditory	Interpretation of fundamental characteristics of sound such as pitch and intensity.
Auditory association	Translation of speech into meaningful ideas.
Primary gustatory	Taste.
Primary olfactory	Smell.
Gnostic	Receives signals from all other sensory areas, enabling an overall impression of sensory experience. Transmits impulses to other areas to enable the appropriate responses.
Primary motor	Voluntary motor pathways to specific skeletal muscles.
Premotor	Control of complex sequences of motor actions such as writing or playing a piano.
Frontal eye field	Controls voluntary eye scanning movements such as are used during reading.
Motor speech (Broca's)	Translation of thoughts into speech. Generation of speech by controlling the actions of the larynx, throat, mouth and muscles responsible for breathing. This area is usually located in the left cerebral hemisphere.

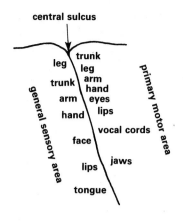

central sulcus

leg
trunk
leg
arm
trunk
arm
hand
hand
eyes
lips
face
vocal cords
lips
jaws
tongue

general sensory area

primary motor area

from which motor pathways originate. Voluntary motor activity is controlled by a strip of cerebral cortex just anterior to the central sulcus. Different parts of the body are controlled by specific parts of the voluntary motor area. Similarly, sensory information from particular parts of the body is interpreted in specific parts of the **general (somesthetic) sensory area** just posterior to the central sulcus (Fig 18.20). The functions of the main regions of the cerebral cortex are summarised in Table 18.4.

Fig 18.20 Localised regions of the general sensory and primary motor areas

The enormous numbers of neurones in the cerebral cortex are constantly active. The electrical activity accompanying transmission of impulses in the brain can be measured by placing electrodes on the scalp (Fig 18.21). Rhythmic waves are produced and these are displayed as an **electroencephalogram (EEG)**. Disturbances may produce EEG patterns which are abnormal and are of value in the diagnosis of some brain diseases (Fig 18.22).

Fig 18.21 Electrodes on the scalp and face of a patient prior to taking an electroencephalogram (Alexander Tsiaras/Science Photo Library)

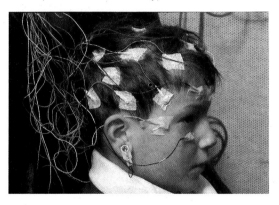

Fig 18.22 EEG trace showing 'spike and wave' activity from an individual suffering from petit-mal epilepsy (courtesy Mrs C. Dunn, Dept. of Neurophysiology, Queen's Medical Centre, Nottingham)

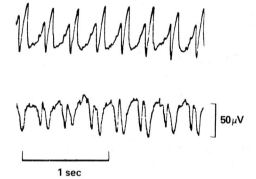

50 μV

1 sec

SUMMARY

The nervous system is an integrated assembly of neurones connected into many pathways. Divisions include central nervous system, (CNS; brain and spinal cord) and peripheral system (somatic and autonomic nerves). Neuronal pathways display several patterns, e.g. diverging, converging and reverberating.

Many body functions are controlled by nerve reflexes; fast, automatic responses to stimuli, not necessarily involving the conscious levels of the brain. The simplest reflexes involve receptors, receptor (effector) neurones, spinal cord, effector neurones and effectors (e.g. muscles).

Spinal cord contains central grey matter (transverse nerve fibres) surrounded by white matter (longitudinal fibres). Complex reflexes can involve longitudinal conduction to different levels of the body and crossing over to the opposite side of the body. Inhibitory synapses can switch off a pathway while another is switched on, e.g. activation of antagonistic skeletal muscles. Reflexes can be either inherited (innate) or learned (conditioned). ▶

> The autonomic system has two divisions, sympathetic and parasympathetic.
> Sympathetic nerves are adrenergic, i.e. they secrete noradrenaline at the neuroeffector junctions. Parasympathetic nerves are cholinergic, i.e. they secrete acetylcholine at the neuroeffector junctions. All secrete acetylcholine at the synapses. Many organs receive dual innervation; activity of the organ is determined by the balance between the two autonomic divisions.

Hindbrain:

Medulla oblongata contains vital reflex centres; cardiac, respiratory and vasomotor; also non-vital centres; swallowing, vomiting, hiccupping, coughing and sneezing.

Cerebellum is concerned with the control of subconscious skeletal movements involved in posture and balance; essential for smooth, precise movements, e.g. writing.

Midbrain:

Midbrain contains the reticular formation; activates conscious awareness in the forebrain; controls wakefulness and sleep.

Forebrain:

Thalamus directs impulses to appropriate parts of the cerebrum; some sensory awareness.

Hypothalamus contains reflex centres, e.g. thermoregulation, feeding and satiety, thirst and osmoregulation centres. Many pituitary hormones originate here.

Cerebrum: largest part of the brain; cortex highly folded due to disproportionate growth of the surface (cortex). Specific functional regions include sensory, motor, speech, gustatory, auditory and visual areas. Association areas enable integration of information to take place.

QUESTIONS

1 (a) Name the regions of the mammalian brain or associated structures concerned with the following.
 (i) Regulation of the heart beat
 (ii) Reflex adjustment of posture
 (iii) Regulation of body temperature
 (iv) Release of antidiuretic hormone (ADH) into the blood
 (v) Memory
(b) Indicate how images falling on the retina are interpreted by the mammalian brain. (L)

2 The diagram below is of a sagittal section through a human brain.

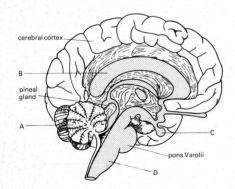

For each of the labels **A** to **D** give (i) the name and (ii) one function of the region you have identified. (O)

3 (a) Make a large, fully-labelled diagram to show the pathways involved in a simple reflex action.
(b) Describe the sequence of events which occur from the detection of the stimulus to the making of a response.
(c) Explain the importance of such pathways in the lives of animals. (L)

4 The diagram below represents a transverse section through the spinal cord of a mammal.

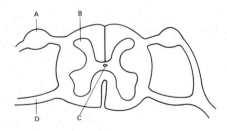

(a) Name the structures labelled **A–D**.
(b) Copy the diagram and draw in and label fully, on the left hand side, the neurones involved in a reflex response such as withdrawal of the hand from a hot object. Indicate also on the diagram the position of the receptor and effector organs and the direction in which the impulse travels.
(c) Indicate *three* ways in which the spinal cord is protected in the body of a mammal. (L)

19 The eye and ear

19.1 The eye   358

   19.1.1 Structure of the eye   358
   19.1.2 Accommodation (focusing)   360
   19.1.3 Defects of the eye   361
   19.1.4 Photoreception   362
   19.1.5 Perception of distance and
           size   366

19.2 The ear   367

   19.2.1 Sound   368
   19.2.2 Structure of the ear   370
   19.2.3 Hearing   371
   19.2.4 Balance   373
   19.2.5 Defects of the ear   375

   Summary   376

   Questions   377

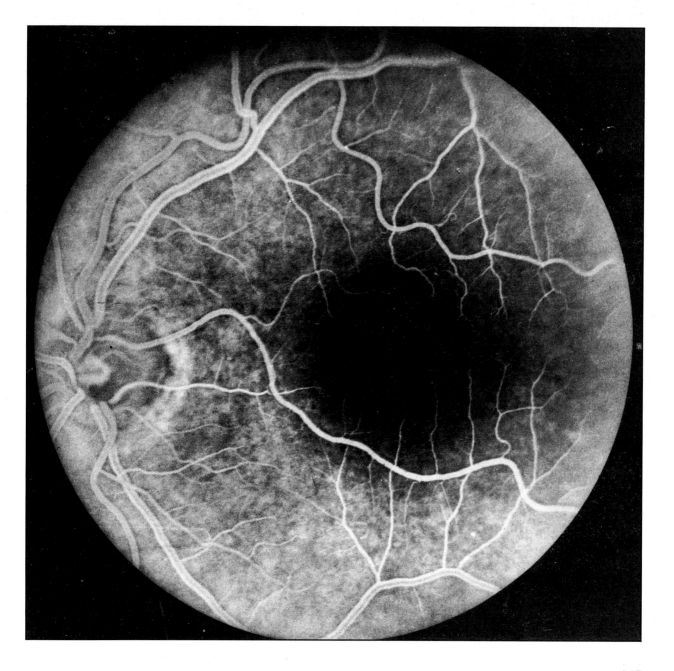

19 The eye and ear

The detection of stimuli depends on the conversion of the stimuli by **receptors** into impulses in the nervous system. The body has a variety of receptors which feed impulses into the nervous system in response to an equal variety of stimuli. Different receptors are sensitive to heat, light, sound, touch, stretch, spatial orientation and to chemicals.

The simplest receptors are single nerve cells which respond directly to a stimulus. Receptors in the skin are of this type. Some receptors consist of groups of sensitive neurones such as the cardiac and respiratory centres in the brain (Chapters 8, 9 and 18). Other receptors are grouped in complex **sense organs**. The structure of sense organs causes the stimulus to be channelled into a receptive region of the organ.

The action potentials produced in activated receptors are called **generator potentials**. They lead to the production of impulses in receptor neurones connected to the receptors. The way in which impulses are generated by receptor cells is important in discriminating between strong and weak stimuli. Measurement of action potentials (Chapter 17) in receptor neurones shows no significant difference between one impulse and another. However, a strong stimulus usually causes impulses to be generated rapidly in receptor neurones. The frequency of impulses generated by weak stimuli is lower. The appropriate sensory region of the cerebral cortex of the brain interprets the intensity of a stimulus according to the frequency with which it receives impulses.

Among the most important sense organs are the **eyes** and **ears**.

19.1 The eye

We depend greatly on vision to sense our environment. While reading this page, your conscious awareness is almost totally stimulated by visual information.

19.1.1 Structure of the eye

The eyes are spherical structures with a wall consisting of three layers, the outer **sclera**, the middle **choroid** layer and the inner **retina** (Fig 19.1). The sclera is a tough, fibrous coating which protects the delicate inner layers. It is white in appearance except at the front where there is a transparent area called the **cornea**. The cornea allows light into the eye. Tears secreted by the **lachrymal glands** lubricate the exposed surface of the eye, including the **conjunctiva** which covers the cornea except in the centre. The watery secretion helps to prevent abrasion of the eye's surface by dust particles and helps combat infection of the eye. Periodic closure of the eyes, blinking, clears away debris.

Fig 19.1 (a) Photomicrograph of a longitudinal section of the eye from a monkey, × 4

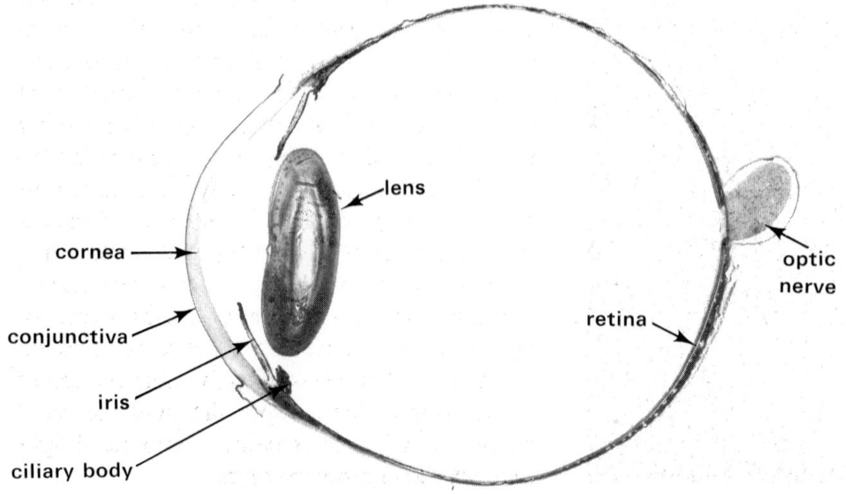

358

Fig 19.1 (b) Diagram of a vertical section through the eye

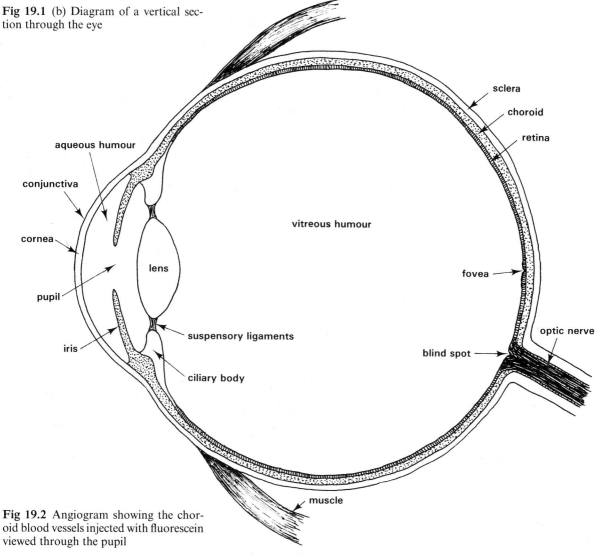

sclera

choroid

retina

aqueous humour

conjunctiva

cornea

vitreous humour

pupil

fovea

lens

iris

suspensory ligaments

optic nerve

blind spot

ciliary body

muscle

Fig 19.2 Angiogram showing the choroid blood vessels injected with fluorescein viewed through the pupil

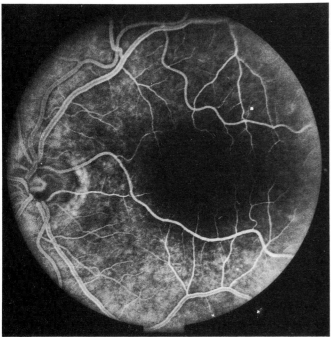

Inside the sclera is the choroid layer which contains numerous blood vessels (Fig 19.2). At the front of the eye it is modified as the **iris** containing pigments which give the eye its colour. The iris also contains radial bands and a ring of circular smooth muscle. Contraction of the radial muscles and relaxation of the circular muscles cause dilation of an aperture called the **pupil**, in the centre of the iris. Constriction of the pupil occurs when the radial muscles of the iris relax and the circular ones contract. Variation in pupil size is controlled by autonomic reflexes and is usually a response to change in the intensity of light entering the eye. In bright light the pupil constricts and prevents excessive illumination of the interior of the eye. In dim light the pupil dilates, allowing the maximum amount of light to reach the photoreceptor cells.

The photoreceptors are in the retina situated immediately inside the choroid layer. Fibres of receptor neurones lead from the retina at the back of the eye as the **optic nerve** which transmits impulses generated in the retina to the brain. The retina develops in the embryo as an outgrowth of the brain, and can be regarded as a modified part of the central nervous system.

Suspended in the fluid inside the eye and just behind the pupil is a biconvex, crystalline **lens**. It is held in position by **suspensory ligaments** attached to a ring of smooth muscle called the **ciliary body**.

19.1.2 Accommodation (focusing)

Light rays entering the eye are redirected or **refracted**. Refraction occurs at three surfaces of the eye before the light reaches the retina. The first of the refracting surfaces is the cornea, then the front surface of the lens and finally the rear surface of the lens (Fig 19.3).

Between the cornea and the lens is a colourless, watery fluid called **aqueous humour**. At the back of the eye between the lens and the retina is the **vitreous humour** made of a gelatinous mucoprotein. The humours are transparent so that transmission of light through the cavities of the eye to the retina is not normally impeded.

Fig 19.3 The three main refracting surfaces in the eye

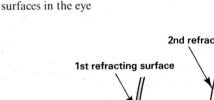

2nd refracting surface

1st refracting surface

3rd refracting surface

object

image

cornea   lens

In the normal eye light rays are refracted sufficiently to be brought to a point on the retina. In this way the object viewed is brought into focus and a clear image is formed. The **image** is **inverted** by the lens (Fig 19.4). However, objects are perceived the right way up because of the way in which the brain interprets images. Experiments have been performed in which human volunteers wore special spectacles, the lenses of which produced upright images on the retina. For a while the subjects in the experiment were confronted with an upside-down world. However, after a few days they became used to the situation and their perception became adjusted so that once more they perceived things the right way up. When the spectacles were removed they again experienced a period of seeing things upside down before normal perception returned. Thus the mechanism of image interpretation, situated in the brain's cerebral cortex, is somewhat flexible.

Fig 19.4 Light from all points on an object are focused in such a way that the image on the retina is inverted. Only two light rays are shown.

The image on our retina is upside down. How do we see the world the right way up?

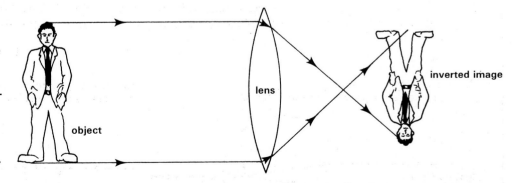

lens

inverted image

object

Light rays from an object near the eye strike the cornea and lens at an acute angle depending on the object's size. If the same object is moved further away from the eye the angle is less acute. Consequently the degree of refraction necessary to **focus** light rays on the retina is greater for close objects than for distant objects. Changes in the degree of refraction of light are achieved by altering the curvature of the lens surfaces (Fig 19.5). The ciliary body contains a ring of involuntary muscles; contraction reduces the tension of the suspensory ligaments which hold the lens in place, and relaxation increases their tension. It is the tension of the ligaments applied to the lens which determines the shape of the lens. When the tension is increased the lens is pulled into a flattened shape suitable for focusing distant objects. When the tension is decreased, the lens becomes a more spherical shape suitable for focusing near objects.

Fig 19.5 Focusing (accommodation)

(a) Light from a distant object is focused on the retina by a flattened lens

(b) Light from a near object is focused on the retina by a near-spherical lens

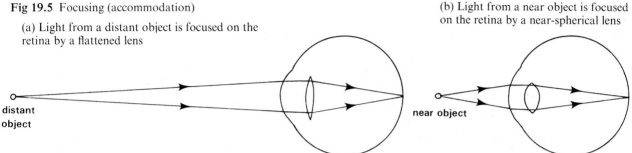

19.1.3 Defects of the eye

There are several abnormalities of the eye's focusing mechanism. The commonest are **myopia (short-sightedness), hypermetropia (long-sightedness)**, and **astigmatism**.

1 Myopia

Myopia results if the lens curvature is too great or the entire eyeball becomes elongated (Fig 19.6(a)). Light rays entering the eye are refracted more than is necessary. Consequently light is focused in front of the retina. By the time the light stimulates the retina it has diverged from the focal point of the lens. The image perceived is thus blurred. The condition is called short-sightedness as objects near the eye are less out of focus than those further away. This is because the light rays from near objects require greater refraction to be focused on the retina than rays from distant objects. Since the lens in a myopic eye refracts light excessively, distant objects appear more blurred than near ones. Myopia can be corrected by placing a **concave lens** in front of the eye. The surface of the concave lens refracts light rays in such a way that the rays diverge slightly from their original path. The lens of the myopic eye now refracts the diverged light rays in to focus on the retina (Fig 19.6(b)).

(continued on p. 362)

Fig 19.6 Myopia
(a) Light is focused at a point in front of the retina

(b) Light may be focused on the retina by placing a concave lens in front of the eye

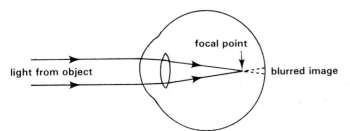

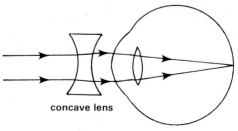

2 Hypermetropia

Hypermetropia results when the curvature of the eye's lens is not great enough. Light rays are not refracted enough and would thus be focused behind the retina (Fig 19.7(a)). The condition is called long-sightedness because distant objects are less out of focus than near ones. This happens because light rays from distant objects require less refraction than rays from near objects. Correction of hypermetropia requires placing a **convex lens** in front of the eye. The lens converges light rays before they enter the eye so that the eye's lens focuses the light correctly on the retina (Fig 19.7(b)).

Fig 19.7 Hypermetropia
(a) Light would be focused at a point behind the retina

(b) Light may be focused on the retina by placing a convex lens in front of the eye

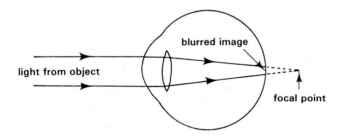

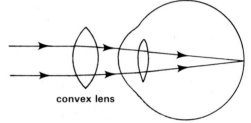

3 Astigmatism

Astigmatism occurs if either the cornea or lens is distorted. One part of the focusing mechanism then refracts light rays too much, another not enough. Usually most of the imaged perceived is out of focus. Light rays from parts of the object are focused in front of the retina, as in myopia. Rays from other parts would be focused behind the retina, as in hypermetropia. Astigmatism can be corrected by placing a lens in front of the eye. The curvature of this lens varies from one part to another to compensate for the eye's deficiencies.

What causes short- and long-sightedness? How can these eye defects be corrected?

19.1.4 Photoreception

The transmission of nerve impulses to the brain in response to stimulation of **photoreceptors** in the retina by light is the function of the optic nerves. In human eyes there are more than 100 million of the two types of photoreceptors called **rods** and **cones**.

1 Rods and cones

Rods are sensitive to different **intensities** of light. Most mammals have only rods. Cones are sensitive to different **wavelengths** of light and enable us to see things in colour. The assumption that an angry bull will charge a red object is somewhat misplaced, since the retina of a bull's eye has no cones. The bull, if it is annoyed is just as likely to charge an object of some other colour.

2 Distribution of rods and cones

The arrangement of photoreceptors in the retina is such that light has to travel through several layers of neurones which are not sensitive to light before reaching the rods and cones: the retina is **inverted** (Fig 19.8).

Fig 19.8 Main cellular components of the retina

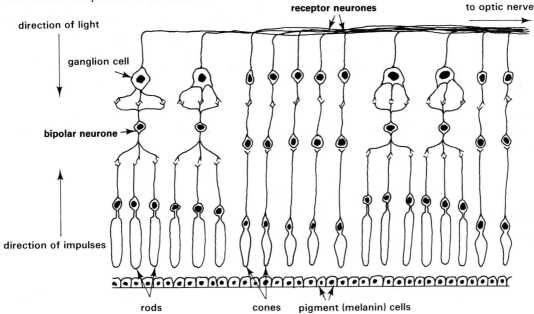

Outside the photoreceptors is a layer of cells containing the black pigment melanin. Melanin is not sensitive to light but absorbs light rays which would otherwise pass through the retina. In this way the formation of hazy images caused by reflection of light to other parts of the retina is prevented. Some mammals have a reflective layer called the **tapetum** in the retina. The reflective 'cat's eyes' along the centre of roads create a similar effect to the reflective tapetum of real cat's eyes. A tapetum is common in nocturnal mammals and enables the maximum use of what little light is available at night.

Many synapses link the photoreceptors with receptor neurones. Impulses generated by the photoreceptors are first transmitted to a small region where the neurones converge to project through the retina into the optic nerve. This region is called the **blind spot**, since no photoreceptors are located there (Fig 19.1). It is possible to demonstrate that there are blind spots in your eyes by referring to the circle and cross illustrated below. Close your right eye and hold the page about 30 cm away from your open left eye. Keeping your left eye focused on the circle, the cross should be visible but slightly less clear. Now move the page slowly towards you and notice that at a distance of about 15 cm from your left eye the cross disappears. At this point light from the cross falls on the blind spot of your left eye. If you now move the page even nearer to your left eye the cross should reappear.

+   ○

In this simple demonstration the circle appears clearer than anything around it in the field of vision. The reason for this is that when you direct your eye at the circle the light from it is focused on to a region of the retina called the **fovea**. Only cones are found in the foveas of the eyes of animals which have colour vision. Each cone forms a synapse with very few receptor neurones (sometimes only one). Consequently the signals sent to the brain

Fig 19.9 Cones often synapse with single receptor neurones whereas usually many rods synapse with a single receptor neurone. Cones are therefore important in the discrimination of close points in the image

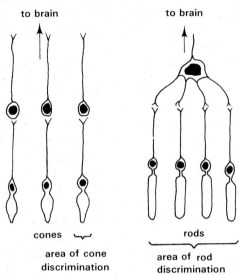

from each cone come from a small area of the retina on which a small part of the image is focused. The cones can thus discriminate between two points of the image which are close to one another. For this reason cones are said to have high **visual acuity**. On the other hand, many rods synapse with each receptor neurone. The signals transmitted to the brain from the rods therefore come from a relatively large area of the image (Fig 15.9). The rods are distributed throughout the retina but are absent from the fovea. The cones in the fovea are much longer and thinner compared with those in other parts of the retina. It enables many cones to be packed together in the fovea. The part of the image focused on the fovea therefore appears much clearer than the rest of the image.

Which small part of your retina gives you the best vision? Which part of your retina cannot 'see' at all?

---

3 Photosensitive pigments

The functioning of rods and cones depends on **photosensitive pigments**. Electron microscopy has revealed the intricate subcellular structure of rods (Fig 19.10). The outer segments of rods contain a great number of membranous, disc-like **lamellae**. The lamellae contain a photosensitive pigment called **rhodopsin**. Cones contain a similar pigment sometimes called **iodopsin**.

Fig 19.10 (a) Electronmicrograph of parts of several rod cells from the retina of a cat, × 5800

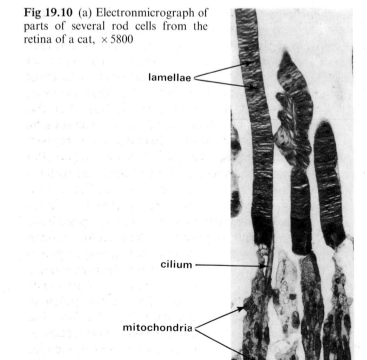

Fig 19.10 (b) Diagram of rod cell structure

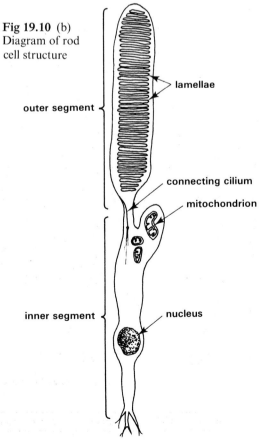

i. Rhodopsin consists of a protein called **opsin** attached to **retinal**, a derivative of vitamin A (Chapter 15). When exposed to light rhodopsin is split into opsin and retinal. The retinal molecules change shape and produce a generator potential which is transmitted from the rod as an impulse to a receptor neurone. Strangely, the initial stimulus causes **hyperpolarisation** in the receptor membrane. The usual effect of stimulation is depolarisation (Chapter 17).

Rhodopsin, split in this way, has to be resynthesised to maintain the rod's ability to respond to light. Resynthesis of rhodopsin requires energy and the rods have many mitochondria which make ATP for this purpose. However, rhodopsin resynthesis takes time. It is a common experience to suffer a brief period of poor vision after going from a well-lit room into darkness. When exposed to bright light rhodopsin is broken down rapidly and the reserve of rhodopsin in the rods is low. The eyes are **light-adapted**. If the retina is then exposed to dim light the rods show little response, so vision is poor. The period required to get used to the dark is the time taken for enough rhodopsin to be resynthesised. When the retina is sufficiently sensitive for us to see in dim light the eyes are **dark-adapted**.

A brief period of poor vision may also be experienced when you go from a very dark to a brightly lit room. When dark-adapted, the retina can work in dim light. Exposure of a dark-adapted retina to bright light overloads the photoreception mechanism. Light rays, even from the darker areas of an object in view, stimulate the rods which are now rich in rhodopsin. The eyes are light-adapted when excess rhodopsin is broken down and the retina once more adjusts to working in bright light. Prolonged exposure to very intense light, however, can reduce sensitivity of the retina too much. The rate of rhodopsin resynthesis may then be unable to keep pace with its breakdown. **Snow blindness** is caused by such an effect.

Fig 19.11 (a) Absorption spectra showing the wavelengths of light (perceived as colours) absorbed most by the iodopsin in the three main types of cones (after Marks, Dobelle, MacNichol, Brown and Wald)

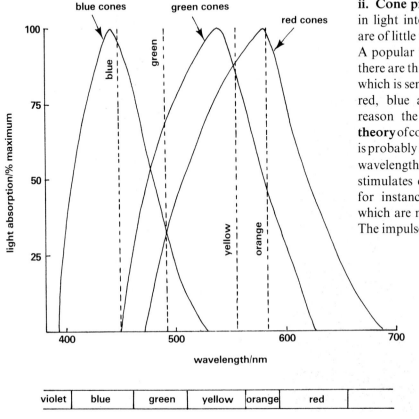

ii. Cone pigments are less sensitive to changes in light intensity than rhodopsin, so the cones are of little value in helping us to see in dim light. A popular theory of colour vision suggests that there are three variants of cone pigments, each of which is sensitive to light of the primary colours, red, blue and green (Fig 19.11(a)). For this reason the theory is called the **trichromatic theory** of colour vision. Each type of cone pigment is probably located in different cones. Light with a wavelength between those of the primary colours stimulates combinations of cones. Yellow light for instance, simultaneously stimulates cones which are most sensitive to red and green light. The impulses generated in the receptor neurones in response to generator potentials in the cones are interpreted by the brain as the appropriate intermediate colour, yellow in this case (Fig 19.11(b)). The interpretation or **perception** of colour pictures seen by our eyes is a complex function of the brain. It is located in the occipital lobes of the cerebral cortex (Chapter 18).

365

Fig 19.11 (b) Different colours are perceived in the brain from the sensory information received from the cones. Signals from a combination of different cones produce the sensation of intermediate colours based on red, green and blue detected by the three primary cones

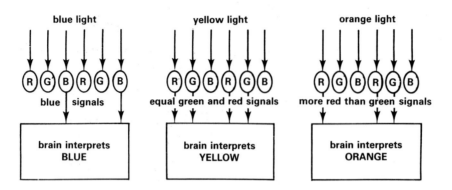

Deficiency of one or more of the three primary colour cones results in **colour blindness**. A glance at Fig 19.11(a) shows that there is some overlap in colour sensitivity between the three types of cones. For example, a green cone is sensitive to red light but its sensitivity to red is far less than that of a red cone. Absence of red cones therefore, means that it is still possible to perceive green, yellow, orange and red.

However, the brain cannot distinguish satisfactorily between these colours because there are no impulses from red cones with which to contrast impulses from green cones. A similar effect occurs when green cones are absent. The condition is called **red-green colour blindness**. More rarely, blue cones may be absent causing **blue weakness**.

The exact mechanism of colour vision is not known. Many people challenge the simplicity of the trichromatic theory and several alternative theories have been proposed in recent years.

19.1.5 Perception of distance and size

The region of the environment from which each eye collects light is called the **visual field**. Since both our eyes point forwards, there is an overlap between the visual fields of each eye. This is called **binocular vision**. Most of the image perceived by the visual cortex in the brain results from the integration of information from both eyes. Furthermore, signals from both eyes are transmitted to each half of the visual cortex. This is because approximately half the receptor neurones from each eye cross to the other side in the brain (Fig 19.12).

Subtle differences appear between the images from the two eyes because each eye is looking at the environment from a slightly different position. Comparison of the two images in the visual cortex enables us to perceive the shapes, textures, distances and relative movements of objects.

You can appreciate the effect by closing your left eye. Hold one of your fingers vertically about 10 cm in front of your open right eye and line it up with some distant object such as the vertical part of a door frame. Now

Fig 19.12 The pathway for sensory impulses from the eyes to the visual cortex in the brain. About half the fibres from each eye cross over to the opposite side in the optic chiasma (modified from Tortora and Anagnostakos)

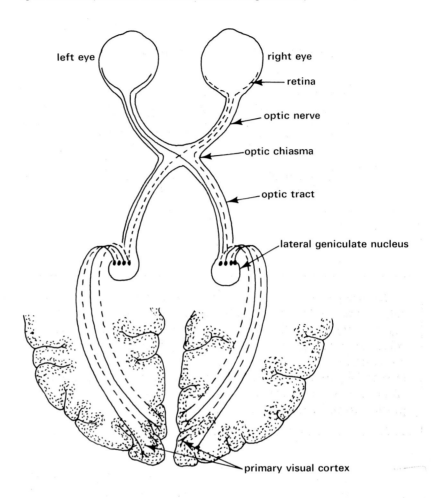

open your left eye and at the same time close your right eye. Your finger and the distant object will now be misaligned. The effect is called **stereoscopic vision**. It is only significant for objects nearer than about 70 m. Perception at greater distance depends partly on **memory**. We are familiar with most of the objects around us. The smaller they appear, the more distant we assume they are and vice versa. Another effect is called the **moving parallax**. When we scan objects in the visual field, near objects move across the image to a greater extent than distant objects.

Visual perception is extremely complex. The brain relies on a variety of sensory information which is integrated in such a way as to provide a complex picture of our surroundings.

19.2 The ear

The ability to generate sounds, as well as to receive and interpret them is valuable as a means of communication. Sound waves trigger off the transmission of receptor nerve impulses from the ears to the cerebral cortex of the brain.

The ears also perform another important function. They transmit to the brain information about the head's relative position in space.

19.2.1 Sound

Sound is all about us. We make many uses of sound in our everyday lives. We use some sounds, such as sirens, to give warning of danger and others, such as door bells, to attract attention. We sing and make music. Most important of all, we communicate with each other by the spoken word. All these activities require the listener not only to hear the sounds but to discriminate between different sounds in order to obtain meaning from them. So, what is **sound**?

Sound is a series of disturbances of the medium through which it travels. For example, if we pluck a guitar string we cause movement of the air molecules in contact with the string (Fig 19.13). As the string moves into the surrounding air it **compresses** the air molecules. As the string moves back towards its original position, the compressed air molecules have more space to occupy. The air pressure increases and decreases each time the string moves back and forth.

Fig 19.13 Generation of sound by a vibrating string

(a) regions of compression and rarefaction of air molecules

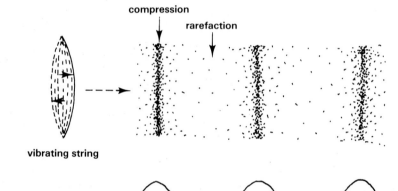

(b) sound waves – fluctuating air pressure changes caused by the vibrating string

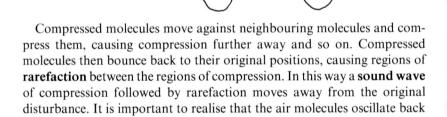

Compressed molecules move against neighbouring molecules and compress them, causing compression further away and so on. Compressed molecules then bounce back to their original positions, causing regions of **rarefaction** between the regions of compression. In this way a **sound wave** of compression followed by rarefaction moves away from the original disturbance. It is important to realise that the air molecules oscillate back and forth but the waves of compression they cause move away to neighbouring molecules.

The **frequency** or, as we perceive it, the **pitch** of a sound is determined by the time interval between each wave. This is the **wavelength** (Fig 19.14).

Fig 19.14 Two sound waves of the same amplitude:

(a) one of relatively low frequency

(b) one of relatively high frequency

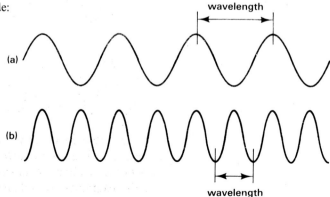

The greater the frequency, that is the shorter the wavelength, the higher the pitch of the sound we hear. Our ears are sensitive to sounds of frequencies between approximately 20 and 23 000 Hz. **Hertz** (Hz) is the unit used to express the number of sound waves as **cycles per second**. We are most sensitive to sounds of frequencies between about 1000 and 4000 Hz.

The frequency of a sound produced by a moving object, such as a train, depends on the speed of the object and the direction in which it is moving relative to the listener. You have probably experienced the sudden change in frequency (pitch) of a sound coming from a fast-moving car as it passes you. The frequency drops as the car goes away. It is called the **Doppler effect** (Fig 19.15).

Fig 19.15 The Doppler effect

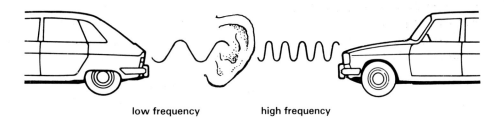

low frequency   high frequency

Wavelength, frequency and velocity of sound are related by the equation:

$$\lambda = c/f$$

where λ = wavelength,
            c = velocity, and
            f = frequency

From this we can see that sound of high frequency has a short wavelength and vice versa (Fig 19.14).

Fig 19.16 Two sound waves of the same frequency: (a) one of relatively low intensity (amplitude), (b) one of relatively high intensity. Compare with Fig 19.14

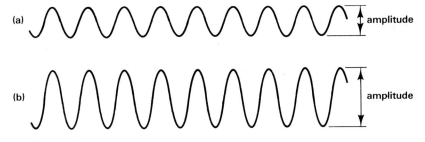

(a)   amplitude

(b)   amplitude

The **intensity** or, as we perceive it, the **loudness** of sound is determined by the **amplitude** of each sound wave; that is the difference in pressure between maximum compression and rarefaction (Fig 19.16). In the guitar string example of sound production, the distance the string moves back and

Why does the pitch of the sound made by an approaching train suddenly drop as the train passes and goes away?

forth determines the intensity of the sound produced. Consequently if we pluck a string forcefully we hear a relatively loud sound. The intensity of sound is often expressed in **decibels, dB**. The intensities of two sounds are compared and expressed as a logarithm:

$$dB = 10 \times \log (I/I_0)$$

where I = new intensity, and I_0 = original (reference) intensity.

The reference level (original intensity) must be specified. In terms of hearing, a sound of 0 dB is said to be one that can just be heard by a perfect ear. It is equivalent to $0.00002 Nm^{-2}$. This is the **hearing threshold level, HTL**. Expressing sound intensity in dB gives convenient numbers over the very wide range of intensities to which our ears are sensitive (Table 19.1).

Table 19.1 Approximate intensities of sounds commonly encountered in everyday life

10 dB = 10 × increase (10^1) above 0 dB reference level
20 dB = 100 × increase (10^2)
60 dB = 1 000 000 × increase (10^6)

hearing threshold level	0 dB
whisper	30 dB
normal talking	60 dB
shouting	90 dB
near pain threshold	120 dB

The sound in a noisy factory at 120 dB is 1 000 000 000 000 (i.e. 10^{12}) times louder than a whisper.

19.2.2 Structure of the ear

The ear is divided into three main regions, the **outer ear, middle ear** and **inner ear** (Fig 19.17). The outer ear channels sound waves from the surrounding air into the middle ear where the energy of sound waves is converted to mechanical vibrations. In the inner ear nerve impulses are generated in response to vibrations received from the middle ear and to changes in position of the head.

Fig 19.17 Structure of the ear

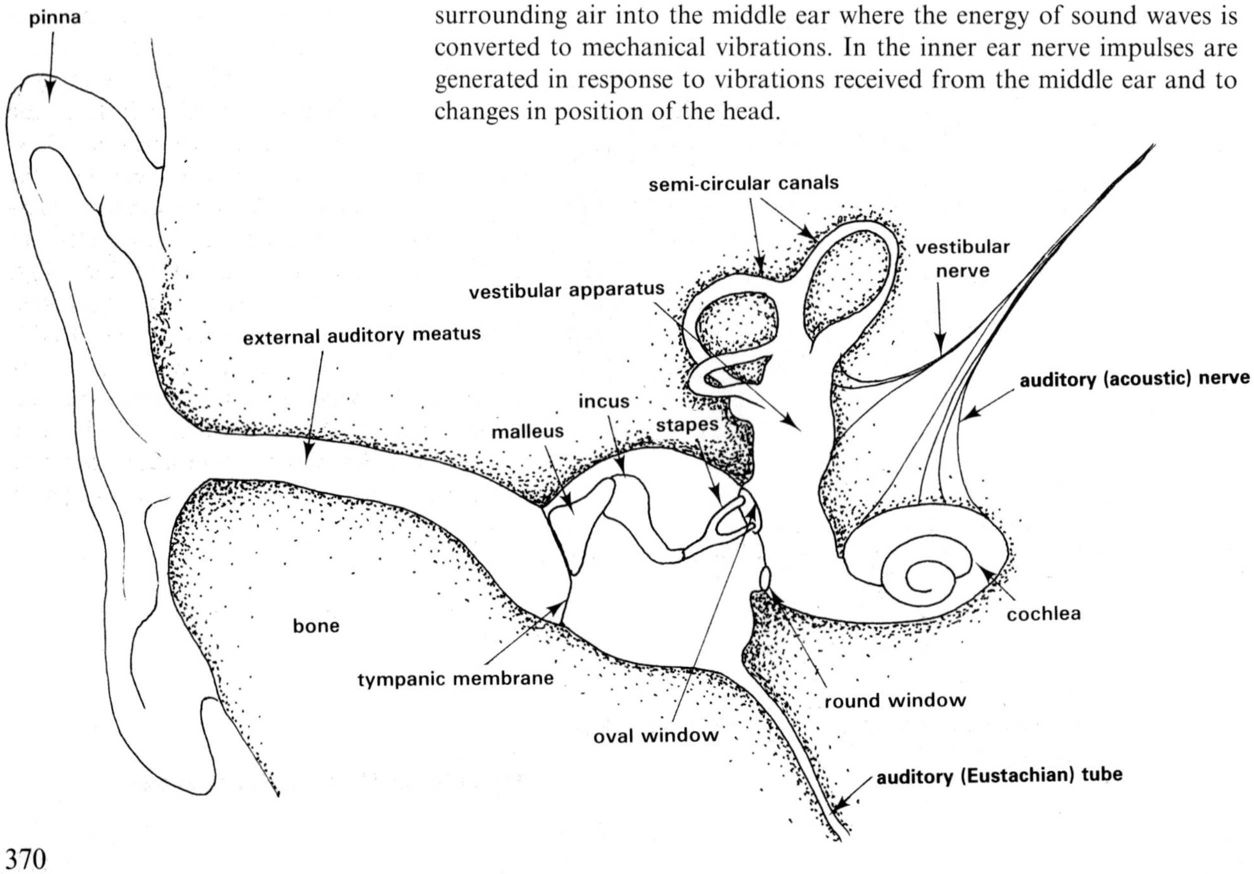

19.2.3 Hearing

Sound waves enter each ear by a short tube called the **external auditory meatus** (plural meati). The **pinnae** on each side of the head may help in directing sound waves into the meati. At the inner end of each meatus is an elastic membrane called the **typanic membrane** or **eardrum**.

Bridging the air-filled middle ear are three small bony **ossicles** held in place by muscles and ligaments. The ossicles articulate freely with each other and are the **malleus** (hammer), **incus** (anvil) and **stapes** (stirrup). Sound waves vibrate the

Fig 19.18 (a) TS diagram of structure of the cochlea

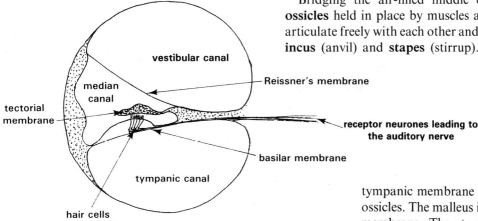

Fig 19.18 (b) TS photomicrograph of cochlea (guinea pig), × 70

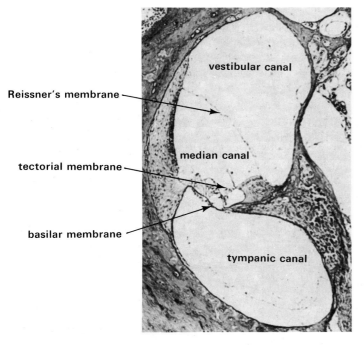

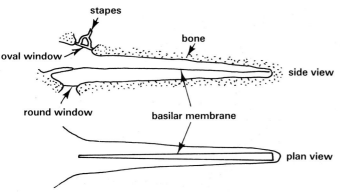

Fig 19.18 (c) 'Unrolled' view of cochlea

tympanic membrane which in turn vibrates the ossicles. The malleus is attached to the tympanic membrane. The stapes is attached to another membrane called the **oval window** which is part of the inner ear. The oval window is less than 5 % of the area of the tympanic membrane. Consequently, vibrations of the tympanic membrane are amplified about 20 times in the oval window. Amplification makes it easier for vibrations to pass through the dense fluid in the inner ear.

An air-filled canal called the **auditory tube (Eustachian tube)** connects the middle ear with the pharynx. The air pressure in the atmosphere and in the middle ear is usually the same. Should there be a sudden large increase in external air pressure there is a possibility that the eardrums would burst. The danger of this happening is usually avoided because air taken in from the outside enters the auditory tubes during swallowing. In this way, air pressure on either side of the eardrums is equalised. The oval window transmits the vibrations of the middle ear ossicles into a coiled, fluid-filled tube called the **cochlea** (Figs 19.17 and 19.18). It contains three longitudinal canals separated from each other by two flexible membranes. The upper **vestibular canal** is connected to the oval window. Between the vestibular canal and the **median canal** is Reissner's membrane. The **basilar membrane** separates the median canal from the lower **tympanic canal**.

371

Vibrations of the oval window generate pressure waves in the fluid filling the vestibular canal. The pressure waves pass to the median canal and vibrate the basilar membrane. The tympanic canal is connected to a circular membrane called the **round window** just beneath the oval window. This arrangement allows the pressure waves to transmit through the cochlear fluid. Since liquids are not compressible, when the oval window is pushed inwards the round window pushes outwards and vice versa (Fig 19.19).

Fig 19.19 Vibrations of the stapes move the oval window back and forth. The oscillations of the oval window push and pull on the cochlear fluids, causing vibrations of the cochlear membranes. Note the effect of the vibrations on the round window.

(a) Effect when the oval window is 'pushed'

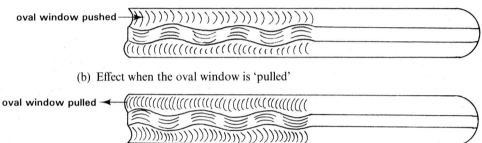

(b) Effect when the oval window is 'pulled'

At the tip of the cochlea, called the **apex**, the fluid in the vestibular canal is continuous with that in the tympanic canal through a narrow channel called the **helicotrema**. The helicotrema is thought to allow sudden, large pressure changes to transmit directly to the round window without undue distortion of, and possible damage to the cochlear membranes.

The sensory region of the cochlea is called the **organ of Corti** (Fig 19.18(a)). It contains many **hair cells** rooted in the basilar membrane. The hair cells are arranged in outer and inner ranks. Short hairs called **stereocilia** project into the fluid of the median canal. Some of them are attached to the **tectorial membrane**. Vibrations of the basilar and tectorial membranes cause the stereocilia to distort, resulting in the generation of impulses in the hair cells (Fig 19.20).

Fig 19.20 Organ of Corti. The median canal contains endolymph. The vestibular and tympanic canals contain perilymph. The fluids differ chemically. Endolymph is similar to cytoplasm. When the stereocilia distort, the hair cells depolarise.

The hair cells are connected by synapses to receptor neurones which transmit the impulses to the brain along the **auditory (acoustic) nerve**. It is part of cranial nerve VIII and leads to the **primary auditory areas**. They are regions of the cerebral cortex where the sensory impulses from the ears are interpreted as sound (Fig 19.21).

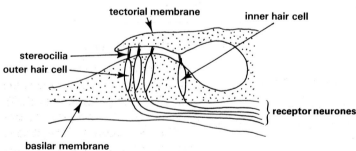

Fig 19.21 The right primary auditory area. Different regions of the area receive impulses from specific parts of the organs of Corti

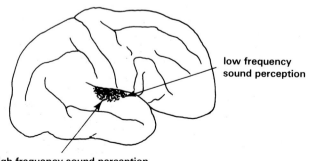

Vibrations of the basilar membrane are crucial to the conversion of sound waves into nerve impulses. The basilar membrane is about 2.5 times wider at the apex than at its base between the oval and round windows. However, the cochlea tapers from base to apex (Fig 19.18(c)). Sound of short wavelength, that is of high frequency, vibrates a relatively short portion of the basilar membrane. Only the hair cells nearest the oval window are stimulated. The impulses arriving at the brain are interpreted as high-pitched sound. A longer portion of the basilar membrane is vibrated by sound of longer wavelength, that is lower frequency. Hair cells are stimulated further along the basilar membrane. The brain interprets the impulses from the hair cells as a low-pitched sound. Sounds of intermediate wavelength stimulate the basilar membrane, mostly in the middle regions (Fig 19.22).

Fig 19.22 Vibrations of the basilar membrane in response to two pure tones of 500 Hz and 2000 Hz frequency. Each numbered line represents the position of the membrane at a particular instant of the vibration cycle (from Rosenberg after Eldredge)

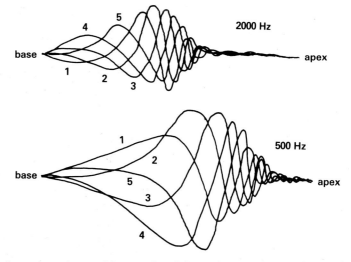

The intensity of sound is translated from the intensity with which the appropriate region of the basilar membrane is vibrated. The greater the amplitude of basilar membrane movement, the more impulses are generated per second in the hair cells and vice versa. Many impulses per second are interpreted in the brain as loud sounds, fewer impulses per second as soft sounds. Interpretation, that is **perception** of sound is thus a function of the brain.

How does the cochlea in the inner ear enable us to distinguish between sounds of different intensities and frequencies?

19.2.4 Balance

Just above the cochlea and connected to it by a short tube is the **vestibular apparatus** (Fig 19.23). It consists of two lymph-filled sacs called the **saccule** and the **utricle**. Projecting from the top of the utricle are three **semicircular canals**, also containing lymph.

Fig 19.23 Vestibular apparatus

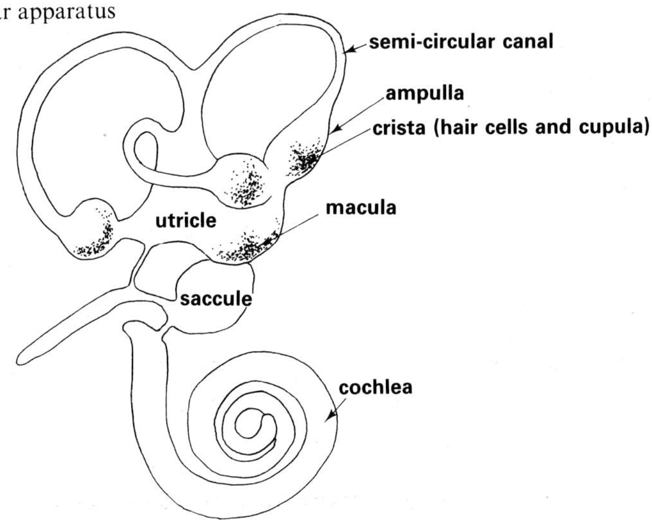

semi-circular canal

ampulla

crista (hair cells and cupula)

macula

utricle

saccule

cochlea

373

The saccule and utricle contain receptors called **maculae** which are sensitive to gravity. Small hairs project from the receptor cells into the lymph. The hairs are attached to calcium carbonate granules called **otoliths** (Fig 19.24). Gravity causes the otoliths to distort the sensory hairs in a direction determined by the position of the head. In response to the distortion, nerve impulses pass along the **vestibular nerve** to the brain. If the head is moved to a different position, the otoliths distort the sensory hairs in a different direction. Information about the head's new position is interpreted in the brain.

Fig 19.24 (a) Structure of the macula from the utricle

Fig 19.24 (b) The otolith deflects the sensory 'hairs' in a direction dictated by gravity and the relative position of the head

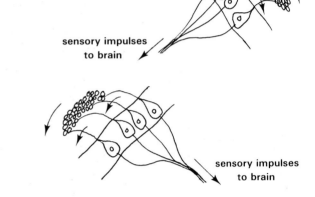

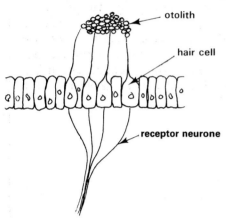

The semicircular canals provide information about head movements, rather than the position of the head when it is stationary. The end of each canal is enlarged to form a **ampulla**. Each contains a receptor, consisting of hair cells similar to those of the maculae. The ampullary hairs, however, project into a gelatinous mass called a **cupula** which is suspended in the lymph inside the ampulla (Fig 19.25(a)).

Any movement of the head of course moves the semicircular canals in the same direction. The lymph inside the canals, however, lags behind and pushes the cupulae in the opposite direction (Fig 19.25(b)). As a result, the hairs projecting into the cupulae are bent and nerve impulses are sent to the brain along the vestibular nerve. There are three semicircular canals and each is arranged at a right angle to the others. Consequently, at least one cupula is stimulated by lymph movements, whatever direction the head is moved in.

Information about the orientation and movement of the head is vital, especially when the whole body is moved. Visual information from the eyes also contributes to the brain's awareness of the head's spatial position. The information is used by the brain to coordinate movement and posture of the body.

Fig 19.25 (a) Structure of the receptors in the ampullae of the semi-circular canals

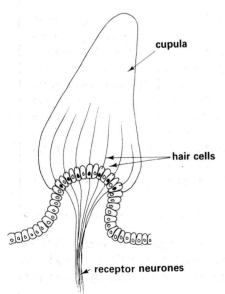

Fig 19.25 (b) Movements of the fluid inside the semi-circular canals, caused by movements of the head, displace the gelatinous cupula and stimulate the sensory 'hairs'

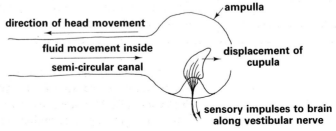

19.2.5 Defects of the ear

Several defects of the ear lead to hearing loss or even deafness. **Conductive hearing loss** is caused by inability of the outer and middle ear to conduct vibrations properly to the cochlea. One of the commonest causes of conductive hearing loss is blockage of the external auditory meatus with wax secreted from **ceruminous** glands in the skin lining the meatus. In some people wax accumulates in the meatus and hardens, sometimes pressing against the eardrum. Normal hearing is usually restored after the hardened wax is removed with a special syringe.

Another cause of conductive hearing loss, is a perforated eardrum. Perforation can be caused by infection in the middle ear or by mechanical injury resulting from a nearby explosion or a sudden blow to the head. Injury to the head can also cause the ossicles of the middle ear to become disconnected from one another, thus breaking the conductive path to the cochlea. Patients with a conductive defect which does not respond fully to treatment may be helped with a **hearing aid**. It is a device which amplifies sound waves before they enter the inner ear (Fig 19.26).

Fig 19.26 'Behind-the-ear' hearing aid (from DHSS booklet: General Guidance for Hearing Aid Users)

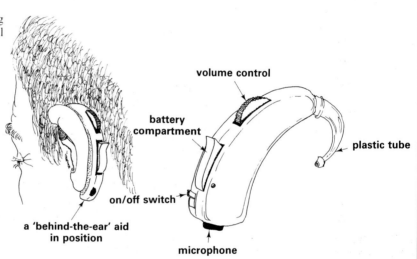

volume control

battery compartment

plastic tube

on/off switch

a 'behind-the-ear' aid in position

microphone

Malfunction of the cochlea and acoustic nerve can be the cause of hearing loss, even though vibrations are conducted perfectly into the inner ear. Such hearing loss is called **sensorineural (perceptive) hearing loss**. Sensorineural hearing loss can be inherited, though it is often acquired. Acquired forms of the condition can result from infection, head injury, blast from explosions or exposure to excessive noise. Much concern has been expressed in recent years about the possible harmful effects on hearing of the very high noise levels in disco's, at airports and in places where noisy machinery is used. So serious is the problem that people employed in extremely noisy places are encouraged to wear ear muffs to protect their hearing from permanent damage.

How can loud music at a disco damage our ears?

Although the eye and ear are extremely complex sense organs, it is the interpretation of sensory information which ultimately limits their use. Processing of information from them occurs in the brain by mechanisms which are, as yet, poorly understood.

SUMMARY

Receptors are cells which convert a stimulus into nerve impulses. Many receptors are housed in sense organs which channel the stimulus to the receptors.

Eyes: three layered wall; sclera, choroid and retina. Transparent cornea (covered by conjunctiva) at the front allows light into the eye. Iris diaphragm determines how much light enters the eye. Lens focuses light from objects to make sharp images on the retina. The image is inverted. Impulses are transmitted to the occipital cortex of the cerebrum along the optic nerves.

Focusing (accommodation) is brought about by reflex control of ciliary muscles. The crystalline lens is flattened (distant object focusing) or allowed to become rounded by elastic recoil (near object focusing). If the eyeball becomes elongated (or lens too rounded) light is focused before it impinges on the retina; image is blurred; myopia (short sightedness).
If the lens is too flat, light is not yet focused when it impinges on the retina; hypermetropia (long sightedness).
If the lens is distorted, so is the image; astigmatism.
Defects can be corrected by use of artificial lenses in front of eyes.

Photoreception occurs in rod and cone cells of the retina. Cones are mostly in middle of retina (centre of the image; fovea).
No photoreceptors where the optic nerve leaves the retina (blind spot). Visual acuity is the ability to discriminate between two points close together in the image. Visual acuity is greatest at the fovea; cones connected to visual cortex individually, rods in clusters.

Photosensitive pigments in rods and cones cause generator potentials and impulses when activated by light. Rhodopsin in rods consists of opsin (protein) and retinal (derivative of vitamin A). In bright light, most rhodopsin is broken down; eyes are not very sensitive; light adapted. In sudden dark, little is seen until more rhodopsin is synthesised. In dim light, little rhodopsin is broken down; eyes are very sensitive; dark adapted. Sudden bright light swamps the retina.

There are three kinds of cone pigments, each sensitive to a different range of light wavelengths (perceived as colours); red, green and blue. Trichromatic theory of colour vision; different combinations of cones are activated by different light wavelengths. Deficiencies of cones lead to colour blindness.

Distance perception is largely a function of binocular vision; each eye sees a slightly different view of the environment.

Ears: outer ear (pinna and external auditory meatus) and middle ear conduct sound to the inner ear where receptors are located. Inner ear is concerned with hearing (cochlea) and balance and head movements (vestibular apparatus). Middle ear is connected to the pharynx by auditory (eustachian) tube; equalises middle ear pressure with that of atmosphere.

Hearing: Vibration of air (sound) causes vibration of the tympanic membrane (eardrum) which causes movement of middle ear ossicles; malleus, incus and stapes. These vibrate the oval window. Movements of fluid in the cochlea (allowed by vibration of the round window) cause vibration of the tectorial and basilar membranes. Consequent deflection of cilia of hair cells in the organ of Corti generate impulses in the acoustic nerve. They are transmitted to the temporal lobes of the cerebral cortex where sound is perceived. Discrimination of sound frequencies (pitch) is enabled by the basilar membrane; furthest part vibrates most in low frequencies, nearest part vibrates most in high frequencies. Sound intensity (loudness) discrimination is by the number of impulses generated per second in the hair cells; amplitude of the basilar membrane's vibration.

Vestibular apparatus: spatial orientation of head; maculae in saccule and utricle respond to gravity; otolith granules deflect cilia in hair cells. Head movement: fluid in the semi-circular canals flows according to head movements. It deflects cupulae in the ampullae of the canals. Each canal is at right angles to the others, providing sensitivity in all planes.

QUESTIONS

1 The diagram below represents the retina of the human eye as seen in section.

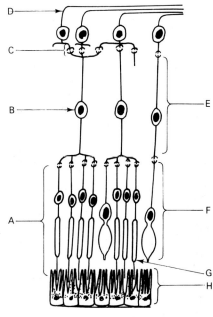

(a) Name the parts **A** to **H**.
(b) Account for the following.
(i) Rod cells produce an indistinct image.
(ii) Rod cells are concerned with night vision.
(iii) Cone cells are capable of colour perception.
(iv) Visual acuteness is greatest at the region of the fovea.
(v) When a person enters a dimly lit room from bright sunlight, the room at first seems dark but gradually objects become visible. (L)

2 The diagram represents a vertical section through the eye of the mammal.

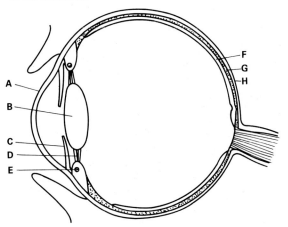

(a) Name the structures labelled **A, B, C, D, E, F, G,** and **H**.
(b)
(i) Explain what is meant by the term *accommodation* of the eye.
(ii) How does an eye viewing a near object accommodate to viewing a distant object?
(c) How is the size of the pupil adjusted as a mammal moves from a dim light to a bright light?

(d)
(i) Explain what you understand by the term *stereoscopic vision.
(ii) Give an example of a group of animals that have effective stereoscopic vision. (C)

3 (a) Make a large, clearly-labelled diagram of the *inner* ear of a mammal.
(b) Describe how each of the following is achieved in a mammal:
(i) hearing,
(ii) balance. (L)

4 Describe how each of the following is achieved in a mammal
(a) the control of light entering the eye,
(b) focusing on a distant object,
(c) colour perception. (L)

5 The diagram shows part of the retina of a human eye.
(a) Explain how
(i) light energy stimulates a rod cell;
(ii) an axon in the optic nerve transmits a nerve impulse.

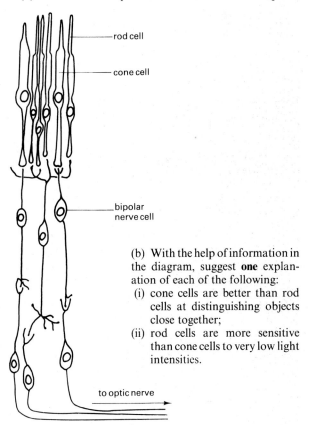

(b) With the help of information in the diagram, suggest **one** explanation of each of the following:
(i) cone cells are better than rod cells at distinguishing objects close together;
(ii) rod cells are more sensitive than cone cells to very low light intensities.

(c) Explain the role of the retina in colour vision.
(AEB 1986)

6 This question refers to the mammal.
Explain the role of
(a) the cochlea in the detection of sound and the discrimination of volume and pitch,
(b) the retina in the detection of colour and the discrimination of colour. (JMB)

377

7 Photograph 2.5II shows a transverse section of part of the cochlea of the inner ear of a mammal.

(a) Identify parts 1 to 4.

(b) Outline the sequence of events from the time when sound waves reach the tympanum to the time when resulting nerve impulses are received by the auditory centres of the brain.

(c) How is the ear able to discriminate between sounds of
 (i) different pitch (frequency)?
(ii) different amplitude (loudness)? (AEB 1986)

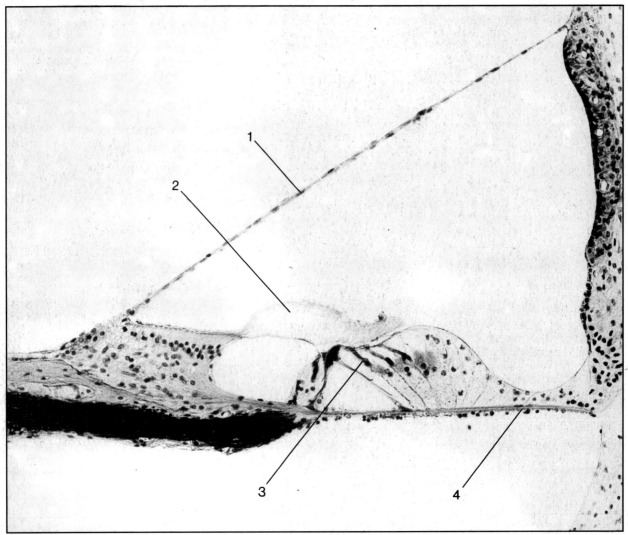

Gene Cox

20 Thermoregulation

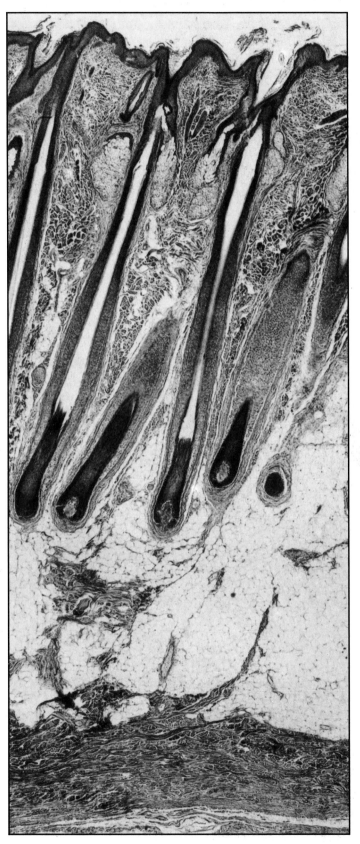

20.1	Heat production	380
	20.1.1 Measuring metabolic rate	380
	20.1.2 Distribution of body heat	381
20.2	Heat loss	381
	20.2.1 Radiation	381
	20.2.2 Evaporation	382
	20.2.3 Conduction	382
	20.2.4 Surface area and body volume	382
20.3	Thermoregulation	384
	20.3.1 The thermoregulation centres	384
	20.3.2 Regulating heat production	384
	20.3.3 Regulating heat loss	385
	20.3.4 Hibernation	390
Summary		391
Questions		391

20 Thermoregulation

One of the many aspects of homeostasis is regulation of body temperature. It is a characteristic feature of birds and mammals, including humans. These animals are called **endotherms**. They maintain a relatively high and constant body temperature which enables them to lead active lives even when the temperature of their surroundings is low. All other animals are called **ectotherms**. Their body temperature varies with temperature fluctuation of the environment (Fig 20.1). Regulation of body temperature is called **thermoregulation** and is brought about by balancing heat production in the body with heat loss to the environment.

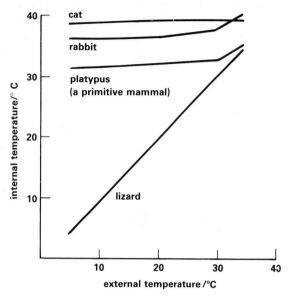

Fig 20.1 Effects of environmental temperature on the body temperatures of three mammals (endotherms) and a lizard (ectotherm)

20.1 Heat production

The main source of heat in the body is tissue respiration (Chapter 5). Over 50 % of the energy released in respiration is heat energy. The rate at which heat is produced is proportional to the metabolic rate which increases greatly during exercise. An adult human produces about $250\,kJ\,h^{-1}$ of heat energy when resting. Heat production increases to about $1000\,kJ\,h^{-1}$ during moderate exercise while the rate may go up to $8000\,kJ\,h^{-1}$ in a few minutes of intense exercise.

20.1.1 Measuring metabolic rate

If we fast, energy is released when food reserves in the body such as fat and glycogen are respired. At rest, nearly all the energy released in respiration is eventually given off as heat. Measurement of the heat given off by a resting, fasting individual gives an indication of the energy required to maintain the body's vital functions.

The rate at which the body respires to provide this amount of energy is called the **basal metabolic rate (BMR)**. The BMR of a human can be measured using a human calorimeter which is an insulated room containing water-filled pipes. The temperature of the water entering and leaving the room through the pipes for a given time is recorded. So is the temperature of the air in the room. Heat lost from the body causes a rise in temperature of the water and air. From the data the BMR can be calculated.

A much quicker way of calculating the BMR involves using a **spirometer** (Figs 8.7 and 8.8). A spirometer allows the volume of oxygen taken in by the subject in a given period of time to be measured. For every $1\,dm^3$ oxygen used in respiration, approximately $20.17\,kJ$ of heat energy are released. This is the average of the amount of heat energy released from the oxidation of carbohydrates, lipids and proteins. If a subject takes in $1.5\,dm^3$ oxygen in 5 minutes, which is equivalent to $18\,dm^3\,h^{-1}$, the rate at which heat energy is released is:

$$18 \times 20.17 = 363.06\,kJ\,h^{-1}$$

Thus the BMR $= 363.06\,kJ\,h^{-1}$

What is the basal metabolic rate of a person who consumes $1.48\,dm^3$ oxygen in 4 minutes? Express the answer in units of $kJ\,h^{-1}$.

The figure is generally adjusted to take into account differences in surface area of the body (section 20.2.4).

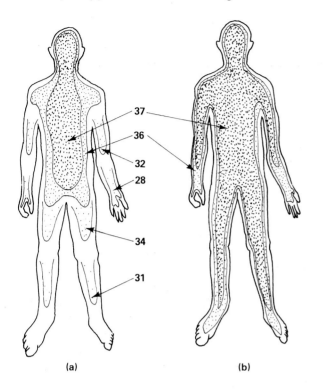

Fig 20.2 Distribution of heat on the surface of the body (a) in a cold, and (b) in a hot environment. Figures are in °C

37
36
32
28
34
31

(a)   (b)

20.1.2 Distribution of body heat

Heat is produced unevenly in the body. Skeletal muscle generates a lot of heat during exercise. Another important producer of heat is the liver (Chapter 15). As blood flows through the skeletal muscles and liver it absorbs heat and distributes it to parts of the body where little heat is produced (Fig 20.2). Blood moving through the circulatory system has kinetic energy which is converted to heat energy when the blood meets resistance, mainly in the arterioles. When blood flows near the body surface, heat is lost through the skin.

20.2 Heat loss

We lose body heat in three main ways, **radiation, evaporation** and **conduction**.

Fig 20.3 Approximate relative heat loss by radiation (wavy line), evaporation (broken line) and conduction (straight line)

20.2.1 Radiation

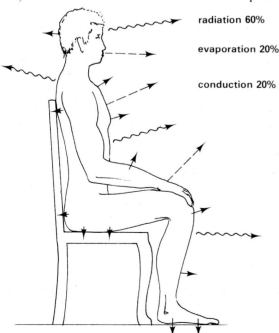

radiation 60%

evaporation 20%

conduction 20%

Radiation is the emission of heat from a body to the surroundings. Except in tropical regions, body temperature is generally higher than the air temperature. Heat is therefore radiated to the surrounding air more rapidly than it is gained by the same means. It has been estimated that 60 % of the total heat loss from a naked man sitting in a room kept at 33 °C is by radiation (Fig 20.3).

Measurements of heat coming from the body as infra-red radiation have been used in the detection of certain types of tumour. The patient is kept in a room of constant temperature, generally between 18 and 20 °C, to minimise fluctuations in skin temperature. Photographs are taken with a special camera which is sensitive to infra-red radiation coming from the body. Hot areas of the skin show up as light patches on the picture. Cool areas shown up as dark patches. Hot spots occur in areas where blood flow and metabolism are increased, which may be diseased areas of the body. The technique is called **thermography**.

20.2.2 Evaporation

When water evaporates from a moist surface, heat energy is taken from the surface. The energy used is the **latent heat of vaporisation**. Our skin contains **sudorific sweat glands** which secrete **sweat** onto the body surface when the body becomes overheated. Sweat is mainly water and when it evaporates the skin is cooled. Evaporation of body water also occurs from the moist linings of our nasal cavities, mouth, trachea and the extensive internal surface of the lungs. In very hot climates the surroundings often have a temperature higher than body temperature. In such conditions the body gains heat from the environment. Profuse sweating and subsequent evaporation of water from the skin's surface prevents overheating. A person can secrete up to 4 dm3 of sweat per hour in very hot, dry conditions. Dehydration of the body can then become a critical problem.

Even in a moderately warm room when sweating is virtually nil, about 20 % of a person's total heat loss is due to evaporation. This is because, even without sweating, water vapour escapes through the skin. Some of the heat loss is also due to evaporation from the linings of the mouth, nose, trachea and lungs (Fig 20.3).

Many environmental factors affect the rate of evaporation and hence heat loss from the skin. Most notable are the relative humidity of the air in contact with the skin, the air temperature and air movement. They control the rate of evaporation from the body.

20.2.3 Conduction

Heat passes from a warm object to a cooler one in direct contact with it by conduction. You have probably noticed how warm a seat becomes after sitting on it for some time. The greater the temperature difference between the body and other objects touching it, the greater is the rate of conductive heat loss. A cool breeze passing over the skin removes heat by conduction to the air. About 20 % of the total heat lost by the person illustrated in Fig 20.3 can be due to conduction.

20.2.4 Surface area and body volume

Heat is released in the body mainly in respiration. The amount of heat released depends on the **volume** of the body. Because most of the heat is lost through the skin, the amount of heat lost depends on the **surface area** of the skin.

As an animal grows, its volume increases in three dimensions whereas its surface area increases only in two dimensions. Consequently, relative to its volume its surface area increases at a slower rate. In terms of heat exchange, a bulky animal has a larger volume of tissues in which heat is released, but relative to this, a smaller surface through which heat is lost to the environment. The **surface area to volume (SA/V) ratio** is the area of skin per unit of body volume. A deer of volume $150\,000\,cm^3$ and $19\,000\,cm^2$ surface area has a SA/V ratio of:

$$\frac{19\,000}{150\,000} = 0.127\,cm^2\,cm^{-3}$$

In comparison, a squirrel of $625\,cm^3$ volume and $550\,cm^2$ surface area has a SA/V ratio of:

$$\frac{550}{625} = 0{\cdot}88\,cm^2\,cm^{-3}$$

In other words, each cm^3 volume of the squirrel has about seven times the

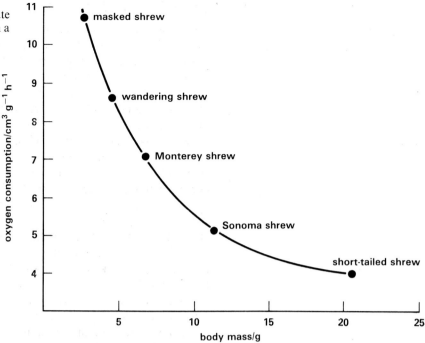

Fig 20.4 Relationship of metabolic rate (oxygen consumption) and body size in a variety of shrews (after Pearson)

skin area available for heat loss compared with the bulkier deer.

Such considerations highlight one of the factors which limits the extremes of size in terrestrial mammals. Bulky animals face overheating while small ones lose heat rapidly. Experimenting with small mammals, Pearson in 1957 obtained data for the metabolic rate of a variety of shrews. He related the metabolic rate to body mass (Fig 20.4). The rate of metabolism controls the rate of heat production. It is, therefore, not surprising that the smaller shrews which have a greater relative surface area through which heat is lost show the highest metabolic rate. However, there is a limit to which increased metabolism can compensate for heat loss. It is estimated that a mammal smaller in size than the smallest species of shrew would be unable to make energy quickly enough from its food to make good the rate at which heat is lost through its body surface.

Why are there no free-living mammals smaller than shrews?

Since metabolic rate and surface area can be related in this way, measurements of BMR are usually expressed in units which take account of surface area. Expressing BMR in this way allows comparison to be made between animals of different sizes. In humans, for example, the BMR is expressed in $kJ\,m^{-2}\,h^{-1}$. The surface area of a human is difficult to measure directly but it can be calculated from measurements of height and mass (Fig 20.5).

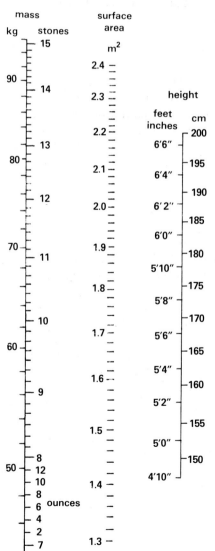

Fig 20.5 Nomogram for determining the surface (skin) area from body mass and height

20.3 Thermoregulation

Regulation of body temperature is brought about by balancing heat production and heat loss.

20.3.1 The thermoregulation centres

Thermoregulation takes place in response to changes in body temperature, but the changes must first be detected. This function is performed by specific parts of the hypothalamus in the forebrain (Chapter 18). In the hypothalamus there are two **thermoregulation centres**. The **cold centre** responds to blood which has a temperature less than normal by triggering off responses which increase heat production and decrease heat loss. The **heat centre** responds to blood which has a temperature higher than normal by triggering off responses which reduce heat production and increase heat loss (Fig 20.6).

Fig 20.6 Action of thermoregulation centres in the hypothalamus

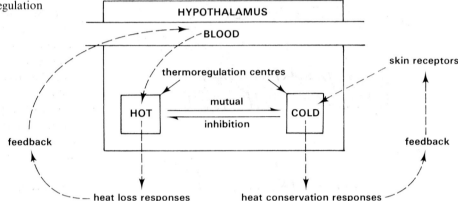

The heat centre acts like a thermostat. It switches on heat loss mechanisms when the temperature of the blood is higher than normal. Conversely it switches on heat conservation mechanisms when the temperature of the blood is lower than normal. The cold centre is inhibited by the heat centre, but becomes active when receptors in the skin signal that the environment is getting cooler.

In this way the body temperature is kept constant, even though the environmental temperature varies. The cold centre also receives information about potential body temperature changes. The information travels along receptor neurones from receptors in the skin which are sensitive to temperature changes outside the body. It is an early warning system which enables the thermoregulation centres to trigger appropriate responses before external changes alter the internal body temperature too much.

20.3.2 Regulating heat production

Heat is generated when respiratory substrates taken in as food are oxidised in respiration (Chapter 5). Heat generation in the body is thus ultimately limited by the intake of food. Assuming this to be adequate, other factors become important. Among these is production of the hormone **thyroxine** which controls the BMR (Chapter 21). Increased thyroxine output can double BMR, but the response takes several days to come into effect. A similar but more immediate response is brought about by **adrenaline** made in the adrenal glands and **noradrenaline** secreted by sympathetic nerve endings (Chapters 18 and 21). When the environmental temperature is very low, considerable heat can be generated by **shivering**. Shivering is very rapid alternate contraction and relaxation of the skeletal muscles.

20.3.3 Regulating heat loss

There are several means by which we can alter the rate at which heat is lost from the body. Nearly all involve the **skin**. The skin acts as a physical barrier which prevents excessive loss of heat and water from the body and stops foreign matter and pathogenic microbes gaining access to the underlying tissues and organs. The skin has a thin outer layer called the **epidermis**, one of the functions of which is to replace cells which are constantly lost from the surface. It consists of **stratified epithelium**. The cells lying next to the basement membrane divide constantly and comprise the **germinal layer**. As the products of cell division are pushed upwards towards the epithelial surface, they become flattened. Eventually the cells degenerate and are sloughed off into the surroundings to be replaced by more cells from underneath. The degenerated surface cells become filled with a protein called **keratin**. It enables the epithelium to resist mechanical abrasion and protects the underlying tissues from desiccation (Fig 20.7(b)).

Below the epidermis is a much thicker layer, the **dermis**. The dermis contains blood vessels and a variety of receptors sensitive to heat, cold, pressure, touch and pain. Also in the dermis are the **sudorific glands** which produce **sweat**. The sudorific glands are coiled structures, each with a duct leading through the epidermis to the surface.

Hairs originate in the dermis and grow in pits called **follicles**. The bases of the follicles are attached to small **erector-pili muscles**, the contraction of which can raise the hairs projecting from the surface. Beneath the dermis are fat deposits contained in **adipose tissue** (Fig 20.7).

Fig 20.7 (a) Photomicrograph of a vertical section through the skin from the scalp, ×20

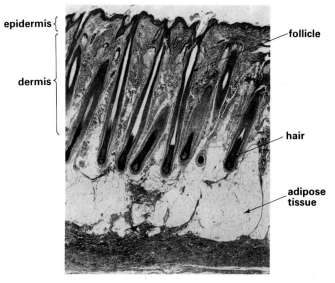

Fig 20.7 (b) Diagram showing some of the components of skin

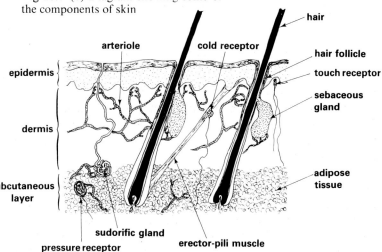

Fig 20.7 (c) Photomicrograph of a thin section of stratified epithelium in skin, ×150

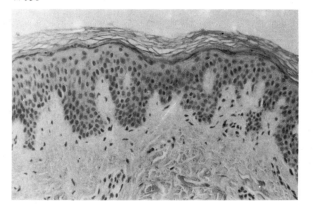

Fig 20.7 (d) Photomicrograph of a thin section of adipose connective tissue, ×400

385

1 Adipose tissue

Fat is a poor conductor of heat and adipose tissue is an effective thermal insulator. The adipose tissue in the skin which insulates against most heat loss makes up **white fat**. Some adipose tissue also makes up **brown fat**. Its cells contain many mitochondria and have a high metabolic rate. Brown fat can be an important generator of body heat. Thin people often have relatively large amounts of brown fat. People who have less brown fat are often overweight and have extensive reserves of white fat. It may be that the body deposits more heat-insulating white fat to compensate for the lack of heat production if little brown fat is present.

2 Cutaneous blood vessels

Heat is taken into the dermis by blood in numerous arterioles and capillaries. When the volume of blood flowing through the dermis is high, much heat is lost through the epidermis. This happens when the arterioles in the skin are in a state of **vasodilation** (Figs 20.8(a) and 20.10).

Fig 20.8 (a) Vasodilation in the skin leading to heat loss through the epidermis

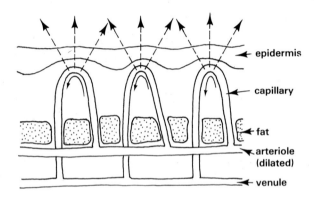

Fig 20.8 (b) Vasoconstriction in the skin reducing epidermal heat loss to a minimum

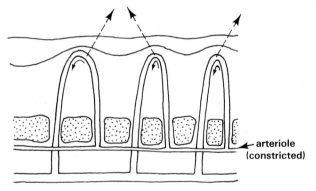

Sympathetic stimulation of the arterioles by nerve impulses from the cold centre of the brain causes **vasoconstriction**, thereby reducing heat loss. Blood flow in human skin can be reduced by vasoconstriction to about one hundredth of the volume which flows when the arterioles are dilated (Fig 20.8(b)). Prolonged constriction of the arterioles resulting from long exposure to intense cold deprives the dermis of the oxygen and nutrients it requires to maintain its metabolic functions. The tissues in the skin may then die or degenerate. This is the cause of **frostbite**.

3 Hair

Most mammals are covered with **hair** which helps thermoregulation by keeping a layer of air next to the skin. Air is a relatively poor conductor of heat. Since it is kept in contact with the skin and is not readily replaced by cold air from the surroundings, the temperature gradient between the skin and the trapped air is relatively small. Consequently, although some of the heat in the trapped air is lost to the atmosphere, the total heat loss from the skin in reduced (Fig 20.9(a)). The volume of warm air trapped by the hair depends on whether the hairs are erect or flat. In cold conditions the hairs stand on end, providing maximum heat conservation. In warm conditions the hairs lie flat on the skin, allowing maximum heat loss (Fig 20.9(b)). When the atmosphere is hotter than the body, the air trapped under the hair prevents excessive inward heat transfer (Fig 20.9(c)). This is particularly important to mammals living in hot climates.

Fig 20.9 Variation in heat exchange with the atmosphere caused by elevation and flattening of hair on the skin in:

(a) a relatively cold environment

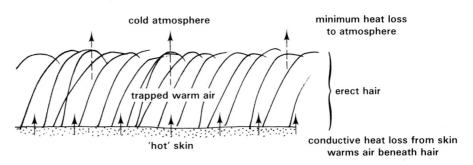

(b) a relatively warm environment

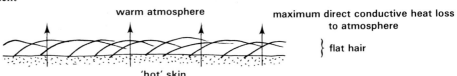

(c) a relatively hot environment

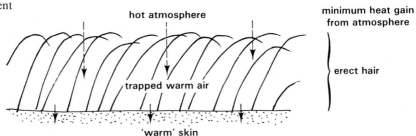

Our bodies do not have abundant hair. We rely on clothes to trap air above our skin, thus keeping us warm in cold weather. Nonetheless, the erector-pili muscles still pull on the hair follicles in our skin. This causes **goose-pimples** in cold weather.

4 Sweat

Evaporation of **sweat** from the skin's surface is a means of increasing heat loss. When it is necessary to conserve heat, sweating stops. Activity of the sweat glands is controlled by autonomic nerves. The impulses come from the thermoregulation centres of the brain.

Benzinger in 1961, experimenting on humans, found that after his subjects had ingested ice, there were strong correlations between fluctuations in evaporation from the body, skin temperature and body temperature (Fig 20.10). Ice in the gut removes heat from the blood. Within a few minutes, a lower blood temperature is detected in the cold centre of the hypothalamus. Evaporation from the skin decreases almost at once, hence the skin temperature rises. Within minutes the internal body temperature begins to rise, having initially dropped by only 0.35 °C. After about twenty-five minutes the internal body temperature returns to normal.

How do our bodies prevent overheating when we enter a very warm environment? How do they respond to the cold?

Fig 20.10 Relationship between evaporative heat loss, skin and hypothalamus temperatures in humans following experimental ice meals (after Benzinger)

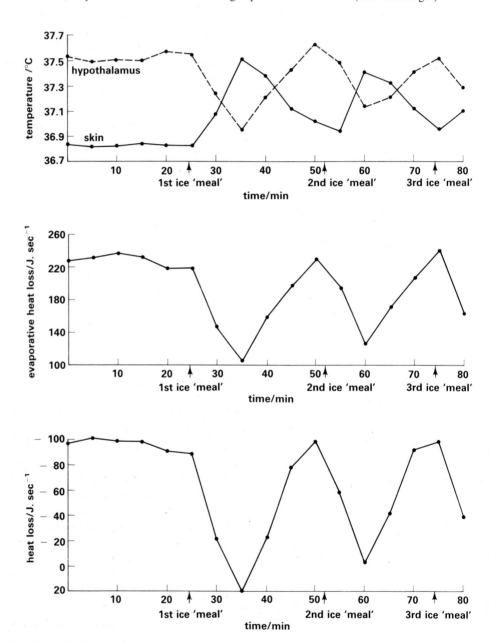

Camels and many other mammals living in the desert conserve water at the expense of sweating. Because it sweats so little a camel's temperature can fluctuate by as much as 6 °C.

The temperature of a human does not fluctuate as much as this (Fig 20.11). However, body temperature often rises following an infection. In these circumstances the heat centre responds to blood which is much hotter than normal, by triggering off thermoregulatory mechanisms at about 40 °C compared with 37 °C, the normal body temperature. The maximum upper limit which the human body can survive is about 45 °C and the lower limit is about 24 °C. Above 45 °C, metabolic reactions occur even more quickly, contributing further to body heat. The temperature of the body thus continues to increase indefinitely, causing enzyme denaturation and permanent tissue damage. This is an example of a **positive feedback** mechanism which clearly does not contribute to homeostasis. Below 24 °C the body's heat-generating mechanisms fail to work, causing **hypothermia**, a lowering of body temperature (Fig 20.12). Hypothermia is a common cause of loss of consciousness and death in old people who cannot keep themselves warm during cold weather.

Fig 20.11 Average daily fluctuations in human body temperature

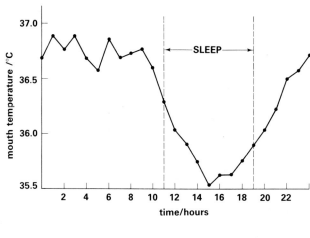

Fig 20.12 Critical body temperatures

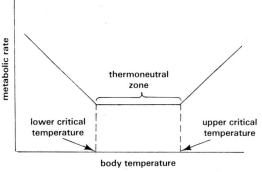

Many mammals, such as dogs, have few sweat glands. Nevertheless evaporation is an important means of heat loss in dogs. In hot weather dogs **pant**. Panting is rapid breathing usually with the mouth open. Breathing is shallow and air in the mouth, trachea and bronchi is exchanged at a faster rate than normal. The rapid exchange of air is accompanied by increased evaporation. Lung ventilation is not significantly increased, an important point as this could change the balance of blood gases (Chapter 8).

Camels and many other mammals living in the desert sweat very little, if at all. They conserve water at the expense of sweating. Because of this, a camel's body temperature can fluctuate by as much as 6 °C. Such a fluctuation is **tolerated** by a camel yet would be fatal to other mammals.

Other desert mammals such as the kangaroo rat avoid the heat by sleeping in burrows during the day. They are active at night, when the desert can be very cold and less likely to encourage evaporative losses.

Fig 20.13 summarises the mechanisms of thermoregulation in humans.

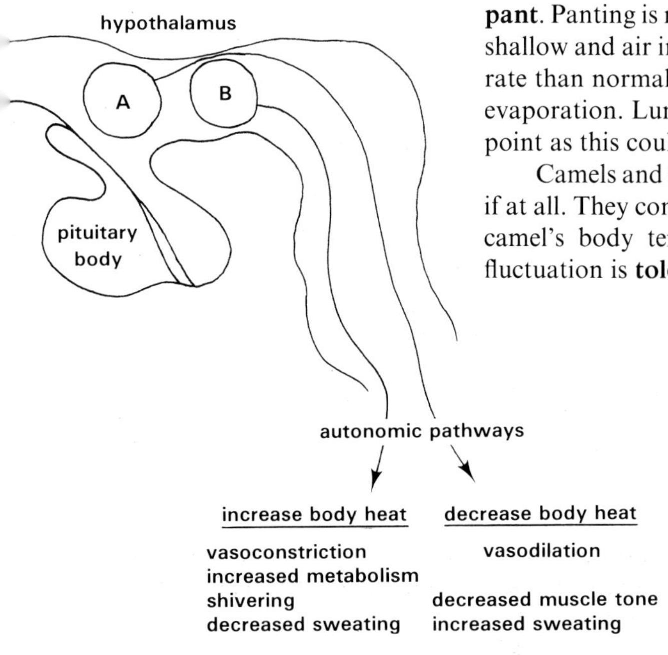

Fig 20.13 (a) Summary of the main thermoregulatory mechanisms (A = hot centre, B = cold centre)

Fig 20.13 (b) Pathways involved in responses to cold

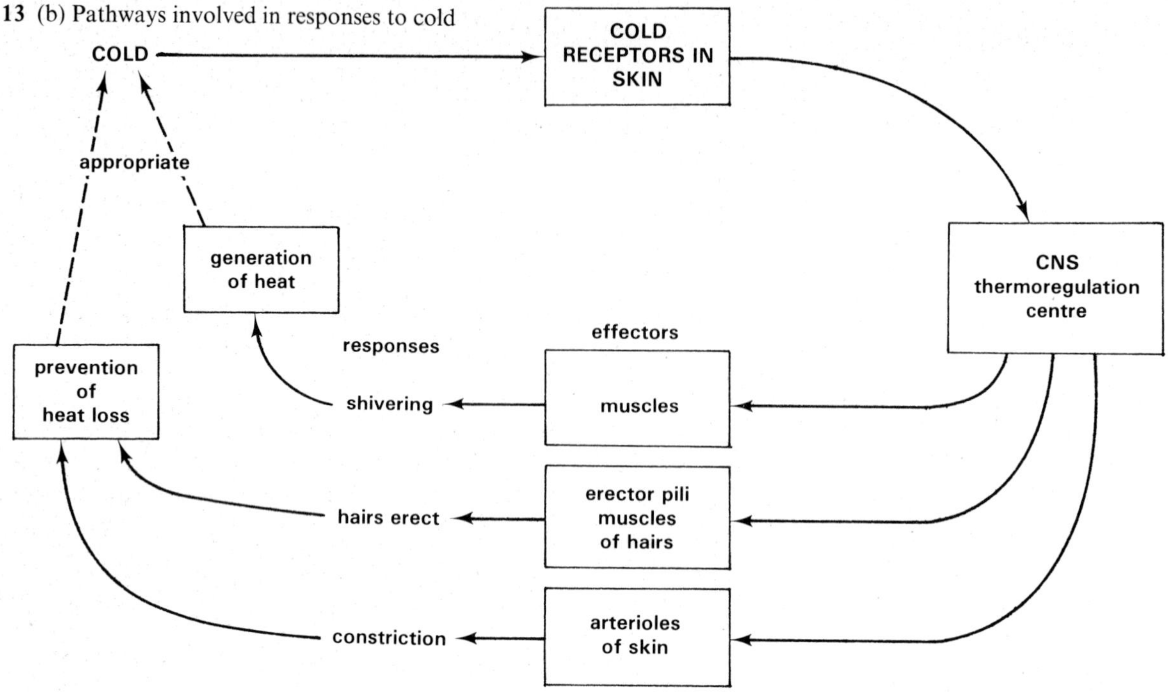

20.3.4 Hibernation

Many endothermic animals become inactive in cold weather and enter a prolonged sleep. In this condition, called **hibernation**, the metabolic rate, oxygen consumption, heart and breathing rates are very low. The body temperature falls to only a few degrees above that of the surroundings, yet it is still kept fairly constant (Table 20.1). Hibernation is a means of surviving long periods of cold weather when food is scarce. Reserves of food in the body provide the energy needed to maintain the slow rate of metabolism during hibernation. When the climate becomes warmer, hibernating animals become aroused very rapidly. In some species, the oxygen consumption necessary to increase their metabolic rate can increase several thousand times in a few hours. Some large mammals, including grizzly bears, have a period of hibernation sometimes called **winter sleep**. The body temperature remains high, even when the surroundings are intensely cold. It is thought that hibernation of some mammals is brought on by a peptide hormone secreted by the hypothalamus. During hibernation, urination and defaecation stop, so helping to conserve body heat.

Several desert-living mammals have a form of hibernation which is of value in surviving long periods of drought rather than cold. Hibernation of this kind is called **aestivation**. During severe drought food is not likely to be plentiful. The reduced rate of breathing during aestivation is of survival value, since even a moderate rate of breathing results in considerable water loss which could lead to desiccation (Chapter 8).

Table 20.1 Comparison of some body functions in active and hibernating marmots (from Hughes 1965)

Functions	Active	Hibernating
body temperature/°C	34–39	3–8
basal metabolism/kJ m–2day–1	1714	113
heart rate/beats min–1	80	4–5
breathing rate/ventilations min–1	25–30	0.2

SUMMARY

Birds and mammals are endotherms; they can regulate their internal body temperature by physiological means (thermoregulation). Other animals are ectotherms; their body temperature fluctuates with that of the surroundings. Thermoregulation is brought about by balancing heat production and heat loss.

The main source of body heat is tissue respiration. Basal metabolic rate, BMR, is the rate at which the body provides energy for survival in resting conditions. BMR can be calculated from the body's heat loss or from the volume of oxygen consumed in respiration.

Heat is transported around the body by the blood. Heat is unevenly distributed. In cold weather there can be up to a 10°C temperature gradient along the limbs in humans.

Heat is lost from the body by radiation, evaporation and conduction. Heat loss through the skin is affected by the surface area/volume ratio. Small mammals lose heat most easily because each unit volume of respiring tissue has a relatively large skin area: limit to smallest size of birds and mammals. Large mammals lose heat less easily.

Thermoregulation centres in the hypothalamus:
Cold centre, activated by skin receptors is sensitive to a drop in temperature.
Hot centre is activated by blood temperature.
The centres are mutually inhibiting.
Activated centres bring about heat retention or heat loss responses.

Thyroxine controls BMR. Shivering generates heat in the muscles.

Skin controls heat loss:
Adipose tissue; brown fat generates heat by respiration. White fat insulates against heat loss. Blood may be directed beneath adipose tissue by vasoconstriction; heat loss is decreased.
Vasodilation brings blood to the surface; heat loss is increased.
Hair: contraction of erector-pili muscles traps warm air on skin surface beneath erect hair; heat loss is decreased.
Relaxation of erector-pili muscles; hair lies flat on skin; heat loss is increased.
Evaporation of water from body increases heat loss (latent heat of vaporisation); sweat is lost from sudorific (sweat) glands. Dry skin; heat loss is decreased.
Some mammals have no sweat glands, e.g. dogs; panting substitutes for sweating. Camels tolerate relatively wide temperature fluctuation and so save water in the dry desert.

In very cold weather many mammals hibernate. BMR, oxygen consumption, heart and breathing rates all become low. Body temperature falls to within a few degrees of the surroundings.
Some large mammals enter a winter sleep. Body temperature remains high.
Some desert-living mammals aestivate; similar to hibernation.

QUESTIONS

1 (a) (i) Define ectothermy.
(ii) Give *one* advantage and *one* disadvantage of ectothermy.
(b) Describe, and explain, *two* ways in which structures in the skin of a mammal respond to a rapid fall in temperature.
(c) The temperature of the open sea is usually lower than that of mammalian blood. Suggest *two* ways in which the skin of a whale might differ from that of a typical land mammal, and give a reason for each answer.   (L)

2 What are the advantages of a constant body temperature to a mammal? Give an account of the physiological and behavioural control of body temperature in mammals.
(L)

3 (a) An Arabian camel, storing fat mainly in its hump, weighs 400 kg and lives in deserts where the temperature by day is often 40 °C. The fat in the camel's hump weighs 40 kg and is a source of metabolic water. The table shows the day temperature, oxygen content and water content of desert air and air expired from the camel's lungs.

	Temperature (°C)	Oxygen content ($cm^3 dm^{-3}$)	Water content ($mg\, dm^{-3}$)
Desert air	40	200	5
Expired air	37	160	44

During aerobic respiration 1 g fat requires $2\,dm^3$ oxygen

and yields 1.07 g water.
(i) How do these figures show that the fat in the camel's hump cannot be its only source of water at 40 °C?
(ii) A 70 kg human contains 14 kg fat. Compare the proportion and distribution of fat in the camel and the human. Suggest why the differences are advantageous to each mammal.
(b) Mammals can produce urine more concentrated than blood plasma.
(i) Make a large labelled diagram of a mammalian nephron.
(ii) By means of annotations (notes) on your diagram, explain how the filtrate is concentrated as it passes along the nephron to the pelvis of the kidney.
(iii) The concentration of urine produced by a camel can be twice that of a human. Suggest how the structure of the camel kidney may differ from that of the human kidney and how this difference can account for the camel's greater urine concentration.   (AEB 1985)

4 The table shows (in arbitrary units) heat losses ($-$) and heat gains ($+$) by a naked man at rest at different environmental temperatures. All other environmental conditions are constant.

| Environmental temperature (°C) | Heat loss ($-$) and heat gain ($+$) in arbitrary units | | |
| | Skin surface | | Body core |
	By radiation and convection	By evaporation	
20.0	-160	-20.0	-120
22.5	-135	-22.5	-85
25.0	-110	-25.0	-50
27.5	-85	-27.5	-20
30.0	-55	-30.0	0
32.5	-25	-60.0	$+5$
35.0	$+5$	-100.0	$+5$
37.5	$+40$	-140.0	$+5$
40.0	$+80$	-180.0	0

(a) On the same axes plot graphs of the heat losses and heat gains by radiation and convection at the skin surface, by evaporation at the skin surface, and by body core against environmental temperature.
(b) Describe and explain the trends in heat loss and heat gain by
(i) radiation and convection at the skin surface,
(ii) evaporation at the skin surface,
(iii) the body core.

(c) Explain why heat losses at the skin surface do not result in a similar loss of body-core heat.
(d) Explain how the data above support the view that humans evolved in tropical regions and were only able later to spread to temperate and cold climates. (AEB 1986)

5 The figure which follows shows the average sweating rates of a man under various conditions in the desert. The results were obtained by accurately weighing the man during 4-hour periods when the air temperature was 38 °C. The results are expressed as rate of water loss in grams per hour.

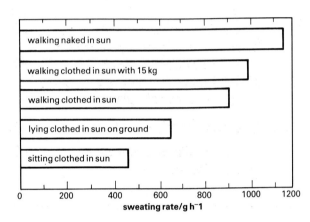

Comment, with reference to the data, on the differences in sweating rates of the man under the following contrasting conditions. In each case, explain the physical principles which account for the differences.
(a) Walking naked in the sun and walking clothed in the sun.
(b) Walking clothed in the sun and walking clothed in the sun carrying 15 kg.
(c) Sitting clothed in the sun and lying clothed in the sun on the ground.   (C)

6 (a) What are the benefits and the costs of maintaining a constant, high body temperature in mammals?
(b) Describe the responses of a mammal to cold temperatures, in both the short term and the long term.
(c) Why is prolonged exposure to cold frequently fatal?   (C)

21 Endocrine system

21.1 Thyroid and parathyroid
     glands   395
     21.1.1 Thyroid gland   395
     21.1.2 Control of thyroid activity   396
     21.1.3 Measurement of thyroid
            activity   397
     21.1.4 Malfunction of the thyroid
            gland   398
     21.1.5 Parathyroid glands   399
     21.1.6 Role of the thyroid gland
            in calcium control   400
     21.1.7 Malfunction of the
            parathyroid glands   401

21.2 Control of blood glucose   401
     21.2.1 Insulin and glucagon   402
     21.2.2 Cortisol   404

21.3 Control of blood sodium
     and potassium   405
     21.3.1 Aldosterone   405

21.4 The adrenal medulla   406
     21.4.1 The medullary hormones   406

21.5 The pituitary body   407
     21.5.1 The anterior pituitary   407
     21.5.2 The posterior pituitary   410

Summary   411

Questions   412

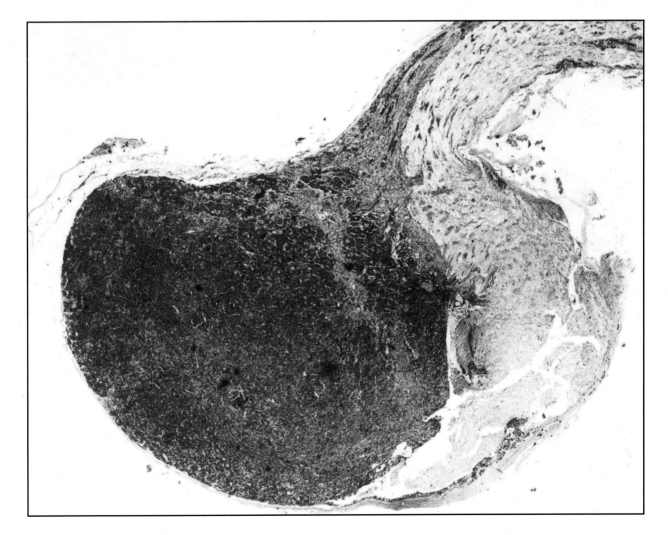

21 Endocrine system

Many functions of the body are coordinated and controlled by **hormones** which are produced by **endocrine glands**. Endocrine glands are ductless and secrete hormones directly into the body fluids, mainly the blood. Reference has already been made in Chapter 15 to hormones which regulate secretion of digestive fluids. Those which control reproduction are described in Chapter 25. The endocrine glands are distributed throughout the body (Fig 21.1). Because they are transported in the blood, hormones can control body functions taking place some distance from the endocrine glands. This distinguishes endocrine glands from exocrine glands which have ducts to channel their products into nearby regions without using blood for transport (Fig 21.2).

Fig 21.1 Sites of the main endocrine glands of a man

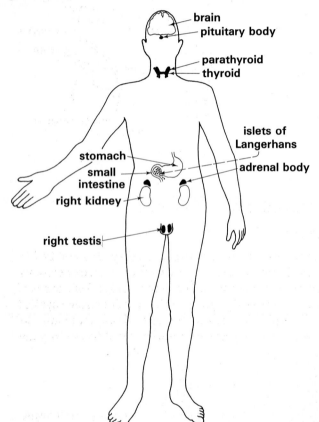

Fig 21.2 Basic construction of (a) an exocrine gland such as a digestive gland in the gut and (b) an endocrine gland

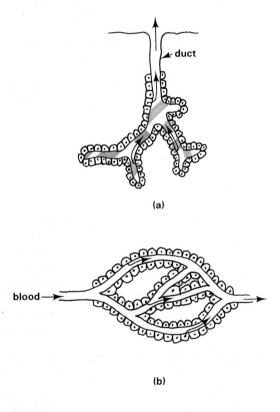

Although only small amounts of hormones are found in blood, their effects on the body tissues are very great. Once in the blood, hormones are bound to plasma proteins which carry them to their sites of activity called **target organs**. Activity of most endocrine glands is controlled by the nervous system (Chapter 18).

The cells of the body that are affected by a particular hormone are called the hormone's **target cells**. Target cells respond to a hormone because they possess **receptors** which are sensitive to the hormone. Receptors are proteins. They are found in the target cells' membranes, cytoplasm and nucleus. Because hormone–receptor binding is specific, hormones only cause a response in their particular target cells.

One mechanism involves receptors in the target cell membrane. When a hormone binds with a membrane receptor, a substance called cyclic AMP (cyclic adenosine monophosphate) is produced from ATP (Chapter 5). Cyclic AMP triggers a chain of reactions which results in the activation of certain enzymes, secretion of cell products, protein synthesis or a change in membrane permeability (Fig 21.3(a)).

Another mechanism involves intracellular receptors. When the hormone enters the target cell, a hormone–receptor complex is formed. The complex activates certain genes. Specific protein (enzyme) synthesis is thus stimulated and the function of the cells is modified (Fig 21.3(b)).

Fig 21.3 Mechanisms of hormone action

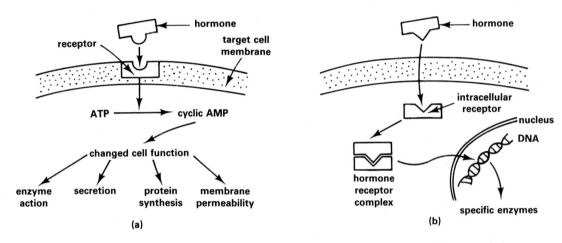

21.1 Thyroid and parathyroid glands

21.1.1 Thyroid gland

The **thyroid gland** is in the neck close to the larynx (Figs 21.1 and 21.11). Under the microscope the thyroid gland is seen to be composed of globular groups of cells called **follicles** (Fig 21.4). The closely packed follicles are of cuboidal epithelium and are held together by connective tissue supplied with blood vessels. The epithelial cells secrete thyroid hormones into the cavities of the follicles where they are temporarily stored before they are taken into the blood.

Fig 21.4 (a) Photomicrograph of a section of thyroid gland, ×400

Fig 21.4 (b) Cuboidal epithelium

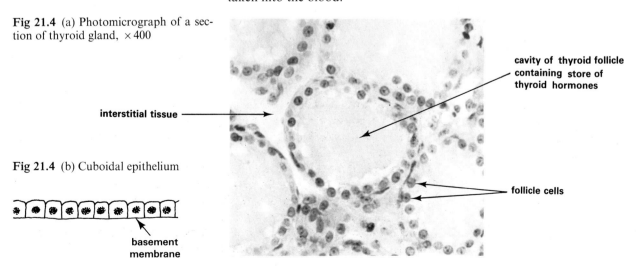

The thyroid gland continuously removes **iodine** from the blood. The iodine comes from the diet. In the thyroid gland iodine combines with the amino acid tyrosine, to produce mono- and diiodotyrosine, then tetra- and finally triiodothyronine (Fig 21.5). **Triiodothyronine (T_3)** and **tetraiodothyronine (T_4)** in a ratio of about 1:9 in humans are the thyroid hormones. T_4 is usually called **thyroxine**.

The thyroid hormones affect all tissues of the body and have two main functions. First, they regulate growth and development by controlling the growth and differentiation of cells. It is thought that the hormones modify the activity of appropriate genes. The second main function of thyroid hormones is to control the basal metabolic rate (BMR).

Fig 21.5 Pathway of thyroxine production

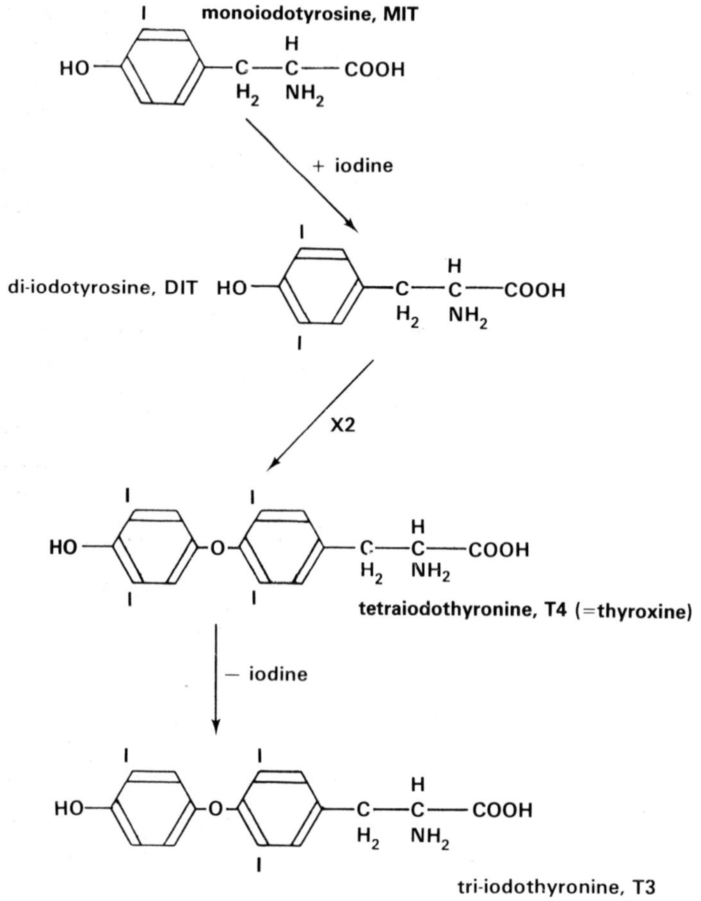

monoiodotyrosine, MIT

+ iodine

di-iodotyrosine, DIT

X2

tetraiodothyronine, T4 (=thyroxine)

− iodine

tri-iodothyronine, T3

Fig 21.6 Mechanism controlling thyroid activity

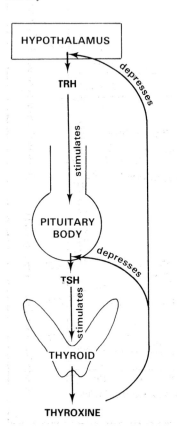

HYPOTHALAMUS

depresses

TRH

stimulates

depresses

PITUITARY BODY

TSH

stimulates

THYROID

THYROXINE

21.1.2 Control of thyroid activity

As with several other endocrine glands, growth and secretion of the thyroid gland are controlled by the **pituitary body** (Figs 21.1 and 21.28). The pituitary body secretes a number of **tropins (trophic hormones)** which affect other endocrine glands (section 21.5.1). The tropins include **thyroid stimulating hormone, TSH (thyrotropin)**. Growth of the thyroid gland and its output of hormones is stimulated by TSH. Secretion of TSH is in turn suppressed by the thyroid hormones. The interaction is a **negative feedback control** mechanism. Another hormone called **thyrotropin releasing hormone, TRH (thyroliberin)** controls the release of thyrotropin. It is made in the **hypothalamus**, the part of the brain immediately above the pituitary body (Fig 21.6). Such control ensures that activity of the thyroid gland meets the body's needs at any time. It is a homeostatic device which helps maintain the stability of the body's internal environment. Other comparable examples are described in the remainder of the chapter.

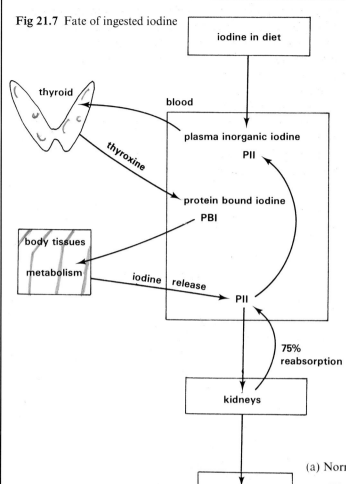

Fig 21.7 Fate of ingested iodine

iodine in diet

thyroid

blood

thyroxine

plasma inorganic iodine
PII

protein bound iodine
PBI

body tissues

metabolism

iodine release

PII

75% reabsorption

kidneys

iodine in urine

21.1.3 Measurement of thyroid activity

Thyroid activity can be measured with the use of **radioactive isotopes (radionuclides)** of iodine. Sodium iodide containing ^{131}I is given orally in doses which produce harmless levels of radiation to the patient. The radio-iodine circulates in the plasma as **plasma inorganic iodine (PII)**. Much of the iodine in PII is normally absorbed by the thyroid gland and used to make thyroid hormones. The hormones are secreted into the blood where the radio-iodine is carried as **protein bound iodine (PBI)**. When the hormones have performed their functions in the tissues, iodine is released back into the blood to be absorbed once more by the thyroid gland. Some of the body's total iodine pool including the radionuclides is excreted in the urine (Fig 21.7). There is a well-defined relationship between the amounts of radio-iodine in the plasma, thyroid gland and urine following a dose of the radionuclide (Fig 21.8(a)). Consequently, measurements of the radiation in the thyroid gland, samples of blood and urine are useful in detecting abnormal activity of the thyroid gland (Fig 21.8(b)).

Fig 21.8 Relationship between the levels of iodine in the thyroid gland, blood and urine after oral administration of radioactive iodine (after Greig, Boyle and Boyle)

(a) Normal subject

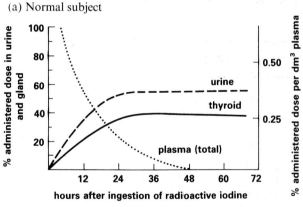

(b) Thyrotoxic subject

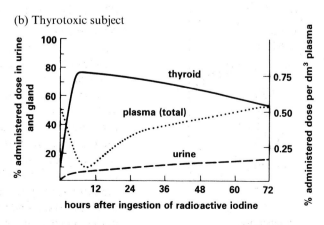

An alternative to the iodine uptake studies described above is the use of a radionuclide to produce an image of the thyroid gland. An isotope of technetium, $^{99}Tc^m$, is administered by intravenous injection. The radionuclide is absorbed by the thyroid gland. About 30 minutes later, the gamma radiation emitted from the thyroid gland is used to produce an image with a gamma camera. Images obtained in this way are useful in assessing an overactive thyroid before and after treatment. They can also be used to detect regions of the thyroid, called **nodules**, which secrete thyroid hormones independently of the TSH control mechanism (Fig 21.9).

Fig 21.9 Thyroid scans, glands shown in outline: (a) normal, (b) cyst in the left lobe results in no uptake of radionuclide; the right lobe is unaffected

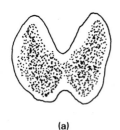

(a)

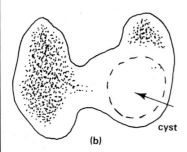

(b)

cyst

21.1.4 Malfunction of the thyroid gland

Overactivity of the thyroid gland is called **hyperthyroidism**. The condition can result from a failure of the thyrotropin–thyroid control mechanism, resulting in increased thyroxin output. Another cause of hyperthyroidism is uncontrolled secretion of thyroid hormones by a thyroid tumour. Either way, an increase in basal metabolism occurs with increased heart rate, extreme irritability and loss of body mass. Removal or destruction of part of the thyroid gland may be necessary to check the condition. A common side-effect of an overactive thyroid gland is the secretion by the pituitary body of **exophthalmos-producing substance**. It causes excessive growth of the tissues immediately behind the eyes. Consequently protrusion of the eyes called **exophthalmos** often accompanies hyperthyroidism (Fig 21.10).

Underactivity of the thyroid gland is called **hypothyroidism**. In adults hypothyroidism is the cause of a condition called **myxoedema**. The effects on basal metabolism are the opposite to those of hyperthyroidism. The BMR slows down, a smaller proportion of the energy-rich ingredients in the diet are respired, hence body mass increases. Mental activity also slows down so the patient is less alert than normal. At one time hypothyroidism was common in parts of Derbyshire and the Swiss Alps. Here the local soil and water supply are deficient in iodine and the population was thus deprived of an essential requirement for the synthesis of thyroid hormones. Addition of traces of iodine to table salt has largely overcome the problem.

The abnormal growth and development which accompanies an underactive thyroid gland is particularly distressing in infants. **Cretinism** is the name given to the effects of thyroid deficiency in children. The main symptoms are retardation in mental, physical and sexual development.

Both hyper- and hypothyroidism usually result in excessive growth of the thyroid gland. The enlarged thyroid gland is called a **goitre** which causes swelling of the neck (Fig 21.10).

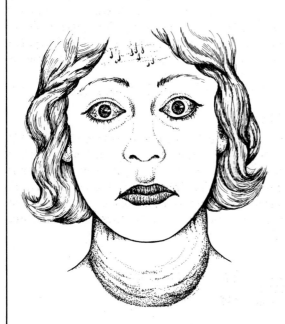

Fig 21.10 Facial characteristics of a person with hyperthyroidism. Note the protruding eyes and goitre

21.1.5 Parathyroid glands

The **parathyroid glands** are four small oval bodies embedded in the thyroid gland (Fig 21.11). The parathyroid glands make a polypeptide hormone called **parathormone (parathyrin)** which has a profound effect on the calcium ion content of the blood and other body fluids. Parathormone elevates the amount of calcium in the blood which, by negative feedback control, regulates parathormone output (Fig 21.12).

Fig 21.11 Position of the parathormone glands

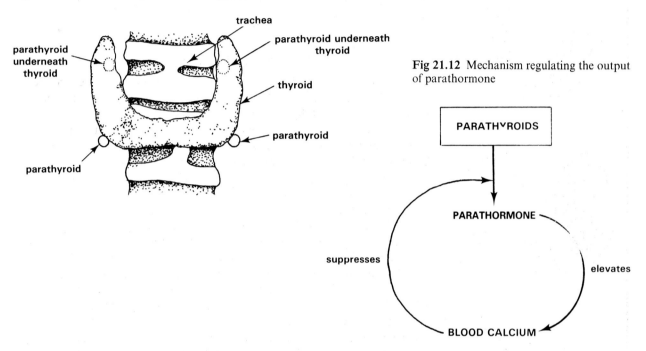

Fig 21.12 Mechanism regulating the output of parathormone

Calcium is an important constituent of body fluids. It participates in many body functions including coagulation of the blood (Chapter 10), nerve and muscle activity (Chapter 17) and teeth and bone formation (Chapter 16). These functions depend not just on the presence of calcium ions but on a precise concentration of calcium ions. It is therefore important that the amount of calcium is regulated within narrow limits. Calcium is one of the most precisely regulated constituents of the body. The concentration of calcium ions in normal human blood serum ranges between 9 and 11 mg per 100 cm^3.

Parathormone affects blood calcium in several ways, as follows:

1 Release of calcium from bone

Bone contains calcium salts, especially calcium phosphate (Chapter 16). Calcium and phosphate(v) ions are continuously released from bones into the tissue fluids and redeposited in bones. Clearly, the composition and structure of bone depends on regulating this reversible process (Fig 21.13). By stimulating the **release of calcium** and phosphate ions from bone, parathormone helps maintain normal concentrations of calcium and phosphate ions in blood.

Fig 21.13 Dynamic equilibrium between the calcium and phosphate in bone and that in the tissue fluid and blood

2 Reabsorption of calcium from the urine

Calcium and phosphate ions are filtered from the blood in the nephrons. Parathormone promotes **reabsorption of calcium** ions from the filtrate in the proximal convoluted tubules at the expense of phosphate ions which are excreted (Chapter 11). Parathormone thus elevates the concentration of calcium ions in the blood while lowering the concentration of blood phosphate ions (Fig 21.14). The inverse relationship between calcium and phosphate ions in blood

$$[Ca^{2+}] \propto \frac{1}{[PO_4{}^{3-}]}$$

prevents the accumulation of unwanted calcium phosphate in the blood and tissues.

Fig 21.14 Summary of the effects of parathormone and calcitonin on blood calcium and phosphate

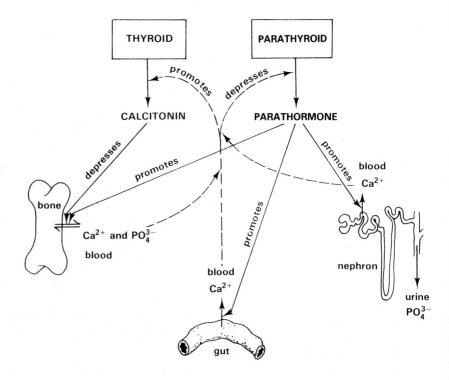

3 Absorption of calcium from the gut

Parathormone is also thought to promote **absorption of calcium** ions from the gut (Chapter 15), thus helping to maintain the normal concentration of calcium ions in the blood.

21.1.6 Role of the thyroid gland in calcium control

Another hormone called **calcitonin** also affects the calcium ion concentration of blood. Calcitonin is secreted from **C cells** in the thyroid gland. Its effect is to lower the concentration of calcium ions in blood by causing the deposition of calcium phosphate in bone. Thus it is the combined effects of parathormone and calcitonin which regulate calcium and phosphate ions to the concentrations necessary for many physiological functions to take place normally (Fig 21.14).

21.1.7 Malfunction of the parathyroid glands

Underactivity of the parathyroid glands, **hypoparathyroidism**, results in a lowering of the concentration of calcium ions in blood, **hypocalcaemia**. If the concentration of calcium ions in blood serum drops below 7 mg per 100 cm³ a condition called **tetany** results. Tetany is characterised by increased excitability of the nervous system. Muscular activity becomes spasmodic and uncontrolled. Surgical removal of the parathyroid glands results in tetany within a rew days. Subsequent injection of parathyroid extract quickly corrects the condition but the effect is shortlived. Subsequent maintenance of normal concentrations of calcium ions in the blood requires continued treatment with parathormone (Fig 21.15)

Fig 21.15 Effects of removing the parathyroids on the concentration of blood calcium

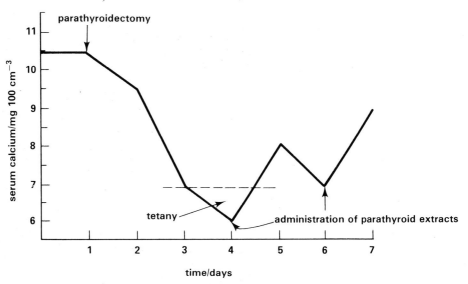

Overactivity of the parathyroid glands, **hyperparathyroidism**, can result in breakdown of bone structure and consequent elevation of the concentration of calcium ions in blood, **hypercalcaemia**. Recent work indicates that the decalcifying activity of parathormone increases in women after the menopause. Before the menopause, output of parathormone is inhibited by oestrogen hormones secreted by the ovaries (Chapter 25). After the menopause oestrogen production stops. It explains why the bones of older women are often fragile and prone to fracture.

Why are skeletal fractures and breaks more likely in older women than in men of the same age?

21.2 Control of blood glucose

Carbohydrate is the main source of energy released in respiration (Chapter 5). Nearly all the carbohydrate absorbed from the gut circulates in the blood as **glucose** and it is this sugar which is the main respiratory substrate. It is therefore vital that the blood maintains a constant and adequate supply of glucose to all the tissues. It is equally important, however, that the concentration of glucose in blood and tissue fluids is not excessive. Abnormally high concentrations of glucose in the tissue fluid would draw water from cells by osmosis (Chapter 1). Furthermore, the water content of the body as a whole can be affected by the concentration of glucose in the blood. Normally the concentration of glucose in blood is sufficiently low for all the glucose in the renal filtrate to be reabsorbed into the blood in the kidneys (Chapter 11). Thus glucose is not usually excreted in urine. Since reabsorption of water from the nephrons depends on the water potential gradient between the renal filtrate and the blood, any glucose remaining in the filtrate would lower the gradient and reduce water reabsorption. Thus as well as tissue dehydration, a high concentration of glucose in blood could cause body dehydration.

Mechanisms exist which regulate the concentration of glucose in the blood so that the tissues' metabolic needs are met without adversely affecting the body's water content. A number of hormones help to control the concentration of glucose in blood.

21.2.1 Insulin and glucagon

The pancreas is an exocrine gland which secretes pancreatic juice into the gut (Chapter 15). It is situated in a loop of the small intestine just below the stomach (Fig 21.1). Embedded in the pancreas, and sometimes also in the wall of the small intestine, is a large number of microscopic patches of endocrine tissue called the **islets of Langerhans** (Fig 21.16). Histochemical studies show that the islets contain two types of cells called α- and β-cells. They are responsible for the production and secretion of two hormones called **glucagon** and **insulin** respectively. The hormones are discharged directly into the islet blood capillaries. The rest of the pancreas and its exocrine ducts are not involved in hormone production or secretion. Experimentally tying off the pancreatic ducts prevents secretion of pancreatic digestive enzymes but has no effect on hormone secretion.

Fig 21.16 Photomicrograph of two islets of Langerhans seen in a thin section of pancreas, × 100

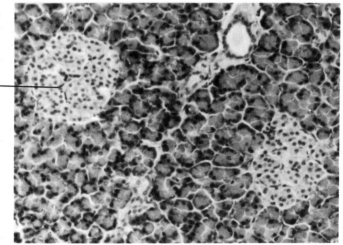

islet—

Fig 21.17 Effects of insulin and glucagon on the levels of blood glucose

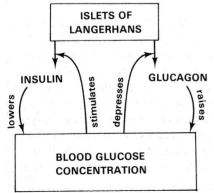

Insulin lowers the concentration of blood glucose while glucagon raises it. Output of the pancreatic hormones is regulated by negative feedback control based on the concentration of glucose in blood (Fig 21.17).

The functions of insulin

Insulin performs two main functions. The first is to regulate the concentration of glucose in the blood. Glucose is used by the tissues at varying rates dependent on the rate of metabolic activity. More glucose is used during exercise than when the body is at rest. The body's input of glucose varies depending on what has been eaten. The concentration of glucose in the blood draining the gut can double after a meal. The problem then is to reconcile the variable input and respiration of glucose with a relatively constant concentration of glucose in the blood. In dealing with this problem the liver, insulin and glucagon play vital roles.

Blood leaving the gut contains the absorbed products of digestion and passes through the hepatic portal system into the liver (Chapter 15). The liver cells contain enzymes which, under the control of insulin, promote the synthesis of **glycogen**, a polymer of glucose. It is as glycogen that much of the glucose absorbed from the gut is stored in the liver. Some glycogen is also stored in the skeletal muscles. Insulin therefore prevents any undue rise in the concentration of glucose in the blood. Glucose removed from the blood for respiration is replaced from the glycogen stores. Glucagon has the opposite effect to insulin. It stimulates the conversion of glycogen to glucose. The balance between the effects of the two hormones results in regulation of the concentration of glucose in blood (Fig 21.18). The negative feedback relationship between the hormones and glucose in blood ensures that glucose is released from the glycogen stores at a rate sufficient to match its uptake from the blood by respiring tissues.

Fig 21.18 Summary of the fate of glucose in the body

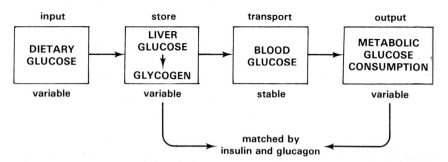

The second main function of insulin is to speed up the rate of entry of glucose into respiring cells. Glucose is taken into living cells by **active absorption**. Insulin greatly increases the rate of glucose absorption, possibly by triggering a membrane carrier mechanism or by acting as a carrier itself. In a normal healthy human a concentration of about 3 to 5 mmol glucose per dm^3 of blood is sufficient to meet the requirements of respiring cells. In the absence of insulin, this concentration would have to increase by between ten and twenty times for glucose to enter the cells at the same rate by diffusion only.

Malfunction of the islets of Langerhans

Underactivity of the islets of Langerhans results in reduced secretion of insulin and is the cause of **diabetes mellitus**. This should not be confused with diabetes insipidus (Chapter 11). In diabetes mellitus the glucose concentration in blood rises, **hyperglycaemia**, and glucose exceeds the maximum concentration which can be totally reabsorbed from the renal filtrate in the kidneys. Consequently glucose is excreted in the urine, a condition called **glycosuria**. The presence of glucose in the urine disturbs the water potential gradient which normally results in water reabsorption from the nephrons. Large volumes of dilute urine are thus produced, a condition called **diuresis**. Diuresis is dangerous because it may bring about dehydration of the body. Since the breakdown of glycogen is uninhibited in diabetes mellitus the stores of glycogen in the liver and muscles are quickly used up. Body fats and proteins are then used as respiratory substrates, causing a rapid loss of body mass.

The condition can be rectified by regular doses of insulin and by eating a carefully controlled diet. A clinical test once used in hospitals to assess whether insulin production is normal involves measurement of **glucose tolerance**. Patients are made to fast for several hours before ingesting 50 g of glucose in $150\,cm^3$ water. The concentration of glucose in the patient's blood is measured immediately and at 30-minute intervals over a period of two to three hours. If necessary the urine is also analysed for glucose. The ▶

concentration of glucose in the blood is plotted against time and a glucose tolerance curve is obtained (Fig 21.19). In a healthy individual there is a slight rise in blood glucose concentration after the glucose drink. Insulin then brings the concentration of glucose in the blood down to its original value within about two hours and no glucose is excreted in the urine.

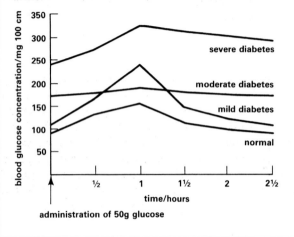

blood glucose concentration/mg 100 cm

350
300
250 — severe diabetes
200
150 — moderate diabetes
100 — mild diabetes
50 — normal

½ 1 1½ 2 2½
time/hours

administration of 50g glucose

Fig 21.19 Glucose tolerance curves

The type of diabetes described above is called Insulin Dependent Diabetes. It results from a deficiency of insulin production and must be treated by periodic insulin injection. Non-Insulin Dependent Diabetes is a commoner type of diabetes. Insulin secretion is not inhibited, but the membrane receptors in the target cells are insensitive to insulin. Consequently, glucose entry into the cells is less and it accumulates in the blood, hypoglycaemia. Non-Insulin Dependent Diabetes is usually treated by a careful selection of diet in which glucose intake is strictly limited.

During late pregnancy the renal threshold for glucose may be lower than normal. Thus, even though the concentration of glucose in the blood is normal and only slightly rises after a meal, the rise may be sufficient for some glucose to be excreted in the urine. Glycosuria during pregnancy therefore does not necessarily mean that the mother has diabetes mellitus.

21.2.2 Cortisol

Fig 21.20 Photomicrograph of a transverse section of an adrenal body from a cat, × 11

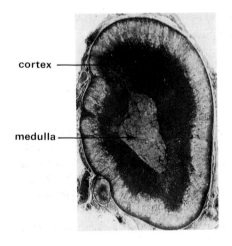

cortex

medulla

The reserves of glycogen in the body are limited and in certain conditions fats and proteins can be converted to glucose to supply metabolic demands. The conversion is influenced by a number of steroid hormones called **glucocorticoids** secreted by the **adrenal bodies**, a pair of glands lying close to the kidneys (Fig 21.1). Each adrenal body is divided into an inner **medulla** and an outer **cortex** (Fig 21.20). It is in the cortex that glucocorticoids are made, the most abundant of which is **cortisol** (Fig 21.21).

Cortisol stimulates the conversion of fats and proteins to glucose and is thus involved in regulating the glucose concentration in blood. Output of cortisol is controlled by a tropin called **adrenocorticotrophic hormone, ACTH (corticotropin)** from the pituitary body. Secretion of ACTH is affected by the concentration of cortisol in the blood, another example of a negative feedback control mechanism (Fig 21.22).

Fig 21.21 Molecular structure of cortisol

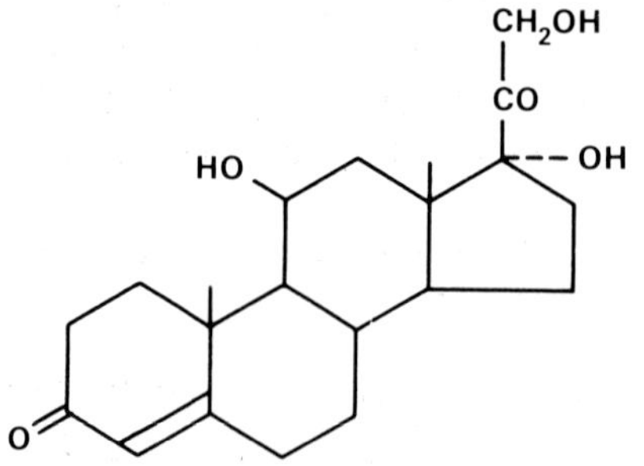

CH_2OH
CO
HO — — OH
O

Another important function of cortisol is its ability to suppress autolysis of damaged body cells by lysosomes (Chapter 6). Cortisol-like drugs are valuable in promoting the repair of body organs damaged in degenerative diseases such as arthritis. One of the effects of adrenaline, secreted by the medulla of the adrenal glands, is to increase output of ACTH (section 21.4.1). This partly explains the increased secretion of cortisol which normally accompanies mental and physical stress.

21.3 Control of blood sodium and potassium

Fig 21.22 Mechanism regulating cortisol production

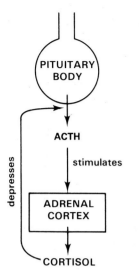

Sodium and potassium salts, especially chlorides, participate in a variety of body functions, particularly nerve and muscle activity. Consequently the maintenance of stable concentrations of sodium and potassium ions in the blood is vital for normal physiological activity (Chapter 17). Furthermore, because sodium and potassium salts are soluble in water they affect the water potential of the body fluids. The concentrations of sodium and potassium ions in the body are regulated by **mineralocorticoid** hormones made in the adrenal cortex (Fig 21.20).

21.3.1 Aldosterone

The most important mineralocorticoid hormone is **aldosterone** (Fig 21.23). This steroid is strikingly similar in molecular structure to cortisol (Fig 21.21). Nevertheless their functions are very different. Aldosterone promotes reabsorption of sodium ions (Na^+) into the blood from the filtrate in the nephrons (Chapter 11). Chloride ions (Cl^-) usually accompany the sodium ions, probably because of electrostatic attraction. Uptake of sodium ions suppresses reabsorption of potassium ions (K^+). Output of aldosterone is inhibited by a high concentration of sodium ions in the blood, yet another example of a feedback mechanism which assists homeostasis (Fig 21.25).

Fig 21.23 Molecular structure of aldosterone

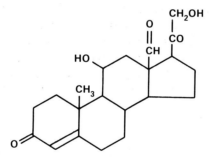

How can glucose appear in the urine of a pregnant woman who does not suffer from diabetes mellitus?

Suppressed output of aldosterone can cause excessive excretion of water by the kidneys resulting in a fall in the blood volume. One effect of the fall is the secretion into the blood of the enzyme **renin** from juxtaglomerular cells in the kidney (Fig 21.24). Renin activates the conversion of a plasma protein called proangiotensin into an active derivative **angiotensin**.

Fig 21.24 Juxtaglomerular cells form part of the afferent arterioles in the kidneys (see Fig 11.3)

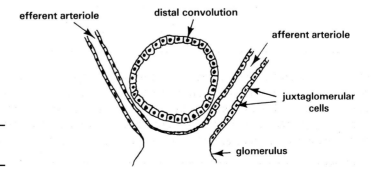

What makes us feel thirsty when our bodies are short of water?

Angiotensin stimulates secretion of aldosterone, thereby aiding water reabsorption by the kidneys. Angiotensin also stimulates the **thirst centre** in the hypothalamus of the brain to trigger off impulses which create the sensation of thirst (Fig 21.25). These are appropriate responses to too little water in the body fluids, and provided water is consumed, the body's state of hydration is rectified.

Fig 21.25 Effect of aldosterone on water reabsorption by the nephrons. Angiotensin stimulates the thirst centre in the brain as well as aldosterone secretion

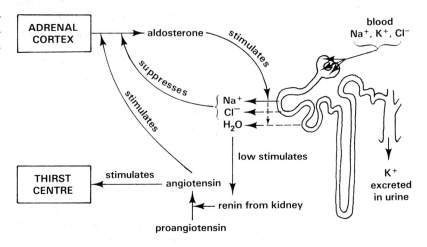

21.4 The adrenal medulla

The centre of each adrenal gland, the **medulla**, is derived from the same embryonic tissue which gives rise to sympathetic nerve ganglia. For this reason the medulla of the adrenal glands can be regarded as a modified part of the sympathetic nervous system. Sympathetic nerves, as part of the autonomic system, play an important role in regulating many body functions (Chapter 18). Medullary hormones have similar effects to stimulation of sympathetic nerves.

21.4.1 The medullary hormones

Two hormones are made in the adrenal medulla, **adrenaline** and **noradrenaline** (Fig 21.26). Noradrenaline is the neuroeffector transmitter secreted by sympathetic nerve endings. However, adrenaline is the main secretion of the medulla. The effects of both hormones are similar and, like sympathetic activity generally, they prepare the body to expend a lot of energy quickly.

Fig 21.26 Structure of (a) noradrenaline and (b) adrenaline. These hormones belong to a group of chemicals called catecholamines

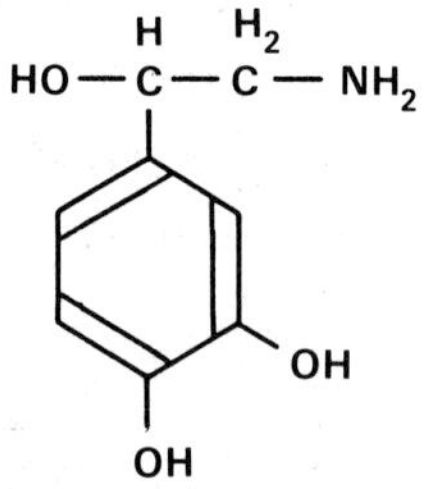

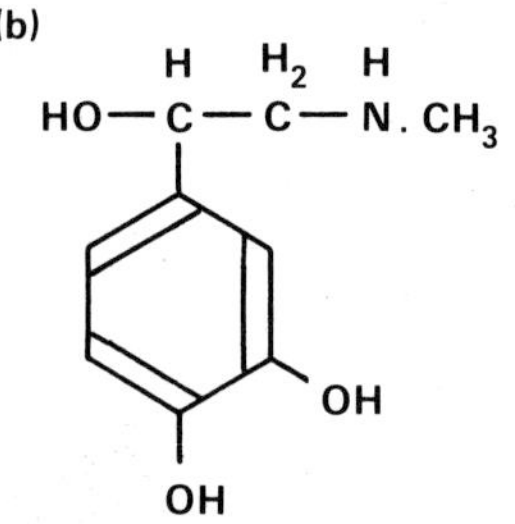

In fact, stimulation of the sympathetic nerves activates the medulla and promotes secretion of adrenaline. Normally the adrenal medulla secretes small amounts of adrenaline and noradrenaline into the blood. However, following increased activity of the sympathetic nerves which usually accompanies physical and mental stress, larger quantities are released. Because the medullary hormones prepare mammals to run away from or to face an enemy, they are sometimes called the **flight or fight** hormones. In

How do the 'flight or fight' hormones prepare our bodies to expend a lot of energy quickly in order to respond to a sudden danger?

many respects this is an appropriate description. The effects of adrenaline and noradrenaline can be summarised as follows.

1 Effects on the gut and respiratory system

The smooth muscle of the gastro-intestinal tract relaxes and the bronchi become dilated. The thorax is enlarged because the diphragm can now be pushed down further into the abdomen. The net result is that volumes of air larger than normal can be drawn in and out of the lungs, thereby increasing the rate of oxygen uptake by the blood.

2 Effects on the cardiovascular system

The heart rate and the power of cardiac contractions increase, with consequent rise of blood pressure. Arterioles in the skin and gut become constricted while the vessels supplying the skeletal muscles dilate. These effects, together with increased ventilation of the lungs, ensure an adequate supply of oxygenated blood to organs such as muscles which produce energy for movement.

3 Effects on blood glucose

Glycogen stored in liver and skeletal muscles is converted to glucose, thus providing the source of energy required for increased muscular activity. Adrenaline stimulates secretion of ACTH and hence cortisol. Some of the extra glucose may therefore come from fats and proteins (section 21.2.2).

4 Effects on the nervous system

Adrenaline increases sensitivity of the nervous system, thereby increasing the speed with which the body may react to environmental stimuli. This is of obvious advantage in flight or fight. The various effects of adrenaline are illustrated in Fig 21.27.

Fig 21.27 Summary of the effects of adrenaline

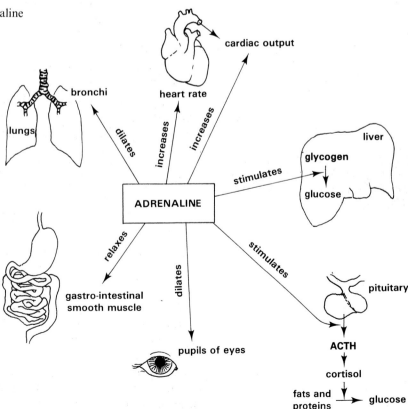

Why is the pituitary body sometimes called the 'master gland' of the endocrine system?

407

21.5 The pituitary body

Projecting downward from the base of the forebrain and almost totally enclosed by bone is the **pituitary body**, also called the **hypophysis**. It consists of two functional glands called the **anterior pituitary**, derived from the embryonic pharynx, and the **posterior pituitary**, derived from the hypothalamus in the forebrain (Figs 21.28 and 18.15).

Fig 21.28 Photomicrograph of a vertical section through the pituitary body, × 10

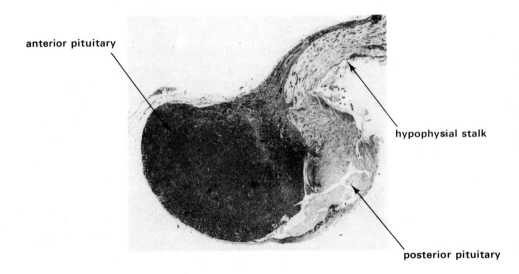

anterior pituitary

hypophysial stalk

posterior pituitary

21.5.1 The anterior pituitary

The anterior pituitary secretes several hormones including the various **tropins** referred to earlier in the chapter. **TSH** controls the secretion of thyroid hormones while **ACTH** controls the output of cortisol from the adrenal cortex. Other tropins affecting mainly the testes or ovaries are called **gonadotrophic hormones** and are also produced by the anterior pituitary. The activities of **follicle stimulating hormone** and **luteinising hormone** are described in Chapter 25.

Another hormone made in the anterior pituitary is **somatotrophin**. It causes the release of the growth hormone **somatomedin** from the liver. The main function of somatomedin is to promote the growth of the body's tissues and organs by stimulating the synthesis of macromolecules, especially proteins. The precise mechanism of the stimulation is unknown but it may involve promoting the absorption of vital nutrients, especially amino acids by body cells from tissue fluid, and activation of the genes which control growth.

Secretion of somatotrophin occurs throughout life but diminishes after the growing period. The way in which its secretion is controlled is unknown.

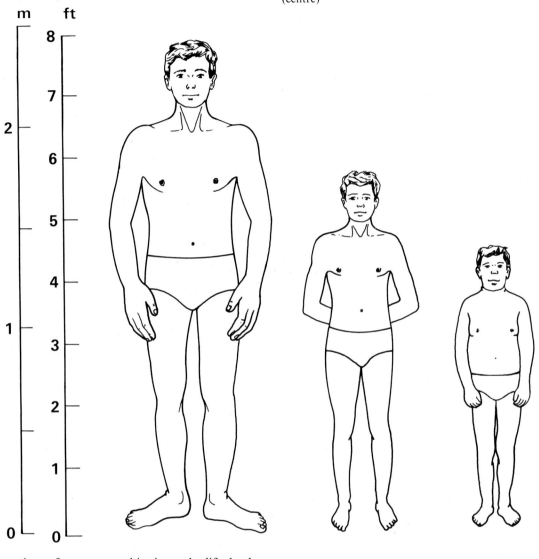

Fig 21.29 Diagram showing gigantism (left) and dwarfism (right) contrasted with normal growth (centre)

Oversecretion of somatotrophin in early life leads to **gigantism**, while undersecretion causes **dwarfism** (Fig 21.29). Abnormally high output of somatotrophin during adulthood when normal growth is complete leads to **acromegaly**. Many of the internal organs become enlarged, as do the hands and feet. The most striking characteristic is the lower jaw which often grows to protrude foward rather noticeably (Fig 21.30). Body height does not usually increase since after adolescence the limb bones do not normally grow further.

Fig 21.30 Facial characteristics of acromegaly. Note the protruding jaw and forehead

Secretion by the anterior pituitary is controlled partly by a variety of substances from the hypothalamus. The secretions are carried from the hypothalamus into the anterior pituitary in blood vessels (Fig 21.31). One secretion is **thyrotropin releasing hormone, TRH**. TRH stimulates the release of **TSH** which in turn stimulates secretion of the thyroid hormones (Fig 21.6). Other liberin hormones made in the hypothalamus control the release of the rest of the hormones made in the anterior pituitary.

Fig 21.31 The hypothalamic-hypophysial portal system responsible for transporting tropins, made in the hypothalamus, into the anterior pituitary

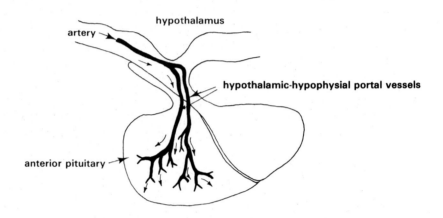

21.5.2 The posterior pituitary

The posterior pituitary produces two polypeptide hormones, **antidiuretic hormone, ADH (vasopressin)** and **oxytocin**. Both are made by neurons in the hypothalamus and migrate in the axons of neurones to the posterior pituitary where they are stored. When the neurones connecting the hypothalamus and posterior pituitary are stimulated, ADH and oxytocin are released into the blood (Fig 21.32).

Fig 21.32 Hypothalamic-hypophysial neurons responsible for transporting ADH and oxytocin from the hypothalamus into the posterior pituitary

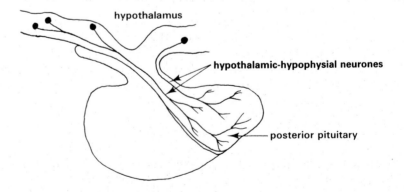

Oxytocin affects smooth muscle and is particularly important in females at the end of pregnancy when it brings about rhythmical contraction of the uterine wall during birth. This function and its role in controlling the ejection of milk from the mammary glands are described in Chapter 25. ADH plays an important role in the regulation of water reabsorption by the renal nephrons (Chapter 11).

SUMMARY

Endocrine (ductless) glands secrete hormones directly into the blood; they are secreted in minute quantities; effects are distant from site of secretion. Target cells possess receptors sensitive to specific hormones. Receptors found in target cells' membranes, cytoplasm and nucleus; Response can be a change in membrane permeability or the secretion of a specific protein (enzyme).

Thyroid gland contains follicles which store thyroxine; maintains basal metabolic rate, BMR, and controls growth. Pituitary secretes thyroid stimulating hormone, TSH; stimulates thyroid growth and secretion; by negative feedback thyroxine suppresses TSH output.
Thyroid releasing hormone from hypothalamus stimulates TSH output.
Hyperthyroidism (over-activity); swollen thyroid called a goitre; increase in BMR; exophthalmos producing substance from pituitary; protruding eyes (exophthalmos).
Hypothyroidism (under-activity); cretinism in children; myxoedema in adults; decrease in BMR; abnormal growth.

Parathyroids next to thyroid; secrete parathormone which elevates blood calcium concentration; by negative feedback blood calcium suppresses parathormone output.
Parathormone releases calcium and phosphate from bone; stimulates calcium reabsorption from nephrons (phosphate is excreted); stimulates calcium absorption from the gut.
Calcitonin from thyroid has opposite effect to parathormone.
Hyperparathyroidism causes excess breakdown of bone; suppressed by oestrogen hormones.
Hypoparathyroidism causes lowering of blood calcium; leads to tetany.

Islets of Langerhans in the pancreas secrete insulin (β-cells) and glucagon (α-cells).
Insulin lowers the concentration of blood glucose; conversion of glycogen to glucose in the liver is suppressed. By negative feedback blood glucose stimulates insulin output. Glucagon has the opposite effect.
Insulin also stimulates the entry of glucose into cells.
Poor insulin secretion leads to diabetes mellitus; raised blood glucose (hyperglycaemia); glucose may be excreted in the urine (glycosuria); leads to excess water loss in urine (diuresis). Cells dehydrate by osmosis.

Adrenal cortex secretes steroids:
Cortisol stimulates the conversion of fats and proteins to glucose in emergencies. ACTH from the pituitary stimulates cortisol output. By negative feedback cortisol suppresses ACTH output. Aldosterone stimulates reabsorption of sodium and chloride ions and (by osmosis) water from nephrons. By negative feedback sodium chloride suppresses aldosterone output.
Low water content of blood stimulates kidneys to secrete renin; blood proangiotensin converted to angiotensin; this stimulates thirst centre.

Adrenal medulla is part of the sympathetic nervous system; secretes adrenaline and noradrenaline in times of stress; 'flight or fight' hormones prepare body to expend a lot of energy quickly; increased cardiac and respiratory activity, relaxed gut, liver glycogen converted to glucose, pituitary secretes ACTH heightened sensitivity in nervous system.

Anterior pituitary:
Tropins include TSH, ACTH and gonadotrophic hormones. Direct-acting hormones include somatotrophin; stimulates release of somatomedin from liver; promotes growth.

Posterior pituitary; attached to hypothalamus:
Antidiuretic hormone controls water reabsorption in nephrons.
Oxytocin causes uterine wall (smooth muscle) contractions at birth; expels milk from lactating mammary glands.

411

QUESTIONS

1 Two human subjects of similar build, **A** and **B**, were each given 60 g of glucose by mouth after an 8-hour overnight fast. Their blood glucose was then monitored at intervals and the results are shown in the graphs.

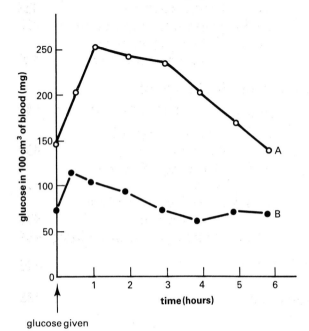

glucose given

(a) Why did the subjects' blood glucose rise after they took oral glucose?
(b) (i) Which hormone would bring about the observed drop in the blood-glucose level of subject **B**?
  (ii) Give **two** effects of this hormone which result in a reduced blood-glucose level.
(c) Give **two** differences between the blood-glucose levels of subjects **A** and **B**. Explain each difference. (AEB 1986)

2 Discuss the role of hormones in the life of mammals. (L)

3 (a) Define the term *hormone*.
(b) Describe how hormones released by the hypothalamus and pituitary gland affect the functions of the thyroid gland, the ovaries and the kidney.
(c) Suggest reasons why hormonal rather than nervous stimuli are used to control these processes. (C)

4 (a) What is meant by the term *homeostasis*?
(b) Explain the role of the endocrine system in mammalian homeostasis. (C)

5 Three patients A, B and C were starved for 12 hours and then each was given 50 g of glucose in 150 cm^3 of water. The blood glucose concentration was measured for each patient immediately and then at 30 minute intervals for a period of $2\frac{1}{2}$ hours. The table below summarises the results from the three patients.

Time after ingestion of glucose in hours	Blood glucose concentration in mg per 100 cm^3		
	A	B	C
0	90	105	240
$\frac{1}{2}$	132	165	275
1	155	240	325
$1\frac{1}{2}$	110	145	310
2	95	120	300
$2\frac{1}{2}$	90	105	290

(a) Plot a graph of these results.
(b) Comment on and explain the results obtained from all three patients one hour after the ingestion of the glucose.
(c) Comment on the results for the three patients in the period 1 to $2\frac{1}{2}$ hours after the ingestion of the glucose.
(d) Give an interpretation of the results for each patient, with reasons to support your explanation. (L)

22 Reproduction in flowering plants

22.1 Sexual reproduction 414

    22.1.1 Flower structure 414
    22.1.2 Development of pollen and ovules 416
    22.1.3 Fertilization and its consequences 420

22.2 The life cycle of a flowering plant 423

22.3 Pollination 424

    22.3.1 Mechanisms for ensuring outbreeding 424
    22.3.2 Pollination mechanisms 426

22.4 Asexual reproduction 428

    22.4.1 Parthenogenesis 428
    22.4.2 Vegetative propagation 428
    22.4.3 The consequence of asexual reproduction 429

22.5 The physiology of angiosperm reproduction 430

    22.5.1 The physiology of flowering 430
    22.5.2 Physiology of pollen growth 432
    22.5.3 Physiology of seed and fruit formation 433
    22.5.4 Physiology of fruit ripening 434

Summary 435

Questions 435

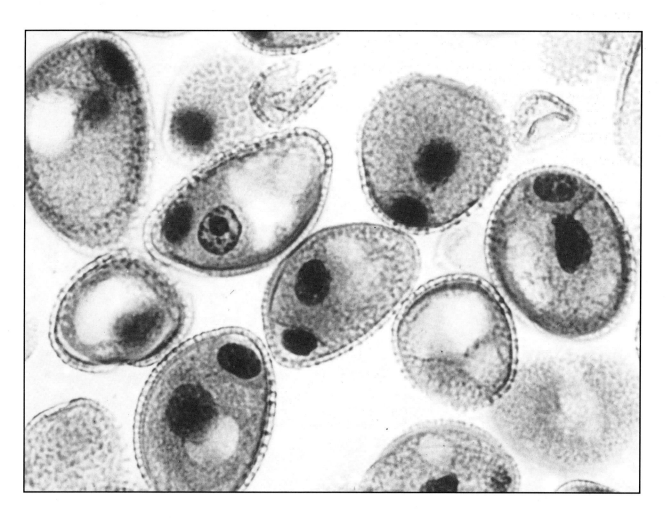

22 Reproduction in flowering plants

Reproduction is the formation of new individuals. It is the means whereby successive generations are produced resembling their parents to a greater or lesser extent. The transmission of hereditary material from parent to offspring is an essential feature of reproduction. In flowering plants the transmission can occur in one of two main ways. In **sexual reproduction** male and female sex cells called **gametes** are produced. The gametes come together and nuclear fusion takes place. The genome of a sexually produced organism is a mixture of genes from two sources. In **asexual reproduction** offspring are formed in a number of ways but fusion of nuclear material does not take place. Offspring produced asexually nearly always have an exact replica of the genome of their parent. Many species of flowering plants reproduce both sexually and asexually.

Man is highly dependent on the end products of flowering plant reproduction. Seeds and fruits, tubers and bulbs figure prominently in our diet and in the diets of our domesticated stock.

22.1 Sexual reproduction

In flowering plants male gametes develop in germinated **pollen grains**, female gametes in **ovules**. Pollen grains and ovules are produced in **flowers**. Most species are **hermaphrodite**, forming both kinds of gametes on the same plant, often in the same flower.

22.1.1 Flower structure

Following a period of vegetative growth, most flowering plants produce flowers. Annual plants flower in their first season of growth, biennials in their second season. After flowering both types die leaving seeds to germinate the following year. Perennial species generally flower every year when established. Some angiosperms, for example the tulip, produce large **solitary flowers** but others such as the dandelion have small flowers massed together in an **inflorescence**.

Flowers develop from apical meristems which previously gave rise to leafy shoots. Some of the factors which regulate the switch from a vegetative to a reproductive function are described in section 22.5. As a flower bud opens it is possible to see four kinds of floral organs in most flowers. On the outside are the **sepals**, usually green in colour. Collectively the sepals of a flower are called the **calyx**. Immediately inside the calyx are the **petals**, collectively called the **corolla**. The petals of many species of flowering plants are coloured. Yellow, blue and red are the most common colours in wild flowers. Next inwards are the **stamens**, each consisting of a **filament** and an **anther** in which pollen is produced. The term **androecium** is used to describe the collection of stamens in a flower. At the centre of the flower are found one or more **carpels**, collectively known as the **gynoecium** or **ovary**. Each carpel or group of carpels contains one or more ovules and has a **stigma** and a **style**.

The apex of the flower stalk from which the floral organs arise is called the **receptacle**. The numbers, size, colour and arrangement of the various organs vary from one species to another, although the flowers of closely related species are very similar in appearance. Fig 22.1 shows a range of flower structures of a few common British species.

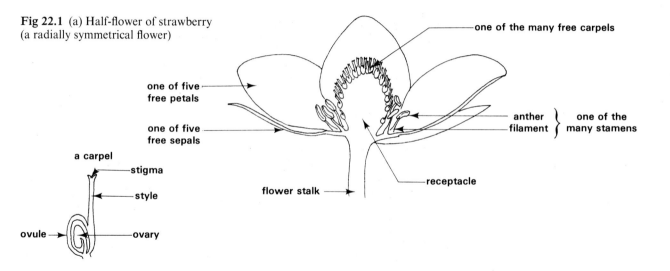

Fig 22.1 (a) Half-flower of strawberry (a radially symmetrical flower)

one of the many free carpels

one of five free petals

one of five free sepals

anther
filament } one of the many stamens

a carpel

stigma

style

ovule → ovary

receptacle

flower stalk →

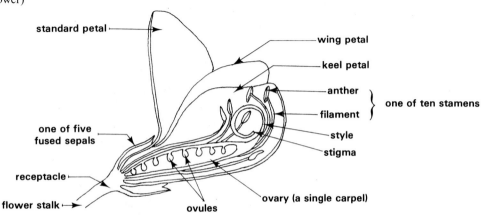

Fig 22.1 (b) Half-flower of broom (a bilaterally symmetrical flower)

standard petal

wing petal

keel petal

anther
filament } one of ten stamens

one of five fused sepals

style

stigma

receptacle

flower stalk →

ovary (a single carpel)

ovules

What is the essential difference between a radially and a bilaterally symmetrical flower?

Fig 22.1 (c) Flowers of dandelion

inflorescence of many small flowers (florets)

a floret

stigma

style →

corolla of five fused petals

tube of five fused anthers through which the style grows

calyx (hairs)

ovary of two fused carpels containing one ovule

22.1.2 Development of pollen and ovules

Sexual reproduction cannot occur without the formation of **gametes**. In flowering plants male gametes are formed from **pollen grains**. Female gametes are formed in the **ovules**.

1 Pollen formation

Fig 22.2 Development of anther and pollen
(a) TS developing anther

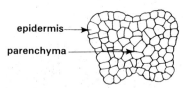

epidermis

parenchyma

Fig 22.2 (b) transverse section through a young anther of lily, × 40

Pollen production takes place in the anthers (Fig 22.2). Early in their development anthers are elongate masses of parenchyma tissue enclosed in an epidermis. In transverse section a young anther is seen to have two or four lobes according to the species. A row of cells extending the whole length of each lobe becomes organised into a central mass of **pollen mother cells** surrounded by several layers of flattened cells. The layer immediately surrounding the pollen mother cells is called the **tapetum**.

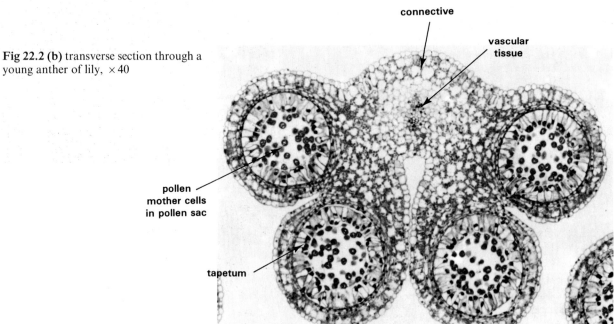

connective

vascular tissue

pollen mother cells in pollen sac

tapetum

Soon the mother cells undergo meiosis to produce tetrads of haploid **pollen grains**, also called **microspores**. Each pollen grain secretes a thick wall and its haploid nucleus divides by mitosis to form a **generative cell** and a **tube nucleus** (Fig 22.3).

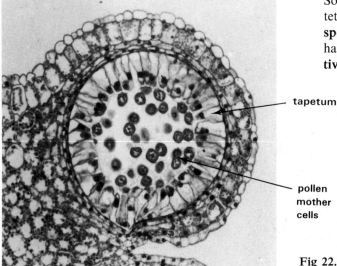

tapetum

pollen mother cells

Fig 22.3 Development of pollen grains of lily (contd. on p. 417)
(a) Pollen mother cell stage, × 70

416

Fig 22.3 (b) Tetrad stage, × 70

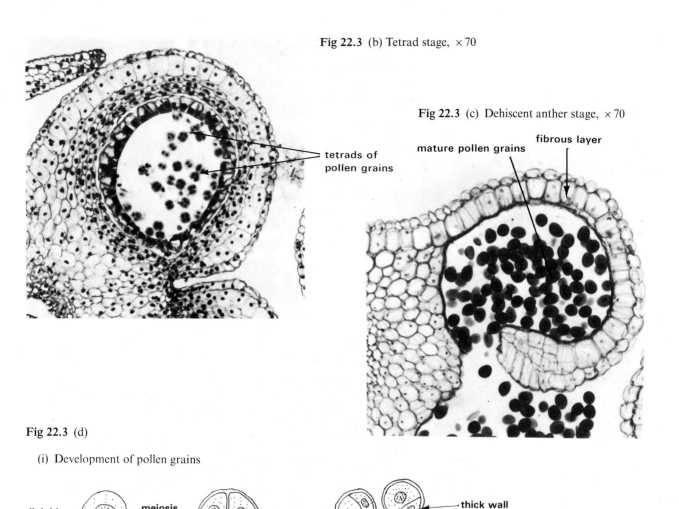

tetrads of
pollen grains

Fig 22.3 (b) Tetrad stage, × 70

Fig 22.3 (c) Dehiscent anther stage, × 70

mature pollen grains

fibrous layer

Fig 22.3 (d)

(i) Development of pollen grains

diploid
nucleus

microspore
mother
cell

meiosis

tetrad of
pollen grains,
each with a
haploid nucleus

thick wall

generative cell

tube nucleus

mature
pollen
grains

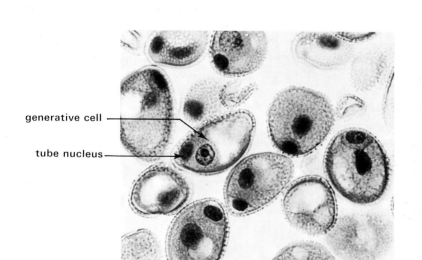

generative cell

tube nucleus

(ii) section through
mature pollen grains,
× 400

(iii) scanning electronmicrograph of a
pollen grain

While these changes are going on the anther grows considerably in size, and depending on the species, two or four pollen sacs become prominent. The cells around each sac become thickened with bands of lignin forming a **fibrous layer**. On drying out the fibrous layer shrinks causing the pollen sacs to rupture. Usually they open by splitting lengthwise down each side of the anther (Fig 22.4), though in some species pores or valves are formed instead. At this stage of its development the anther is **dehisced**.

Fig 22.4 TS dehisced anther

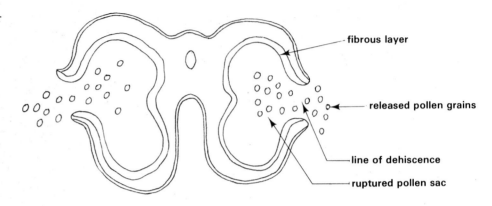

- fibrous layer
- released pollen grains
- line of dehiscence
- ruptured pollen sac

2 Ovule formation

Ovules are formed inside the carpels. Depending on the species, one or more ovules develop in each carpel. In all species, however, an ovule grows from a pad of carpellary tissue called the **placenta** to which it is attached by a short stalk called a **funicle**. At first the young ovule is just a bulge of parenchyma tissue called the **nucellus**. As it enlarges the nucellus is enclosed by two layers called **integuments** which grow up from the funicle. When fully developed the integuments completely surround the nucellus except for a tiny pore called the **micropyle**. Meanwhile, near the tip of the nucellus a cell divides by mitosis to form an outer **parietal cell** and an inner **megaspore mother cell**. The latter divides by meiosis to form a row of four haploid **megaspores**. Only the innermost megaspore usually develops further. It grows considerably in size and its nucleus undergoes three mitotic divisions. The eight haploid nuclei which are formed become arranged in a characteristic fashion (Fig 19.3). The resulting structure is called an **embryo sac** (Fig 22.5).

Fig 22.5 Development of an ovule of strawberry. Longitudinal sections through ovules at various stages of development (contd. on p. 419)

(a) young ovule in carpel

- stigma
- style
- ovule
- ovary
- funicle
- placenta

(b) appearance of megaspore mother cell

- megaspore mother cell
- parietal cell
- integuments
- nucellus

(c) formation of embryo sac

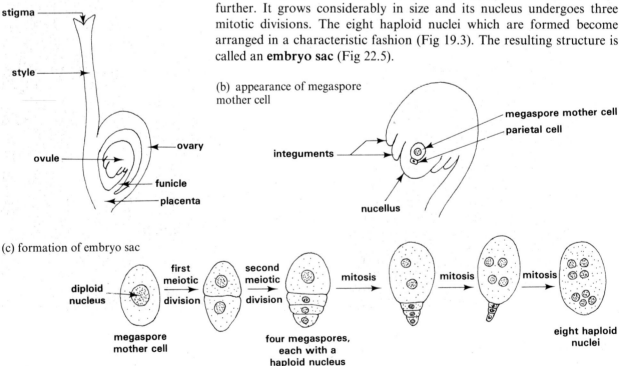

diploid nucleus — megaspore mother cell — first meiotic division — second meiotic division — four megaspores, each with a haploid nucleus — mitosis — mitosis — mitosis — eight haploid nuclei

(d) mature ovule

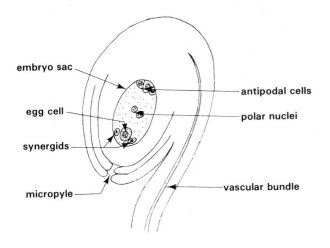

embryo sac

egg cell

synergids

micropyle

antipodal cells

polar nuclei

vascular bundle

(e) megaspore mother cell stage, ×250

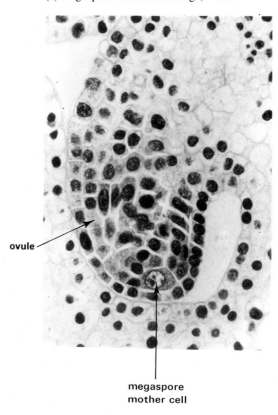

ovule

megaspore
mother cell

(f) embryo sac stage, ×125

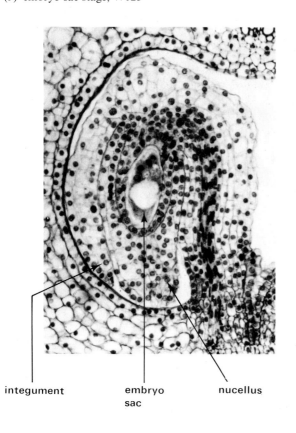

integument   embryo
sac

nucellus

(g) mature embryo sac, ×300, eight
nuclei stage

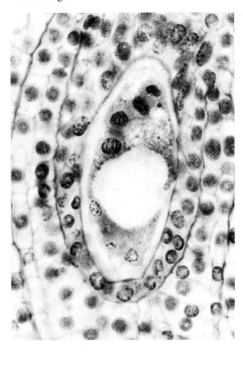

22.1.3 Fertilisation and its consequences

If fertilisation is to occur the male and female sexual cells must be brought together. The first step in bringing the gametes near to each other is **pollination** (section (22.3). Wind and insects are the main pollinating agents. They carry pollen from the anthers to the stigmas.

The **egg cell** of each embryo sac is a **female gamete**. Following the transfer of pollen to the stigma a **pollen tube** is formed from each pollen grain. As the tube grows down into the style the tube nucleus is situated at its tip. The generative cell then enters the tube and divides by mitosis to form two cells which are **male gametes** (Fig 22.6). In some species the male

Fig 22.6 (a) Germinating pollen grains

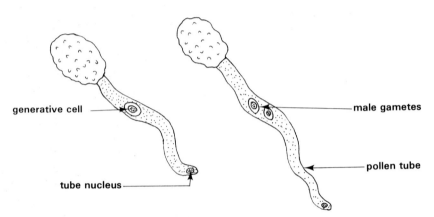

Fig 22.6 (b) Pollen grains germinating on the stigma of evening primrose, × 400

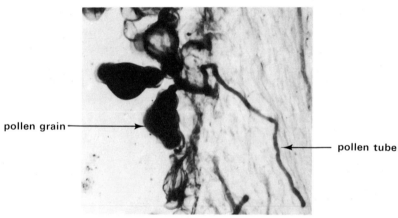

Fig 22.7 Fertilisation

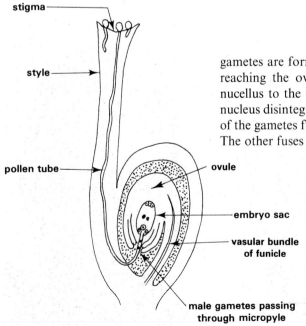

gametes are formed before the generative cell enters the pollen tube. On reaching the ovule the pollen tube grows through the micropyle and nucellus to the embryo sac where its tip opens out (Fig 22.7). The tube nucleus disintegrates and the two male gametes enter the embryo sac. One of the gametes fuses with the nucleus of the egg to form a **diploid zygote**. The other fuses with the diploid nucleus formed by the fusion of the two polar nuclei. The result is a **triploid fusion nucleus**. This **double fertilisation** is unique to flowering plants. Following fertilisation the ovules develop into **seeds** and the ovary of the flower into a **fruit**.

1 Seed formation

The subsequent events leading to seed production vary from one species to another. Essentially what happens is that the zygote develops into an **embryo plant** which is nourished as it grows by **endosperm** formed from the triploid fusion nucleus.

Fig 22.8 (a) Main stages in the development of an embryo of shepherd's purse

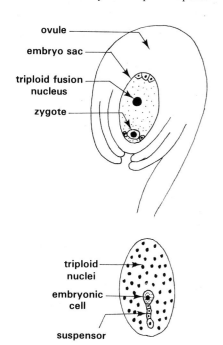

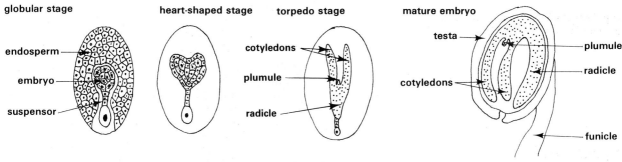

In shepherd's purse, mitosis of the zygote produces a short chain of cells. The first division is at right angles to the long axis of the ovule. The cell furthest away from the micropyle eventually becomes the embryo. The rest of the chain called the **suspensor** elongates, pushing the **embryonic cell** deeper into the embryo sac (Fig 22.8). Successive divisions of the triploid fusion nucleus produce a large number of triploid nuclei lying in the embryo sac. The nuclei may or may not become separated by cell walls forming a tissue called **endosperm** which surrounds the developing embryo. The embryonic cell divides to form a multicellular embryo with its **radicle**, the embryonic root, pointing towards the micropyle. In shepherd's purse the long axis of the mature embryo is bent. The two cotyledons with the **plumule**, the embryonic shoot, tucked between them point in the same direction as the radicle. During its development the embryo goes through globular, heart-shaped and torpedo stages. As the embryo grows the endosperm gradually disappears. Ultimately no endosperm remains so the seeds of shepherd's purse are **non-endospermic**. However, in some flowering plants such as grasses, some of the endosperm is still present when the embryo is fully developed. These seeds are described as **endospermic (**Chapter 23).

During the final stage in the formation of a seed some of the integument cells usually become lignified, forming a tough protective seed coat called the **testa** perforated only by the micropyle. Gradual dehydration of the entire seed takes place so that eventually as little as 5–10 % of its mass is water. The embryo is now in a state of **dormancy** (Chapter 23).

Fig 22.8 (b) Stages in the development of an embryo of shepherd's purse

(i) Globular stage, × 150   (ii) Torpedo stage, × 100   (iii) Mature embryo, × 100

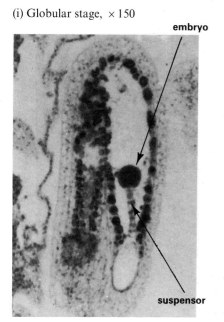

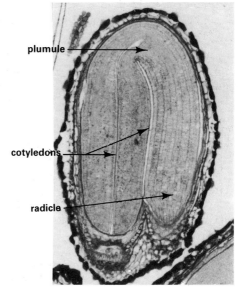

2 Fruit formation and seed dispersal

While the seeds are developing, changes take place in the surrounding carpels. Growth of the carpels keeps pace with enlargement of the developing seeds. At maturity the seeds are completely enclosed and protected by carpellary tissue called the **pericarp**. The gynoecium has now developed into a fruit, the structure of which depends on the number and arrangement of carpels in the flower and on the changes taking place after fertilisation.

In many species the pericarp dries out at the same time as the seeds. The **dry fruits** which result often open by valves or pores through which the seeds are released when the fruit is shaken by wind. The pods of leguminous plants such as gorse and broom open by splitting along the dorsal and ventral edges and the seeds are shot out (Fig 22.9(a)). In small-seeded dry

Fig 22.9 Fruits

(a) Broom

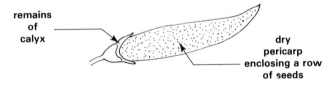

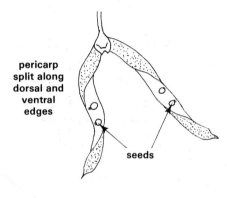

fruits the pericarp often has an extension enabling the fruit to remain airborne for some time after its release from the parent. Dandelion fruits, for example, have a parachute-like pappus (Fig 22.9(b)). In the dandelion and in many other flowering plants the seeds are not released. The pericarp splits open when the seeds germinate in the soil.

Some flowering plant species produce fleshy, **succulent fruits**. The fleshiness is sometimes due to growth of the pericarp as in plums and cherries. In other species another part of the flower enlarges instead. In the strawberry for example the receptacle swells considerably after fertilisation, pushing the tiny one-seeded carpels apart so that they are eventually scattered over the surface of the fruit (Fig 22.9(c)). Fleshy fruits are often part of the diets of many animals. Although the seeds may be eaten they come to no harm unless crushed by the animals' teeth. This is because they are protected by the tough pericarp which is undigested in its passage through the animal's gut. By the time the animal has egested the seeds in its faeces they have probably been carried some distance from the parent plant.

(b) Dandelion

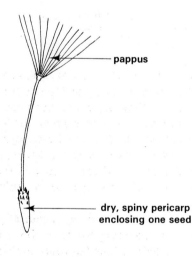

(c) Strawberry

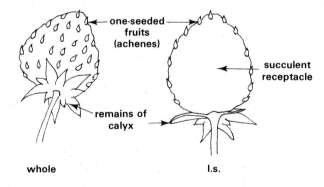

There are many other patterns of fruit development but the end-product is generally structured so that the seeds are self-dispersed or are dispersed by wind, animals, or, occasionally, water. Dispersal helps to avoid overcrowding and competition which occurs if a large number of seeds germinate in a confined area. It also provides opportunities for species to colonise territories where previously they have not grown and where they may thrive.

22.2 The life cycle of a flowering plant

The reproductive function of flowers involves the formation of two kinds of spores. Pollen grains are sometimes called **microspores** and the anthers in which they are produced can be called **microsporangia**. The larger **megaspores** are formed in the ovules or **megasporangia**. Thus a diploid spore-forming flowering plant is a **sporophyte** and produces two types of haploid spores. On developing further the spores give rise to gamete-bearing structures called **gametophytes**. The germinated pollen grain is a **male gametophyte** from which two male gametes arise whilst an embryo-sac is a **female gametophyte** in which a female gamete is formed. The life cycle is completed when an embryo sporophyte plant is produced in the seed following fertilisation (Fig 22.10).

Fig 22.10 Life-cycle of a flowering plant

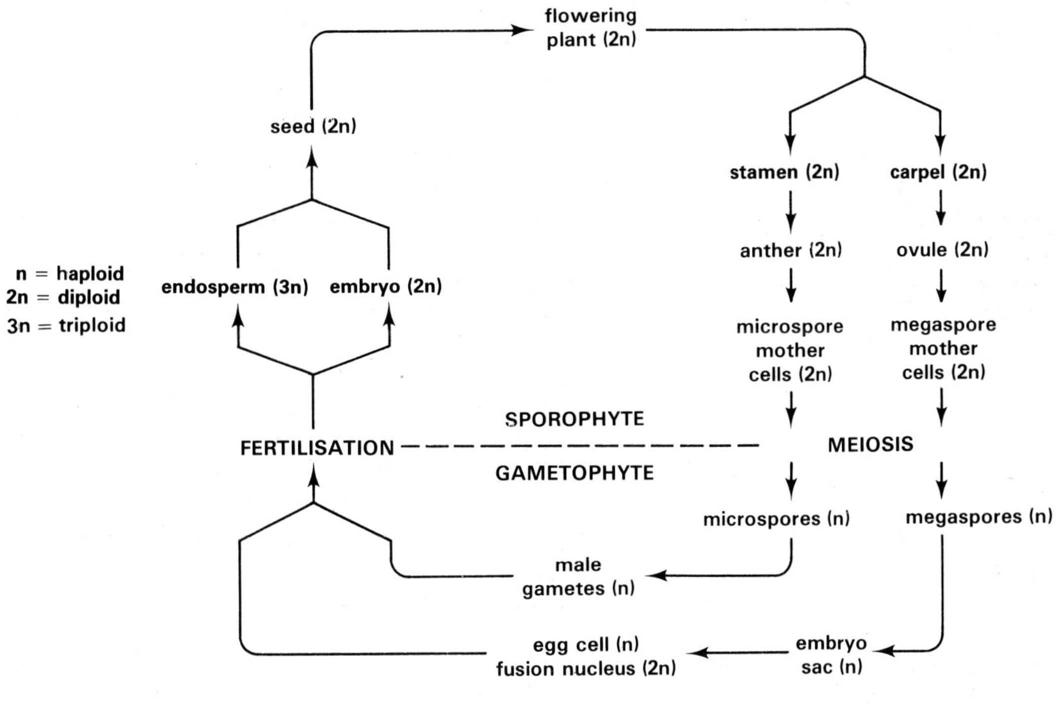

In the life cycle of a flowering plant there is thus an **alternation of generations**. Such an alternation of sporophyte and gametophyte is also found in other groups of terrestrial plants. In flowering plants, however, the gametophytes are vestigial and rely entirely on the sporophyte for their development. The situation is rather different in ferns where the sporophyte and gametophyte are independent plants and where the gametophyte has distinctive sexual organs. It is also different from the life cycle of a moss where the gametophyte is the free-living plant on which the sporophyte develops (Chapter 28).

A feature of flowering plant reproduction is the development of the young sporophyte inside the carpel of the parent sporophyte. There the young sporophyte is nourished and protected and thus has ideal conditions in which to develop. Because of this the chances that the next generation will be produced are very good. The development of seeds may be one of the reasons for the success of flowering plants compared with other groups of land plants.

22.3 Pollination

The transfer of ripe pollen from an anther to the stigma of a flower of the same species is called **pollination**. **Self-pollination** occurs when pollen is carried from an anther to the stigma of the same flower or to the stigma of another flower on the same plant. When pollen is transferred from one plant to another of the same species, **cross-pollination** has taken place.

The two forms of pollination have very different genetic consequences. Self-pollination leads to **self-fertilisation**, cross-pollination to **cross-fertilisation**. Self-fertilised species depend on random assortment and crossing over in meiosis leading to pollen grain and embryo sac production and on mutation to bring about variation in the genomes of male and female gametes. Self-fertilised species therefore display less **genetic variation** than cross-fertilised species which are produced from gametes from two different individuals. Plants which are self-pollinated are called **inbreeders** while cross-pollinated plants are called **outbreeders**.

Inbreeding has its virtues because it can preserve particularly good genomes which may be suited to a relatively stable environment. Outbreeding is of greater evolutionary significance because it continually produces a variety of genomes. In the struggle for survival some genomes are more successful than others. Although a few species never outbreed it is interesting, in view of its potential evolutionary advantage, to find that most flowering plants do. In fact many flowering plant species have evolved a variety of mechanisms to ensure that cross-fertilisation usually takes place.

22.3.1 Mechanisms for ensuring outbreeding

When the stamens and ovules of flowers do not ripen at the same time pollen has to come from another plant of the same species if fertilisation is to occur. In the wood-sage the anthers discharge their pollen before the style has completed its growth and the stigma is receptive. The condition is called **protandry** (Fig 22.11(a)). It contrasts with the behaviour of the rib-wort plantain in which the styles protrude from the flowers and the stigmas are ready to be pollinated before the stamens have grown. The latter

Fig 22.11 (a) Protandrous flowers of wood-sage

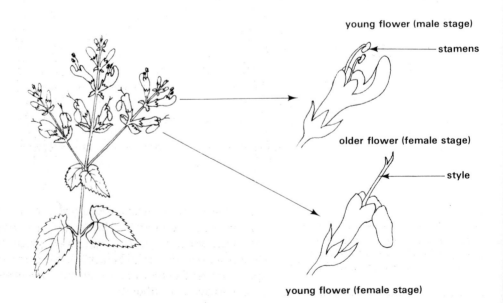

young flower (male stage)

stamens

older flower (female stage)

style

young flower (female stage)

condition is called **protogyny** (Fig 22.11(b)). Occasionally flowers of species which are normally hermaphrodite fail to produce pollen as a result of mutation of the gene or genes which regulate pollen formation. Such **male sterile** flowers have to cross pollinate if they are to produce seed.

Some angiosperms such as willows and poplars have separate plants for the production of pollen and ovules. They have no option but to cross-pollinate.

Fig 22.11 (b) Protogynous flowers of ribwort plantain

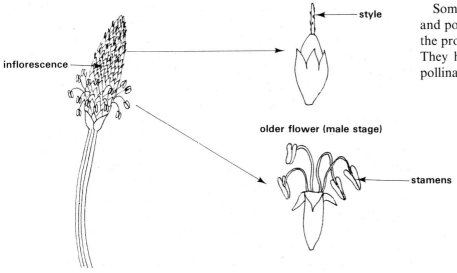
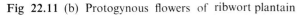

Fig 22.11 (b) Protogynous flowers of ribwort plantain

inflorescence

style

older flower (male stage)

stamens

The primrose is usually cross-pollinated by bees and butterflies. How are flower structure and insect mouthparts complementary in achieving pollen transfer between pin- and thrum-eyed varieties?

The primrose has two types of flower, pin-eyed and thrum-eyed (Fig 22.12). As you can see, the structure of the pin-eyed flowers makes self-pollination very unlikely. Self-pollination can readily occur in the thrum-eyed flower when pollen falls down the corolla tube. However, self-pollination in both pin- and thrum-eyed flowers does not result in self-fertilisation. This is because pollen will not germinate on the stigma of the flower in which it was produced.

Fig 22.12 Flowers of primrose

pin-eyed flower

thrum-eyed flower

stamens at top of corolla tube

stigma

stamens in corolla tube

style

calyx

ovary

A favourite variety of apple grown in many gardens in the UK is Cox's orange pippin. Which varieties of apple act as pollinators for it?

Like the primrose, many other flowers, although producing both pollen and ovules, are **self-sterile**. When transferred to the stigma, pollen will not germinate unless it comes from another plant of the same species. Self-sterility is a genetic trait (Chapter 26). It is the reason why many varieties of apple, pear and cherry will not set fruit unless another variety is growing nearby to act as a **pollinator**.

22.3.2 Pollination mechanisms

A number of agencies bring about pollination, the two most common being **wind** and **insects**.

1 Wind-pollinated flowers

The flowers of **wind-pollinated** species often show a complete lack of petals and sepals. They have no means of attracting insects (Fig 22.13(a)). The flowers are often borne in inflorescences which either appear before the leaves as in trees such as ash, and shrubs such as hazel, or well above the leaves as in grasses and some plantains. In this way the leaves do not interfere with air movements around the flowers.

Fig 22.13 (a) Wind-pollinated flowers of rye-grass

Fig 22.13 (b) Mature inflorescences of rye-grass. Notice the pendulous stamens

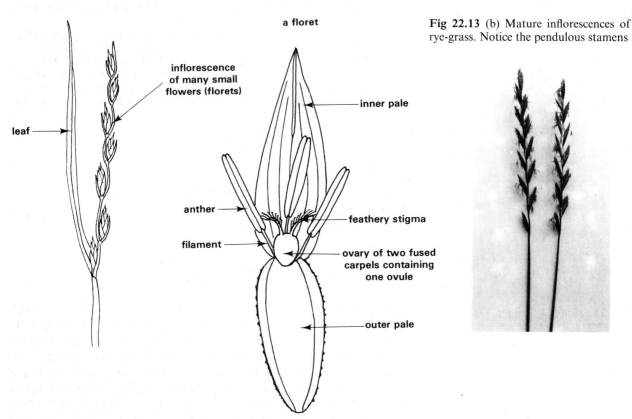

leaf

inflorescence of many small flowers (florets)

a floret

inner pale

anther

feathery stigma

filament

ovary of two fused carpels containing one ovule

outer pale

The long, feathery and often sticky stigmas which protrude into the air improve the chances of pollination. The anthers are often hinged on to the filaments and hang out of the flower where the slightest air movement shakes out the pollen (Fig 22.13(b)). The pollen grains are extremely light, often with air bladders, so that they remain suspended for long periods in the atmosphere. Many people suffer from hay-fever and similar allergies during the summer when the density of airborne pollen is high. Grass flowers make a large contribution to the pollen count.

How and for what purposes are pollen counts made? Where are the figures published?

2 Insect-pollinated flowers

One of the most remarkable facets of plant biology is the way in which certain insects and flowers of many species have evolved to their mutual advantage. **Insect-pollinated** flowers offer food. While taking it insects transfer pollen. Insects, like all animals, must have a balanced diet and many flowers provide carbohydrate, protein and lipids. The carbohydrate is found in **nectar**, a sugary solution secreted by special glands called nectaries which are often situated at the base of the petals. Pollen contains the other main dietary ingredients. Some insect-pollinated flowers such as poppies offer pollen only. Nectar and pollen have very little smell so are unlikely to be found by insects, except by chance. Flowers pollinated by insects have powerful insect attractants. Two senses which are highly developed in insects are smell and sight. It is not surprising therefore that insect-pollinated flowers are often highly scented. They are often relatively large and so easily seen. If small, they are grouped into conspicuous inflorescences. They are also brightly coloured. Significantly, the most common flower colours, yellow, blue and red, are those to which insects respond most readily.

The mouth-parts of the insects involved, such as bees and wasps, are adapted for sucking up nectar. Their legs are often modified to collect pollen (Fig 22.14). Often the insects are social creatures and in co-operating

Fig 22.14 Head and legs of worker bee

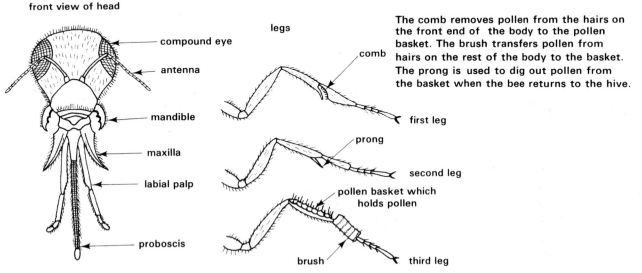

front view of head

compound eye

antenna

mandible

maxilla

labial palp

proboscis

The proboscis is used as a suction tube to draw nectar into the mouth.

legs

comb

The comb removes pollen from the hairs on the front end of the body to the pollen basket. The brush transfers pollen from hairs on the rest of the body to the basket. The prong is used to dig out pollen from the basket when the bee returns to the hive.

first leg

prong

second leg

pollen basket which holds pollen

brush

third leg

in their search for food they increase the chances of pollination. Although pollen is collected as food by insects, insect-pollinated flowers are generally so structured that some pollen is transferred to the stigma by a foraging bee or wasp. This condition is seen at its extreme in bilaterally symmetrical flowers. The structure of such flowers ensures that insects stand only in one position while feeding. Consequently pollen is dusted on to a specific site on the insect's body. In the white dead nettle it is the insect's back which receives the pollen. In gorse and broom it is its belly. Often the heavier bees operate a lever mechanism which brings the anthers down on to the body (Fig 22.15). When visiting another flower of the same species the pollen is then at just the right position to make contact with the stigma.

Fig 22.15 Pollen transfer in sage

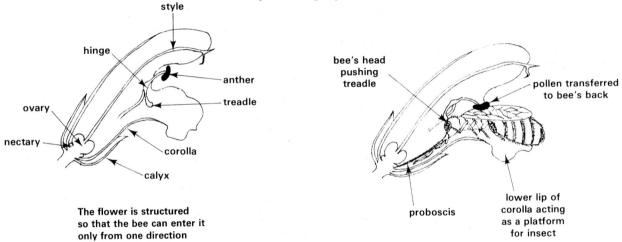

style

hinge

ovary

nectary

anther

treadle

corolla

calyx

The flower is structured so that the bee can enter it only from one direction

bee's head pushing treadle

pollen transferred to bee's back

proboscis

lower lip of corolla acting as a platform for insect

22.4 Asexual reproduction

Some plants which produce flowers can produce seeds without sexual fusion taking place as previously described. The production of seeds without fertilisation is called **parthenogenesis**. It occurs in a number of ways.

22.4.1 Parthenogenesis

Diploid parthenogenesis gives rise to seeds with normal diploid embryos. In the dandelion the megaspore mother cell invariably does not undergo meiosis and develops into an embryo without fertilisation. In the black-berry, embryos develop from a diploid cell in the nucellus.

Haploid parthenogensis results in the formation of seeds with haploid embryos. This occurs when an unfertilised egg cell grows into an embryo, or more rarely when a male gamete on reaching the embryo sac does so. Examples of this form of reproduction have been reported in the thorn apple and the evening primrose. However, because a complete set of chromosomes is lost the plants which grow from the seeds are sterile.

22.4.2 Vegetative propagation

Many angiosperms can produce offspring by means other than seed formation. In these plants progeny are formed by proliferation of part of the vegetative body of the parent. Species which have **vegetative propagation** usually flower too. The garden strawberry, for example, flowers in early summer and produces fruit. Later, lateral shoots from axillary buds on the parent plant grow over the soil surface. At the tip of each **runner** is a terminal bud (Fig 22.16). Along the length of the runner are scale leaves in the axils of which are axillary buds. One or more of the axillary buds produces roots which grow into the soil, and leaves grow up into the air. When the internodes of the runner decay several new plants are independent of the parent.

Fig 22.16 Vegetative propagation in strawberry

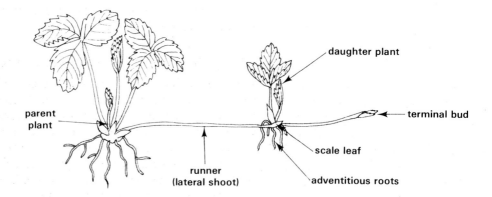

Many other methods of vegetative propagation are seen in wild and cultivated plants. Often propagation is coupled with a means of survival in adverse seasons. **Rhizomes, corms, bulbs** and **tubers**, for example, are frequently swollen with reserves of food which are used to establish new growth when the weather is suitable (Fig 22.17). As growth proceeds, lateral buds develop into daughter organs which later give rise to new plants.

Fig 22.17 Organs of perennation and vegetative propagation

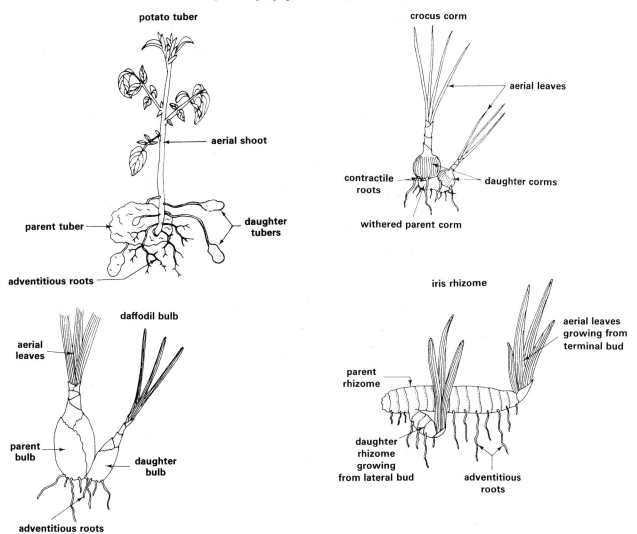

The various forms of asexual reproduction observed in flowering plants are often collectively called **apomixis**.

22.4.3 The consequences of asexual reproduction

In section 22.3 the differences between inbreeding and outbreeding were described and the genetic consequences were explained. All forms of asexual reproduction are examples of **inbreeding**. They are a means whereby genomes are passed with little or no change from parent to progeny. Asexual reproduction gives less scope for genetic variation than self-fertilisation where some change in the genomes of gametes will result from random assortment and crossing-over in meiosis. There is a tendency therefore for asexual reproduction to give rise to progeny of genetic uniformity. Such a group of plants is called a **clone**. Even so, the ability of many cultivated plants to reproduce by vegetative propagation is of great advantage to humans. It ensures that the genomes of useful varieties of plants are kept unchanged. Nevertheless occasional mutants, called **sports** by gardeners, appear among plants which propagate vegetatively.

Very few species of wild plants reproduce by asexual methods only. However, plants which can produce asexually and sexually are more versatile than those which rely entirely on sexual reproduction. It means that progeny can be produced even when seed production fails.

In recent times **tip propagation** has been used to asexually propagate plants such as orchids which are difficult to rear by natural means. The technique involves aseptic removal of growing shoot apices and transferring them to a sterile jelly containing the required nutrients and hormones for the tips to grow into juvenile plants. In this way the risk of disease being carried from one generation to the next is greatly minimised.

22.5 The physiology of angiosperm reproduction

Earlier in the chapter an orderly sequence of events beginning with the development of flowers and ending with the formation of seed-containing fruits is described. We will now examine some of the reasons for the changes. What causes a vegetative shoot to produce flowers and what controls the development of fruits and seeds? Why does a pollen tube grow towards an embryo sac when a pollen grain terminates on the stigma?

22.5.1 The physiology of flowering

The flowers of wild and cultivated plants grown outdoors appear at a particular time of the year. Daffodils, snowdrops, crocuses and hyacinths produce flowering shoots in early spring. Grasses and roses flower in summer, chrysanthemums and Michaelmas daisies flower in the autumn. This suggests that climate controls the times of flowering in different species. Among the more obvious seasonal changes of climate in temperate zones are **day length** and **temperature**.

1 Effect of day length

The number of hours of light each day varies from one time of the year to another. In midwinter the sun rises at about 8 am in southern England and dusk occurs at about 4.30 pm. This contrasts with midsummer when dawn is at 5 am and sunset at 10 pm. British vegetation is thus exposed to little more than 8 hours of sunlight each day in winter and as much as 17 hours in summer. From December to June the number of daylight hours gradually increases while from June to December they gradually decrease.

The effect of day length in controlling the time of flowering was discovered in America in the 1920s. It was found that a variety of tobacco flowered much more readily if the plants were given as little as five hours light a day compared with longer periods of light. The light had to be above a certain intensity for it to be effective. Since then it has been shown that many other species respond in the same way. They flower after exposure to several days when the number of light hours or **photoperiod** is small. Other species flower only after exposure to a number of days with long photoperiods. A third group are unaffected by day length. Thus we have three categories of **photoperiodic response**:

i. Short-day plants which flower after a number of photoperiods of less than twelve hours, as occurs in Britain during autumn and winter. This group of plants includes chrysanthemums, rice and cotton.

ii. Long-day plants which flower only after a succession of long photoperiods of more than twelve hours. In this group are many important crop plants such as wheat, clover and lettuce.

iii. Day-neutral plants which produce flowers regardless of the length of photoperiod to which they are exposed. They include tomato, cucumber and dandelion.

The number of photoperiods required to induce the formation of flowers varies from one species to another. The cocklebur requires only one **inductive photoperiod**. Most plants need about ten consecutive photoperiods of suitable length. It may then take several days or even weeks for vegetative apices to change into floral apices. A plant must be at a certain stage in its development when it receives the photoperiod treatment if eventually it is to flower. In some species the plant can be at the seedling stage. In others several mature leaves must be present.

Commercial growers use time switched lighting systems to regulate day length indoors. In this way they can get plants to flower out of season. Visit a local florist, nursery or garden centre at several times in a year and make a list of plants in flower at times of the year other than when they flower naturally outdoors.

The length of the photoperiod is less critical than the length of the **dark period** in a daily cycle. If the photoperiods are interrupted with short periods of darkness, flowering still follows. However, if the dark period is interrupted by as little as one minute's exposure to light, flowering is prevented. Red light is most effective in this respect. Yet the effect of red-light treatment can be overcome if the plant is immediately exposed to infra-red light. This suggests that a flower-inducing substance is made in darkness and is broken down by red light. Flowering plants produce a pigment, **phytochrome**, which is affected in this way (Chapter 23). Phytochrome may therefore play a part in regulating flowering. In what way is still unknown, but phytochrome may stimulate production of gibberellic acid (GA) or may inhibit the production of abscissic acid. GA has been shown to enhance flowering when applied to long-day plants grown in short days whereas abscissic acid prevents such plants from flowering, even when raised in long days. Grafting experiments strongly indicate that hormones are involved in regulating flowering and that the leaves perceive the stimulus of day length (Fig 22.18).

Fig 22.18 Some aspects of photoperiodism in the chrysanthemum

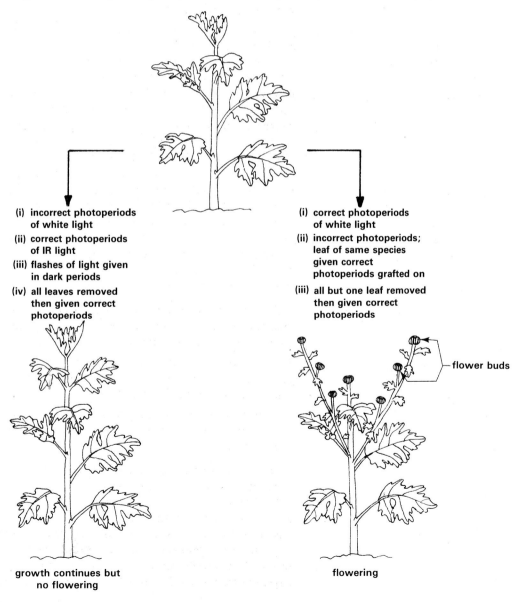

(i) incorrect photoperiods of white light

(ii) correct photoperiods of IR light

(iii) flashes of light given in dark periods

(iv) all leaves removed then given correct photoperiods

growth continues but no flowering

(i) correct photoperiods of white light

(ii) incorrect photoperiods; leaf of same species given correct photoperiods grafted on

(iii) all but one leaf removed then given correct photoperiods

flower buds

flowering

431

2 Effect of temperature

As well as a suitable photoperiod treatment some plants require an appropriate temperature treatment if they are to flower. Winter varieties of wheat and barley sown in autumn in Britain will flower the following year, but if sown in spring they do not flower unless the seedlings are exposed to a temperature of between 1 and 5 °C for several days. This shows that cold weather has an effect in regulating flowering of some species. Low temperature treatment induces flowering in many biennial species. In the first year of growth biennial plants such as the carrot and parsnip grow vegetatively and build up a reserve of food which is stored in their swollen taproots. Next year, if the plants are left out of doors, a flowering shoot is produced using the stored food (Fig 22.19). However, if they are kept over winter in a warm greenhouse they continue to grow vegetatively.

The cold-requirement for initiation of flowers is called **vernalisation**. How exposure to low temperatures promotes flowering is still poorly understood. It is likely, however, that it changes the hormone balance in the plant, possibly by promoting gibberellic acid production. Biennial plants treated with gibberellic acid will flower without vernalisation.

It is possible to investigate the involvement of hormones in flowering using tissue culture techniques. When floral apices are grown in a culture medium, floral organs differentiate if auxin, gibberellic acid and cytokinin are added (Chapter 24). The cucumber has been much studied in this respect. Application of auxin to male flower buds causes them to produce female flowers. Gibberellic acid has the opposite effect. Studies of this kind suggest that protrandry, protogyny and the existence of separate male and female flowers on the same plant may be due to changes in hormone balance at different stages of growth.

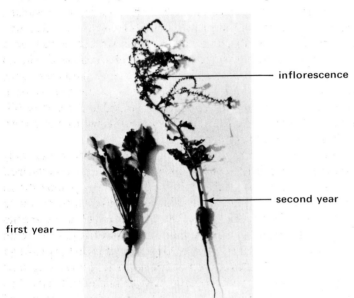

Fig 22.19 A beeetroot plant in the first and second year of its growth

inflorescence

second year

first year

22.5.2 Physiology of pollen growth

A lot has yet to be learned about the reason why pollen tubes transfer male gametes with such accuracy into embryo sacs. The very fact that the tips of pollen tubes so frequently find the micropyle suggests that the nucellus or embryo sac secretes substances to which the pollen tube is sensitive. Growth of the tube is probably an example of **positive chemotropism**. However, there has been little success in identifying the substance or substances involved. More is known of the conditions which encourage germination of pollen grains. Whereas the pollen of some species will germinate in pure water, others need the presence of sugars, amino acids and other ingredients all of which are present in stigmatic and stylar tissues. The required concentrations of these substances is often quite high. Up to 15 % sucrose is needed in some cases. The trace element boron also stimulates pollen tube growth (Chapter 14).

Early growth of the pollen tube uses materials stored in the pollen grain. There is evidence that digestion of stylar and carpellary tissues by pectinase and cellulase enzymes takes place when the pollen tube grows further. Even so, pollen may fail to germinate despite successful pollination. Incorrect amounts of stimulants in the stigma and style, secretion of inhibitory materials by the stigma, style and ovules and inability of the pollen tube to produce enzymes which help the pollen tube penetrate the stigmatic surface are among the reasons for this.

A better understanding of the physiology of pollen tube growth may enable plant breeders to break down sterility barriers between varieties of flowering plant species. By bringing together genetic material from two different sources some of the hybrid progeny would be superior to existing varieties. This could mean higher yields of crop and go some way to alleviate the existing world food shortage.

22.5.3 Physiology of seed and fruit formation

Some information of the conditions required for embryo growth has been obtained by analysing the endosperm which surrounds the embryo during its natural development. The coconut produces a large volume of liquid endosperm, commonly called coconut 'milk'. The milk is rich in sugars, minerals and plant hormones such as cytokinins and auxin. Recently young embryos of shepherd's purse have been grown successfully in a synthetic mixture of sucrose, various minerals, indole-acetic acid, kinetin and adenine. As the embryos develop they become less reliant on organic substances in their surroundings. Older embryos continue to develop if supplied with carbon dioxide, water, minerals and light. Other than this and the fact that embryos respire actively as they grow, little is known about the physiology of embryo development.

Simultaneous rapid growth of ovules and pericarp occur immediately after fertilisation. This indicates that growth of the various parts of the fruit are synchronised. Expansion of the fruit is stimulated by pollination in some species (Fig 22.20). If auxin or gibberellic acid is sprayed on to unpollinated flowers, fruit will form. Yet the pollen of most species contains only a little auxin. It appears that pollination triggers off auxin synthesis in the ovules and carpels so that growth of seed and pericarp go 'hand in hand' (Fig 22.21). Production of hormones in the growing fruit also inhibits the development of an abscission layer in the fruit stalk. The ovaries of unpollinated flowers usually fall away like the leaves of deciduous trees in autumn.

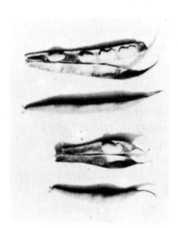

Fig 22.20 Effect of seed growth on growth of pericarp. In the small pod only one ovule was fertilised. Several ovules were fertilised in the larger pod.

Fig 22.21 Growth of a plum

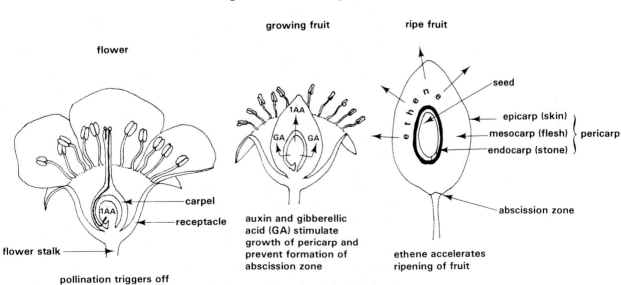

433

22.5.4 Physiology of fruit ripening

The pericarps of dry fruits quickly lose water when the growth is finished. Drying is usually accompanied by lignification of the pericarp, forming a structure which helps disperse the seeds and protects them until they germinate. The ripening of fleshy fruits is accompanied by many chemical changes in the swollen pericarps or receptacles, making them attractive and palatable to animals. Unripe fruits are generally sour because they contain high concentrations of carboxylic acids. On ripening, the acids together with starch are converted to sweet-tasting sugars. Simultaneously the fruit becomes softer as pectinase and cellulase enzymes digest cell wall materials such as pectin and cellulose. The extent to which this occurs varies between species. In ripe plums for instance there is a lot of unchanged pectin while in cherries there is little. For this reason jam made from plums usually sets readily, the pectin acting as a gelatinising agent, but cherry jam does not set unless extra pectin is added.

Pigment changes are also common. Green unripe fleshy fruit become red, yellow and orange as chlorophyll is broken down and xanthophylls and carotenes predominate. Volatile aromatic oils may be synthesised giving the fruit a pleasant odour. The changes make the fruit attractive to some animals which help in dispersal of seeds inside the fruit.

The many chemical conversions use energy from ATP made during respiration. Ripening fruit thus show a marked increase in respiratory rate. They also exhibit an increased output of ethene (ethylene) gas which accelerates ripening.

SUMMARY

The organs of sexual reproduction in flowering plants are flowers. They are concerned with producing pollen from which male gametes arise and ovules in which female gametes are formed. Fertilisation results in the formation of seeds, each containing an embryo plant with a store of food. Because the developing embryos are attached to the parent plant and are enclosed in a protective tissue, the pericarp, their chance of reaching maturity is high.

Various strategies have evolved which enhance the probability of fertilisation in flowering plants. They are concerned with ensuring that pollen is brought into contact with the stigma where it will germinate to deliver male gametes to the ovules. Wind and insects are particularly effective in bringing about cross-pollination which leads to cross-fertilisation, hence genetic variability among seeds. Variation is of evolutionary significance; it is the raw material on which natural selection works to preserve better genomes.

The life cycle of flowering plants which reproduce sexually involves an alternation of generations. In contrast to other groups of plants such as the bryophytes the gametophyte generation is suppressed, the sporophyte assuming a dominant role.

Many species of flowering plants can reproduce asexually too. Corms, bulbs and tubers are examples of organs of asexual reproduction. They also serve as a means of survival in periods of unfavourable weather. Asexual reproduction generally gives rise to offspring which are genetically and hence phenotypically similar to their parents. This is of advantage as it preserves genomes which are suited to a given environment. Many ornamental plants and some important crops are raised by asexual methods of reproduction.

The numbers of hours darkness in a day is a vital factor in determining the time of year different species come into flower. In this context there are three categories of plants – short day, long day and day neutral. Some species also require a period of cold weather before they will flower. Experimental evidence suggests that hormones may be involved in regulating the flowering response.

QUESTIONS

1 The flowers of the primrose, *Primula vulgaris*, are polymorphic. The two most common forms occur on separate plants and are termed 'pin-eyed' and 'thrum-eyed'. In most localities, pin-eyed and thrum-eyed plants are found in equal proportions. The flowers are pollinated by insects, such as bees, which have a long proboscis or tongue. The bee pushes its head into the mouth of the corolla tube and the proboscis is extended to collect the nectar at the base of the flower.

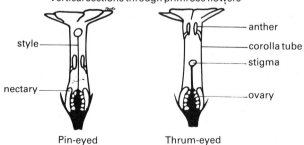

Vertical sections through primrose flowers

Pin-eyed   Thrum-eyed

(a) Explain how the polymorphism is likely to *increase* the variability of the offspring of primrose plants.

Pin-eyed and thrum-eyed primroses were pollinated by hand. In addition, measurements were made of the cross-sectional area of the styles of pin-eyed and thrum-eyed flowers. The results are given in the tables below.

Pollen source (anther)	Stigma pollinated	Mean number of seeds per pollinated flower
Pin	Thrum	106.1
Thrum	Pin	64.9

	Mean cross-sectional area of styles (μm^2)
Pin	127.5
Thrum	284.8

(b) Assuming that in all cases pollination involved ample quantities of pollen and that the two types of pollen germinated equally successfully, suggest **two** hypotheses which could account for the larger number of seeds set in thrum-eyed flowers.

Thrum-eyed and pin-eyed flowers were then self-pollinated, with the following results.

Cross	Mean number of seeds per pollinated flower
Thrum × Thrum	3.6
Pin × Pin	12.5

(c) From these results it was suggested that a chemical

produced by the stigma of one type of flower is needed to stimulate the germination of pollen from the other kind of flower. Briefly outline how you could test this hypothesis.

In the inheritance of the primrose polymorphism, the thrum-eyed characteristic (S) is dominant to pin-eyed (s).

(d) (i) From all the information given in this question, state the *most* likely genotype of primrose plants with thrum-eyed flowers.

(ii) Explain, giving **one** reason only, how you reached your answer to (d) (i). (Nuffield/JMB)

2 (a) The diagram shows a half flower.

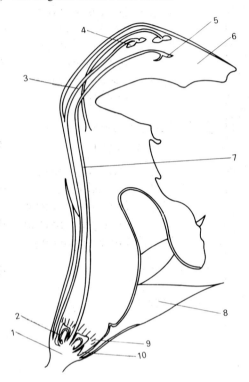

(i) Name parts **1** to **10**.

(ii) Describe **five** visible features in the flower in the diagram which you would expect to be different or absent in wind-pollinated flowers.

(b) Explain where each of the following **four** parts of an angiosperm flower develops:

 (i) microspore,

 (ii) megaspore,

 (iii) male gamete,

 (iv) female gamete.

(c) *Zostera* is a marine angiosperm whose flowers are *submerged*. Give **two** ways in which you would expect its pollen grains to differ from those of terrestrial insect-pollinated angiosperms. (AEB 1986)

3 (a) Describe the particular features of flowers which represent adaptations to pollination by (i) wind and (ii) insects.

(b) (i) Describe the events from pollination up to and including fertilisation, giving the factors which regulate pollen germination and pollen-tube growth.

(ii) Explain how in some instances fruit development may fail to occur after pollination has taken place and in other instances fruit development can take place without pollination. (JMB)

4 A biologist was investigating the reproductive biology of two closely related species of flowering plant referred to as Species **A** and Species **B**.

One of the studies concerned pollen production. For this, the ripe anthers were sectioned and examined microscopically. Pollen dimensions were also noted. Figure 1 shows a typical half anther of each species and gives some data about the pollen.

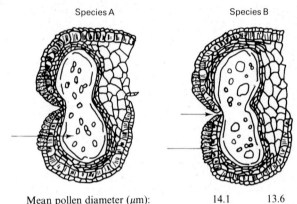

	Species A	Species B
Mean pollen diameter (μm):	14.1	13.6
Standard deviation of the mean (μm):	1.2	4.9

(a) On the diagrams label the tissues or regions shown by arrows.

(b) What do you understand by the term standard deviation and why, in this case, is it useful to have this information?

(c) On examining root tip squash preparations the biologist noted that the chromosome number of each species was

 Species **A** 14

 Species **B** 21

Relate this information to the variation in size of the pollen grains in Species **B** as compared with Species **A**.

(d) Next, the biologist made aqueous extracts of the stigmas of each species and placed pollen grains from each in both extracts. Some of the pollen grains germinated (Figure 2).

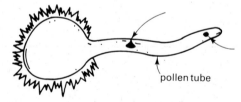

pollen tube

(i) Label the two nuclei shown in Figure 2.

(ii) Explain whether you consider this pollen to be wind- or insect-distributed.

(e) Data concerning the percentage germination of the grains in the stigma extracts are shown in the table.

	Percentage germination of pollen in the stigma extract of the species shown	
	Stigma extract from A	Stigma extract from B
Pollen from Species A	95	0
Pollen from Species B	0	0

Briefly suggest explanations for the results found for (i) pollen from Species A, (ii) pollen from Species B. (JMB)

23 Seed structure and germination

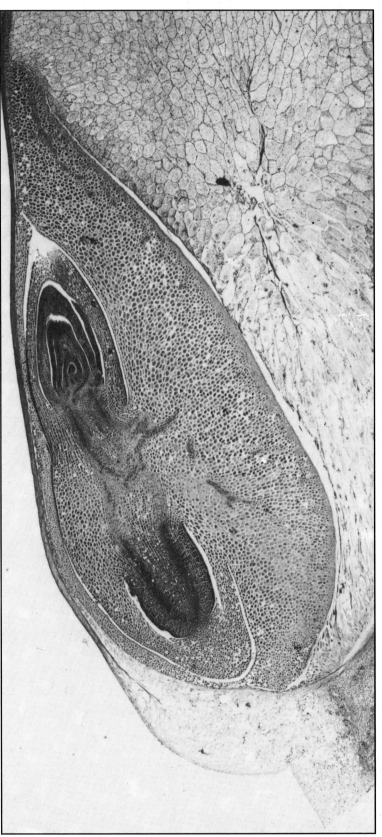

23.1	Seed structure	438
23.2	Morphological aspects of seed germination	438
	23.2.1 Germination of broad bean	438
	23.2.2 Germination of sunflower	440
	23.2.3 Germination of maize	441
23.3	Physiology of germination	442
	23.3.1 Mobilization of stored food	442
	23.3.2 Factors affecting seed germination	444
	23.3.3 Seed dormancy	448
Summary		449
Questions		450

23 Seed structure and germination

Few species of flowering plants fail to produce seeds. Indeed, the evolutionary success of flowering plants can to some extent be attributed to seed production. New genomes arise when seeds are produced. Seeds also provide a resting stage which can be maintained for long periods. It is not uncommon for seeds to germinate after being stored for up to fifty years. The longest life span for seeds recorded to date is that of the Arctic lupin, *Lupinus arcticus*. Seeds of this species, discovered in frozen silt at Miller Creek, Yukon, Canada, in 1954, were successfully germinated. Radiocarbon dating methods showed the seeds to be 10 000 to 15 000 years old. The longevity of seeds is due mainly to their extremely low metabolic rate and to the presence of a reserve of energy in the form of stored food. Humans have long known of the food reserves in seeds and a great deal of time and effort has been spent in improving and cultivating crops such as cereals and legumes which produce edible seeds.

Although biologists are a long way from a complete understanding of many aspects of seed germination we shall see that the physiology of seeds is largely controlled by their environment.

23.1 Seed structure

A **seed** is an embryo plant together with its store of food. The whole structure is encased in a seed coat called the **testa**. The **embryo** consists of an immature shoot, the **plumule**, and an undeveloped root, the **radicle**. In the seeds of monocotyledonous plants such as grasses and maize the embryo is attached to a single seed leaf called the **cotyledon**. There are two seed leaves in dicotyledonous plants. The food reserve is stored in swollen cotyledons in sunflower and bean seeds. In maize, food is stored in a separate tissue called **endosperm** (Figs 23.1, 23.2 and 23.3).

23.2 Morphological aspects of seed germination

Germination is the first step in the development of a mature plant from the embryo of the seed. The appearance of a seedling plant marks the end of germination. However, the way in which this happens differs from one species to another. The various patterns of germination can best be appreciated by studying the germination of seeds of several species.

23.2.1 Germination of broad bean

The seeds of the broad bean develop as a single row inside an elongated pod-shaped **pericarp**. Each seed is attached to the pod by a stalk, called the **funicle**, at the base of which is a tiny hole, the **micropyle**. When a seed is detached from its funicle a scar called the **hilum** is seen at one end of the testa. Each seed has two very large cotyledons packed with starch and protein. The embryo is tucked between the cotyledons to which it is attached by two short stalks (Fig 23.1(a)).

Fig 23.1 (a) Seed of broad bean

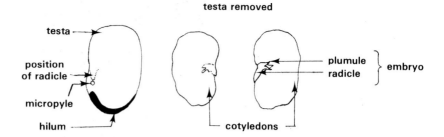

testa removed

If placed in moist soil at a suitable temperature the seed quickly swells as it absorbs water. Shortly afterwards the radicle elongates and pushes its way through the micropyle, bursting open the testa as it does so. As the radicle extends down into the soil a number of lateral roots appear, forming a branched root system. While the radicle emerges the base of the plumule called the **epicotyl** also elongates and becomes hook-shaped as it grows between the cotyledons into the surrounding soil. Eventually the plumule reaches the soil surface where the epicotyl straightens and grows vertically. The embryonic leaves expand and develop into the large green compound leaves typical of the broad bean.

The cotyledons gradually shrink as the food reserves are used up. Such a pattern of germination where the cotyledons remain below the surface of the soil is called **hypogeal** (hypo = below; ge = earth) (Fig 23.1(b)).

Fig 23.1 (b) Germination of broad bean seed

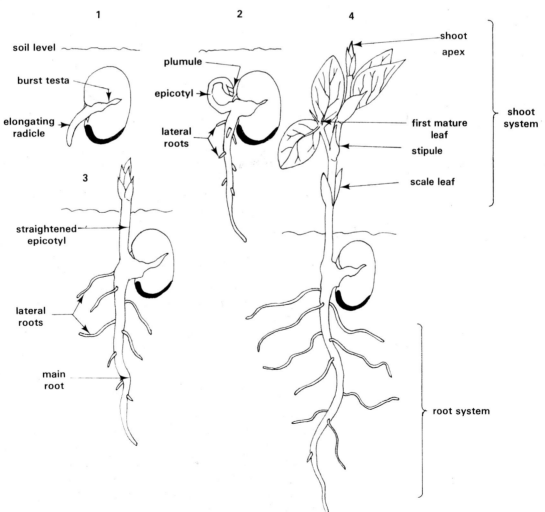

Fig 23.2 (a) Seed of sunflower

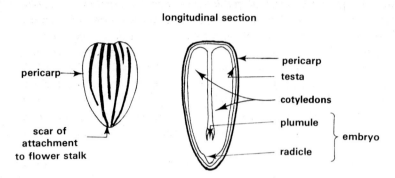

longitudinal section

pericarp →

scar of
attachment
to flower stalk

pericarp
testa

cotyledons

plumule

radicle

} embryo

23.2.2 Germination of sunflower

Each sunflower seed is surrounded by a testa and a leathery pericarp and is therefore a one-seeded fruit called an **achene**. At the tapering end of the achene there is a small scar showing where it was attached to the flower stalk during its development. As in the broad bean there is no endosperm. The food reserve mainly of protein and oil is stored in the swollen cotyledons (Fig 23.2(a)).

Fig 23.2 (b) Germination of sunflower seed

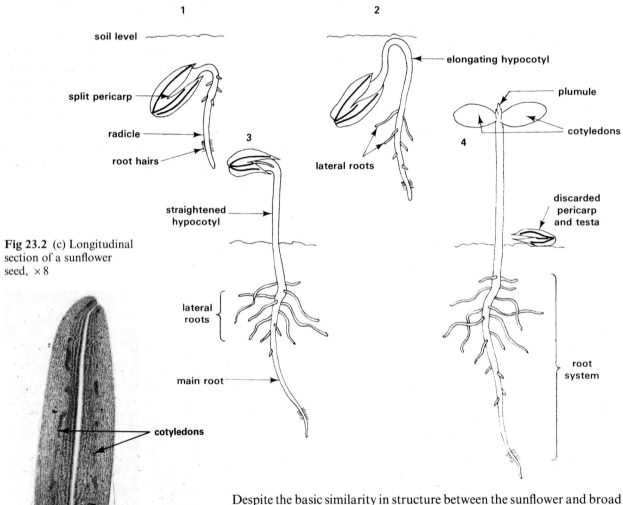

soil level

split pericarp

radicle

root hairs

straightened
hypocotyl

lateral
roots

main root

elongating hypocotyl

lateral roots

plumule

cotyledons

discarded
pericarp
and testa

root
system

Fig 23.2 (c) Longitudinal
section of a sunflower
seed, ×8

cotyledons

plumule

radicle

Despite the basic similarity in structure between the sunflower and broad bean seeds there are differences in the patterns of germination. After the absorption of water and subsequent emergence of the radicle, the **hypocotyl**, the part of the embryo between the cotyledon stalks and the radicle, begins to elongate. As it grows it becomes curved and the cotyledons with the tiny plumule tucked between them are dragged above soil level where the hypocotyl straightens. The cotyledons, by now shrunken because their food reserves have been used up, become green and start to photosynthesise. At this stage the plumule elongates and the first simple mature leaves appear. This type of germination where the cotyledons appear above soil level is called **epigeal** germination (epi = above) (Fig 23.2(b)).

Fig 23.3 (a) Seed of maize

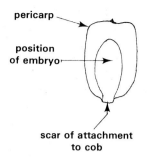

pericarp

position of embryo

scar of attachment to cob

23.2.3 Germination of maize

A maize seed is enclosed in a tough pericarp which is fused with the testa. This type of one-seeded fruit is called a **caryopsis** and develops with several hundred others on a cob. At the tapered base of the fruit is a scar showing its point of attachment to the cob. On one of the broader sides of the fruit a lighter oval area can be seen under which the embryo is found. The radicle is enclosed in a hollow tube called the **coleorhiza**. The plumule is surrounded by a similar structure called the **coleoptile**. There is a small triangular cotyledon, the **scutellum**, to one side of which the embryo is attached. At its other side the scutellum is fused to the endosperm which stores starch, protein and oil (Fig 23.3(a)).

Soon after water uptake the radicle emerges, piercing the coleorhiza as it grows. Shortly afterwards the coleoptile appears, enclosing the plumule as it grows upwards through the soil. Growth of the coleoptile is mainly from its base called the **mesocotyl**. Once through the soil surface the coleoptile soon stops growing and the first long, strap-shaped leaf bursts through its tip. Adventitious roots grow from the mesocotyl and form a fibrous root system. Because the cotyledon remains below the soil level the germination of maize is **hypogeal** (Fig 23.3(b)).

Fig 23.3 (b) Germination of maize seed

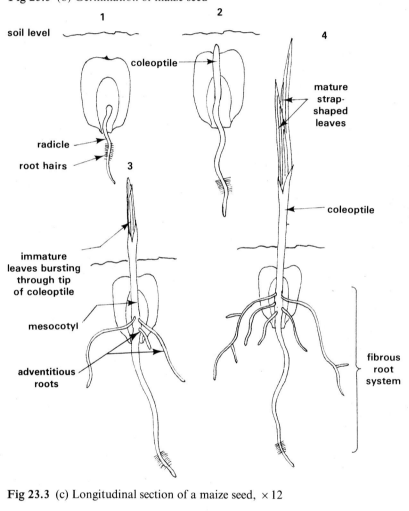

soil level

1

2

4

coleoptile

mature strap-shaped leaves

radicle

root hairs

3

coleoptile

immature leaves bursting through tip of coleoptile

mesocotyl

adventitious roots

fibrous root system

Fig 23.3 (c) Longitudinal section of a maize seed, × 12

longitudinal section

fused pericarp and testa

endosperm

cotyledon (scutellum)

coleoptile

plumule

radicle

coleorhiza

embryo { plumule radicle

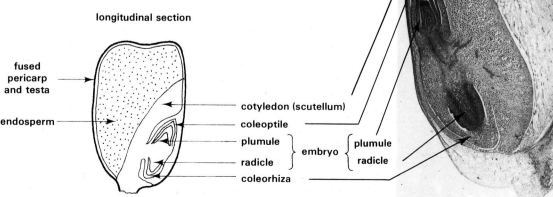

endosperm

plumule

radicle

23.3 Physiology of germination

The structural changes described so far are the consequences of physiological and metabolic changes occurring inside the cells and tissues of the germinating seed. The first changes are mainly the result of **cell expansion** following the uptake of water (section 23.3.2). Later changes involve the **growth** of new cells and tissues at the apices of the radicle and plumule. Growth requires raw materials and energy and initially it occurs at the expense of energy-rich molecules such as starch and lipids stored in the seed. Once the leaves of the young plant have expanded the seedling photosynthesises.

23.3.1 Mobilization of stored food

Table 23.1 gives an indication of the main materials stored in a variety of seeds. Because of their richness in **carbohydrates, lipids** and **proteins**, seeds feature prominently in the diets of many animals including man. Food reserves in seeds are insoluble in water and cannot be transported in the seedling. They must be broken down into relatively simple, soluble substances which dissolve in water to be moved to the growing apices of the plumule and radicle. As in the mammalian gut (Chapter 15), hydrolytic enzymes catalyse the breakdown of proteins, lipids and polysaccharides such as starch.

Table 23.1 Chemical composition of seeds

Species	Percentage dry mass		
	Carbohydrates	Lipids	Proteins
maize	50–75	5	10
wheat	60–75	2	13
rice	65–70	2	10
broad bean	57	2	36
sunflower	2	45–50	25
peanut	12–33	40–50	20–30
pea	34–46	2	20

1 Carbohydrates

The hydrolysis of **starch** into the soluble disaccharide sugar **maltose** is catalysed by a complex of enzymes called **amylase.** In seeds germinating under natural conditions maltose is further hydrolysed by the enzyme **maltase** to **glucose** which is converted to **sucrose** for transport to the growing apices of the embryo. At the apices sucrose is used for the synthesis of cellulose, hemicelluloses and pectic compounds, the main components of plant **cell walls**. Some sucrose is **respired** to provide energy for growth.

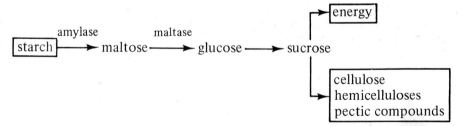

Commercial application

The conversion of starch to maltose is used to produce malt for the brewing industry. Barley seeds are germinated for a few days and the sprouted grain then dried and ground into powder. When later steeped in water the maltose dissolves and provides the source of energy for yeast cells in the fermentation stage of brewing.

2 Proteins

Proteins are hydrolysed to polypeptides and **amino acids** by **peptidase enzymes**. Some amino acids are moved in solution to the embryo. Most are transported as **amides**. At the growing points of the plumule and radicle the amides are de-aminated and the amino acids are used to synthesise structural and enzymic **proteins**.

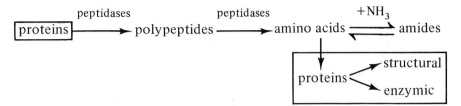

3 Lipids

Fats and oils are first hydrolysed to **glycerol** and **fatty acids** by **lipase** enzymes. The fatty acids may be oxidised to release energy, converted to sucrose ready for transport or used in membrane synthesis. Glycerol, which is only a small part of a lipid molecule, is also converted to transportable sugars.

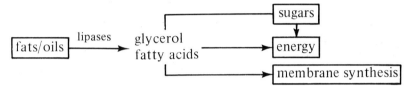

The consequences of some of the changes described above are shown for wheat seeds in Table 23.2. The general picture is a gradual depletion of reserves in the food storage areas and an increase in dry mass of the embryo (Fig 23.4).

Table 23.2 Changes in dry mass of maize seedlings during germination

| Time/days | Whole seedling | Dry mass/mg g^{-1} | |
		Endosperm	Embryo
0	225	200	2
1	210	189	3
2	208	188	5
3	206	155	5
4	175	115	15
5	155	84	23

Fig 23.4 Changes in dry mass of embryo and endosperm during germination (after Stokes (1952))

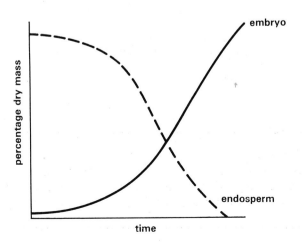

What causes such changes to occur? Studies with cereal grains show that the secretion of hydrolytic enzymes is triggered off by the hormone **gibberellic acid (GA)** made by the embryo (Fig 23.5(a)). In cereals the release of enzymes takes place from the **aleurone layer** surrounding the endosperm. After absorption of water by the grain there is a marked rise in embryonic production of GA which diffuses into the endosperm where food reserves are hydrolysed (Fig 23.5(b)). In some seeds the hydrolytic enzymes are found in lysosomes in the food-storing cells. Other hormones, notably **cytokinin** and **indolacetic acid (IAA)** promote cell division and enlargement at the growing apices of the embryo. IAA also controls differentiation of vascular tissue in the developing shoot and root of the seedling (Chapter 24).

Fig 23.5 (a) Investigation of the effect of the embryo on the release of the starch-hydrolysing enzyme amylase by the aleurone layer of barley seeds (after Black (1972))

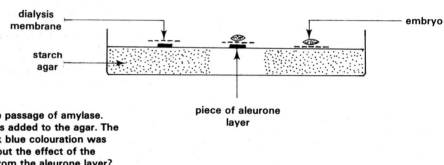

The dialysis membrane prevents the passage of amylase. After a few days iodine solution was added to the agar. The stippled area indicates where a dark blue colouration was observed. What does this tell us about the effect of the embryo on the release of amylase from the aleurone layer?

Fig 23.5 (b) Summary of the physiology of food mobilization in a barley seed (after Black (1972))

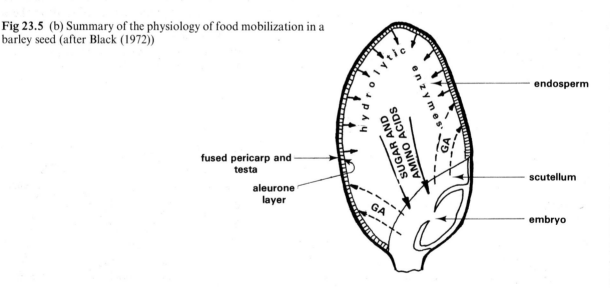

23.3.2 Factors affecting seed germination

If a gardener or farmer sows seeds at the wrong time of the year, few if any will germinate. **Environmental factors** can thus influence seed germination. Instructions concerning the best time of year to plant seeds are usually given on seed packets. For many outdoor crops of vegetables the most suitable time for sowing seeds in Britain is spring and early summer when the temperature is higher than in winter. In preparing a seed bed an experienced gardener ensures that the soil has a good crumb structure and is moist but not waterlogged. Too much water fills pore spaces in soil which are normally filled with air which contains oxygen. Thus a suitable **temperature, water** and **oxygen** have to be available for successful seed germination. Some seeds also require **light** before they germinate.

1 Temperature

Many of the metabolic changes which take place in germination are catalysed by **enzymes**. The effect of temperature on enzyme activity is described in Chapter 3. For the seeds of each species of flowering plant there is an **optimum temperature** at which the germination rate is at its highest, providing no other factor is limiting germination (Fig 23.6). The fact that seeds of most plants germinate early in the year if sown in a warm greenhouse is used by nurserymen in preparing trays of bedding plants for sale in early summer.

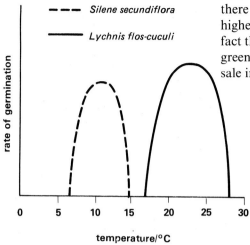

Fig 23.6 Effect of temperature on the rate of seed germination (after Thompson (1970))

--- Silene secundiflora

—— Lychnis flos-cuculi

S. secundiflora inhabits the Mediterranean coastal area. The seedlings of this species avoid the hot, dry summer weather because the seeds germinate in the cooler, moister winter months. This is reflected in its relatively low optimum germination temperature.
L. flos-cuculi is widespread in temperate deciduous forest areas including Britain. How does the higher optimum germination temperature for this species reflect the climate of its habitat?

The seeds of many species such as red fescue grass, peach and apple germinate only if they have been subjected to a period of low temperature after they have absorbed water. A temperature between 0–5 °C is normally sufficient, the period of time varying from a few days to several weeks (Fig 23.7). **Chilling** reduces the abscissic acid content of the seed coat and enables the embryo to make gibberellic acid (Fig 23.8). In Britain the cold weather of winter provides the chilling requirement of seeds in the wild.

In this country gardeners often chill seed-containing fleshy fruits such as holly berries and rose hips by placing layers of them in boxes of sand and keeping them outdoors throughout winter. What other purposes does this technique called **stratification** have?

Fig 23.7 Effect of chilling on the rate of germination of apple seeds (after Luckwill (1952))

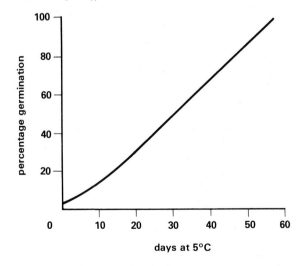

After the required period of chilling the seeds were germinated at 25°C. The graph shows the percentage of seeds germinating after 16 days at 25°C.

Fig 23.8 Effect of chilling on hormone concentrations in the seeds of sugar maple (after Wareing (1973))

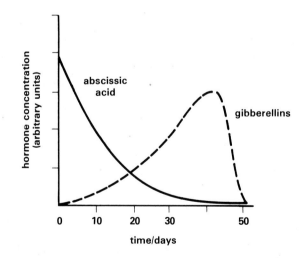

2 Water

The water content of many types of seeds is as little as 5–10 % of their dry mass, far too low for rapid **metabolic activity**. Water is also needed for **vacuolation** of cells and for **transport** during germination.

It is normal practice to store seeds in an atmosphere which keeps them dry and prevents germination. When placed in water, seeds quickly swell enormously, often doubling their mass in a few hours. The initial absorption of water is largely by **imbibition**, the attraction between hydrophilic colloids such as membrane-bound proteins in the seed and water molecules (Chapter 1). Later the roots absorb water by **osmosis**.

In natural circumstances the water content of many seeds does not remain low for long. Most seeds are dispersed as soon as they are formed and on reaching the soil they absorb water. Even so, few seeds germinate at this stage. It is therefore evident that factors other than water content normally control germination.

3 Oxygen

The synthesis of new cellular components at the apices of plumule and radicle requires energy. Energy is supplied by ATP produced in **respiration** from stored energy-rich compounds such as carbohydrates and lipids. The manufacture of ATP occurs most efficiently when oxygen is available and aerobic respiration can occur (Chapter 5). As germination proceeds there is rapid intake of oxygen through the seed coat (Fig 23.9).

Fig 23.9 Intake of oxygen by dormant and non-dormant seeds (after Chen and Varner (1970))

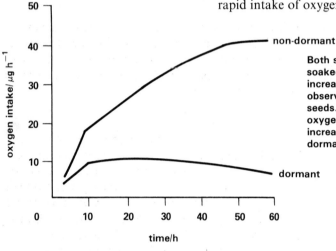

Both sets of seeds were first soaked in water. An initial increase in oxygen intake is observed even in dormant seeds. However the rate of oxygen intake continues to increase only in the non-dormant seeds.

Some of the energy in the reserves of food is liberated as heat, so it is possible to detect a rise in temperature of germinating seeds, particularly if kept in bulk. This is a problem farmers frequently encounter when grain with too high a water content has been stored. In these conditions the seeds begin to germinate, and **self-heating** occurs as heat energy is released. The temperature reached is so high that enzyme denaturation occurs and the embryos die. Self-heated grain is useless as seed for a potential crop and its value as fodder is considerably reduced because much of the food reserve has been spent.

Fig 23.10 Changes in respiratory quotients of seeds of pea and castor oil during germination

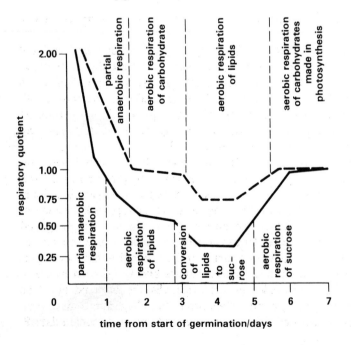

Carbon dioxide is a product of respiration. The ratio of carbon dioxide given off to oxygen used in a given period of time is called the **respiratory quotient, RQ** (Chapter 5). Fig 23.10 shows the respiratory quotients of two types of seeds at various stages in germination. The quotients give an indication of the type of substrate oxidised and whether aerobic respiration, anaerobic respiration, or both is taking place.

446

The testa of some seeds is relatively impermeable to oxygen so they may respire anaerobically at the start of germination. Large, bulky seeds such as broad beans have a small surface area to volume ratio which may result in an inadequate supply of oxygen to respiring cells and consequently some anaerobic respiration. For these reasons the RQ can be greater than 1.0 during the first 24–48 hours of germination. The RQ of most seeds then falls to 0·7. This is because lipids are now respired aerobically. If large quantities of lipids are stored, as in castor-oil seeds, the RQ at this stage may drop to as little as 0.35 owing to the conversion of lipids to sucrose.

$$C_{16}H_{40}O_6 + 11O_2 \longrightarrow C_{12}H_{22}O_{11} + 4CO_2 + 9H_2O$$
$$\text{lipid} \qquad\qquad\qquad \text{sucrose}$$

$$RQ = \frac{4}{11} = 0.35$$

In carbohydrate-rich seeds the RQ soon climbs to 1.0 as aerobic respiration of sugar gets under way. When the seedling eventually photosynthesises the RQ is 1.0 as in the mature plant. Aerobic respiration of carbohydrates made in photosynthesis now takes place. At this stage RQ measurements have to be made in darkness. Why is this so?

4 Light

Although the seeds of many plants can germinate in total darkness, others such as tobacco and some varieties of lettuce germinate only after exposure to **light**. The time of exposure varies from several seconds to a few days. Light-treatment is effective only if the seeds have absorbed water. The light need only be of low intensity and it is the red part of the spectrum which is required. Infra-red light inhibits germination, even though the seeds have previously been subjected to red light. The cause of this response is the effect of light on the pigment **phytochrome**. Phytochrome exists in two forms, P_r which absorbs mainly red light and P_{ir} which absorbs mainly infra-red light (Fig 23.11). Both P_r and P_{ir} are present in seeds. On absorbing red light, P_r is converted to P_{ir} while P_{ir} is reconverted to P_r when it absorbs infra-red light.

$$P_r \underset{\text{infra-red}}{\overset{\text{red}}{\rightleftharpoons}} P_{ir}$$

When a seed contains more than a certain amount of P_{ir} germination can proceed.

Sunlight, to which seeds are naturally exposed, contains both red and infra-red light. The proportion varies from time to time and place to place. Generally there is sufficient red light for the formation of enough P_{ir} to trigger off germination. The effect of P_{ir} accumulation is not known with certainty but it may stimulate the production of gibberellic acid or inhibit production of abscissic acid. Seeds which can germinate in darkness seem to have sufficient P_{ir} and thus do not require light.

Many light-dependent seeds are small. If they germinated too deeply in soil the embryonic shoot system would not reach the soil surface. Their dependence on light ensures that they germinate only when at or near the soil surface where there is a reasonable chance of the seedlings' aerial system becoming established. Ploughing and digging bring the light-dependent seeds of weeds to the soil surface where their dormancy is broken. Another ecological aspect of light-dependence is seen in deciduous forests and woodlands where in summer the canopy of leaves on the trees absorbs much red light. The light reaching the ground is then proportionally richer than normal daylight in infra-red light. In these conditions seeds accumulate P_r

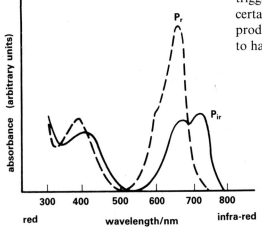

Fig 23.11 Absorption spectra of phytochromes (after Black 1972))

447

which inhibits germination. Germination of the seeds of woodlands herbs therefore normally takes place in spring before the leaf canopy appears. In the absence of shade, intensity of light in the wood is then also favourable for seedling growth.

23.3.3 Seed dormancy

Even when provided with water, a suitable temperature, oxygen and light, the seeds of many species fail to germinate and remain dormant. There are a number of reasons for **seed dormancy**.

Fig 23.12 Effect of gibberellic acid on germination of hazel seeds (after Jarvis (1968))

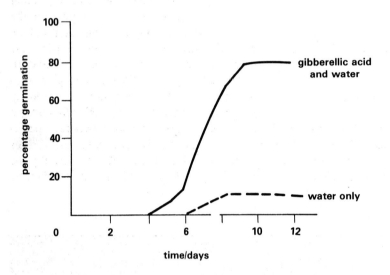

time/days

1 Immaturity of the embryo

The embryo may be **structurally immature** when the seeds are dispersed from the parent plant. This has been shown to be so in the ash and the wood anemone. In some species such as hazel the embryo is unable to synthesise gibberellic acid. This means that the hydrolytic enzymes needed for mobilising food reserves are not produced. This is an example of **physiological immaturity**. Fig 23.12 shows the effect of gibberellic acid in stimulating germination of dormant hazel seeds.

2 Impermeability of the seed coat

The seed coat may be impermeable to oxygen and water or is too hard for the swelling embryo to burst through. This is so in many leguminous species such as lupin, clover and sweet pea. If the testa is chipped or removed, dormancy is broken if the seeds are given conditions suitable for germination. In bulky fruits such as apples the thick wall of the fruit restricts the diffusion of oxygen to the seeds. In natural conditions the seed coat and fruit wall are gradually broken down by microbial activity in the soil.

Gardeners often chip hard seed coats with a knife or rub them on a rough surface such as a file. It is recommended that this be done on the opposite side of the seed to the embryo. Why is this so?

Fig 23.13 Effect of abscissic acid on germination of rye grass seeds (after Sumner and Lyon (1967))

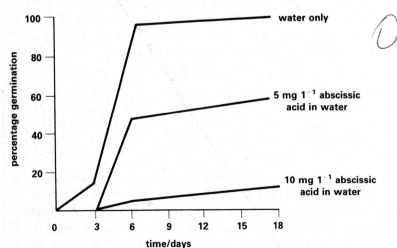

time/days

3 Germination inhibitors

Seeds of plants with succulent fruits such as tomatoes develop surrounded by water and nutrients, conditions which seem ideal for germination. They fail to germinate because of **inhibitor** chemicals in the sap of the fruit. In other species such as birch and rose the inhibitor is in the testa and can be removed by immersing the seeds in several changes of water. In natural conditions rainfall probably slowly leaches out the inhibitor. The most important natural inhibitor so far discovered is **abscissic acid** (Fig 23.13).

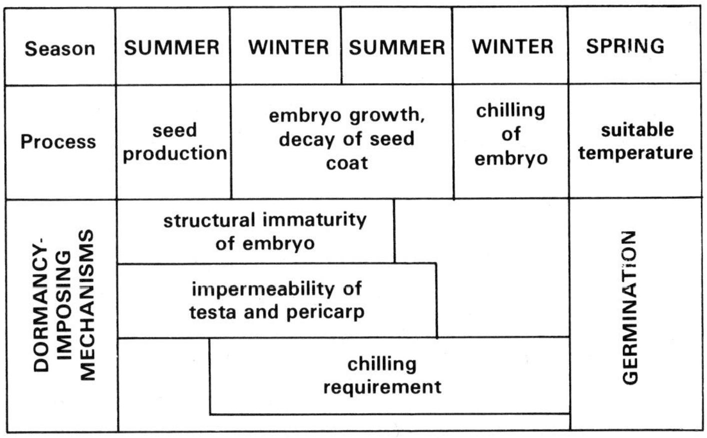

Although the various factors responsible for seed dormancy have been considered separately it is often a combination of factors which generally keeps the seeds of most species dormant. Fig 23.14 shows the combination of dormancy-imposing mechanisms which operate in the European ash.

Fig 23.14 Combination of dormancy-imposing mechanisms in the European ash (after Villiers (1975))

Season	SUMMER	WINTER	SUMMER	WINTER	SPRING
Process	seed production	embryo growth, decay of seed coat		chilling of embryo	suitable temperature
DORMANCY-IMPOSING MECHANISMS	structural immaturity of embryo				GERMINATION
	impermeability of testa and pericarp				
		chilling requirement			

The few aspects of dormancy considered here demonstrate the subtle interactions between seed and environment and show how germination is controlled in the wild so that it usually takes place when there is a reasonable chance that successful growth will follow. In Britain most flowering plants produce their seeds in summer. If germination took place immediately the seeds are dispersed, large numbers of seedlings would compete for the available space and light during autumn. Few would become established properly before the cold weather of winter. Because germination is delayed until the following spring or summer, subsequent growth is much more likely to be successful with a long spell of favourable weather about to begin. The reasons for dormancy vary between species and the time required for inhibitory factors to disappear also varies. Thus germination is usually spread out over a long period of time whereby the effects of competition between seedlings is diminished and adverse climatic conditions are avoided.

SUMMARY

Seeds are embryo plants and their food store produced by angiosperms and gymnosperms. The embryo comprises an embryonic shoot called the plumule and an embryonic root, the radicle. In monocotyledons the embryo is attached to a single seed leaf, the cotyledon. There are two seed leaves in dicotyledons. The embryo's store of food may be in the cotyledons or in a separate tissue called endosperm.

Germination is the first step in the development of a new plant from a seed. Hypogeal germination is where the cotyledons stay under the soil surface as in the broadbean and maize. In contrast, the cotyledons emerge above the soil in epigeal germination as shown by the sunflower.

The water content of seeds is too low for rapid metabolic activity. Hence water uptake by imbibition is an essential pre-requisite of germination during which the metabolic rate is extremely high. Water is the medium in which enzymic reactions are catalysed and is a reactant in the hydrolysis of food reserves. It is also the solvent in which the products of hydrolysis are transported to the sites of growth in the embryo. Secretion of hydrolytic enzymes in the seed is triggered by the hormone gibberellic acid.

For seeds to germinate they need a suitable temperature and access to oxygen gas as well as water. Some also require sunlight. The many enzyme-catalysed reactions of germination require an optimum temperature for them to proceed efficiently. Access to oxygen gas permits aerobic respiration to supply ATP at a rate commensurate with the energy requirements of germination. For those seeds in need of light, red light is most effective in promoting the production of P_{ir}, a form of phytochrome which most seeds already have in sufficient amounts to stimulate germination.

Internal factors sometimes cause seeds to remain dormant. Structural and physiological immaturity of the embryo, impermeability of the seed coat and the presence of inhibitors, notably abscissic acid are of significance in this context. Interaction between the seed and its environment determines the time of year when it will germinate.

QUESTIONS

1 (a) The diagram shows the appearance of a barley grain cut lengthwise.

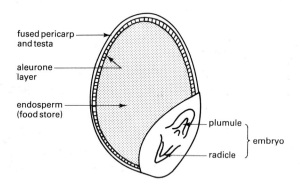

fused pericarp and testa

aleurone layer

endosperm (food store)

plumule

radicle

embryo

Which **one** of the parts shown suggests that the grain is **not** a seed?

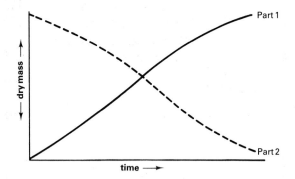

Part 1

Part 2

dry mass

time

(b) The graph shows changes in the dry mass of two different parts of the grain during germination. Use the labels from the diagram above to identify these parts.

(c) (i) During the first stage of germination, barley grains imbibe water. If you were provided with about 25 grains, list the steps you would take in order to determine the mean mass of water absorbed per unit mass of dry grain in unit time.

(ii) Write the symbols for the units you would use to express this value. (WJEC)

2 It is a common practice to soak seeds in water before planting.

(a) Suggest **three** reasons why soaking seeds in water before planting may stimulate them to germinate more rapidly.

(b) The table indicates the effect that soaking pea seeds for different periods of time has on the percentage germination and the subsequent dry mass of the roots, shoots and cotyledons. The dry mass measurements were made 14 days after the period of soaking.

Soaking time/days	1	2	3	4
% germination	99	98	75	66
Dry mass (mg seedling^{-1})				
root	19.9	14.5	12.0	12.3
shoot	25.0	21.7	16.6	13.7
cotyledons	58.8	70.3	91.6	119.0

(i) By reference to the figures in the table, state how prolonged soaking affected (1) the seeds and their ability to germinate, and (2) development of the seedlings.

(ii) Suggest a possible explanation of these effects. (C)

3 The table shows the effects of keeping 10 mm long segments of oat radicles and coleoptiles in a culture solution containing IAA (an auxin) in the dark for 24 hours.

Culture solution (IAA in parts per million)	Mean segment length after 24 hours (mm)	
	Radicle	Coleoptile
2 % sucrose only	11.5	12.0
2 % sucrose + 0.0001 ppm IAA	14.3	12.5
2 % sucrose + 0.01 ppm IAA	12.2	15.8
2 % sucrose + 1.0 ppm IAA	11.0	16.7
2 % sucrose + 100.0 ppm IAA	11.0	16.8

(a) Why were the radicle and coleoptile segments kept in the dark throughout the experiment?

(b) Why was sucrose included in the culture solutions?

(c) Why was IAA omitted from one solution?

(d) What can you conclude from these results about the effect of IAA on the growth of radicle and coleoptile segments?

(e) If an oat seedling is placed on its side in the dark, the radicle will grow downwards and the coleoptile upwards. With the help of results from the table above, explain this phenomenon. (AEB 1987)

4 The relationship between seed fresh mass in the lupin, *Lupinus texensis*, and percentage seed germination, percentage seedling survival and seedling fresh mass is shown in the table.

Seed fresh mass class /mg	Percentage germination	Percentage of seedlings surviving to 2 leaf stage	Mean seedling fresh mass 5 weeks after germination/mg
below 16	41.9	84.6	24.3
16–25	90.2	96.8	44.2
26–35	95.6	98.1	60.7
36–45	97.5	100.0	86.4
above 45	100.0	100.0	106.4

(a) With reference to the figures in the table indicate the relationships between

(i) seed fresh mass and percentage seed germination,

(ii) seed fresh mass and percentage seedling survival,

(iii) seed fresh mass and seedling fresh mass.

(b) State the evidence provided by the figures that the seedlings produced from large seeds grow more rapidly than the seedlings produced from small seeds. Suggest an explanation for the more rapid growth. (C)

5 Under natural conditions some seeds with a hard testa germinate only after one or two winters.

(a) Suggest **three** factors that might be responsible for breaking the dormancy of these seeds.

(b) Design an experiment to test **one** of these factors. (AEB 1984)

24 Growth and development of flowering plants

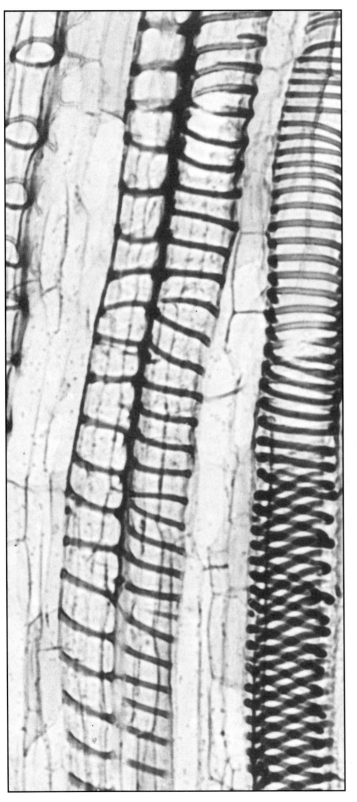

24.1 Morphological and anatomical aspects of growth and development 453

   24.1.1 Primary growth and development 453

   24.1.2 Secondary growth and development 463

24.2 Physiological aspects of growth and development 467

   24.2.1 External factors 467

   24.2.2 Internal factors 469

   24.2.3 Practical uses of plant hormones 475

Summary 477

Questions 477

24 Growth and development of flowering plants

Growth, which can be thought of as an increase in mass, and **development**, an increase in complexity, go hand in hand in all living organisms. In flowering plants growth and development begin when a seed germinates and ends when a mature plant has grown. Trees can take many years to reach maturity yet a few weeks is sufficient for some annual plants. It is well known that plants kept in greenhouses grow more quickly than similar plants growing outdoors. It is also known that oak trees and not dandelions appear if we sow acorns. This is because **environment** and **heredity** control the pattern and the rate of growth and development.

In recent times, a good deal of effort has been made in an attempt to unravel the physiological basis of plant growth. It is now apparent that growth and development of plants are the consequences of subtle interactions between many **external** and **internal factors**. With this knowledge it may soon be possible for humans to control plant growth more effectively than at present. As a result there could be significant improvements in the production of food, timber and other essential plant products.

24.1 Morphological and anatomical aspects of growth and development

In the previous chapter, seed germination and the subsequent establishment of a small immature plant called a seedling is described. How does a seedling grow and develop into a much larger and more complex mature plant? Mature plants have many more cells than seedlings. It is therefore evident that somatic cell division, mitosis (Chapter 7), is an integral part of growth. But the adult plant is more than a large number of identical cells. It has a variety of cell types, each type adapted to perform specific physiological and mechanical roles. The changes which take place to produce several types of cells is called **differentiation**.

24.1.1 Primary growth and development

In seedling plants mitosis is mainly confined to groups of cells at the tips of radicles and plumules. These areas of somatic cell division are called **apical meristems**. They produce the cells which differentiate into the tissues of the **primary plant body**.

1 The primary shoot system

A thin longitudinal section cut through the middle of a shoot tip reveals, at the apex, a mass of closely packed cells with conspicuous nuclei. The thin cell walls consist largely of pectic substances and the cytoplasm lacks any large vacuoles. Some of the cells are undergoing mitosis (Fig 24.1). This is the apical meristem of the shoot.

i. The apical meristem of the shoot   It is usually possible to distinguish an outer zone of cells called the **tunica** and an inner mass of cells called the **corpus** in the apical meristem of a shoot. In the tunica, which can be one to several cells thick, cross walls are laid down at right angles to the surface of the shoot apex during mitosis. Its main task is to enlarge the surface area of the shoot. The outermost layer of cells derived from the tunica later differentiates into the **epidermis** of the primary shoot system. Division of the corpus cells occurs in several planes. An internal mass of cells is thus produced, the size of which keeps pace with the ever-increasing surface layer formed by the tunica. Cells derived from the corpus differentiate into the tissues of the **cortex**, **vascular bundles** and **pith** of the primary stem.

Fig 24.1 (a) Longitudinal section through the shoot apex of a dicotyledonous plant, × 30

Fig 24.1 (b) Close-up view of apex, × 160

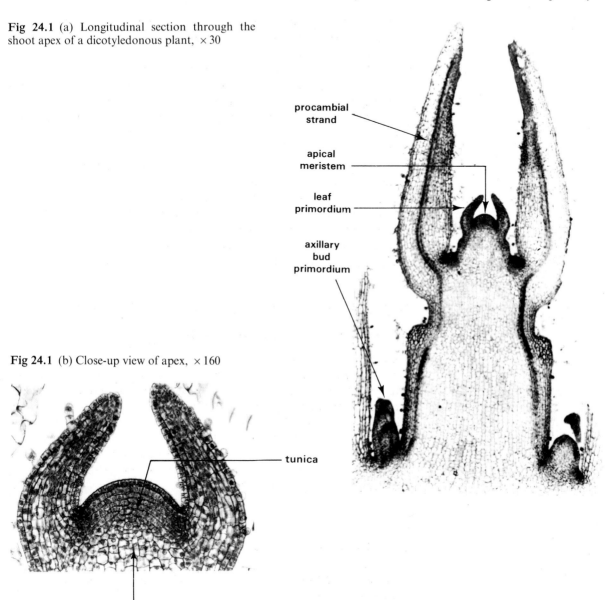

Both tunica and corpus take part in the formation of swellings which arise at regular intervals on the surface of the shoot apex. The swellings are called **leaf primordia**. Their growth is so rapid that the earlier formed primordia soon enclose the apical meristem at the tip of the shoot. Mature leaves gradually differentiate from the older primordia, new primordia constantly being formed at the tip of the shoot as apical growth continues. The spatial arrangement of the leaf primordia determines the positions of mature leaves on the shoot. Paired primordia for example give rise to pairs of leaves on the mature stem. **Axillary bud primordia** which have the potential to grow into lateral shoots are generally seen in the axils of developing leaves near the shoot apex (Fig 24.2). The tunica and corpus again produce the axillary bud primordia. The superficial origin of lateral shoots contrasts with the deep-seated origin of lateral roots (Fig 24.8).

Fig 24.2 Differentiation at the shoot apex of a dicotyledonous plant

(a) Longitudinal section

(b) Transverse sections

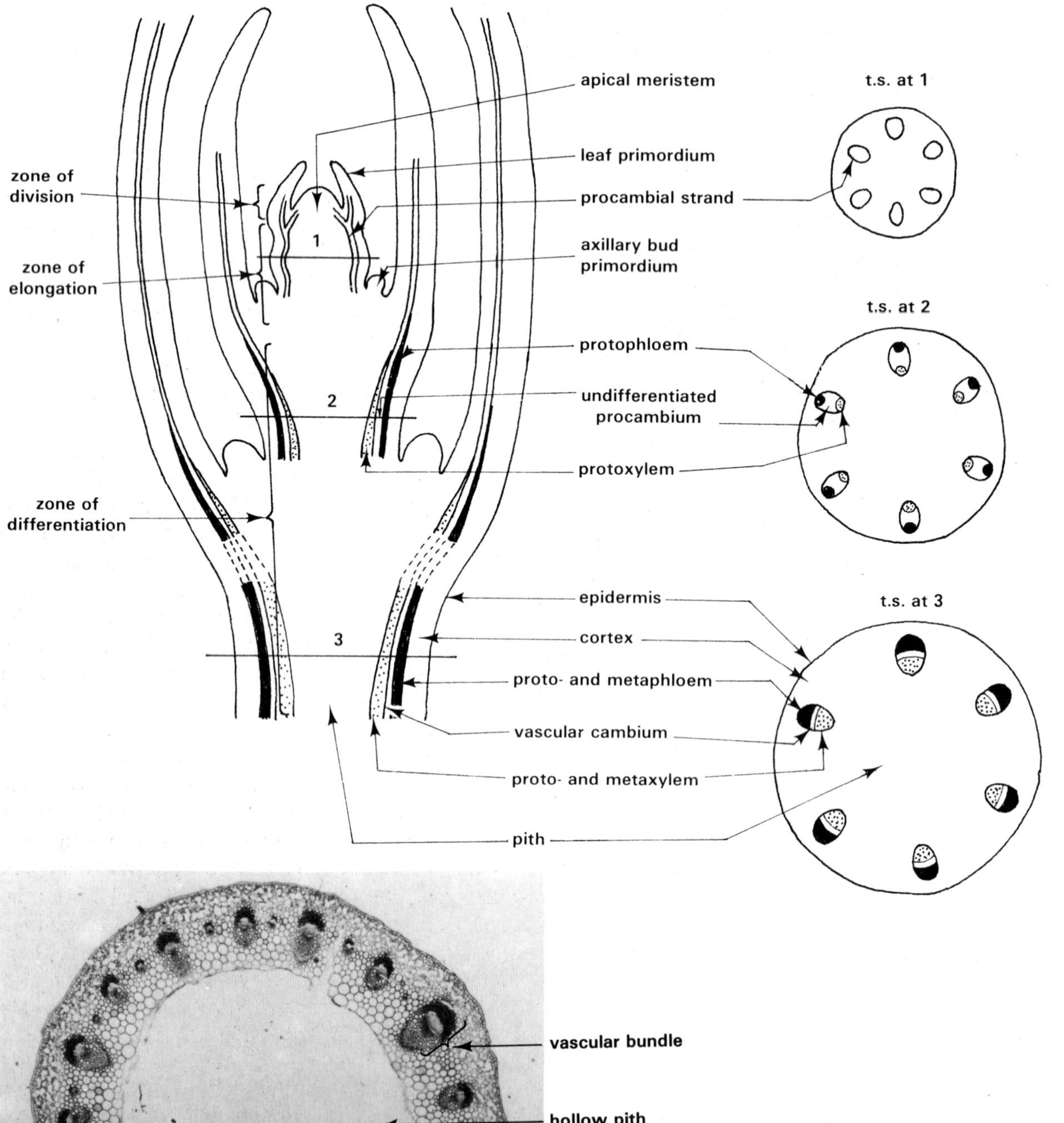

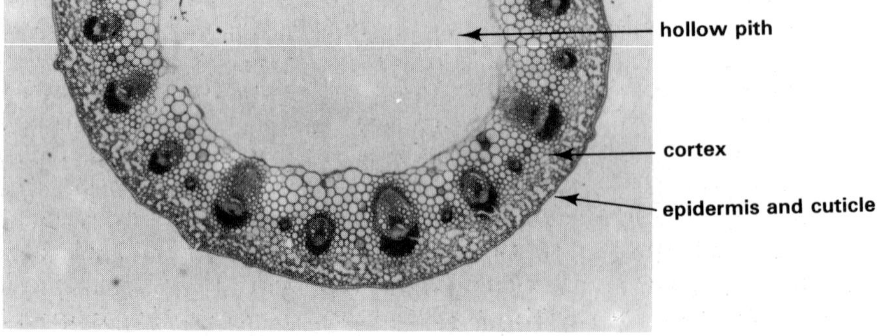

Fig 24.2 (c) Transverse section of a mature buttercup stem, ×20

Fig 24.2 (d) Enlarged view of one of the vascular bundles, × 120

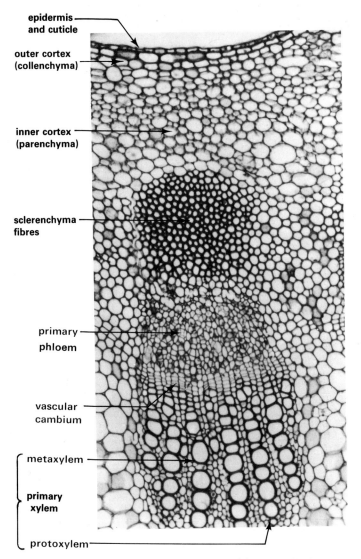

epidermis and cuticle

outer cortex (collenchyma)

inner cortex (parenchyma)

sclerenchyma fibres

primary phloem

vascular cambium

metaxylem

primary xylem

protoxylem

ii. Tissue differentiation in the stem Just behind the apical meristem the first signs of **tissue differentiation** are seen. Conspicuous vacuoles appear in the cytoplasm of cells cut off from the tunica and corpus. The effect is to cause considerable increase in length and girth of the cells. **Vacuolation** accounts mainly for elongation of the shoot at this stage. However, there is an increase in the protein and nucleic acid content of the cells. Cellulose is also laid down in the expanding cell walls.

Not all of the apically-derived cells undergo the changes outlined above. Several strands of cells extending backwards from the apex remain narrow with pointed ends. Although they elongate they do not develop large vacuoles. They also retain the power to divide even when the surrounding cells can no longer do so. Thus a number of discrete bundles of tissue called **procambial strands** appear among the vacuolate ground tissue. The positions of the strands are determined by the leaf primordia, one strand developing below each primordium.

The outermost cells of the procambial strands differentiate into the first formed cells of primary phloem called **protophloem**. First there is an unequal longitudinal division of each procambial cell. The smaller of the derivatives differentiates into a companion cell. At the same time the larger cell differentiates into a sieve-tube element. Sieve tubes transport organic substances in solution to all parts of the plant including the growing shoot apex (Chapter 13). Shortly afterwards the first formed water-carrying tissue called **protoxylem** differentiates from the innermost cells of each procambial strand. The main changes in protoxylem differentiation are loss of the protoplast, breakdown of the end walls of adjacent cells and secretion of a lignified secondary wall as annular or spiral bands inside the primary cell wall (Fig 24.3(a)). The primary walls of protoxylem cells can continue growing for some time keeping pace with elongation of surrounding tissues. The secondary wall gives the necessary support to keep the lumen of each xylem element open for water transport.

While the changes outlined above are taking place the shoot apex continues to grow upwards. The part of the shoot where protophloem and protoxylem have formed now stops elongating and further differentiation occurs in the procambial strands. The procambial cells immediately inside the protophloem differentiate into **metaphloem** sieve-tubes and companion cells. The procambial cells immediately inside the protoxylem differentiate into **metaxylem** cells. Again in the formation of metaxylem cells there is a loss of end walls and primary walls become covered internally by scalariform, reticulate or pitted, lignified secondary walls (Fig 24.3(b)). This form of wall thickening gives extra rigidity compared with protoxylem cells and reflects the cessation of elongation in this part of the shoot.

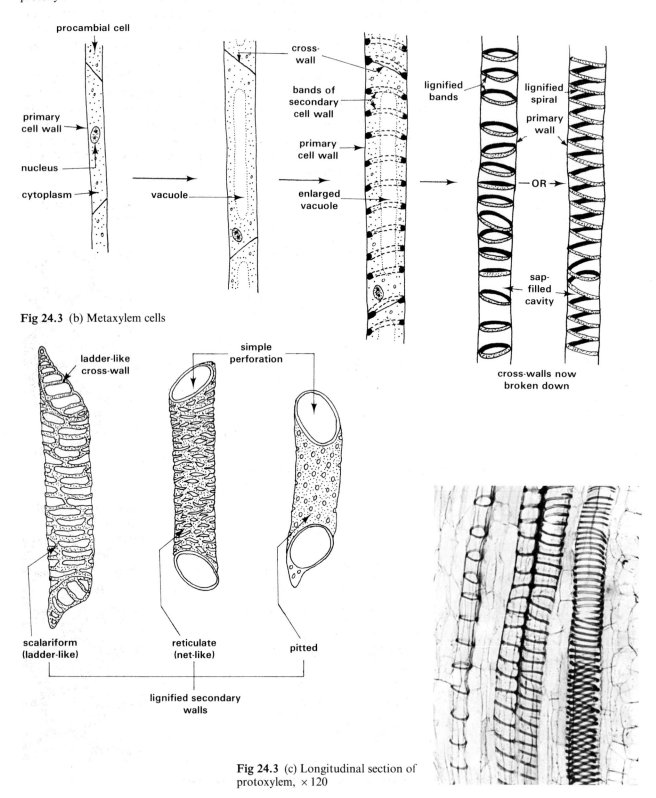

Fig 24.3 (a) Differentiation of protoxylem

procambial cell

primary cell wall

nucleus

cytoplasm

vacuole

cross-wall

bands of secondary cell wall

primary cell wall

enlarged vacuole

lignified bands

lignified spiral

primary wall

OR

sap-filled cavity

cross-walls now broken down

Fig 24.3 (b) Metaxylem cells

ladder-like cross-wall

simple perforation

scalariform (ladder-like)

reticulate (net-like)

pitted

lignified secondary walls

Fig 24.3 (c) Longitudinal section of protoxylem, ×120

The cells of much of the ground tissue of the stem, the cortex and pith, show less obvious changes as they differentiate into a tissue called **parenchyma**. The main changes are enlargement in girth and length, the development of thin cellulose walls and extensive vacuolation of the cytoplasm. The changes prepare the parenchyma cells for the storage of large quantities of sap which helps maintain turgidity of the young shoot. Starch grains too are frequently stored in parenchyma. In the outer cortex the cell walls often become thickened with stiff, yet plastic, deposits of cellulose forming a supporting tissue called **collenchyma**. Parenchyma and collenchyma cells retain the ability to divide and often contain chloroplasts. Some cortical cells, particularly those immediately around each vascular bundle, become extremely elongated and have tapering end walls. They develop extremely thick, lignified secondary walls. These are **sclerenchyma fibres**, the ends of which dovetail together and form a very efficient supporting tissue (Fig 24.4). The epidermal cells derived from the tunica secrete a waxy material called **cutin** over their outer surface making the epidermis waterproof. Some epidermal cells become modified into **guard cells** of **stomata**.

Fig 24.4 (a) Parenchyma

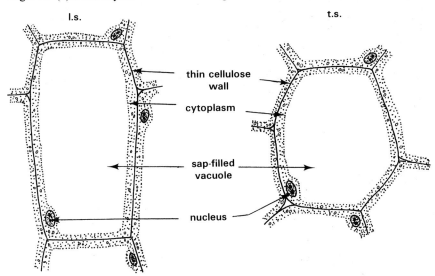

l.s.　　　　　　　　　　t.s.

thin cellulose wall

cytoplasm

sap-filled vacuole

nucleus

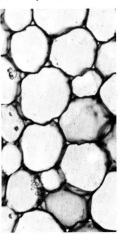

Transverse section of parenchyma, ×300

Parenchyma cells near the surface of the stem usually contain chloroplasts

Fig 24.4 (b) Collenchyma

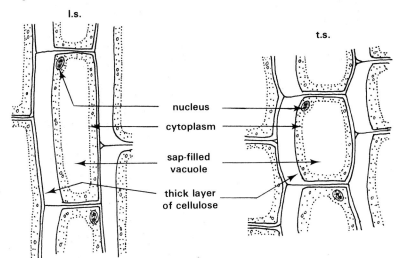

l.s.

t.s.

nucleus

cytoplasm

sap-filled vacuole

thick layer of cellulose

Transverse section of collenchyma, ×300

The position of the thick deposit of cellulose varies from one type of collenchyma to another

457

Fig 24.4 (c) Sclerenchyma fibres

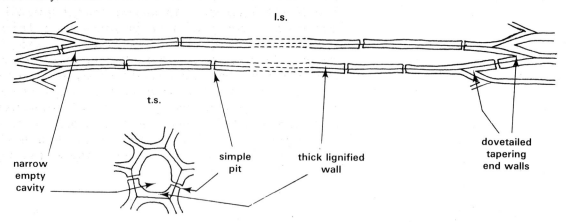

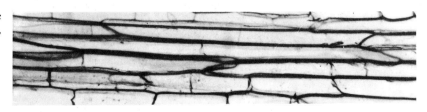

Sclerenchyma: transverse and longitudinal sections, both, × 300

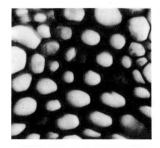

There are several obvious ways in which the structure of a stem is related to its functions. The presence of stomata on its surface enables it to exchange gases with the surrounding air. Chloroplasts in its outer tissues trap solar energy in photosynthesis. Turgidity of its parenchyma and thickening of the cell walls of its collenchyma serve to keep it erect. This role is supplemented by the stiff lignified walls of its sclerenchyma and xylem elements. The distribution of the vascular bundles is where maximum stress occurs when a stem is bent. Civil engineers have copied the design in deciding where best to place steel rods in reinforced concrete pillars.

iii. Development of leaves and lateral shoots A leaf primordium grows initially into a peg-like outgrowth with an **adaxial meristem** on its lower surface. Division of the cells of the adaxial meristem produces a thicker peg containing procambial strands continuous with those in the stem. The peg later differentiates into the tissues of the leaf midrib. **Marginal meristems** appear on either side of the peg and divide to produce cells which differentiate into the epidermis, mesophyll and vascular tissue of the leaf blade (Fig 24.5). The spatial arrangement or **phyllotaxis** of leaves on the mature stem is therefore determined by the positions of the leaf primordia at the stem apex. In Fig 24.5 it can be seen that the developing leaves are so close to each other at the shoot apex that they overlap and form a terminal **bud**. Growth and differentiation of the primary stem separates the older leaves as elongation of the **internodes**, the length of stem between successive leaves, occurs. Spacing of the mature leaves allows efficient gas exchange and light absorption, important for photosynthesis, the main physiological function of leaves. Lateral buds in the axils of leaves may grow into side shoots. Development of side shoots is the same as described earlier for the main shoot.

Fig 24.5 (a) Stages in the early development of a leaf

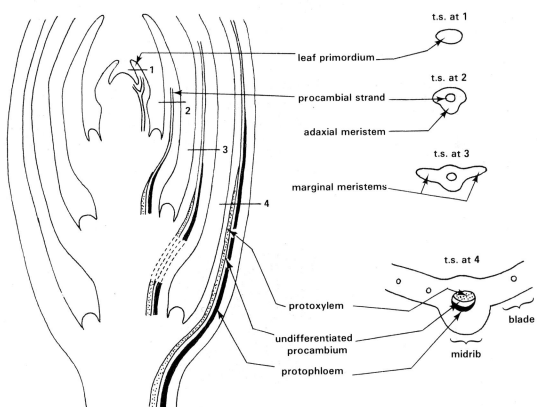

Fig 24.5 (b) Transverse section of a mature leaf, ×30

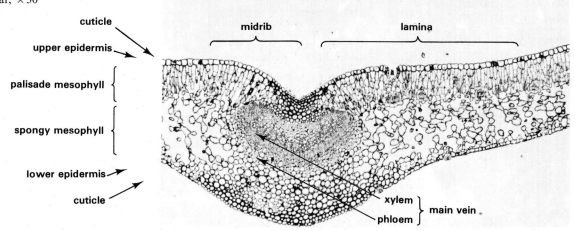

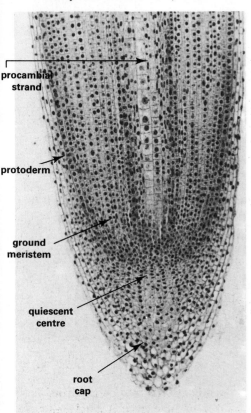

2 The primary root system

i. The apical meristem of the root The most obvious difference in appearance between longitudinal sections of stem and root apices is the absence of bulges comparable to leaf and bud primordia on the root apex. The root apex is also covered by a **root cap** (Fig 24.6). There is, however, a marked similarity in appearance and behaviour of the apical cells which constantly divide by mitosis. In most roots it is possible to distinguish a number of zones of cells at the apex. The outermost zone is called the **protoderm**. It produces cells which differentiate into the root epidermis and root cap. Inside the protoderm is the **ground meristem**, the derivatives of which differentiate into the root cortex. Just behind the root apex a single **procambial strand** can be seen at the centre of the root. Some roots have an additional meristematic layer, the **calyptrogen**, which gives rise to the cells of the root cap. The meristematic zones radiate from a clump of cells called the **quiescent centre** situated immediately behind the root cap. The significance of the quiescent centre is not as yet fully understood. Its cells divide slowly and it is probably the site from which the other meristematic layers arise (Fig 24.7).

ii. Tissue differentiation in the root Differentiation of vascular tissue begins near the root apex. Several strands of sieve-tube elements and companion cells appear near the outside of the procambial strand. Shortly afterwards a similar number of strands of protoxylem cells alternating with the primary phloem strands differentiate. Metaxylem cells differentiate last of all at the centre of the procambial strand. The outermost procambial cells undergo little change and retain their ability to divide. They become the **pericycle** which may later produce lateral roots.

Fig 24.7 Differentiation at the root apex of a dicotyledonous plant

(a) Longitudinal section   (b) Transverse sections

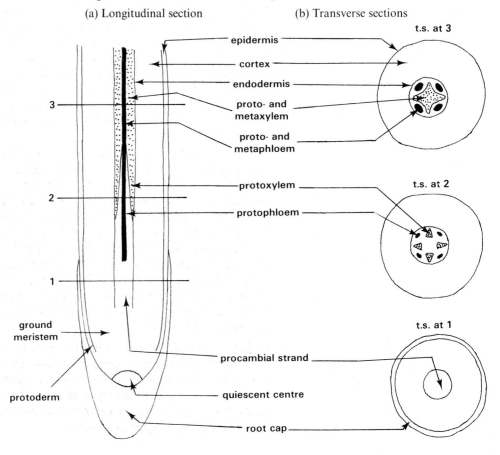

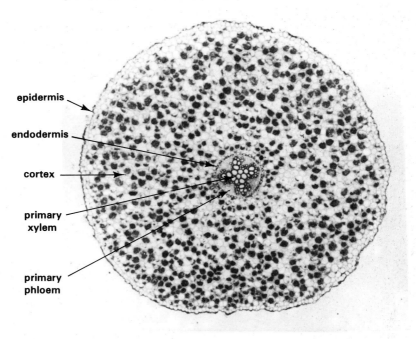

Fig 24.7 (c) Transverse section of a mature primary root of a dicotyledonous plant, × 70

epidermis

endodermis

cortex

primary xylem

primary phloem

The derivatives of the ground meristem differentiate into parenchyma cells of the **root cortex** which are used mainly for the storage of starch and the transport of water and dissolved minerals from the soil solution to the root vascular tissue. The cortical cells of the root do not usually develop chloroplasts as do the outer cortical cells of the stem. The innermost layer of the root cortex differentiates into a special band of cells called the **endodermis**. Early in development the endodermal cells secrete a suberised **Casparian strip** on their radial wall. Later, all but the outer of the endodermal cell walls become thickened and suberised (Fig 12.17). The endodermis is thought to control the passage of absorbed water and dissolved minerals from the cortex into the root vascular tissue (Chapters 12 and 14).

The root epidermis derived from the protoderm is very different from that of the shoot system. There is no cuticle as this would prevent one of the most important tasks of the root, the absorption of water and dissolved minerals. Just behind the root apex the epidermal cells develop tubular extensions called **root hairs** which vastly enlarge the root surface area through which absorption from the soil can occur. Root cap cells develop thick gelatinous walls of pectic substances which lubricate and protect the root apex as it constantly probes its way through the soil.

The lack of specialised supporting tissues in the root reflects the density of the surrounding soil. It contrasts with the lack of support given by air to stems where mechanical tissues such as collenchyma and sclerenchyma are usually well developed.

iii. Lateral root formation   The formation of **lateral roots** shows two main points of difference compared with the formation of lateral shoots. Firstly, lateral roots normally arise some distance from the root apex. Secondly, they originate deeply in the root from the pericycle. For this reason the origin of lateral roots is described as **endogenous** whereas lateral shoots are exogenous in origin.

Mitosis of the pericycle cells produces bulges of cells (Fig 24.8). The bulges are the beginnings of lateral roots. They normally arise opposite the groups of protoxylem cells. As they grow, each of the bulges develops an apical meristem and a root cap. The growing lateral roots, helped by enzyme secretion, force their way through the cortex of the main root. Eventually the lateral roots emerge through the ruptured epidermis. By this time a procambial strand has usually differentiated in each lateral root. Further differentiation takes place as in the main root, and the vascular tissues of the lateral roots become joined to the xylem and phloem of the main root.

Fig 24.8 Stages in the early development of a lateral root (a) Pericycle cells begin to divide

(b) Lateral root primordium formed

(c) Tissue differentiation

Fig 24.8 (d) Transverse section of a dicotyledonous root showing the early growth of a lateral root, × 30

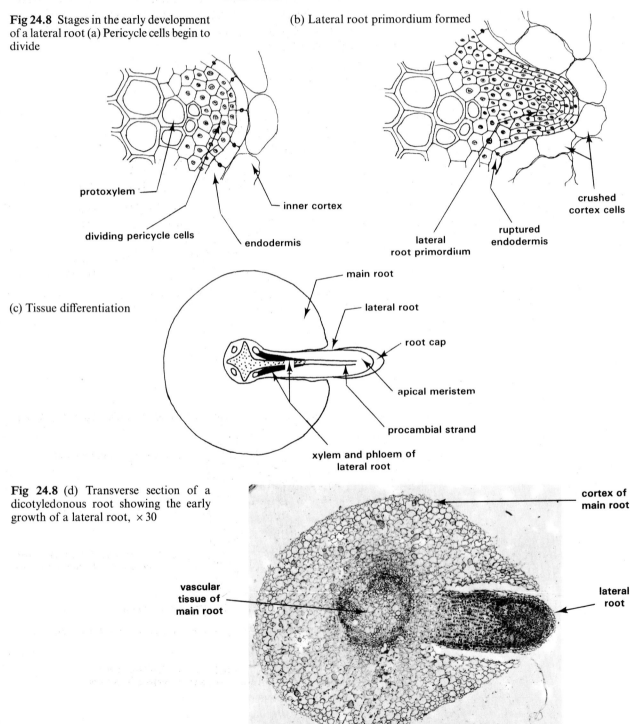

24.1.2 Secondary growth and development

Whereas some plants do not develop further than the primary stage, others add secondary tissues to the primary body. The **secondary plant body** is derived from **lateral meristems**, the most important of which are the **vascular cambium** and **cork cambium**.

1 The vascular cambium

In plants which undergo secondary growth not all the procambial cells in the primary body differentiate into vascular tissue. What remains is called **vascular cambium** and is situated between the primary xylem and phloem of stems and roots. The vascular cambium consists of two types of cells, **ray and fusiform initials** (Fig 24.9), and usually begins its activity during the first year of the plant's life.

Fig 24.9 The vascular cambium

(a) Tangential LS

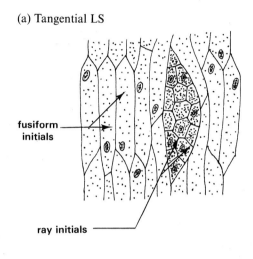

(b) Relative positions of ray and fusiform initials

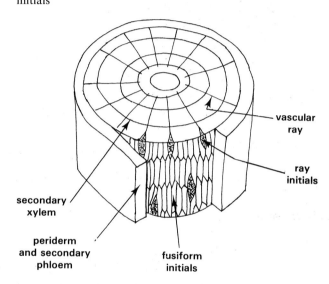

(c) Stages in the differentiation of secondary xylem and phloem (modified from Ray, 1963)

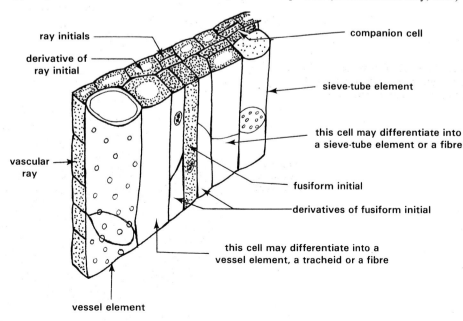

In the stem of a dicotyledonous plant the vascular cambium is at first confined to the vascular bundles. However, a complete cylinder of vascular cambium is formed as parenchyma cells between the bundles become meristematic. The main plane of division of the ray and fusiform initials is tangential. Derivatives cut off from the fusiform initials to the inside of the cylinder of vascular cambium **differentiate** into xylem vessel elements, tracheids and fibres. Those cut off to the outside become sieve-tube elements, companion cells and phloem fibres. Derivatives of the ray initials differentiate into parenchyma cells which cross the secondary xylem and phloem as radial **vascular rays**. Occasionally the ray and fusiform initials divide by radial walls so that the circumference of the cylinder of vascular cambium increases, accommodating the growing amount of secondary xylem it encloses (Fig 24.10).

Fig 24.10 Stages in the secondary growth of a dicotyledonous stem as seen in transverse section

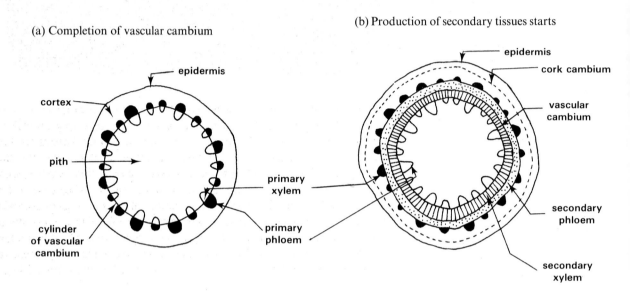

(a) Completion of vascular cambium

(b) Production of secondary tissues starts

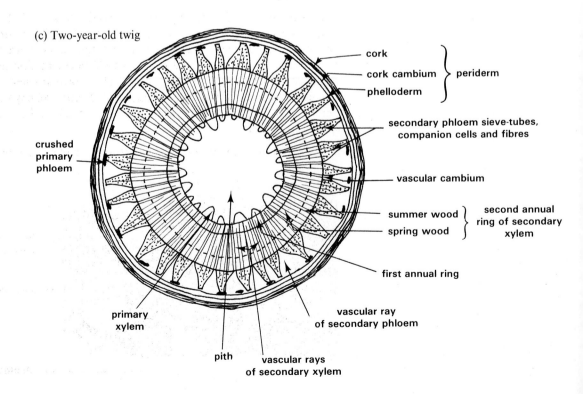

(c) Two-year-old twig

Fig 24.11 Transverse section through a 5-year-old twig of lime, × 15

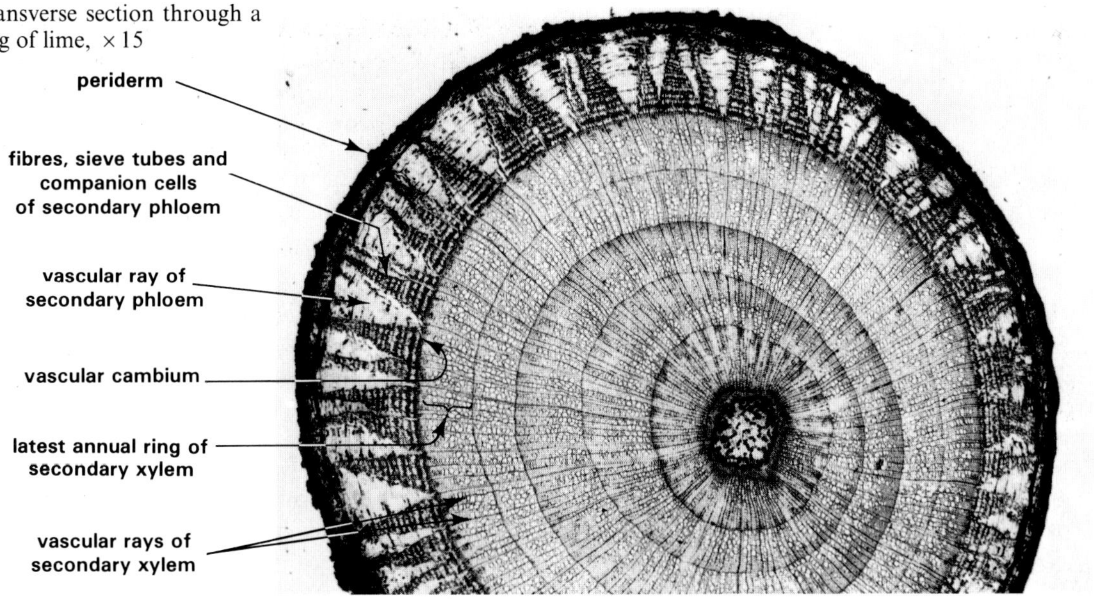

periderm

fibres, sieve tubes and companion cells of secondary phloem

vascular ray of secondary phloem

vascular cambium

latest annual ring of secondary xylem

vascular rays of secondary xylem

The vascular cambium is seasonal in its activity. It is dormant in winter and active in spring and summer. A cylinder of secondary xylem called an **annual ring** is added each year to the existing plant body. Vessels produced early in the growing season are generally larger in diameter than those arising later. Each annual ring thus consists of two distinct regions, **spring wood** on the inside and **summer wood** on the outside. The age of a shrub or tree can be found by counting the number of annual rings of secondary xylem in a cross-section of the main stem or trunk. There is no obvious layering in the secondary phloem. Here the vascular ray cells often undergo further division by radial walls so that wide, wedge-shaped phloem rays separate groups of sieve-tubes, companion cells and fibres (Fig 24.11).

2 The cork cambium

The increase in girth resulting from the addition of secondary vascular tissue imposes a strain on the outer tissues of the stem. If the primary cortex and epidermis remained, the stem would split open with the risk of desiccation. This is prevented by the development of tissues called the **periderm** which replace the outer tissues of the primary stem.

A cylinder of **cork cambium** arises, usually from parenchyma cells in the outer cortex (Fig 24.12). In transverse section the cells of the cork cambium

Fig 24.12 Transverse section through the periderm of elder, × 360

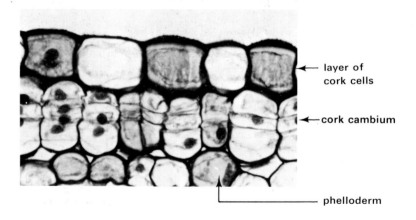

layer of cork cells

cork cambium

phelloderm

Fig 24.13 Transverse section through a lenticel of elder, × 80

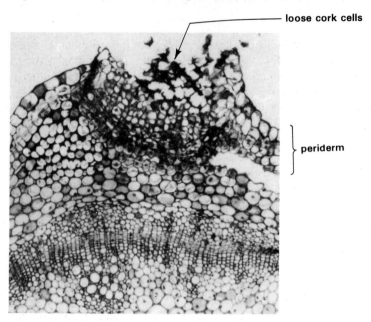

loose cork cells

periderm

are uniformly rectangular in shape. As in the vascular cambium, mitotic division in a tangential plane cuts off derivatives to the outside and to the inside. The outer derivatives secrete a layer of waxy material called suberin in their secondary wall after which their protoplasts disappear. They are now **cork cells** and form an effective waterproof barrier which prevents dehydration of internal tissues. Here and there the cork cells are loosely arranged, forming **lenticels** which allow gas exchange between the outer tissues and the atmosphere (Fig 24.13). The small number of derivatives cut off internally undergo little change and remain as a narrow band of uniform cells called **phelloderm**. The cells of the cork cambium now and again divide by radial walls to enlarge the circumference of the meristem, thereby keeping pace with the increase in girth caused by the activity of the vascular cambium.

Commercial application

Cork from the cork oak, *Quercus ruber* is used for making bottle stoppers, liners for bottle caps and floor tiles. A layer of cork several centimetres thick develops on the trunk and main branches of this species. The cork is carefully levered off after making a vertical split in the bark, leaving the cork cambium undamaged to grow again. Repeat harvests are made every ten years or so. Bottle stoppers are cut such that the lenticels run across them rather than lengthwise. Why is this so?

What structural features of cork make it suited to its role in bottle stoppers?

Fig 24.14 Intercalary meristems in a grass shoot

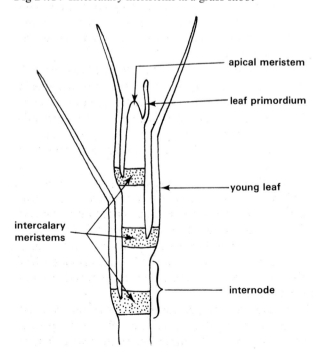

apical meristem

leaf primordium

young leaf

intercalary meristems

internode

A comparable pattern of secondary growth also happens in the roots of many dicotyledonous plants. The vascular cambium originates from strands of undifferentiated procambium lying between the primary xylem and phloem. It then develops into a cylinder from which annual rings of secondary vascular tissue are produced as in the stem. The cork cambium often originates from the pericycle so that early in its secondary growth the root becomes temporarily much thinner when the endodermis, cortex and epidermis are lost.

Few monocotyledonous species undergo secondary growth. However, in plants of the grass family shoot growth is not entirely dependent on the apical meristem. In grasses, tissues at the bases of the internodes remain meristematic even though the rest of the internode is fully differentiated. Each internode thus has its own growth zone called an **intercalary meristem** (Fig 24.14). The activity of the intercalary meristem is one of the reasons why grass shoots regenerate rapidly after being severely damaged in haymaking, lawnmowing and animal grazing.

24.2 Physiological aspects of growth and development

A seedling becomes established using the reserve of food stored in the seed (Chapter 24). When the food reserve has been used up continued growth and development depend on the young plant obtaining alternative sources of energy and raw materials. The capture of energy in sunlight and its conversion to chemical bond energy is described in Chapters 5 and 13. The roles of minerals in plant metabolism are dealt with in Chapter 14. There are, however, many other physiological aspects of plant growth. What for example causes roots to grow down into the soil and shoots to grow up into the air? What controls the orderly sequence of cell division and differentiation at the apices of roots and shoots? Why do deciduous trees shed their leaves in autumn, why do buds spurt into growth in spring and summer and why is the vascular cambium seasonal in its activity?

Climate plays some part in controlling plant growth. The metabolism of green plants occurs efficiently only if such factors as temperature, light intensity and water availability are at an optimum. Nevertheless many species of flowering plants stop growing even when environmental factors are ideal. This suggests that there is an inbuilt growth control mechanism which causes plants to enter a period of dormancy at certain times of the year. Thus **external** and **internal factors** must be examined if we are to find a physiological explanation for the morphological and anatomical changes described earlier.

24.2.1 External factors

1 Light

When the **compensation point** (Chapter 13) has been passed a green plant begins to increase in dry mass as it stores photosynthesised energy-rich substances, mainly carbohydrates like starch. Some of the energy released from the food reserves as high energy phosphate bonds of ATP is used in growth for building up the many complex constituents of the protoplasm and walls of new cells. The **intensity of light** largely determines the time required to reach the compensation point each day. If a plant cannot build up sufficient food reserves in a day to provide for energy-requiring processes such as growth, it cannot remain active for long and must either die or undergo a period of dormancy.

In a temperate climate the length of time a plant is subjected to light each day varies from season to season. Flowering plants respond to changes in **day length**. Growth takes place in seasons of favourable climate and stops when winter approaches. This response is called **vegetative photoperiodism**. It accounts for leaf fall and bud dormancy in deciduous trees and the production of storage organs, such as bulbs, tubers, corms and taproots by herbaceous perennials.

The pattern of growth is affected by the **wavelength of light** to which green plants are exposed. Kept in total darkness shoot systems soon begin to show signs of **etiolation**. The internodes become abnormally long and the leaves remain embryonic and do not develop chlorophyll (Fig 21.15). Red light alone can prevent etiolation. Red light converts the inactive form of phytochrome P_r to the active form P_{ir} (Chapter 23). P_{ir} promotes leaf expansion and inhibits internode elongation and the normal growth pattern of the shoot is established.

Both vegetative photoperiodism and prevention of etiolation are good examples of the ways in which light controls the balance of growth promoters and inhibitors in plants. The **direction** from which light is

Fig 24.15 Etiolation in pea seedlings

(a) Etiolated seedling grown without light

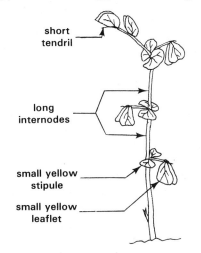

short tendril

long internodes

small yellow stipule

small yellow leaflet

(b) Seedling grown in normal daylight

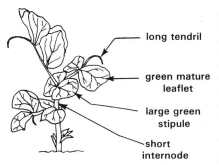

long tendril

green mature leaflet

large green stipule

short internode

received by shoots and roots also affects their direction of growth by altering the distribution of growth promotors (section 24.2.2).

2 Temperature

Because growth depends on biochemical reactions catalysed by enzymes, it is affected by **temperature**. In the tropics the average yearly temperature is high enough to allow plants to grow for most of the time. In temperate zones growth is seasonal and often stops altogether in the winter because of low temperatures. In winter annual plants survive as **seeds**, herbaceous perennials as **bulbs, corms, tubers** and **rhizomes** or by the production of some other kind of fleshy storage organ. Deciduous shrubs and trees develop **dormant winter buds**.

Exposure to a period of **low temperature** is essential to break the dormancy of perennating organs and winter buds (Fig 24.16). As with seeds, the effect of chilling is thought to reduce the amount of growth inhibitors. This ensures that growth normally occurs when the more suitable weather conditions of spring and summer are about to follow.

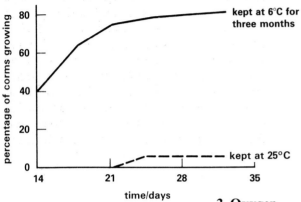

Fig 24.16 Effect of chilling in breaking the dormancy of *Gladiolus* corms (after Ginzburg (1973))

3 Oxygen

Higher plants can produce sufficient energy for growth only if **oxygen** is available for aerobic respiration. Root growth of most plants is inhibited in waterlogged soils where oxygen is scarce. Flowering plants which live totally or partly immersed in water are called hydrophytes. They often have special **aerating tissue** which stores oxygen made in photosynthesis (Fig 31.14). The storage of oxygen helps hydrophytes to overcome the relative scarcity of dissolved oxygen in water.

4 Carbon dioxide

Carbon dioxide is essential for photosynthesis. If growth is to occur beyond the seedling stage carbon dioxide must be available in sufficient quantities.

5 Water availability

Elongation of cells during the primary growth of roots and shoots is mainly due to vacuolation. **Water** is essential for **vacuolation** and the subsequent maintenance of **turgidity** of young growing organs. The direction of root growth too is affected by water availability (section 24.2.2).

The synthesis of new cellular components, an essential feature of growth, takes place in water like all enzyme-catalysed reactions. It is in water that the raw materials for growth are **absorbed** and **circulated** inside plants.

6 Mineral elements

Minerals are required for many facets of plant metabolism on which growth depends (Chapter 14). However high concentrations of minerals such as copper are toxic to plants and inhibit growth.

7 Gravity

Gravity affects the distribution of growth promoters at root and shoot apices thus influencing the direction in which they grow (section 24.2.2). However, unlike the other factors listed above, gravity does not effect the amount of growth.

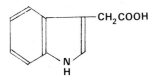

Fig 24.17 Indoleacetic acid

24.2.2 Internal factors

Plants contain **growth promoters** and **inhibitors**. Because an understanding of the ways in which they work could enable man to control plant growth more efficiently, much effort has been put into investigations concerning these substances which are generally called **plant hormones**. The more important plant hormones so far identified are **auxins**, **gibberellins**, **cytokinins** and **abscissic acid**.

Fig 24.18 Polar transport of IAA

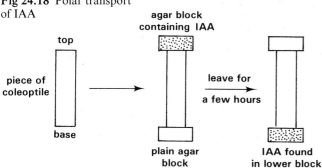

1 Auxins

Indoleacetic acid (IAA) is the main **auxin** made in plants (Fig 24.17). Produced at the apices of shoots from the amino acid tryptophan, IAA is transported backwards to the zones of elongation in roots and shoots, where it regulates **cell extension** and the **differentiation** of vascular tissues. Auxin transport is uni-directional or polar (Fig 24.18). In coleoptiles IAA is carried in parenchyma tissue but in mature roots and shoots the main transporting tissue is probably the phloem.

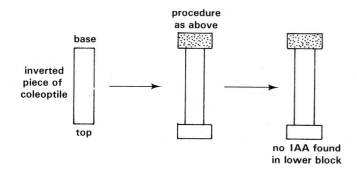

Fig 24.19 Demonstration of the role of IAA in apical dominance

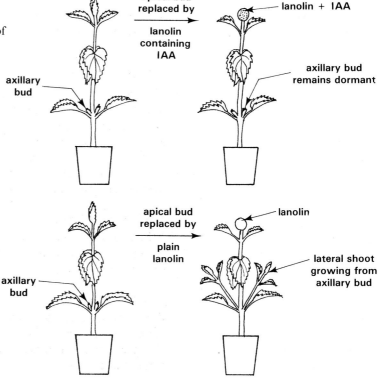

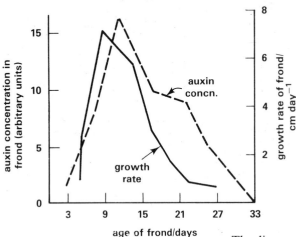

The IAA made by apical buds usually inhibits lateral buds from growing into side shoots, a phenomenon called **apical dominance** (Fig 24.19). Leaf expansion is IAA-controlled (Fig 24.20) as is the growth of fruits and seeds. The positions of procambial strands at shoot apices too are determined by IAA made in leaf primordia (section 24.1.1). IAA released from the expanding buds of woody plants during spring triggers the vascular cambium into activity. IAA also stimulates translocation of organic substances in the phloem. It is therefore apparent that the general shape of a plant, its orderly internal growth and development and its rate of growth are very much affected by auxins.

Fig 24.21 (a) Effect of light on distribution of IAA at the apex of an oat coleoptile

The discovery of auxins resulted from investigations into the tendency of shoot systems to grow towards a source of light (Fig 24.21) a phenomenon called **positive phototropism**. Although there is still much to be understood of the underlying mechanism, auxins almost certainly play a key role. The effect of unilateral light on a shoot apex is to cause more auxin to be present on the shaded side. Thus more IAA is transported to the elongating zone on the shaded side of the shoot. The result is that the shaded side grows more rapidly and the shoot bends towards the source of light (Fig 24.22).

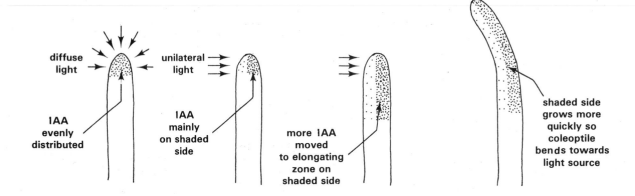

diffuse light — IAA evenly distributed

unilateral light — IAA mainly on shaded side

more IAA moved to elongating zone on shaded side

shaded side grows more quickly so coleoptile bends towards light source

Fig 24.21 (b) Phototropic response of the shoots of cress seedlings (Biofotos)

Fig 24.22 Some important landmarks in the discovery of auxins (after Wareing and Philips (1970))

(a) Darwin (1880)

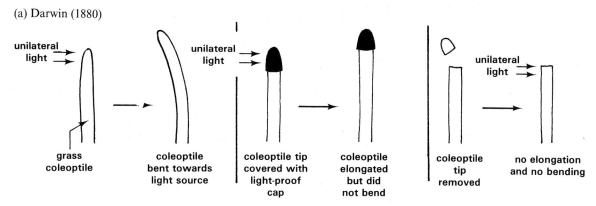

Darwin's results show that it is the tip of the coleoptile which detects the stimulus of light.

(b) Boysen-Jensen (1913)

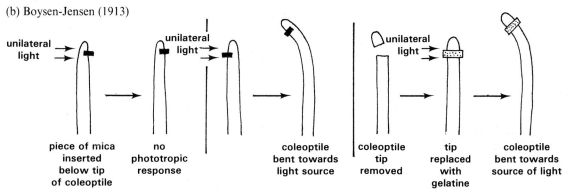

These results suggest that a stimulus for growth coming from the tip cannot pass through a solid barrier such as mica but can pass through an aqueous medium.

(c) Paàl (1919)

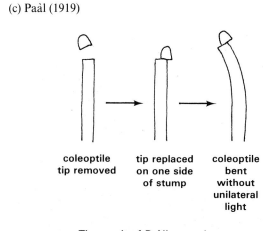

coleoptile tip removed   tip replaced on one side of stump   coleoptile bent without unilateral light

The result of Paàl's experiment suggests that a response from one side of the tip brings about bending of the coleoptile.

(d) Went (1928)

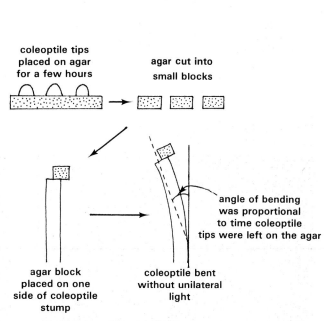

coleoptile tips placed on agar for a few hours   agar cut into small blocks

angle of bending was proportional to time coleoptile tips were left on the agar

agar block placed on one side of coleoptile stump   coleoptile bent without unilateral light

Went's observations indicate that a growth-promoting substance had passed into the agar blocks.

Roots are usually **negatively phototropic** and grow away from a light source. The effect of unilateral light on auxin distribution at the root apex is similar to that at a shoot apex. What then causes the shaded side of a root treated in this way to grow less quickly than the side exposed to light? Experiments have shown that whereas relatively high concentrations of auxin stimulate shoot growth, roots grow quickest when given relatively low concentrations of IAA (Fig 24.23). Thus the small amount of IAA transported to the elongating zone on the illuminated side of a root causes more rapid growth than on the shaded side so the root gradually bends away from the source of light.

Fig 24.23 Effect of IAA on root and shoot growth (after Audus (1959))

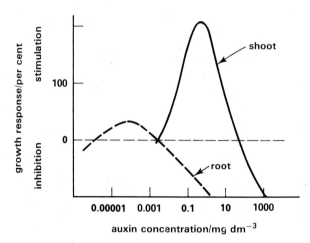

Having explained the phototropic response in terms of changes in auxin distribution it is much more difficult to determine why uni-directional light has such an effect. Blue light of wavelength 440–460 nm is more effective than any other in stimulating a phototropic response (Fig 24.24). The **action spectrum** for phototropism also has a minor peak at 360 nm and resembles combined absorption spectra for carotene and riboflavin. If, or how, these substances could affect auxin distribution is not yet known. Little is known too of the ways in which IAA affects cell growth. Some workers think that auxins soften the cell wall substances causing rapid cell enlargement by vacuolation as water is absorbed. Auxins may also stimulate cell growth by helping the unloading of metabolites from phloem sieve-tubes.

Fig 24.24 Action spectrum for phototropism in oat coleoptiles

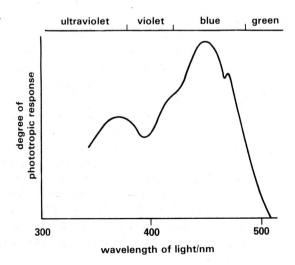

Roots and shoots also react to a unilateral stimulus of gravity. Shoots are usually **negatively geotropic**, growing away from a gravitational pull, while roots show **positive geotropism**. Changes in the normal pattern of auxin distribution are again thought to account for the responses (Fig 24.25). Why gravity should bring about such changes is still not at all clear. Roots are also **positively hydrotropic**, growing towards a source of water, but again the reasons for this are still something of a mystery.

Fig 24.25 Geotropism in pea seedling

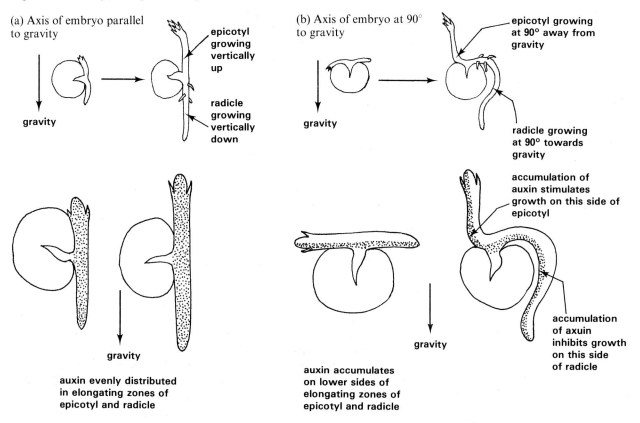

(a) Axis of embryo parallel to gravity

epicotyl growing vertically up

radicle growing vertically down

gravity

auxin evenly distributed in elongating zones of epicotyl and radicle

(b) Axis of embryo at 90° to gravity

epicotyl growing at 90° away from gravity

radicle growing at 90° towards gravity

accumulation of auxin stimulates growth on this side of epicotyl

accumulation of axuin inhibits growth on this side of radicle

gravity

auxin accumulates on lower sides of elongating zones of epicotyl and radicle

2 Gibberellins

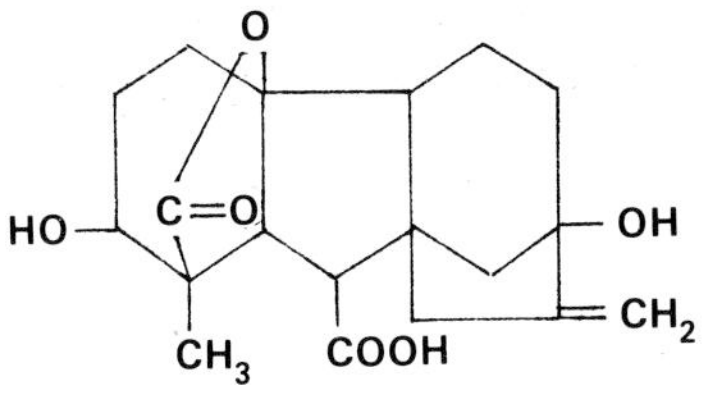

Gibberellins take their name from the fungal parasite *Gibberella fujikuroi* which causes a disease of rice plants known as 'foolish seedling'. Seedlings which are attacked by the fungus develop very long internodes so that the shoot systems are much taller than those of healthy plants. By the late 1920s Japanese workers showed that a growth-promoting substance called gibberellin was present in a liquid medium in which the fungus had been grown in pure culture. However, the chemical nature of gibberellin was unknown until the 1950s when British and American biochemists showed it to be **gibberellic acid** (Fig 24.26).

Fig 24.26 Gibberellic acid

Since then a number of closely related substances having similar effects have been isolated from higher plants. They are synthesised at the apices of roots and shoots but little is known of the way in which they are transported. The best known effect of gibberellic acid (GA) is to bring about **internode elongation**. GA also stimulates enzyme production during seed germination and plays a role in **cell division** and **tissue differentiation**. When applied to woody stems GA promotes division of the vascular cambium, but the derivatives fail to differentiate into mature xylem elements unless IAA is added. However GA on its own brings about differentiation of mature phloem elements. This suggests that the order in which primary xylem and phloem appear in root and shoot may be governed by the ratio of IAA:GA passing back from the apical meristems.

Fig 24.27 Zeatin

3 Cytokinins

Interest in the **cytokinins** began in the 1950s when **kinetin** was found to affect the growth of plant tissue in sterile culture media. Kinetin is DNA which has been autoclaved in an acid solution. Kinetin closely resembles adenine, one of the nitrogenous bases of DNA and RNA, but it has not yet been isolated from higher plants. **Zeatin** is one of several similar compounds having comparable effects which has been extracted from a number of plants in recent years (Fig 24.27).

Cytokinins are frequently found in xylem sap and are thought to be synthesised in the roots from where they are carried to other parts of the plant in the transpiration stream. Production of cytokinins is at its highest when a plant is growing rapidly and falls off as ageing begins.

Much of what is known about cytokinins comes from tissue culture experiments. Small samples of living cells called **explants** are taken from plant organs and aseptically transferred to sterile growth media containing sugar, minerals, vitamins and hormones. Growth of the transplanted tissue follows, but the pattern of growth depends on the composition of the medium, in particular the plant hormones present. With no cytokinin but IAA present the tissue grows into a mass of similar large multinucleate cells. When a small amount of cytokinin is added, a **callus** of uninucleate cells appears indicating that cytokinins have an effect on cell division. By changing the ratio of IAA to cytokinin the first signs of **organ differentiation** are seen. Roots appear on the explant if the ratio of auxin to cytokinin is high while shoots appear if the ratio is low (Fig 24.28). With no hormones, growth of the explant does not take place.

Fig 24.28 (a) Root formation from callus tissue of cowpea, ×4

Fig 24.28 (b) Shoot formation from callus tissue of tobacco, ×2.5

If these findings reflect what happens in intact plants then cytokinins evidently play a significant role in various aspects of plant growth. Particularly important are the ways in which cytokinins and other plant hormones interact in controlling plant growth.

4 Abscissic acid

Plant cells contain many substances which inhibit some process or another. One of the best known compounds which inhibits growth is **abscissic acid** (Fig 24.29). Abscissic acid (ABA) was discovered in 1964 and has also been

Fig 24.29 Abscissic acid

called abscissin and dormin. The quantity of ABA in the leaves of deciduous trees increases in the shorter days of autumn. The accumulation of ABA slows down the growth rate of shoots and stimulates the formation of dormant **winter buds**. Soon leaf fall, **abscission**, occurs (Fig 24.30). In this way the plant is prepared for winter conditions. The quantity of ABA in the shoot system falls after a period of cold weather. Dormancy of winter buds is then broken and growth follows in spring and summer. Once again it can be seen that hormone balance is controlled by climate and growth generally occurs only when a spell of good weather is about to follow.

Fig 24.30 (a) Longitudinal section through an abscission zone of sycamore, × 10

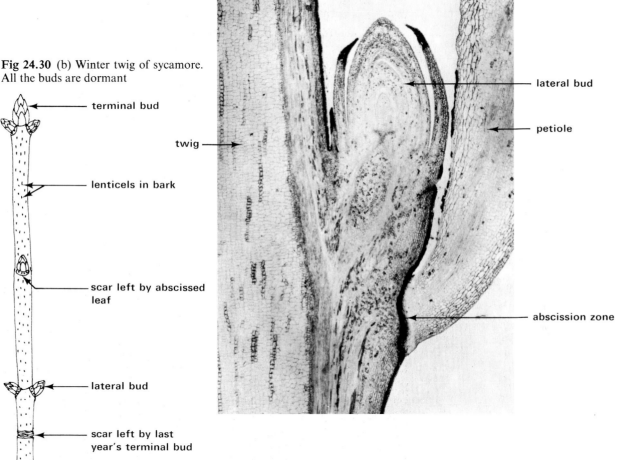

Fig 24.30 (b) Winter twig of sycamore. All the buds are dormant

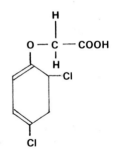

- terminal bud
- lenticels in bark
- scar left by abscissed leaf
- lateral bud
- scar left by last year's terminal bud

- lateral bud
- petiole
- abscission zone
- twig

24.2.3 Applications of plant hormones in agriculture and horticulture

With the explosion of information obtained about growth-regulating substances over the past twenty years or so it is not surprising that some of the knowledge has been put to practical use. There are many synthetic compounds with auxin-like effects now in common use in agriculture and horticulture. They include various **phenoxyacetic acids** such as 2,4D (Fig 24.31) which stimulates the formation of adventitious roots at the bases of

Fig 24.31 2,4-dichlorophenoxy-acetic acid

cut shoots (Fig 24.32) and is an invaluable aid to horticulturalists in accelerating the **striking of cuttings**. Many species of ornamental plants such as carnations and chrysanthemums are propagated by means of cuttings. Synthetic auxins also encourage the formation of wound tissue which knits together parts of plants after **grafting**.

Fig 24.32 Effect of synthetic auxin on the striking of chrysanthemum cuttings

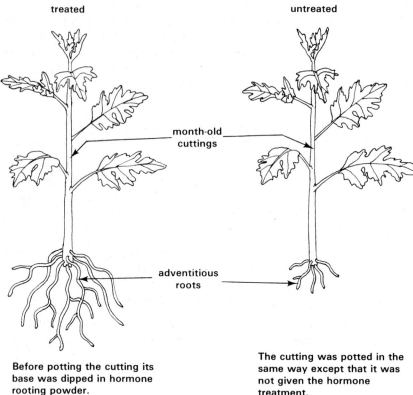

treated   untreated

month-old
cuttings

adventitious
roots

Before potting the cutting its base was dipped in hormone rooting powder.

The cutting was potted in the same way except that it was not given the hormone treatment.

When sprayed on to fruit trees after flowering the phenoxyacetic acids prevent young fruit from dropping off. Large increases of apple and pear crops have been achieved in this way (Chapter 22).

Perhaps the most widespread use of synthetic auxins has been as **weed killers** applied to fields of cereal crops, pasture and lawns. The roots and shoots of the same plant respond differently to a particular concentration of auxins (section 24.2.2). It has also been found that different groups of plants vary in their reaction to applications of similar concentrations of synthetic auxin. Dicotyledonous plants such as dandelions and thistles are more sensitive to 2,4D than are monocotyledons such as grasses, wheat and maize. When sprayed on to an area of mixed species, for example a weedy lawn or a field of wheat infested with thistles, the dicotyledons are stimulated into abnormal growth. The internodes become grossly elongated, sieve-tubes become blocked and cells with unbalanced chromosome numbers arise. Distortion of the root system also occurs with the net result that the growth pattern is thrown completely out of control and death quickly follows. Meanwhile the monocotyledons are unaffected and are thus freed from unwanted competitors.

Despite the many practical benefits which result from their uses, synthetic auxins can cause serious ecological problems if used indiscriminately. Some are toxic to insects and have been known to kill large numbers of useful insects such as bees with considerable economic loss. Many weeds are the food plants of pollinating insects which starve when weeds are destroyed. Finally some weeds have evolved auxin-resistant varieties which are virtually impossible to eradicate by treatment with synthetic auxins.

SUMMARY

Growth is an increase in mass, development an increase in complexity. They go hand in hand from the time a seed germinates until a mature plant is formed. The pattern and rate of growth and development are influenced by heredity and by the environment.

Mitosis and differentiation are integral features of growth and development. Primary growth is brought about by the activity of the apical meristems at the tips of roots and shoots. Herbaceous flowering plants have no other growth pattern. In contrast woody plants have lateral meristems, the vascular and cork cambium, which produce additional tissues called the secondary plant body. It is added to the primary body derived from the apical meristems. Grasses have intercalary meristems which replace parts of the shoot system removed in grazing and grass-cutting.

Differentiation is the change in structure of cells cut off from the meristems which produces the variety of cells and tissues seen in complex plants. Each type of cell is suited to one or more physiological or mechanical roles.

The intensity and number of hours of sunlight in a day, temperature, availability of oxygen gas, concentration of carbon dioxide in the air, water and mineral availability are the more important external factors which affect the rate of growth. The direction of light and gravity also determines the direction in which roots and shoots grow.

Auxins play a key role in tropisms, the direction of growth in response to unilateral stimuli such as light and gravity. Auxins also regulate enlargement of cells and leaves, differentiation of vascular tissue and activity of the vascular cambium. Other plant hormones which interact with auxins in promoting cell division, growth and differentiation are gibberellins and cytokinins.

The fall of leaves which signals cessation of growth in deciduous trees in the autumn is triggered by yet another hormone called abscissic acid. The balance of hormones and hence the rate and pattern of growth and development is affected by climatic factors, especially temperature.

QUESTIONS

1 The drawings show cells from two different plant tissues as seen in transverse section.

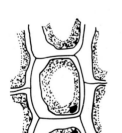

Cell P Collenchyma (×400)   Cell Q Sclerenchyma (×400)

(a) (i) List **three** structural differences, visible in the drawings, between the two cells.
(ii) State **one** further difference which could be shown by the use of a suitable stain.
(b) Make an accurate **scale** drawing of cell **Q** as it would appear in longitudinal section at a magnification of × **200**.
(c) Indicate clearly, by drawing on the outlines provided, where you would expect to find tissues P and Q in a typical dicotyledonous plant. (Do not draw individual cells.)

Label the regions P and Q respectively.

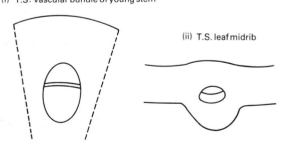

(i) T.S. vascular bundle of young stem

(ii) T.S. leaf midrib

(d) (i) What is the function of the two tissues in plants?
(ii) Name **two** tissues which perform a similar function in mammals.
(iii) Suggest how this function is performed in the flower stalk of a bluebell, which does not contain tissues **P** and **Q**.   (WJEC)

2 For a young stem of a flowering plant give an illustrated account of:
(a) the formation of a cross wall following nuclear division in a cell of the apical meristem;
(b) the differentiation from procambial cells of (i) a sieve

tube and companion cells and (ii) a pitted xylem vessel;
(c) secondary thickening resulting in the formation of new xylem and phloem. (O)

3 (a) The diagrams **A** to **D** illustrate the effect of one-sided illumination on oat coleoptiles. In each, an oat coleoptile is shown **before** and **after** the period of illumination.

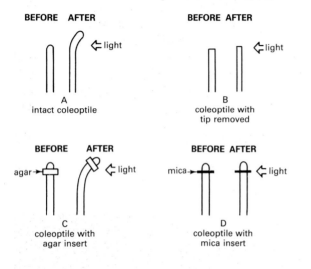

(i) Copy diagram **C** and add an arrow labelled **X** to indicate the site which is sensitive to light, and an arrow labelled **Y** to indicate the site of visible response.
(ii) Explain the difference in response between (1) **A** and **B**, and (2) **C** and **D**.
(b) (i) Further experiments were set up as shown in the diagrams below. Copy the diagrams and draw the expected appearance of the coleoptiles after exposure to the light.

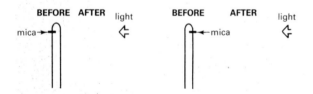

(ii) Explain the results.
(c) A coleoptile tip was removed and placed on blocks of agar. After a period of illumination from one side, the blocks were placed on one side
of decapitated coleoptiles.
The coleoptiles curved as
shown on the right.

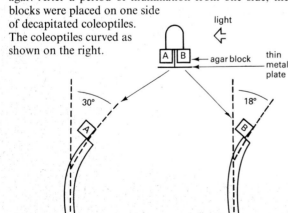

(i) Which block contained the highest concentration of

auxin?
(ii) Which **one** of the following hypotheses is supported by the evidence: (1) light destroys auxin, (2) light causes auxin to move away from light? Give one reason for your choice.
(d) State how the hormones produced by the thyroid gland of a mammal and auxins
(i) are similar in effect,
(ii) differ in operation. (WJEC)

4 Two leaf blades were removed from a healthy intact plant (see the figure). One of the petioles was left exposed and the other was treated with a hormone. Two weeks later a longitudinal section through the petioles and stem was taken. The tissue distribution of this section is shown in the figure.

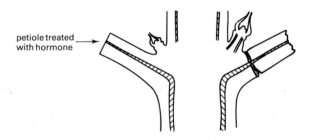

(a) (i) Describe **three** differences between the treated and untreated parts of the plant which are apparent from the figure.
(ii) Copy the figure and on it mark clearly with crosses the position of **two** areas where actively dividing cells would be present.
(b) (i) What method should be used to apply the hormone?
(ii) Describe a suitable control for this investigation.
(c) The effect of two hormones, **A** and **B**, on abscission of leaves in similar plants was measured. The results are shown in the graph.

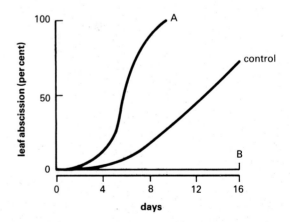

(i) From the results shown in the graph, which of the two hormones do you think was applied to the cut petiole of the plant shown in the figure? Explain your answer.
(ii) Give the name of a hormone which could be hormone **A**. (JMB)

25 Mammalian reproduction

25.1 The male reproductive system — 480

25.1.1 Structure of the testes — 481
25.1.2 Spermatogenesis — 481
25.1.3 Male hormones — 483

25.2 The female reproductive system — 483

25.2.1 Structure of the ovaries and oogenesis — 483
25.2.2 Hormonal control of ovulation — 485
25.2.3 Other effects of female hormones — 487

25.3 Reproduction — 487

25.3.1 Copulation — 487
25.3.2 Fertilization and implantation of the embryo — 488
25.3.3 The placenta — 489
25.3.4 Birth and lactation — 491
25.3.5 Birth — 492

Summary — 493

Questions — 494

25 Mammalian reproduction

Whereas most flowering plants are hermaphrodite all mammals are **unisexual**. This means that mammalian gametes are produced in separate male and female animals and **cross-fertilisation** is inevitable. Consequently genetic variation among the offspring of mammals is ensured.

For sexual reproduction to occur successfully gametes have to be mature at the right time. Males and females have to be in the same locality and attracted to each other. Fertilisation must occur and the resulting embryo must be protected and provided with food for it to survive and develop properly.

Reproduction in mammals is a carefully regulated process although the controlling mechanisms vary in different species. An important feature of mammalian reproduction is that it occurs in such a way that the effects of chance and adverse environmental conditions are minimised. A successful outcome to sexual reproduction is therefore more likely in mammals than it is in many other groups of animals. Mammalian gametes are made in sex organs called **gonads**. The **testes** are the male gonads, the **ovaries** the female gonads. **Gametogenesis** is the process which gives rise to male and female **gametes**.

25.1 The male reproductive system

The production and maturation of **sperm**, the male gametes, takes place in a pair of compact testes. The testes originate in the abdomen from embryonic tissue which also gives rise to the urinary system. Indeed the urethra, through which urine passes to the outside, also provides the route for sperm to leave the male reproductive system (Fig 25.1). As well as making sperm, the testes produce male sex hormones.

Fig 25.1 Male reproductive organs (from the side)

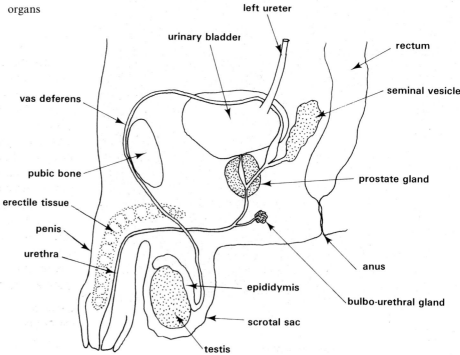

25.1.1 Structure of the testes

As they mature the testes descend from the abdomen into two sacs of skin called **scrotal sacs**. Under the microscope a thin section of a testis is seen as thousands of very fine, highly coiled **seminiferous tubules** in which sperm are produced. The seminiferous tubules are continuous with other tubules called the **vasa efferentia, epididymis** and **vasa deferentia** (Fig 25.2). The sperm travel through the system of tubules to the **seminal vesicles** where they are stored before passing out of the body through the urethra.

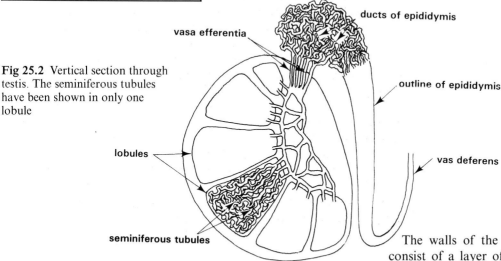

Fig 25.2 Vertical section through testis. The seminiferous tubules have been shown in only one lobule

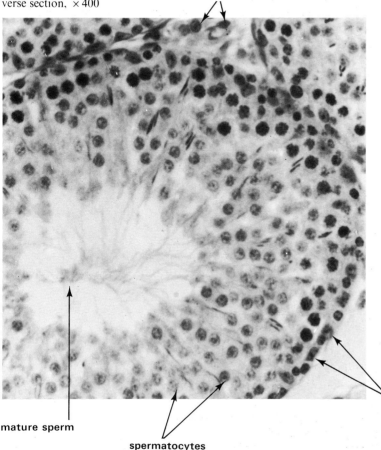

Fig 25.3 Photomicrograph of a seminiferous tubule in transverse section, × 400

The walls of the seminiferous tubules consist of a layer of cells called the **germinal epithelium** from which sperm originate (Fig 25.3). Sperm at different stages of development are found inside the seminiferous tubules. As they mature the sperm become attached to the relatively large **Sertoli cells**. The tubules are held together by connective tissue which contains **interstitial (Leydig) cells**, nerve fibres, blood and lymphatic capillaries.

25.1.2 Spermatogenesis

Spermatogenesis is the process whereby sperm are made. The production of sperm begins with divisions of the **primordial germ cells** which make up the germinal epithelium. The first divisions are mitotic (Chapter 7) and produce many genetically identical cells called **spermatogonia**. Next the spermatogonia increase in size and mass to become **primary spermatocytes**. The primary spermatocytes then divide by meiosis (Chapter 7). The first meiotic division produces **secondary spermatocytes**, two from each primary spermatocyte. The second meiotic division gives rise to **spermatids**, two from each secondary spermatocyte (Fig 25.4). The final stage of spermatogenesis is differentiation of sperm from the spermatids. Maturing sperm cluster around the Sertoli cells from which they are thought to obtain materials essential for differentiation.

Fig 25.4 Spermatogenesis

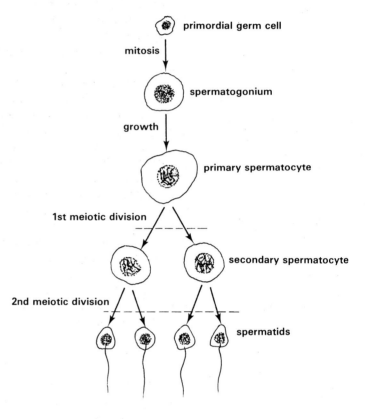

Fig 25.5 (a) Diagram of a sperm

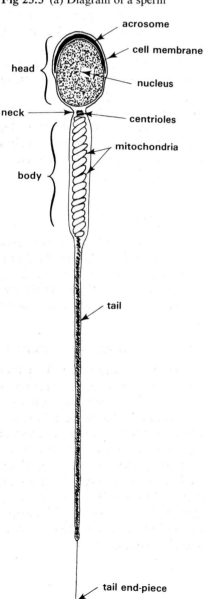

A mature sperm has a **head** which contains a nucleus. Attached to the head is a long **tail** (flagellum). At the base of the tail where it joins the sperm head, there are many mitochondria (Fig 25.5). They provide energy for contraction of protein filaments in the flagellum which consequently undulates. It provides the propulsive force for sperm to swim. Sperm can swim at a rate of 1–4 mm a minute.

Spermatogenesis is inhibited by heat. Because the testes are located in the scrotal sacs where the temperature is lower than in the abdomen, production of sperm is normally encouraged.

Fig 25.5 (b) Photomicrograph of some human sperm, × 1000

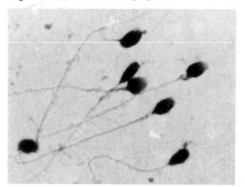

Fig 25.6 Mechanism controlling testosterone secretion

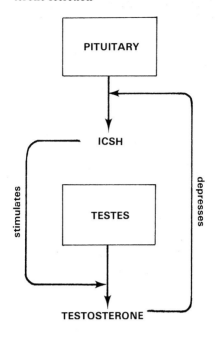

Fig 25.7 Structure of testosterone

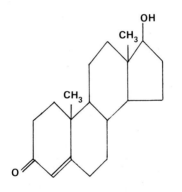

25.1.3 Male hormones

Spermatogenesis is stimulated by a gonadotrophic hormone called **follicle stimulating hormone, FSH** produced by the anterior pituitary (Chapter 21). In both male and female, FSH stimulates gametogenesis.

Another gonadotrophic hormone made in the anterior pituitary is **luteinising hormone, LH**. In the male, LH is alternatively called **interstitial cell stimulating hormone (ICSH)**. As the name suggests ICSH stimulates the interstitial cells to secrete a male sex hormone called **testosterone**. ICSH and testosterone interact in a negative feedback control mechanism so that the output of testosterone is kept relatively constant (Fig 25.6).

Male sex hormones are collectively called **androgens**. The functions of androgens are twofold. First, they regulate the development of the male **accessory sex organs**. They include the vasa efferentia, epididymes, vasa deferentia, penis and scrotal sacs. The second main function of androgens is to control the development and maintenance of the **secondary sexual characteristics** which in men include growth of facial and pubic hair, a deep voice and general muscular development. Androgens may also be partly responsible for a number of behavioural characteristics. Removal of the testes, castration, does not usually lead to loss of secondary sexual characteristics as androgens are steroids (Fig 25.7) and following castration there is increased output from the adrenal cortex of steroids similar to androgens (Chapter 21).

The development of male secondary sexual characteristics and the production of sperm begin in the human at a period of life called **adolescence**. The point at which sexual maturity is reached is called **puberty**. The testes also produce female sex hormones; their function in males is not clear.

25.2 The female reproductive system

Female gametes are called **ova**. Their production and maturation take place in the **ovaries**. The ovaries also produce female sex hormones. A pair of ovaries lies in the abdominal cavity. The ova they release are sucked by ciliary action into funnels from where they move into the **oviducts** and **uterus**. If an ovum is not fertilised it passes into the **vagina** and to the exterior through the **vulva** (Fig 25.8).

25.2.1 Structure of the ovaries and oogenesis

Spermatogenesis in the testes (section 25.1.2) is paralleled in the ovaries by **oogenesis**, the production of ova. Unlike the testes, ovaries do not contain tubules but consist mainly of connective tissue. Like the testes, however, ovaries have a germinal epithelium. Ovarian germinal epithelium is on the outside of the ovaries. **Primordial germ cells** in the germinal epithelium divide by mitosis to form many **oogonia**. The oogonia become surrounded by **follicle cells**, also derived by mitosis from the germinal epithelium. The follicle cells and enclosed oogonia are called **primary follicles**. They migrate to the centre of the ovary.

Fig 25.8 Female reproductive organs (ventral view)

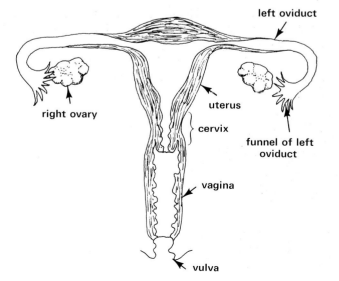

At birth there are up to 400 000 primary follicles in a human ovary, where they remain dormant until puberty. Hormones from the pituitary gland then start the process of oogenesis. In a woman only about 400 primary follicles normally mature over a reproductive lifetime of about 30–40 years. The end of this period, when follicle development stops, is called the **menopause**. The remaining primary follicles degenerate into cyst-like **atretic follicles** and remain in the ovaries.

Ova usually develop one at a time. Each oogonium grows into a **primary oocyte** which by this stage is surrounded by several layers of follicle cells. Between the dividing follicle cells, cavities appear, filled with follicular liquid. The follicle becomes surrounded by two layers derived from the ovarian connective tissue. An outer fibrous **theca externa** encloses an inner vascular **theca interna**. The mature follicle is called an **ovarian follicle** and can usually be seen bulging at the surface of the ovary (Fig 25.9). During development, a follicle grows from about 0.05 mm to about 12 mm in diameter.

Fig 25.9 (a) Photomicrograph of a thin section of ovary from a rat, × 50

Fig 25.9 (b) Photomicrograph of ciliated columnar epithelium in an oviduct, × 1000

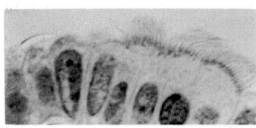

Fig 25.10 Oogenesis. Compare with Fig 25.4

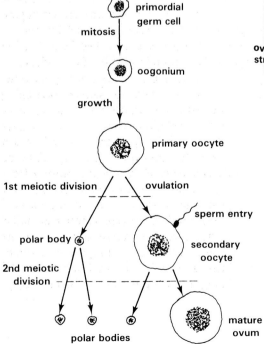

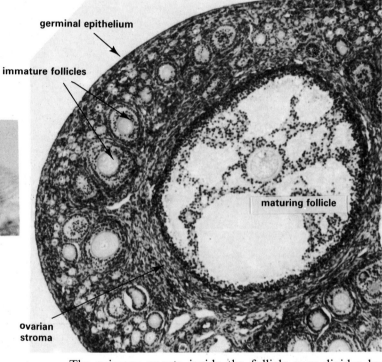

The primary oocyte inside the follicle now divides by meiosis. The first meiotic division, which occurs just before ovulation, produces a **secondary oocyte** and attached to it, a small cell called a **polar body**. **Ovulation** is the release of the secondary oocyte from the ovary into the abdominal cavity from where it is drawn into the oviduct. The second meiotic division usually occurs in the oviduct, just after a sperm nucleus enters the secondary oocyte. The secondary oocyte then produces the ovum and another small polar body. The second meiotic division gives rise to two more polar bodies from the polar body produced in the first meiotic division. The polar bodies have no known function (Fig 25.10).

During oogenesis, only one functional ovum arises from each primary oocyte, whereas spermatogenesis produces four functional sperm from each primary spermatocyte (section 25.1.2).

25.2.3 Hormonal control of ovulation

As in the male, gametogenesis in the female mammal is influenced by hormones from the anterior pituitary, particularly **FSH**. Unlike spermatogenesis, however, oogenesis is not a continuous process. It is a characteristic of mammals that the embryos develop inside the mother's reproductive tract for a period prior to birth. This period is called the **gestation period**. Since there is limited space in the mother for embryo development, there is a limit to the number of embryos which can develop at a time. The number of embryos carried by a mother during the gestation period is usually determined by the number of ova released at a time from the ovaries. In women only one ovum normally leaves the ovaries at each ovulation.

For many mammals living in the wild, it is advantageous to mother and offspring if birth occurs when food is plentiful and the weather is mild. In many species this is ensured by the seasonal secretion of FSH. Secretion of FSH in **seasonal reproducers** is influenced by day-length. In some species such as ferrets, FSH output is stepped up in spring when the number of hours of light in a day increases. In sheep, production of FSH increases in autumn when the day-length shortens. Other mechanisms such as **delayed implantation** can also determine the time of year when birth takes place.

Secretion of FSH in some species is triggered off by the nervous stimulation of copulation. The mechanism is called **induced ovulation** and occurs, for example, in rabbits. Induced ovulation ensures that ripe ova are provided immediately after sperm have been introduced into the female tract. Since fertilisation is then likely to occur it is not surprising that rabbits have gained a reputation as prolific breeders.

Many mammals produce ova at regular intervals by a mechanism in which FSH interacts with ovarian hormones to produce a rhythm of activity. The rhythm is called the **oestrus** or **menstrual cycle**. This internal method of controlling ovulation is largely independent of external influences.

Since the ovarian cycle is a sequence of events which repeats itself in a cyclical fashion, there is no fixed beginning or ending. For convenience of description we shall begin with the onset of follicle development.

Follicle development is stimulated by FSH from the anterior pituitary. FSH also causes the ovary to secrete a female sex hormone, an **oestrogen**. The most prominent oestrogen is called **estradiol**. Estradiol has several effects in the female body, some of which are described later (section 25.2.4). Among these is a negative feedback effect on FSH (Fig 25.11).

Fig 25.11 Mechanism controlling estradiol secretion

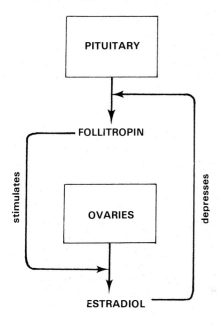

Fig 25.12 Structures of (a) estradiol and (b) progesterone. Compare with Fig 25.7

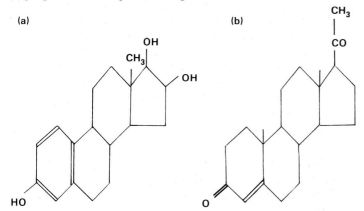

Another effect of estradiol is to cause the anterior pituitary to secrete **luteinising hormone, LH** which brings about ovulation. Ovulation usually occurs 14 days after a follicle starts to form. LH has a second function which is to stimulate the development of the **corpus luteum** from the remains of the ovarian follicle. As it grows, the corpus luteum secretes a hormone called **progesterone** which is similar in structure to estradiol and testosterone (Figs 25.12 and 25.7). Progesterone stimulates growth of the **endometrium**, the lining of the uterus, and its blood supply.

485

Fig **25.13** Mechanism controlling progesterone secretion

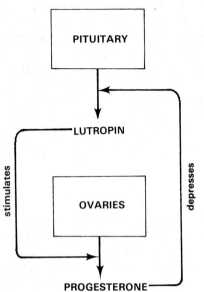

Progesterone also inhibits the release of FSH and LH from the anterior pituitary (Fig 25.13). What happens next depends on whether or not the ovum has been fertilised and implanted in the uterine wall.

If implantation occurs, then a new source of progesterone develops, the **placenta**. Continued production of progesterone inhibits secretion of FSH, so that the development of new follicles during gestation is prevented. This is an important aspect of pregnancy. If follicles continued to appear and ovulation occurred after the first conception, it might happen that other embryos would be conceived later on. A number of embryos at different stages of development would then be present in the uterus, and a severe physiological burden would be placed on the mother. In addition, relatively violent muscular contractions of the uterine wall occur at birth. Embryos which are not ready to be born are unlikely to survive the contractions.

Maintaining progesterone secretion during pregnancy is important for another reason. Progesterone stimulates growth and improves the blood supply of the endometrium. Maintenance of the endometrium is necessary for successful implantation of a fertilised ovum and subsequent attachment of the placenta.

Fig 25.14 summarises the hormonal interactions which regulate the ovarian cycle and their effects on the ovary and uterus.

Fig **25.14** Summary of the main factors involved in the ovarian cycle

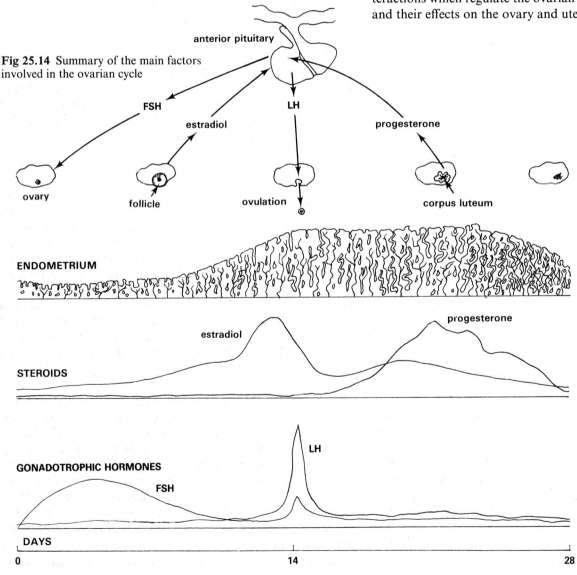

What prevents the continuing production of ova in the ovaries if an ovum is fertilised? Why is this important? What enables ova production to restart in the mother after the birth of her child?

Unwanted pregnancies can be prevented if a woman takes the contraceptive pill. The pill contains chemicals which have effects similar to estradiol and progesterone. In particular, they prevent secretion of FSH from the anterior pituitary. Consequently no follicles develop in the ovaries and fertilisation is impossible.

If fertilisation does not occur, the lack of LH from the anterior pituitary causes the corpus luteum to degenerate, and progesterone production by the ovary then stops. Growth of the endometrium which is promoted by progesterone cannot now be maintained. Consequently, the endometrium breaks down and passes through the vagina to the outside as the **menstrual flow**. Menstruation normally lasts for about four or five days.

Lack of progesterone in the blood also means that secretion of FSH from the anterior pituitary can begin once more. The cycle restarts with the development of a new follicle. About 14 days elapse between ovulation and the start of a new cycle. With ovulation marking the mid-point, the ovarian cycle takes about 28 days to complete. The length of the cycle varies from woman to woman, and may be different for the same person at different times.

25.2.4 Other effects of female hormones

Like androgens in males, oestrogens in females regulate the development of the accessory sex organs. These organs include the **oviducts**, **uterus**, **vagina** and **clitoris**. Oestrogens also control the development and maintenance of the secondary sexual characteristics which in women include growth of pubic hair, development of the mammary glands and a relatively broad pelvic girdle.

25.3 Reproduction

Fertilisation and subsequent development of the embryo occur inside the mother's body. Internal fertilisation requires the introduction of sperm from the male into the female's reproductive tract where the ova are found after ovulation. The transfer of sperm from male to female takes place in an elaborate pattern of behaviour called **copulation**.

25.3.1 Copulation

Before copulation the male and female usually interact so that each partner becomes sexually aroused. Sexual excitation of the male results in erection of the penis. The penis is made mainly of spongy **erectile tissue**. Entry of blood into the cavities of the tissue causes **erection**. Appropriate parts of the male's parasympathetic nervous system are activated in response to stimuli from the female, triggering off the sequence of events leading to erection. When erect, the penis can be pushed into the vagina, the wall of which also contains erectile tissue. Appropriate excitation of the female causes this tissue to become gorged with blood, enabling the penis to be gripped as it is pushed back and forth during copulation. The to and fro motion of the penis in the vagina is helped by mucus secretions. In the male

It has been known for many years that the ovaries secrete androgens as well as oestrogens and that testes secrete oestrogens as well as androgens. The roles of androgens in females and oestrogens in males remain a mystery. However, an important effect of oestrogens in both sexes has been discovered. After the menopause, the skeleton of women can diminish in mass by as much as 2% each year. Oestrogens inhibit the action of **parathormone** which stimulates the absorption of calcium phosphate from bone into blood and tissue fluid (Chapter 21). After the menopause, when oestrogen secretion stops, output of parathormone is uninhibited and bone deterioration sets in. Women who have oestrogen treatment during and after the menopause have significantly fewer bone fractures compared with women of a similar age who have not received oestrogen. Men have no such problem of bone deterioration because their testes secrete oestrogens as well as androgens throughout life.

the secretions come from the **bulbo-urethral gland** at the upper end of the urethra and from mucus glands along the length of the urethra (Fig 25.1). In the female mucus comes from glands inside the vagina.

When sexual excitation reaches a peak in the male, sympathetic nervous stimulation of the genitalia brings about **ejaculation**. Ejaculation is caused by rhythmic constriction of the genital ducts, beginning in the testes and progressing to the penis. The peristaltic wave pushes sperm from the testes to the urethra of the penis and into the vagina. The ejaculated fluid is called **semen**. Semen consists of sperm suspended in mucus and a milky fluid from the **prostate gland**. The prostate fluid stimulates sperm to swim. Most of the ejaculated fluid comes from the seminal vesicles. In man, approximately 350 000 000 sperm are normally present in each ejaculation.

During copulation the female also becomes excited. The to and fro motions of the penis rub against the erect clitoris. When sexual excitement reaches a peak in the female, peristalsis occurs in her reproductive tract. The direction of peristalsis is opposite to that in the male and it may help to carry sperm into the oviducts. In this way sperm are propelled efficiently to the regions where ova are likely to be found.

25.3.2 Fertilisation and implantation of the embryo

Following ovulation, secondary oocytes are sucked from the abdominal cavity into the oviducts by the action of cilia lining the ducts. Oocytes die if they are not fertilised between 8 and 24 hours after ovulation. By this time they have been wafted only a short distance down the oviducts. Consequently fertilisation normally takes place in the upper regions of the oviducts.

1 Fertilisation

Fertilisation involves entry of the sperm nucleus into the secondary oocyte. The oocyte is usually surrounded by many small cells called the **corona radiata**. The corona is dispersed by the enzyme **hyaluronidase**, secreted by the sperm when they arrive near an oocyte. Although an oocyte is approached by many thousands of sperm, only one usually enters it. The oocyte's membrane now becomes impermeable to the entry of other sperm. Entry of additional sperm is undesirable as it could produce a zygote with more chromosomes than normal. This could have adverse consequences for the embryo.

When the sperm enters, the secondary oocyte undergoes the second meiotic division to become the ovum proper. Only at this point are both gametes haploid. Fusion of the sperm and ovum nuclei produces a **diploid zygote**. It is from the zygote that the embryo develops. Sometimes the

---

What are 'test tube babies'?

In some women the tubes are blocked and cannot be opened. Until recently such women had no hope of having their own children. However, research has developed a technique whereby an ovum is removed from the woman, fertilised externally and the resulting embryo placed in its mother's uterus. The technique, which was first performed successfully with humans in 1978 in Britain, produces what are popularly called **test-tube babies**. The procedure is now so well perfected that it is possible to determine the sex of the embryo before it is placed in the mother. Consequently, such mothers can avoid having children who would probably inherit serious defects.

Fig 25.15 Photomicrograph of a 15-day human embryo embedded in the uterine wall, ×30

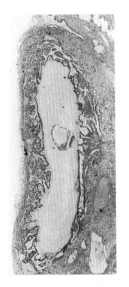

zygote divides into two separate cells from which two individuals develop. These individuals are called **monozygotic twins**. Monozygotic twins, coming from the same sperm and ovum, are genetically identical. Small differences appear as the twins develop. **Dizygotic twins** result from two different zygotes produced from two ova fertilised at the same time. Consequently dizygotic twins are genetically different from one another.

2 Implantation

The zygote undergoes several divisions, producting a ball of 16 to 32 cells by the time it reaches the uterus. It is in this form that the embryo is **implanted** in the endometrium (Fig 25.15). Peptidase enzymes from **trophoblast cells** on the outside of the embryo digest part of the endometrium, making space for the embryo. The products of digestion are absorbed by the trophoblast and are used as nutrients by the embryo. Meanwhile the trophoblast, along with surrounding endometrial cells, divide to form the placenta and fetal membranes.

25.3.3 The placenta

Considerable growth and development of the embryo, now called the **fetus**, takes place inside the uterus. Growth and development require the provision of oxygen and nutrients and the removal of waste materials. Oxygen and nutrients are supplied to the fetus by the mother's blood which also serves to remove fetal wastes. However, the fetal blood and the mother's blood do not mix. The two circulations come near to each other in the **placenta** (Fig 25.16). Materials are exchanged between the blood of the mother and fetus across the placenta almost entirely by diffusion. The arrangement of the blood vascular systems of the mother and fetus is described in Chapter 9.

Fig 25.16 Ultrasonic B-scan showing a fetus inside the uterus of a pregnant woman

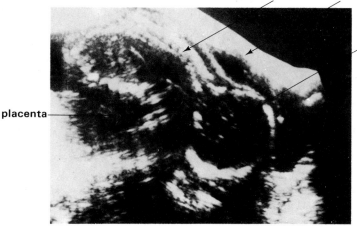

arched back of fetus   uterine wall

head of fetus

placenta

The placenta is formed by growth of some of the **fetal membranes** and part of the uterine wall in which the fetus is implanted. At an early stage several fetal membranes develop. They are the **yolk sac, allantois, amnion** and **chorion** (Fig 25.17). The yolk sac is an extension of the embryo's hind

Fig 25.17 Development of the embryonic membranes in the mammal

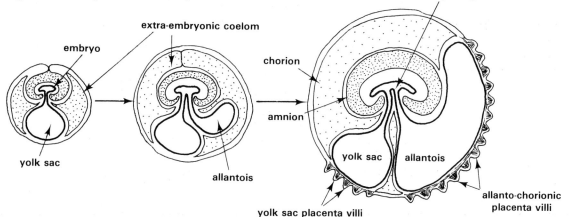

embryonic gut

extra-embryonic coelom

embryo

chorion

amnion

yolk sac

allantois

yolk sac   allantois

yolk sac placenta villi

allanto-chorionic placenta villi

gut. In birds and reptiles yolk is the source of nutrients for the embryo. This is also the case in primitive mammals such as the duck-billed platypus which lay reptile-like eggs. In marsupial mammals also, the yolk sac provides nutrients for the embryo. In placental mammals, however, the yolk sac is relatively small. Nevertheless it is important in some mammals because antibodies pass through it into the embryo from the uterus. The antibodies provide the developing fetus with passive immunity (Chapter 10). Like the yolk sac, the allantois projects from the embryonic gut.

In reptiles, birds and egg-laying mammals, the allantois acts as a urinary bladder, collecting wastes. The allantois is important in these animals, since the embryo, encased in its shell, has no means of disposing of solid wastes. In these animals too the allantois acts like a lung. Respiratory gases are exchanged across the allantoic membrane between the allantoic fluid and the atmosphere (Fig 25.18). The amnion and chorion are membranes which grow to enclose the whole embryo. They are derived from the outermost layers of the embryo which by now lies in a fluid-filled cavity called the amniotic cavity. The chorion has a great number of **villi** which are attached to the uterine wall. The blood vessels of the allantois project into the chorionic villi and it is this vascular structure which becomes the **placenta** (Fig 25.19).

Fig 25.18 Embryonic membranes of a reptilian egg

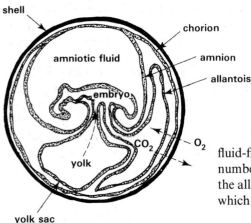

Fig 25.19 Blood vessels in the placenta

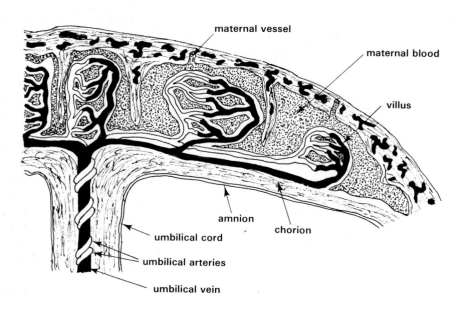

As well as acting as a route by which materials are exchanged between maternal and fetal blood, the placenta performs other vital functions. Among its other functions is the secretion of hormones. By about the second month of pregnancy in women the chorion secretes **chorionic gonadotrophic hormone**. This hormone is present in the urine of pregnant women and its detection is the basis of modern pregnancy tests. Small latex beads which are coated with an antibody specific to **human chorionic gonadotrophic hormone (HCG)** are added to a sample of urine. If the urine contains HCG the beads clump together due to the binding of HCG with the antibody. Some tests are more complicated and work by preventing the clumping of latex beads (Fig 10.16).

The chorionic villi provide a large surface area across which materials can be exchanged between maternal and fetal blood. Exchange of materials is further aided by the pattern of blood flow in the placental vessels. The maternal and fetal vessels are arranged in the placenta in such a way that maternal blood is close to the blood of the fetus (Fig 25.20). Only substances of low relative molar mass can cross the placental barrier. Essential metabolites such as oxygen, sugars, amino acids, minerals, water and vitamins are transported from the mother's blood into that of the fetus. Antibodies which provide passive immunity in the fetus also pass from the mother across the placenta. Carbon dioxide, urea and other waste products made by the fetus move in the opposite direction. Most substances pass through the placenta by diffusion, though there is evidence that some sugars are moved by active transport.

Estradiol and progesterone are also secreted from the placenta. These hormones prevent output of FSH from the anterior pituitary, thus stopping the development of new follicles in the ovaries. They also maintain the endometrium during pregnancy. Relatively high concentrations of progesterone in the uterus also keep the smooth muscle of the uterus relaxed, so protecting the developing embryos from premature birth.

25.3.4 Birth and lactation

During pregnancy, estradiol and progesterone from the placenta bring about growth of the milk ducts and secretory tissue in the mammary glands. **Lactation** is the production of milk by the mammary glands. It is stimulated by the hormone **prolactin** from the anterior pituitary body (Chapter 21) under the influence of **prolactin releasing factor, PRF,** from the hypothalamus. During pregnancy PRF output rises. However, lactation is inhibited by another hypothalamic hormone called **prolactin inhibiting factor, PIF**. Estradiol and progesterone, initially from the corpus luteum and later from the placenta, stimulate PIF output. After birth when the placenta is removed, PIF secretion subsides and PRF allows prolactin release. During breast-feeding the sucking action of the infant's mouth stimulates the nipple area. Sensory nerve impulses are transmitted to the hypothalamus, resulting in PRF secretion and hence prolactin release. The impulses also cause release of **oxytocin** from the posterior pituitary body (Chapter 21). Oxytocin encourages contraction of smooth muscles lining the mammary ducts. Milk is forcibly expelled into the infant's mouth (Fig 25.20).

Fig 25.20 (a) Mammary gland, internal structure

muscle

fat

lobules

mammary duct

nipple

lactiferous duct

rib

intercostal muscles

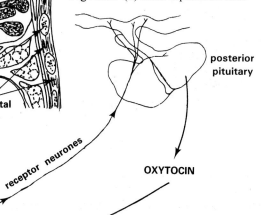

Fig 25.20 (b) Milk expulsion reflex

posterior pituitary

receptor neurones

OXYTOCIN

tactile stimulation

milk

mammary gland

What physiological effects does pregnancy have on a woman? How do the effects help the fetus to grow?

True **milk** appears a few days after birth. Just before then a fluid called **colostrum** is produced by the mammary glands. Colostrum contains no fat and little lactose compared with milk. It is in colostrum, however, that some kinds of maternal antibodies are transmitted to the baby. They provide important passive immunity against a variety of diseases in early infancy.

Fig 25.21 Stages of birth viewed from the side. (a) Dilation of cervix, (b) expulsion, (c) delivery of the placenta (afterbirth) (modified from Tortora and Anagnostakos)

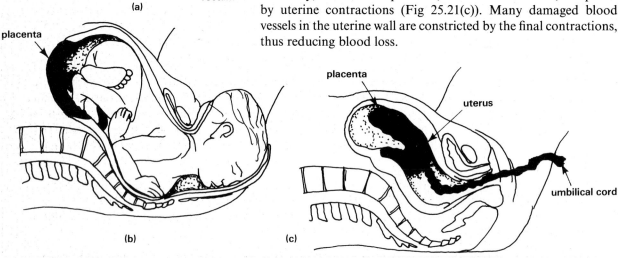

(a)

placenta

(b)

placenta

uterus

umbilical cord

(c)

25.3.5 Birth

The time during which the fetus grows and develops inside the uterus is called **gestation** and usually lasts about nine months. At the end of this period a series of events called **labour** result in **parturition**, that is, **birth** of the baby.

The onset of labour is brought about by several hormones. Among them is oxytocin from the posterior lobe of the pituitary gland. Artificial induction of labour may be brought about by administration of oxytocin into the mother's blood. Oxytocin stimulates rhythmic contractions of the uterine wall. Contractions begin at the top of the uterus and travel dowards in waves. They become regular and produce **labour pains**. The amniotic sac usually ruptures early in labour, releasing amniotic fluid, called 'breaking of the waters'. As labour progresses, the contractions become more and more frequent and intense. The uterine cervix dilates. The baby is pushed downwards, usually head first (Figs 25.21(a) and (b)). Complete expulsion of the baby from the mother is called **delivery** (Fig 25.22).

Finally, the detached placenta, called the **afterbirth**, is expelled by uterine contractions (Fig 25.21(c)). Many damaged blood vessels in the uterine wall are constricted by the final contractions, thus reducing blood loss.

Fig 25.22 Delivery (courtesy Camilla Jessel, PAPS)

What mechanisms in the body of a pregnant woman bring about the birth of her child?

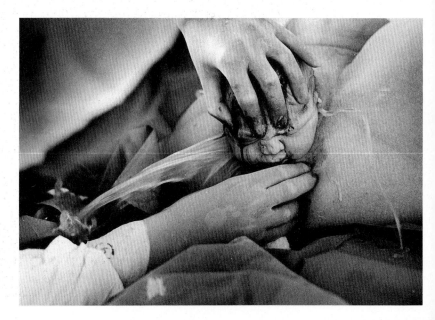

One of the most important changes which occurs in the baby at birth is the establishment of breathing. The lungs must be filled with air during the first minute or so of life outside the uterus. An important factor is the presence of a surface active agent (surfactant) in the baby's lungs. It is a lipoprotein which lowers the surface tension of the fluid lining the alveoli, thus making it easier to expand the lungs at inhalation (Chapter 8). Failure of the lungs of very premature new-born babies to inflate properly is sometimes due to insufficient surfactant production. Such babies suffer from respiratory distress syndrome. They are kept in intensive care until they produce enough surfactant, when their breathing becomes normal.

SUMMARY

Sperm (male gametes) are produced by spermatogenesis in seminiferous tubules of testes under the influence of FSH from pituitary. Primordial germ cells divide by mitosis to form spermatogonia; increase in size to form primary spermatocytes. Meiosis results in secondary spermatocytes (1st meiotic division) and finally spermatids (2nd meiotic division). Maturing sperm cluster around Sertoli cells. Sperm; head (nucleus), neck (mitochondria) and tail (flagellum).

Luteinising hormone, LH (also called Interstitial Cell Stimulating Hormone, ICSH) from pituitary; stimulates interstitial cells in testes to secrete testosterone (a steroid). By negative feedback testosterone suppresses LH output. Testosterone is main male sex hormone (androgen); responsible for accessary sex organs (male genitalia other than testes) and secondary sexual characteristics.

Ova (female gametes) produced by oogenesis in ovaries. Primordial germ cells divide by mitosis to form oogonia; become enclosed in primary follicles. Ova develop (usually one at a time in women). Oogonia grow into primary oocytes; mature into ovarian follicles. Each primary oocyte divides by meiosis to form a secondary oocyte (1st meiotic division) and a polar body. Secondary oocyte released from ovary at ovulation and taken into oviduct. Second meiotic division occurs (just after sperm nucleus enters it) to form an ovum and another polar body.

Follicle development and ovulation controlled by follicle stimulating hormone (FSH) from pituitary. Many mammals reproduce seasonally; FSH output may be influenced by day-length. FSH output may be brought about by coitus; induced ovulation, e.g. in rabbits. In some primates, including women, ovulation is controlled internally; ovarian cycle.

FSH stimulates development of a follicle and secretion from ovary of an oestrogen (female sex hormone) e.g. estradiol. By negative feedback estradiol suppresses FSH output. Estradiol also stimulates pituitary to secrete lutropin, LH. LH brings about ovulation (after 14 days from start of follicle development in women); also stimulates development, from remains of follicle, of corpus luteum.

The corpus luteum secretes progesterone; stimulates growth of endometrium (wall of uterus) ready for implantation of embryo. Progesterone also suppresses LH output by negative feedback. If fertilisation occurs, the placenta develops and secretes progesterone. If fertilisation does not occur, corpus luteum atrophies (LH suppression); endometrium sloughs away as the menstrual flow (after 28 days in women); cycle begins again with FSH secretion.

Female sex hormones also stimulate the development of accessory sex organs and secondary sexual characteristics.

Oocytes die if they are not fertilised within 8–24 hours of ovulation; fertilisation occurs in oviducts. Sperm head enters secondary oocyte which undergoes second meiotic division. Fusion of nuclei restores diploid state; zygote is formed. Zygote divides several times to form a ball of 16–32 cells; implants in uterine wall. The placenta develops partly to provide an alternative source of progesterone to the corpus luteum; also to enable fetal and maternal blood to come close enough together to allow exchange of metabolites.   ▶

> ► Mammary glands produce milk (lactation) under influence of prolactin from anterior pituitary (stimulated by Prolactin Releasing Factor, PRF, from hypothalamus). Prolactin Inhibiting Factor, PIF (hypothalamus) inhibits milk production. At birth PIF subsides. Breastfeeding brings about a neurohormonal reflex (involving oxytocin) which ejects milk into the baby's mouth.
>
> Birth is brought about by oxytocin, which stimulates uterine contractions; placenta expelled as afterbirth.

QUESTIONS

1 (a) Give an account of the production of gametes by the mammalian ovary.
(b) Explain the role of hormones in the sexual cycle of a female mammal.
(c) How does the role of the hormones change
 (i) in the early stages of pregnancy
 (ii) in the final stages of pregnancy? (L)

2 The drawing below represents a mammalian spermatozoon.

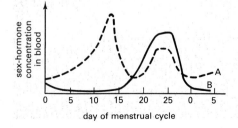

(a) Name the parts labelled **A–F**.
(b) Explain briefly the functions of the parts **A**, **B** and **D**.
(c) What events occur in the egg immediately following the entry of the spermatozoon? (L)

3 The table below refers to mammalian hormones concerned in the menstrual cycle.
(a) Complete the table by writing the name of the appropriate hormone in each case.

Secreted by ovary	Causes ovulation	Suppresses secretion of FSH	Secreted by pituitary	Name of hormone
×	✓	×	✓	
✓	×	×	×	
✓	×	✓	×	
×	×	×	✓	

(b) Name *two* hormones involved in birth. For each state *one* effect.
(c) What is the role of hormones in lactation? (L)

4 The graphs **A, B, C** and **D** show the changes in the concentrations in the blood of four hormones associated with a menstrual cycle.

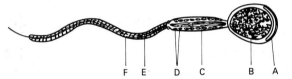

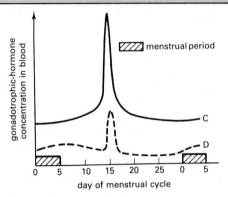

day of menstrual cycle

(a) The table shows the four main hormones associated with the menstrual cycle.

Name of hormone	Graph (A, B, C, or D)	Site of production	Effect
follicle-stimulating hormone (FSH)			
luteinising hormone (LH)			
oestradiol (oestrogen)			
progesterone			

(i) Copy the table and complete it by giving the graph (**A, B, C** or **D**) by which each hormone is represented and by givine **one** site of production and **one** effect of each hormone in a mature female.
(ii) What changes occur in the levels of oestradiol and progesterone in the maternal blood in the early stages of pregnancy?
(b)
(i) Explain negative feedback. What is its significance in biological systems?
(ii) Explain how negative feedback operates in the control of the hormones of the menstrual cycle. (AEB 1986)

5 The figure below illustrates the changes taking place during the human oestrous cycle.

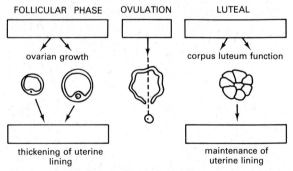

In each of the boxes shown in the figure write the name of the hormone, or hormones controlling the stage in the human oestrous cycle. (O)

494

26 Genetics

26.1 Autosomal inheritance 496

   26.1.1 Monohybrid inheritance 497
   26.1.2 Dihybrid inheritance 500
   26.1.3 Polyhybrid inheritance 502
   26.1.4 Co-dominance and
         incomplete dominance 503
   26.1.5 Multiple alleles 503
   26.1.6 Epistasis and gene
         interaction 504
   26.1.7 Linkage 505
   26.1.8 Chromosome maps 507
   26.1.9 Autosome abnormalities 509

26.2 Non-autosomal inheritance 510

   26.2.1 Inheritance of sex 510
   26.2.2 Sex linkage 511

26.3 Inheritance of quantitative
     characteristics 512

26.4 Population genetics 514

   26.4.1 Inbreeding and
         outbreeding 514
   26.4.2 Clones 515
   26.4.3 The Hardy–Weinberg
         equation 515
   26.4.4 Natural selection 517

26.5 Detection of carriers 517

26.6 Karyotyping 518

26.7 Barr bodies, Y chromosomes
     and white cell drumsticks 520

26.8 Genetic counselling 520

Summary 522

Questions 522

(Bruce Coleman: Kim Taylor)

26 Genetics

Genetics is the science of **inheritance**, the way in which characteristics are passed from parent to offspring. All organisms closely resemble their parents. Progeny created asexually are likely to resemble their parents more closely than those arrived at by sexual reproduction. This happens because mitotic nuclear division, which gives rise to progeny asexually, reproduces the genetic material with little or no variation. Meiotic nuclear division is an integral part of the life cycle of every creature which reproduces sexually and allows scope for genetic variation (Chapter 7).

It is desirable to be acquainted with some commonly used genetic terms before studying the inheritance of characteristics. A **gene** is a part of a molecule of DNA which codes the synthesis of a polypeptide. Protein molecules consist of one or more polypeptides (Chapter 2). The characteristics of organisms are largely determined by the types of protein, especially enzymes, they are able to synthesise. The combination of genes an organism has is called its **genotype** or **genome** whereas its traits are called its **phenotype**. Interaction of the genotype and environment determine the phenotype. Each of the characteristics of an organism is usually determined by two **alleles**. An individual in which both alleles are similar is described as **homozygous** or pure-breeding for that characteristic. Alleles of **heterozygous** individuals are dissimilar. Organisms can be homozygous for some traits and heterozygous for others. Alleles are thus alternative forms of a gene.

26.1 Autosomal inheritance

The chromosomes of somatic cells exist as homologous pairs (Chapter 7). In many organisms, including humans, one pair are the sex chromosomes, the rest are called **autosomes**. Gregor Johann Mendel (Fig 26.1), an Austrian monk, was the first to discover how genes carried by autosomes are inherited. He collected the seeds from several varieties of the garden pea and devised a procedure for following the passage of characteristics from parents to their offspring. In 1865 Mendel presented the results of his studies to the local Natural History society. A year later his findings were published in the Annual Proceedings of the Society. They lay neglected until the beginning of this century when several geneticists extended his

Fig 26.1 (a) Gregor Johann Mendel (Ann Ronan Picture Library)

Fig 26.1 (b) The monastery at Brno where Mendel worked

Fig 26.2 Inheritance of height in the garden pea

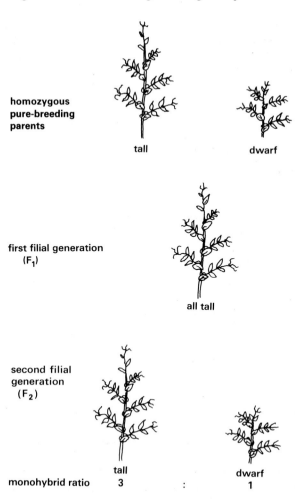

homozygous
pure-breeding
parents

tall   dwarf

first filial generation
(F₁)

all tall

second filial
generation
(F₂)

tall   dwarf
monohybrid ratio   3   :   1

methods to other species and confirmed his conclusions. At the time Mendel was alive, little was known of human inheritance. For instance, it was not until 1901 that Garrod worked out the way in which the human characteristic alkaptonuria (Chapter 4) is inherited. A few years later, Landsteiner's discovery of ABO blood groups led to an explanation of how we inherit our blood group.

26.1.1 Monohybrid inheritance

Mendel's first experiments were designed to follow the inheritance of one well-defined characteristic. His procedure was to cross-pollinate pure-breeding varieties of the garden pea, collect the seeds and sow them the following year. The plants which grew were then allowed to self-pollinate, and the seeds they produced were again collected. The phenotypes of the plants which grew from the seeds were carefully noted. One of the traits Mendel worked with was the height of the plants. In one investigation he crossed a pure-breeding tall variety with a pure-breeding dwarf variety. All the progeny (F₁) were tall but when self-pollinated they produced F₂ plants in which there was a ratio of 3 tall : 1 dwarf (Fig 26.2). This is known as a **monohybrid ratio**. A third of the tall F₂ plants were pure-breeding. The remainder, on selfing, produced a mixture of tall and dwarf progeny, showing that they were heterozygous (hybrids). The tall characteristic, which the F₁ plants showed, Mendel termed **dominant**. Its contrasting trait, dwarfness, was termed **recessive**.

The observations led Mendel to formulate his First Law of Inheritance which states:

Of a pair of contrasting characteristics, only one can be represented in the gametes.

Fig 26.3 Genetic explanation of inheritance of height in the garden pea

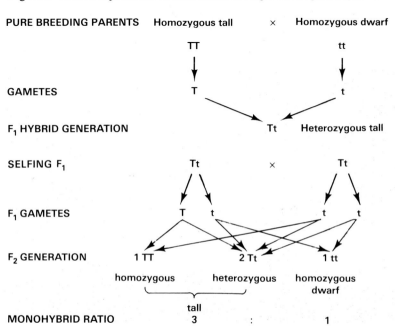

PURE BREEDING PARENTS   Homozygous tall   ×   Homozygous dwarf

TT   tt

GAMETES   T   t

F₁ HYBRID GENERATION   Tt   Heterozygous tall

SELFING F₁   Tt   ×   Tt

F₁ GAMETES   T   t   t   t

F₂ GENERATION   1 TT   2 Tt   1 tt

homozygous   heterozygous   homozygous
dwarf

tall

MONOHYBRID RATIO   3   :   1

The law can be more readily understood if symbols are used to represent the alleles which determine the characteristics. Using **T** as the allele for tallness and **t** for dwarfness the experiment may be depicted as shown in Fig 26.3. The fact that plants of genotype **Tt** are hybrids can be detected by crossing them with a pure-breeding recessive.

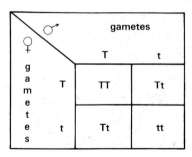

		gametes	
♀ ♂		T	t
gametes	T	TT	Tt
	t	Tt	tt

Application

These days it is common practice for gardeners to grow F_1 hybrids from seed obtained by crossing two related varieties. Hybrid tomatoes, cucumbers and geraniums, to name but a few, combine the better traits of both parents, are more vigorous and give better yields.

This is called a **back-cross** or **test-cross** (Fig 26.4). The 1 : 1 ratio obtained is termed a **back-cross ratio**.

Fig 20.4 A back-cross

PARENTS
Hybrid × Homozygous dwarf
Tt
tt
GAMETES
T
t
t
T
t
t
Tt
tt
BACK-CROSS RATIO
Tt tall
1
:
tt dwarf
1

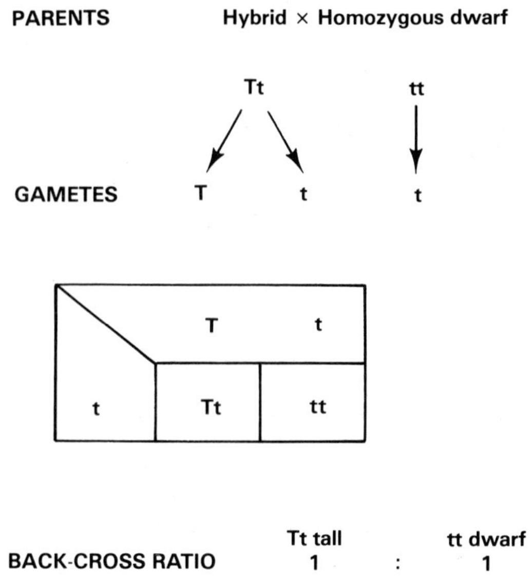

What ratio would result in a backcross if the tall plants were homozygous?

Mendel observed that seven other characteristics of the garden pea were inherited in a comparable way.

An example of a human trait which follows the same pattern of inheritance is **Huntington's chorea (HC)**. The condition was first described by a Dr Huntington in the USA in 1872. People with HC gradually lose muscular coordination and develop severe mental deterioration. The symptoms do not usually begin to appear until the affected person is about thirty-five years of age. By this time many people, unaware that they have the disease, will have passed the dominant allele for HC to their children (Fig 26.5). At the moment there is no way of detecting carriers of the allele. The use of gene probes (Chapter 4) may offer a solution in the near future.

mother
(with HC)
father
(normal)
Parents
Children
daughter
with HC
son
with HC
normal
daughter
normal
son

Fig 26.5 Family pedigree showing inheritance of Huntington's chorea (after Clarke 1970)

Throughout this chapter the symbols ○ and □ are used to denote females and males respectively in family pedigrees.

Using appropriate symbols, produce a genetic diagram similar to that shown in Fig 26.3 to explain the inheritance of HC in this family.

Other human characteristics determined by dominant alleles are **brachydactyly** and **polydactyly**. People with brachydactyly have a very short bone in the middle of each finger. Consequently each finger is only a little longer than the thumb. In the homozygous state the gene causes severe defects of many bones, and death occurs shortly before or after birth. Thus all surviving individuals with brachydactyly are heterozygous. Polydactylous people have six fingers on each hand and six toes on each foot. The

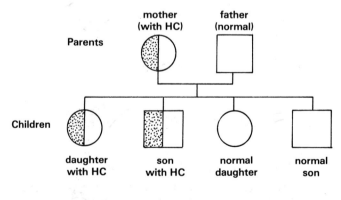

Fig 26.6 A polydactylous hand and foot

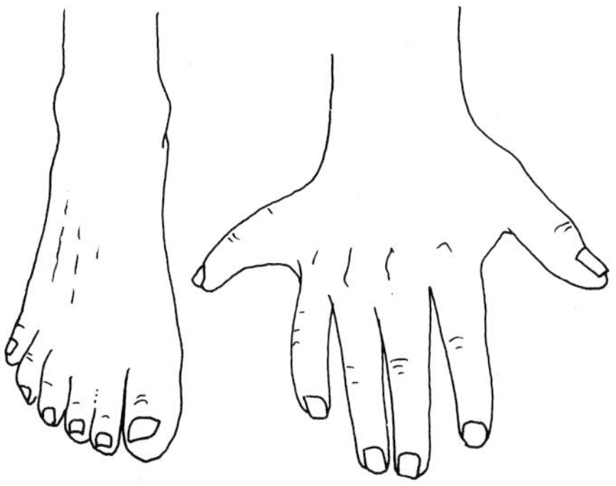

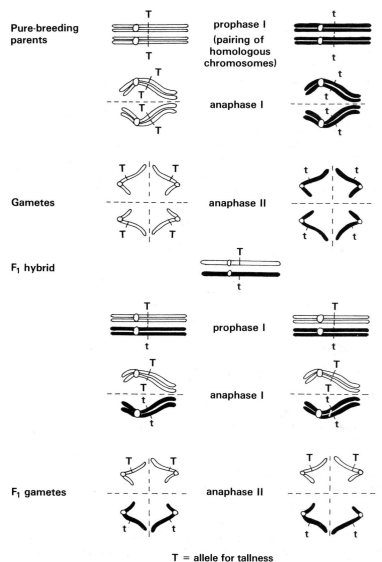

Is the individual whose hand and foot are shown in Fig 26.6 likely to be homozygous or heterozygous for polydactyly?

Fig 26.8 Segregation of alleles in meiosis

extra toes and fingers of individuals who are homozygous for polydactyly are of normal size (Fig 26.6). Heterozygotes have much smaller extra digits and occasionally they are not externally visible.

Fig 26.7 Family pedigree showing inheritance of FCD (After Clarke 1970)

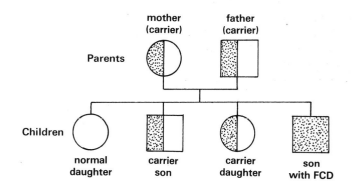

Albinism and **fibrocystic disease of the pancreas (FCD)** are caused by recessive alleles. Hence people with these conditions are homozygous recessive. The family pedigree in Fig 26.7 shows how FCD was inherited by a child whose parents were carriers. Again use appropriate symbols as in Fig 26.3 to produce a genetic diagram to explain the inheritance of FCD in this family.

What are the symptoms of FCD and how are they alleviated?

Chromosome behaviour and monohybrid inheritance

Organisms which reproduce sexually form gametes. They are formed by meiotic nuclear division which cause them to have a haploid chromosome number (Chapter 7).

Mendel's First Law of Inheritance is also called the **Law of Segregation** because alleles become segregated in meiosis. A study of the behaviour of chromosomes in meiosis shows how segregation occurs (Fig 26.8).

T = allele for tallness
t = allele for dwarfness

26.1.2 Dihybrid inheritance

Mendel also studied the simultaneous inheritance of two characteristics. In one experiment he traced the inheritance of seed colour and texture in the garden pea. He first crossed a pure-breeding variety having round and yellow seeds with another pure-breeding variety having wrinkled and green seeds. All the F_1 plants had round, yellow seeds, showing these to be dominant traits. When self-pollinated, the plants which grew from the F_1 seeds produced four kinds of seeds: round and yellow, round and green, wrinkled and yellow, wrinkled and green in a ratio of $9:3:3:1$ respectively (Fig 26.9). This is a **dihybrid ratio**. The results led Mendel to formulate a Second Law of Inheritance which states:

> **Each of a pair of contrasting characteristics segregates independently of those of any other pair.**

Once more, it is easier to understand the law using symbols to represent the alleles (Fig 26.10). Mendel observed that several other pairs of contrasting characteristics of the garden pea were inherited in the same way.

Fig 26.9 Inheritance of seed colour and texture in the garden pea

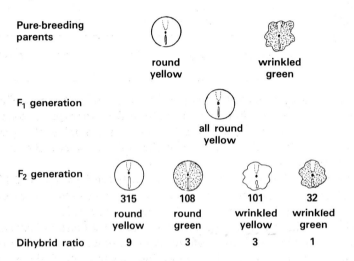

Pure-breeding parents	round yellow	wrinkled green		
F_1 generation		all round yellow		
F_2 generation	315 round yellow	108 round green	101 wrinkled yellow	32 wrinkled green
Dihybrid ratio	9	3	3	1

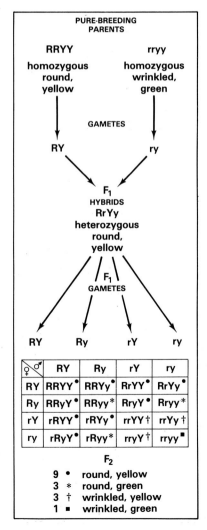

Fig 26.10 Genetic explanation of inheritance of seed colour and texture

The chi-square (χ^2) test

In the experiment just described the data Mendel obtained were:

315 round, yellow   101 wrinkled, yellow
108 round, green   32 wrinkled, green

We can work out the figures we might expect if the data is to show an exact 9:3:3:1 ratio. This is done by adding up the total number of seeds, dividing by 16, then multiplying by 9,3,3 and 1 respectively. The χ^2 value for the data is then computed using the equation:

$$\chi^2 = \sum \frac{(O-E)^2}{E} \text{ where } \sum = \text{sum of (Table 26.1)}$$

Table 26.1 Calculation of χ^2 value

	round yellow	round green	wrinkled yellow	wrinkled green	Total
Observed (O)	315	108	101	32	556
Expected (E)	313	104	104	35	556
$O-E$	+2	+4	−3	−3	–
$(O-E)^2$	4	16	9	9	–
$\dfrac{(O-E)^2}{E}$	0.001	0.154	0.086	0.257	$\chi^2 = 0.498$

Using a χ^2 statistical table it is now possible to determine the probability, P, of obtaining purely by chance such a deviation from the expected data. The relationship between χ^2 and P is a function of the number of degrees of freedom (df). If $n=$ the number of classes of phenotypes, the df value $=n-1$. In this case, as there are 4 classes, $df=3$.

Now when $\chi^2 = 0.498$ and $df=3$, P lies between 0.95 and 0.90 (Table 26.2). Here we can expect a deviation of the order seen in the observed data to occur by chance between 95 and 90 times out of a 100.

Whenever the _P value is greater_ than 0.05, the deviation is not considered to be statistically significant from the expected data. However if P is less than 0.05 but greater than 0.01 the deviation is significantly different, and if P is less than 0.01 the deviation is very significantly different from what is expected.

Table 26.2 χ^2 table (From _Fisher and Yates Statistical Tables_, Oliver & Boyd Ltd, Edinburgh)

df	$P=0.99$	0.95	0.90	0.50	0.30	0.01
1	0.000157	0.00393	0.0158	0.455	1.074	6.635
3	0.155000	0.38200	0.5840	2.366	3.665	11.345

Assignment

The following data were obtained in the F_2 generation by Bateson and Punnett when they investigated the inheritance of pollen shape and flower colour in the sweet pea:

long pollen, purple flower   round, purple   red, long   red, round
226   95   97   1

Calculate the χ^2 value and determine whether or not the deviation is very significant from the expected ratio.

Chromosome behaviour and dihybrid inheritance

A study of the behaviour of chromosomes in meiosis reveals how independent segregation occurs (Fig 26.11). The alleles which determine the two pairs of contrasting characteristics are located on different pairs of homologous autosomes. Because the chromosomes of one pair separate independently of the other pair, the alleles segregate independently. Mendel's Second Law is also known as the **Law of Independent Assortment**.

Fig 26.11 Independent segregation of two pairs of alleles in meiosis

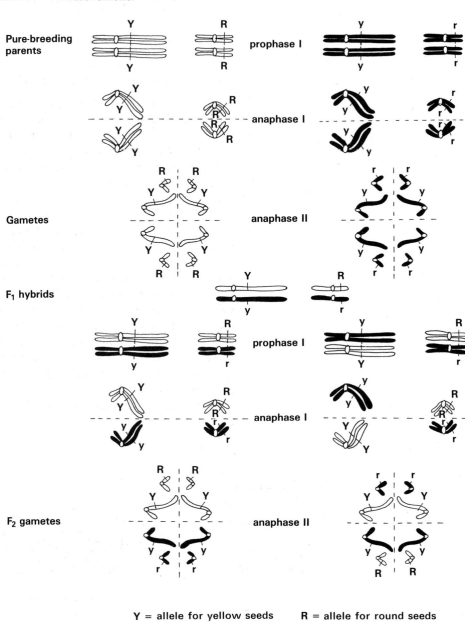

Y = allele for yellow seeds   R = allele for round seeds

y = allele for green seeds   r = allele for wrinkled seeds

26.1.3 Polyhybrid inheritance

Tracing the simultaneous inheritance of more than two pairs of contrasting characteristics is more difficult. Nevertheless, the principles of segregation and independent assortment still apply. Thus there is an increase in the variety of combinations of genes in the gametes produced by hybrid F_1 individuals, giving rise to an even greater variety of genotypes in the F_2

generation (Table 26.3). Many quantitative characteristics show this pattern of inheritance (section 26.3).

Table 26.3 Relationship between number of alleles and F_2 genotypes

Pairs of alleles	No. of gene combinations in F_1 gametes	No. of genotypes in F_2 generation
1	$2^1 = 2$	$3^1 = 3$
2	$2^2 = 4$	$3^2 = 9$
3	$2^3 = 8$	$3^3 = 27$
n	2^n	3^n

26.1.4 Co-dominance and incomplete dominance

Complete dominance is not always observed in an allelic pair. For example, there is a pair of alleles which express themselves equally in people of blood group AB (section 26.1.5). Such alleles are **co-dominant**.

In other instances an allele may show **incomplete dominance** over its partner. A typical example is seen in people who have the **sickle cell trait** (Chapter 10). The condition is caused by an allele which codes the production of haemoglobin S. Individuals who are heterozygous produce equal amounts of haemoglobin S and normal haemoglobin A. They have a relatively mild form of anaemia. Homozygotes produce only haemoglobin S and have a severe form of anaemia. It is therefore apparent that the allele for producing normal haemoglobin A does not completely suppress the allele which regulates production of haemoglobin S. Fig 26.12 shows how children with sickle cell disease could be among the progeny of parents who have sickle cell trait. Another example of incomplete dominance is seen in people who suffer from **thalassaemia** (Mediterranean anaemia). The condition is caused by an allele **Hb^B** which is only partly dominated by the allele **Hb^A** which controls production of normal haemoglobin A. It means that people who are heterozygous for the condition **Hb^{A}Hb^B** have a mild form of anaemia (thalassaemia minor). Homozygotes **Hb^{B}Hb^B** suffer from very severe anaemia (thalassaemia major) and usually die in childhood.

Fig 26.12 Inheritance of sickle cell disease

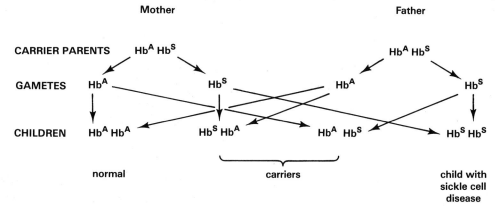

26.1.5 Multiple alleles

The patterns of inheritance described so far involve two alternative forms of a gene. Many of the characteristics of most organisms are determined by such alleles. However, some traits are determined by several forms of an allele known as **multiple alleles**. A good example is seen in the inheritance of blood groups of the **ABO system**. As described in Chapter 10 the grouping is based on the presence or absence of A and B antigens on red

503

blood cells. The alleles I^A and I^B which code production of A and B antigens respectively are co-dominant. Both are dominant to the allele i which does not code for antigen production (Table 26.4).

Table 26.4 Genotypes of A, B, AB and O blood groups

Group	Antigens	Genotypes
A	AA or AO	$I^A I^A$ or $I^A i$
B	BB or BO	$I^B I^B$ or $I^B i$
AB	AB	$I^A I^B$
O	none	ii

If a mother's blood group is B and her child's is AB, which blood group(s) could the father have?

If a mother and her baby belong to group O, the father could be of group O, A or B but not AB (Fig 26.13). Evidence of this kind is sometimes used in legal cases where the paternity of a child is in question.

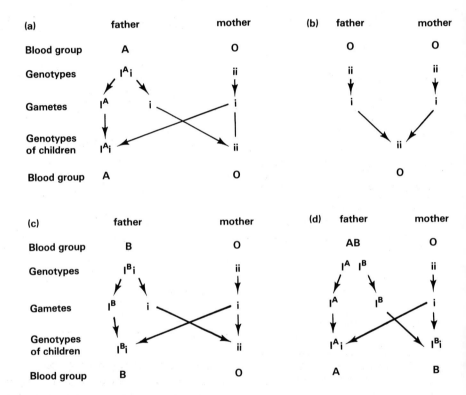

Fig 26.13 Inheritance of ABO blood groups

26.1.6 Epistasis and gene interaction

Epistatic genes prevent the action of other genes. The recessive gene for albinism inhibits the activity of several genes which determine the amount of melanin in the eyes, hair and skin. In a similar way the recessive gene for blue eye colour is epistatic to genes which control the amount of melanin in the iris.

Alleles from different genes sometimes **interact** to give rise to unusual phenotypic ratios. Each of the common breeds of fowl has one of four comb shapes – rose, pea, walnut or single. Fowls with rose and pea combs are true-breeding, but when crossed give rise to F_1 fowls with walnut combs. If the F_1 birds are selfed, the phenotypic ratio among the F_2 fowls is

9 walnut : 3 rose : 3 pea : 1 single (Fig 26.14). The 9 : 3 : 3 : 1 ratio indicates that this is a case of dihybrid inheritance. However, it is an unusual case of dihybrid inheritance, because:

 (i) the F_1 fowls have a comb form which neither of the parents had, and
 (ii) a fourth comb form, single, appears in the F_2 generation.

The genes are inherited in typical Mendelian fashion. A walnut comb is produced when the alleles for pea (P) and rose (R) interact. A pea comb arises when the P allele is present without the R allele. When the R allele is present without the P allele, a rose comb appears. In the absence of both R and P alleles, a single comb develops:

Fig 26.14 Inheritance of comb form in fowls

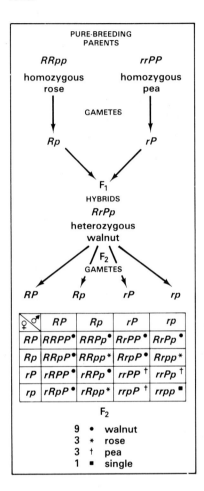

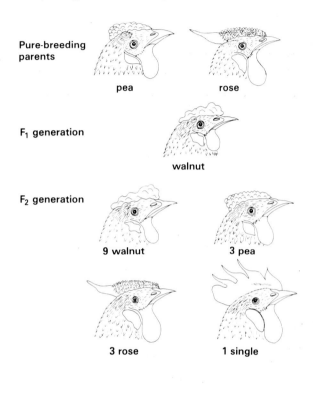

Gene interaction is known to occur in humans. Two alleles determine the type of haptoglobin, a plasma protein, in human blood. People who are homozygous for the two alleles have genotypes Hp^1Hp^1 or Hp^2Hp^2 and have different varieties of haptoglobin. In heterozygotes, the alleles Hp^1 and Hp^2 interact to form a third variety of haptoglobin.

26.1.7 Linkage

Up to now we have considered patterns of inheritance in which the genes are located on separate autosomes and thus segregate independently. However, each chromosome carries a number of genes. Genes which are on the same chromosomes are said to be **linked** and are inherited *en bloc*.

A well-known case of linkage in humans involves the genes for **ABO blood groups** and the **nail-patella syndrome (NP)**. People with the syndrome have small, discoloured nails especially on the thumbs, index fingers and the first and second toes. The patella is missing or is small and pushed to one side of the knee. The syndrome is caused by a dominant allele

Fig 26.14 Inheritance of comb form in fowls

How would you detect whether or not:
(a) the fowls with walnut combs are heterozygous for both alleles, and
(b) the fowls with single combs are homozygous for both alleles?

505

which is on the same chromosome as the allele for ABO blood groups (Fig 26.15). Most people with NP syndrome belong to the two most common blood groups, A and O (Table 26.5). Examination of the pedigrees of families in which the syndrome has occurred clearly reveals that the ABO and NP genes are linked (Fig 26.16).

Table 26.5 Relationship between frequencies of ABO blood groups and NP syndrome

Blood group	A	O	B	AB
% normal	49.4	38.4	8.5	3.7
% with NP	38.1	44.6	14.4	2.9

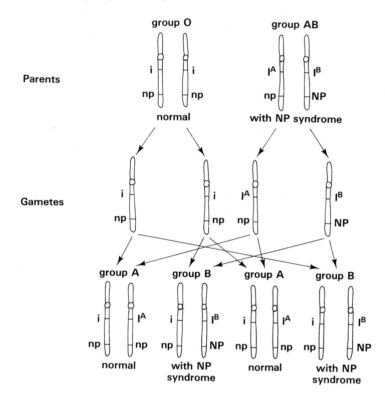

Fig 26.15 Inheritance of linked genes for ABO blood group and NP syndrome

Fig 26.16 Pedigree of a family showing linkage between ABO blood group and NP syndrome

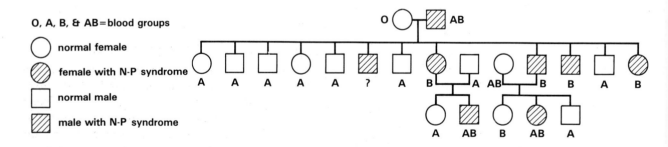

Linked genes can be separated when crossing-over occurs during meiosis (Fig 26.17). Crossing-over produces a relatively small number of gametes containing **recombinant genes**. Linkage contrasts markedly with independent assortment which gives rise to equal numbers of each combination of genes in the gametes. Consequently most of the F_2 progeny arising in the inheritance of linked genes have the characteristics of one or other of their parents. The few which have traits from both parents are produced from gametes containing recombinant genes.

Fig 26.17 Separation of linked genes during crossing-over

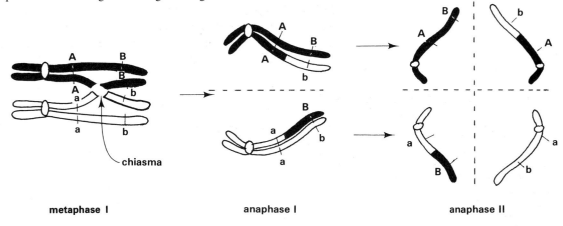

metaphase I   anaphase I   anaphase II

Fig 26.18 Inheritance of closely linked genes

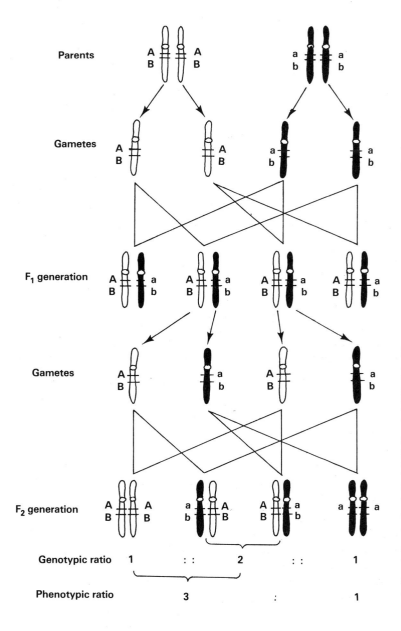

In general terms, the closer a pair of linked genes are on a chromosome, the less is the probability that a chiasma will form between them. Thus a very small proportion of recombinants indicates close linkage. The absence of recombinants tells us that the genes are so close that they are not separated by crossing-over. When this is the case, the ratio of progeny in the F_2 generation is 3:1 (Fig 26.18). Compare this with the 9:3:3:1 dihybrid ratio which occurs when the genes are not linked.

26.1.8 Chromosome maps

In the previous section we saw that the gametes produced by an individual who is heterozygous for linked genes are less likely to contain recombinant than parental genes. The frequency of the recombination is constant and specific for any pair of linked genes. T. H. Morgan was the first to suggest that the frequency is proportional to the distance along the chromosome between the linked genes in question. Thus the further apart they are, the greater is the chance that a chiasma will form between them when homologous chromosomes pair and the greater is the probability that they will be separated.

Fig 26.19 *Drosophila* (Bruce Coleman: Kim Taylor)

Assignment

Draw a series of genetic diagrams summarising Morgan's experiment.

To test this assumption, Morgan and his colleagues carried out a number of breeding programmes involving linked genes in the fruit fly *Drosophila* (Fig 26.19). One of the investigations first involved crossing pure-breeding grey-bodied flies having normal eye colour with pure-breeding black-bodied flies having purple eyes. All the F_1 flies had grey bodies and normal eye colour. The F_1 females were then back-crossed with males having black bodies and purple eyes. The researchers obtained the following results:

grey, normal 250   The proportion of recombinants was
black, purple 234
black, normal 15 $\dfrac{15+16}{515} \times 100 = 6.02\,\%$.
grey, purple 16
—
515

Such a **cross-over value** indicated that the genes for black body and purple eye-colour were about 6 map units apart on one of the autosomes.

Next, pure-breeding flies having grey bodies and long wings were crossed with pure-breeding flies having black bodies and vestigial wings. The F_1 female flies were again back-crossed as described in the previous experiments. The progeny showed the following characteristics:

grey, long   192   The cross-over value was
black, vestigial 170
black, long   43 $\dfrac{43+39}{444} \times 100 = 18.5\,\%$.
grey, vestigial   39
—
444

It indicated that the genes for black body and vestigial wings were 18.5 map units apart on the same autosome. The linear order of the genes could thus be:

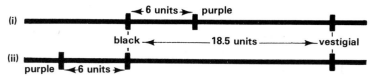

If it is the former, then the cross-over value for purple and vestigial should be about $18.5 - 6.0 = 12.5\,\%$; if the latter, the cross-over value should be $18.5 + 6.0 = 24.5\,\%$.

To find out which of the two orders is correct, Morgan and his colleagues carried out a further experiment in which they crossed pure-breeding normal-eyed, long-winged flies with pure-breeding flies having purple eyes and vestigial wings. Once more the F_1 flies were back-crossed as in the two previous investigations. The results were:

normal, long   230   The cross-over value was
purple, vestigial 200
purple, long   32 $\dfrac{32+28}{490} = 12.2\,\%$.
normal, vestigial 28
—
490

It was therefore evident that the linear order of the genes on the homologous autosomes was:

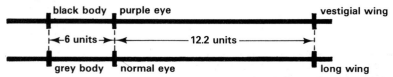

Similar procedures have been used to map the chromosomes of a wide variety of living organisms. Gene probes are currently being used to produce maps of human chromosomes (Chapter 4).

26.1.9 Autosome abnormalities

Fig 26.20(a) shows the chromosomes of a human female. They can be arranged in homologous pairs. The chromosomes of the first twenty-two pairs are called **autosomes**, the twenty-third pair are the sex chromosomes, People are sometimes born with autosomes which are unusual in number or structure. Failure of a pair of homologous autosomes to separate at anaphase of the first meiotic nuclear division occurs fairly often. It gives rise to gametes having one less or one more autosome than normal. If fertilised by a normal gamete, the offspring will have 45 or 47 chromosomes respectively, rather than the usual number of 46. Such a condition is called **polysomy**. Loss of an autosome is lethal in humans. **Down's syndrome** is a condition in which individuals usually have an extra twenty-first autosome (Fig 26.20(b)). The extra autosome often becomes joined (translocated) to one of the fourteenth pair of autosomes. The chromosome number then appears normal. However, the translocation can be detected from the unusual structure of one of the fourteenth autosomes.

People with Down's syndrome have characteristic facial features (Fig 26.20(c)). They are mentally retarded and have a relatively short life expectancy. Although some live into their forties, others die much younger from congenital heart disease or from common infections. Even so, they are loving, trusting individuals and often serve to strengthen the bonds within families.

Fig 26.20 (a) Chromosomes of a normal female

Fig 26.20 (b) Chromosomes of a female with Down's syndrome

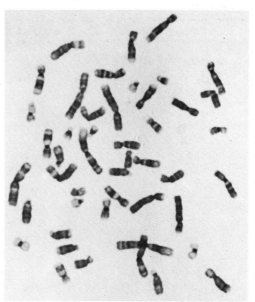

Fig 26.20 (c) A child with Down's syndrome (Science Photo Library: Richard Hutchings)

How many pairs of chromosomes are there in Fig 26.20(a)? Is there any difference in the number of chromosomes shown in Fig 26.20(b)?

Very occasionally all the homologous pairs of chromosomes fail to separate in meiosis. This produces gametes with a diploid chromosome number. If fertilised by a normal gamete, a triploid offspring with an extra set of chromosomes arises. The condition is called **polyploidy**. Only a few instances of triploid humans have been recorded and they all died shortly after birth. However, polyploidy is quite common among plants. Polyploid crop plants are usually more vigorous than their diploid parents and therefore produce higher yields. For example, bread wheat is a hexaploid with a diploid number of 42 chromosomes whereas its ancestors had a diploid number of 14 (Fig 27.27).

26.2 Non-autosomal inheritance

26.2.1 Inheritance of sex

In a male mammal the sex chromosomes are different in size and shape and are called **X** and **Y chromosomes**. Females have two X chromosomes (Fig 26.21). Note that the longer arm of the X chromosome has no counterpart in the Y chromosome.

In the first nuclear division of meiosis, the sex chromosomes pair and then segregate, just as homologous pairs of autosomes do (Chapter 7). As a result, half of the sperm of humans normally have 22 autosomes and an X chromosome; the other half have 22 autosomes and a Y chromosome. For this reason the male is known as the **heterogametic** sex. All the ova of humans normally have 22 autosomes and an X chromosome. This is why the female is called the **homogametic** sex.

There is a 50 % chance that an ovum is fertilised by a sperm carrying an X chromosome. The zygote so formed will be XX and develop into a female. There is also a 50 % chance that the sperm carries a Y chromosome, and the XY zygote so produced will develop into a male (Fig 26.22).

Many plants have a similar mechanism of sex inheritance. However in birds, butterflies, most reptiles, some fish and amphibians the female is heterogametic and the male homogametic.

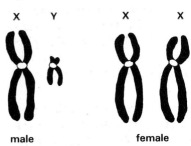

Fig 26.21 Sex chromosomes

X   Y   X   X

male   female

Fig 26.22 Inheritance of sex

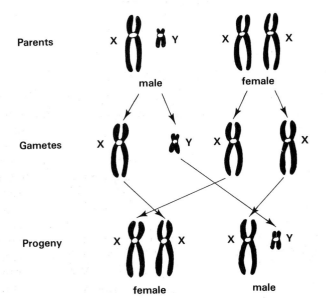

Sometimes the sex chromosomes fail to separate at anaphase of meiosis. When this happens, some ova develop with an excess of X chromosomes and others with none, or some sperm develop with an excess of X or Y chromosomes. Possible outcomes in humans are as follows.

i. An XX ovum is fertilised by a Y sperm. The XXY zygote produced develops into a male with **Klinefelter's syndrome**. During early life the child may be normal. At puberty, however, the testes will have grown to only about half the expected size and fail to produce sperm. Though obviously infertile, males with the syndrome are capable of sexual intercourse, their ejaculate coming from the prostate and other accessory sex glands (Chapter 25). They also tend to grow taller than average, have high-pitched voices and develop small breasts. About 75 % of individuals with Klinefelter's syndrome are XXY. Others are usually XXXXY, having developed from an ovum containing four X chromosomes fertilised by a Y sperm. Such an ovum would arise if the pair of X chromosomes failed to separate at anaphase in both meiotic divisions.

ii. A normal ovum is fertilised by a sperm containing two Y chromosomes. A sperm of this kind is produced if the Y chromosome failed to split at anaphase II of meiosis. The zygote is XYY, and often gives rise to tall aggressive male offspring. Recent studies have shown that an unusually high proportion of male inmates of institutions for the criminally insane are in this category.

iii. An XX ovum is fertilised by an X sperm. The XXX zygote would develop into a female with **triple-X syndrome**. Many individuals with this condition are physically and mentally normal and fertile. Others do not menstruate and are sterile.

iv. An ovum with no X chromosomes is fertilised by a Y sperm. Human zygotes of the YO type have never survived. It indicates that the X chromosomes carry genes which are essential for life.

v. An ovum lacking an X chromosome is fertilised by an X sperm. The XO zygote produced develops into a female with **Turner's syndrome**. At birth the child has a thick fold of skin on both sides of the neck causing the neck to appear very wide when viewed from the front or back. Other than this the girl appears normal until puberty when the secondary sexual characteristics fail to appear (Chapter 25). Consequently the sex organs remain child-like, the ovaries do not produce ova and there is no menstruation or breast development.

26.2.2 Sex linkage

As well as carrying genes which regulate sexual development, the sex chromosomes, especially the X chromosome, carry genes which determine other traits. Such genes are said to be **sex-linked**. If they occur on the longer arms of the X chromosome, their presence will be apparent in males even when the alleles are recessive. This is because there are no corresponding alleles on the Y chromosome. In females the other X chromosome may carry dominant alleles which would mask recessive alleles on this part of the X chromosome (Fig 26.23).

Fig 26.23 Sex-linked genes

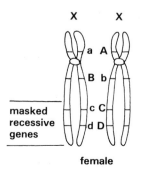

Fig 26.24 Inheritance of haemophilia (after Pedder and Wynne 1972)

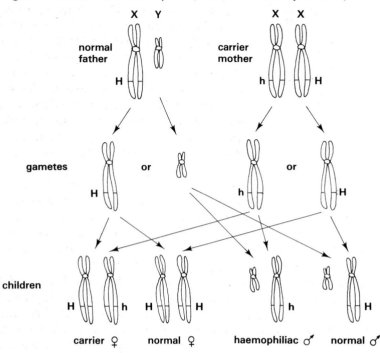

H = gene for normal clotting of blood
h = gene for haemophilia

Two well-known examples of sex-linked traits in humans are **haemophilia** and **red–green colour-blindness**. Both conditions are caused by recessive alleles located on the longer arms of the X chromosome. For females to have haemophilia or red–green colour-blindness the alleles must be present on both X chromosomes. Males who have these traits carry the alleles only on their single X chromosome. For this reason haemophiliac and colour-blind males are more common than females. Women who have the recessive alleles on one X chromosome, and the dominant alleles on the other for blood clotting and colour vision, are called **carriers**. They do not have the diseases but can transmit the diseases to their children (Fig 26.24).

Perhaps the most publicised pedigree showing the inheritance of haemophilia is that of the descendants of Queen Victoria (Fig 26.25). She was a carrier of the defective gene which was inherited by the Prussian, Russian and Spanish royal families.

Fig 26.25 Pedigree of descendants of Queen Victoria (after Pedder and Wynne 1972)

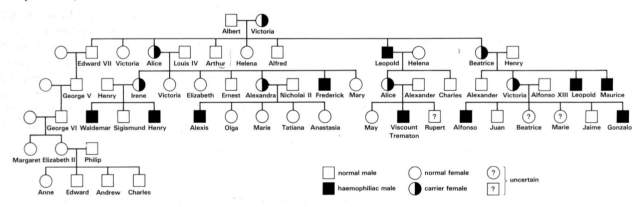

Fig 26.26 A frequency distribution curve for height of men in South East England (from Carter 1962)

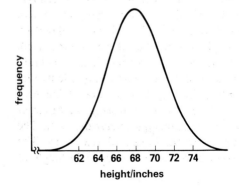

26.3 Inheritance of quantitative characteristics

The examples of inheritance dealt with so far have been concerned with **discontinuous variation**. These are clear-cut alternatives of a given trait such as pigmentation in contrast to albinism. Other traits show there is a range of variation. A typical example of a human characteristic showing **continuous variation** is height. Within a population of humans there is a range of height. Quantitative characteristics of this kind are usually determined by several genes.

If the frequency of height of a large number of humans of a given age group is plotted graphically, a **frequency distribution curve** is obtained (Fig 26.26).

The observations can be explained as follows. Let us assume that three

Table 26.6 Polygenic inheritance

Gametes	ABC	ABc	AbC	aBC	Abc	aBc	abC	abc
ABC	12	11	11	11	10	10	10	9
ABc	11	10	10	10	9	9	9	8
AbC	11	10	10	10	9	9	9	8
aBC	11	10	10	10	9	9	9	8
Abc	10	9	9	9	8	8	8	7
aBc	10	9	9	9	8	8	8	7
abC	10	9	9	9	8	8	8	7
abc	9	8	8	8	7	7	7	6

pairs of alleles **Aa, Bb** and **Cc** control height. Let us also assume that each dominant allele allows for two units of height, while each recessive allele allows for only one unit. The tallest individuals will have the genotype **AABBCC** (12 units) and the shortest **aabbcc** (6 units). Heterozygotes for all three genes **AaBbCc** will have a height of nine units, exactly half way between the extremes. The distribution of height among the progeny of heterozygotes can be worked out as in Table 26.6. In reality, height in humans is controlled by more genes than shown in this model.

When displayed as a graph, the data appear as in Fig 26.27. It is a **normal distribution curve**. Note the similarity between this curve and the frequency distribution curve. The resemblance supports the notion that continuously variable characteristics are mainly determined by the additive effects of several genes. This is called **polygenic inheritance**.

Pigmentation of the skin and eyes of humans is also determined by polygenes. Skin and eye colour depend on the amount of melanin produced in the Malpighian layer and iris respectively. Melanin is synthesised only if a person has a dominant allele to code its production. However, the amount of melanin produced is regulated by polygenes. As a consequence, a range of skin colour from black to almost white is possible, whereas eye colour can be dark, medium or light brown, hazel, green or blue. People with blue eyes have no pigment in the iris but there is melanin in the choroid layer of their eyes. Albinos are double recessive and are thus unable to produce any melanin. For this reason their skin is white. The eyes of albinos are pink because of the blood capillaries in the iris.

Another human trait determined by polygenes is the **Rhesus blood group**. The Rhesus group to which a person belongs depends on three dominant alleles **C, D, E** and their recessive alternatives **c, d** and **e**. With the exception of **d**, the alleles can regulate production of red cell antigens. However, only the **D** antigen is usually important in blood transfusion and pregnancy (Chapter 10). The most common genotype for Rhesus group is **CcDdee** (Table 26.7). Double dominant (**DD**) and heterozygous (**Dd**) individuals are Rhesus positive, whereas double recessives (**dd**) are Rhesus negative.

Fig 26.27 A normal distribution curve for stature, assuming that the alleles with the properties stated in the text determine this characteristic

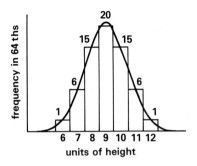

Table 26.7 Frequency of some genotypes for Rhesus group (after Emery 1975)

Genotype	Frequency/%	Phenotype
CcDdee	33	Rhesus positive
CCDDee	17	Rhesus positive
ccddee	15	Rhesus negative

Mean and Standard Deviation

Every organism that reproduces sexually is genetically unique. For this reason the biological data obtained for a group of organisms of the same species is variable. Consider the range of heights of your colleagues; some are on the short side, others on the tall side whereas many are above average. The average is the **mean** ($\bar{x}$) of a set of data. It is calculated by adding up the data (Σx) and dividing by the number of pieces of data (n). Hence if you needed to know the mean height of ten adult men, you would measure the height of each man, add up all the heights and divide by ten. Stating a mean value saves having to reproduce a lot of data. However it gives no idea of the spread of data. One way of expressing the range of values in a set of data is to calculate the **standard deviation** (SD). The larger the SD the wider is the spread of the data and vice versa. The example on page 514 shows how to calculate the SD value for the height of a group of ten men.

Man no	Height (m)	Deviation (d) from mean	d^2
1	1.6	−0.2	+0.04
2	1.9	+0.1	+0.01
3	1.8	0.0	0.00
4	2.1	+0.3	+0.09
5	1.7	−0.1	+0.01
6	1.8	0.0	0.00
7	2.0	+0.2	+0.04
8	1.9	+0.1	+0.01
9	1.5	−0.3	+0.09
10	1.7	−0.1	+0.01
$\Sigma x = 18.0$			$\Sigma d^2 = 0.30$

$$\text{Mean} = \frac{\Sigma x}{n}$$

$$= \frac{18.0}{10} = 1.8\,\text{m}$$

$$\text{SD} = \sqrt{\frac{\Sigma d^2}{n}}$$

$$= \sqrt{\frac{0.30}{10}}$$

$$= \sqrt{0.03}$$

$$= 0.173$$

26.4 Population genetics

A population is a group of individuals living in the same locality. No two individuals are exactly alike. Even identical twins are not the same, despite the fact that they are derived from the same zygote and thus have the same genotype. They are different because they did not develop in identical environments. One of the twins may have been in a more favourable position to obtain nourishment from the placenta. Consequently it could grow more rapidly and would be larger at birth. Hence the full genetic potential of an individual is realised only in an ideal environment. **Environmental differences** are the cause of some of the variation observed among humans. However, much of the variation is due to **genetic differences**.

26.4.1 Inbreeding and outbreeding

The genes of parents are passed in their gametes to the next generation. Thus all the genes of offspring come from a common pool of parental genes. Many of the individuals in a population have many alleles in common. They also have some alleles which are different from those of other individuals. Gametes are formed by meiotic nuclear division which brings about recombination of parental genes. Provided mating is at random, genetic variation constantly occurs among the individuals in each generation. This is called **outbreeding**. Inbreeding limits the amount of genetic variation because fertilisation is not at random. It also increases the proportion of homozygous individuals in a population (Fig 26.28).

Fig 26.28 Effect of successive inbreeding on the proportion of homozygotes in a population

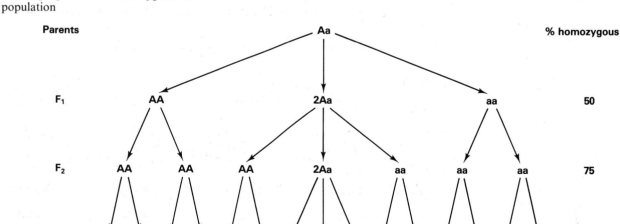

In most organisms an increase in homozygosity causes a reduction in vigour of the population. Such a loss of vigour is unlikely to occur among human populations, because religious and cultural barriers prevent marriage between close relatives. Another effect of **inbreeding** is to create genetically distinct lines of pure-breeding individuals. Some are homozygous dominant, others homozygous recessive for a given gene. The two lines can be distinguished phenotypically and maintained separately. Inbreeding is practised when it is desirable to have a large number of genetically uniform organisms for a specific purpose. For example, if a newly produced drug is to be tested on experimental animals to find out whether or not it is suitable for use on humans, it is important that the test animals should be as identical as possible to one another. The variation in reaction to the drug is much less than occurs in a batch of random-bred animals and the results are much more reliable.

In the wild, occasional inbreeding among organisms which normally outbreed, perhaps as a consequence of physical isolation, may result in a new line which is unable to breed with the parent stock. This could be a way in which new species arise.

26.4.2 Clones

Organisms which reproduce asexually give rise to progeny by mitotic nuclear division which normally produces exact copies of the parental genotype (Chapter 7). Bacteria and protozoa which reproduce by fission and flowering plants which multiply by parthenogenesis or by vegetative propagation usually have identical genotypes too. Populations of individuals having the same genotypes are called **clones**. Cloning is useful because it preserves genotypes which may be very well suited to a particular environment. Gardeners frequently maintain clones of desirable varieties of plants by vegetative propagation (Chapter 22).

26.4.3 The Hardy–Weinberg equation

As described earlier in the chapter each characteristic of an individual is usually determined by a gene pair, the alleles of which often occur in dominant and recessive forms. If the genotypes of two parents for a given characteristic are known, it is possible to calculate the probability of their having offspring with a particular genotype. For example, if the father is of blood group A with a genotype $I^A i$, and the mother is of blood group B with a genotype $I^B i$, the probability that they will have a child of blood group O is 0.25 or 25 % (Fig 26.29). It is worked out by multiplying the probability of sperm having the allele i by the probability of ova having the same allele.

Fig 26.29 Determining the probability of the genotypes of children

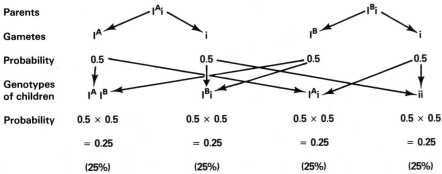

Table 26.8 Determination of frequencies of genotypes of progeny in a population producing gametes of known frequencies

	♂ gametes	
♀ gametes	A(p)	a(q)
A(p)	AA(p^2)	AA(pq)
a(q)	Aa(pq)	aa(q^2)

In a population the probability that a sperm may contain such an allele may be different from the probability for an ovum. Nevertheless, the same principle applies. When a population breeds at random it contains homozygous dominant, homozygous recessive and heterozygous individuals. Consider an allele **A** and its recessive form **a**. Individuals of genotypes **AA**, **Aa** and **aa** are present in the population and contribute their genes to a large pool of gametes. If the probability that a gamete contains **A** is p and the probability that it contains **a** is q, the proportions of genotypes among the progeny can be calculated as shown in Table 26.8.

Because there is 100 % probability that a gamete contains **A** or **a** then $p + q = 100$ % or 1.0. There is also a 100 % probability that the progeny are of the genotypes **AA + 2Aa + aa**. Thus:

$$p^2 + 2pq + q^2 = 1.0 \text{ or } 100 \%$$

Table 26.9 Frequencies of tasters and non-tasters of PTC among parents in Ohio, USA (after Snyder 1932)

Phenotype	Percentage	Frequency
taster	71.2	0.712
non-taster	28.8	0.288
total	100	1.000

This is called the **Hardy–Weinberg equation**. It has been verified with a number of human traits, including our ability or otherwise to taste phenylthiocarbamate (PTC). Some people are unable to taste PTC, others experience a bitter taste when a small amount of it is placed on the tip of the tongue. The allele for tasting (**T**) is dominant to that for non-tasting (**t**). Non-tasters must therefore have the genotype **tt**, whereas tasters could be **TT** or **Tt**. In the early 1930s Snyder tested a large number of North American parents for their ability to taste PTC. His results are summarised in Table 26.9. Applying the Hardy–Weinberg equation to the findings:

$$p^2 + 2pq = 0.712$$
$$\text{and } q^2 = 0.288$$
$$\text{Therefore } q = \sqrt{0.288}$$
$$= 0.537$$

$$\text{Now } p + q = 1.0$$
$$\text{Thus } p + 0.537 = 1.0$$
$$\text{and } p = 1.0 - 0.537$$
$$= 0.463$$

The probability of someone having the genotype **TT** (p^2) is therefore:

$$(0.463)^2 = 0.214 \text{ or } 21.4\%$$

The probability of someone having the genotype **Tt** ($2pq$) is:

$$2(0.463 \times 0.537) = 0.497 \text{ or } 49.7 \%$$

In this way Snyder calculated the proportions of the various genotypes among the parents. He also detected the same ratio of tasters and non-tasters among their children. It was therefore apparent that there had been no change in the proportions of individuals of the three genotypes in successive generations. This is known as the **Hardy–Weinberg equilibrium**. It is disturbed when:

i. mating is non-random,

ii. the alleles in question mutate,

iii. migration occurs between populations, and

iv. there is natural selection against one of the genotypes (section 26.4.4).

The gene pool of many human populations is constantly changing because people frequently move within their own country or they emigrate to another country. Such movement eventually leads to intermarriage of people of different cultures and races. The resulting exchange of genes between such populations is called **gene flow**. However, in many parts of the world small communities are isolated from the general population for cultural, political or religious reasons.

Should a gene mutation arise in an isolated population, it is likely to increase in frequency because of inbreeding. This is known as the **founder principle** and obviously disturbs the Hardy–Weinberg equilibrium. The equilibrium may also be changed in isolated populations as a consequence of **random genetic drift**. If for example allele **X** is at high frequency in the general population, whereas allele **x** is rare, the recessive allele is still retained because of random mating. In a small isolated population, only a few individuals would carry the **x** allele. If they do not have offspring the recessive allele will disappear from the gene pool of the next generation. Such genetic drift is thought to account for the high frequency of blood group A among Blackfeet Indians in contrast to the majority of North American Indians who are mainly blood group O.

26.4.4 Natural selection

Instances occur when individuals with a particular genotype do not reproduce. For example, the genotypes for haemoglobin production in some African populations are **SS** (normal), **Ss** (sickle-cell trait) and **ss** (sickle-cell disease). Those with **sickle cell disease** do not normally survive to reproductive age. Hence their genes are not passed on to the next generation. On the contrary, people who are homozygous dominant are more susceptible to some forms of malaria than are heterozygotes. The loss of **s** alleles to the next generation is thus to some extent counterbalanced by the loss of **S** alleles in people dying of malaria before they produce children. For these reasons the heterozygotes are the main contributors of genes to each new generation. As a consequence, the harmful **s** allele is retained at a high frequency in such populations.

New genes also constantly arise by mutation (Chapter 4). However, a mutated gene is only retained in a population if it is advantageous or neutral in terms of survival. An increase in frequency of a harmful gene is usually prevented because individuals having such genes may not live to produce children. Preserving two or more forms of a gene in a population by means other than mutation is called **genetic polymorphism**. Sickle-cell anaemia and ABO blood groups are examples of how alternative forms of a gene may be maintained at relatively high frequencies. The allele for sickle-cell anaemia is advantageous in helping heterozygotes to resist malaria. The multiple alleles for the ABO blood groups appear to be neutral so far as survival is concerned.

26.5 Detection of carriers

Carriers are individuals who are heterozygous for an undesirable recessive allele. Parents are often unaware that they carry a defective allele until they have an impaired child because it is homozygous recessive.

Simple biochemical tests can sometimes be used to detect carriers. Blood can be tested to determine whether or not parents carry the gene for **acatalasia**. The red blood cells of normal people contain the enzyme catalase which breaks down hydrogen peroxide into water and oxygen. However, homozygotes have twice as much catalase activity as heterozygotes. The activity is determined by mixing a known volume of blood with a known amount of hydrogen peroxide and measuring the frothing (oxygen production) which occurs in a given time. Homozygous recessives lack catalase altogether. Biochemical tests on samples of blood can also be used to detect carriers of **phenylketonuria** and **galactosaemia** (Chapter 4). People who carry the gene for **haemophilia** can be identified by testing the clotting of their blood.

In some instances it is possible to detect whether or not a fetus will develop into a defective child. The technique most often used for this purpose is **amniocentesis** (Fig 26.30). A small sample of amniotic fluid is withdrawn from around the developing fetus through a fine hollow needle which is pushed through the mother's abdominal wall and uterus. The fluid contains urine which the fetus has passed. Amniotic fluid can be analysed to see if it contains unusual constituents such as homogentisic acid, a sign that the fetus has **alkaptonuria** (Chapter 4). Also in the fluid are fetal skin cells. They can be grown in a cuture medium and tested to see if important enzymes are missing. An alternative technique called **chorionic villus sampling** is currently being tried out. A sample of embryonic tissue is taken from the chorionic villi (Chapter 25) by suction using a narrow tube inserted into the womb via the vagina. This procedure can be carried out at 10–12 weeks of pregnancy compared with 16–18 weeks for amniocentesis.

What are the potential benefits of using chorionic villus sampling as a screening procedure?

Fig 26.30 Amniocentesis

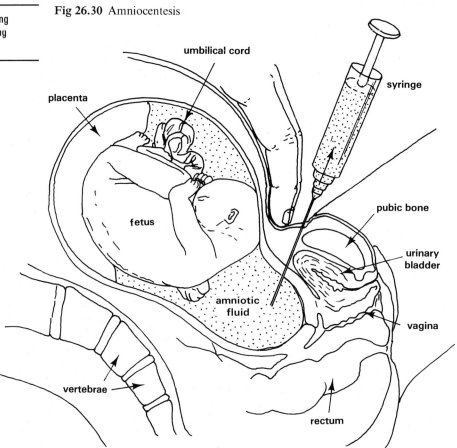

26.6 Karyotyping

Unusual numbers of chromosomes or abnormal chromosomes may be detected by allowing cells to divide in culture and examining the chromosomes microscopically. The technique is called **karyotyping**. In the 1950s Tijo and Levan developed a technique which clearly demonstrated that the diploid chromosome number in humans is 46. The technique which is still universally used in **cytogenetic laboratories**, is as follows.

A sample of amniotic fluid is centrifuged to sediment the fetal cells (Fig 26.31). The cells are sucked into a pipette and are then aseptically transferred to a nutrient solution containing **phytohaemagglutinin** which stimulates the cells to divide by mitosis. After about three days at 37 °C, several mitotic divisions will have occurred. Some **colchicine** is now added which prevents the formation of the spindle fibres. Cell division is thus stopped at metaphase when the chromosomes are shortest and thickest. Next, some weak saline solution is added to the medium, causing the cells to absorb water by osmosis. They swell up and the chromosomes spread out. Some of the cells are placed on a glass slide and viewed with a light microscope. The chromosomes are photographed and the images of individual chromosomes cut out of the picture. They are now paired in decreasing order of size and numbered 1–23, the last pair being the sex chromosomes. The arrangement of chromosomes is a **karyotype** (Fig 26.32).

Fig 26.31 Preparation of a karyotype

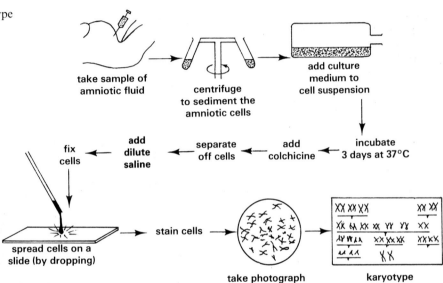

The technique of karyotyping makes it possible to detect fetuses with abnormal chromosome numbers at a relatively early stage in gestation. A decision can then be taken whether or not to terminate the pregnancy. Karyotyping of fetal cells also makes it possible to tell the sex of a child well before it is born. The same technique can be applied to white blood cells to diagnose chromosome abnormalities after birth.

Fig 26.32 Karyotype of a male with Down's syndrome. Note the extra chromosome 21.

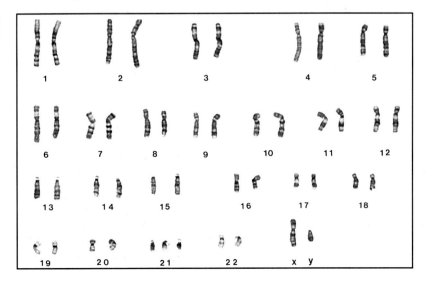

26.7 Barr bodies, Y chromosomes and white cell drumsticks

In 1949 Murray Barr reported the presence of a small semicircular body lying just inside the nuclear membrane of normal female cells to which a stain specific for DNA had been applied (Fig 26.33). The **Barr body**, as it is now called, does not occur in the cells of normal males. This is because at least two X chromosomes must be present for the body to appear. Barr body studies can be used to determine the sex of a fetus, applying the stain to fetal cells obtained by amniocentesis. It is another way of finding out the sex of a child before it is born. Smears of buccal epithelium can also be examined using this technique when there is doubt as to the sex of athletes participating in major events such as the Olympic Games.

Fig 26.33 Nucleus showing Barr body

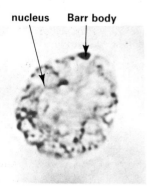

nucleus   Barr body

These days a piece of amnion is also routinely examined for Barr bodies at birth. The reason for doing so is to identify children who may have an unusual complement of sex chromosomes. The number of Barr bodies in a cell is one fewer than the number of X chromosomes. Thus normal females have one Barr body and normal males have none. However, triple-X females or XXXY males will have two Barr bodies. If the baby is male and a Barr body is present then Klinefelter's syndrome is suspected. The absence of a Barr body in a female child would indicate that it may have Turner's syndrome.

Fluorescent microscopical techniques can be used to examine cells for Y chromosomes. The cells are stained at interphase with acridine dyes and viewed with an ultraviolet microscope. Y chromosomes take up the dye and emit a green light (fluorescence), enabling their number to be determined. The test is now performed routinely at birth on cells from the umbilical cord. It is also performed on buccal smears of female athletes where their sex is in doubt.

Polymorphonuclear granulocytes of normal females have a body shaped like a **drumstick** protruding from one of the lobes of the nucleus (Chapter 10). Drumsticks do not occur in such cells of normal males. The number of drumsticks is one less than the number of X chromosomes. Examining blood smears microscopically for drumsticks is therefore another way of finding out the sex of a child and of detecting unusual numbers of X chromosomes.

26.8 Genetic counselling

With the aid of the techniques described in the previous three sections it is now possible to counsel parents about the possibility of genetic defects in their unborn children. The parents are then able to decide whether or not to have a pregnancy terminated. Decisions of this kind are not easy to arrive at. Some people believe it ethically and morally wrong to prevent any fetus from reaching maturity. Others are of the opinion that children with very serious abnormalities should not be born because of the suffering they may have to endure.

Fig 26.34 Predicting the probability of fibrocystic disease

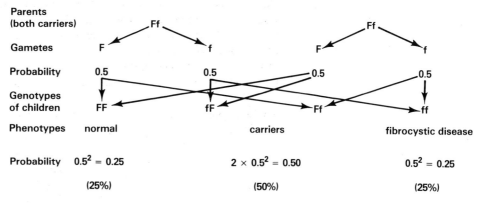

F = allele for normality
f = fibrocystic disease

Genetic counselling is available to families with a history of genetic abnormalities and those without. For example, both parents may be carriers of a defective allele. The first child they produce may be a homozygous recessive for a condition such as fibrocystic disease of the pancreas. It is then possible to determine the probability of further children having the defect. In this instance it would be 25 % (Fig 26.34). In a similar way it is possible to predict that of the sons born to a normal man and a woman who is carrier for haemophilia, half will be normal and half will be haemophiliacs (Fig 26.35).

Some abnormalities in babies are not inherited. The embryo may be affected by **teratogenic agents** while inside the uterus. For example, a woman who has rubella (German measles) in early pregnancy is very likely to have a child with defective sight, hearing or heart function. Drugs taken by pregnant women, as thalidomide was in the 1960s, may also seriously impair fetal development.

Fig 26.35 Predicting the probability of haemophilia

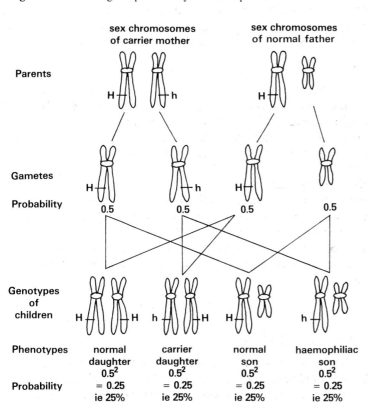

SUMMARY

The foundations of genetics, the study of inheritance, were established by Gregor Mendel in the middle of the last century. He formulated two laws to explain how the characteristics of the garden pea are inherited. It has since been shown that Mendel's laws are applicable to many of the traits inherited by most kinds of organisms including humans.

Most inherited traits are determined by alternative forms of genes called alleles, many of which exist as dominant and recessive forms. A few genes occur in multiple forms, others are co-dominant. The combination of genes an organism has is its genotype. Interaction of the genotype and the environment determines the phenotype, the traits displayed by the organism. The behaviour of chromosomes in meiosis offers an explanation of how the alleles they carry are distributed to gametes. One pair of chromosomes are the sex chromosomes; the remainder are autosomes.

Genes which control different traits and are on the same chromosome are linked. Such traits give rise to unusual phenotypic ratios. Linked genes provide a means of mapping the location of alleles on chromosomes.

The number of chromosomes in a nucleus is characteristic for each species. An abnormal number of chromosomes is often harmful. Down's syndrome is an example of a human condition in which there is an extra autosome.

The sex chromosomes carry the genes which determine sex and sex-linked traits. Female mammals are homogametic having two X chromosomes. Males are heterogametic with one X and one Y sex chromosome. Abnormal numbers of sex chromosomes in humans gives rise to such conditions as Klinefelter's and Turner's syndromes in which sexual development is impaired. Haemophilia and red–green colourblindness are examples of human sex-linked traits.

Discontinuous variation is a clear cut alternative of a trait, for example, red flower colour as opposed to white in the garden pea. In contrast continuous variation is where a trait shows a gradation of forms such as height in humans. It is attributed to the additive effect of many genes, polygenic inheritance, as opposed to a pair of alleles as in discontinuous variation.

In large closed populations where mating occurs at random, where alleles do not mutate and where natural selection does not operate against a given genotype, the frequency of the various genotypes remains constant from generation to generation. This is called the Hardy–Weinberg equilibrium and provides a means of calculating gene frequencies in populations. The equilibrium is disturbed by gene flow when mating occurs with individuals from an adjacent population.

Carriers are individuals who have an undesirable recessive allele but do not display the trait associated with it. Detection of carriers is of importance in medical genetics whereby people can be counselled of the probability of having an impaired child. Karyotyping, determining the chromosome complement, as well as the detection of Barr bodies, Y chromosomes and white cell drumsticks are also valuable aids in this context.

QUESTIONS

1 (a) The nitrogenous bases in nucleic acids are adenine (A), cytosine (C), guanine (G), thymine (T) and uracil (U). In part of a molecule of deoxyribonucleic acid (DNA), the sequence of bases is CTT. Using the letters A, C, G, T and U where appropriate, write down the base sequence in:
  (i) The corresponding part of the complementary strand of the DNA molecule.
  (ii) The codon of messenger ribonucleic acid (mRNA) transcribed by the triplet CTT.
  (iii) The anti-codon of the transfer ribonucleic acid (tRNA) molecule used to translate the codon.
(b) A change in the DNA molecule caused the base sequence in the triplet to change from CTT to CAT.
  (i) Write down the anti-codon required to translate the change.
  (ii) State **two** factors which could have caused such a change.
  (iii) What name is given to this kind of change?
(c) A change of the kind described in (b) causes some humans to produce sickle cell haemoglobin (HbS). People can be homozygous or heterozygous for the recessive sickle cell allele. Heterozygotes synthesise approximately equal amounts of HbS and normal haemoglobin and have a mild form of anaemia. Homozygotes produce HbS only

and die of sickle cell disease before puberty. In Central Africa, a population of 500 tribesmen sampled in each of several generations showed a constant 2:1 ratio of normal to heterozygous individuals.
   (i) State the probable cause of early deaths of people with sickle cell disease.
   (ii) Using the Hardy–Weinberg equation ($p^2 + 2pq + q^2 = 1$), calculate the proportion of HbS and normal alleles in the population gene pool.
   (d) Indicate **one** way in which the data in (c) differ from the usual assumptions on which the Hardy-Weinberg equilibrium is based. (WJEC)

2 (a) Which one of the following populations is in the Hardy–Weinberg equilibrium?

	Genotypic frequencies		
	AA	Aa	aa
Population I	0.4300	0.4810	0.0890
Population II	0.6400	0.3200	0.0400
Population III	0.4225	0.4550	0.1225
Population IV	0.0025	0.1970	0.8005

   (b) Give the frequency of the A allele in this population.
   (c) Give the frequency of the a allele in this population.
   (d) If this is a large inbreeding population, without mutations, show how you would determine the genotype frequencies in the next generation. (O)

3 In a genetics investigation using the fruit-fly *Drosophila melanogaster*, the following experiments were performed.

Experiment 1
Male flies showing the two recessive characteristics claret eyes (cl) and vestigial wings (vg) were mated with female flies which were true breeding for wild type eyes (cl^+) and wild type wings (vg^+).
   The female offspring of this first cross were then mated to their male parents. The results of this second cross were as follows:

Characteristics	Number
Claret eyes, vestigial wings	72
Claret eyes, wild type wings	80
Wild type eyes, vestigial wings	76
Wild type eyes, wild type wings	84

Experiment 2
Male flies showing the two recessive characteristics purple eyes and vestigial wings were mated to female flies which were true breeding for wild type eyes and wild type wings.
   The female offspring of this first cross were then mated to their male parents and the following results were observed:

Characteristics	Number
Purple eyes, vestigial wings	128
Purple eyes, wild type wings	21
Wild type eyes, vestigial wings	17
Wild type eyes, wild type wings	136

   (a) (i) Using suitable genetic cross diagrams (and the symbols given) explain fully the reasons for the results of the two crosses obtained in Experiment 1.
   Give the genotypes and phenotypes of the offspring of the first and second crosses.
   (ii) What deductions can be made about the positions of the claret eye allele and the vestigial wing allele on the *Drosophila* chromosomes?
   (b) (i) Explain why the results observed in the second cross of Experiment 2 are different from those observed in Experiment 1.
   (ii) What can be deduced about the relative positions of the purple eye allele and the vestigial wing allele on the *Drosophila* chromosomes?
   (c) Comment on the significance of these results in relation to Mendel's Second Law of Inheritance (the law of 'Independent Assortment'). (O)

4 The following statement is to be found on certain seed packets:
'Seed saved from F_1 hybrids will not breed true'
   (a) What is meant by (i) F_1 *hybrids*, (ii) *breed true*?
   (b) Explain why F_1 hybrids will not breed true if they are self-fertilised.
   (c) Give **one** advantage to the grower of having F_1 hybrids. (AEB 1987)

5 In a genetics experiment, tomato plants which were heterozygous for the recessive alleles non-pigmented (green) stem and smooth ('entire') leaflets were allowed to interbreed.
   Three hundred and twenty seedlings were produced as a result of the cross and these were sorted into four different phenotype groups and counted.
   The results were:

Pigmented (purple) stem, serrated ('cut') leaflets	190
Pigmented (purple) stem, smooth ('entire') leaflets	53
Non-pigmented (green) stem, serrated ('cut') leaflets	63
Non-pigmented (green) stem, smooth ('entire') leaflets	14
TOTAL	320

   (a) Copy the table and fill in the necessary figures that would be required in order to calculate the χ^2 value.
   Assume that the inheritance is of simple Mendelian type giving a 9:3:3:1 ratio of phenotypes.
   Purple (**P**) is dominant to green (**p**) and serrated (**S**) is dominant to smooth (**s**).

	P/S	P/s	p/S	p/s
Observed numbers				
Expected numbers				
O–E				
$(O–E)^2$				
$\dfrac{(O–E)^2}{E}$				

$$\chi^2 = 3.32$$

   (b) In this example the number of degrees of freedom is 3. Below is an extract from χ^2 distribution tables giving the probability levels for 3 degrees of freedom.

Prob. = 0.99   0.95   0.05   0.001
χ^2 = 0.115   0.352   7.82   11.35

(i) Explain briefly how the experimenter could interpret this χ^2 value of 3.32 and what conclusions could be drawn.
(ii) If a χ^2 value of greater than 7.82 (probability is less than 0·05) had been found, what conclusion would you have drawn?   (O)

6 The diagram shows part of a family tree of a mammal. Several members have the condition called albinism where there is a lack of normal pigment in the body.

(a) Using the information in the family tree, explain why it is unlikely that albinism is controlled by any of the following (in each case give one reason):
 (i) an autosomal dominant gene.
 (ii) a sex-linked (X-linked) recessive gene.
 (iii) a gene on the Y chromosome.
(b) Assume that pigment levels in this mammal are controlled by a single autosomal gene which has two alleles. The albino condition is produced by the recessive allele, a, while the normal pigment level is controlled by the

dominant allele, A. What are the possible genotypes of the following members of the family tree?
 (i) number 2 (in generation I),
 (ii) number 3 (in generation II),
 (iii) number 12 (in generation III).

(c) The same family was scored for inability to distinguish red and green colours, a condition called colour-blindness. This is controlled by a sex-linked gene; individuals with normal colour vision carry at least one B allele of this gene, while colour-blind individuals do not carry any B alleles. Using the B and b, give the likely genotypes of:
 (i) a female homozygous for normal pigmentation and normal colour vision;
 (ii) an albino male with normal colour vision;
 (iii) a male with normal pigmentation whose mother was colour-blind.

(d) Derive and give the phenotypic ratio produced in the F_2 of the following mating:

Female	×	Male
Colour-blind and homozygous for normal pigmentation		Normal colour vision and albino
		(JMB)

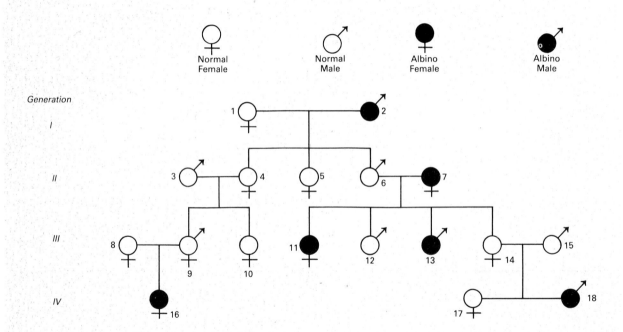

524

27 The origin and evolution of life

27.1 The origin of life   526

  27.1.1 Monomer synthesis   526
  27.1.2 Polymer synthesis   527
  27.1.3 Polymer isolation and replication   527
  27.1.4 The first cells   528

27.2 Evidence of evolution   529

  27.2.1 Fossils   529
  27.2.2 Affinities   539
  27.2.3 Geographical distribution   541

27.3 How evolution came about   542

  27.3.1 Lamarckism   542
  27.3.2 Darwinism   542
  27.3.3 Variation   543
  27.3.4 Selection   545
  27.3.5 Isolating mechanisms   547
  27.3.6 The formation of species   549

Summary   550

Questions   550

(Bruce Coleman: Kim Taylor)

27 The origin and evolution of life

The earth is populated by an enormous variety of plant and animal life (Chapters 28–30). Some organisms clearly resemble others and there is little difficulty in accepting the fact that they are closely related. Others are so different in appearance that it would seem that they are unrelated. Evolution implies that all living organisms, however unalike, arose from common ancestors. They are the descendants of previous forms of life and in this sense they are related to each other.

For centuries people have wondered how life originated and how the diversity of living organisms evolved. Scientists can still only speculate about the origin of life. Much more is known about how life evolved, mainly due to the work of Charles Darwin and Alfred Russel Wallace in the nineteenth century and to the findings of twentieth-century geneticists and biochemists.

27.1 The origin of life

Radionuclide methods of dating (section 27.2.1) indicate that the earth was formed about 4600 million years ago, whereas the oldest fossils are only 3200 million years old. The first 1400 million years of the earth's history are thus something of a mystery. There are many theories as to how life began. Modern astronomers such as Sir Fred Hoyle have recently produced evidence to suggest that microbial life may have begun in outer space and was carried to earth in the dust of comets and on meteorites.

The British biochemist J. B. S. Haldane appears to have been the first to propose that life originated on earth when the atmosphere was devoid of oxygen gas. In 1929 he suggested that an atmosphere lacking oxygen would have no ozone layer which today stops the penetration of most of the ultraviolet radiation from the sun. Ultraviolet light is assumed to have provided energy for organic molecules to be made from atmospheric gases such as carbon dioxide, ammonia and water vapour. The organic molecules gradually accumulated in the oceans, in Haldane's words, as a 'dilute soup'. One of the difficulties in accepting Haldane's theory is that ultraviolet light can disrupt bonds in organic molecules as well as providing energy for synthesis.

A few years later, the Russian biochemist A. I. Oparin came up with similar ideas as to how life began. There was little interest at the time in Haldane's and Oparin's suggestions. However, with present-day knowledge it is reasonable to assume that the main steps in the evolution of life were probably:

1. **Synthesis of organic monomer molecules** such as amino acids, nitrogenous bases and sugars.
2. **Formation of organic polymers** such as proteins and nucleic acids.
3. **Isolation of the polymers** from the non-living environment followed by their **replication**.

27.1.1 Monomer synthesis

An indication of the kinds of organic **monomers** that can be synthesised from simple gases came in the early 1950s. Harold Urey and Stanley Miller

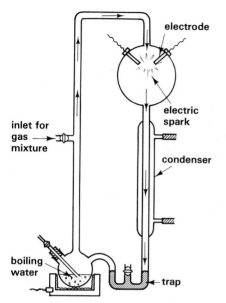

in the USA carried out a series of laboratory experiments in which they subjected a variety of gas mixtures to ultraviolet light and an electric spark (Fig 27.1). The electric spark simulated lightning flashes. These may have been an additional or alternative source of energy for the synthesis of organic molecules in the earth's early atmosphere. Urey and Miller observed that gas mixtures such as hydrogen, methane, ammonia and water vapour gave rise to aldehydes, amino acids and carboxylic acids. Other mixtures – for example, carbon dioxide, carbon monoxide, nitrogen and water vapour – gave similar results.

Amino acids were possibly formed in the following way, known as the **Strecker synthesis**:

In the atmosphere:

$$R.CHO + NH_3 \quad H_2O \rightarrow R.CH.NH + HCN \rightarrow R.CH.NH_2CN$$

aldehyde + ammonia $\longrightarrow$ imine $\longrightarrow$ aminonitrile

In the ocean:

$$R.CH.NH_2CN + 2H_2O - NH_3 \rightarrow R.CH.NH_2COOH$$

aminonitrile + water $\longrightarrow$ amino acid

Several of the amino acids formed in the experiments performed by Urey and Miller are found in the proteins of living organisms. In the early 1960s, using other gas mixtures, Oró and Kimball made purine and pyrimidine bases. Pentose sugars can be made from methanal (formaldehyde) in the **formose reaction**:

$$5H.CHO \rightarrow C_5H_{10}O_5 - O \rightarrow C_5H_{10}O_4$$

methanal $\longrightarrow$ ribose $\longrightarrow$ deoxyribose

These are among the monomers of which nucleic acids are synthesised (Chapter 4).

27.1.2 Polymer synthesis

When amino acids join by peptide linkages to form proteins, water is one of the products of the reaction (Chapter 2). The same happens when nitrogenous bases and pentose sugars join to make nucleic acids. They are both examples of **polymerisation** by condensation. It is difficult to see how condensation reactions could readily occur in the earth's primaeval oceans. Here water was abundant and would have kept the monomers apart. Bernal has suggested that the amount of water in the vicinity of the monomers may have been reduced by their adsorption on to clay or mica particles. Katchalsky has shown experimentally that polypeptides will form on clay minerals. Polymers formed in such a way would have their monomers arranged at random, unlike proteins and nucleic acids where their sequence is specific (Chapter 4). Presumably simple polynucleotides able to replicate and code the synthesis of enzyme molecules would have to be formed to achieve specific sequencing.

27.1.3 Polymer isolation and replication

In living cells, the energy required to build polymers from monomers comes from ATP (Chapter 5). Before life evolved, it may have come from ultraviolet light.

Oparin in the Soviet Union and Fox in the USA have closely studied some of the ways in which natural polymers may have become separated from their surroundings. The tendency of biological polymers to form polymer-rich droplets called **coacervates** has been examined by Oparin. He reports that coacervates are readily formed from aqueous suspensions of proteins and polysaccharides, and proteins and nucleic acids. Some coacervates are unstable and quickly sediment from suspension. Such coacervates can be made stable by adding enzymes to the suspensions. The enzyme molecules concentrate in the coacervates, where they are able to catalyse biochemical reactions. One such enzyme is phosphorylase which catalyses the polymerisation of glucose phosphate to form starch (Fig 27.2). As starch accumulates, the coacervates grow and eventually break up into smaller coacervates which continue to grow in the same way. Could this be the way in which biological polymers were first replicated?

Fig 27.2 Polymerisation of starch in a coacervate

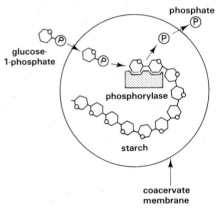

In addition to condensation reactions, Oparin has shown that coacervates can also bring about electron transport. For example, coacervates containing dehydrogenase enzymes will reduce redox dyes such as methyl red when an electron carrier such as NADH is added to the suspension. Furthermore, he has succeeded in getting coacervates containing chlorophyll to develop a reducing power when exposed to visible light. This is comparable to the way in which chloroplasts reduce $NADP^+$ in photosynthesis (Chapter 5).

Fox has worked mainly with protein coacervates. When heated in water at 130–180 °C, they aggregate spontaneously into spheres 1–2 μm in diameter. The spheres have a boundary which resembles the structure of a unit membrane, though it lacks the double phospholipid layer (Chapter 6). They grow by absorbing protein dissolved in the surrounding water and split in the same way as bacteria multiply by fission (Chapter 28). In addition, they are able to catalyse a variety of biochemical reactions, for example esterification, glucose oxidation and peroxidase activity.

The experiments of Oparin and Fox show that biological polymers may form stable systems capable of catalysing biochemical reactions. Such systems may well have been the forerunners of living cells.

27.1.4 The first cells

The first cells to evolve were possibly **unicellular organisms**. Because oxygen gas was absent in the earth's oceans at the time, they probably acquired energy by fermenting some of the organic monomers synthesised from atmospheric gases. In many ways they were probably similar to anaerobic bacteria living today. The capacity of the earth to support **consumer organisms** of this kind was limited by the rate at which monomers were formed. When photosynthetic organisms evolved, the earth's capacity to support life became greatly enhanced. Such **primary producers** probably used hydrogen sulphide as a source of hydrogen for reducing carbon dioxide. This form of photosynthesis is used by some bacteria today. At a later stage, organisms able to use water instead of hydrogen sulphide in photosynthesis appeared. They were probably similar to blue-green algae (Chapter 28). It is these organisms which are found as the oldest fossils in rocks at least 3200 million years old.

The release of oxygen as a photosynthetic product resulted in a gradual but dramatic change in the composition of the earth's atmosphere. An important consequence of the change was the formation of the ozone layer in the upper atmosphere which reduced the penetration of ultraviolet light from the sun. It greatly limited the non-biological production of organic matter. From this time on, synthesis of organic molecules was brought about mainly by photosynthesis. Another significant outcome was that organisms began to evolve aerobic methods of respiration, using photosynthetically produced oxygen. Aerobic respiration is much more efficient at energy conversion than anaerobic respiration (Chapter 5).

27.2 Evidence of evolution

The reason why it is believed that the enormous variety of life has evolved from simple ancestors is based on three main sources of evidence. They are the evidence from **fossils**, from **affinities** between organisms and from the **geographical distribution** of organisms.

27.2.1 Fossils

A **fossil** is any buried object which points to the existence of prehistoric life. Fossils are formed in various ways. One of the commonest is where a hard part of an animal, such as the shell or skeleton, immersed in water, has become infiltrated or replaced by minerals causing it to become **petrified**, changed to stone. In other instances, the body of an organism has become embedded in mud which hardened to form a **mould** when the body subsequently decayed. When the mould later filled with minerals brought in by water, a stony **cast** was formed, similar in shape to the organism. Casts of the burrows of mud-dwelling invertebrates have been formed in this way. Alternatively, an **impression** of the shape of the body or its surface structure may be left in material which later changed into rock (Fig 27.3).

Fig 27.3 Various types of fossils

cast of an ammonite

shell of a gastropod

tooth of a shark

impression of fern leaves

cast of a trilobite

Most fossils are found in **sedimentary rocks** formed from mud and silt deposited in lakes and on the sea bed millions of years ago. Earth

Fig 27.4 Rock strata at Aberystwyth, Wales

movement later raised many sedimentary rocks above sea level. Often the rocks are in **strata**, the oldest at the bottom and most recently formed at the top (Fig 27.4). Some indication of the course of evolution can be obtained by comparing the sorts of fossils in the strata.

It is also possible to age the rocks by radionuclide (radioisotope) dating. Nearly all elements exist as several nuclides (isotopes), that is as atoms having slightly different atomic masses e.g. C^{12} and C^{14}. Some nuclides are unstable and decay by releasing sub-atomic α- and β-particles until they become more stable. Some examples are as follows:

$$\text{uranium}^{238} \rightarrow \text{lead}^{206}$$
$$\text{potassium}^{40} \rightarrow \text{calcium}^{40} + \text{argon}^{39}$$
$$\text{carbon}^{14} \rightarrow \text{carbon}^{12}$$

The time required for half a given quantity of an unstable nuclide to break down into a stable form is its **half-life**. The half-lives of many unstable nuclides have been determined. It is thus possible to estimate accurately the age of a fossil by measuring the ratio of unstable:stable nuclides, for example $U^{238}:Pb^{206}$ in the rock in which the fossil was found. If the ratios of several nuclides are measured at the same time, the dating is cross-checked and thus reliable. The sequence in which organisms evolved can thus be determined. It usually coincides with the order suggested from rock strata studies.

Other information of the kinds of living organisms which existed in past times comes from **amber**, the hardened resin of primitive seed-bearing trees, in which whole insects have been trapped. The **frozen bodies** of mammoths and woolly rhinoceroses found in Siberia, and the **pollen-grains** trapped in peat bogs, have also provided valuable information.

It is important to realise that fossils provide only a sketchy history of previous life. Many fossils have been found purely by chance when rocks have been disturbed in mining, quarrying and in the construction of roads, bridges and buildings. Many more lie as yet undiscovered. Fossils of soft-bodied organisms are not particularly plentiful. This is presumably because the organisms quickly decayed after death. For some forms of life such as fungi there is hardly any fossil evidence.

1 The first fossils

Rocks over 600 millions years old belong to the **Precambrian** geological period. Until quite recently there was no fossil evidence of life as far back as this. In the 1950s, two American geologists, Barghoorn and Tyler, reported the presence of fossil micro-organisms called **stromatolites** in Precambrian rocks near Lake Superior, Ontario. The fossils which are over 3000 million years old resemble bacteria and blue-green algae. Living stromatolites have since been found in the coastal waters of Western Australia. They are layered communities of bacteria and blue-green algae (Fig 27.5).

Fig 27.5 (a) General view of stromatolites (Bruce Coleman: Jan Taylor)

Fig 27.5 (b) Section of a fossil stromatolite (B&B)

Living blue-green algae are able to photosynthesise and some can fix atmospheric nitrogen. If the stromatolite microbes were also able to do this, they may have had a profound influence on the course of evolution. By releasing oxygen gas, a product of photosynthesis, they may have created the conditions in which aerobic organisms first evolved. Before this time the earth's atmosphere is thought to have contained very little oxygen gas.

2 Animal and plant fossils

The oldest animal and plant fossils suggest that the earliest forms of life were simple and probably aquatic. Fossil burrows have been found in rocks of the late Precambrian period, less than 700 million years old. Living invertebrate animals which form similar burrows are worm-like creatures which use their hydrostatic skeleton to bore through mud and soil (Chapter 29). It is likely that the fossil burrows were formed by animals of this kind. Precambrian rocks dated as 580–680 million years old contain fossils which resemble **jellyfish** and **annelids**. Fragments of fossil invertebrate skeletons have also been found in rocks of the late Precambrian period. However, it is impossible to say to which group of invertebrates they belong. These earliest animals probably lived near the bottom of the oceans where they fed on particles of organic debris suspended in the water or deposited on the mud.

By the end of the **Cambrian** period, a wide variety of invertebrate groups had evolved. Some later became extinct, others have living descendants. One of the groups resembled **sponges**, another looked like **molluscs**. The plant life which existed at the time resembled **algae**. Some were single cells, others multicellular. They enriched the oceans with oxygen and detritus which may have stimulated subsequent evolution of a range of consumer organisms. During the **Ordovician** period, the first **bryophytes** appeared, as did agnathans, jawless fish which fed on particles filtered from water passing through their gills.

In the **Silurian** period, the first **jawed fishes** evolved, their jaws enabling these animals to use a choice of food. It may explain the greater variety of fish existing in the succeeding **Devonian** period. At this time the first vascular land plants had appeared. They were **psilophytes**, a group of extinct pteridophytes (Fig 27.6) which most likely grew in soil covered with a shallow layer of water. As plants began to colonise land, invertebrates such as annelid worms and **arthropods** probably followed.

Amphibians were well-established by the **Carboniferous** period. Many were fairly large, omnivorous animals which fed on the evolving terrestrial plant and animal life. They used water for fertilisation and embryo development as do living members of this group (Chapter 30). Later amphibians were smaller and consequently less noticeable to predatory vertebrates which were beginning to emerge. It was from the smaller types that present-day amphibians probably descended. At this time, vascular land plants were abundant. They included many of the larger extinct **pteridophytes** such as the giant horsetails and tree ferns (Fig 27.7). Primitive **spermatophytes** called the pteridosperms had also evolved by this time. Some of them secreted resin which attracted terrestrial insects. Some of the insects have become fossilised in amber formed from the resin.

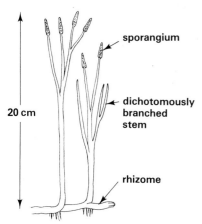

Fig 27.6 *Rhynia*, a psilophyte

sporangium

dichotomously branched stem

20 cm

rhizome

Psilophytes were the earliest vascular terrestrial plants. None survive today

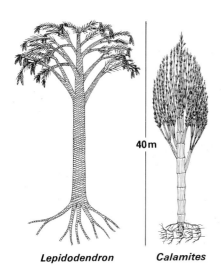

40 m

Lepidodendron   *Calamites*

Fig 27.7 Tree-sized pteridophytes, now extinct

Fig 27.8 A pelycosaur, Dimetrodon, one of the earliest known reptiles

Speculate on the possible functions of the dorsal fin of *Dimetrodon* (Fig 27.8).

It has been suggested that a creature as large as a *Brontosaurus* would have to spend much of its life almost submerged in water. What is the reason behind such thinking?

What ideas have been proposed to explain the extinction of the dinosaurs?

The earliest known **reptiles** appeared in the **Permian** period. Like living reptiles, they probably did not depend on water for reproduction and were able to colonise the land quickly. Here they preyed on and competed with the larger amphibians. Most were lizard-like and some had extremely large dorsal fins (Fig 27.8). In the **Triassic** period, they were replaced by the **dinosaurs** which survived well into the **Cretaceous** period, over 150 million years later. Dinosaurs were the most successful of terrestrial vertebrates to evolve up to this time and they exploited every ecological niche. Some, such as the pterodactyls were the first known vertebrates to fly

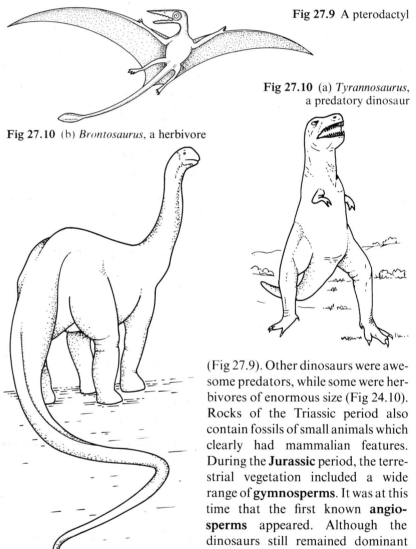

Fig 27.9 A pterodactyl

Fig 27.10 (a) *Tyrannosaurus*, a predatory dinosaur

Fig 27.10 (b) *Brontosaurus*, a herbivore

(Fig 27.9). Other dinosaurs were awesome predators, while some were herbivores of enormous size (Fig 24.10). Rocks of the Triassic period also contain fossils of small animals which clearly had mammalian features. During the **Jurassic** period, the terrestrial vegetation included a wide range of **gymnosperms**. It was at this time that the first known **angiosperms** appeared. Although the dinosaurs still remained dominant among the terrestrial vertebrates, **birds** and **egg-laying mammals** began to emerge.

The dinosaurs became extinct in the **Cretaceous** period. By then a variety of **placental mammals** had evolved, including what is though to be man's ancestors. The diversity of angiosperms also increased at this time.

The variety of **pollinating insects** which evolved in the subsequent **Tertiary** period reflects the fact that angiosperms became the dominant vegetation on land at the time. Simultaneously there was an increase in the variety of birds and terrestrial mammals.

The major phases of evolution described to date are summarised in Table 27.1.

Table 27.1 Summary of the main phases of evolution from the fossil record

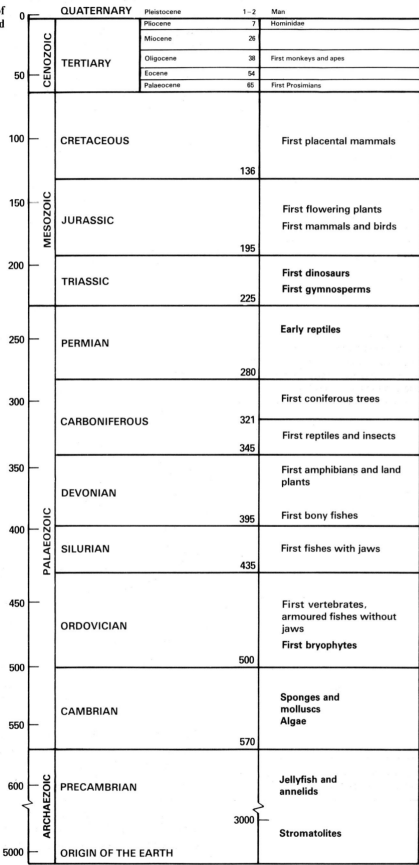

10⁶ Years	Era	Period	Epoch	Beginning of Period (millions of years BP)	Biological features
0	CENOZOIC	QUATERNARY	Pleistocene	1–2	Man
		TERTIARY	Pliocene	7	Hominidae
			Miocene	26	
			Oligocene	38	First monkeys and apes
50			Eocene	54	
			Palaeocene	65	First Prosimians
100	MESOZOIC	CRETACEOUS			First placental mammals
150				136	
		JURASSIC			First flowering plants First mammals and birds
200		TRIASSIC		195	First dinosaurs First gymnosperms
				225	
250	PALAEOZOIC	PERMIAN			Early reptiles
300		CARBONIFEROUS		280	First coniferous trees
				321	First reptiles and insects
350		DEVONIAN		345	First amphibians and land plants
400		SILURIAN		395	First bony fishes
				435	First fishes with jaws
450		ORDOVICIAN			First vertebrates, armoured fishes without jaws First bryophytes
500		CAMBRIAN		500	Sponges and molluscs Algae
550				570	
600	ARCHAEOZOIC	PRECAMBRIAN			Jellyfish and annelids
3000					Stromatolites
5000		ORIGIN OF THE EARTH			

Fig 27.11 Bush baby, a prosimian (after J Z Young 1976)

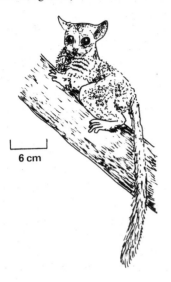

Fig 27.11 Bush baby, a prosimian (after J Z Young 1976)

6 cm

Fig 27.12 *Ramapithecus* (after R A Leakey 1981)

What is the reason for linking an upright gait with an upwardly directed foramen magnum?

27.2.2 Human evolution

Man is a member of the class **Mammalia** which taxonomists divide into about twenty orders (Chapter 30). The order to which man belongs is the **Primates** which has two sub-orders, the **Prosimii** and **Anthropoidea**. The prosimians include the earliest fossil primates as well as a few living representatives such as the lemur, loris and bush baby (Fig 27.11). They are probably similar to what our ancestors were like 60–70 million years ago, small tree-dwelling creatures. At the beginning of the Eocene epoch these early primates disappeared. At that time the earth's climate was generally warmer than it is now. The widespread tropical forests were the habitats of many early prosimians, some of which were similar to lemurs. About 70 distinct species have been described. It is from these creatures that anthropoids probably evolved. The evidence as to how this evolution took place is incomplete.

1 Early anthropoids

Fragments of several kinds of anthropoids have been discovered dating from the middle of the Oligocene epoch, about 35 million years ago. It appears that by then there was a clear distinction between ape and monkey lines of descent. From then on monkeys evolved along a separate path. The stock from which apes and man evolved thus probably diverged from that of monkeys 35–25 million years ago.

During the Miocene epoch the climate became generally cooler and drier. Open grassy plains replaced large tracts of forest and stimulated the evolution of horses and other fast-running herbivores. Fossils of the teeth and jaws of Miocene apes have been found. Some were ancestors of gibbons, others of apes. A third group of the genus *Ramapithecus* are more like humans with rounded jawbones, small canines and spade-like incisors. A dentition of this kind would allow the jawbone to be moved from side to side in a grinding action. The diet probably included grass seeds collected from open grassland. Seed collection was probably performed by the hands suggesting that *Ramapithecus* was bipedal (Fig 27.12). This is only speculation because no other parts of the skeleton have been found. Palaeontologists do not agree on when the ape and human lines diverged. Traditionally the lines are thought to have been distinct since late Miocene times, ten million years ago. However, some evidence indicates a more recent separation. Most typically human characteristics evolved within the Pleistocene epoch.

2 *Australopithecus*

Several human-like fossils from the early Pleistocene epoch have been discovered this century in various parts of Africa. Some taxonomists put them all into the genus *Australopithecus* with two species, *A. africanus* and *A. robustus*. Fossils of *A. africanus* were found at Taungs in the Transvaal, South Africa by Raymond Dart in 1925, and by Louis Leakey at Olduvai Gorge, Tanzania in the early 1960s. *A. africanus* has a flattened face with a forward projecting jawbone (Fig 27.13(a)). The foreamen magnum pointed upwards, a feature taken by some to indicate an upright gait. The teeth were generally like those of humans except the premolars and molars which were bigger. Large numbers of bones of various animals including hares, gazelle and wildebeest are frequently present in caves where fossils of *A. africanus* have been unearthed. This suggests that its diet may have included the flesh of herbivores which lived on nearby open grassland. Alternatively it may be that *A. africanus* was among the prey of carnivores such as leopards and hyenas which used the caves as lairs. One of the fossils of *A. africanus* had a cranial volume of about 800 cm^3 and was named

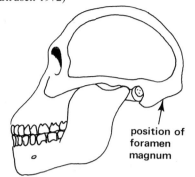

Fig 27.13 (a) Skull of *Australopithecus africanus* (from E J Clegg 1978 after Birdsell 1972)

position of foramen magnum

Fig 27.13 (b) Skull of *Australopithecus robustus* (from E J Clegg 1978 after Howells 1964)

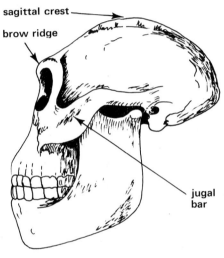

sagittal crest

brow ridge

jugal bar

Fig 27.14 Pelvic bones of *Australopithecus* and man (from J Z Young 1976 after Broom and Robinson 1955). The ilium of *Australopithecus* resembles that of man but the ischium is much longer

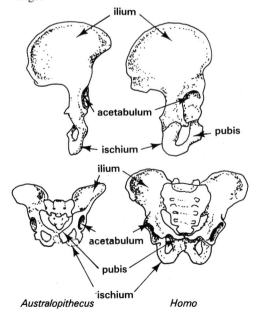

ilium

acetabulum

ischium

pubis

ilium

acetabulum

pubis

ischium

Australopithecus   *Homo*

Homo habilis by Leakey. Habiline means skilful and refers to the ability of this early primate to make simple tools. They were nothing more than stones which it may have used to throw at prey.

In 1938 Robert Broom found fragments of fossil crania and teeth of *A. robustus* in a cave at Kroomdraar, South Africa. A complete cranium of this species was discovered by Mary Leakey at Olduvai Gorge in 1959. The popular press christened him 'nutcracker man'. The skull of this species is heavy with a prominent ridge at each brow (Fig 27.13(b)). Its cheek teeth were also larger and the cranium had a prominent sagittal crest to which a large temporalis muscle was probably attached. The jugal bar was also thick indicating that *A. robustus* had a large masseter muscle. The facts strongly suggest that vegetable material figured prominently in its diet. This is supported by scanning electron microscope studies of its teeth. They show that the tooth-wear pattern is similar to that of wild chimpanzees whose food is mainly hard, thick-skinned fruits. A diet of this kind is of low quality and has to be eaten in large quantities. It also has to be thoroughly chewed to help its digestion. The evolution of large grinding cheek teeth and massive jaw muscles in *A. robustus* enabled it to cope with such a diet. Neither species of *Australopithecus* had a diastema, a gap between the incisor and premolar teeth, seen in apes.

Something is also known of the pelvic girdle of *Australopithecus* (Fig 27.14). The ilium is short and broad as in modern man. However, it has only a small area of articulation with the sacrum and its acetabulum is simpler than that of modern man. The ischium is relatively long, hence the hamstring muscle was probably inserted a long way from the acetabulum. The arrangement would enable the hindlimb to be drawn backwards and straightened, making a bipedal gait possible.

A third australopithecine species called *A. afarensis* was unearthed from river sediments in the Afar Triangle, Ethiopia in the early 1970s. The fossils are up to 3 million years old. Nearly 40 % of the bones of one skeleton, nicknamed Lucy, have been pieced together. Owen Lovejoy examined the bones in great detail and commented:

'They look incredibly primitive above the neck and incredibly modern below. The knee looks very much like a modern human joint; the pelvis is fully adpated for upright walking and the foot, although a curious mixture of ancient and modern is adequately structured for bipedalism.'

3 Homo

The following is an abbreviated form of Le Gros Clark's definition of *Homo*:

'A genus of the family Hominidae, distinguished mainly by a large cranial capacity with a mean value of more than 1100 cm^3, mental eminence variably developed; dental arcade evenly rounded with no diastema, first lower premolar bicuspid, molars variable in size with a relative reduction of the last molar; canines relatively small with no overlapping after the initial stages of wear; limb skeleton adapted for fully erect posture and gait.'

The scientific evidence suggests that man gradually evolved to such a state. Some human characteristics such as bipedalism were evident in *Australopithecus*. Others, for example cranial capacity, were not. The cranial volume of modern man ranges from 1000 to 2000 cm^3, with a mean of 1400 cm^3, compared with an estimated range of 400–600 cm^3 for *Australopithecus*. Hence, various human

With reference to Le Gros Clark's definition, suggest an order in which the characteristic features of *Homo* may have evolved.

characteristics evolved at different times in different populations. This phenomenon is called **mosaic evolution**.

Fossils have been found showing that creatures anatomically intermediate between *Australopithecus* and modern man existed. They support the idea that man's evolution was gradual.

A. *Homo erectus*

The first fossils of *H. erectus* were discovered in the early 1890s by Dubois in Java, hence this species is often called Java man. Subsequently other specimens have been located in China (Peking man), Germany, East Africa, South Africa, Algeria and Morocco. The remains are mainly bones of the skull, teeth, a mandible and a femur. The skull was thick with a cranial volume ranging from 775 to 1225 cm³ (mean 980 cm³). The forehead was less steep than in modern man and had prominent ridges at the brow (Fig 27.15). *H. erectus* had a heavier jaw and larger teeth than present-day man. However, the teeth were typically human in appearance and had wear patterns similar to those of modern man indicating that *H. erectus* was probably omnivorous. East African fossils of this kind are about 5 hundred thousand years old, those from Java are dated as 7 hundred thousand years old. The Chinese specimens are younger, dating from 3.5 to 5 hundred thousand years ago. They were discovered in caves alongside heaps of charcoal and ashes, suggesting *H. erectus* used fire. Also in the caves were the remains of deer and rhinoceros, together with simple tools. They were mainly sharp-edged stone flakes 2–3 cm long and were probably used to cut up carcasses of large animals. The flakes were knapped from stones, the remnants of which look like crude axeheads. As far as can be judged the rest of the skeleton of *H. erectus* was very much like our own.

Our limited knowledge of *H. erectus* tells us they lived throughout Africa, Asia and Europe in the early to middle Pleistocene epoch from about a million to 2.5 hundred thousand years ago. Their distribution extended much further north than *Australopithecus*. This may be related to the ability of *H. erectus* to use fire and so keep warm in cooler climates.

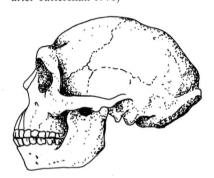

Fig 27.15 Reconstructed skull of *Homo erectus* from Peking (from E J Clegg 1978 after Tattershall 1970)

B. *Homo sapiens neanderthalensis*

Fossils of *H. sapiens* were discovered in 1856 in the Neander Valley near Dusseldorf, West Germany. They are referred to as Neanderthal man, *H. sapiens neanderthalensis* and are no more than 50 thousand years old. Other sites where similar remains have been found are China, Gibraltar, North and South Africa. An almost complete skull and most of a skeleton were dug up at La Chapelle-aux-Saints, France in 1908.

Neanderthal man had a large skull with a cranial volume ranging from 1300 to 1600 cm³. There is a prominent brow ridge across the slanted forehead. The jawbone and teeth are larger than those of modern man. Points of muscle attachment on the stout limb bones are well marked, suggesting a muscular build. The average height was about 1.67 metres (5 feet 8 inches). Neanderthal features first evolved at a time when the world was in a warm interglacial period. They were fully developed by the last Ice Age about 70 thousand years ago. Bulky bodies and short limbs are suited to cold conditions as there is less surface area per unit volume from which to lose body heat. They also used fire. *H. sapiens neanderthalensis* was thus adapted to cope with low temperatures.

Neanderthal man made a much wider variety of tools than his prede-

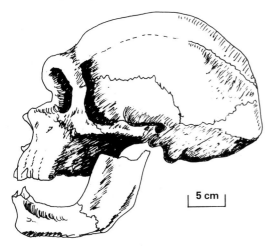

Fig 27.16 Skulls of (a) a classic, *Homo sapiens neanderthalensis*, and (b) a progressive neanderthal, *H. sapiens steinheimensis* (from J Z Young 1976)

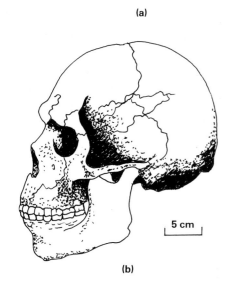

(a)

(b)

cessors. Many of them were made from stone, including hand axes and knives. Others were of bone. Their production required skilful trimming and fine control of the hands. They also point to the Neanderthals as meat eaters, preying on herbivores such as reindeer and horses.

As in all populations, fossils of Neanderthal man show considerable variation. Some palaentologists divide them into 'classical' and 'progressive' types (Fig 27.16). The progressive type is more like modern man. He had a more rounded cranium, a high forehead and smaller brow ridge. Skulls of this description have been found at Swanscombe, Kent and Steinheim in Germany. They are dated at about 250 thousand years old and are thought by some people to be related to progressive Neanderthals. Others believe them to be more like *Homo erectus* who lived at that time. More recent finds of Neanderthals have been made in caves at Mount Carmel, Israel. In one there are the remains of the classical type, in another nearby are fosssils of the progressive type. Carbon dating gives an age of about 45 thousand years. It appears that both types were in existence for many thousands of years. The progressive type is more like modern man, and is thought to be our ancestor.

C. *Homo sapiens sapiens*

All human remains from about 35 thousand years ago to the present are modern man, *Homo sapiens sapiens*. He first appeared during the last glacial period when the climate placed great stress on man's capacity for survival. Even in southern France and Spain the summer temperature probably averaged only 16 °C. One of the most famous finds of early modern man was made at Cro-Magnon in the Dordogne, France. In all respects he is typical of present-day man. Other similar remains have come to light at Combe-Capelle, France and Lautsch, Czechoslovakia.

Many of the typical features of modern man reflect the arboreal existence of our early ancestors. One of the ways our predecessors probably moved in trees was to swing from branch to branch. This requires great mobility of the upper limb, especially at the shoulder joint, and the ability to flex the phalanges of the hands. Being able to twist the forearm also helps. Moving about in this way places great demands on motor coordination. The ability to judge distances is also important. Having eyes at the front of the head enables fields of view to overlap to give stereoscopic vision which permits distances to be judged with precision.

The food of our tree-dwelling forebears was probably mainly leaves and fruit. The evolution of a herbivorous dentition enabled our ancestors to deal with a diet of this kind. Our teeth retain many of these features. The incisors have flat, sharp upper edges suited to cutting leaves and nibbling fruit. The canines are reduced in size and the molars and premolars provide grinding surfaces on which plant food is reduced to pieces small enough to be swallowed.

Bipedal locomotion evolved when our ancestors began to colonise the expanding grassy plains during the Miocene epoch. Our long-armed predecessors probably used their knuckles to support their bodies as they moved on their hind legs. A full striding bipedal gait requires the gluteus maximus muscle to be used as an extensor of the hip, thus permitting the trunk to be raised into an upright position. Moving about on hindlimbs gave our ancestors a better view of their surroundings and freed the hands which could thus be used for grasping food.

Colonisation of the plains probably brought with it a change to an omnivorous diet. Lizards, birds and small mammals were killed at first,

Fig 27.17 (a) The power grip, and (b) the precision grip of the human hand (from J Z Young 1976 after Napier 1956)

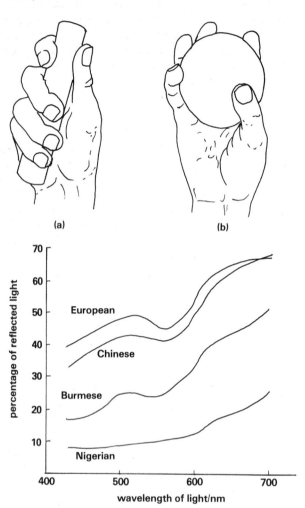

(a)   (b)

large mammals later. As the hands evolved, instruments for killing prey such as spears, bows and arrows, and simple tools for cutting up carcasses were made. An important feature in this respect was the lengthening of the thumb so that it could be opposed to the fingers (Fig 27.17). This made it possible for the hands to grip and to manipulate objects with precision, thus facilitating a hunting way of life. To strike prey would require weapons to be held with a power grip. Making and using simple tools such as stone flakes for butchering carcasses could only occur if the hands could be used precisely to manipulate objects.

Another human feature which may well have evolved at the same time as the hunting habit was loss of body hair. It may be that the exertions involved in hunting, especially in a hot climate, placed considerable stress on the body's cooling system. Loss of body hair would facilitate heat loss. Evaporation of sweat from large numbers of sweat glands in the skin helps maintain an even body temperature during exertion (Chapter 20). Pigmentation of the skin is an advantage in sunny climates. A dark skin absorbs ultraviolet light (Fig 27.18), lowering the risk of sunburn. In cloudy, temperate regions, lack of skin pigment is of some advantage. Light coloured skin permits UV light to penetrate deeper into the skin where ergosterol is converted to vitamin D. Dark skinned individuals living in such conditions may therefore suffer from rickets if their diet is inadequate in vitamin D (Chapter 15). A summary of the main stages of man's evolution is given in Table 27.2.

Fig 27.18 Skin reflectance curves for individuals of four different racial groups (From E J Clegg 1978)

Table 27.2 Main stages of man's evolution (from J Z Young 1976 after Campbell 1964)

		Years BP × 10^3						
		100	200	300	400	500	1000	1500
EPOCH	Upper Pleistocene		Middle Pleistocene			Lower Pleistocene		
FOSSILS	*H. s. sapiens* *H. s. neanderthalensis*	*H. erectus*		*H. erectus*		*A. robustus* *A. africanus* *A. ? robustus*	*A. ? africanus* (= *H. habilis*)	
COUNTRY AREA	Asia, Europe Africa, Israel	Morocco		China, Hungary		Germany Java Tanzania (Olduvai)	S. Africa (Taungs) Tanzania (Olduvai)	Tanzania (Olduvai)

27.2.3 Affinities

All living organisms have common features. Some have features which are virtually identical to those of another organism. In other words, they have a close **affinity** to each other and for this reason they are classified in the same group. Among the features which indicate that living organisms are related by descent are **anatomical, physiological, embryological, immunological, biochemical** and **behavioural** affinities. The relationship is more reliably established when a number of the features are common.

1 Anatomical affinities

When the anatomy of various animals, which superficially appear to be unrelated, is compared, similarities may be evident which suggest a common ancestry. During the course of evolution, vertebrate limbs have become adapted to serve a variety of purposes. For example, the forelimbs of a whale are used as paddles, the forelimbs of birds for flight and the hands of humans for grasping and manipulating objects. Even so, the limbs of all vertebrates develop from the same embryonic tissues and fit a common plan known as the **pentadactyl limb** (Fig 27.19). For this reason they are called **homologous structures**.

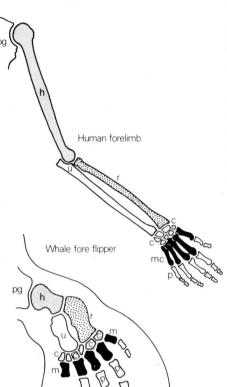

Fig 27.19 Forelimbs of (a) man, (b) whale and (c) bird

Human forelimb

Whale fore flipper

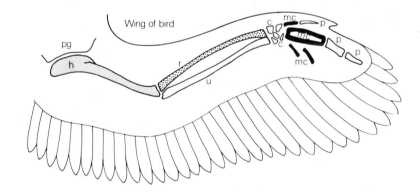

pg = pectoral girdle   h = humerus   r = radius   u = ulna
c = carpals   mc = metacarpals   p = phalanges

Wing of bird

Insect mouthparts vary in function and appear to be quite different in structure. The mouthparts of locusts are used for biting vegetation, those of butterflies and bees for sucking nectar, whilst those of fleas are used for piercing skin and siphoning blood. Yet they all have a common structural plan (Chapter 29).

Some animals have **vestigial structures** which in related creatures are fully developed. For instance, whales have vestigial pelvic girdles but lack hind limbs. This suggests that whales in common with other vertebrates have evolved from the same ancestors. Humans have a vestigial appendix suggesting that our forebears were herbivores.

The examples given above show how structures of similar origin have become adapted for different roles in various creatures. They illustrate **adaptive radiation** and are examples of **divergent evolution**. There are instances where structures of quite different origin have evolved to carry out similar functions. The wings of both insects and birds are organs used for flight, yet they do not have the same origin. They are **analogous structures**, which illustrates that **convergent evolution** can also occur.

2 Immunological affinities

In 1904, George Nuttall devised a procedure based on immunological methods for determining the relationship between animals. When, for example, the serum of a human is injected into the blood of an experimental animal, the latter forms **antibodies** against the proteins in the serum (Chapter 10). If the serum from the experimental animal is then mixed with more human serum, the antibodies combine with those serum proteins which caused their production to form a precipitate. Should the antibody-containing serum from the experimental animal be mixed instead with serum from another animal, the amount of precipitate formed is a measure of the affinity between human and the other animal. The more precipitate, the nearer the relationship (Table 27.3). Such methods suggest that humans are more closely related to the chimpanzee than to other primates.

Table 27.3 Precipitin reactions with human serum (from Marshall 1978)

Serum	Percentage precipitation
human	100
gorilla	64
baboon	29
deer	7
kangaroo	0

3 Biochemical affinities

Between man and chimpanzee there is only one difference in the sequence of amino acids in the α-globin and two differences in the β-globin fractions of **haemoglobin**. This contrasts with seventeen differences between human and horse haemoglobin. It points to man and chimpanzee having a more recent common ancestor than other mammals.

Another protein which has been studied in this respect is **cytochrome c**, a respiratory co-enzyme (Chapter 5). The amino acid sequence of cytochrome c has been determined for organisms as diverse as bacteria, yeasts, invertebrates and vertebrates. The variations are caused by slight differences in the nucleotide sequence of DNA which codes the production of the co-enzyme (Chapter 4). In this way it is possible to compute the minimum number of nucleotide differences between the genes which code the synthesis of cytochrome c in various organisms. A comparison of the differences show how closely the organisms are related (Fig 27.20). The results agree well with the fossil record.

4 Embryological affinities

Examination of the **embryos** of vertebrates, which as adults are very different, reveals similarities in development. The shapes and positions of the brain, eyes, nostrils, limbs and tail are comparable in animals as remotely related as fish, amphibians, reptiles, birds and mammals. At some stage in development, each of the embryos has gill clefts to which deoxygenated blood is pumped from the heart. With the exception of fish, the gill clefts later degenerate.

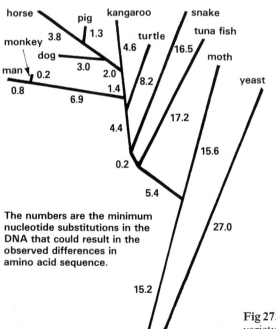

The numbers are the minimum nucleotide substitutions in the DNA that could result in the observed differences in amino acid sequence.

Fig 27.20 Computer predicted relationship between a variety of organisms based on differences in the amino acid sequence of their cytochrome c (after Ayala 1978)

5 Behavioural affinities

Organisms whose patterns of **behaviour** are similar are more likely to be closely related than those whose behaviour is very different.

Jane Goodall in Tanzania observed many aspects of chimpanzee behaviour. Apart from man, chimpanzees are the most habitual users of simple tools. They use sticks for attacking, exploring, poking and teasing. They collect ants and termites using small twigs and blades of grass. They clean themselves with leaves and use stones to crack nuts. The play behaviour of adult and juvenile chimpanzees resembles that of humans. Before it is weaned, a young chimpanzee is also held by its parents in a way similar to that in which human babies are nursed (Fig 27.21). The observations further substantiate the claim that chimpanzees and humans have common ancestors.

Fig 27.21 Chimpanzee suckling its young (London Zoological Society)

27.2.4 Geographical distribution

Each of the main land masses of the world has its characteristic plant and animal life. Marsupials are confined to Australia and South America, elephants to Africa and India.

Wallace was one of the first naturalists to survey the world-wide distribution of birds and mammals. He noted that there are six major land masses separated from each other by natural barriers such as mountain ranges and oceans. The number of species native to Africa, Australia and South America is about twice as great as in Asia, Europe and North America (Fig 27.22). It is thought that the first mammals evolved in Asia and migrated into Africa and Europe. They were non-placental types. From here they spread into Australia and the Americas using bridges of land which at that time connected the continents (Fig 27.23). Movement of the land masses by the processes of **plate tectonics**, caused Australia and South America to become cut off in the Cretaceous period when placental mammals were just beginning to evolve. Plate tectonics explains why marsupial mammals are indigenous to these areas. Africa and North America became separated from Europe later on, by which time many species of placental mammals had evolved. They are more successful than marsupials, which is why they are more prevalent in Asia, Africa, North America and Europe. Because Asia did not move apart from Africa, migration is still possible between these land masses. Hence elephants are native to both Africa and Asia.

Fig 27.22 World-wide distribution of species of mammals and birds (from Marshall 1978). The figures show the percentage of the world's total number of species endemic to each area

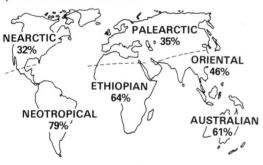

Fig 27.23 Changes in positions of major land masses with time as a result of plate tectonics

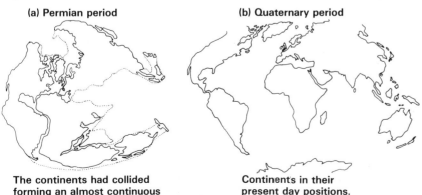

(a) Permian period

The continents had collided forming an almost continuous mass of land.

(b) Quaternary period

Continents in their present day positions.

27.3 How evolution came about

There have been many attempts to explain how life evolved into the enormous variety of living organisms which now inhabit the earth. The theories of Lamarck, Darwin and Wallace are particularly important.

27.3.1 Lamarckism

Jean-Baptiste Lamarck (1774–1829), a French naturalist, was the first modern biologist to propose how the variety of living organisms has evolved. His theory was based on two main premises:

i That organisms acquired new features because of an inner need.
ii That the acquired features are inherited by future generations.

Lamarck's ideas are often illustrated with reference to the evolution of the giraffe's long neck. According to Lamarck, the ancestor of the giraffe was a short-necked creature whose inner need to eat the leaves of trees caused it to stretch higher so that its neck became progressively longer. The offspring of these animals inherited the longer neck so that they could feed in the same way. It is the second part of the theory which is incorrect. Living organisms may acquire new characteristics during their lifetime. For example, a blacksmith acquires large biceps because of the nature of his work. However, the children of a blacksmith have biceps which are normal in size. It is thus evident that acquired characteristics are not inherited.

27.3.2 Darwinism

Fig 27.24 Charles Darwin (Ann Ronan Picture Library)

Charles Darwin (1809–1882) (Fig 27.24) was well acquainted with the variety of life. Between 1831 and 1836 he worked as a naturalist on HMS Beagle on its voyage around the world. He made careful records of the diversity and distribution of living organisms, particularly in South America, the Pacific Islands and Australia. Darwin was also a keen geologist. He made an extensive collection of fossils from various parts of the world.

After many years of observation, Darwin's theory of evolution was presented in a paper to the Linnaean Society of London in 1858. At the same meeting, a paper written by **Alfred Russel Wallace**, an English naturalist who had recently explored the East Indies and Australia, was also read. Darwin and Wallace had arrived independently at the same conclusions. Darwin's theory of evolution, illustrated by numerous examples, was published in 1859 in the *Origin of Species by Means of Natural Selection*. It was based on the following observations and deductions:

i. Most organisms have the potential to produce very large numbers of progeny. For example, in a single season a female cod lays about 85×10^6 eggs, a frog 1–2×10^3 eggs, an orchid produces $1 \cdot 7 \times 10^6$ seeds. There must be intense **competition** because the numbers of adult organisms remain fairly stable. Some degree of differential death occurs either among gametes or zygotes, or during development. Very few sperm produce by man fertilise ova, and relatively few ova are fertilised, hence there is a high death rate among human gametes.

ii. All organisms, even members of the same species, **vary**. You have only to look around at your colleagues to notice that this is so!

iii. The progeny which are most likely to survive competition are those having the combination of features which enable them to cope best with their environment. This is called **survival of the fittest by natural**

selection. Fitness can be defined as the capacity of an individual to survive and the probability of it producing viable offspring.

iv. The features favoured by natural selection are **inherited**.

Darwin's interpretation as to how the long neck of the giraffe was evolved would go something like this:

Among the ancestors of the giraffe were animals with a range of neck lengths. Those with long necks could browse on the leaves of trees. In this way they could also avoid the competition with other herbivores which ate vegetation growing nearer the ground. The short-necked ancestors succumbed to the competition and became extinct. The long-necked variety survived and gave rise to long-necked progeny which are the ancestors of the modern giraffe.

27.3.3 Variation

Darwin had no idea of the origin and causes of variation. Nor did he know how characteristics were inherited. Mendel's work, though published in Darwin's lifetime, was neglected until 1900, well after Darwin's death. Explanations of how variations come about and are passed on to future generations are incorporated into an extension of Darwin's theory called **Neo-Darwinism**.

Genetic variability in gametes can arise by random orientation and crossing over between chromosomes during meiosis (Chapter 7). Hence the offspring of organisms which reproduce sexually always differ from each other. This is partly why humans vary in many characteristics such as height, colour of skin, hair and eyes, blood pressure, resistance to disease and intelligence.

Inheritable variation is of two kinds, **continuous** and **discontinuous**. Continuous variation is caused by the random assortment of many genes which affect some human characteristics such as height, skin and eye colour. Discontinuous variation occurs where a characteristic is determined usually by a single gene (Chapter 26). The discovery of the nucleic acids and their functions have made it possible to explain the causes of discontinuous variation (Chapter 4). Changes in the genetic code are known as **gene mutations**. All genes mutate, but some do so more frequently than others. **Chromosome mutations** are caused by an alteration in the number of chromosomes or by fragmentation of chromosomes. Mutations may occur in any body cell but are inherited only when they appear in gametes. Somatic mutations may occur in any of our body cells, giving rise to local effects, for example, a white lock in a head of dark hair.

Gradual accumulation of somatic mutations may cause the changes occurring in our bodies as we age. Physical and chemical agents are **mutagenic**. The main physical agent is high energy radiation such as UV light, **a**, **b** and **c** rays. Mutagenic chemicals include methanal and mustard gas. If they are to be effective in bringing about evolutionary change, mutations must not affect the viability of individuals before reproductive age. Unsuitable mutations may cause death and may not therefore be inherited.

1 Gene mutations

Proteins perform numerous functions in living organisms (Chapter 2), so they determine many of an organism's characteristics. The role played by a protein is largely determined largely by its primary structure, the sequence of amino acids in its molecule. The primary structure affects the three-dimensional structure, the **conformation** of the protein molecule. It is the conformation which determines the protein's function. For example, the type of substrate which binds to an enzyme depends on the conformation of the enzyme's active centre. If the primary structure of the enzyme is abnormal, its conformation may be so altered that substrate molecules cannot bind to the active centre.

The sequence of amino acids in proteins is precisely regulated by the genetic code (Chapter 4). Any change in the code, a **gene mutation**, causes abnormal proteins to be made. Sometimes the change is harmless. It may even be beneficial. If this is so, the new gene may confer an advantage on its possessor. As a consequence, it may increase in frequency in a population, especially if it is a dominant gene. Usually gene mutations are harmful, because the abnormal proteins which are synthesised are unable to perform their normal functions. Individuals with such mutations may not survive to reproductive age, so the mutated gene is not passed to the next generation. Gene mutations are the cause of haemophilia and sickle-cell anaemia in humans. Interestingly, people who are heterozygous for sickle-cell anaemia, are resistant to malaria, so it is more likely that they will survive to childbearing age than normal individuals in countries where malaria is endemic. It is for this reason that the gene for sickle-cell anaemia is frequent among people who live in tropical countries.

2 Chromosome mutations

During meiosis, one or more chromosomes may break. A variety of things can happen to fragments of broken chromosomes:

  i. they may become lost (**deletion**),

 ii. they may become attached to the end of another chromosome (**translocation**),

iii. they may become turned round and rejoin the chromosome from which they became detached (**inversion**),

 iv. they may become inserted into another chromosome (**duplication**).

Clearly, these changes may alter the genomes of gametes.

Exposure to ultraviolet light is one of the causes of chromosome fragmentation. Nucleic acids strongly absorb UV light. New varieties of moulds such as *Penicillium* are produced by irradiation with UV light. Many new strains of crop plants have been developed by exposing their seeds to γ radiation.

Sometimes a pair of homologous chromosomes fails to separate in meiosis, giving rise to gametes with one chromosome less and one chromosome more than normal. The progeny formed from such gametes are called **polysomics**, having fewer or more chromosomes than normal. Down's syndrome in humans is caused by an extra chromosome (Chapter 26). In other instances, whole sets of homologous chromosoms do not separate in meiosis, so that diploid gametes are produced. Fusion with a normal haploid gamete gives rise to progeny with a triploid chromosome number. This condition, called polyploidy, is more often observed in plants than in animals. Many modern varieties of crop plants are **polyploids** (section 27.3.4). Polypoloids are usually more vigorous than their diploid counterparts.

An example of how polyploidy can give rise to new species of plants is seen in the cord grass, *Spartina*. In 1870, a new species of cord grass was observed growing in Southampton Water. It was named *Spartina townsendii* but has since been renamed *S. anglica*. Before this time two species of cord grass, *S. maritima* and *S. alterniflora*, were known in the locality. It is thought that *S. anglica* is a polyploid hybrid of the other two species (Fig 27.25). Its vigorous growth is of great help in building up mud flats which can be reclaimed from the sea to provide useful farmland (Chapter 32). In the 1920s a few plants of *S. anglica* were planted in the upper reaches of the Dyfi estuary, West Wales. Today the new species is established in over 400 hectares of mud in the estuary, a clear illustration of its remarkable powers of colonisation.

Fig 27.25 Evolution of a new species of cord grass *Spartina anglica*

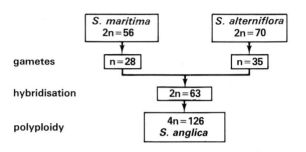

27.3.4 Selection

Darwin saw the important role of selection as an evolutionary force. In *The Origin of Species by Natural Selection* he described many examples of both **natural** and **artificial selection**.

1 Natural selection

A number of experimental and field studies carried out this century have provided an insight into how selection works in natural conditions. One of the first studies was of the peppered moth, *Biston betularia*. Collections of preserved peppered moths show that until almost the middle of the nineteenth century the only variety known had the speckled, light colour of pepper dust. In 1845, a black moth of the same species was caught in Manchester. By the end of the eighteen-hundreds the black variety made up 98 % of the population of the moth in the Manchester area.

The change in frequency of the gene for black colour coincided with the spread of heavy industry in the locality. Before the Industrial Revolution the air in and around Manchester was clean. The bark of trees growing in the area was covered with lichens. Against such a background the speckled variety of the moth is camouflaged (Fig 27.26). Insectivorous birds presumably had difficulty in seeing it, so the moth had a good chance of surviving and reproducing. The soot and smoke generated by heavy industry polluted the air and was deposited on the bark of trees. Many of the lichens were killed by the high concentrations of sulphur dioxide in the atmosphere. Against a dark background the speckled moth would be seen easily and taken by insect-eating birds. In contrast, the black form would be less conspicuous. It would be less likely to be taken as part of the diet of insectivorous birds. Consequently its survival was favoured. It was for this reason that the gene for black colour gradually became more frequent in subsequent generations.

Fig 27.26 Non-melanic and melanic varieties of the peppered moth (Heather Angel)

To prove that natural selection was the cause of the changed gene frequency, Dr H B D Kettlewell bred large stocks of the two varieties of the peppered moth, which were marked and then released in two areas:

i. in polluted Birmingham where over 90 % of the indigenous peppered moth population was the black variety, and
ii. in an unpolluted rural area of Dorset where none of the peppered moths was black.

He observed and filmed insectivorous birds such as robins, redstarts and hedge sparrows feeding in the two localities. The birds were particularly severe on the variety of moth which was not camouflaged by its

545

Table 27.4 Natural selection of the peppered moth in two localities (from Marshall 1978)

	Non-melanic	Melanic
Dorset, 1955 (unpolluted)		
released	496	473
recaptured	62	30
% recaptured	12.5	6.3
Birmingham, 1953 (polluted)		
released	137	447
recaptured	18	123
% recaptured	13.1	27.5

Explain how such resistant populations have evolved.

What factors are taken into account in selecting cattle or pigs for meat production these days?

Fig 27.27 Evolution of bread wheat, *Triticum aestivum*

background. Kettlewell also recaptured the surviving marked moths (Table 27.4). The experimental field-work confirmed that **natural selection** was the cause of the increase in frequency of the gene for blackness among peppered moths in industrialised areas. This is an example of **industrial melanism**. It has since been observed among many other types of moth in Britain, Europe and the USA.

Many other examples of natural selection have since come to light. A number of them have important implications in terms of human health and survival. For example, strains of malarial mosquitoes resistant to insecticides, especially DDT, have evolved in some parts of the world. For this reason malaria has returned to countries from where it had been eradicated in recent years. The evolution of populations of rats which are resistant to the rat poison warfarin is another instance of natural selection. Antibiotic resistance in bacteria is yet another example.

2 Artificial selection

For centuries man has selected livestock and crop plants because they have desirable features such as resistance to disease and high yield. Where this happens it is called **artificial selection**. In *The Origin of Species* Charles Darwin quoted many examples of how livestock had been improved by artifical selection.

A well-documented example of the influence of man in selection is seen in the evolution of wheat. In wild wheats the ripe ear is brittle and breaks up to release each seed-containing spikelet separately. It brings about efficient dispersal of the seeds. This type of wheat is of little use for cultivation because the ear breaks off when the grain is harvested. Over the past 10 000 years or so, man has selected non-brittle varieties of wheat for agriculture. One of the first was Einkorn wheat, *Triticum monococcum*, from Iran with a diploid number (2*n*) of 14 (genome AA). It was widely cultivated in Neolithic times. Einkorn wheat is, however, a low-yielding variety, so it was later replaced by Emmer wheat, *Triticum dicoccum*, a tetraploid with a chromosome number of 28 (genome AABB). Half of its chromosomes (AA) are thought to have come from Einkorn wheat. The other half (BB) probably came from a grass *Agropyron*. Emmer wheat was grown for thousands of years by the Ancient Egyptians. It is still cultivated on a small scale in a few parts of India and Russia. One of its disadvantages is that it is difficult to remove its seeds from the chaff in threshing. For this reason 'naked' wheats have been selected in more recent times. Bread wheat, *Triticum aestivum* (2*n* = 42, genome AABBDD) is the most widely grown naked wheat today. It is a hexaploid, probably a hybrid of Emmer wheat and the grass *Aegilops* (Fig 27.27).

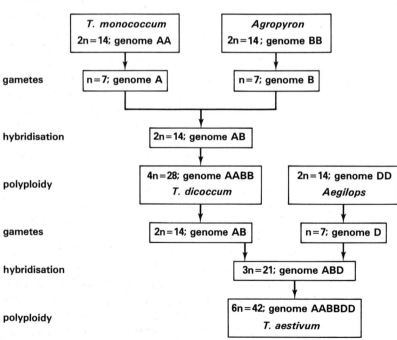

New varieties of *T. aestivum* are continually bred and selected with the foremost aim of maximising crop yield. Among the factors which are artificially selected in this context are number of spikelets per plant, number of grains per spikelet, grain size, disease resistance and hardiness to climatic conditions. The same principles apply to the development of new strains of other cereals such as barley. Genetic engineering techniques are currently being applied to design new varieties of crops which will grow in areas which are presently unsuited to arable farming. Such genetic engineering techniques are discussed more fully in Chapter 34.

27.3.5 Isolating mechanisms

A **species** can be defined as

> **a natural population, the individuals of which are actually or potentially capable of breeding with each other to produce fertile progeny, and which do not interbreed with members of other species.**

Not all individuals in a natural population have the same chance of breeding with one another. Every species is thus made up of a number of breeding sub-units called **demes**. When demes become isolated, they may in time lose their ability to breed with other demes of the same species. They may then be thought of as new species. There are several ways in which isolation can occur.

Fig 27.28 Darwin's finches

Insect-eater

Warbler finch

Mainly insect-eaters, but also some seeds eaten

Large insectivorous tree finch Medium insectivorous tree finch Small insectivorous tree finch

Tool-using finch Mangrove finch

Seed-eater

Vegetarian tree finch

Mainly seed-eaters, but also some insects eaten

Large ground finch Medium ground finch Small ground finch

Sharp-beaked ground finch Cactus ground finch Large cactus ground finch

1 Geographical isolation

One of the most striking examples of how new species can arise when demes become spatially separated was observed by Darwin. On his journey in HMS Beagle, Darwin visited the volcanic islands of the Galapagos archipelago, west of Ecuador. He noticed that there were just twenty-four species of birds on the islands, most of which he had not seen on the South American mainland. Fourteen of the species were finches belonging to four genera. In his account of the journey in *Voyage of the Beagle*, Darwin referred to the finches as follows:

> 'One might really fancy that from the original paucity of birds on this archipelago, one species had been taken and modified to different ends.'

The species of Darwin's finches can be distinguished by their feeding habits and the shapes of their beaks (Fig 27.28). The single species of finch on the adjacent mainland is a ground-dwelling, seed-eating bird. However, on the Galapagos Islands the finches eat a variety of foods. Seed crushers have rather heavy beaks; insect eaters have slender beaks for probing crevices where their prey may reside. One insectivorous species holds the spines of cacti in its beak to spear grubs hidden beneath the bark of trees. Several species have taken to eating the flesh of cacti which is pierced and cut using sharp, stout beaks. Subsequent studies of the anatomy and reproductive

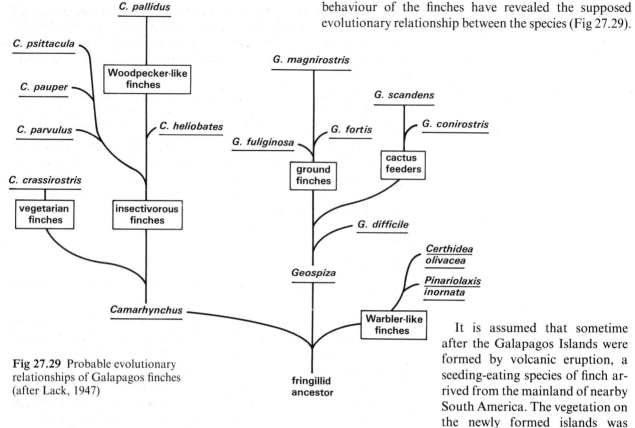

behaviour of the finches have revealed the supposed evolutionary relationship between the species (Fig 27.29).

Fig 27.29 Probable evolutionary relationships of Galapagos finches (after Lack, 1947)

It is assumed that sometime after the Galapagos Islands were formed by volcanic eruption, a seeding-eating species of finch arrived from the mainland of nearby South America. The vegetation on the newly formed islands was sparse, so seeds were scarce and the finches were forced to adopt new eating habits. This is another example of **adaptive radiation** and is partly attributable to a lack of competition for the same food source. In time the demes of finches became sufficiently different in their general habits that they no longer interbred. As a result, fourteen new species gradually evolved from the single species which first colonised the islands.

Comparable patterns of evolution have been observed in island populations in Britain. The mainland, Hebridean and Shetland forms of the British wren are slightly different in colour and wing length and do not interbreed. At the moment they are considered to be varieties of the same species. The Orkney vole exists as five sub-species on different islands off the coast of Scotland.

It is likely that geographical isolation has played a part in human evolution, different races adapting to different climates.

2 Reproductive isolation

Even when related demes live in the same locality, they do not always breed with each other. The causes of this are twofold:

A. Pre-fertilisation

i. Seasonal barriers Many flowering plants and mammals reproduce seasonally because of different responses to daylength (Chapters 22 and 25). It is probable that some species have evolved from demes which reproduce at a different time of the year from other demes in a population. For example, viable hybrids of pine can be obtained by crossing *Pinus radiata* and *Pinus attenuata*. However, they do not normally occur naturally because they produce pollen at slightly different times of the year.

ii. Behavioural barriers The plumage of male finches living on the Galapagos Islands is so similar that it is difficult to identify the species to which they belong. However, female finches have no problems in finding a male of the same species. During courtship the male passes food to the female in his beak. It is the shape of the male's beak which enables the female to recognise her mate. In this way interbreeding between demes does not happen.

iii. Mechanical barriers The genitalia of male and female animals in related demes may be different in size. It is often assumed that this may prevent copulation. However, this has not been confirmed by observation.

iv. Physiological barriers Many related demes of flowering plants are reproductively mature at the same time and there are no spatial, behavioural or mechanical barriers. Pollination may occur, yet they do not interbreed. The reason is that the stigma and style of the recipient flower does not provide suitable physiological conditions for germination of pollen grains from a related deme. Such conditions are genetically determined.

B. Post-fertilisation

Some demes intermate, but do not produce viable progeny. Among the reasons for this are:

i. Hybrid inviability Fertilisation occurs when cross-mating takes place between many species. However the embryos do not fully develop hence the progeny are still-born. The reasons for this are not fully known.

ii. Hybrid sterility Mules are produced by crossing a horse with a donkey. Because the chromosomes from the parents do not pair at meiosis, the gametes of mules are non-viable. It is for this reason that mules are sterile. In this way the genes of horses and donkeys remain isolated.

27.3.6 The formation of species

Once a deme is isolated, mutation and interbreeding will bring about a change in gene frequency in the population. Coupled with random genetic drift (section 26.4.3), this alters the gene pool of the deme such that some degree of reproductive incompatibility exists between it and other demes. After prolonged isolation total sterility may develop, so that the isolated deme becomes a new species.

The stages in the formation of a new species are as follows:

i. Variety – a recognisable type within a deme,

ii. Race – members of partially isolated demes,

iii. Sub-species – partially fertile with original species,

iv. Species – infertile with original species.

The process is not irreversible. Where it has not gone to the final stage, interbreeding may take place if the barriers preventing it are removed. There are many human demes differing in gene pool and in appearance. Viable offspring are produced when people of any two demes mate. Hence there is only one human species *Homo sapiens*. The various demes are races called *Homo sapiens sapiens*.

Evolution is a controversial subject of which much is still to be learned. Darwin compared it with trying to unravel the story of a novel of which only one page was available with half the lines missing and in each line half the words missing. Shortly after the publication of *The Origin of Species*, Darwin was severely criticised by members of the Church who favoured the idea of Special Creation. However, in the hundred years since Darwin's death many new and important biological and geological discoveries have served to confirm his ideas of how the variety of life we see today has come into being.

SUMMARY

Evolution is the process whereby the diversity of living organisms we see today arose from common ancestors. The way in which life originated is a subject of speculation among scientists. Some believe that life began in outer space and was carried to earth on the dust of comets and on meteorites. Others are of the opinion that life originated in the oceans. Many biologically important molecules are polymers such as proteins and nucleic acids. Thus any theory of the origins of life has to explain how the monomers of such molecules were formed, how they polymerised, how the polymers became isolated from their surroundings and how they replicated to form exact copies of themselves as happens in present day forms of life.

Evidence of evolution comes from several sources. Fossils provide an insight of the kinds of plants and animals inhabiting the earth in the distant and recent past. The age of fossils can be determined with reasonable accuracy using radionuclide dating techniques. In this way the sequence in which the major forms of life evolved has been established. Some fossils are remarkably similar to living forms still in existance. Others, such as the skeletons of dinosaurs have no modern counterparts.

Whereas fossils give a clear picture of the evolution of some creatures such as the horse, the fossil record in general is very patchy. Much has yet to be learned about our immediate ancestors who first appeared only 35 000 years ago.

The close resemblance in anatomy of organs in quite unrelated animals suggests a common line of descent. One of the clearest examples of an anatomical affinity is the pentadactyl limb which has a common structural plan in such diverse organs as the human arm and hand, the flipper of a whale and the wing of a bird. Immunological, biochemical, embryological and behavioural affinities strengthen the case that present day forms of life had common ancestors. The geographical distribution of plants and animals today also supports such a notion.

Darwinism is the most widely accepted theory of how evolution came about. First proposed by Charles Darwin in the middle of the last century it is based on four premises – variation and competition among living organisms, survival of the fittest by natural selection and inheritance of desirable traits. In this century the theory has been extended to Neo-Darwinism which includes genetic explanations of how variation comes about and is inherited.

Man has played a part in the creation of new species by artificial selection. Many of our modern crop plants have evolved in this way. In wild populations isolation plays a key role in evolution. Geographical isolation has resulted in the characteristic forms of life on remote islands. Darwin's finches on the Galapagos Islands are proof of the significance of spatial isolation. Even where geographical barriers do not exist, sub-groups of a population can become isolated by behavioural and reproductive barriers.

QUESTIONS

1 (a) What is a fossil?

(b) Indicate *two* ways in which fossils may be formed.

(c) Briefly describe a method used for estimating the age of a fossil.

(d) Describe an example of fossil evidence which has helped establish the possible evolutionary development of a group of organisms.

(e) Suggest *two* possible reasons for gaps in the fossil record.

(f) Describe briefly *one* further type of supporting evidence for evolution other than fossils, giving a *named* example. (L)

2 Write accounts of **two** of the following:

(a) the ways in which heritable variation arises;

(b) the ways in which genetic isolation may occur;

(c) the theory of Natural Selection. (WJEC)

3 (a) Explain what you understand by chromosome mutation and gene mutation.

(b) How does the behaviour of chromosomes during the first meiotic division

(i) differ from their behaviour during mitosis,

(ii) result in genetic variation? (Nuffield/JMB)

4 (a) What is a species?

(b) Describe how new species may arise. Illustrate your answer with **named** examples.

(c) Explain why closely-related species may be unable to interbreed successfully. (C)

5 'A mutagen causes a dominant mutation in an organism and results in a new phenotype . Natural selection of the new form leads to a change in the gene pool and isolation leads to eventual speciation .'

Explain the processes involved in this sequence of events, making clear the meaning of the underlined words. (JMB)

28 Variety of life: Micro-organisms and plants

28.1 Nomenclature and classification   552

28.2 Viruses   553
    28.2.1 Structure   553
    28.2.2 Replication   554
    28.2.3 Importance to humans   554

28.3 Kingdom Monera: Class Bacteria   555
    28.3.1 Structure   555
    28.3.2 Nutrition   556
    28.3.3 Respiration   556
    28.3.4 Reproduction   557
    28.3.5 Endospores   557
    28.3.6 Importance of bacteria to humans   557

28.4 Kingdom Monera: Class Cyanobacteria   558

28.5 Kingdom Protoctista: Phylum Protozoa   559
    28.5.1 Rhizopoda   559
    28.5.2 Flagellata   560
    28.5.3 Ciliata   560
    28.5.4 Sporozoa   562

28.6 Kingdom Protoctista: Phylum Euglenoidea   563

28.7 Kingdom Protoctista: Phylum Chlorophyta   563

28.8 Kingdom Protoctista: Phylum Phaeophyta and Rhodophyta   565

28.9 Kingdom Fungi   566
    28.9.1 Structure   566
    28.9.2 Nutrition   567
    28.9.3 Reproduction   567
    28.9.4 Importance of fungi to humans   570

28.10 Kingdom Plantae: Phylum Bryophyta   570
    28.10.1 Class Hepaticae   571
    28.10.2 Class Musci   572

28.11 Kingdom Plantae: Phylum Tracheophyta   573
    28.11.1 Class Lycopsida   573
    28.11.2 Class Sphenopsida   574
    28.11.3 Class Pteropsida   575
    28.11.4 Class Gymnospermae   577
    28.11.5 Class Angiospermae   580

Summary   581

Questions   582

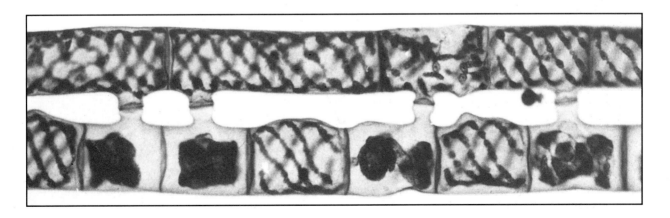

28 Variety of life: Micro-organisms and plants

Biological classification tries to make order and sense out of the great variety of living things. Biologists are constantly revising classifications and there is much argument and controversy on the subject of taxonomy. The following outline of micro-organisms and the plant kingdom and that of the animals in chapters 29 and 30 is a widely used system of classification with which most biologists are familiar.

28.1 Nomenclature and classification

Naming things is one of the fundamental functions of language. It enables us to identify objects in our surroundings and to tell other people about the objects we see. Stone-Age Man identified animals that were important to him. He made elaborate cave paintings of the animals he hunted and the predators he avoided. The Ancient Greeks went a stage further. They divided things into large groups which had common features. Animals were distinguished from plants.

A problem with naming things, however, is the language used. An organism may have several different names, even in the same language, let alone the names given it by people who speak another language. For example, *man* in English may also be called *people* or *humans* or even *folk*. To overcome this problem in biology, Latin has been used to name living things.

The system of naming organisms used by biologists today was devised in the eighteenth century by Carolus Linnaeus, Professor of Medicine at the university of Uppsala in Sweden from 1741 to 1778. Linnaeus published several editions of *Systema naturae*. In his writings he used a **binomial system** which gives animals and plants two names. The first is the **generic name**, that is, the **genus** to which the organism belongs. A genus is a small group of closely related organisms. The second name is the **specific name**. A **species** may be defined as a group of very closely related organisms which interbreed and produce fertile offspring. *Homo sapiens* (L) is the binomial name for humans. Including (L) after the specific name means that Linnaeus first described the organism by that name.

As well as naming organisms, biologists classify them. The branch of biology concerned with classifying organisms is **taxonomy**. Organisms can be classified merely for reference and filing. For example, all flowers with yellow petals can be grouped together. This is **artificial** classification. Alternatively, the system most useful to biologists is a **natural** classification in which closely related organisms are collected into groups called **taxa** (singular **taxon**). The largest of the taxa are kingdoms. There are five kingdoms, **Monera**, **Protoctista**, **Fungi**, **Animalia** and **Plantae**. They are subdivided into **phyla**. Phyla are sub-divided into **classes, orders** and **families**. Generally the smaller the group, the more closely related are the organisms in it. By closely related we mean organisms which have many fundamental features in common.

When choosing **diagnostic features** for classification, biologists look for features that are characteristic of the whole group. Superficial features such as colour are of little use in natural classifications. Crocuses, for example

How do biologists organise the great variety of living organisms?

exist in white, yellow and blue-flowered forms yet they all belong to the same species. We must also beware of **analogous** structures that adapt basically different organisms to a similar habit. Bats, birds and butterflies all have wings, but are not closely related. However, bats and birds have wings constructed of the same **homologous** structures: they are pentadactyl limbs (Figs 27.12 and 30.32). In contrast, a butterfly's wing has an entirely different origin.

28.2 Viruses

Table 28.1 Viruses of importance to humans

Disease of	Name	Transmission	Cause
humans	influenza	coughs, sneezes	influenza virus A, B, C
	poliomyelitis	faeces	poliovirus
	common cold	coughs, sneezes	rhinovirus
	yellow fever	mosquitoes	togavirus B
rabbit	myxomatosis	fleas	myxomavirus
dog, humans	rabies	saliva	rabies virus
tobacco	tobacco mosaic	touch	tobacco mosaic virus
sugar beet	beet yellows	aphids	beet yellows virus

Viruses do not fit into any of the five kingdoms. All **viruses** are obligate parasites which depend absolutely on host cells for reproduction. The only criterion of life that they have in common with other living organisms is their ability to reproduce.

Viruses were first discovered in the 1890s when the Russian scientist Iwanowski showed that the organism which caused tobacco mosaic disease could pass through a filter capable of trapping bacteria. The term virus was used by Beijerink in 1898 to describe the cause of the disease. About the turn of this century, it was realised that yellow fever in humans was caused by a virus. **Bacteriophages** were discovered in 1915. They are viruses which parasitise bacteria.

28.2.1 Structure

Viruses are very small, ranging in size between 10 and 300 nm in diameter. It is for this reason that they were not seen until the electron microscope became available in the 1940s. An individual virus is called a **virion**. At the centre of each virion is DNA, RNA or hybrid nucleic acid. The acid is usually surrounded by a protein coat called a **capsid**, which is normally made up of protein globues called **capsomeres**. Some viruses have an outer **envelope**, composed of polysaccharide or lipoprotein (Fig 28.1).

Fig 28.1 Electronmicrographs showing a variety of viruses:
(a) tobacco mosaic virus TMV, × 185 000
(b) influenza virus, × 230 000
(c) bacteriophage, × 200 000

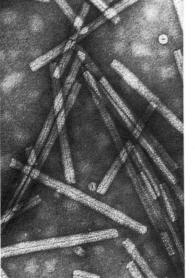

(a)

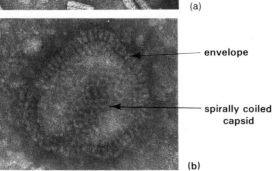

— envelope

— spirally coiled capsid

(b)

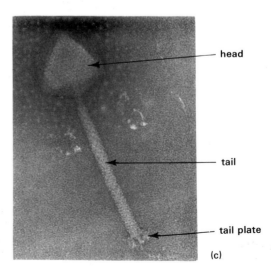

— head

— tail

— tail plate

(c)

553

28.2.2 Replication

For viruses to reproduce, their nucleic acid must get into host cells. Bacteriophages have short tails which they use to inject their nucleic acid into the host cell (Fig 28.2). Other viruses may enter their hosts by phagocytosis and pinocytosis. Insects and nematodes are vectors of many viruses. They are passed into the host in the vectors' mouthparts which pierce the host's tissues when the vector feeds. Mosquotoes are the vectors of yellow fever virus; aphids and nematodes transmit viruses which infect plants. After infection of the host, a **latent period** follows. During this time the viral nucleic acid represses the host cell's metabolism. The viral nucleic acid replicates and large numbers of identical virions are formed. The host cell eventually bursts and the new virions are released into the surroundings. Further infection of more host cells is now possible. These events are called the **lytic cycle** (Fig 28.3).

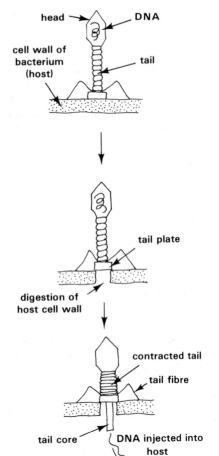

Fig 28.2 Injection of viral nucleic acid into host cell by bacteriophage

Fig 28.3 Lytic cycle

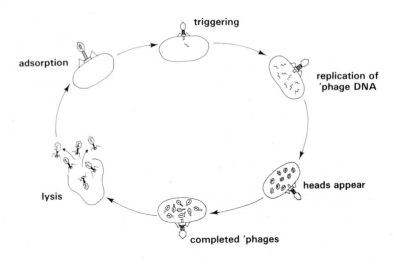

28.2.3 Importance to humans

The most obvious impact of viruses is their role as the cause of diseases of humans, livestock and crops (Table 28.1). This has been put to good effect in pest control. Probably the best-known case of the use of viruses in this respect is the **biological control** of rabbits by myxoma virus. Fleas which are the vectors of the virus were introduced into rabbit burrows in Britain in the 1950s. The wild rabbit population soon succumbed to **myxomatosis**. Within a few years, rabbits were scarce in the British countryside. Today numbers are beginning to recover, mainly from the breeding activities of rabbits which are resistant to the disease.

28.3 Kingdom Monera: Class Bacteria

Fig 28.4 (a) Structure of a bacterium based on electronmicrographs from Williams and Shaw, 1982)

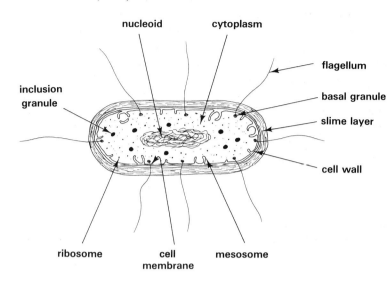

The Monera are **prokaryotic** organisms. This is, their nuclei are not surrounded by a membrane envelope.

Bacteria are microscopic organisms first seen in 1683 by van Leeuwenhoek. Very little was known about bacteria, however, until the work of Louis Pasteur and Robert Koch in the last half of the nineteenth century. They were the first to isolate and grow bacteria experimentally. Today more than 1500 species of bacteria have been described.

28.3.1 Structure

Bacteria are larger than viruses and range in size from 0.5 to 1.5 μm in diameter. They can be seen with the aid of a light microscope. Bacteria exist as single cells or as colonies (Fig 28.4).

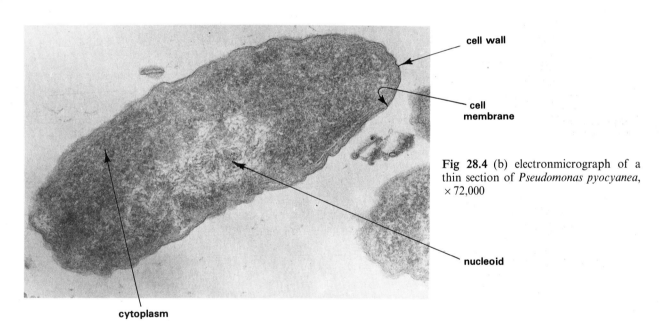

Fig 28.4 (b) electronmicrograph of a thin section of *Pseudomonas pyocyanea*, $\times 72{,}000$

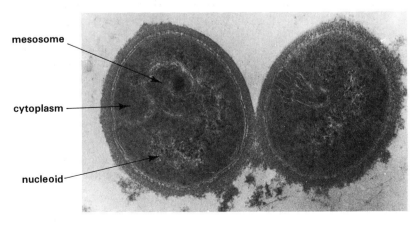

Fig 28.4 (c) electronmicrograph of thin sections through two cells of *Micrococcus lysodeikticus* immediately after binary fission, $\times 47\,000$ (see Fig 28.7)

Each cell is surrounded by a **wall** of mucopeptide which maintains the shape of the cell and allows interchange of materials with the surroundings. Outside the cell wall there is often a gelatinous **capsule** which helps protect the cell from desiccation. Just inside the cell wall is a partially permeable **cell membrane**. Folds of the plasma membrane, called **mesosomes**, project into the cytoplasm. Mesosomes contain a variety of enzymes, including those used in respiration and cell wall synthesis. Mitochondria are not found in bacteria, but the cytoplasm does contain **ribosomes**. In species capable of photosynthesis, small vesicles called **chromatophores** containing light-absorbing pigments may be seen in the cytoplasm. Bacteria do not have a nucleus. DNA is concentrated in a region called the **nucleoid** which is not bounded by a membrane. Chromosomes are not visible during cell division.

Many kinds of bacteria move when suspended in liquid. Motility is achieved using **flagella** (Fig 28.5), by flexing or by a corkscrew action. Bacterial flagella are hollow tubes of a contractile protein called flagellin.

When seen with the aid of a microscope, bacterial cells appear in a variety of different shapes (Fig 28.6). There are four basic forms. A rod-shaped cell is called a **bacillus**, a spherical cell a **coccus**, a comma-shaped cell a **vibrio** and a spiral form is a **spirillum**.

Fig 28.5 Bacterial flagella: (a) monotrichous e.g. *Pseudomonas*, (b) lophotrichous e.g. *Spirillum*, (c) peritrichous e.g. *Salmonella*

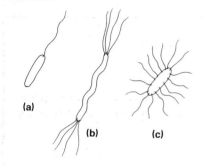

Fig 28.6 Shapes of bacteria: (a) bacilli, (b) diplococci, (c) streptococci, (d) staphylococci, (e) vibrio, (f) spirilla

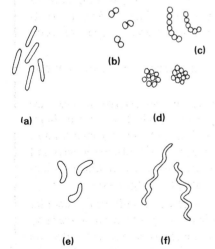

28.3.2 Nutrition

Most bacteria are **heterotrophic**, feeding as **saprophytes** or as **parasites**. Like animals, they require organic nutrients such as carbohydrates as a source of energy. Fastidious species need a wide range of organic nutrients, including vitamins and amino acids. Non-fastidious species can manage on simple sugars and minerals. Some bacteria are **photo-autotrophic**. Carbon dioxide is their source of carbon, and sunlight their source of energy. Photo-autotrophic bacteria contain the light-absorbing pigment **bacteriochlorophyll**. It is similar to chlorophyll, but mostly absorbs light of wavelengths 730–900 nm. Most photo-autotrophic bacteria use hydrogen sulphide to reduce carbon dioxide. Green plants use water for this purpose (Chapter 5). **Chemo-autotrophic** bacteria also use carbon dioxide. Their energy for carbon dioxide fixation is provided from the oxidation of simple inorganic compounds. They include the nitrifying and sulphur-oxidising bacteria (Table 28.2 and Chapter 31).

Table 28.2 Variety of energy sources used by chemo-autotrophic bacteria

1 $2NH_4^+ + 3O_2 \rightarrow 2NO_2^- + 4H^+ + 2H_2O + \text{energy}$	(e.g. *Nitrosomonas*)
2 $2NO_2^- + O_2 \rightarrow 2NO_3^- + \text{energy}$	(e.g. *Nitrobacter*)
3 $2S + 3O_2 + 2H_2O \rightarrow 2H_2SO_4 + \text{energy}$	(e.g. *Thiobacillus*)

28.3.3 Respiration

Most species of bacteria are **facultative anaerobes**. They respire aerobically when oxygen gas is present. In the absence of gaseous oxygen, they obtain energy by fermentation. Fermentative bacteria are used in the dairy industry. They ferment the milk sugar lactose into lactic acid in the production of cheese and yoghurt. A few kinds of bacteria are **strict anaerobes** and die or form dormant spores when exposed to oxygen gas. They include the free-living nitrogen-fixing bacterium *Clostridium pasteurianum*.

28.3.4 Reproduction

Bacteria reproduce by an asexual process called **binary fission**. Prior to fission, the bacterial DNA replicates. The nucleoid then divides in two, each part containing a copy of the DNA. Finally the cell wall grows in to produce two normally identical cells (Fig 28.7). Given favourable conditions, fission takes as little as twelve minutes in some species.

Fig 28.7 Binary fission

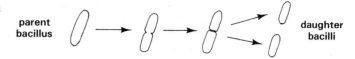

28.3.5 Endospores

Fig 28.8 Endospores in (a) *Bacillus* sp., (b) *Clostridium* sp.

(a)   (b)

In adverse conditions, a few species form **endospores**. The thick wall of the endospore protects the cell inside from dessication, from certain chemicals such as disinfectants and from extremes of temperature (Fig 28.8). When favourable conditions return, an endospore germinates to release a young vegetative cell.

28.3.6 Importance of bacteria to humans

1 Pathogenic bacteria

A relatively small number of species of bacteria are the cause of many serious diseases of humans and domesticated animals. A few species are also the cause of plant diseases.

A. Mammalian diseases

Bacterial diseases of mammals are transmitted in four main ways:

i. Through the air Airborne bacterial diseases include diphtheria, tuberculosis, bacterial pneumonia and whooping cough.

ii. In food and water Waterborne bacterial diseases include typhoid fever, gastroenteritis, bacterial dysentery and cholera. Among foodborne bacterial diseases are botulism and several types of food poisoning.

iii. By sexual contact Syphilis and gonorrheoa are two common venereal diseases.

iv. Through the skin Erysipelas, impetigo and puerperal fever are diseases caused by species of *Streptococcus* which enter the body through wounds or abrasions of the skin and mucous membranes. Various species of *Staphylococcus* are the cause of boils, carbuncles and abscesses. Tetanus and gas gangrene are caused by species of *Clostridium* which enter the body tissues through damaged skin.

Vectors are sometimes involved in transmitting pathogenic bacteria. The bacterium which causes bubonic plague is transmitted to humans by the rat flea. The fleas normally feed on the blood of rats, but when rats are scarce they attack humans. When their mouthparts penetrate the skin, a drop of saliva is injected into the wound. The saliva may contain the bacteria which cause bubonic plague.

B. Plant diseases

Galls, wilts, cankers, scab and spot diseases of plants can be caused by bacterial. **Galls** are disorganised swollen masses of cells which, as they

▶

How do bacteria affect our lives?

grow, may cut off the flow of water and nutrients, resulting in the death of the plant. **Wilts** are caused by slime-producing bacteria which plug the xylem vessels, causing the shoot system to droop. **Cankers** are areas of tissue decay, often beginning in the xylem and spreading into surrounding tissue. Species of *Streptomyces* are responsible for potato scab in which lesions develop in the skin of potato tubers. **Spot diseases** normally occur on the leaves and are caused by species of *Pseudomonas* and *Xanthomonas*.

2 Gut bacteria

The diet of herbivores contains a large amount of cellulose. No mammal produces an enzyme which can break down cellulose. Herbivores rely on bacteria in the gut to perform this task. In non-ruminant animals, such as the horse and the rabbit, the bacteria in the **caecum** ferment cellulose into fatty acids such as ethanoic, butyric and proprionic acids. In ruminants such as the cow and goat the fermentative bacteria are located in a large compartment of the stomach called the **rumen**. The fatty acids are absorbed and account for a significant amount of the energy-rich molecules metabolised in the liver of these animals. Gut bacteria are important in other ways for the growth of mammals, including humans. They produce vitamins of the B group and vitamin K. In ruminants, bacteria from the rumen are killed in the abomasum (true stomach) and provide an important source of protein for the animal to digest.

3 Bacteria which improve soil fertility

The breakdown of dead organic matter to form humus in soils is mainly due to the activities of saprophytic bacteria and fungi called **decomposers**. Chemo-autotrophic bacteria such as the **nitrifiers** can convert some of the products of decomposition into minerals which are essential for the growth of higher plants. **Nitrogen-fixing bacteria** are also found in most soils. They can increase the nitrogen content of soils, thereby promoting soil fertility. See Chapter 34.

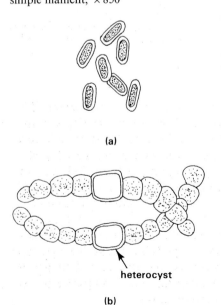

Fig 28.9 Two cyanobacteria (a) *Gloeothece*, a unicell, × 750, (b) *Nostoc*, a simple filament, × 850

(a)

heterocyst

(b)

28.4 Kingdom Monera: Class Cyanobacteria

The cyanobacteria are the **blue–green algae**. Fossil evidence suggests that the first living organisms on earth were very like the blue–green algae (Chapter 27). One of the characteristics of blue–green algae is that they are pigmented. The pigments include **chlorophyll, carotene, xanthophyll, phycocyanin** (blue) and **phycoerythrin** (red). Various combinations of pigments result in the wide range of colours seen among the cyanobacteria. Most, however, are blue–green.

Blue–green algae have a relatively simple structure, being either **unicellular** or simple **filaments** (Fig 28.9). The cells or filaments are usually surrounded by a sticky film of mucilage. The cell wall is made of mucopeptide similar to that of bacteria. Also like bacteria, blue–green algae are **prokaryotic**. Nuclear material is not enclosed in a nuclear envelope. The photosynthetic pigments are located in cytoplasmic lamellae. Food reserves such as cyanophycin granules, oil and glycogen are stored in the cytoplasm. There are no mitochondria. Gas vacuoles in the cytoplasm of some aquatic species enable the cells to float, placing them in the best position for light absorption. Some filamentous species form **heterocysts**, thick-walled cells with a clear protoplast.

Blue–green algae do not reproduce sexually. Asexual production is by **fission** similar to that seen in bacteria or by fragmentation of filaments.

Fig 28.10 Some rhizopods (Biophoto Associates):

(a) *Amoeba*, ×150, showing ingested *Paramecium*

(b) *Actinophrys*, ×500. Large black circles are contractile vacuoles.

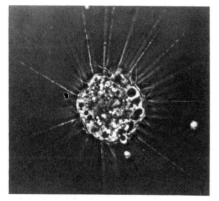

(c) *Difflugia*, ×250. Pseudopodia are just visible projecting through the shell

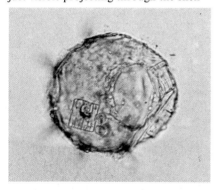

(d) *Entamoeba* in human intestine, ×500

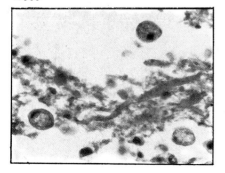

Some species form thick-walled resistant spores called **akinetes**.

Many cyanobacteria are found in **symbiotic** relationship with higher plants. Others are the algal component of lichens (Chapter 31). As well as producing carbohydrate by photosynthesis, many cyanobacteria can also use atmospheric nitrogen to make amino acids and proteins. Nitrogen fixation takes place in heterocysts. These properties enable blue–green algae to colonise regions hostile to other organisms. Cyanobacteria are often the first to appear on bare ground.

28.5 Kingdom Protoctista: Phylum Protozoa

The Protoctista are **Eukaryotic** organisms. That is, their nuclei are surrounded by a membrane envelope. Protozoans are micro-organisms whose bodies consist of a single cell and are thus **unicellular**. Some biologists see the protozoan body as not being divided into cells. From this point of view, protozoans are sometimes described as **acellular**. The phylum Protozoa is divided into four classes: Rhizopoda, Flagellata, Ciliata and Sporozoa.

28.5.1 Rhizopoda

The rhizopods are protozoans which have cytoplasmic extensions called **pseudopodia** (Fig 28.10). *Amoeba*'s body changes shape constantly because pseudopodia are formed. When the body touches a solid surface, the cytoplasm changes viscosity, becoming a liquid **sol** at the rear of the body. The **plasmasol** flows forward into the leading edge of the animal as it advances. When the plasmasol enters the pseudopodia, it becomes a more viscous **plasmagel**. Each pseudopodium extends further as more plasmasol flows in from behind (Fig 28.11). There is no permanent front or

Fig 28.11 Changes occurring in the cytoplasm during amoeboid movement

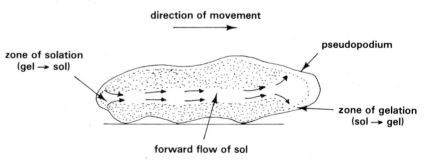

direction of movement

zone of solation (gel → sol)

pseudopodium

zone of gelation (sol → gel)

forward flow of sol

rear. *Amoeba* can stop suddenly and reverse its direction of movement. This often happens if it meets an obstruction. The use of pseudopodia for locomotion is called **amoeboid movement**. Amoeboid movement is also seen in some of the white blood cells of multicellular animals. Pseudopodia are also used for capturing prey. The pseudopodia are extended over the prey, which are engulfed into food vacuoles in the cytoplasm. The process is called **phagocytosis**. The pseudopodia of some rhizopods are supported by thin rods such as the silica spicules of *Actinophrys*. In others the entire body is surrounded and supported by a shell through which the pseudopodia project. An example of a shelled rhizopods is *Difflugia*. Two groups of shelled rhizopods, the *Foraminifera* and the *Radiolaria*, are abundant in the sea. When they die, their shells sink to the sea bed, forming thick **oozes** which have given rise to limestone and siliceous rocks.

Reproduction in the rhizopods is mainly by **binary fission** in which mitosis occurs. The parasitic *Entamoeba histolytica* produces resistant cysts which are egested in the faeces of an infected host and bring about the spread of amoebic dysentery. Sexual reproduction occurs in many rhizopods, but not *Amoeba*. Most rhizopods are free-living in the sea, freshwater ponds and lakes or in thin films of water in soil and on vegetation. In freshwater species, water is constantly taken in by osmosis. It accumulates in one or more vacuoles which increase in size as the water enters. At intervals they expel the water to the outside through the body surface. Because they pulsate doing this, they are called **contractile vacuoles**.

28.5.2 Flagellata

Protozoans which have one or several **flagella** belong to the Flagellata. Flagella are whip-like organelles, relatively long compared with cilia, but with the same ultrastructure (Fig 28.12). They are used in locomotion. *Trypanosoma* is an example of an organism with a flagellum (Fig 28.13).

What methods do different protozoans use in locomotion?

Fig 28.12 Electronmicrograph of a transverse section of cilia, × 50 000. Cilia contain longitudinal fibrils, two in the centre surrounded by nine outer pairs. Contraction of the fibrils causes bending of the cilia

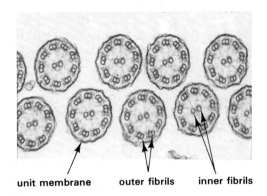

unit membrane   outer fibrils   inner fibrils

Fig 28.13 *Trypanosoma*, × 2000 (Science Photo Library: John Durham)

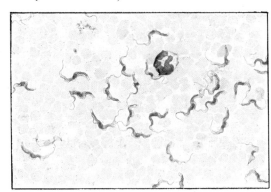

28.5.3 Ciliata

Paramecium and a variety of stalked protozoa such as *Vorticella* are **ciliates** (Fig 28.14). Their main characteristics are the presence of numerous **cilia** and two nuclei, a small **micronucleus** and a large **meganucleus**.

Fig 28.14 Two ciliates:

(a) *Paramecium*, × 200 (Bruce Coleman: Frieder Sauer)

contractile vacuoles

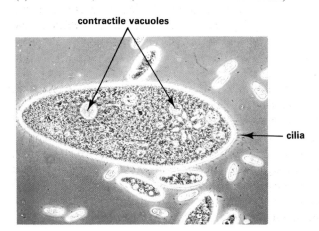

cilia

(b) *Vorticella*, × 150 (Bruce Coleman: Frieder Sauer)

cilia

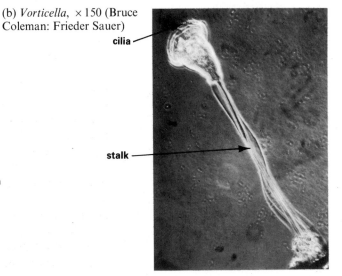

stalk

The cilia may be all over the body as in *Paramecium* or in patches as in *Vorticella*. Each cilium is attached to a **kinetosome** or **basal body** in the pellicle (Fig 28.15). The kinetosomes are linked in rows by fine threads called **kinetodesmata** arranged longitudinally under the pellicle. Just beneath the pellicle in many ciliates are numerous sacs called **trichocysts**. They contain threads which may be discharged outwardly and used to capture prey, for defence against predators or the attachment of the body to a solid surface.

Fig 28.15 *Paramecium*:

(a) part of the pellicle (after Vickerman and Cox)

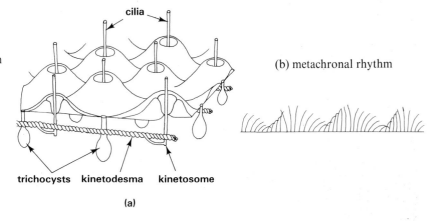

(b) metachronal rhythm

trichocysts   kinetodesma   kinetosome

(a)

Fig 28.16 Path of food vacuoles in *Paramecium*

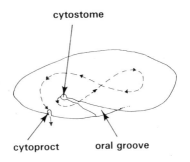

cytostome

cytoproct   oral groove

Fig 28.17 Binary fission in *Paramecium* (compare with Fig 29.5)

Fig 28.18 Conjugation in *Paramecium*, × 160

Cilia are often used for locomotion. In *Paramecium*, the cilia are arranged in tracts across the pellicle. Each cilium beats rhythmically and slightly out of phase with the adjacent cilia. This is called **metachronal rhythm** and creates a thrust against the surrounding water. *Paramecium* usually moves in a straight path, the body rotating about the longitudinal axis. If an individual meets an obstruction, reversal of the metachronal rhythm brings about a change of direction.

Cilia may also help in the capture of food. *Paramecium* has an **oral groove** on one side of its body. Small food particles such as bacteria are sucked down the oral groove by the beating of cilia. The food is engulfed into the cytoplasm by phagocytosis at the **cytostome** at the end of the oral groove. Food vacuoles produced in this way circulate in definite paths in the cytoplasm. Digestion occurs within the vacuoles from which the products are absorbed into the cytoplasm. Eventually, the unused contents of the vacuoles are discharged to the outside through the pellicle at the **cytoproct** (Fig 28.16). The stalked ciliates such as *Vorticella* live attached by contractile stalks to solid surfaces such as larger animals, rocks or aquatic weeds. The cell is capped by rings of cilia. Rhythmic beating of the cilia helps to direct food into the oral groove. **Contractile vacuoles** regulate the water potential of freshwater ciliates.

Ciliates reproduce both asexually and sexually. The asexual process of **binary fission** involves mitosis of the micronucleus, fission of other organelles and transverse division of the body (Fig 28.17). Sexual reproduction occurs by **conjugation**, a complex process resulting in genetic variation among the offspring because of nuclear transfer and meiosis (Fig 28.18). Self-fertilisation, called **autogamy**, occurs in some ciliates. This is a process in which conjugation occurs in a single individual.

28.5.4 Sporozoa

All the members of this class are parasitic. An example is *Plasmodium*, several species of which cause malaria in humans. The main characteristics of the Sporozoa result from their parasitic way of life. They include a complex life cycle, a relatively simple body form and an absence of locomotory organelles. The name Sporozoa refers to the production of **spores** which often occurs in sporozoan life cycles.

Plasmodium will be described as an important example of the group. Three species of *Plasmodium* infect humans. *P. vivax* (Fig 28.19) causes benign tertian malaria, *P. falciparum* malignant tertian malaria and *P. malariae* quartan malaria. All species are dispersed by the mosquito *Anopheles* which acts as a **vector** and **secondary host**.

Fig 28.19 (a) Life cycle of *Plasmodium vivax*

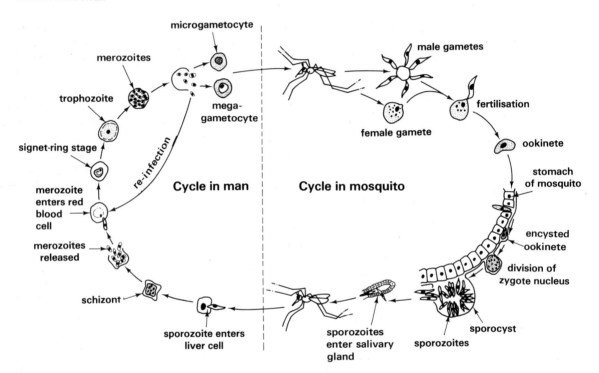

Fig 28.19 (b) Stages in the life cycle of *Plasmodium* (from J I Williams and M Shaw 1982)

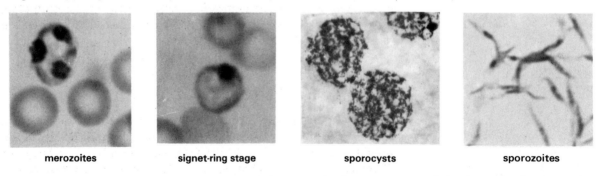

merozoites   signet-ring stage   sporocysts   sporozoites

Malaria is the most common cause of death among humans after malnutrition. In the 1940s over three million people, mainly Asians and Africans, died of malaria annually. Today the figure has fallen to about two million. The reduction has come about by a combination of measures which are based on a knowledge of the life cycle of *Plasmodium*

▶

and its vector. Denying the anopheline mosquito access to humans prevents transmission of the disease. Spraying of human dwellings with insecticides such as DDT has proved effective in this respect. However, there is evidence that strains of *Anopheles* resistant to DDT have evolved in some countries where malaria is on the increase. The vector has an aquatic larval and pupal stage. Draining off, oiling or poisoning swamps and marshes reduces the mosquito's breeding territory. This approach has been very effective in recent years in China.

Fig 28.20 (a) *Euglena* (b) flagellate swimming in *Euglena* (c) Euglenoid movement

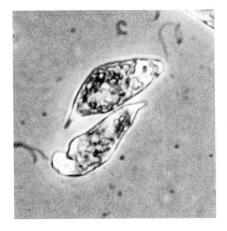

28.6 Kingdom Protoctista: Phylum Euglenoidea

Euglenoidea are flagellate unicells. They contain chlorohyll and are capable of photosynthesis. An example, is *Euglena* (Fig 28.20 (a)). It has a single locomotory flagellum. A second very short flagellum is found at the base of the main flagellum inside a cup-shaped reservoir at the anterior end of the body. Waves of contraction spread along the locomotory flagellum from the base to the tip. Spiral movements of the flagellum pull the animal through the water in a helical path (Fig 28.20 (b)). Locomotion may also be brought about more slowly by pulsation of the whole body, called **euglenoid movement**, in which contractile plates in the flexible **pellicle** are used (Fig 28.20 (c)). *Euglena* is phototactic, moving towards light to photosynthesise. A photoreceptor near the base of the long flagellum and a **stigma** in the adjacent cytoplasm are involved in the phototactic response. Contractile vacuoles regulate the water potential of freshwater euglenoids.

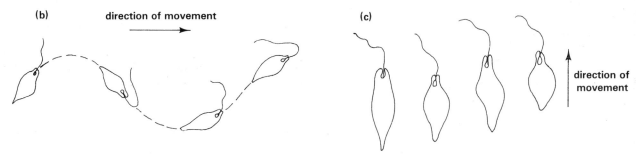

(b) direction of movement →

(c) direction of movement ↑

Reproduction is by binary fission. In adverse conditions the flagellum may withdraw and a resistant cyst is secreted around the protoplast. Fission may occur inside cysts. Several new individuals then emerge when the cysts burst in favourable conditions.

Fig 28.21 (a) *Chlamydomonas*, a unicellular chlorophyte, × 1000

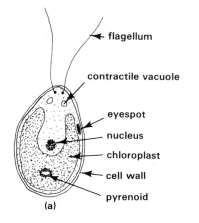

flagellum

contractile vacuole

eyespot

nucleus

chloroplast

cell wall

pyrenoid

(a)

28.7 Kingdom Protoctista: Phylum Chlorophyta

The chlorophytes are the **green algae**. They are nearly all aquatic plants, a few are terrestrial. Green algae include **unicells, colonies, filaments** and **thalloid** forms (Fig 28.21). A thallus is a multicellular plant body not differentiated into roots, stems and leaves. The cells are **eukaryotic**, having extensive internal membranes, including a nuclear envelope. **Chlorophylls *a*** and **b**, **carotene** and **xanthophyll** are found in chloroplasts. **Starch** and **oil** are the main food reserves and **cellulose** is the main component of the cell wall. **Zoospore** formation is the usual method of axesual reproduction. The **sex organs are unicellular** and the male gametes are motile. In many green algae the female gametes are motile too.

Fig 28.21 (b) *Spirogyra*, a filament, × 150 (Biofotos), (c) *Ulva*, a thallus, × $\frac{1}{2}$

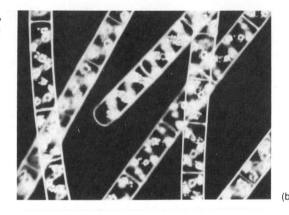

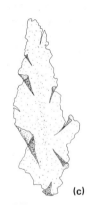

(b)　　　　(c)

Fig 28.22 *Spirogyra*, × 500

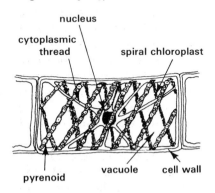

An example is *Spirogyra*, which grows as unbranched filaments made of cylindrical cells. The cellulose cell wall is enclosed in a layer of mucilage. Large fluid-filled vacuoles occupy much of the intracellular space. One or more spiral chloroplasts lie in the cytoplasm (Fig. 28.22). The haploid nucleus is suspended at the centre of the cell by cytoplasmic threads.

Cell division causes the filaments to grow in length, after which they fragment into short chains of cells. **Fragmentation** is a form of asexual reproduction.

Sexual reproduction is by **conjugation**. Two or more filaments come to lie next to each other with their cells aligned. Short bulges grow from adjacent cells and elongate, gradually pushing the filaments apart. The cross walls between the bulges break down and the filaments are now joined by **conjugation tubes** (Fig 28.23(a)). The filaments are therefore **gametophytes**. Because the gametes are similar in size, *Spirogyra* is **isogamous**.

Fig 28.23 *Spirogyra*: (a) conjugation, × 300; (b) zygospore formation, × 300

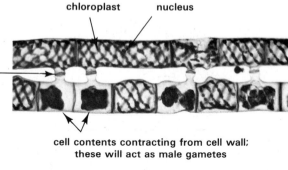

(a)

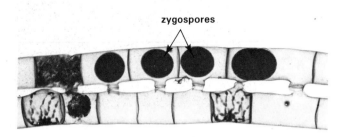

(b)

Fig 28.24 Haplontic life-cycle of *Spirogyra*

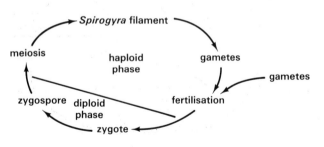

The protoplasts of one filament become amoeboid and pass through the conjugation tubes into the adjacent filament. The nuclei of the gametes fuse to form diploid zygotes which develop a resistant wall to become **zygospores** (Fig 28.23(b)). Zygospores are released when the parent filaments decay. The zygospore nucleus divides by meiosis. Three of the resulting four nuclei degenerate. In favourable conditions when the zygospore germinates, a haploid cell emerges. It divides by mitosis to form a new filament (Fig 28.24).

Fig 28.25 The bladder wrack, *Fucus ves-iculosus*: (a) vegetative thallus, $\times \frac{1}{3}$, (b) female conceptacle, $\times 75$, (c) male conceptacle, $\times 75$, (d) thallus showing bladders and receptacles (Geoscience)

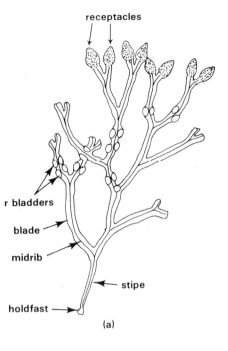

28.8 Kingdom Protoctista: Phyla Phaeophyta and Rhodophyta

The phaeophytes are the **brown algae.** They include **filamentous** forms and the **thalloid** macroscopic marine seaweeds. Their yellow or brown colour is caused by **xanthophyll** and **fucoxanthin** although **chlorophylls** *a* and *c* are present too. The pigments are located in chloroplasts. **Laminarin**, a soluble polymer of glucose, is the main food reserve. Cell walls have an inner layer of cellulose and an outer gelatinous layer made of **algin**. The cells are **eukaryotic**. Most brown algae produce **unicellular sex organs**. Many also reproduce asexually by forming **zoospores**.

An example of a phaeophyte is the **wrack** *Fucus* (Fig 28.25). The **thallus** is attached to a solid substrate by a **holdfast**. Between the holdfast and the extensive flattened **blade** of the thallus is a short **stipe**. The blade has a tough leathery texture and is branched **dichotomously**. A thickened **midrib** is present in each blade. *F. vesiculosus* has air-filled bladders which provide buoyancy for the thallus when submerged by the tide. At the tip of each branch there are swollen **receptacles** containing pits called **conceptacles**. The sex organs are inside the conceptacles. In *F. vesiculosus*, male and female conceptacles are in separate plants. Conceptacles open to the outside by pores called **ostioles**. The male organs are **antheridia**, which produce the flagellate male gametes called **antherozoids**. The female organs are **oogonia**. They produce the non-motile female gametes called **oospheres**. *Fucus* is therefore **oogamous**. The thallus is the gametophyte stage of the life cycle. Unlike the chlorophytes, the gametophyte of *Fucus* is diploid. Gametes are produced by meiosis, so the antherozoids and oospheres are haploid. Fertilisation occurs in the surrounding water. The diploid zygote develops into a new diploid thallus and the cycle is completed (Fig 28.26). *Fucus* does not reproduce asexually.

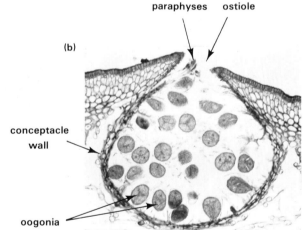

Fig 28.26 Life cycle of *Fucus vesiculosus*

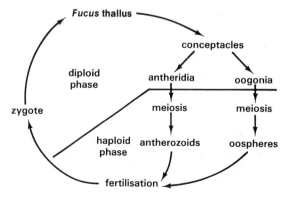

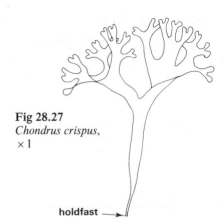

The rhodophytes are the **red algae**, nearly all of which are marine and **thalloid**. Their colour is caused by the presence of relatively large amounts of **phycoerythrin**. An example is *Chondrus* (Fig 28.27).

28.9 Kingdom Fungi

Fungi range in size from microscopic forms to the bracket fungi, toadstools and puffballs which are easily seen with the unaided eye (Fig 28.28).

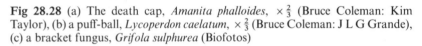

Fig 28.27 *Chondrus crispus*, ×1

(c)

Fig 28.28 (a) The death cap, *Amanita phalloides*, ×$\frac{2}{3}$ (Bruce Coleman: Kim Taylor), (b) a puff-ball, *Lycoperdon caelatum*, ×$\frac{2}{3}$ (Bruce Coleman: J L G Grande), (c) a bracket fungus, *Grifola sulphurea* (Biofotos)

(a)

(b)

28.9.1 Structure

Some fungi are **unicellular**, such as the yeast *Saccharomyces*. More often they exist as a **mycelium** made of a network of branching multicellular threads called **hyphae** (Fig 28.29). Fungal hyphae are bounded by a fibrous **cell wall** made largely of glucans, mannans and chitin. The protoplast inside contains a **nucleus**, or more often many nuclei, each bounded by a nuclear envelope. Chromosomes appear when the nucleus divides. Also present are **mitochondria** and **endoplasmic reticulum**. All these structures distinguish fungi from bacteria.

Fig 28.29 Yeast, *Saccharomyces* sp: (a) budding, ×1000 (Biophoto Associates), (b) aseptate hyphae, (c) septate hyphae

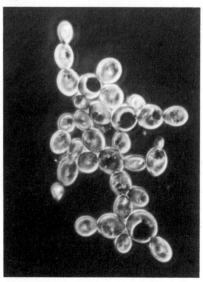

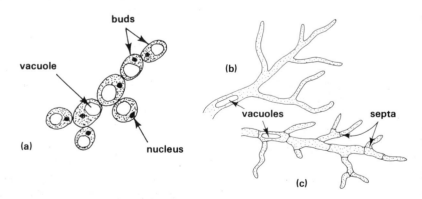

Fungi are divided into four phyla on the basis of hyphal structure and mode of reproduction (Table 28.3). Hyphae may be either septate or aseptate. **Septate** hyphae have cross-walls called **septa** which divide the protoplast into cells containing one or more nuclei. Septa are absent from **aseptate** hyphae, where a single **coenocytic** protoplast containing very many nuclei is present in the mycelium. A mycelium grows by apical elongation and branching of the hyphae.

Table 28.3 Classification of fungi

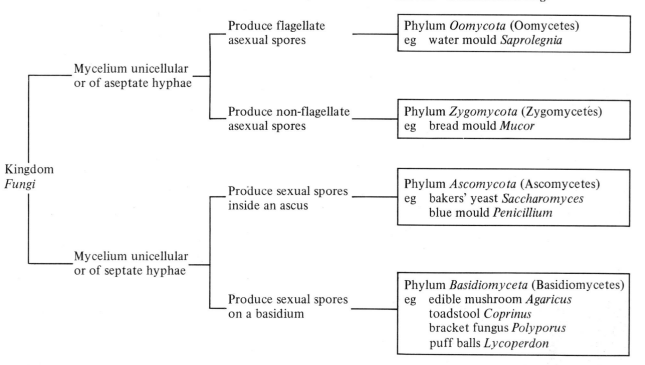

	Produce flagellate asexual spores	Phylum *Oomycota* (Oomycetes) eg water mould *Saprolegnia*
Mycelium unicellular or of aseptate hyphae	Produce non-flagellate asexual spores	Phylum *Zygomycota* (Zygomycetes) eg bread mould *Mucor*
Kingdom *Fungi*	Produce sexual spores inside an ascus	Phylum *Ascomycota* (Ascomycetes) eg bakers' yeast *Saccharomyces* blue mould *Penicillium*
Mycelium unicellular or of septate hyphae	Produce sexual spores on a basidium	Phylum *Basidiomyceta* (Basidiomycetes) eg edible mushroom *Agaricus* toadstool *Coprinus* bracket fungus *Polyporus* puff balls *Lycoperdon*

Fig 28.30 Asexual reproduction in fungi (part): (a) *Saprolegnia*, × 200, (b) *Mucor*, × 200

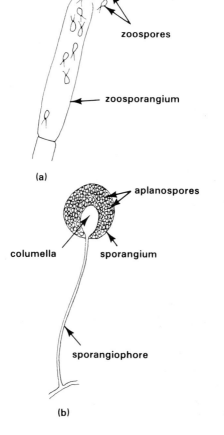

28.9.2 Nutrition

A major difference between fungi and higher plants is the absence of chlorophyll in fungi. All fungi are **heterotrophic**, depending on organic matter as a source of energy and nutrients. **Saprophytic** fungi such as *Mucor* use dead organic matter. Digestive enzymes are secreted by the growing tips of hyphae. The extra-cellular enzymes convert the organic substrate on which the fungus is growing into soluble products which are absorbed by the hyphae.

Parasitic fungi such as *Phytophthora* absorb soluble organic matter from the cells of their host using haustoria (Chapter 31). The fungi of lichens and mycorrhizae are **symbionts** obtaining organic matter from their partners.

28.9.3 Reproduction

Fungi produce **spores** when they reproduce. The method of spore production may be asexual or sexual and varies from phylum to phylum. In favourable conditions, spores germinate to form a new mycelium.

1 Asexual spores

These are formed after mitotic nuclear division. They normally give rise to progeny which are similar to the parent.

1. Oomycetes, such as *Saprolegnia*, are aquatic and produce flagellate **zoospores** in **sporangia** (Fig 28.30(a)). The flagella allow dispersal by swimming through water. In the appropriate conditions, the zoospores germinate and new hyphae emerge. **Zygomycetes** such as the pin-mould *Mucor* produce aerial sporangia. They are borne on vertical branches of the mycelium called **sporangiophores** (Fig 28.30(b)). The spores are uni-cellular but without flagella and are called **aplanospores**. Aplanospores are released from the sporangia into the air and are dispersed by air currents.

567

Fig 28.30 Asexual reproduction in fungi (contd.)

(c) *Erysiphe*, ×400

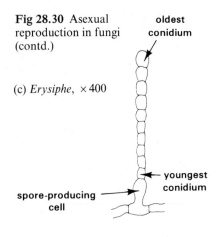

oldest conidium

youngest conidium

spore-producing cell

Biophoto Associates

iii. Ascomycetes produce asexual spores called **conidia**. Conidia are often formed in chains from special hyphae called **conidiophores** (Fig 28.30(c)–(e)). Yeasts such as *Saccharomyces* reproduce asexually by **budding**.

iv. Basidiomycetes Asexual spore production is not often seen in this phylum. When formed, the asexual spores are similar to conidia.

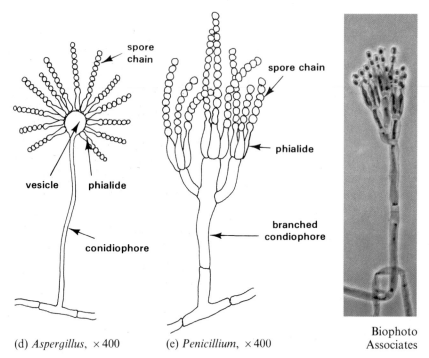

spore chain

spore chain

phialide

vesicle

phialide

branched condiophore

conidiophore

(d) *Aspergillus*, ×400   (e) *Penicillium*, ×400

Biophoto Associates

2 Sexual reproduction

This is common in fungi and involves the fusion of nuclei followed by meiosis. It results in genetic variation among the offspring.

i. Oomycetes produce gametes in unicellular sexual organs known as **gametangia**. The male gametangium is called an **antheridium**, the female an **oogonium**. Most aquatic oomycete fungi are **homothallic**, male and female gametangia arising on the same mycelium. The self-fertilised eggs develop into thick-walled **oospores** (Fig 28.31(a)).

ii. Zygomycetes *Mucor* is **heterothallic**, and cross-fertilisation occurs between two genetically different mycelia. Gametangia arise from adjacent mycelia on hyphal branches called **suspensors**. The mycelia and gametangia are identical to look at and are usually described as + and − rather than male and female. Gametangia join to form a **zygospore** in which fusion takes place between + and − nuclei (Fig 28.31(b) & (c)). Meiosis follows, restoring the normal haploid chromosome number in each nucleus. After a period of rest, the zygospore germinates. A **promycelium** grows from the zygospore and supports a sporangium. The sporangium contains either + or − asexual spores.

Fig 28.31 (a) Sexual reproduction in *Saprolegnia*

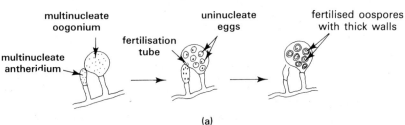

multinucleate oogonium

multinucleate antheridium

fertilisation tube

uninucleate eggs

fertilised oospores with thick walls

(a)

568

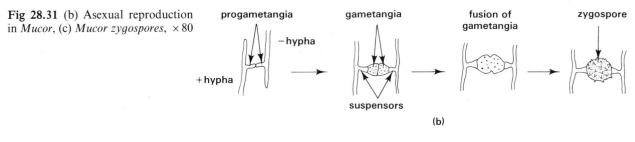

Fig 28.31 (b) Asexual reproduction in *Mucor*, (c) *Mucor zygospores*, ×80

progametangia   gametangia   fusion of gametangia   zygospore

−hypha

+hypha

suspensors

(b)

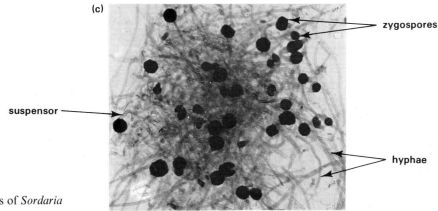

(c)

zygospores

suspensor

hyphae

Fig 28.32 Ascospores of *Sordaria fimicola*, ×150

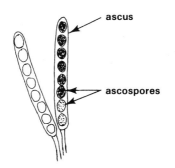

ascus

ascospores

iii. Ascomycetes are heterothallic. The sexual organs are unicellular, a male **antheridium** and a female **ascogonium**. The male nuclei pass into the ascogonium through a fertilisation tube. Nuclear fusion takes place and **ascospores** develop inside **asci** (Fig 28.32). They may be surrounded by a protective **ascocarp** made of sterile hyphae from the surrounding mycelium. The ascospores are shot out of the ascocarp or released when the ascocarp decays.

iv. Basidiomycetes are heterothallic, but there are no sexual organs. **Conjugation** occurs between adjacent hyphae from different mycelia. A new **secondary mycelium** develops containing nuclei from each of the parent mycelia. The secondary mycelium gives rise to a **basidiocarp** (Fig 28.33) consisting of a vertical **stipe** supporting a cap-shaped **pileus**. Beneath the pileus is a number of radiating gills called **lamellae** which bear spore-producing **basidia**. Nuclear fusion followed by meiosis occurs in the basidia. The resulting **basidiospores** are released and may be dispersed by air currents, water, insects or other animals.

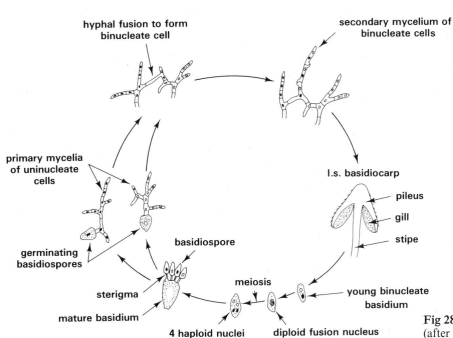

hyphal fusion to form binucleate cell

secondary mycelium of binucleate cells

primary mycelia of uninucleate cells

l.s. basidiocarp

pileus
gill
stipe

germinating basidiospores

basidiospore

sterigma

meiosis

young binucleate basidium

mature basidium

4 haploid nuclei   diploid fusion nucleus

Fig 28.33 Life cycle of *Coprinus lagopus* (after Williams and Shaw 1982)

569

28.9.4 Importance of fungi to humans

1 Parasitic fungi

Some fungi are plant and animal **pathogens**. Ringworm, thrush and athlete's foot are examples of human diseases caused by fungi. A much larger number of fungi are plant pathogens. One of the most notorious of the parasitic fungi is *Phytophthora infestans*, the cause of late blight of potatoes. In the 1840s this fungus ravaged the European potato crop, leaving millions of starving peasants in its wake. Between 1841 and 1851, the population of Ireland alone was thus reduced, by starvation and emigration, by one and a half million people.

2 Saprophytic fungi

Many **saprophytic** fungi, together with bacteria, decompose plant and animals remains to form **humus** in the soil. Humus is rich in minerals and is able to hold moisture. It is therefore a valuable component of all fertile soils. Plant remains contribute most of the material from which humus is formed (Chapter 32).

Not all saprophytic fungi benefit humans. Prepared food readily becomes mouldy, making it unpalatable. Mouldy food may be lethal if eaten, because some saprophytic fungi release poisons called **myco-toxins** on to the substrate on which they are growing. Saprophytic fungi attack the buildings in which we live. Dry rot of timber, for example, is caused by the basidiomycete fungus *Merulius lacrymans*. Other saprophytic fungi destroy wallpaper and fabrics in damp houses. See chapter 34.

Fig 28.34 *Marchantia polymorpha*:

(a) whole plant showing gemma cups, ×2

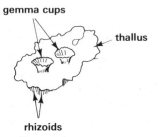

gemma cups

thallus

rhizoids

Geoscience

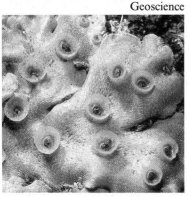

(b) vertical section through gemma cup, ×35

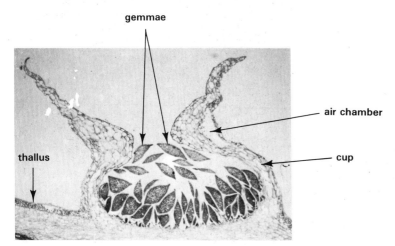

gemmae

thallus

air chamber

cup

28.10 Kingdom Plantae: Phylum Bryophyta

Plants are multicellular, eukaryotic and photosynthetic organisms whose cell walls contain cellulose.

The bryophytes are **liverworts** and **mosses**. They are green plants, nearly all terrestrial. Most are restricted to damp regions because they need water for fertilisation and they lack a waterproof cuticle. Bryophytes have a **heteromorphic alternation of generations** in which the dominant generation of the life cycle is the haploid **gametophyte**. The gametophyte of many liverworts is thalloid. In mosses and many liverworts, the gametophyte is differentiated into stem and leaves. Fine threads called **rhizoids** hold the gametophyte to the substrate. There is no vascular tissue.

Many liverworts and some mosses reproduce asexually by means of **gemmae**. Gemmae are multicellular, microscopic structures, often produced in **gemma cups** (Fig 28.34). Raindrops falling into the cups scatter the gemmae, which may germinate to produce new gametophyte plants. Bryophytes are **oogamous**, because they produce small flagellated male gametes and large non-motile female gametes. The **sex organs are multicellular**. The diploid **sporophyte** never exists as an independent plant and remains attached to the gametophyte.

The bryophytes are divided into two classes, *Hepaticae* and *Musci*.

28.10.1 Class Hepaticae

Fig 28.35 A thalloid liverwort, *Pellia epiphylla*, ×3

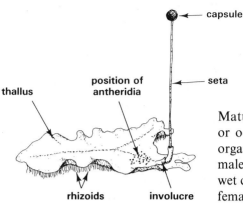

- capsule
- seta
- position of antheridia
- thallus
- rhizoids
- involucre

The Hepaticae are the **liverworts**. They are usually found in moist environments such as rocks moist with freshwater and on wet soil.

Pellia is a **thalloid** liverworth (Fig 28.35). The thallus is flattened and dichotomously branched. It is attached to the substrate by numerous **unicellular rhizoids** on the underside of the thallus. The thallus is the gametophyte phase of the life cycle. The sex organs develop on the upper surface of the thallus near the tip of the branches. The female organs, called **archegonia**, are in small pockets covered by a flap of tissue called an involucre. Mature archegonia each contain a haploid female gamete called an egg cell or oosphere. A little behind the archegonia are pits in which the male organs, **antheridia**, develop. The antheridia produce unicellular haploid male gametes called antherozoids. The antherozoids are flagellated and, in wet conditions, swim and seek out an oosphere. They are attracted to the female gametes by sucrose secreted from the necks of the archegonia. The movement of unicells towards a chemical stimulus is called **chemotaxis**. A diploid zygote produced by fertilisation divides and forms a multicellular embryo which becomes the sporophyte. As it develops, the **sporogonium** stays inside the remains of the archegonium, the wall of which grows to form a **calyptra**. The mature sporogonium consists of a **foot, seta** and a large **capsule** in which diploid spore mother cells undergo meiosis to produce haploid spores. The sporophyte rests in this form over winter. During the following spring, the spore capsule is pushed upwards by elongation of the seta (Fig 28.36). Growth of the sporophyte is entirely dependent on nutrients absorbed from the gametophyte by the foot of the sporogonium. The capsule wall splits into four parts as it dries out (Fig 28.37). Within the spore mass lie many **elaters**, long cells with spirally-thickened walls. The elaters twist as they dry out, helping to dislodge the spores which are dispersed by air currents. New haploid thalli arise from germinated spores.

An example of a **leafy** liverwort is *Lophocolea* (Fig 28.38). It has an erect stem bearing two rows of trough-shaped leaves.

Fig 28.36 *Pellia* sporogonium, ×50

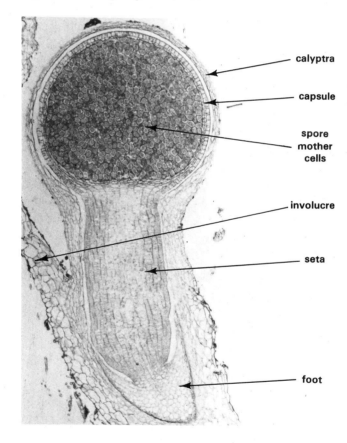

- calyptra
- capsule
- spore mother cells
- involucre
- seta
- foot

Fig 28.37 *Pellia*: (a) spore release from bursting capsule, (b) elater

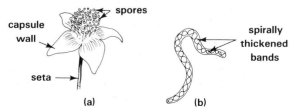

- spores
- capsule wall
- seta
- (a)
- spirally thickened bands
- (b)

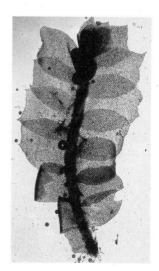

Fig 28.38 A leafy liverwort, *Lophocolea*, ×8 (Biophoto Associates)

28.10.2 Class Musci

The Musci are the **mosses**. Like liverworts, most mosses inhabit damp places. However, mosses are more tolerant of desiccation and can survive in fairly dry places such as sand dunes and on dry stone walls. Each plant has a simple **stem** attached to the substrate by **multicellular rhizoids**. Simple **leaves** are usually arranged spirally along the stem. Antheridia and archegonia, usually on separate plants, arise from the growing tips of the stems. As in liverworts, the sporophyte is attached to the parent gametophyte plant.

Fig 28.39 (a) *Funaria hygrometrica*, × 2, (b) *Polytrichum* sp., × ½

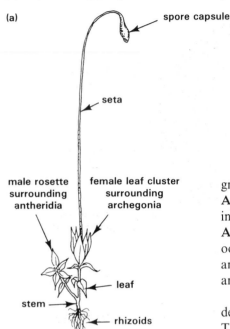

(a)

- spore capsule
- seta
- male rosette surrounding antheridia
- female leaf cluster surrounding archegonia
- leaf
- stem
- rhizoids

(b)

An example of a moss is *Funaria* (Fig 28.39). Each plant is erect and grows to a few centimetres in height, often with a branching stem. **Antheridia** arise at the tip of the main stem. Repeated mitotic divisions inside the antheridia produce many haploid biflagellated antherozoids. **Archegonia** develop at the tip of a side-branch. There is a single haploid oosphere inside each archegonium. The antherozoids are chemotactic and are attracted to the oospheres by proteins secreted from the necks of archegonia. On fertilisation, a diploid zygote is formed.

Usually only one zygote develops further at each stem apex. The development results in a **sporogonium** with a long seta bearing a capsule. The sporogonium is attached to the gametophyte by a **foot**, through which water and nutrients are absorbed. However, the sporophyte is not totally dependent on the gametophyte. The **seta** and **capsule** become green and make organic materials by photosynthesis. The developing sporogonium is enclosed at first by the remains of the archegonium. As the seta elongates, the capsule is carried upwards. A cap called the **calyptra**, the remnant of the archegonium, covers the capsule (Fig 28.40). The capsule contains a

Fig 28.40 Funaria: (a) archegonia, × 20, (b) developing sporophyte, (c) antherozoids

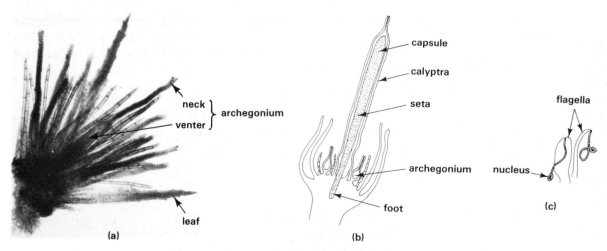

(a)

- neck } archegonium
- venter }
- leaf

(b)

- capsule
- calyptra
- seta
- archegonium
- foot

(c)

- flagella
- nucleus

central tissue mass called the **columella**, surrounded by sporogenous tissue which produces the spores. Spore formation involves meiosis, and the haploid spores eventually occupy most of the space inside the capsule. The capsule is topped by the **peristome** covered by an outer domed cap, the

Fig 28.41 *Funaria*: (a) VS capsule, × 15, (b) peristome teeth, × 30

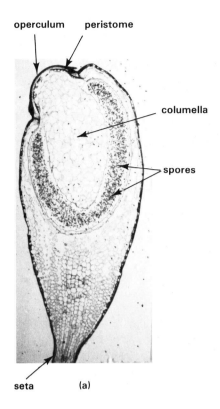

operculum   peristome

columella

spores

seta   (a)

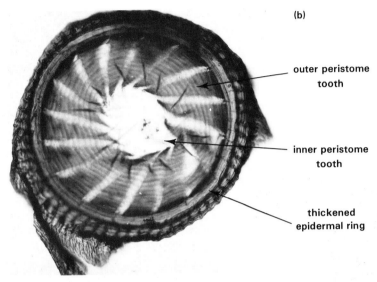

(b)

outer peristome
tooth

inner peristome
tooth

thickened
epidermal ring

operculum (Fig 28.41). As the spores mature, the columella disappears and the capsule dries out. The calyptra and operculum fall away, exposing the teeth of the peristome to the air. The peristome teeth are hygroscopic, bending open when dry and closing back when wet. When open, the ripe spores fall out, because the spore capsule is usually held upside down. Spores are dispersed by air currents. In suitable conditions, the spores germinate to produce a branching filament called the **protonema**. Haploid moss plants develop from 'buds' on the protonema and the life cycle is completed.

28.11 Kingdom Plantae: Phylum Tracheophyta

The tracheophytes are the **vascular plants**. They include **clubmosses, horsetails** and **ferns, gymnosperms** and **flowering plants**. The diploid sporophyte is the dominant phase of the life cycle and is independent of the haploid gametophyte. Unlike the algae and bryophytes, tracheophytes have **vascular tissue** containing **tracheids or xylem vessels** and **sieve cells**. The sporophyte plant is differentiated into **roots, stems** and **leaves**; the shoot system is covered by a **waterproof cuticle**. These features represent a greater adaptation to terrestrial life than is seen in the bryophytes. In the Carboniferous period, about three hundred million years ago, tracheophytes were the dominant plants on land. Their fossilised remains are useful to us as coal.

28.11.1 Class Lycopsida

The Lycopsida include the **clubmoss** *Selaginella* (Fig 28.42). *S. selaginoides* grows in upland pastures in the north and west of Britain. Most other

Fig 28.42 Clubmoss, *Selaginella kraussiana*: (a) strobilus, × 70, (b) whole plant, × 1, (c) antherozoid

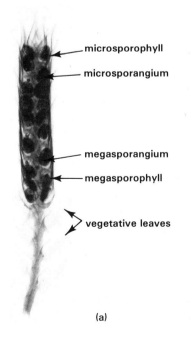

microsporophyll

microsporangium

megasporangium

megasporophyll

vegetative leaves

(a)

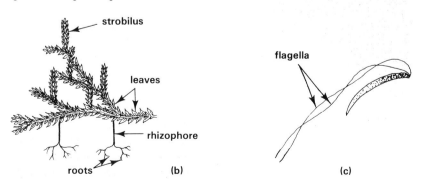

strobilus

leaves

rhizophore

roots   (b)

flagella

(c)

573

species are tropical. In *S. kraussiana*, the sporophyte plant has a prostrate branching stem. Four rows of leaves, two small and two large, project laterally and the stem is anchored to the soil by roots. The roots develop from rhizophores which grow down from the stem.

In a mature **sporophyte**, cones called **strobili** develop at the ends of the shoots. Each strobilus carries modified leaves called **sporophylls**, in the axils of which develop the sporangia. Two kinds of sporangia are formed in the same strobilus. Each **microsporangium** produces many small **microspores**. Four large **megaspores** usually develop inside each **megasporangium**. For this reason *Selaginella* is **heterosporous**. Meiosis occurs in the production of spores and each spore is haploid. When ripe, the sporangia split open and the spores are released into the air. A **male gametophyte** which produces many biflagellated antherozoids develops inside each microspore. The antherozoids are released by rupture of the microspore wall. A **female gametophyte** develops inside each megaspore. This is made up of a cellular prothallus, on which archegonia develop at one end. An oosphere appears in each archegonium. As the archegonia develop, the megaspore wall splits, exposing the apex of the gametophyte. Fertilisation of the oosphere by an antherozoid produces a diploid zygote which develops into a new sporophyte plant to complete the life cycle.

28.11.2 Class Sphenopsida

The Sphenopsida are represented by a single living genus, *Equisetum*, the **horsetail** (Fig 28.43). The stem of the **sporophyte** is a root-like rhizome

Fig 28.43 Common horsetail, *Equisetum arvense*:

(a) sterile aerial branch, $\times \frac{1}{2}$ (Biofotos)

(b) fertile aerial branch, $\times 1$ (Biofotos)

strobilus

leaf

node

(c) spores with coiled and uncoiled elaters

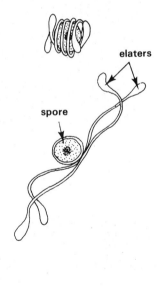

elaters

spore

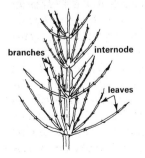

branches
internode
leaves

which grows horizontally in the soil. Whorls of small scale-like leaves arise at nodes on the rhizome. Adventitious roots also arise at the nodes where the rhozome may be branched. In *E. arvense*, two kinds of aerial shoots grow from the rhizome. Sterile shoots are green and much-branched at the nodes. Photosynthesis occurs in the stem, the whorls of leaves at each node being scale-like and colourless. Fertile shoots are colourless, unbranched and have a single **strobilus** at the apex. The strobilus is made up of several whorls of sporangiophores. Each consists of a stalk with a disc on the outside which supports the **sporangia**. Meiosis occurs in the production of spores. Each haploid spore secretes a wall containing four coiled strips. The strips are hygroscopic and spring open when dry, thus helping to free the spores.

Spores are short-lived and germinate in suitable conditions to produce the **gametophyte** generation. Each gametophyte is a small green lobed thallus less than a centimetre across, attached to the soil by rhizoids. Antheridia and archegonia develop on the thallus, and zygotes are formed when oospheres are fertilised by motile antherozoids. The zygotes develop into new sporophyte plants.

28.11.3 Class Pteropsida

The Pteropsida are the **ferns** such as *Dryopteris* (Fig 28.44). The **sporophyte** plant has an underground rhizome. A tangled mass of adventitious roots grows from the rhizome, which is usually covered by the bases of old leaves from previous years' growth. The apex of the rhizome projects above the surface of the soil. From it arise compound leaves called **fronds**. Each frond has a main axis called the **rachis**, from which arise numerous lateral **pinnae**. Each pinna is divided into many green **pinnules**. Fronds develop over two years. They begin as coiled structures and gradually unroll as growth proceeds.

Fig 28.44 The male fern, *Dryopteris filix-mas*: (a) whole plant, (b) two fronds

(a)

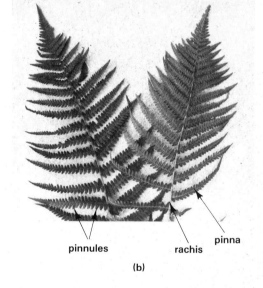

pinnules rachis pinna

(b)

Fig 28.45 *Dryopteris*, longitudinal section sorus, × 150

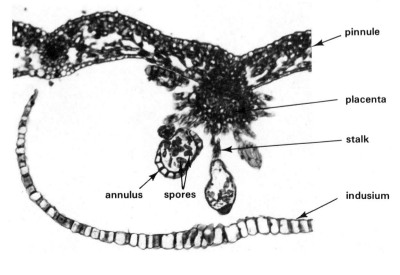

pinnule

placenta

stalk

indusium

annulus spores

The pinnules of mature fronds have **sporangia** on their lower surface. Sporangia are in clusters called **sori** and are covered by a protective flap of tissue called the **indusium** (Fig 28.45). The indusium degenerates as the sori develop, exposing the ripe sporangia to the air. Each mature sporangium consists of a flat spore-containing capsule attached by a stalk to the **placenta**. A row of cells with partially thickened walls extends three-quarters of the way around the capsule. These cells make up the **an-**

575

Fig 28.46 *Dryopteris* sporangium: (a) mature, (b) dehisced with annulus bent back, (c) annulus recoil propelling spores into the air

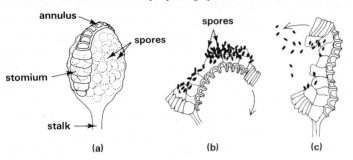

nulus (Fig 28.46(a)). Next to the annulus is a row of large thin-walled cells which make up the **stomium**. Dehiscence of a sporangium occurs when the annulus cells dry out and shrink. The thinner-walled stomium ruptures and the annulus springs back to reveal the haploid spores. As the annulus cells dry out more, their ability to hold the annulus in its sprung position fails. The annulus suddenly recoils and many of the spores are catapulted away to be dispersed by air currents (Fig 28.46(b) and (c)).

After a short dormant period, the haploid spores germinate to produce the **gametophyte** generation. The gametophyte is a flat heart-shaped **prothallus** about a centimetre long. It is green and attached to the soil by **rhizoids** (Fig 28.47). Sex organs develop on the underside of the prothallus.

Fig 28.47 *Dryopteris*: (a) photomicrograph of prothallus, ×20, (b) antheridium, ×600, (c) antherozoid, ×1000, (d) archegonium, ×650

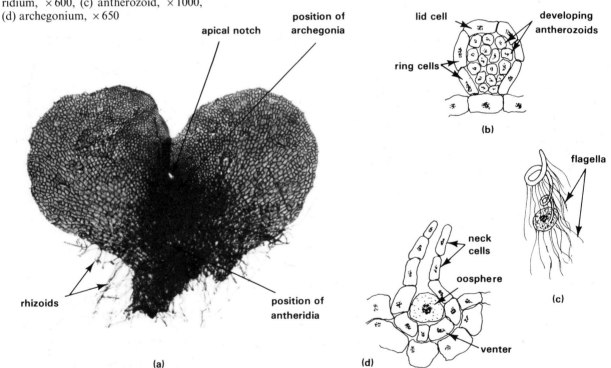

Spherical antheridia are found among the rhizoids. Antherozoids are coiled and flagellated. Archegonia develop some distance away from the antheridia on the central cushion near the apical notch of the prothallus. Each archegonium has a cup-shaped **venter** from which projects a short **neck**. The venter encloses a single oosphere. Chemotaxis by the antherozoids brings them towards the archegonia; the attractant is malic acid, secreted by the necks of the archegonia. Antheroids swim down the archegonial necks and fuse with the oospheres to form diploid zygotes. Usually only one zygote develops further on each prothallus to give rise to a new sporophyte generation (Fig 28.48).

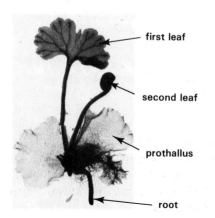

Fig 28.48 *Dryopteris*, developing sporophyte, ×15

28.11.4 Class Gymnospermae

Why have gymnosperms and angiosperms come to dominate the earth's terrestrial vegetation?

The **gymnosperms** and **angiosperms** dominate the earth's terrestrial vegetation. Except in a few genera, there are no motile antherozoids such as are found in the bryophytes and tracheophytes. Consequently, gymnosperms and angiosperms do not require water for fertilisation. The male gametes are haploid nuclei inside resistant **pollen**. **Seeds** contain and protect the embryo (Chapter 22). The main phase of the life cycle is the diploid **sporophyte** (Fig 28.49), usually with well-developed roots, stem and leaves. Efficient **vascular tissue** is present and the aerial parts of the sporophyte are protected from desiccation by a **waterproof cuticle**. The development of **woody tissue** in many species enables them to grow to considerable size. These features are largely responsible for the relative success of gymnosperms and angiosperms as terrestrial plants.

Fig 28.49 Life cycles of (a) a bryophyte, (b) a fern, and (c) an angiosperm

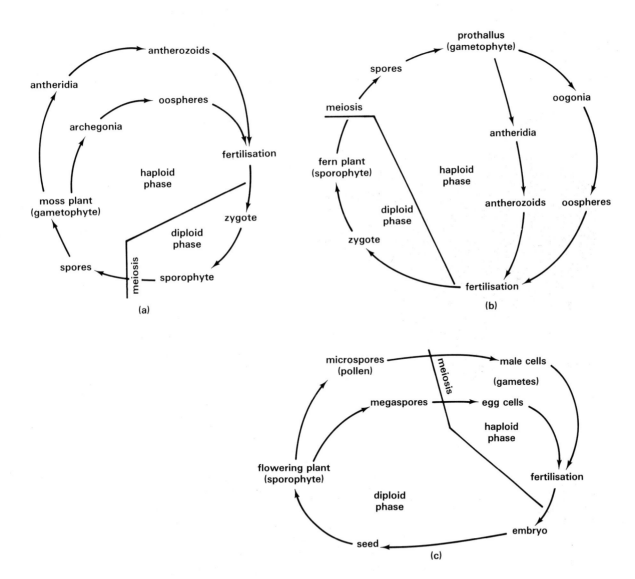

The gymnosperms contain **cycads** and **conifers**. The seeds of gymnosperms are naked and not enclosed in a pericarp as in angiosperms. **Tracheids** are the water-transporting cells, and **sieve cells** carry organic materials. Secondary vascular tissue is always present. The gymnosperms contain several orders, including the Coniferales.

Fig 28.50 The Scots pine, *Pinus sylvestris*, ×$\frac{1}{120}$

1 Coniferales

The Scots pin, *Pinus sylvestris* (Fig 28.50) is a conifer. It is a tall tree, growing to a height of 30 m or more. Underground it has a branched system of roots which arise from a vertical tap root. Old roots are woody and give sturdy anchorage. As a pine tree grows, it loses its lower branches, leaving the bottom of the shoot as a straight trunk. A pyramidal mass of branches grows from the upper part of the trunk. Photosynthetic leaves are restricted to lateral dwarf or spur shoots which are eventually shed from the tree. The dwarf shoot develops from buds in the axils of brown scale-like leaves on the main shoot. Each spur shoot consists of a short stem, a series of overlapping deciduous bud scales and two needle-like leaves (Fig 28.51).

The thick cuticle of the leaves helps prevent drying out in windy and in arid conditions. The leaves also tolerate frost.

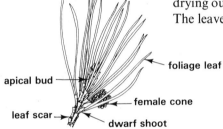

Fig 28.51 *Pinus* twig, growing tip

Pinus is **monoecious** and **heterosporous**, bearing male and female cones on the same tree. **Male cones** take the place of dwarf shoots and arise in clusters on young long shoots, just behind the terminal buds. A male cone has many **microsporophylls**, each bearing two **microsporangia** on the underside. Haploid **microspores (pollen grains)** are formed by meiosis. When mature, they each contain an incomplete male gametophyte and have two air-filled sacs. When they have shed their pollen, male cones wither and drop off the tree (Fig 28.52).

Female cones take the place of long shoots. On the axis of the cone are small spirally-arranged **ovuliferous scales**, under each of which is a **bract scale** (Fig 28.53). Each ovuliferous scale bears two **ovules** on its upper surface.

Fig 28.52 *Pinus*: (a) cluster of male cones, (b) longitudinal section of male cone, ×5, (c) pollen grains, ×80

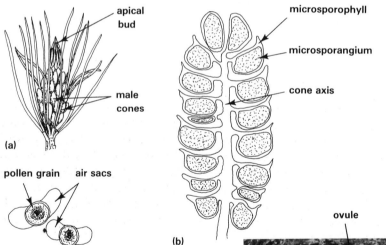

Fig 28.53 *Pinus* female cone: (a) beginning of second year; (b) two ovules, ×10

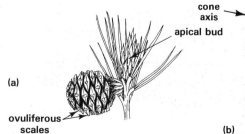

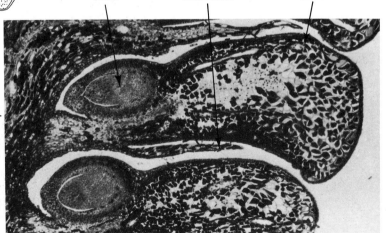

Ovules contain a mass of tissue, called the **nucellus**, which is equivalent to a **megasporangium**. It is covered by an **integument**. At one end is a small pore, the **micropyle.** A single haploid **megaspore** develops inside each ovule.

Pollination, the transfer of pollen grains from male cones to female cones, is by wind. The air sacs keep the pollen grains suspended for long periods in the air. Pollen is drawn into the ovule through the micropyle. After pollination, the development of the male and female gametophytes proceeds slowly and is interrupted by winter. In the following spring, the **female gametophyte** (prothallus) becomes cellular, and three or four **archegonia** are formed (Fig 28.54). The **male gametophyte** completes its development, and the male gametes are carried through the nucellus to the archegonia by growth of the **pollen tube** (Fig 28.55).

Fig 28.55 *Pinus*, development of pollen grain: (a) just before shedding, (b) at the time of shedding, (c) the following spring, (d) just before fertilisation

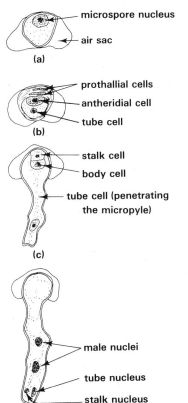

Fig 28.54 *Pinus*, longitudinal section of ovule at fertilisation stage (from photomicrographs), × 5

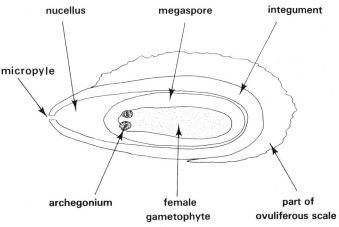

Fertilisation occurs when the male nucleus from the pollen fuses with the female nucleus in the oosphere. A diploid zygote is formed. Usually only one zygote develops in each ovule to produce an **embryo**. The integument becomes a hard testa and protects the embryo inside the **seed**. Each seed has a thin papery wing, part of the ovuliferous scale (Fig 28.56). Maturing of the seeds ready for release from the female cones takes a further year. Female cones therefore require three years to complete their seed-producing function. Ripe seeds are dispersed by the wind and germination is epigeal (section 23.2).

Fig 28.56 *Pinus*: (a) female cone, third year, (b) seed, (c) longitudinal section of embryo, × 7

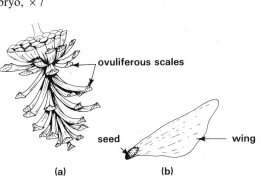

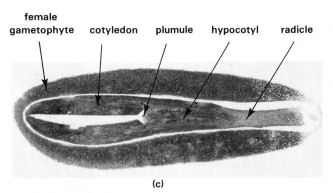

28.11.5 Class Angiospermae

The angiosperms are the **flowering plants**. The sexual reproductive organs are found in **flowers** and the seeds are enclosed in a **pericarp**. The ovules with two integuments are borne within a structure called the **carpel**. The female gametophyte is an **embryo sac** lacking archegonia. It develops from a megaspore produced by the nucellus of the ovule. The microspores (pollen) which are produced in anthers germinate on a special region of the carpel called the **stigma**. Pollination is by wind, water, insects and sometimes birds. **Double fertilisation** occurs, producing a zygote and usually a triploid endosperm nucleus. The vascular system consists of **vessels** and **tracheids** and **sieve tubes** with **companion cells**. These elements are more effective at transportation than the tracheids and sieve cells of gymnosperms. Secondary vascular tissue is often present.

For details of angiosperm anatomy and reproduction refer to Chapters 12, 13 and 22. There are two subclasses, Monocotyledones and Dicotyledones.

1 Monocotyledones

These produce seeds with a **single cotyledon**. The vascular tissue in the leaves is arranged in **parallel veins** (Fig 28.57). In the stem, the vascular bundles are scattered (Fig 28.58). The floral parts are usually arranged in **threes** or multiples of three (Fig 28.59). Secondary growth (Chapter 24) occurs in only a few species. For this reason, monocotyledonous plants are mainly **herbaceous**. Some perennate by forming bulbs, corms and rhizomes, which are also a means of propagation (Chapter 22).

Among the monocotyledonous plants most important to man are the cereals, such as barley, oats, maize, rice and wheat, and the grasses. Grasses make up most of the vegetation in pastureland.

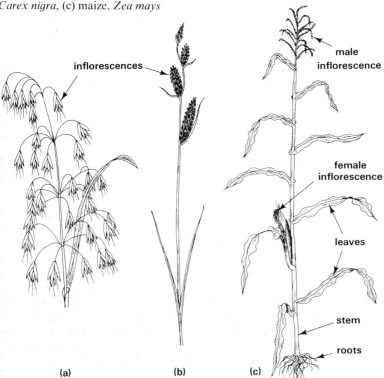

Fig 28.57 Some monocotyledons: (a) wild oat, *Avena fatua*, (b) common sedge, *Carex nigra*, (c) maize, *Zea mays*

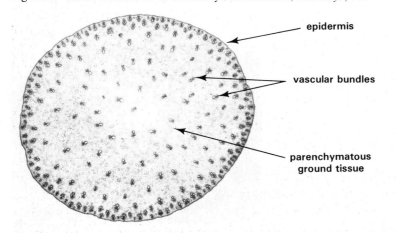

Fig 28.58 Transverse section of monocotyledonous stem, *Zea mays*, × 5

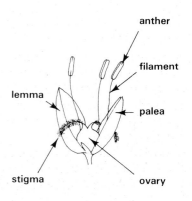

Fig 28.59 Grass floret (after Gill and Vear)

Fig 28.60 Some dicotyledons: (a) common buttercup, *Ranunculus acris* (Bruce Coleman: Dennis Green), (b) dandelion, *Taraxacum officinale* (Biofotos), (c) oak *Quercus robur*, $\times \frac{1}{100}$ (Bruce Coleman: Jane Burton)

(a)

(b)

(c)

2 Dicotyledones

These produce seeds with **two cotyledons** (Fig 28.52). The vascular tissue in the leaves is net-like. In the stem, the vascular bundles form a circle between cortex and pith. The floral parts are generally in **fours** or **fives** or in multiples of these numbers.

Secondary growth occurs in many dicotyledons, which grow as **shrubs** and **trees**. The climax vegetation in most parts of Britain comprises dicotyledonous deciduous trees, such as elm and oak, and shrubs such as the hawthorn. In autumn they lose their leaves, and their woody twigs survive the winter. There are many herbaceous dicotyledonous species including **annuals** which overwinter as seeds, and **perennials** which survive unfavourable conditions as tubers and swollen tap roots. Biennial and perennial dicotyledons such as beetroot, carrots, parsnips and potatoes are an important part of the diet of humans, as are the seeds of some annuals such as peas and beans.

Gymnosperms and angiosperms make up the bulk of the earth's terrestrial vegetation. Among the reasons for their success is their enormous variety of habit, independence of water for fertilisation and the ability of many species to tolerate and survive desiccation. For these reasons there are few terrestrial habitats where flowering plants are not found.

SUMMARY

The binomial system of nomenclature (after Linnaeus) includes generic and specific names. Classifications are: artificial (for convenience and filing); and natural (closely related organisms are grouped).

Viruses: obligate parasites and very small – 10–300 nm diameter. Nucleic acid is surrounded by a protein capsid. Reproduction occurs in host cells which eventually burst – lytic cycle.
Importance: many viruses cause diseases in humans, livestock and crops.

Kingdom Monera
Class Bacteria: single cells or colonies; the cells are surrounded by a wall, often with a capsule outside; inside the wall are: plasma membrane, mesosomes, ribosomes, nucleoid (DNA) prokaryotic. Flagella allow motility in many bacteria. There are four basic forms: bacilli, cocci, vibrio and spirilla. Most bacteria and heterotrophic; some are photo-autotrophic and some are chemo-autotrophic. Reproduction is by binary fission. Endospores are produced in some species.
Importance: Many bacteria are pathogenic; gut bacteria aid digestion in animals (especially herbivores); some bacteria help the breakdown of dead organic matter in soil; some are important in food spoilage. Industrial uses include dairy products, antibiotics, vinegar, amino acids and proteins, retting fibres, silage and sewage.
Class Cyanobacteria (blue-green algae); similar to the first organisms on earth. Similar to bacteria; prokaryotic cells.

Kingdom Protoctista, Phylum Protozoa: unicellular (acellular) animals. Class Rhizopoda; pseudopodia, amoeboid movement, phagocytosis. Osmoregulation by means of contractile vacuoles; reproduce by binary fission, e.g. *Amoeba*.
Class Flagellata: possess flagella. Some are parasitic, e.g. *Trypanosoma*.
Class Ciliata: possess numerous cilia; micro- and mega-nucleus. Sexual reproduction is by conjugation, e.g. *Paramecium*.
Class Sporozoa: all parasitic; complex life cycles often involving more than one host, e.g. *Plasmodium*.
Phylum Euglenoidea; Flagellate unicells capable of photosynthesis e.g. *Euglena*
Phylum Chlorophyta (green algae); eukaryotic cells, e.g. *Spirogyra*

▶

▶ Phylum Phaeophyta (brown algae); include seaweeds. Life cycle of sporophyte and gametophyte alternation of generations, e.g. *Fucus*.
Phylum Rhodophyta (red algae), Bacillariophyta (diatoms) and Euglenophyta (unicellular flagellates).

Kingdom Fungi: some are unicellular (yeasts), most exist as mycelia (network of hyphae). They are bounded by cell wall and contain nuclei (eukaryotic), mitochondria and endoplasmic reticulum. Reproduction is by sexually or asexually produced spores.
Importance: Some fungi are pathogenic; many help the breakdown of dead organic matter in the soil; some are edible. Industrial uses include the production of single cell protein, brewing and wine making, antibiotics, plant hormones, vitamins, steroids and enzymes.

Kingdom Plantae: Phylum Bryophyta; mostly terrestrial green plants inhabiting damp places; no vasc. tissue. Heteromorphic altern'n of generations (gametophyte dominant). Asex. reprod. by gemmae.
Class Hepaticae (liverworts): usually a thalloid gametophyte; rhizoids, e.g. *Pellia*.
Class Musci (mosses): gametophyte differentiated into stem, leaves and rhizoids, e.g. *Funaria*.

Phylum Tracheophyta: heteromorphic alternation of generations (sporophyte dominant); vascular tissue; roots, stem and leaves; waterproof cuticle on shoot system; terrestrial adaptations.
Lycopsida (clubmosses, e,g. *Selaginella*), Sphenopsida (horsetails, e.g. *Equisetum*) and Pteropsida (ferns, e.g. *Dryopteris*).

Gymnosperms and Angiosperms: dominate terrestrial vegetation; do not require water for fertilisation. Male gametes inside resistant pollen; seeds contain and protect the embryo. Sporophyte is the dominant part of the life cycle. Vascular tissue; waterproof cuticle on aerial parts; woody tissue in some species.

Gymnospermae (cycads and conifers, e.g. *Pinus*).

Angiospermae (flowering plants) include –
Monocotyledones: seed with single cotyledon; vascular tissue in parallel veins in leaves, scattered in stem; floral parts in three or multiples of three; few produce wood; grasses and cereals, e.g. *Avena*.
Dicotyledones; seeds with two cotyledons; vascular tissue is net-like in leaves, and arranged in a circle in the stem; floral parts in fours or fives (or multiples of these); many produce wood (secondary thickening), herbeceous plants, shrubs and trees.

QUESTIONS

1 Give an account of the structures, life-cycles and pathogenicity of viruses. (JMB)

2 (a) List the Divisions of the Plant Kingdom, indicating the main characteristics of each Division.
(b) Explain why the Spermatophyta are often referred to as the 'higher plants'. (C)

3 (a) The diagram below summarises the life cycle of a fern. Complete the diagram by writing the most appropriate word in the boxes.

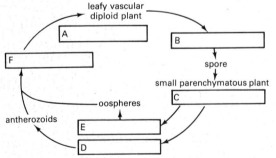

(b) State *three* ways in which water is essential for reproduction in a fern.
(c) Describe the mechanism by which fern spores are discharged and dispersed.
(d) How does spore production in ferns contribute to genetic variation in ferns? (L)

4 Copy the diagram of a fern life cycle and place the following terms in the appropriate positions: gametophyte, sporophyte, meiosis
(AEB 1987)

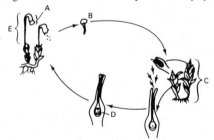

5 The diagram illustrates the life cycle of a bryophyte.

(a) Name the structures labelled A–E.
(b) State clearly at which stage in the life cycle meiosis occurs.
(c) (i) What is meant by the term 'Alternation of Generations' in relation to this life cycle?
(ii) State *two* ways in which the alternation of generations shown by this cycle differs from that found in flowering plants. (L)

29 Variety of life: Non-chordate animals

29.1 Phylum Cnidaria 584

    29.1.1 Cell differentiation in cnidarians 584
    29.1.2 Variety of cnidarians 586

29.2 Phylum Platyhelminthes 589

    29.2.1 Turbellaria 589
    29.2.2 Trematoda 590
    29.2.3 Cestoda 592

29.3 Phylum Nematoda 593

29.4 Phylum Annelida 594

    29.4.1 Polychaeta 595
    29.4.2 Oligochaeta 596
    29.4.3 Hirudinea 598

29.5 Phylum Arthropoda 598

    29.5.1 Crustacea 600
    29.5.2 Myriapoda 602
    29.5.3 Insecta 602
    29.5.4 Arachnida 606

29.6 Phylum Mollusca 606

    29.6.1 Gastropoda 607
    29.6.2 Bivalvia 607

29.7 Phylum Echinodermata 609

    29.7.1 Asteroidea 610
    29.7.2 Echinoidea 610

Summary 611

Questions 612

Biofotos

29 Variety of life: Non-chordate animals

Non-chordate animals do not have a dorsal notochord, which is characteristic of the chordate animals. Non-chordate animals are often called **invertebrates**, as they do not have vertebrae either. However, some chordate animals are also invertebrate (Chapter 30).

29.1 Phylum Cnidaria

The Cnidaria are commonly called **coelenterates**. *Hydra*, *Obelia*, sea anemones, corals and jellyfish are cnidarians. They are aquatic. The body wall is of two layers of cells enclosing a central water-filled cavity called the **enteron**. Cnidarians exist in two basic forms, the **hydroid** and the **medusa** (Fig 29.1). Each form has a single opening to the outside surrounded by tentacles. Through the opening food is brought into the enteron. Wastes are discharged from the enteron through the same opening. Coelenterates are said to be **diploblastic**, because the body is derived from two tissue layers. The outer layer is called the **ectoderm** and is separated from the **endoderm** by a jelly-like **mesogloea**. The diploblastic state is sometimes called the **tissue level of development** (Fig 29.2). The ectoderm and endoderm consist of a variety of cells which perform specific functions. However, the tissues are not aggregated into organs. All coelenterates are **radially symmetrical** (Fig 29.13).

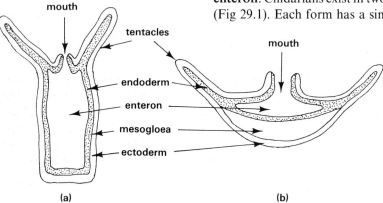

Fig 29.1 Cnidarian body forms: (a) hydroid, (b) medusoid

29.1.1 Cell differentiation in cnidarians

1 Muscle cells

These cells are not like those of mammals (Chapter 17). However, they do contain contractile fibres. In the ectoderm, the fibres are arranged along the axis of the body. In endodermal muscle cells, they are arranged in a circle around the body (Fig 29.18). The longitudinal and circular fibres act

Fig 29.2 The main cell types in the cnidarian body wall

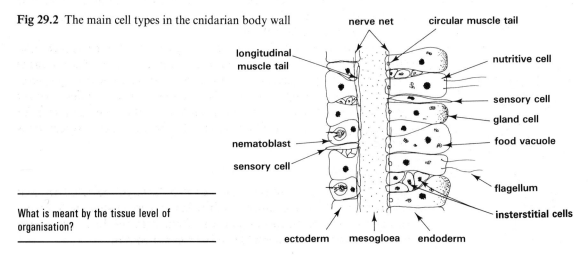

What is meant by the tissue level of organisation?

584

Fig 29.3 Cnidarian movement: (a) pulsation in a jellyfish (Bruce Coleman: Frieder Sauer), (b) somersaulting in *Hydra* (Bruce Coleman: Kim Taylor).

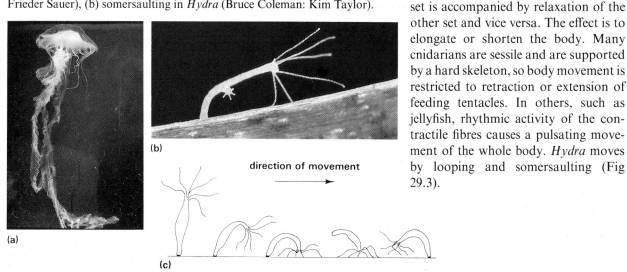

(a)

(b)

direction of movement

(c)

antagonistically. Contraction of one set is accompanied by relaxation of the other set and vice versa. The effect is to elongate or shorten the body. Many cnidarians are sessile and are supported by a hard skeleton, so body movement is restricted to retraction or extension of feeding tentacles. In others, such as jellyfish, rhythmic activity of the contractile fibres causes a pulsating movement of the whole body. *Hydra* moves by looping and somersaulting (Fig 29.3).

Fig 29.4 Nematocysts: (a) coiled, (b) discharged (note the thread turns inside out)

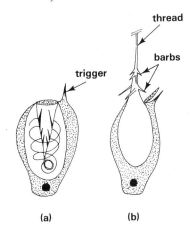

thread

barbs

trigger

(a)   (b)

2 Nematoblasts

These are also called **cnidoblasts** and are found in groups in the ectoderm, especially in the tentacles. Nematoblasts are used to capture prey. They contain a coiled and often barbed thread inside a structure called a **nematocyst**. The thread uncoils and quickly projects from the cell when it is triggered. The triggering device is on the outer surface of the nematoblast and may be activated by touch and chemical stimulation (Fig 29.4).

3 Interstitial cells

These are small undifferentiated cells which may develop into any of the other kinds of cells in the cnidarian body wall.

4 Gland cells

These are found in the ectoderm at the base of cnidarians such as *Hydra* and sea anemones. They secrete a sticky substance by which the animal adheres to the substrate. Gland cells in the endoderm secrete a juice of hydrolytic enzymes into the enteron. Some digestion of food occurs in the enteron which for this reason is often called the **gastrovascular cavity**.

5 Nutritive cells

These are muscle cells in the endoderm. They form pseudopodia which engulf food particles from the enteron. Digestion is intracellular. Nutritive cells may also have flagella, the beating of which causes water to enter and leave the enteron through the mouth.

Fig 29.5 Nerve net in *Hydra*

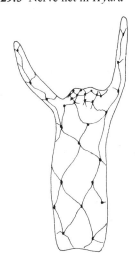

Some cnidarians, especially reef-building corals, have **symbiotic** unicellular green or brown **algae** in the endoderm layer. Sugars made photosynthetically by the algae are used by its partner as an energy source. When food is scarce, up to 50 % of the cnidarian's energy comes from the algae. Amino acids made by the algae are also used by the cnidarians for protein synthesis. The benefit derived by the algae from the association is obscure.

6 Sensory cells

These connect to a **nerve net** which extends through the ectoderm and endoderm (Fig 29.5). It enables the cnidarian to sense its surroundings and respond to stimuli in an appropriate way. Responses may be to attack prey or avoid danger by retracting the body from a predator.

29.1.2 Variety of cnidarians

The Cnidaria are divided into three classes: Hydrozoa, Anthozoa and Scyphozoa.

1 Hydrozoa

These are single, free-moving organisms such as *Hydra*, or more usually colonial, sessile organisms such as *Obelia*. The main body form is the **hydroid**.

Hydra reproduces sexually and asexually. In sexual reproduction, sperms are made in **testes** and ova in **ovaries**. The sexual organs are ectodermal in origin and often arise on the same individual. *Hydra* is thus **hermaphrodite**. However, the gonads mature at different times and cross-fertilisation is usual between different individuals. In asexual reproduction, a new *Hydra* is formed by **budding** on an established hydroid. When full-grown, the junction between the bud and parent is constricted and the new *Hydra* is released (Fig 29.6). Budding involves mitosis of interstitial cells followed by cell differentiation. While it is developing, the enteron of the bud and parent are joined.

Fig 29.6 *Hydra*: (a) showing testes, × 80, (b) budding, × 120

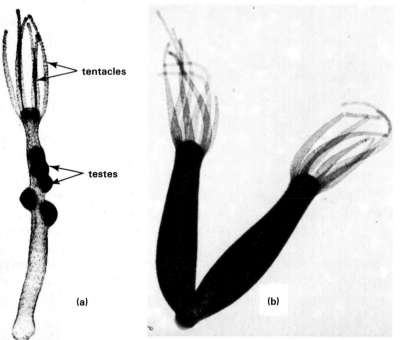

Obelia is a colonial organism made of very many hydroids joined together by a system of branching tubes called the **coenosarc**. The colony is surrounded by a chitinous exoskeleton called the **perisarc**. At the ends of the colony's branches, hydroids, called **hydranths** or **polyps**, project into the surrounding water. Their main function is food capture (Fig 29.7).

Fig 29.7 *Obelia*: (a) colony, × 25, (b) medusa, × 15

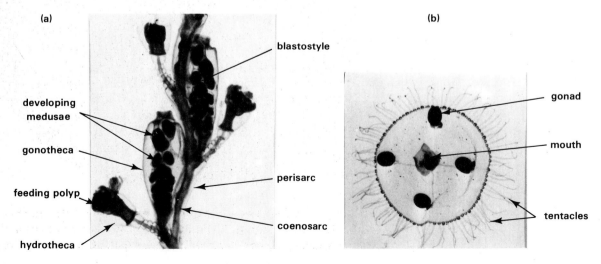

Asexual reproduction in *Obelia* involves the budding of medusae from reproductive polyps called **blastostyles**. Medusae are small jellyfish-like structures which break free from the parent colony. Each has gonads, from which arise the gametes used in sexual reproduction. The product of fertilisation is a **planula larva** which settles and grows into a new colony.

2 Anthozoa (Actinozoa)

These are **hydroids** which do not have a medusoid stage in the life cycle (Fig 29.8). **Sea anemones** and **corals** are anthozoans. The sea anemone *Actinia* lives on the British shoreline between high and low tide marks. Its mouth is ringed by numerous tentacles covered in nematocysts which are used for food capture. In the rim of the mouth are ciliated grooves.

Fig 29.8 (a) Sea anemone (Biofotos), (b) LS body plan, (c) life cycle

(a)

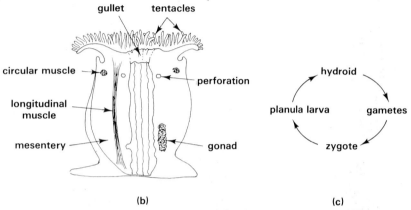

(b)

(c)

Rhythmic beating of the cilia creates movement of water which enters and leaves the enteron through the mouth. The endoderm grows into the enteron to form folds called **mesenteries** (Fig 29.9). They provide a large internal surface area for the absorption of food from the enteron. *Actinia* retracts its whole body when longitudinal muscle fibres in the mesenteries contract. When this happens, only the ectoderm is exposed to the drying air between tides.

Fig 29.9 *Actinia*, TS through gullet, ×6

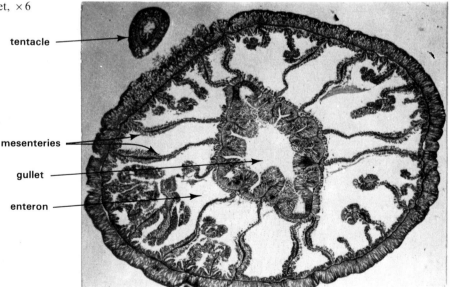

The gonads are found on the mesenteries of separate male and female animals. Gametes are released to the outside and fertilisation occurs in the sea, or occasionally in the enteron of females. A planula larva arises from

587

Fig 29.10 Underwater view of corals (Bruce Coleman: Bill Wood)

the zygote and is the **dispersal phase** of the life cycle. When the larva settles, perhaps miles away from its parents, a new hydroid develops, providing the surroundings are suitable.

Corals are colonial organisms which secrete a hard supportive and protective skeleton of limestone (Fig 29.10). Corals living in the same area often form a **reef**. A reef is built up by new corals growing on layers of old skeletons left by past colonies. The Great Barrier Reef of Australia stretches for more than 2000 kilometres (Fig 29.11). An **atoll** is a circular coral reef which developed around the coast of an island that has since sunk beneath the surface of the sea.

Fig 29.11 Great Barrier Reef, Australia (Bruce Coleman: Fritz Prenzel)

3 Scyphozoa

These are **medusoid** cnidarians, the hydroid stage being absent from the life cycle. They include **jellyfish** such as *Aurelia* which swims in the surface waters of the sea around Britain (Fig 29.12). The bell-shaped medusa has a ring of small

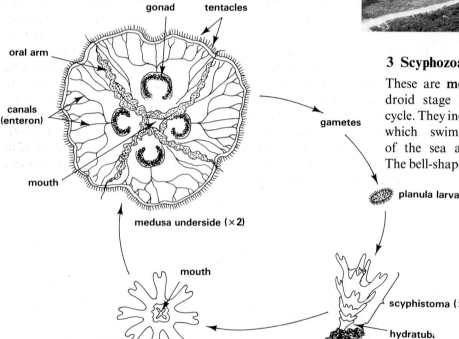

gonad tentacles

oral arm

canals (enteron)

mouth

medusa underside (×2)

gametes

planula larva

scyphistoma (×40)

hydratuba

mouth

young ephyra (×15)

(a)

Fig 29.12 The common jellyfish, *Aurelia*: (a) life cycle, (b) ephyra larva, ×10

The Great Barrier Reef is more than 2000 km long yet it is made up of microscopic animals. How are they organised?

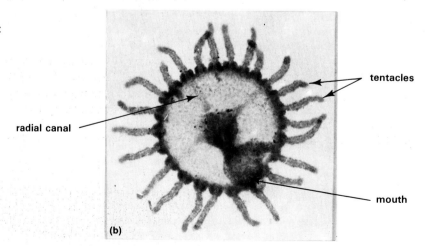

tentacles

radial canal

mouth

(b)

tentacles around the rim. The mouth is at the centre on the underside. Projecting from the mouth are four frills of tissue called **oral arms**, which are used in the capture and ingestion of prey. The mouth leads into a lobed enteron called the **gastric cavity**, which radiates as a system of narrow branched **canals** to the edge of the medusa. In the canals, food and oxygen are brought to all parts of the body. Much of the upper part of the medusa is mesogloea, hence the jelly-like consistency of jellyfish. The sexes are separate and the gonads develop in the gastric cavity. Ova are fertilised in the frills of the oral arms and planula larvae are formed. When the larvae settle, they develop into a small hydroid-like polyp from which small **ephyra larvae** are released by budding into the sea. Each ephyra is a small medusa and develops into an adult jellyfish.

29.2 Phylum Platyhelminthes

Flukes, tapeworms and turbellarians are **flatworms** belonging to the phylum Platyhelminthes. The flatworm body is **triploblastic**, originating from three tissue layers; an inner endoderm, separated from an outer ectoderm by a solid middle layer called the **mesoderm**. Flatworms are **acoelomate**, because there is no body cavity or coelom in the mesoderm. Reproductive, nervous and excretory organs develop in the mesoderm, so the triploblastic condition is sometimes called the **organ level of development**. This is a distinct advance on the body form of cnidarians. Flatworms are **bilaterally symmetrical**, whereas poriferans and cnidarians are radially symmetrical (Fig 29.13). There is a distinct head, often with sense organs and a primitive internal brain. The development of complex structures in the head is called **cephalisation**. Flatworms have no blood system. An extensively branched gut is usually present and delivers food to all parts of the body. As in cnidarians, there is a single gut opening through which food enters and wastes leave the body.

Fig 29.13 Planes of symmetry (a) radial, (b) bilateral

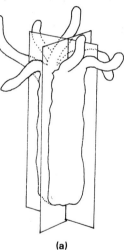

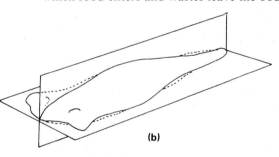

(a)

(b)

The *Platyhelminthes* is divided into three classes: Turbellaria, Trematoda and Cestoda.

29.2.1 Turbellaria

Turbellarians are mainly free-living, aquatic flatworms such as *Dendrocoelum* and *Planaria* (Fig 29.14). They move by an undulating action brought about by antagonistic longitudinal and circular muscles beneath the ectoderm. The ectoderm cells on the underside of the body are ciliated.

Fig 29.14 *Dendrocoelum*, ×7

eye   branched gut   pharynx

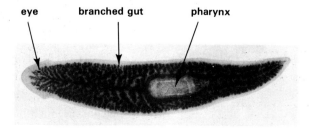

Turbellarians glide over solid surfaces when the cilia beat. The mouth is at the end of a muscular, tubular **pharynx**, which is extended from the ventral surface when feeding (Fig 29.15(a)). *Planaria* feeds on small animals and secretes a digestive juice on to the prey through the extended mouth. A pumping action of the pharynx then sucks small pieces of partly digested food into the gut. Secretory cells lining the gut produce enzymes which complete the digestion of the food. Metabolic wastes are removed by a system of **excretory canals** fed by **flame cells**. The cells have flagella which create a current in the excretory system. The excretory canals open on to the body surface by several pores, through which the wastes are discharged to the outside (Fig 29.15(b) and (c)).

Fig 29.15 Body systems in *Planaria*: (a) digestive, (b) excretory, (c) flame cell, (d) nervous, (e) reproductive

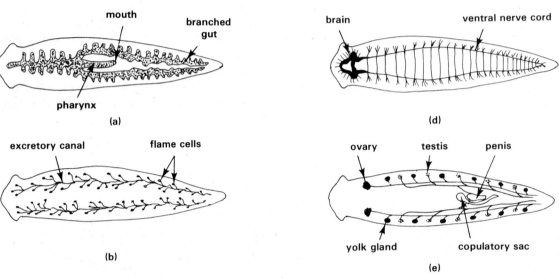

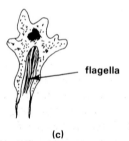

Turbellarians have **eyespots**, usually on the dorsal surface of the head. As far as can be judged, turbellarians cannot perceive clear images. However, the eyespots can distinguish between light and dark and may be able to detect movement. On the leading edge of the head there are groups of **chemoreceptors**. Turbellarians use them to detect the presence and location of food.

The nervous system is more organised than the simple nerve net of cnidarians. There are two longitudinal bundles of nerve fibres making a primitive **nerve cord**, which is positioned ventrally in the mesoderm. In the head, the double nerve cord joins to form a concentration of nervous tissue which acts as a simple **brain** (Fig 29.31(d)).

Although turbellarians are **hermaphrodite**, they engage in copulation which ensures **cross-fertilisation**. There is no larval stage in the life cycle. Some turbellarians reproduce asexually by splitting into two halves, anterior and posterior. Each half then grows into a new adult. Many also show remarkable powers of **regeneration**. If a turbellarian is cut into many pieces, a new individual may grow from each of the severed parts.

29.2.2 Trematoda

The trematodes are the **flukes** such as *Echinostoma*, *Fasciola* and *Schistosoma* (Fig 29.16). Many of the body features of flukes result from their parasitic way of life. Most are **endoparasites** living inside their hosts. There are no external cilia in the adult. A protective **cuticle** covers the outer surface of the body. There are usually one or more **suckers** by which the fluke clings to its host. The mouth leads into a branched gut. Flame cells

Fig 29.16 *Echinostoma revolutum*, × 10

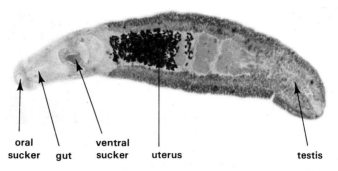

oral sucker gut ventral sucker uterus testis

linked to a system of excretory tubes remove metabolic wastes. There is a well developed nervous system, but no sense organs.

A significant feature of flukes is the well developed reproductive system. Individuals are usually **hermaphrodite**. The life cycle is often complex and may involve more than one host.

Fasciola hepatica is the **liver fluke** which lives in the bile ducts of its primary hosts, sheep and cattle. It has a pair of testes and a single ovary (Fig 29.17). Although the fluke is hermaphrodite, **cross-fertilisation** takes

Fig 29.17 (a) *Fasciola hepatica* (Biofotos), (b) *Fasciola*, reproductive system

(a)

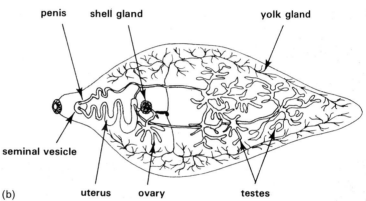

penis   shell gland   yolk gland

seminal vesicle

(b)   uterus   ovary   testes

place. The yolky fertilised eggs have a protective shell and are egested in the faeces of the primary host. In water, each egg develops into a ciliated **miracidium larva**. Miracidia swim by ciliary action and enter the secondary host, usually the freshwater snail *Limnaea truncatula*. The miracidia change into **sporocysts**, from which arise another larval form called **rediae**. Each redia changes into yet another larval form called a **cercaria**, which passes into the pulmonary cavity of the snail and from

Fig 29.18 Life cycle of *Fasciola hepatica*

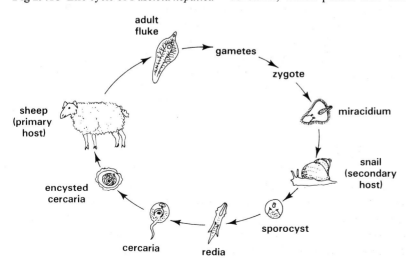

adult fluke
gametes
zygote
miracidium
sheep (primary host)
snail (secondary host)
sporocyst
encysted cercaria
cercaria
redia

there to the outside. The cercariae wriggle on to vegetation and become encased in a cyst. Here they may lie dormant for many months before being eaten by the primary host. In the host's gut the cysts hatch into small flukes which migrate into the body cavity. They find their way to the liver, where they cause considerable damage. The flukes eventually come to lie in the bile ducts, grow to maturity and the life cycle is completed (Fig 29.18).

Schistosoma lives in the abdominal veins of humans. It is called the **blood fluke**. The sexes are separate, the female lying permanently in a groove along the surface of the male's body (Fig 29.19). The life cycle involves several larval stages as in *Fasciola* and the secondary host is also a water snail.

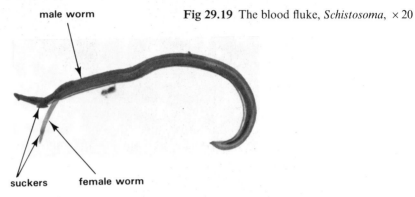

Fig 29.19 The blood fluke, *Schistosoma*, × 20

29.2.3 Cestoda

The cestodes are **tapeworms** such as *Taenia*. All tapeworms are **endoparasites**. Many of their body features reflect the parasitic way of life. There are no sense organs or gut, and the excretory and nervous systems are degenerate. The body is covered by a tough cuticle which protects it from the host's digestive juices. Nearly all the space inside tapeworms is occupied by reproductive organs.

Tapeworms live in the gut of their host, often growing to great lengths. *Taenia solium*, the pork tapeworm of humans, for example, may become over 4 metres long. The head is called the **scolex** and has a ring of **hooks** and **suckers** by which it clings to the inside of the host's intestine (Fig 29.20). Behind the head the tapeworm produces a great number of **proglottides**. As the proglottides mature, they move further away from the scolex and develop reproductive organs. Eggs are fertilised by sperms from the same proglottis. The fertilised eggs are then provided with yolk and a thin shell. Each egg develops into a small six-hooked embryo encased in a resistant **onchosphere**. Ripe proglottides containing onchospheres

Why are tapeworm infestations in humans rare in developed countries?

Fig 29.20 *Taenia*: (a) scolex, × 15, (b) proglottis showing reproductive organs, × 25, (c) gravid proglottis just before shedding, × 20, (d) bladderworm, × 10

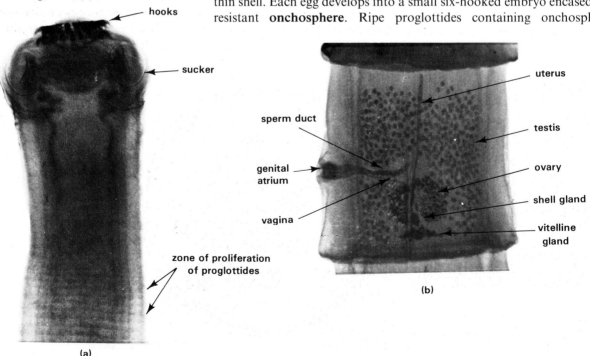

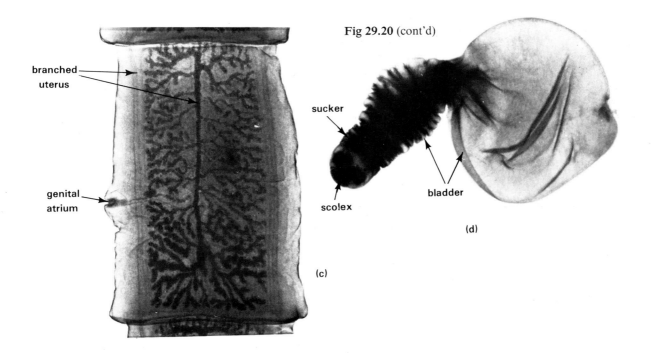

branched
uterus

genital
atrium

(c)

Fig 29.20 (cont'd)

sucker

scolex

bladder

(d)

become detached from the tapeworm and pass to the outside in the host's faeces. If the onchospheres are eaten by a suitable secondary host, in this case a pig, further development takes place. The embryo is released from its case in the host's stomach and bores its way into a blood vessel. Eventually the embryo becomes encysted, usually in the host's muscles, where it develops into a fluid-filled **cysticercus** or **bladderworm**. Should pork infected with bladderworms be eaten without sufficient cooking to kill them, the worm attaches itself to the gut wall and develops into a new adult tapeworm. Pork tapeworms are rare in developed countries where meat is inspected for parasites before it can be sold for human consumption.

29.3 Phylum Nematoda

Fig 29.21 (a) *Ascaris lumbricoides*, entire male (smaller) and female (Trustees of the Wellcome Trust)

The nematodes are **roundworms** such as *Ascaris lumbricoides* which lives in the small intestine of humans (Fig 29.21). Many nematodes are parasites. Most kinds of multicellular plants and animals act as hosts.

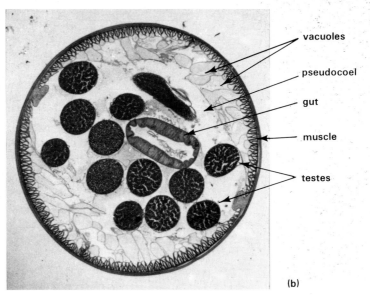

vacuoles

pseudocoel

gut

muscle

testes

Fig 29.21 (b) *Ascaris lumbricoides* TS ×30

(b)

Nematodes are all thread-like animals almost devoid of external features. The body is covered by a cuticle and there are no cilia. The **gut** has two openings, the **mouth** and **anus**. This is an advance on the single opening to the enteron seen in cnidarians and flatworms, because food moves through the gut in one direction, making feeding, digestion and absorption more efficient. Longitudinal muscles beneath the ectoderm enable movement of the body. The absence of circular muscles, however, permits only a back-and-forth thrashing motion, so movement from place to place is relatively inefficient.

The many cells making up the ectoderm have no plasma membranes. For this reason it is a **syncitium**. When viewed microscopically, the ectoderm appears as a mass of separate nuclei. Between the muscles and gut there is a fluid-filled cavity formed from the vacuoles of giant cells in the mesoderm. Consequently, the cavity is not a coelom but is a **pseudocoel.** The excretory system consists of canals which open to the outside by an excretory pore. There are no flame cells. Sensory cells, probably chemoreceptors, are found around the mouth. They may be used to locate food. Nearly all nematodes are **dioecious**. The reproductive organs are found in the pseudocoel.

29.4 Phylum Annelida

The phylum Annelida includes the bristleworms, fanworms, earthworms and leeches. Annelids are **triploblastic**. However, unlike the flatworms and nematodes, the mesoderm is split into two parts by a fluid-filled cavity called the **coelom** (Fig 29.22). It is though that the coelom may have evolved primarily as a space for storing gametes. Another suggestion is that it evolved to store excretory products prior to their elimination from the body. In living annelids, the coelom performs both these functions and is an advance on the acoelomate condition. The body of annelids is divided internally by partitions called **septa** into many co-ordinated compartments. This kind of structural organisation is known as **metameric segmentation**. Annelids have a tubular gut with a mouth and anus at opposite ends of the body. The excretory organs are coiled tubes called **nephridia** arranged one pair per segment. They remove wastes from the coelomic fluid through a ciliated funnel called the nephrostome. The wastes are discharged to the outside through small openings called nephridiopores.

What is a coelom? What main functions are performed by the coelom?

Fig 29.22 Three grades of organisation: (a) diploblastic, (b) triploblastic acoelomate, (c) triploblastic coelomate

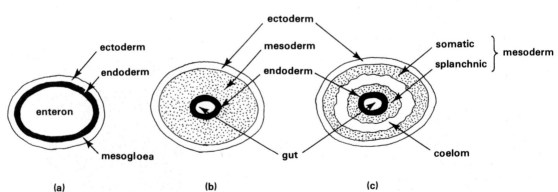

(a) (b) (c)

Annelids show advanced cephalisation. There is a primitive brain which consists of a pair of **dorsal nerve ganglia** at the anterior end.

The circulatory system consists of closed vessels containing blood. Blood moves through the vessels by the pumping action of some of the vessels. Annelid blood often contains respiratory pigments such as **haemoglobin**

(a)

Fig 29.23 *Nereis*: (a) whole animal, × 1 (Biofotos), (b) TS, × 20, (c) head, pharynx extended, (d) VS eye, (e) trochophore larva, (f) parapodium, × 45

and **chlorocruorin** which enable the transport of relatively large volumes of oxygen. Gills are found in some annelids.

Another annelid feature is the presence of **chaetae**, bristle-like structures composed of **chitin**, which project from the body surface. Chaetae are often used in locomotion, which is achieved by the antagonistic action of circular and longitudinal muscles.

The annelids are divided into several classes, including the Polychaeta, Oligochaeta and Hirudinea.

29.4.1 Polychaeta

The polychaetes are so called because they have many chaetae. They include the **bristleworms** and **fanworms**. All are aquatic, mostly marine.

The **ragworm**, *Nereis diversicolor* (Fig 29.23) lives in U-shaped burrows in the mud along the coast of Britain, often in estuaries. The chaetae project from flaps called **parapodia** on the sides of the body (Fig 29.23(b) and (f)). Ragworms use the parapodia to set up water currents in their burrows. The currents bring food and oxygen and eliminate wastes from the burrow. Projecting from the head are sensory **tentacles** and **palps**. They are thought to detect food, which is grabbed by the jagged jaws on the protrusible pharynx (Fig 29.23(d)).

Nereis has well-developed **eyes**. These consist of a cup of photosensitive cells beneath a gelatinous lens. It is doubtful that they can perceive images, but *Nereis* certainly detects movements in its surroundings. Respiratory

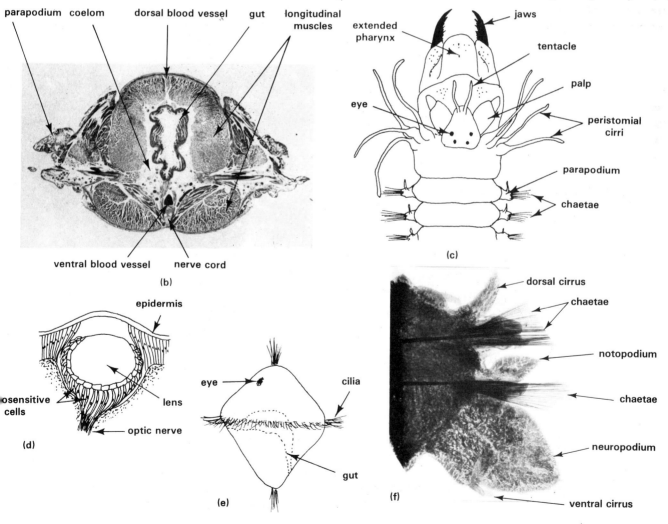

Fig 29.24 Two polychaetes:

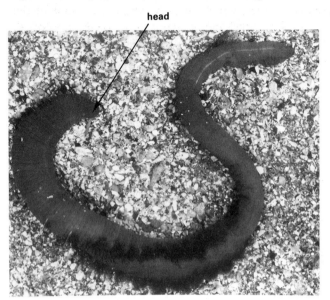

(a) the lugworm, *Arenicola*, × ¾
(Biofotos)

(b) a fanworm, *Sabella*, × ⅓
(Bruce Coleman, Jane Burton)

gas exchange occurs through the body surface. In some polychaetes such as the **lugworm** *Arenicola*, gills are present (Fig 29.24(a)).

Polychaetes are nearly all **dioecious**, and gonads develop in most segments. Very often segments containing ripe gametes break away from the parent worm and fertilisation is external. Ripening and release of gametes from males and females are co-ordinated so that sperms and eggs mix together in the sea. Swarming is seen in some polychaete worms prior to fertilisation. A famous example of swarming is shown by *Leodice viridis*, the Pacific palolo worm. It takes place exactly at dawn on the day of the last quarter of the October–November moon. The rear halves of the worms project from their burrows and break away. They rise to the surface and writhe in large numbers, eventually releasing their gametes for fertilisation to occur. A **trochophore** larva is typical of polychaetes (Fig 29.23(e)).

Fanworms, such as *Sabella*, live in tubes which the worms secrete. Extensive tentacles on the head project into the sea from the mouth of the tube. The action of cilia on the tentacles creates water currents which bring small food particles into the fanworm's mouth (Fig 29.24(b)).

29.4.2 Oligochaeta

The best known of the oligochaetes are the **earthworms** (Fig 29.25(a)). Oligochaetes have few chaetae. In *Lumbricus terrestris*, four pairs of chaetae project from each segment. There are no parapodia. Earthworms are terrestrial, commonly found burrowing in soil. The chaetae are used to grip the sides of the burrow while the action of longitudinal and circular muscles pushes the worm through the soil. During movement, the coelomic fluid acts as a **hydroskeleton**. It is squeezed by the muscles, causing the body to become shortened or lengthened (Fig 29.25(b)).

Excretion is by means of paired **nephridia** (Fig 29.25(c)). There is evidence that substances are selectively reabsorbed into the coelom from the nephridia in much the same way that reabsorption occurs in the nephrons of vertebrate kidneys (Chapter 11). In this way, nephridia help to regulate the chemical composition of the coelomic fluid.

Earthworms ingest soil and detritus which they take into their burrows. Food enters the mouth by the sucking of the muscular pharynx.

Earthworms have no legs. How do they burrow through soil?

596

Blood flows the length of the body towards the head in a contractile dorsal vessel and towards the rear in a ventral vessel. Circular vessels connect the dorsal and ventral vessels in each segment. In some of the anterior segments, thicker-walled, contractile circular vessels are found. They are called **pseudohearts** and help keep the blood circulating.

The reproductive organs are restricted to a few segments in the anterior part of the body (Fig 29.25(e)). Although hermaphrodite, earthworms copulate in pairs so cross-fertilisation takes place. When mating, two individuals come together with their ventral surfaces touching, their bodies pointing in opposite directions. Segments 9 to 11 of one worm lie opposite the **clitellum** of the other (Fig 29.25(f)). The clitellum of each worm secretes **mucus** which ensheaths the two worms in the region of the sex organs. Sperm are ejected from the **sperm ducts** which open on to the

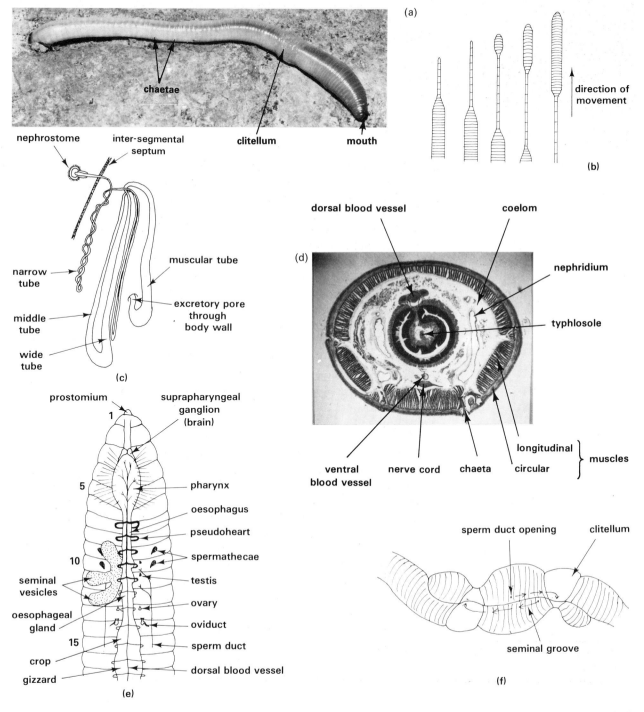

Fig 29.25 Earthworm: (a) whole animal, $\times \frac{2}{3}$, (b) movement involving elongation of anterior segments, anchorage by chaetae to walls of soil burrow and finally drawing posterior part of body up to the front, (c) nephridium, (d) TS intestinal region, $\times 10$, (e) dissected anterior segments displaying reproductive organs; dorsal view with seminal vesicles on the right removed, (f) mating and sperm transfer

597

surface of segment 15. The sperm are then transported forwards in **seminal grooves** on the ventral surface and enter the **spermathecae** on segments 9 and 10 of the mating partner. After sperm have been exchanged, the worms separate. Later, the clitellum secretes a band of mucus which glides forward over the worm. As it passes segment 14, the **oviducts** deposit several ripe eggs into the mucus. Sperm stored from a previous mating are then added to the eggs as the mucus passes the spermathecae. Fertilisation takes place and the band of mucus slips off the worm's head to form a protective **cocoon**. After a period of development, young worms escape from the cocoon. There is no larval stage in the life cycle of oligochaetes.

29.4.3 Hirudinea

Leeches are the best known hirudineans (Fig 29.26). Many leeches are ectoparasites, sucking blood from their hosts. Others are free-living carnivores, preying on small animals. Leeches are dorso-ventrally flattened and the coelom is filled with connective tissue. There are usually two **suckers**, one at each end of the body. They are used to cling to the host, to stones and other solid structures. Leeches have no chaetae. The numerous annulations on the outer surface do not correspond to the 33 internal segments.

Inside the mouth is a group of tooth-like jaws. Parasitic leeches use them to puncture the host's skin before sucking blood. Blood-sucking leeches secrete in their saliva an anticoagulant called **hirudin** which prevents the blood from clotting as they feed.

Although many leeches are parasitic, they have simple sense organs and a well-developed nervous system. Leeches are hermaphrodite, but reproduction involves cross-fertilisation and a cocoon is formed. As in the oligochaetes, there is no larval stage in the life cycle.

Fig 29.26 A leech, *Hirudo*, ventral view × 4 (Biofotos)

posterior
sucker

29.5 Phylum Arthropoda

The Arthropoda includes **crustaceans, spiders, centipedes, millipedes** and the most numerous of all animal groups, the **insects**. Arthropods are bilaterally symmetrical, metamerically segmented, triploblastic, coelomate animals. The body is divided basically into a **head, thorax** and **abdomen**. Projecting from some of the body segments are pairs of **jointed limbs** (arthron = joint: podos = foot).

The outermost layer of cells secretes a thick **cuticle**, which consists largely of chitin and makes up an **exoskeleton**. The cuticle of most terrestrial arthropods is often covered with a waterproof layer of **wax** which prevents excessive water loss. To allow a degree of movement in the body and limbs, the exoskeleton is jointed. Joints are regions where the cuticle is flexible. Internal muscles operate at a joint to cause movement of the limb or body segments (Fig 29.27). The limbs of arthropods are modified to perform a variety of functions including walking, swimming, jumping, feeding, reproduction and, where present, ventilation of the gills.

The exoskeleton is fixed in size and does not grow with the animal. This contrasts with our endoskeleton which increases in size as our bodies grow. When an arthropod grows to the limit imposed by its exoskeleton, the cuticle is moulted, a process called **ecydsis**. A new flexible cuticle is secreted which inflates and hardens to a larger size than the old one. The animal's soft tissues grow until the new cuticle needs replacing. In insects a hormone

Fig 29.27 Comparison of limb joints: (a) arthropod, (b) vertebrate

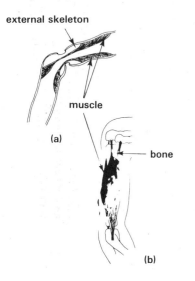

external skeleton

muscle

(a)

bone

(b)

called **ecdysone** regulates ecdysis, which may occur many times in the life cycle (Fig 29.28).

Fig 29.28 Growth curves of: (a) locust, an arthropod, (b) a vertebrate

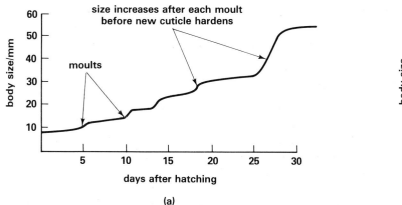

(a)

(b)

Arthropods are more cephalised than annelids. The head often bears complex mouthparts and sense organs. Internally, the brain is relatively well developed. Many arthropods have **compound eyes** consisting of very many photoreceptors called **ommatidia** (Fig 29.29). Since the ommatidia all point in a slightly different direction, each looks at a slightly different part of the environment. It is assumed that the animal perceives a mosaic image of its surroundings.

What are the advantages of an exoskeleton? What constraints does the exoskeleton place on arthropods?

Fig 29.29 (a) Photomicrograph of VS compound eye of a locust, × 15, (b) diagram of an ommatidium, (c) in dimly lit conditions the pigment in the pigment cells may recede enabling more light to reach the photosensitive rhabdome

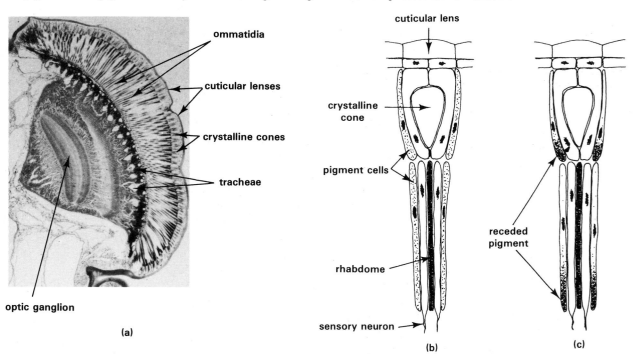

(a)

(b)

(c)

599

Arthropods have an open blood vascular system. Blood circulates in vessels which open into a large mesodermal cavity called the **haemocoel**. The coelom is reduced in size compared with annelids. There is usually a **dorsal tubular heart** which pumps blood forwards towards the head. Blood is collected from the haemocoel through pairs of holes called **ostia** in the sides of the **heart** (Fig 29.30).

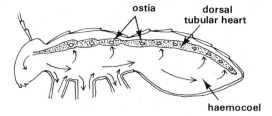

Fig 29.30 Arthropod heart. The arrows indication the direction of blood flow

Fig 29.31 Woodlouse *Porcello*, × 1½ (Biofotos)

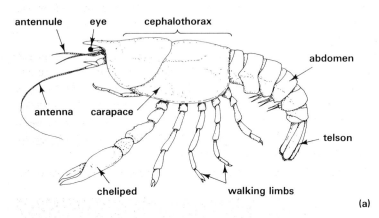

Fig 29.32 Crayfish, *Astacus*: (a) whole animal, (b) TS thorax

The arthropods are divided into several classes including the Crustacea, Myriapoda, Insecta and Arachnida.

29.5.1 Crustacea

Crustaceans include **waterfleas, barnacles, woodlice, crabs, shrimps, prawns, crayfish** and **lobsters**. **Woodlice** are the only truly terrestrial crustaceans (Fig 29.31), the rest being aquatic, mostly marine.

Crustaceans have a thick, often calcareous exoskeleton. The head is often fused with a number of thoracic segments to form a **cephalothorax**. Sometimes a large fold called the **carapace** develops from the posterior edge of the head and grows to cover much of the anterior part of the body (Fig 29.32). The carapace protects the anterior region of the body and often encloses gills (Fig 29.32(b)). In lobsters, crayfish and crabs (Fig 29.33) a current of oxygen-laden water is made to flow over the gills by the wafting

Bruce Coleman: Jane Burton

(a)

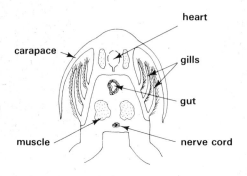

(b)

600

Fig 29.33 Edible crab, *Cancer pagurus*:

(a) adult, ×⅕
(Biofotos)

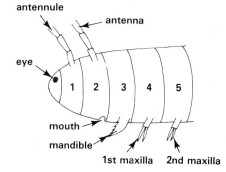

abdomen folded underneath

carapace

chela   eye

Fig 29.34 *Daphnia*, ×25

(b) zooea larva, ×50

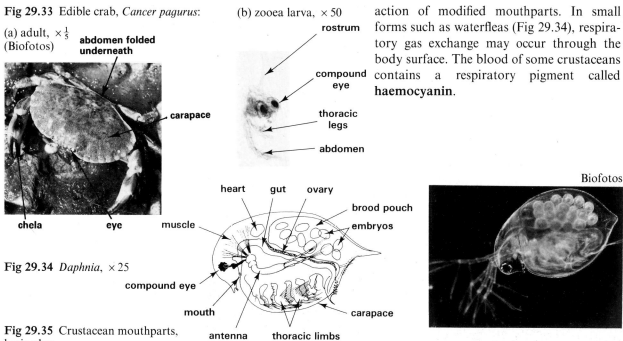

rostrum

compound eye

thoracic legs

abdomen

heart   gut   ovary

brood pouch

embryos

muscle

compound eye

mouth

antenna   thoracic limbs

carapace

Biofotos

action of modified mouthparts. In small forms such as waterfleas (Fig 29.34), respiratory gas exchange may occur through the body surface. The blood of some crustaceans contains a respiratory pigment called **haemocyanin**.

Fig 29.35 Crustacean mouthparts, basic plan

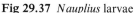

antennule

antenna

eye

1 2 3 4 5

mouth

mandible

1st maxilla   2nd maxilla

Crustaceans have five pairs of appendages on the head (Fig 29.35). There are two pairs of **antennae**. The first part are often small and are called **antennules**. The antennae are tactile sense organs, used to feel the surroundings or food before ingestion. A pair of **mandibles** is used to bite or chew food. Finally there are two pairs of maxillae sometimes called **maxillules** and **maxillae**, which may be used to manipulate food during feeding. In the shore crab *Carcinus*, as many as three pairs of legs are modified to form **maxillipeds** which help in feeding. Many crustaceans have compound eyes.

The basic body plan of some crustaceans is highly modified to suit the requirements of a specialised way of life. An example is the barnacle, *Balanus*, which lives stuck to rocks in the intertidal zone. The body is encased in calcareous plates which open on the top of the body. Food is captured by modified thoracic limbs covered in cirri, which project from the opening and filter small organisms from the sea and transfer them to the mouth (Fig 29.36).

Fig 29.36 Barnacle, *Balanus*, ×2. Note long penis of mating barnacles (Biofotos)

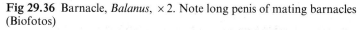

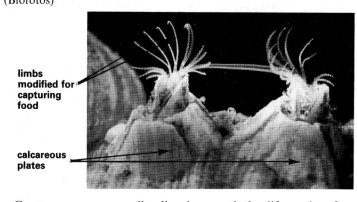

limbs modified for capturing food

calcareous plates

Crustaceans are usually dioecious and the life cycle often includes a larval stage called a **nauplius** (Fig 29.37).

Fig 29.37 *Nauplius* larvae

Fig 29.38 Centipede, *Lithobius*, ×1 (Biofotos)

29.5.2 Myriapoda

The myriapods include **centipedes** and **millipedes** (Fig 29.38). They have elongated, narrow bodies in which the thorax and abdomen are difficult to distinguish externally. The head bears antennae, mandibles and maxillae. There are many pairs of walking legs, one or two pairs per segment (centi = hundred; milli = thousand; podos = foot). Gas exchange occurs through holes called **spiracles** on the side of the body. As in insects the spiracles lead into a system of branched air-tubes called **tracheae**. Another insect-like feature of the myriapods is the presence of **Malpighian tubules** which are used for excretion.

Myriapods are terrestrial animals. The sexes are usually separate and there is no larval stage in the life cycle.

29.5.3 Insecta

The class Insecta includes **ants, aphids, bees, beetles, butterflies, house flies, locusts, mosquitoes** and **moths**. Insects occupy most habitats, however, there are no truly marine insects. There are at least twice as many different kinds of insects as all the other animals put together.

The body of an adult insect is clearly divisible into a **head, thorax** and **abdomen**. A pair of antennae and compound eyes are usually present on the head. The thorax bears three pairs of jointed limbs (Fig 29.39).

Many insects have **wings** and are good fliers. The wings are flat, dorsal extensions of the second and third thoracic segments. They contain air-filled tubes (tracheae) which give strength and rigidity. Gas exchange is by **gills** in aquatic insects and **tracheae** in terrestrial forms. Tracheae are air-filled tubes that open to the air by paired **spiracles** on the sides of each segment. Resting insects rely on diffusion to take in oxygen and to get rid of carbon dioxide. Abdominal movements ventilate the tracheae during activity. The tracheae branch to supply all the major internal organs of the body (Fig 29.40). Gas exchange occurs in the terminal tracheal branches called **tracheoles**. The role of the blood as an oxygen carrier is less important than in non-tracheate arthropods.

Fig 29.39 Insect body, basic plan (from Buchsbaum, after Snodgrass)

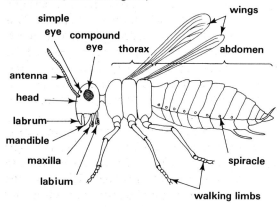

Fig 29.40 (a) Human body louse, *Pediculus humanus corporis* showing tracheal system, ×12, (b) branched tracheae dissected from an insect, ×40

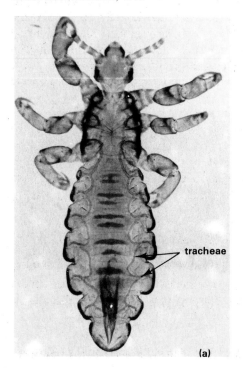

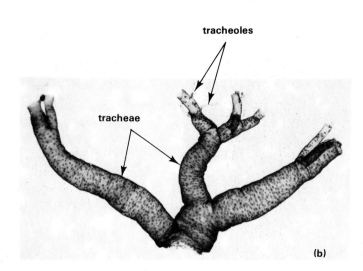

Fig 29.41 Insect gut

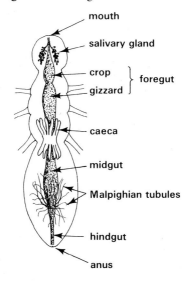

- mouth
- salivary gland
- crop ⎫
- gizzard ⎬ foregut
- caeca
- midgut
- Malpighian tubules
- hindgut
- anus

In what ways are insects particularly well adapted to life on land?

Fig 29.42 Ground beetle, *Cychrus*, × 1½ (Ardea: John Mason)

elytra

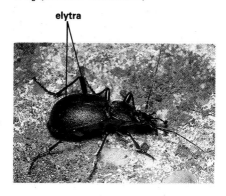

Fig 29.43 The bed bug, *Cimex*, × 10

The gut of insects is divisible into three main parts: fore-, mid- and hindgut (Fig 29.41). At the junction of the foregut and midgut is a ring of finger-like projections called **caeca** which secrete digestive juices. Digestion and absorption take place in the midgut. At the junction of the midgut and hindgut is a mass of narrow excretory tubes called **Malpighian tubules**. They collect metabolic wastes from the haemocoel and discharge them into the gut. Nitrogenous waste in the form of potassium urate is absorbed from the blood by the tubule cells. In the tubules, potassium urate, water and carbon dioxide react to form potassium hydrogencarbonate and **uric acid**. The potassium hydrogencarbonate is reabsorbed into the blood and the uric acid is discharged into the gut. Uric acid is almost insoluble in water which is absorbed into the blood from the Malpighian tubules and the hindgut. The faeces and nitrogenous excretory products are relatively dry. Conservation of water in this way helps insects resist desiccation and is one reason why they are able to live in very dry habitats.

Insects undergo metamorphosis in their life cycles. **Complete metamorphosis** is shown by some species such as the housefly, bee and butterfly (Fig 29.45). Fertilised eggs hatch into **larvae**, which give rise to **pupae**. In a pupa, the larval tissues are metamorphosed into an **imago**, the adult insect. Larval and pupal stages are missing the life cycles of insects such as cockroaches and locusts, which display **incomplete metamorphosis**. Their fertilised eggs hatch into **nymphs**, which are like the adult insect but smaller and sexually immature. After several moults, the nymphs grow into imagos.

Moulting and metamorphosis are regulated by hormones. The **moulting hormone, ecdysone**, is secreted by the **ventral gland** in the head of some insects and by the **prothoracic gland** in others. The gland is stimulated to do so by neurosecretions from the **cerebral ganglion**. Release of the neurosecretions occurs in some species when the abdomen is stretched after feeding. In others it is the act of swallowing. When the imago stage is reached the ecdysone-producing gland degenerates and moulting stops. The role of hormones in incomplete metamorphosis has been carefully worked out in the blood-sucking bug *Rhodnius*. In the first four larval stages **juvenile hormone, neotenin**, is secreted by the **corpus allatum** just behind the cerebral ganglion. Neotenin interacts with ecdysone keeping the soft, juvenile cuticle between moults. When the fifth larval stage is reached secretion of neotenin ceases and the hard, adult cuticle develops. In species which have complete metamorphosis the pupal stage develops when the output of neotenin begins to fall. When it stops completely the imago stage develops.

The class Insecta is divided into many orders, including the following six:

1 Coleoptera

Beetles are in this order (Fig 29.42). The forewings are two horny shields called **elytra** which cover the hindwings. In flight, the elytra project sideways. The mouthparts are modified for chewing solid food.

2 Hemiptera

This includes aphids such as greenfly and bedbugs such as *Cimex* (Fig 29.43). In aphids the mouthparts are adapted for piercing plant shoots and sucking sap from the phloem. Bedbugs suck the blood of mammals.

3 Diptera

Flies and mosquitos are dipterans. The hindwings are a pair of **halteres** which act as stabilisers during flight. The mouthparts are modified for sucking and sometimes for piercing and biting (Fig 29.44).

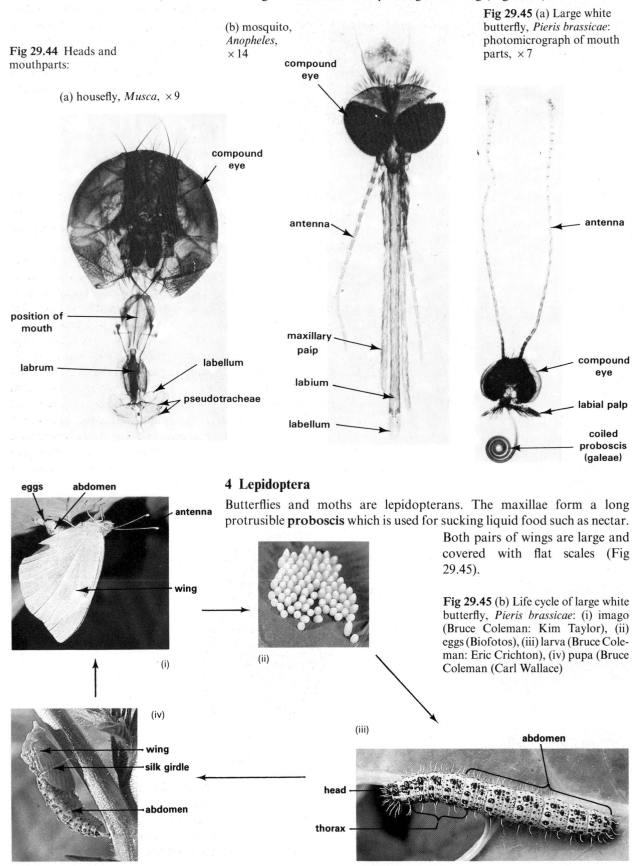

Fig 29.44 Heads and mouthparts:

(a) housefly, *Musca*, ×9

compound eye

position of mouth

labrum

labellum

pseudotracheae

(b) mosquito, *Anopheles*, ×14

compound eye

antenna

maxillary paip

labium

labellum

Fig 29.45 (a) Large white butterfly, *Pieris brassicae*: photomicrograph of mouth parts, ×7

antenna

compound eye

labial palp

coiled proboscis (galeae)

4 Lepidoptera

Butterflies and moths are lepidopterans. The maxillae form a long protrusible **proboscis** which is used for sucking liquid food such as nectar. Both pairs of wings are large and covered with flat scales (Fig 29.45).

Fig 29.45 (b) Life cycle of large white butterfly, *Pieris brassicae*: (i) imago (Bruce Coleman: Kim Taylor), (ii) eggs (Biofotos), (iii) larva (Bruce Coleman: Eric Crichton), (iv) pupa (Bruce Coleman (Carl Wallace)

eggs abdomen

antenna

wing

(i)

(ii)

(iv)

wing

silk girdle

abdomen

(iii)

abdomen

head

thorax

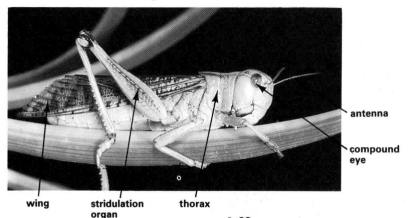

5 Orthoptera

This includes locusts, grasshoppers and crickets. The mouthparts are modified for biting. The third pair of legs are elongated for jumping. On them are ridged **stridulation organs** which are rubbed to produce a chirping sound. It attracts the opposite sex prior to mating (Fig 29.46).

Fig 29.46 Locust, × 1 (Bruce Coleman: John Markham)

Fig 29.47 Honey bees: (a) queen, surrounded by workers (Biophoto Associates), (b) drone (Bruce Coleman: H J Flügel), (c) worker (Bruce Coleman: H J Flügel) (d) hive cells (Biofotos), (e) worker mouthparts, × 9

6 Hymenoptera

Ants, bees and wasps are in this order. They have biting or sucking mouthparts. The two pairs of wings are linked by a row of hooks on the leading edge of the hindwings which engage a groove on the trailing edge of the forewings. The first abdominal segment is fused with the thorax.

Many hymenopterans live in **social colonies** sometimes containing many thousands of individuals. The colony contains several forms or **castes**. In the honey bee, for example, the colony is centred on a single large female called the **queen** (Fig 29.47). Her eggs are fertilised by males called **drones**. Fertilising the queen's eggs is their only function. The queen lays eggs one at a time in hexagonal wax cells in the **hive**. Any unfertilised eggs give rise to drones. Fertilised eggs produce either new queens or sterile females called **workers**. The workers feed the queen and build the wax cells which house the colony's larvae and pupae. Queen larvae are fed a special material called **royal jelly**. Other larvae are fed a mixture of pollen and honey. After moulting, the pupae develop inside a silk **cocoon** from which the adult emerges after a few days. The queen kills new queen larvae with her sting. Sometimes the workers prevent her doing this and a new queen grows to maturity. The old queen may leave the hive with a swarm of workers and establish a new colony elsewhere.

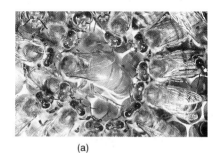

(a)

(b)

(c)

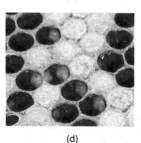

(d)

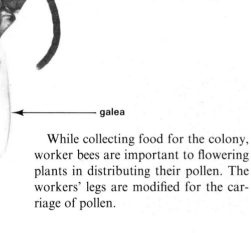

(e)

While collecting food for the colony, worker bees are important to flowering plants in distributing their pollen. The workers' legs are modified for the carriage of pollen.

605

Fig 29.48 Orb web spider, *Argiope*, ×4
(Bruce Coleman: F. Sauer)

walking legs

opisthosoma   prosoma

Fig 29.49 A tick, *Dermacentor*, ×7

29.5.4 Arachnida

The arachnids include **spiders, scorpions, mites** and **ticks**. Most are terrestrial. The body is divided into two parts. The first six segments make up the **prosoma** which bears the head and four pairs of walking legs. The posterior part of the body is called the **opisthosoma** and usually has no appendages (Fig 29.48).

Gas exchange in many arachnids involves the use of **lung books**. These are air-filled chambers containing many flaps of blood-filled tissue called lamellae. The lung books are rather like enclosed gills. In spiders there is also a tracheal system, but it is not as extensive as in insects. Consequently the blood of arachnids is important in the transport of oxygen in the body.

There are no antennae nor mandibles in arachnids. Instead, the mouthparts include paired **chelicerae** and **pedipalps**. The chelicerae each have a sharp claw which may be used to pierce prey. A paralysing poison flows through tubes inside the chelicerae and into the wound. The pedipalps are sensory appendages which may also be used by males to transfer sperms to the female during mating. **Malpighian tubules** are used to excrete nitrogenous wastes. There are no compound eyes, but simple eyes are often present. In spiders, a **web** is produced by three pairs of **spinnerets**, found on the posterior tip of the opisthosoma. Using a web to catch prey is unique to spiders. The sexes are usually separate and the life cycle has no larval stage.

Ticks and mites are ectoparasites found on the skin, usually feeding on the blood of their host. Their bodies are rounded, and the division between prosoma and opisthosoma is lost. They have no spinnerets (Fig 29.49). Scorpions are ferocious predators which attack their prey using the sting at the tip of their arched tails.

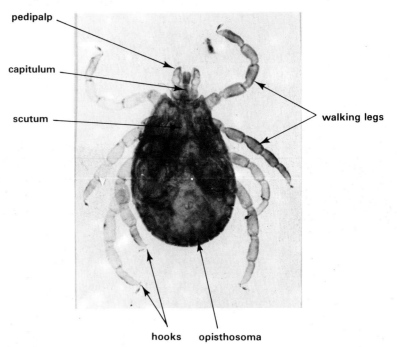

pedipalp

capitulum

scutum

walking legs

hooks   opisthosoma

29.6 Phylum Mollusca

The phylum Mollusca includes **snails, slugs, whelks, winkles, limpets, mussels, octopuses, cuttlefish** and **squids**. Molluscs are coelomate, but the coelom is much reduced. The heart consists of a single ventricle and two lateral atria. Like arthropods, the blood flows in a **haemocoel**.

Fig 29.50 Radula of *Helix*, × 50

Molluscs are not segmented. The body is divided into three parts, the **head, foot** and **visceral hump**. The visceral hump is covered by a soft skin called the **mantle** which usually secretes a hard **calcareous shell**.

Most molluscs are aquatic and exchange respiratory gases by means of gills called **ctenidia** which are found in the **mantle cavity** beneath the shell. In terrestrial forms such as snails, the mantle cavity is air-filled and acts as a simple lung. It has a vascular roof and opens to the air by a small pore called the **pneumostome**. The blood of molluscs contains the oxygen-carrying pigment **haemocyanin**.

In the mouth of many molluscs is a structure called the **radula** which has many rows of small teeth. The radula is used for scraping food off solid surfaces, such as algae from rocks, and for rasping vegetation (Fig 29.50).

The central nervous system consists of cerebral ganglia and a circum-oesophageal ring. Nerve fibres called pedal cords supply the foot; visceral loops pass to the visceral hump.

Molluscs are hermaphrodite, and in marine forms the life cycle includes a **trochophore** larva similar to that seen in polychaetes. It changes into a **veliger larva** from which the adult develops (Fig 29.68).

Molluscs are divided into several classes including the Gastropoda, Bivalvia and Cephalopoda.

Fig 29.51 Veliger larva of a gastropod mollusc

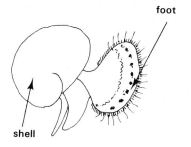

29.6.1 Gastropoda

Snails, slugs, limpets, whelks and winkles are gastropods. They have a distinct head, usually with eyes and prominent **tentacles**. The muscular foot is flattened and brings about a sliding form of movement. In most gastropods the visceral hump is twisted and the shell is coiled to accommodate the twisted body. The visceral hump also displays **torsion**; in addition to being twisted it is turned 180° on the foot (Fig 29.52).

Fig 29.52 (a) Common garden snail, *Helix aspersa*, × $\frac{2}{3}$ (Biofotos), (b) torsion

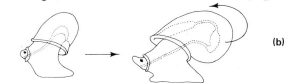

29.6.2 Phylum Bivalvia

This class includes mussels, razorshells, clams and scallops. The body is laterally flattened and **bilaterally symmetrical**. The shell develops as two **valves** which are hinged along one edge. The head is much reduced. Eyes, tentacles and radula are absent. The foot may be extended and used for burrowing in sand or mud (Fig 29.53). Other bivalves such as the edible

Fig 29.53 Razor shell: (a) burrowing in sand, (b) foot extended by increased blood pressure, (c) distal foot expands like an anchor, (d) muscle contraction draws shell down.

direction of movement

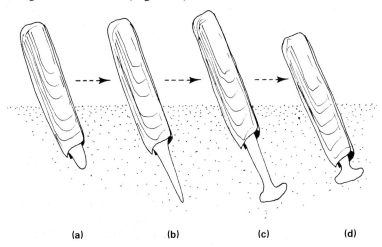

(a)　　　　(b)　　　　(c)　　　　(d)

mussel, *Mytilus edulis*, live attached to rock surfaces by **byssus threads** (Fig 29.54). The mantle cavity contains large gills which have many cilia arranged in tracts on their surfaces. The beating of the cilia causes oxygen and food-carrying water to flow over the gills. **Palps** near the mouth filter food particles out of the current of water.

Fig 29.54 (a) The common mussel, *Mytilus edulis*, ×1 (Biofotos), (b) feeding currents in *Mytilus*, shell removed

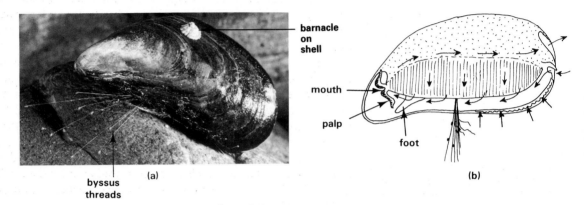

29.6.3 Cephalopoda

Octopuses, squids and **cuttlefish** are cephalopods. The head is well developed and surrounded by a ring of **tentacles**. The nervous system is well developed, and the eyes of cephalopods are remarkably similar to those of vertebrates (Fig 29.55). A **siphon** connects the mantle cavity with

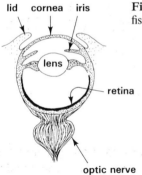

Fig 29.55 Vertical section eye of cuttle-fish, ×10. Compare with Fig 19.1

the sea water outside. When muscle action applies pressure to the water inside the mantle cavity, the water is squirted out through the siphon and the animal is propelled rapidly in the opposite direction. This action is often used as an escape mechanism when danger threatens. As another defence mechanism, some cephalopods secrete a dark fluid called **sepia** from an **ink sac**. Some cephalopods, such as *Nautilus*, have a coiled external shell containing chambers. In the squid and cuttlefish, the shell is internal and greatly reduced, while the shell is absent altogether in the octopus (Fig 29.56).

Fig 29.56 Octopus swimming, ×⅛ (Biofotos)

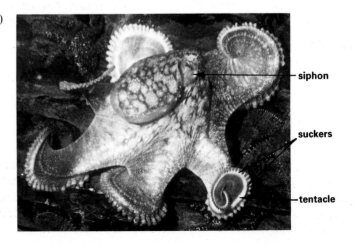

29.7 Phylum Echinodermata

The phylum Echinodermata includes **starfish** and **sea urchins**. One of the most striking features of echinoderms is their **radial symmetry**. This is often **pentamerous**, that is radiating in five directions from the centre. Adult echinoderms are coelomate. They are not segmented and there is no head. The mouth is on the **oral surface** underneath the animal, the anus is on the upper, **aboral surface**. Numerous **spines** usually project from the skin, which contains calcareous plates. Some spines are arranged in groups and may act as pincers. They are called **pedicellariae** and may be used for defence or for gripping and cutting prey. Some have poison glands attached to them.

Another feature peculiar to echinoderms is the **water vascular system**, a complex system of water-filled coelomic tubes which radiate throughout the body. Sea water enters the system through a perforated calcareous plate called the **madreporite** on the aboral surface near to the anus. The action of cilia maintains a current of water in the tubes. From the system, rows of many **tube feet** perforate the skin and project to the outside. The tube feet are kept inflated by the pressure of water in the water vascular system. Muscles in the tube feet enable them to be used to move the whole animal (Fig 29.57). The area covered by the tube feet is called the **ambulacrum** and is on the oral surface.

Fig 29.57 Starfish: (a) water vascular system, (b) transverse section arm. Tube feet are moved by the action of internal muscles.

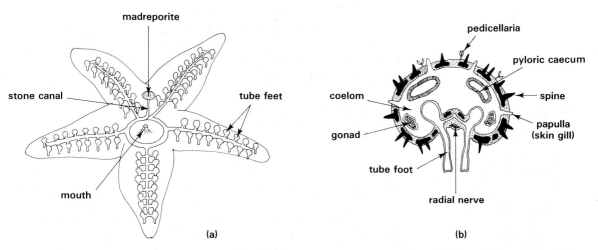

(a)

(b)

The simple nervous system consists of a ring of fibres around the mouth and a series of radiating branches. There is no brain. The blood vascular system is much reduced and probably plays little part in the circulation of materials in the body.

The sexes are separate and fertilisation is external. The life cycle involves a free-swimming, bilaterally symmetrical **pluteus larval** stage (Fig 29.58).

Fig 29.58 *Pluteus* larva, × 70

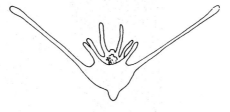

The structure and development of echinoderm larvae is similar to that seen in primitive chordates. For this reason it is thought that echinoderms and chordates evolved from a common stock.

The echinoderms are divided into several classes, including the Asteroidea and Echinoidea.

Why are echinoderms thought to have a common ancestry with chordate animals?

29.7.1 Asteroidea

Starfish such as *Asterias* belong to this class (Fig 29.59). The body is star-shaped with a photosensitive eyespot at the tip of the ambulacral groove on each arm. It is usually raised on the leading arm when the starfish is moving. The calcareous plates are spaced out in the skin, so the arms are flexible. On the oral surface, small thin-walled outgrowths called **papullae** project from the skin. They act as gills and may also be the sites of nitrogenous excretion.

Starfish use their tube feet as suction pads to pull open the shells of bivalve molluscs. The stomach can be everted through the mouth to take in the soft bodies of their prey. Extensions of the gut called **pyloric caeca** project into the arms (Fig 29.57(b)).

Fig 29.59 Common starfish, *Asterias rubens*, × 1 (Biofotos)

madreporite

spines

tube feet

29.7.2 Echinoidea

This includes the sea urchins such as *Echinus* (Fig 29.60). It has no arms. The calcareous plates in the skin are fused to form a spherical shell. Long, movable spines and pedicellariae project from most of the body surface. While feeding, urchins use a complex apparatus called **Aristotle's lantern**. It consists of five calcareous plates called pyramids from which teeth project. Muscle action causes a chewing movement of the pyramids and teeth.

Fig 29.60 Common sea urchin, *Echinus esculentus*, × ½ (Bruce Coleman: Jane Burton)

tube feet

spines

SUMMARY

Phylum Cnidaria: diploblastic body wall (ectoderm and endoderm) encloses enteron; hydroids and medusae; tissue level of organisation (muscle cells, nematoblasts, interstitial, gland, nutritive and sensory cells); radial symmetry.
Class Hydrozoa: e.g. *Obelia*; Class Anthozoa (sea anemones), e.g. *Actinia*; Class Scyphozoa (jellyfish), e.g. *Aurelia*.

Phylum Platyhelminthes (flatworms): triploblastic (ectoderm, mesoderm and endoderm); acoelomate (solid mesoderm); organ level of organisation; bilateral symmetry; cephalisation (development of a head).
Class Turbellaria, e.g. Planaria; mostly free-living; body systems include excretory (flame cells), branched gut, brain and nervous and reproductive systems.
Class Trematoda (flukes), e.g. *Fasciola*; most are endoparasites with suckers for attachment to host; complex life cycle often involving more than one host.
Class Cestoda (tapeworms), e.g. *Taenia*; all endoparasites, scolex with hooks and suckers for attachment to host's gut wall; produce proglottids.

Phylum Nematoda (roundworms): many are parasitic; thread-like bodies; devoid of external features, e.g. *Ascaris*.

Phylum Annelida; triploblastic coelomate (mesoderm split by cavity called the coelom); metameric segmentation. Excretory system contains nephridia; dorsal nerve ganglia (brain) and ventral nerve cord. Blood vessels; pseudohearts; respiratory pigments. Chaetae.
Class Polychaeta, e.g. *Nereis*; marine; parapodia; trochophore larvae.
Class Oligochaeta, e.g. *Lumbricus*; few chaetae; no larvae.
Class Hirudinea (leeches), e.g. *Hirudo*; many are blood-sucking ectoparasites; secrete anticoagulant called hirudin; suckers; no larvae.

Phylum Arthropoda: bilaterally symmetrical, metamerically segmented, triploblastic, coelomate. Body is divided into head, thorax and abdomen; jointed limbs. Exoskeleton; cuticle with waterproof wax in terrestrial forms. Joints moved by internal muscles.
Periodic moulting (ecdysis) allows growth, controlled by hormones. Highly cephalised; head often bears complex mouthparts, compound and simple eyes; well developed brain. Open blood vascular system.

Class Crustacea: all aquatic except woodlice. Exoskeleton often calcareous; cephalothorax; blood of some contains haemocyanin. Two pairs of antennae; nauplius larvae. E.g. *Daphnia, Balanus, Cancer, Astacus, Oniscus*.

Class Myriapoda (centipedes and millipedes): terrestrial; long body with many paired limbs; spiracles and tracheae for gas exchange; excretion by Malpighian tubules; no larvae.

Class Insecta: the most numerous and varied animal group. Many have paired wings; gas exchange by gills (aquatic) or spiracles and tracheae (terrestrial); excretion by Malpighian tubules. Life cycle involves metamorphosis, complete (eggs, larvae, pupae, imago) or incomplete (eggs, nymphs, imago). Many Orders include –
  Coleoptera (beetles): elytra cover wings.
  Diptera (flies): one pair of wings; halteres used as stabilisers in flight, e.g. *Musca*.
  Hemiptera (aphids and bedbugs), e.g. *Cimex*.
  Lepidoptera (butterflies and moths): maxillae form proboscis, e.g. *Pieris*.
  Othoptera (locusts and grasshoppers): stridulation organs; hindlegs adapted for jumping, e.g. *Locusta*.
  Hymenoptera (ants, bees and wasps): social colonies and castes, e.g. *Apis*.
Class Arachnida (spiders, scorpions, mites and ticks): body in two parts, prosoma and opisthosoma; prosoma bears four pairs of jointed legs. Gas exchange by lung books. Mouthparts include chelicerae and pedipalps. Spinnerets on opisthosoma produce web., e.g. *Epeira*. Mites and ticks are ectoparasites. ▶

Phylum Mollusca: coelomate; open blood vascular system (haemocoel); non-segmented; body divided into head, foot and visceral hump (covered by mantle); calcareous shell. Most are aquatic; gas exchange by gills (ctenidia) in mantle cavity (air-breathing snails have simple lung); blood contains haemocyanin. Mouth often contains radula, used for scraping food off hard surfaces such as rock. In marine forms life cycle includes trochophore and then veliger larvae.

Class Gastropoda (snails, slugs, limpets and whelks): head with tentacles. Visceral hump is twisted and turned through 180° on the foot (torsion), e.g. *Helix*.

Class Bivalvia: muscles and clams; body is laterally flattened and bilaterally symmetrical. Shell develops as two hinged valves; head much reduced; eyes, tentacles and radula absent; filter feeders; large gills for gas exchange, e.g. *Mytilus*.

Class Cephalopoda (octopus, squid and cuttlefish); well developed head with a ring of tentacles; eyes similar to those of vertebrates; siphon connects seawater with mantle cavity. E.g. *Sepia*.

Phylum Echinodermata: radially symmetrical (often pentamerous); coelomate; non-segmented; no head; spines and pedicellaria. Water vascular system; water enters by madreporite and inflates tube feet (locomotion). Separate sexes; pluteus larvae.

Class Asteroidea (starfish), e.g. *Asterias*: star-shaped body.

Class Echinoidea (sea urchins), e.g. *Echinus*: no arms; spherical shell; complex feeding apparatus called Aristotle's lantern.

QUESTIONS

1 Copy the table and indicate by ticks (√) those features listed which are characteristic of each of the three phyla.

Feature	Phylum		
	Coelenterata (Cnidaria)	Platyhelminthes	Annelida
Diploblastic			
Triploblastic			
Acoelomate			
Coelomate			
Radially symmetrical			
Bilaterally symmetrical			

(AEB 1987)

2 (a) For each of the specimens **A** honeybee, **B** liverfluke and **C** earthworm state:

(i) the major group (phylum) to which it belongs;
(ii) *one* subgroup to which it belongs.
Write your answer in a table like the one shown below:

Specimen	(i)	(ii)
A		
etc.		

(b) For each of the specimens **A**, **B** and **C**, name *one visible* feature that is characteristic of its major group.
(c) Common on *two visible* structural differences between specimens **B** and **C** that are related to the environments in which they live.
(d) Draw and annotate *two* features of the stickleback which are concerned with locomotion. (L)

3 Discuss the mechanisms, occurrence and importance of amoeboid movement and ciliary activity. (L)

4 (a) Name the phylum to which each of the following belongs: (i) earthworm, (ii) frog, (iii) *Planaria* (a flatworm).
(b) Give **three** diagnostic characteristics of the phylum Coelenterata [Cnidaria]. (AEB 1986)

5 For each of the following organisms, a bryophyte, a homosporous fern, *Hydra* or *Obelia*, and a parasitic platyhelminth:
(a) state where fertilisation takes place;
(b) explain how new individuals become established in places remote from the parent organism;
(c) describe any particular aspect of the life cycle which helps to ensure survival in adverse conditions. (O)

6 For the following examples: the earthworm and *Nereis*,
(a) Make labelled drawings to show the external features of the first 40 segments of the earthworm and the first 8 segments of *Nereis*.
(b) What external features enable you to classify these two examples into the same phylum and into separate classes? (O)

7 (a) Make a large labelled drawing of a lateral view of a honeybee.
(b) (i) List *three* features characteristic of insects and which are shown by both the honeybee and the locust.
(ii) Make a table to show *four* differences between the external features of the honeybee and the locust.
(c) Comment on the ways in which the structure and functions of the limbs of the locust are related to its way of life. (L)

30.1 Class Chondrichthyes
(Elasmobranchii)   614

30.2 Class Osteichthyes   616

30.3 Class Amphibia   618

30.4 Class Reptilia   619

30.5 Class Aves   620

30.6 Class Mammalia   624

Summary   629

Questions   630

Bruce Coleman: Gerald Cubitt

30 Variety of life: Chordate animals

Chordates are **bilaterally symmetrical, coelomate** animals. Their name is derived from the dorsal rod of cells called the **notochord** which extends the length of the body. The notochord is relatively rigid and provides support. Dorsal to the notochord is a **hollow nerve cord**, usually with a well-developed brain at the head. The heart is placed ventrally and pumps blood forward. At some stage in the life cycle, chordates have pairs of **gill clefts** in the sides of the pharynx and a solid **post-anal tail** (Fig 30.1).

Fig 30.1 Chordate body plan

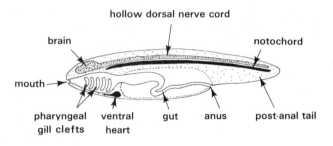

The chordates include several classes, six of which are often collectively called the **vertebrates**.

The vertebrates are chordates in which the notochord is strengthened by cartilage, or more usually bone, forming a **vertebral column** divided into vertebrae. In the head the brain is surrounded by a skeletal case called the **cranium**.

Vertebrate chordates include the classes Chondrichthyes, Osteichthyes, Amphibia, Reptilia, Aves and Mammalia.

30.1 Class Chondrichthyes (Elasmobranchii)

The chondrichthyes are **cartilaginous fish** such as dogfish, sharks, rays and skates (Fig 30.7). Nearly all live in the sea. They have strong jaws lined

Fig 30.2 (a) Mermaid's purse, egg case of lesser spotted dogfish, *Scyliorhinus*, $\times \frac{1}{2}$ (Bruce Coleman: Neville Fox-Davis). (b) Blue shark, *Carcharinus*, $\times \frac{1}{50}$ (Bruce Coleman: Giorgio Gualco). (c) Thornback ray, *Raja*, $\times \frac{1}{10}$ (Biofotos)

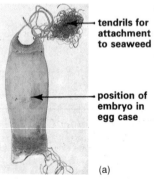

tendrils for attachment to seaweed

position of embryo in egg case

(a)

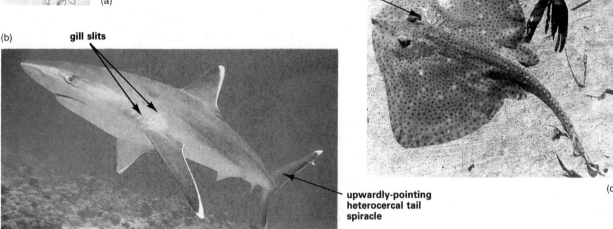

(b)

gill slits

pectoral fin

spiracle

upwardly-pointing heterocercal tail
spiracle

(c)

Fig 30.3 Placoid scales of dogfish, × 10

Fig 30.3 Placoid scales of dogfish, × 10

with teeth and the skeleton is made of cartilage. The skin is thick and covered by many tooth-like **placoid scales** (Fig 30.3). These are thought to be the evolutionary remains of the external armour that once covered their ancestors. The body is **dorso-ventrally flattened**. Beneath the tip of the head is a pair of well-developed olfactory organs. Sharks are well known for their keen sense of smell. The highly developed eyes are similar to those of mammals. Running the length of the body on each side is a **lateral line system**. It consists of a long canal just beneath the skin connected with the sea through numerous perforations in the skin. In the canal are many small hair cells which are sensitive to vibrations in the surrounding water. The lateral line enables cartilaginous fish to detect movements in their surroundings.

Gill clefts, usually five pairs, open at **gill slits** just behind the head. Water taken into the mouth is forced through the gill slits when the floor of the mouth is raised. Blood passing through the gills travels in much the same direction as the sea water. Gas exchange in such a **parallel-flow** arrangement is relatively inefficient (Fig 30.9(b)). A pair of small openings called **spiracles** is found just behind the eyes. In rays and skates, the spiracle allows oxygen-carrying water into the pharynx while the animal is partially buried in sand when water cannot enter the mouth (Fig 30.2(b)).

Cartilaginous fish are powerful swimmers. Alternate contraction and relaxation of **myotomal muscles** flex the body sideways, first one way then the other. The sides of the body and the fin of the **heterocercal tail** push against the water, giving forward thrust. The body is stabilised by the **rigid fins**. They help prevent rolling, pitching and yawing (Fig 30.4). There is no swim bladder. Rays and skates are greatly flattened and mostly live on the sea bottom. Their pectoral fins are very extensive, allowing them to glide over the sea bed in an undulating fashion.

Fig 30.4 Mechanics of swimming in dogfish: (a) lateral flexing, (b) deviations of motion

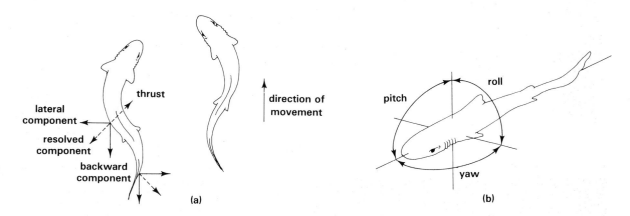

The sexes are separate and reproduction involves copulation. The male has a pair of **claspers** between the pelvic fins. Claspers are used to introduce sperms into the oviducts of the female. After fertilisation, the eggs are enclosed in a tough case and released into the sea (Fig 30.2(c)). Coiled threads trailing from the egg cases become entangled in seaweeds. When sufficiently developed, the young fish break free from the **egg case** and swim away. In some sharks and rays, development of the young takes place in the mother's body.

30.2 Class Osteichthyes

The osteichthyes are **bony fish** (Fig 30.5). They inhabit fresh and brackish water as well as the sea and are by far the most numerous of aquatic chordates. There are about seven times more species of bony fish than cartilaginous species.

Fig 30.5 Brown trout, *Salmo trutta*, $\times \frac{1}{7}$ (Bruce Coleman: Hans Reinhard)

dorsal fin

lateral line

operculum

cordal fin

pectoral fin

pelvic fin

anal fin

The internal skeleton consists of bone and the skin is covered by many thin **bony scales**. Along each side of the body is a **lateral line canal.** There is no spiracle and the gills are covered by a bony flap called the **operculum**. The body is usually **laterally flattened**, the fins are flexible and the tail fin is **homocercal**.

Fig 30.6 Queensland lungfish, *Neoceratodus*: (a) whole animal, $\times \frac{1}{6}$ (Bruce Coleman), (b) position of lungs

(a)

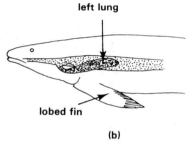

left lung

lobed fin

(b)

One of the unusual features of most bony fish is the **swim bladder**, an air-filled sac close to the pharynx. It is thought that the lungs of air-breathing chordates may have evolved from swim bladders. In **lung fish**, the swim bladder is a paired structure connected to the pharynx and used for gas exchange (Fig 30.6). In most bony fish, the swim bladder is used in maintaining buoyancy. The gas, mainly oxygen, inside the bladder forces the fish upwards towards the surface of the water. The body mass drags the fish downwards under the influence of gravity. The wall of the bladder is highly vascular. Gas may be absorbed from the bladder by the blood or discharged into the bladder from the blood. In this way, bony fish can alter their density, enabling them to swim at an appropriate depth with the minimum of muscular effort (Fig 30.7).

Why are bony fish more numerous than cartilaginous fish?

Mackerel do not have a swim bladder. They swim continuously to maintain their preferred level in the water.

Fig 30.7 Swim (air) bladders in bony fish: (a) position in the herring, (b) maintenance of buoyancy at different depths by inflation or deflation of the swim bladder. Neutral buoyancy is achieved when the fish neither rises nor sinks

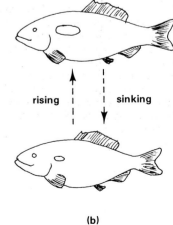

rising   sinking

(b)

swim bladder

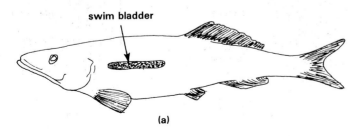

(a)

The **fins** in bony fish are thin sheets of skin supported by rays of bone. The pelvic fins are usually well forward and used with the pectoral fins in diving, rising, turning and braking (Fig 30.8). The rigid fins of cartilaginous fish do not allow the range of movements seen in bony fish. In some bony fish, forward thrust is provided solely by flapping movements of the fins. During fast swimming, myotomal muscles flex the body providing extra thrust. The streamlined shape of many bony fish also aids swimming by reducing friction and drag.

Fig 30.8 Use of the pectoral fins in bony fish: (a) cornering, (b) braking, (c) skeleton

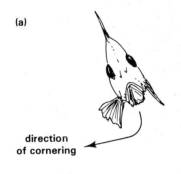

(a)

direction
of cornering

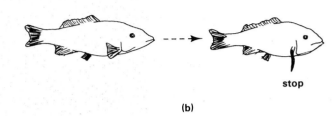

(b)

stop

(c)

Respiratory gas exchange occurs at the **gills**, which are well supplied with blood and supported by bony gill bars. As in cartilaginous fish, there are usually five pairs of gill clefts. Each gill is composed of very many thin plates called **lamellae**. There is thus a considerable surface area for gas exchange, which helps offset the relatively low oxygen content of water, less than 1 % by volume. Water is forced over the gills when the floor of the mouth is raised. The gill lamellae are arranged so that blood and water flow in opposite directions. Gas exchange in such a **counterflow** arrangement is much more efficient than in the parallel flow mechanism seen in cartilaginous fish (Fig 30.9). Osmoregulation in bony fish is described in Chapter 31.

Fig 30.9 (a) Gill structure in bony fish. The lamellae are arranged so that the water flows over them in the opposite direction to that of the blood within them. This counter flow enables maximum gas exchange.
(b) Section through a gill of dogfish. The blood–water counter flow is less efficient than in bony fish because the water flows mostly along the lamellae rather than in between them

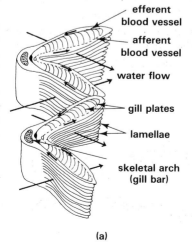

efferent
blood vessel

afferent
blood vessel

water flow

gill plates

lamellae

skeletal arch
(gill bar)

(a)

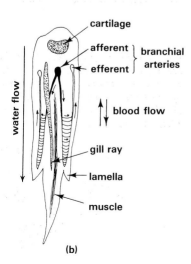

cartilage

afferent } branchial
efferent } arteries

water flow

blood flow

gill ray

lamella

muscle

(b)

Fig 30.10 (a) Common newt, *Triturus vulgaris*, $\times \frac{2}{3}$ (Biofotos), (b) common frog, *Rana temporaria*, $\times \frac{2}{3}$ (Biofotos)

(a)

(b)

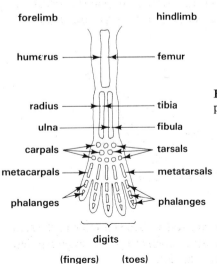

Reproduction in bony fish usually involves external fertilisation. The large numbers of eggs laid by females helps compensate for the losses caused by predators and other hazards. In one season, for example, a female cod may produce as many as ninety million eggs. Reproduction may also involve complex behaviour of mating pairs, nest-building, care of the young, shoaling and long migrations. The salmon, for example, normally returns to the river where it was hatched, after swimming several thousands of miles across the oceans.

30.3 Class Amphibia

The **amphibians** include **frogs, toads, newts** and **salamanders** (Fig 30.10). Amphibians were probably the first chordates to colonise land. They are thought to have evolved at a time when widespread seasonal droughts were common. The theory is that some lobe-finned fish used their fins as primitive legs and crawled to neighbouring ponds when their own pond dried out. The skeleton of amphibians has evolved to support the body on land. The vertebral column acts as a suspension girder from which the thoracic and abdominal organs hang. The force of gravity acting on the body mass is transmitted to the ground through strong **pectoral** and **pelvic girdles** and the limbs. Swimming in amphibians is a fish-like side to side flexing of the body (Fig 30.11). Amphibian skeletons contain bone and

Fig 30.11 (a) The main components of the amphibian skeleton. (b) Fish-like locomotion in an amphibian (compare with Fig 30.9(a))

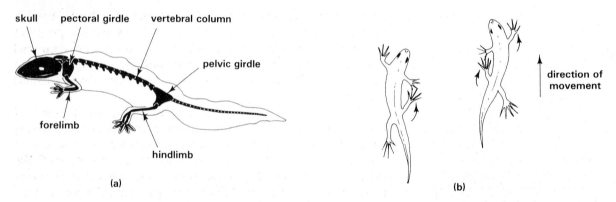

skull pectoral girdle vertebral column

pelvic girdle

forelimb

hindlimb

(a)

direction of movement

(b)

forelimb   hindlimb

humerus — — femur

radius — — tibia

ulna — — fibula

carpals — — tarsals

metacarpals — — metatarsals

phalanges — — phalanges

digits

(fingers)   (toes)

often a lot of cartilage. The bones of the limbs are arranged in a characteristic fashion. The limbs are **pentadactyl**, usually terminating in five digits (Fig 30.12). In the forelimbs of amphibians, the number of digits is usually reduced. The presence of pentadactyl limbs in amphibians, reptiles, birds and mammals points to a common ancestry.

Fig 30.12 Pentadactyl limb, general plan. See Section 16.1.5

About 21 % of the volume of the earth's atmosphere is oxygen. The problem faced by air-breathers is to extract the oxygen from the atmosphere without losing too much water. Respiratory surfaces must be moist to allow oxygen gas to dissolve before the oxygen can diffuse into the blood. As the respiratory surface must also be exposed to the atmosphere, inevitably it loses moisture by evaporation (section 8.1). The lungs of amphibians are relatively simple air-filled sacs (Fig 8.5). Most amphibians use their **smooth moist skin** as an accessory respiratory surface. Larval amphibians use **gills**.

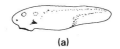
(a)

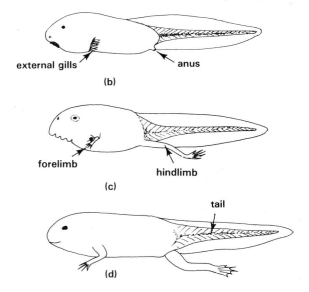
external gills — anus
(b)

forelimb — hindlimb
(c)

tail
(d)

Fig 30.13 Some tadpole stages in the development of the frog: (a) newly-hatched, ×5, (b) 2 weeks, ×8, (c) 7 weeks, ×2, (d) 10 weeks, just before final metamorphosis, ×3

Another important reason why most amphibians live in moist habitats is that they require water for fertilisation and larval development. Amphibian **larvae** are fish-like and usually have gills. The transition from larva to adult involves great changes in body form, known as **metamorphosis** (Fig 30.13). In some amphibians the adult remains aquatic. Others are **neotenous**, the larval body form persisting throughout life (Fig 30.14). A lateral line system is present in aquatic amphibians.

Fig 30.14 The axolotl, *Sirenodon mexicanum*, ×⅔ (Bruce Coleman: Eric Crichton)

external gills

30.4 Class Reptilia

Fig 30.15 Green lizard, *Lacerta viridis*, ×⅔ (Biofotos)

Crocodiles, lizards, snakes and **turtles** are **reptiles** (Fig 30.15). The few present-day reptiles are descendants of a once very successful group of animals, including the **dinosaurs**, which dominated the land about 200 million years ago. Many extinct reptiles were **bipedal**, walking upright on the hind limbs. A few **pterosaurs** even took to the air. Their forelimbs supported flaps of skin rather like the wings of bats.

Most of the features of reptiles make them better suited to life on land than amphibians. The distal parts of the limbs are bent downwards, raising the body clear of the ground (Fig 30.16). Reptiles can thus move on all four limbs without the trunk of the body touching the ground. Pairs of **ribs** project from the vertebrae. They give support to and protect the soft organs in the body cavity. Movements of the ribs, brought about by contractions of **intercostal muscles**, are used to ventilate the **lungs**. The lungs of reptiles have a more complex internal structure than those of amphibians (Fig 8.5). Ingrowths of tissues increase the inner surface area for gas exchange.

Fig 30.16 The gait of (a) amphibians, (b) reptiles

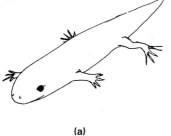
(a)

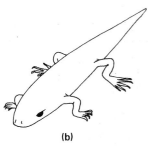
(b)

In what ways are reptiles better adapted to terrestrial life than amphibians?

The **skin is dry** and covered by **horny scales**. Beneath these in the skin there may be **bony plates**. They provide a tough armour in some reptiles. Little evaporation takes place through the skin.

The kidneys of reptiles are capable of producing a concentrated urine and thus help to retain body water. **Uric acid** is the main nitrogenous waste. It requires the minimum of water for its elimination. Marine reptiles have **salt excretion glands** in the orbits of the skull. They discharge tears rich in salts which the reptiles have absorbed from seafood.

Reptiles usually have rows of identical teeth. This **type** of dentition is called **homodont** (homos = same; dens = tooth). **Salivary glands** are also present. They represent an adaptation to the terrestrial way of life and are absent from fish and amphibians. The main function of saliva is to lubricate food when it is taken into the gut. Water performs this function in most aquatic chordates. Eyelids are present, including a third eyelid called the nictitating membrane. In amphibians the lower eyelid is the **nictitating membrane**, the upper lid is fixed. Movements of the eyelids remove abrasive particles from the front of the eye.

Copulation and **internal fertilisation** are features of reptilian reproduction. The embryo develops inside a membraneous fluid-filled compartment called the **amnion** where it is protected from desiccation. Eggs are often laid in nests or buried in sand or soil, where they are hidden from predators. A few reptiles retain the eggs inside the female's body. The **amniote egg** is a feature of reptiles, birds and mammals (Fig 25.18). The embryo obtains food from a **yolk sac** attached to its gut. Wastes are discharged into another sac called the **allantois** which enlarges during development and eventually comes close to the shell.

The shell of most reptilian eggs is leathery in texture and respiratory gases pass through it. The allantois aids in gas exchange between the atmosphere and the embryo. Because the embryos of reptiles are protected from predation the chances of survival are greater than in fish and amphibians which leave the eggs to fend for themselves. The enormous losses of fish and amphibian embryos are offset by the large numbers of eggs produced.

30.5 Class Aves

The members of this class are **birds** (Fig 30.17). One of the main characteristics of birds is that they have **feathers**. These are highly modified skin scales and are used by most birds for **flight**. Feathers develop as outgrowths of the skin. Soft, fluffy **down feathers** are formed in young birds. They are useless for flight but act as heat insulators in much the same way as hair does in mammals (Chapter 20). The down feathers are later

Fig 30.17 (a) Starling, *Sturnus vulgaris*, a passeriform, × ½ (Bruce Coleman: Gordon Langsbury), (b) Imperial eagle, *Aquila heliaca*, a falconiform, × ⅛ (Bruce Coleman: J L G Grande)

(a)

first toe directed
backwards to grip perch

(b)

hooked beak for
piercing flesh

young with
down feathers

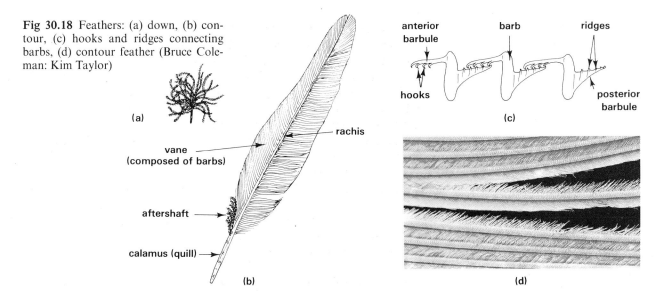

Fig 30.18 Feathers: (a) down, (b) contour, (c) hooks and ridges connecting barbs, (d) contour feather (Bruce Coleman: Kim Taylor)

Birds have been described as 'flying reptiles'. Why is this?

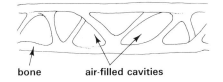

Fig 30.19 Supporting struts inside a pneumatic bone of a bird

---

replaced by **contour feathers** (Fig 30.18). These have a **rachis** which supports numerous **barbs** projecting from it at right angles. The barbs form the **vane** which provides lift and thrust during flight. The barbs are held together by rows of hooks and grooves called **barbules**. They interlock to produce a continuous surface. At the base of the vane there is usually a small tuft called the **aftershaft**. It provides heat insulation in the absence of down feathers. The only gland in the skin of birds is the **preen gland** just above the base of the tail. Oily secretions from the gland are spread over the feathers by the beak during preening. The oil helps to waterproof the feathers. It is especially important in aquatic birds, where it prevents the feathers becoming waterlogged.

The body mass of birds is reduced by the presence of hollow air-filled spaces in certain **pneumatic bones** (Fig 30.19). They include the humerus in the wings, the sternum and the vertebrae. The hollows are crossed by angled struts which give great strength to the bone. The inner design of aircraft wings is similar to that of pneumatic bones.

Many other features of birds are also adaptations to flight. The forelimbs are modified to form **wings**. The feathers on the wings are arranged in definite tracts (Fig 30.20). They provide a surface to the wing which acts as an **aerofoil**. The forces that act on the wing may be resolved into **drag** and **lift** components. Drag pulls down on the wing and is largely created by gravity. Lift pushes up on the wings. When lift forces equal or exceed drag forces, flight is possible. Lift may be increased by tilting the wing at an angle to the oncoming air. Alternatively, lift increases if the velocity of the air

---

Fig 30.20 Wing skeleton showing position of flight feathers. The bastard feathers are extended to prevent stalling during slow flight such as on landing. They are retracted during rapid flight

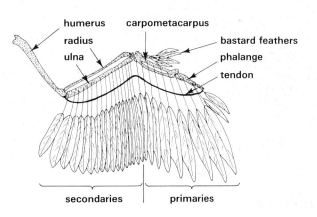

Fig 30.21 Wing aerofoil in section showing air flow, lift and drag forces when (a) flat, and (b) tilted. The air flow over the upper surface is faster than that over the underside. Air pressure on top is thus lowered

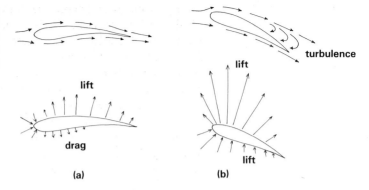

flow increases (Fig 30.21). **Stalling** occurs when lift is not sufficient to keep the bird aloft. Birds may use moving air to **glide** or **soar** simply by holding their wings still and extended at the appropriate angle. Birds of prey often soar in circles at great height while looking out for prey. They use warm upcurrents of air called **thermals** to provide lift. The shape and proportions of the wing are also important in soaring, which requires little energy (Fig 30.22). In **flapping flight**, the wings are forcibly moved up and down.

Fig 30.22 Wings of vulture: (a) soaring upwind, gaining height, (b) upwind on level, (c) downwind, losing height, (d) gliding flight (From J Z Young, after Ahlborn)

Lift is generated as the wings push down on the air. The angle of the wing is changed to minimise drag as the wing is raised (Fig 30.23). Flapping requires a great deal of energy. The relatively large flight muscles are attached to a flat sternal plate called the **keel** (Fig 30.24).

Fig 30.23 Wing flapping in the pigeon showing: (a) downstroke, (b) upstroke (from photographs after Brown)

Fig 30.24 Skeleton of a pigeon

beak   optic ossicles

keel

The tail is much reduced in birds, although the tail feathers may be long. Lowering and fanning out the tail feathers increases lift when the bird is flying slowly while landing.

Birds, like mammals, are **endothermic** (Chapter 20). They regulate their body temperature by physiological means. A relatively high and constant body temperature enables respiration to occur at a rate rapid enough to provide energy for flight. Maintaining a relatively high body temperature also enables birds and mammals to be active even in very cold environments. Animals other than birds and mammals are ectothermic. Their body temperature changes with that of their environment (Fig 20.1). Ectotherms may regulate their body temperature to a certain degree by behavioural means such as basking in the sun, seeking shade or burrowing.

Large volumes of oxygen are needed to provide the energy for flight. Ventilation of the lungs in birds is far more efficient than in other vertebrates. It is assisted by a system of **air sacs** connected to the lungs (Fig 30.25). Some of them connect with the air spaces in the pneumatic bones. When a bird breathes in, any air which remains in the lungs from previous ventilations is sucked into the sacs. The lungs are filled with fresh air, avoiding the dead space that occurs in the lungs of other chordates (section 8.2.1). Ventilation of the lungs and air sacs is brought about by movements of the ribs and sternum. There is no diaphragm. During flight, the action of the flight muscles ventilates the lungs at the same rate as movements of the wings. In some diving aquatic birds such as penguins, the air sacs may provide buoyancy, acting like the swim bladder of bony fish (section 30.2). The lungs of birds are spongy organs with an internal structure similar to that of mammals (Chapter 8).

Fig 30.25 Air sacs in a bird from the side (after Goodrich)

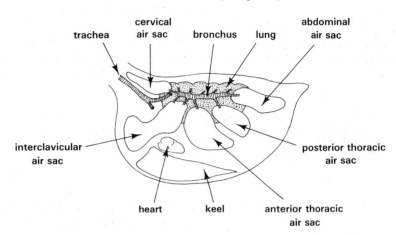

As in all other chordates, the carriage of oxygen in the blood is aided by haemoglobin.

The nervous system is well developed, especially the brain. Birds display more complex behaviour than reptiles. Bird behaviour usually includes the formation of flocks, pairing and mating, seasonal migration, communication between individuals and the rearing of young. Birds are especially vocal and songs are generated by a structure called the **syrinx** in the trachea.

Birds have no teeth. The jaws are drawn out into a horny **beak**. In many birds the beak has evolved to make use of a particular type of food (Fig 30.26).

Fig 30.26 Variety of birds' beaks: (a) cormorant, catching fish, (b) harrier, gripping and ripping flesh, (c) oyster catcher, picking food from mud, (d) hawfinch, crushing seeds, (e) flamingo, filtering water

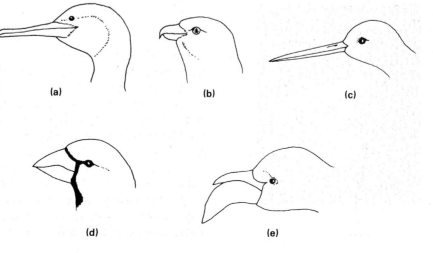

Uric acid is the main nitrogenous excretory material. In marine birds, excess salts are excreted through the nose from **salt glands** behind the eyes.

Birds reproduce in much the same way as reptiles. Copulation followed by internal fertilisation leads to the development of **amniote eggs** (Fig 25.18). Birds' eggs, like those of reptiles, are **cleidoic**. They are protected from desiccation and predators by a tough **calcareous shell**. The eggs are usually laid in a nest, where they are incubated by one or both parents. The young are cared for after hatching.

30.6 Class Mammalia

Mammals include the platypus, kangaroo, antelopes, bats, cattle, cats, dogs, elephants, horses, sheep, shrews, rodents, whales and humans. Features of mammals include:

(i) **hair** and **sweat glands** in the skin which assist **endothermy** (Chapter 20),

(ii) a muscular **diaphragm** separating the thorax from the abdomen (Chapter 8),

(iii) a highly **developed brain** and **intelligent** behaviour involving **learning** and **memory** (Chapter 18),

(iv) **well-developed sense organs** (chapter 16),

(v) **heterodont dentition** and a bony roof to the mouth called the **palate** (Chapter 15),

(vi) development of the embryo inside the mother's body, the offspring being provided with **milk** from well-developed **mammary glands** (Chapter 25).

The limbs of terrestrial mammals are positioned beneath the body, enabling them to travel very rapidly over land. Greyhounds can run at speeds of about $59 \, \text{km h}^{-1}$, while cheetahs have been reported to reach $112 \, \text{km h}^{-1}$ for short periods (Fig 30.27). Some mammals, such as whales and dolphins, are efficient swimmers; others, such as bats, are capable of flight.

Fig 30.27 (a) Skeleton of a cat. (b) Cheetah running at high speed (Bruce Coleman: Gunter Ziesler)

(a)

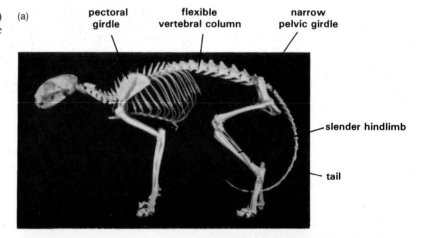

pectoral girdle — flexible vertebral column — narrow pelvic girdle — slender hindlimb — tail

(b)

Fig 30.28 (a) Platypus, *Ornithorhyncus*, ×¼ (Biofotos)

Fig 30.28 (b) Spiny ant-eater, *Tachyglossus*, ×¼ (Bruce Coleman: M R Harris)

Fig 30.29 Three marsupials: (a) opossum, *Marmosa*, ×⅕ (Bruce Coleman: Jane Burton), (b) kangaroo, *Macropus*, ×²⁄₂₅ (Bruce Coleman: John Cancalosi), (c) wombat, *Phascolomis*, ×⅛ (Biofotos)

Mammals are divided into several sub-classes:

1 Prototheria

This has a single order, the Monotremata. There are two **monotremes**, the spiny anteater and the platypus (Fig 30.28). They are primitive mammals and are thought to be the remains of an early mammalian stock. Monotremes have hair and sweat glands, although their body temperature fluctuates more so than in other mammals (Fig 20.1). They care for their young and have simple mammary glands. Among their reptilian features is the laying of **amniote eggs**. They are deposited in a nest by the platypus and carried in a skin pouch by the spiny anteater. The urine ducts and anus have a common opening called the **cloaca**.

2 Theria

Therians are divided into two infra-classes:

A. Metatheria

This includes the order Marsupiala. **Marsupials** are found mainly in Australia. Their evolution there has been due to Australia's geographic isolation from other land masses (Chapter 27). Adaptive radiation of marsupials in Australia is comparable with that of other mammals elsewhere (Fig 30.29). In marsupials, the embryos emerge from their mother's reproductive tract very early and enter a **pouch** on the mother's abdomen. Further development of the offspring takes place in the pouch, where they suck milk from mammary glands. There is no placenta. As in monotremes and reptiles, the urinogenital tract opens into a **cloaca**.

(a)

(b)

(c)

B. Eutheria

These are the **placental mammals**. The embryos of eutherians develop within the mother's uterus, where nutrients are supplied and wastes removed by a **placenta** (Chapter 25). After birth, the young are nourished by milk from **mammary glands**.

Fig 30.30 Common shrew, *Sorex araneus*, ×⅔ (Bruce Coleman: Leonard Lee Rue)

Eutherians are divided into many orders including:

i. Insectivores include shrews, moles and hedgehogs (Fig 30.30). The earliest placental mammals are thought to have been shrew-like insectivores. Most modern insectivores are small and nocturnal. Insects form a large part of their diet, hence the name given to the order. The sense of smell is relatively well developed and the long snout may be pushed into holes or crevices in search of insects. Insectivores have a full mammalian dentition (Fig 15.10). Shrews are among the smallest of all mammals. Small size leads to excessive heat loss (Fig 20.4). Many insectivores overcome this problem by hibernating in winter (section 20.3.4).

Fig 30.31 Black rat, *Rattus rattus*, ×$\frac{1}{4}$. See Fig 15.11(a) (Bruce Coleman: John Markham)

Fig 30.32 Common wolf, *Canis lupus*, ×$\frac{1}{10}$ (Biofotos: Britta Rothawson)

Fig 30.33 Impala, *Aepyceros*, ×$\frac{1}{10}$ (Bruce Coleman: Kim Taylor)

ii. Rodents are the most numerous of all mammals and include beavers, mice, rats, squirrels and voles (Fig 30.31). They are nearly all herbivorous, having a single pair of incisors on each jaw. Open roots allow continuous growth of the incisors, which are used to gnaw food (Fig 15.11(a)). The large caecum contains cellulolytic bacteria which aid the digestion of vegetation (section 15.3.5).

iii. Carnivores include bears, dogs, seals, walruses, wolves and cats such as cheetahs, lions and tigers (Figs 30.27 and 30.32). They are flesh-eaters whose dentition is specialised for gripping, piercing, ripping and tearing flesh (Fig 15.4). The digits have strong claws which are often retractable. Carnivores often form complex social groups usually hunting in packs. Secretions from well-developed anal glands are used frequently for recognition and to mark territory. Some carnivores, such as seals, have become successful swimmers. They are capable of diving to great depths and for long periods in search of food (section 8.4).

iv. Mammals with hoofs are called **ungulates** (ungula = hoof). They are relatively large herbivores, often able to run at great speeds to escape predators. They are often grazing animals and collect into large herds. In many species the pupil of the eye is horizontal, giving a wide angle of vision. The pinnae are often long and mobile. The neck is usually long, so that individuals can keep watch for predators while chewing food. The young are relatively well developed at birth. Their survival often depends on being able to run with the herd within a few minutes of being born.

(a) Artiodactyls are the **even-toed ungulates** such as antelopes, camels, cattle, deer, hippopotamuses, pigs and sheep (Fig 30.33). They have feet with two toes forming a **cloven hoof**. The dentition is modified for grazing, grinding vegetation and cud-chewing (Fig 15.6–8). Cellulolytic bacteria are found in a large multichambered stomach (section 15.3.3).

(b) Perissodactyls are the **odd-toed ungulates** such as horses, zebras and rhinoceroses (Fig 30.34). The digits are reduced to a single long toe. As in artiodactyls, the teeth are modified for grinding food. However, the bacteria necessary for digesting vegetation are not found in the stomach but in the caecum (section 15.3.5).

Fig 30.34 Grant's zebra, *Equus burchelli boehmi*, ×$\frac{1}{15}$ (Bruce Coleman: Frans Lanting)

v. Chiropterans are bats (Fig 30.35). In many respects bats are similar to insectivores. Many bats eat insects, but some are fruit-eaters, while vampire bats drink the blood of other chordates. Bats are the only mammals capable of true flight. Their wings consist of flaps of skin supported by four long fingers, the legs and even the tail. There is a **keel** on the sternum for the attachment of flight muscles. When resting, bats often hang upside down, clinging to something solid with their clawed toes. Bats cool down when not flying and many species hibernate in winter. They have an elaborate **echolocation** system which is used to avoid obstacles during flight and to locate insects.

Fig 30.35 Greater horseshoe bat, *Rhinolophus ferrumequinum*, $\times \frac{1}{3}$: (a) in flight, (b) roosting (Biofotos: G Kinns)

(a)   (b)

Fig 30.36 Blue whale, *Balaenoptera musculus*, $\times \frac{1}{145}$. See Fig 15.12

vi. Cetaceans include dolphins and whales (Fig 30.36). They are marine mammals capable of diving to great depths for long periods (section 8.4).

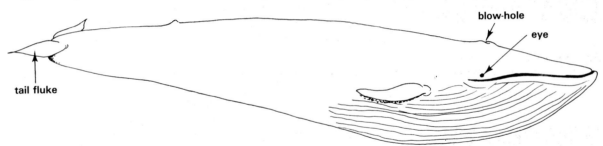

The short neck and a tapered outline enable these creatures to slip easily through water. The forelimbs are modified to form **flippers** which, together with a well-developed dorsal fin, are used as stabilisers while swimming. The hind limbs are absent altogether. The tail is also modified as a fin with horizontal **flukes**. The main thrust in swimming is provided by an up-and-down movement of the body and the tail flukes. It contrasts with the lateral movement of the body and tail in fish. The skin of cetaceans is smooth and there is generally no hair. Heat loss is minimised by a thick layer of fat called **blubber**. The large body mass of whales relative to surface area also helps to reduce heat loss (section 20.2.4).

The blue whale is the largest animal ever seen on earth. Specimens have been known to measure 32 metres long and weigh about 150 tonnes. The nostrils of cetaceans are on the dorsal surface of the head and are used as a **blow-hole** in breathing. Dolphins and **toothed whales** eat fish and other marine animals (Fig 15.12(a)). **Whalebone whales** feed on small organisms such as the shrimp-like krill which form part of the plankton of the sea. Whalebone whales filter the plankton from the sea using a complex system of **baleen plates** as a sieve (Fig 15.12(b)). Dolphins have a well developed **echolocation** system similar to that of bats. Dolphins and whales are also known to communicate with members of their own species by a complex system of sounds. The cetaceans have large brains and are among the most intelligent of mammals.

vii. Primates are divisible into two sub-orders. The *Prosimii* includes the aye-aye, bush babies, lemurs, lorises and tarsiers (Fig 30.37). The *Anthro-*

Fig 30.37 Slender loris, *Loris tardigradus*, $\times \frac{1}{3}$ (Bruce Coleman: Gerald Cubitt)

poidea include apes, marmosets, tamarins, a variety of monkeys and, in the family *Hominidae*, humans (Fig 30.38).

Fig 30.38 Gibbon, *Hylobates*, $\times \frac{1}{8}$ (Bruce Coleman)

Most primates live in trees. They have hands and feet, and sometimes a **prehensile tail** capable of grasping branches. The digits have **nails** instead of claws. The eyes are usually at the front of the head, each eye seeing the same field of view but from a slightly different angle. This is **stereoscopic vision** and makes it possible for the distance of objects to be judged.

Primates often form complex social groups in which communication between individuals may be very detailed. There is considerable care of the young. Egg production by the ovaries is cyclical and involves a complex hormone control (section 25.2). There are usually few offspring. The period of development and growth is very long in primates compared with other mammals, up to 20 years in humans. The brain is relatively large, especially the cerebral hemispheres (Chapter 18). Intelligent behaviour involving teaching, learning, memory, abstraction and prediction occurs in primates to a far greater degree than in other animals. Primates are nearly all **omnivorous**, with an **unspecialised dentition** (Fig 15.9).

Humans are thought to represent the pinnacle of evolution. Is this true?

Two main features of the human body have enabled humans to evolve as the dominant mammal on earth. An inquisitive nature and an ingenuity in solving problems is probably attributable to the **large cerebral hemispheres** of the brain. The other feature is the **manipulative ability of the hands**. The adduction of the thumb and index finger enables humans to handle and shape things with great skill. Manufacture of tools and comfortable houses, the development of agriculture and the domestication of other animals have contributed greatly to human evolution. The organisation of social groups into states, the making of laws and the development of art have all contributed to human **civilisation**. Unfortunately, humans have also developed the capacity to destroy the world as well as improve it.

SUMMARY

Chordates are bilaterally symmetrical coelomates; possess a dorsal notochord and hollow nerve cord; usually a well developed brain; ventral heart; gills in side of pharynx; solid post-anal tail.

Class Chondrichthyes (Cartilaginous Fish), e.g. sharks and rays: Mostly marine; dorso-ventrally flattened bodies; thick skin with placoid scales; lateral line system; spiracle and gill clefts open directly to the outside; heterocercal tail; rigid fins; paired kidneys; claspers in male; internal fertilisation; egg cases.

Class Osteichthyes (Bony Fish): marine and fresh water. Bony scales; lateral line system; gills covered by operculum; usually laterally flattened bodies; homocercal tail. Swim bladder and flexible fins allow considerable manoeuvrability in swimming. Lamellate arrangement and counterflow of blood within and water over the gills provides efficient respiratory gas exchange.

Class Amphibia: frogs, toads, newts and salamanders. Pectoral and pelvic girdles and pentadactyl limbs enable primitive walking and running. Double circulation (systemic and pulmonary); air-breathers using lungs and (often) moist skin for gas exchange. Fish-like larvae have gills; metamorphosis to adults.

Class Reptilia: crocodiles, lizards, snakes and turtles. Ribs moved by intercostal muscles ventilate lungs; dry skin with horny scales; kidneys excrete nitrogenous waste mainly as uric acid; Homodont dentition (all teeth the same); salivary glands. Internal fertilisation is followed by the development of amniote eggs; yolk sac, amnion and allantois; terrestrial adaptations.

Class Aves: birds. Possess feathers; usually capable of flight; forelimbs modified as wings; pneumatic bones containing air-filled cavities. Endothermic; capable of regulating their internal body temperature by physiological means. Amniote eggs with calcareous shell.

Class Mammalia: hair and sweat glands in skin; endothermic; muscular diaphragm separates thorax and abdomen; highly developed brain; intelligent behaviour; heterodont dentition (different types of teeth perform different functions); milk produced by mammary glands; internal development of embryos. Several sub-groups:
  Prototheria: Monotremes; spiny anteater and platypus; amniote egg-laying mammals; reptile-like.
  Metatheria: Marsupials, e.g. kangaroo; mostly restricted to Australia; early embryos migrate to ventral pouch for further development.
  Eutheria: Placental Mammals; embryos develop inside the mother's uterus; attached by a placenta. Many orders include:
    Insectivores, e.g. shrews, moles, hedgehogs.
    Rodents, e.g. mice, rats, squirrels.
    Carnivores, e.g. dogs, cats, seals.
    Ungulates, e.g. deer, cattle, sheep (Artiodactyls, even-toed).
          e.g. horses, zebras, rhinoceroses (Perissodactyls, odd-toed).
    Chiropterans, e.g. bats.
    Cetaceans, e.g. whales and dolphins.
    Primates, e.g. monkeys, apes and humans.

QUESTIONS

1 Examine the drawings of a lion and of a dolphin feeding its offspring. Give **three** characteristics which are found only in mammals and are shown in either of the drawings.

(AEB 1985)

Lion

Dolphin feeding its offspring

2 (a) For each of the examples the garden snail **A**, starfish **B**, snake **C**, bladderwrack **D** and white dead nettle **E** state:

(i) the major group (eg phylum) to which it belongs;
(ii) *one* subgroup to which it belongs.

Your answer should be written in columns as shown.

Specimen	(i)	(ii)
A		
etc		

(b) For each of the specimens **A**, **B** and **C**, comment on *one visible* external structural feature which is concerned with protection.
Write about 20 words for each feature.

(c) (i) Comment on *two visible* features of specimen **D** which are adaptations to its mode of life.
Write about 30 words for each feature.
(ii) For each of the specimens **D** and **E**, state *two visible* features characteristic of its major group. (L)

3 Birds (class Aves) and bats (class Mammalia) are the only two groups of chordates capable of true flight.
(a) Give **three** characteristics which could be used to distinguish birds from bats.
(b) The diagram shows the internal structure of part of a bone in a bird's wing.

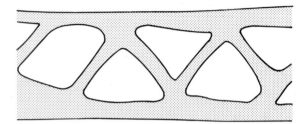

(i) How does this internal structure differ from that of a long bone in a mammalian limb?
(ii) How might this difference be considered an adaptation to flight? (AEB 1987)

4 (a) Make a drawing of a hind limb of a frog. Make annotations on the drawing to show *four* features by which the hind limb differs from the fore limb.
(b) Comment on the ways in which the structure and functions of the limbs of the frog are related to its way of life.
(c) Examine the limbs of a crab and comment on *four visible* structural differences between the limbs of this specimen. (L)

5 (a) Make a drawing of a lateral view of the head region of the bony fish, and label *four* features only.
Make annotations *next to* these four labels to explain the functions of each of these features in the way of life of the bony fish.
(b) (i) Make a labelled drawing of a representative region of hard bone.
(ii) Comment on the functions of hard bone in the life of the bony fish. (L)

31 Ecology I: Distribution of organisms; associations

31.1 Factors affecting the distribution of organisms   632

     31.1.1 Biotic factors   633
     31.1.2 Abiotic factors   637
     31.1.3 The factor complex   643
     31.1.4 Micro-ecology   644

31.2 Associations between organisms   645

     31.2.1 Parasitism   645
     31.2.2 Mutualism   648
     31.2.3 Commensalism   652

Summary   653

Questions   653

Biofotos

31 Ecology I: Distribution of organisms; associations

Ecology is the study of the interaction of living organisms with each other and with their surroundings. Plants and animals inhabit land up to a height of about 6 km above sea level. In water they live as far down as 11 km below the surface. The total volume of the earth in which life permanently exists is called the **biosphere**, in which there are four major **habitats** – marine, estuarine, freshwater and terrestrial. On land there are several **biogeographical zones,** in each of which there is characteristic plant and animal life. They are called **biomes** and include temperate deciduous forest, tropical rain forest and grasslands such as steppe and savanna (Fig 31.1).

Fig 31.1 (a) Distribution of the earth's biomes (from Sands 1978)

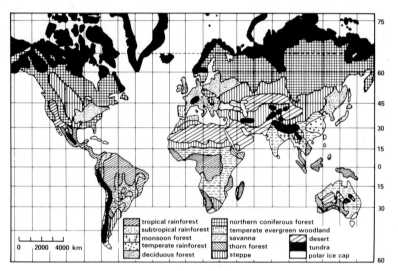

tropical rainforest	northern coniferous forest
subtropical rainforest	temperate evergreen woodland
monsoon forest	savanna
temperate rainforest	thorn forest desert
deciduous forest	steppe tundra polar ice cap

0   2000   4000 km

Fig 31.1 (b) Tropical rain forest (Bruce Coleman: Alain Compost)

Fig 31.1 (c) Tundra (Bruce Coleman: Robert Burton)

31.1 Factors affecting the distribution of organisms

One of the questions constantly asked by ecologists is, 'Why is this organism found in this habitat and not elsewhere?' The factors which determine where an organism lives are either **biotic** (living) or **abiotic** (chemical and physical). Man is one of the most important biotic factors in many habitats.

31.1.1 Biotic factors

The creatures which co-exist in a habitat determine to a large extent each other's success. Living organisms affect each other in a variety of ways. They may **compete** for the same resources. **Grazing** animals may eradicate certain species of plants and encourage the growth of others. Animals live only where their **food source** is available.

1 Competition

Plants compete for light, carbon dioxide, water, minerals, pollinators and sites for spores and seeds to germinate. Animals compete for food, mates, breeding sites and shelter from predators. Intraspecific **competition** occurs between individuals of the same species, interspecific competition between individuals of different species.

i. Intraspecific competition among flowering plants has been studied by sowing seeds at known densities and later measuring the yield of the plants. The total dry mass produced per unit area of soil is the same over a wide range of sowing density (Fig 31.2(a)). This may come about because many seedlings die when seeds germinate close to each other. The few plants that survive then have an adequate share of resources, so they grow quite well. Alternatively, most of the seedlings may survive and grow into mature plants, but each plant has a lesser share of resources so is smaller than normal. In contrast, seed production is reduced when plants are over-crowded, because many fail to flower and set seed (Fig 31.2(b)).

Fig 31.2 Effect of sowing density on (a) yield (dry crop) of clover and (b) seed production by shepherd's purse (after Solomon, *Population Dynamics*, Inst. Biol. 'Studies in Biology', Edward Arnold)

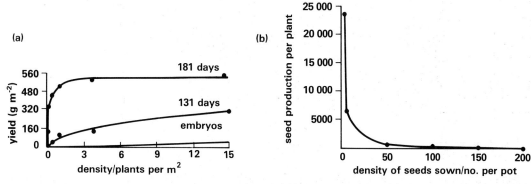

A reduction in reproductive rate is also frequently noted in dense populations of animals. The fruit fly *Drosophila* lays fewer eggs when surrounded by too many of its own kind. The mortality rate of young fish increases at high densities (Fig 31.3). In bird colonies, smaller numbers of fledglings are reared when the adult population density is high. Mammals have fewer live births in crowded conditions. Competition for food is likely to be the main cause of these effects. However, experiments with mammals

Fig 31.3 Effect of population density on (a) egg production by *Drosophila* (from Solomon as above) and (b) mortality rate of brown trout (from Open University 1974)

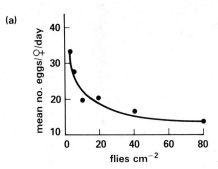
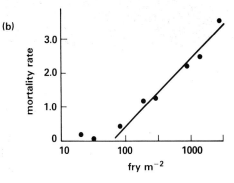

show that fewer young are reared at high densities even when food is plentiful. It is thought that an increase in aggression may in part be the cause of the fall in the rate of reproduction. Rats and hamsters have been seen to cannibalise their young when living in overcrowded conditions.

ii. Interspecific competition among plants is seen wherever crops are grown. Weeds grow in the crop and compete for space, light, water and nutrients. It is often in the early growth of the crop that weeds are very effective competitors. Unless kept in check, they can seriously reduce the yield of the crop (Fig 31.4).

Fig 31.4 Effect of weed density on the yield of field beans (from Hill, *The Biology of Weeds*, Inst. Biol. 'Studies in Biology', Edward Arnold)

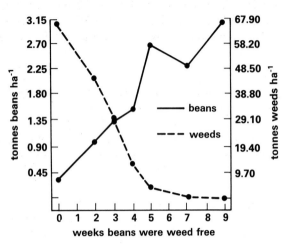

Many kinds of weeds germinate rapidly and grow quickly at the beginning of the crop's growth. By establishing themselves at this stage, they use up space in which a crop plant might have grown. Their shoot systems often intercept light which the crop would otherwise use. When the crop plants reach maturity, they may be taller than the weeds. By this time, however, the weed plants will have flowered and set seed (Fig 31.5).

Fig 31.5 Competition between weed and crop plants

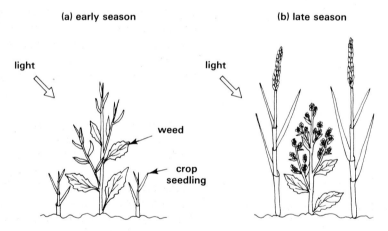

The blackbird and song thrush inhabit parkland and gardens in Britain. Both compete for the same type of food, which includes slugs, earthworms, insects and fruit. They are able to co-exist partly because the thrush eats about five times more animal food than the blackbird. Most of it comes from snails, the shells of which thrushes crack open on large stones called thrushes' anvils. Blackbirds do not have the ability to do this.

The role played by an organism in a community is called its **niche** and includes its diet, the way in which it collects its food, as well as its fate in terms of becoming food for another organism. The examples described above show that inter-specific competition allows populations of different species to co-exist in a given habitat. This is possible because different species can occupy different niches. In contrast, intraspecific competition is more likely to affect the distribution of organisms of the same species as they have similar niches and thus compete for the same resources at the same time.

Fig 31.6 The flour beetle *Tribolium* (Biofotos)

iii. Competitive exclusion The competitive exclusion principle states that two species cannot co-exist if they have the same niche. Competitive exclusion has been studied using two species of flour beetle, *Tribolium castaneum* and *T. confusum* (Fig. 31.6). They can be kept in the laboratory in bottles of flour which is habitat and food for them. The outcome of competition between the two species is governed by temperature and humidity (Table 31.1). At high temperatures (34 °C) and in very humid conditions (70 % RH) *T. castaneum* succeeded better, while at low temperatures (24 °C) and in dry conditions (30 % RH) *T. confusum* did better. Whatever the conditions, only one of the species eventually survived.

Table 31.1 Interspecific competition between flour beetles (after Park 1954)

| Temp. °C | % RH | Percentage remaining | |
		T. castaneum	*T. confusum*
34	70	100	0
34	30	10	90
29	70	86	14
29	30	13	87
24	70	29	71
24	30	0	100

2 Grazing and food source

Cattle, sheep, horses and rabbits are examples of grazers. Their natural habitat is pastureland, where they feed on growing grass and herbs. **Grazing** affects some plant species more than others. Grasses are tolerant of grazing, because their shoots regrow quickly after being nibbled. Some grassland herbs grow so close to the ground that they avoid being grazed to any large extent. Others are unpalatable to livestock, so they are left uneaten.

The effects of grazing by rabbits on grassland vegetation have long been studied. Rabbits were brought into Britain about 850 years ago to be bred for their meat and fur. The number of wild rabbits was fairly small until the middle of the last century. By the 1950s, they were a serious pest to farmers. At this time a lethal viral disease of rabbits, myxomatosis, was introduced in an attempt to reduce the number of wild rabbits. The campaign was very effective and by 1960 the wild rabbit population had severely declined. Observations of the vegetation in grassland areas before and after the outbreak of myxomatosis showed that the absence or reduction in grazing by rabbits had three general effects:

(i) the overall height of the vegetation increased,

(ii) there was a greater variety of dicotyledonous herbs, and

(iii) woody species such as bramble, heather and tree saplings started to grow in the pasture.

However, there were some notable exceptions to these trends. Two dicotyledonous herbs, wild thyme and ragwort (Fig 31.7) virtually disappeared. Wild thyme is a low-growing plant which is not intensively grazed by rabbits. In tall vegetation, however, it is a poor competitor, and gradually died out. Ragwort is a much taller plant, but again is not an effective competitor among tall herbs. The reason why it was abundant before the outbreak of myxomatosis is because rabbits dislike its taste. Where the vegetation is kept short by grazing, the ragwort is able to compete reasonably well.

Fig 31.7 The ragwort *Senecio jacobaea* (ARDEA)

Fig 31.8 Crossbill (Bruce Coleman: Hans Reinhard)

The distribution of many plant-eating animals is confined to places where their food plants grow. Some animals have specialist methods of feeding off plants. The crossbill, for example, has a beak which can prise open pine and spruce cones (Fig 31.8). Not surprisingly, it lives mainly in mature plantations of conifers. Animals able to eat a variety of plant food are more widely distributed.

Such feeding relationships help in the dispersal of many species of plants. The fruits of some flowering plants are hooked and catch in the fur of mammals as they move through vegetation. Succulent fruits are often taken as food by animals. The seeds are later egested unharmed (Fig 31.9). Migratory birds disperse seeds over great distances in this way.

Fig 31.9 Fruits dispersed by animals

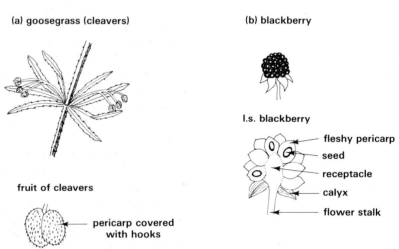

(a) goosegrass (cleavers)

fruit of cleavers

pericarp covered with hooks

(b) blackberry

l.s. blackberry

fleshy pericarp
seed
receptacle
calyx
flower stalk

3 Antibiosis

Plants and animals often deter other creatures by releasing substances into their surroundings. The substances may affect other organisms of the same species or those of other species.

i. Intraspecific antibiosis   Pheromones produced by many vertebrates which live on land are used to mark out their territory. Male rabbits are often seen rubbing their chins on the ground. This activity brings about the release of a pheromone from the submandibular salivary glands. The pheromone warns other bucks that the territory is occupied.

ii. Interspecific antibiosis   Saprophytic soil fungi such as *Penicillium* are well known as producers of **antibiotics** which kill or prevent the growth of many kinds of bacteria. Soil has been shown to contain antibiotics, so it is reasonable to assume that fungi which make antibiotics release them into their natural environment. Here they enable them to compete with saprophytic bacteria.

4 Man

Homo sapiens is undoubtedly the most potent biotic force in the world today. Much of the vegetation on land is not natural. It has been grown there by man in his capacity as a farmer, forester, landscaper and gardener. Man is also a predator and has hunted many animals to the point of extinction. Environmental pollution is largely due to man's activities and influences species distribution locally and worldwide (Chapter 33).

31.1.2 Abiotic factors

Edaphic factors affect soil conditions (Chapter 32); they have a bearing on the distribution of terrestrial plants and hence on animals which use them as food sources and living quarters. **Climatic factors** also determine where many terrestrial species live. Fluctuations in temperature and rainfall, the duration and intensity of solar radiation and air movement, are mong the the factors which contribute to climate. Temperature and solar radiation affect aquatic habitats too, as do water movement, water potential and the availability of oxygen for respiration and carbon dioxide for photosynthesis.

1 Temperature

Fluctuations in temperature in lakes, seas and oceans are relatively small because water has a high specific heat capacity. Radiant heat from the sun does not penetrate water very far, so the temperature of deep waters is fairly low (2–3 °C) but constant. This is partly why aquatic habitats are suitable for many kinds of ectothermic creatures.

Extremes of temperature occur on land. On the fringes of polar areas, the air temperature for much of the year is nearly always below freezing point. Only lichens, mosses, a few flowering plants and one or two endothermic animals such as penguins and polar bears can tolerate such conditions. For similar reasons high mountains have a typical alpine flora and fauna.

In desert areas the daytime temperature is very high and causes water to evaporate very quickly. Here a major problem faced by living organisms is how to resist or tolerate desiccation. Xerophytic plants such as cacti, insects and a few mammals, notably camels, are able to do this.

Although the effects of temperature on living organisms are so apparent, few correlations are known to occur between distribution of species and temperature. One that is established relates to the mosquito *Aedes aegypti* which is confined to the 10 °C isotherm on either side of the equator (Fig 31.10).

Fig 31.10 The geographical distribution of the mosquito *Aedes aegypti* (from King 1980)

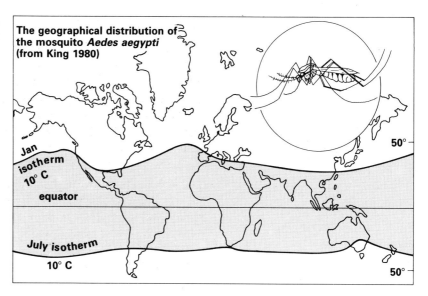

The geographical distribution of the mosquito *Aedes aegypti* (from King 1980)

Jan isotherm 10° C

equator

July isotherm 10° C

50°

50°

2 Solar radiation

Any self-sustaining community must have access to sunlight, because light is essential for photosynthesis (Chapter 5). Although many heterotrophic

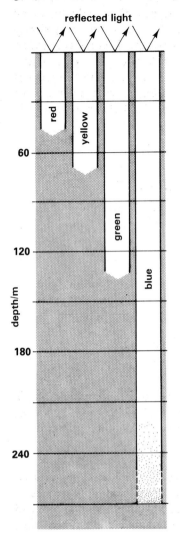

Fig 31.11 Penetration of water by visible light (after Lucas and Critch 1974)

reflected light

depth/m

organisms can and do live in total darkness, they rely on organic molecules made by photo-autotrophs as an energy source.

In the ozone layer of the earth's atmosphere, much of the ultraviolet light from the sun is absorbed. Solar radiation passing through this layer is about 10 % ultraviolet, 45 % visible light and 45 % infra-red. Infra-red radiation is the earth's external source of heat. Clouds of water vapour and dust reduce the amount of radiation arriving at the earth's surface. Latitude, season and the amount of cloud cover greatly affect the intensity of solar radiation reaching a given habitat. It is at its highest in tropical areas which thus have higher mean annual temperatures than elsewhere and consequently higher rates of evaporation.

The shorter wavelengths of light penetrate water deepest. Most of the red light is absorbed in the first 50 m. Only blue light reaches a depth of 250 m. Blue light reflected from the sea bed gives the sea its blue colour when the sun shines. Beyond 250 m, there is total darkness (Fig 31.11). It is for this reason that green algae which require a high intensity of red and blue light for photosynthesis live mainly near the sea surface. Brown algae which can make do with lower intensities of red light live between 10 and 25 m, while red algae which use mainly blue light can grow deeper down. In this way the various kinds of algae avoid competing with each other. As with the atmosphere, the amount of light which penetrates water can be seriously reduced by the presence of suspended particles.

3 Availability of oxygen

Most living organisms are **aerobes** and require oxygen gas for the release of energy in respiration. At sea level, the atmosphere contains about 20.9 % oxygen gas by volume. At high altitudes there is the same proportion of oxygen present, but its partial pressure is less so it is not readily absorbed by living organisms. Given time, mammals produce extra erythrocytes which enable them to carry enough oxygen to their tissues when at high altitudes. Others such as the llama, which normally live at such heights, have a variant of haemoglobin with an exceptional affinity for oxygen (Chapter 8). Even so, there is a limit to the ability of any terrestrial animal to compensate for the lack of oxygen in habitats of this kind. Together with low temperature, it explains why there is very little mountain fauna above a height of about 4.5 km, where the partial pressure of oxygen is roughly half that at sea level.

Oxygen gas is relatively scarce in water. The amount present depends on its solubility, the temperature (Table 31.2), the partial pressure of oxygen in the air, the amount of solutes dissolved in the water and on the rate at which it is consumed by aquatic organisms. Even when fully saturated with oxygen, water contains less than 1 % by volume of the gas. Oxygen is thus much less accessible to aquatic organisms. Some aquatic animals require relatively high concentrations of dissolved oxygen to survive. Others can manage where there is much less (Table 31.3).

Table 31.2 Effect of temperature on concentration of dissolved oxygen in water

Temp. (°C)	Percentage oxygen by volume
10	0.796
15	0.723
20	0.649
25	0.593
30	0.548

Table 31.3 Effect of dissolved oxygen concentration on survival of freshwater fish (after Open University 1972)

Species	Dissolved oxygen (mg dm^{-3})*
rainbow trout	3.7
perch	1.2
roach	0.8
tench	0.7

*The concentrations are the minimum required for 7 days' survival at 16 °C.

Near the surface of a lake or sea where the water is in contact with air, the water may be saturated with oxygen. Photosynthetic plants help to maintain high concentrations of dissolved oxygen. Deeper down, where there are no photo-autotrophs and because oxygen diffuses very slowly through water, much less dissolved oxygen is present. Thermal convection currents help to carry some oxygen downwards from the upper layers. However, in the deepest waters there may be little or no oxygen gas available. Here only invertebrates whose oxygen-carrying pigments become saturated at very low oxygen tensions (Fig 31.12) and anaerobic bacteria can survive.

Fig 31.12 Oxygen dissociation curves for haemoglobin of *Arenicola* and *Nephtys* (from Jones 1955)

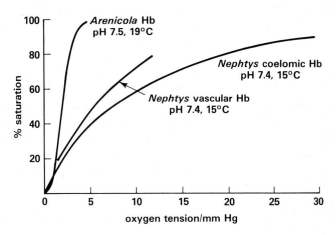

Aquatic habitats may become seriously depeleted of dissolved oxygen when polluted by organic matter. The effect of such pollution on the distribution of river fauna is fully described in Chapter 33.

4 Availability of carbon dioxide

Carbon dioxide is one of the raw materials essential for **photosynthesis**. At sea level, the earth's atmosphere contains about 0.03 % by volume of carbon dioxide. When light intensity is high, it is this small amount of carbon dioxide which limits the rate of photosynthesis. As with oxygen, the partial pressure of carbon dioxide gas at high altitudes is less. It is mainly this, and the lower temperatures, which prevents plants growing at a height of more than 6 km in mountainous areas.

Carbon dioxide is nearly forty times more soluble in water than oxygen. However, because the partial pressure of carbon dioxide in the air is 700 times less, the volume of carbon dioxide dissolved in water is generally small compared with oxygen. Even so, aquatic plants are never short of carbon dioxide for photosynthesis. This is because water contains vast quantities of carbonate and hydrogencarbonate ions which are formed when carbon dioxide reacts with water:

Fig 31.13 Effect of altitude on atmospheric pressure (from Hollingsworth and Bowler, *Principles and Processes of Biology*, Chapman & Hall)

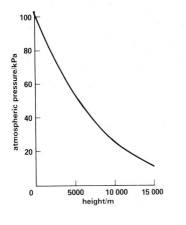

$$CO_2 + H_2O \rightleftharpoons H_2OCO_3 \rightleftharpoons H^+ + HCO_3^- \rightleftharpoons 2H^+ + CO_3^{2-}$$

$$\qquad\qquad\quad \underset{\text{acid}}{\text{carbonic}} \qquad\quad \underset{\text{carbonate}}{\text{hydrogen-}} \qquad\quad \text{carbonate}$$

The ions can be used as a source of carbon dioxide. As dissolved carbon dioxide, HCO_3^- and CO_3^{2-}, sea water contains the equivalent of 4.7 % carbon dioxide, about 150 times more than in the earth's atmosphere.

5 Pressure

At sea level, the pressure of the air is 101.3 kPa (1 atm), but it reduces with altitude (Fig 31.13). Cold and the low partial pressures of oxygen and

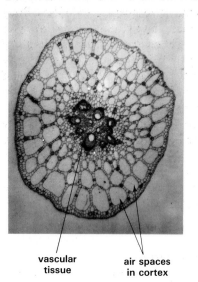

Fig 31.14 Transverse section stem of pondweed *Potamogeton* (B&B photos)

vascular tissue   air spaces in cortex

carbon dioxide prevent organisms living permanently above a height of about 6 km.

In water, the pressure increases by about 100 kPa for each 10 m in depth. At 11 km below the oceans' surface, the greatest depth at which aquatic organisms live permanently, the pressure is over 1000 times that at sea level. Simple forms of life such as bacteria appear to be unaffected by such a high pressure. Other aquatic organisms avoid pressures of this magnitude by floating at an appropriate depth. Bony fishes have **swim bladders** which enable them to adjust the depth and hence the pressure at which they swim. Whales and seals store large amounts of **blubber** which gives them buoyancy. Freshwater plants have **aerating tissue** which retains oxygen and carbon dioxide (Fig 31.14). Seaweeds often have buoyant **air-bladders**. The lift provided by such adaptations ensures that many aquatic plants float near the surface where light, oxygen and carbon dioxide are more plentiful.

6 Water potential

Water is vital for many of the activities of living organisms (Chapter 1). In most terrestrial habitats, the water potential of the air is much less than that of living organisms. Consequently, organisms which live on land constantly lose water by **evaporation**. Species which are unable to minimise water loss due to evaporation thrive best in places where the percentage relative humidity of the air is high. For this reason, liverworts and many species of moss are mainly confined to shaded moist habitats such as woodlands, whereas woodlice congregate under the bark of fallen trees. The external surfaces of most land-dwelling organisms are waterproofed and evaporation of water into the air is minimised. Even so when the stomata of terrestrial plants are opened to take in carbon dioxide for photosynthesis, water vapour is lost by **transpiration**. Animals which live on land conserve water by reabsorption in the renal system and hind gut. Insects, birds and mammals minimise water loss by producing solid or semi-solid faeces. Nevertheless they continually lose water vapour through their respiratory surfaces.

In aquatic habitats there are three possibilities:

  (i) the water potential of an organism is less than its surroundings,
 (ii) the water potential is greater than its surroundings,
(iii) the water potential is the same as its surroundings.

Fresh water contains only small amounts of dissolved solutes (Table 31.4) so usually has a higher water potential than the organisms which live in it. It means that freshwater plants and animals constantly absorb water by **osmosis**. The pressure potential of plant cells limits the intake of water.

Table 31.4 Comparison of ionic composition of fresh water and freshwater organisms (after Hollingsworth and Bowler 1972)

	Concentration (mmol dm^{-3})				
	Na$^+$	K$^+$	Ca^{2+}	Mg^{2+}	Cl$^-$
fresh water	0.5	0.06	2.7	0.4	0.9
crayfish blood	142	4	8	4	140
Nitella sap	86	59	19	22	107

Freshwater animals have to eliminate the excess water. In simple animals such as protozoans, **contractile vacuoles** serve this purpose. More complex animals such as fish have elaborate osmoregulatory devices. Bony fish which inhabit ponds, streams, rivers and freshwater lakes absorb water by osmosis through their gills. The regulate their water potential by producing large volumes of dilute urine. Their renal tubules are very efficient in reabsorbing mineral ions from the renal filtrate. Any loss of minerals in the urine is made good partly by the fishes' diet and also by the active absorption of salts by the **chloride cells** of the gills.

Sea water is relatively rich in dissolved solutes and has a lower water potential than the body fluids of marine bony fish. As a consequence they constantly lose water by osmosis through their gills. They drink sea water to make good the loss of water. Excess sodium, potassium and chloride ions are excreted through the chloride cells of the gills (Fig 31.15). The low glomerular filtration rate of the kidneys ensures little water is passed in their urine.

Fig 31.15 Osmoregulation in fish (a) freshwater bony fish (trout) (b) a marine bony fish (cod)

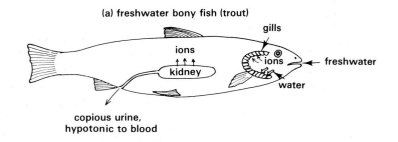

(a) freshwater bony fish (trout)

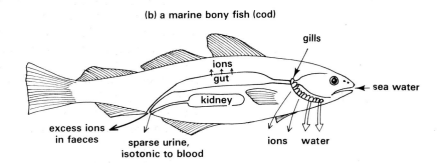

(b) a marine bony fish (cod)

Most marine animals maintain their body fluids at the same water potential as sea water by regulating their intake of solutes. Aquatic animals which live in estuaries where the sea and river waters mix are often able to do this over a wide range of water potential (Fig 31.16). They are said to be **euryhaline**. The majority of marine animals have limited powers at doing this and are said to be **stenohaline**. Migratory fish such as the salmon and eel change their method of osmoregulation when they move from the sea to rivers, and vice versa. The ability to adapt to changes of water potential enables such creatures to exploit a wider range of habitat than those which cannot do so.

How do the differences reflect the habitats of the animals?

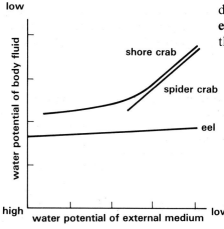

Fig 31.16 Effect of water potential of the external medium on the water potential of the body fluid of three aquatic animals (after Ramsay 1968)

The distribution of some species of plants is also influenced by water potential. Salt marsh plants such as the glasswort *Salicornia* (Chapter 32) can tolerate the low water potential of estuarine mud which is constantly washed by the tide. Inland species are unable to grow in such an environment, either because the water potential is too low or because they cannot cope with the very high salt content of the mud.

7 Water movement

It is in fast-flowing streams and rivers, in wave action at the surface of lakes and seas, on sea shores and in estuaries that **water movement** is most apparent. Turbulent water is likely to be well oxygenated, as it is constantly brought into contact with air. Still, deep water is often deficient in oxygen.

Fig 31.17 River crowfoot *Ranunculus fluitans*

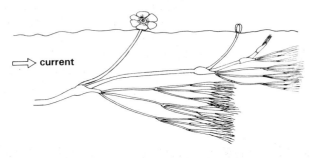

current

Fig 31.18 (a) Ramshorn snail (b) larva of caddis fly

(a)

(b)

Living organisms cope with the mechanical effect of moving water in a variety of ways. Submerged aquatic plants have tough but flexible bodies which enable them to bend with the current without being damaged (Fig 31.17). Attachment organs such as the **holdfast** of seaweeds prevent aquatic plants from being washed away in moving water.

The ways in which aquatic animals withstand or avoid moving water are just as varied. Many estuarine animals such as lugworms and cockles burrow into mud and thus avoid being moved by tidal currents. Shore crabs hide under seaweeds and in rock crevices. Sea anemones, limpets and barnacles are firmly fixed to rock surfaces. In streams and rivers, caddis fly larvae are weighed down by their stony cases, snails by their shells (Fig 31.18)

The larvae of fixed aquatic animals, the spores of non-flowering aquatic plants and the seeds of a few species of flowering plants are **dispersed** by moving water. Many of the larger fish and aquatic mammals are strong swimmers and often migrate long distances against tides and river currents.

8 Air movement

Moving air encourages **evaporation** of water. It is a factor in determining the distribution of terrestrial plants. Windy habitats are often colonised by plants which can minimise transpiration (Chapter 12).

Wind also aids in the **dispersal** of plants and animals. Birds are the masters of the air. They can fly long distances even against strong winds. Insects are not usually very strong fliers and are often blown from place to place. The same applies to plant spores and seeds. Spores are very light and remain suspended in the air for long periods. Seeds sometimes have wing-like extensions, which help their dispersal by moving air (Fig 31.19).

Fig 31.19 Winged fruits

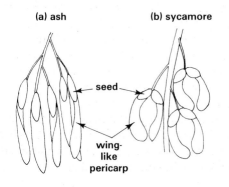

(a) ash   (b) sycamore

seed

wing-like pericarp

31.1.3 The factor complex

Finding a species in a particular habitat does not necessarily prove that it is its optimum environment, only that it can survive there.

It is unusual for a single factor to determine the distribution of a species. More often it is a complex of interacting factors. A study of the distribution of the inhabitants of rocky seashores around Britain helps to illustrate the **factor complex**. Between high and low water marks is the littoral belt, in which brown algae and invertebrates are clearly zoned (Fig 31.20). Tufts of channelled wrack, *Pelvetia canaliculata*, cling to the rocks near the high-water mark on sheltered shores. A little further down the shore are clumps of spiral wrack *Fucus spiralis*. Among the algae and in rock crevices in this **upper zone** is a gastropod mollusc, the winkle, *Littorina saxatilis*. Patches of rock are occupied by the barnacle *Chthamalus stellatus*. Next is a **middle zone** of heavy masses of knotted wrack, *Ascophyllum nodosum*, in which the predatory dog whelk *Nucella lapillus* lurks. Another barnacle, *Balanus balanoides*, and the limpet, *Patella vulgata*, adhere to the rocks in this zone. At the bottom of the shore is a **lower zone** of serrated wrack, *Fucus serratus*, and bladderwrack, *Fucus vesiculosus*.

Fig 31.20 Zonation on a rocky shore (a) channelled wrack and (b) spiral wrack in upper zone (c) barnacles and limpets and (d) knotted wrack in middle zone

(a) (b) (c) (d)

The upper limit of each species on the shore is possibly a measure of its ability to resist desiccation when exposed to air at low tide. Species in the lower zone are exposed for a shorter time than those living higher up the shore. It may be that the species of the upper zone can retain moisture better and are thus able to tolerate longer exposures to drying air.

A more difficult question to answer is what determines the lower limit of each species. Is it a measure of the ability to survive immersion in sea water and buffeting by the waves? Do competition, grazing and predation have a bearing on the zonation?

An insight into the complex of factors which affect the distribution of the two species of barnacles has been obtained by moving rocks on which they live to other zones. In one investigation, rocks covered with *Chthamalus* were moved downshore to the middle zone occupied by *Balanus*. The moved barnacles survived for a long time, showing that they could tolerate the conditions lower down the shore. Eventually, larvae of *Balanus* settled on the stones and, as they grew, *Chthamalus* gradually disappeared. One of the reasons why *Chthamalus* does not do well in the middle zone may therefore be that it does not compete well with *Balanus* for space on the rocks. Predation by the dog whelk and smothering of the rocks by knotted wrack may be contributory factors.

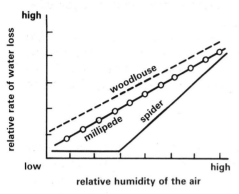

31.1.4 Micro-ecology

The organisms which frequent the same general habitat may occupy a very small part of it, a **microhabitat**, in which conditions are often very different. In a woodland, for example, fallen logs and dead tree trunks provide microhabitats for a variety of small invertebrates. Newly felled logs are soon attacked by the larvae of wood-boring beetles. Woodlice, centipedes and spiders later colonise the space beneath the lifting bark. There the air has a constant high relative humidity and is sheltered from extreme fluctuations in temperature. Cool, humid conditions are favoured by such small animals which, in proportion to their volume, have a large surface area through which water vapour can pass. Spiders and centipedes have a waxy epicuticle which restricts evaporation through the body wall. Woodlice have no such protection (Fig 31.21) and become desiccated if the air is too dry. They spend much of the day hidden beneath the bark but come out at night when the outside temperature falls and humidity rises.

When placed in the open air, woodlice run about at random until they find a humid spot where they rest. This behaviour is called **kinesis**. **Orthokinesis** is the rate of movement in response to the intensity of a stimulus. For example, woodlice rest when the air is saturated with water vapour, move slowly in humid air and quickly in dry air. **Klinokinesis** is the frequency of change of direction or turning and is also dependent on the intensity of a stimulus. In dry air woodlice move around in circles and this may bring them back to more humid conditions. Such reactions enable woodlice to find a microhabitat of their choice.

The preference of woodlice for air of high humidity can be investigated using a choice chamber (Fig 31.22). Between 5 and 10 animals are placed on the gauze and moved around with a soft brush until they are randomly spaced. The lid is then replaced and after 10 minutes the number found on either side of the dish is recorded. Repetition of the procedure provides more data and increases the reliability of the experiment.

What precautions should be taken to ensure that the only variable in the experiment is the humidity of the air?
How can the experiment be modified to investigate the effect of light on the response to humidity?

Fig 31.22 (a) A choice chamber for investigating responses to relative humidity. Each half of the divided dish provides a different relative humidity in the air immediately above it. A range of humidities can be obtained using dilutions of sulphuric acid or potassium hydroxide

Fig 31.22 (b) British woodlice: (a) *Trichoniscus pusillus*, (b) *Philoscia muscorum*, (c) *Oniscus asellus*, (d) *Porcellio scaber*, (e) *Armadillidium vulgare* (Reproduced from *Countryside*, **18**, 105, the Journal of the British Naturalists' Association)

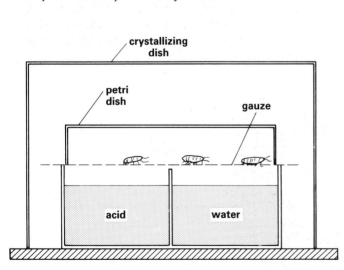

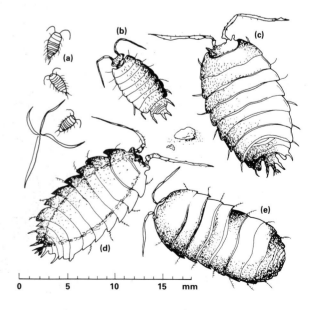

31.2 Associations between organisms

Plants and animals of different species often associate temporarily or permanently with each other. **Associations** are often a means of providing food for one or both of the partners and in the broad sense are regarded as **symbiotic** relationships.

Contact in such partnerships ranges from infrequent, as when a flea occasionally visits its host to suck blood, to permanent, as in the case of a tapeworm in the gut. In obtaining nourishment one of the partners may benefit to the detriment of the other, both may benefit or there may be no apparent overall benefit to either.

31.2.1 Parasitism

A **parasite** is an organism which gets part or all of its food from another species called the **host**. The activities of parasites are usually harmful and they are often the cause of their hosts' death. However, hosts are sometimes carriers of parasites without coming to any serious harm. Many parasites live permanently on or in their hosts. Others visit their hosts only to feed. Whereas the majority are **obligate parasites**, a few such as *Pythium*, a fungus which causes **damping off** of seedlings, are **facultative parasites**. After killing seedlings, the fungus lives as a saprophyte on their dead remains.

Parasites belong to a wide range of living organisms. Those which are most successful are protozoans, platyhelminths, nematodes, arthropods, fungi, viruses and bacteria. The general characteristics of these groups of plants and animals are described in Chapters 28–29. Parasitic micro-organisms are often called **pathogens**.

Parasites are sometimes thought of as **degenerate organisms**. This is because they often lack some of the features of their free-living relatives. Flatworm parasites, for example, often have no digestive system or sensory organs. They also have limited powers of movement and co-ordination. Parasitic flowering plants have no roots and may lack chloroplasts. However, it is probably better to think of parasites as **specialists**. Their morphology, biochemistry, physiology, methods of reproduction and dispersal are specially adapted for a parasitic existence.

1 Habitat

Ectoparasites live on the surface of the host's body. The skin, mouth, nostrils and gills of animal hosts and the epidermis of plant hosts are the usual places where they are to be found. Ectoparasites of animals often cling to their hosts using claws, suckers or hooks. Ectoparasites of plants have attachment organs called **haustoria** (Fig 31.23).

Fig 31.23 Haustoria of dodder, *Cuscuta*

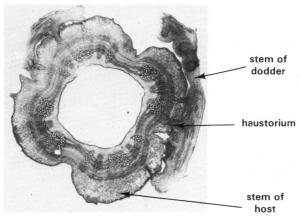

stem of dodder

haustorium

stem of host

Endoparasites live inside their hosts' bodies. Many endoparasites of plants have a preference for young, active tissues such as the mesophyll of leaves. Endoparasites of animals frequently inhabit the digestive tract or body bluids of their hosts. Gut parasites often have hooks and suckers which attach them to the gut wall. In this way their movement along the digestive tract by peristalsis is prevented.

2 Access to host tissues

A **vector** is an animal which carries parasites from one host to another. **Blood-sucking insects** are the most notorious vectors of a range of human parasites. Millions of people are infected annually with yellow fever and malaria, which are transmitted by mosquitoes. The tsetse fly is the vector of sleeping sickness, black flies carry river blindness, while typhus is transmitted by the body louse. **Sap-sucking insects** such as aphids are the vectors of many viral diseases of crops. Enormous sums of money are spent each year on insecticides in an attempt to prevent insect vectors coming into contact with humans and crops. Stopping the vector from breeding is often an effective alternative strategy.

Other parasites enter their hosts through damaged outer tissues or through natural openings such as stomata, the nasal openings and the mouth. Some can even penetrate unbroken skin. An example is the fluke *Schistosoma* which is passed in a larval form to man from water snails. The the disease caused by the fluke is schistosomiasis (bilharzia). It is prevalent in countries where crops are irrigated, as in rice-growing. Molluscicides can be used to kill the snails, thus denying the fluke access to human tissues.

Parasites are often very **invasive** when they enter a host. They may secrete **enzymes** enabling them to penetrate healthy tissues. Fungal parasites invade plant tissues by secreting pectinase and cellulase enzymes. Some pathogenic bacteria attack hyaluronic acid, the material which cements animal cells to each other, by secreting the enzyme hyaluronidase. *Schistosoma* secretes the same enzyme to penetrate human skin. Once in the body, it invades body organs, especially the liver, through blood and lymph capillaries.

3 Host reaction

Microbial parasites of animals trigger the host's **immune system** into action (Chapter 10). However, the parasite often has time to become established before the system comes into action. Phagocytosis also helps to ward off infections caused by microbes. Defence mechanisms of these kinds are ineffective against larger parasites such as helminths and nematodes. It may be expected that the host's digestive enzymes would attack gut parasites. However, these parasites have very tough outer coverings which resist damage by gut enzymes.

Plants do not have an immune system. Nevertheless, they can resist infection in a variety of ways. When infected, many species of plants produce chemicals called **phytoalexins** which protect healthy tissues from attack by the parasite. Others have high concentrations of toxic substances, notably phenols, in their external tissues.

4 Feeding

Many parasites have very exacting nutritional requirements. This partly explains why a parasite often has a specific host or can live in only a limited range of hosts.

Microbial parasites such as viruses, bacteria and protozoa absorb soluble materials through their body surfaces by active transport and diffusion from the body fluids and cells of the host. Insect and nematode parasites frequently have mouthparts adapted for piercing the host's outer tissues and sucking blood or phloem sap. Gut parasites take in food which has been made soluble by the digestive enzymes of the hosts. Many endoparasites have a relatively large surface area through which nutrients are absorbed.

The haustoria of parasitic plants also provide a large area of contact with the host. In addition, they make the membranes of the host cells permeable to nutrients, allowing food to pass into the parasite. Some parasitic flowering plants, such as mistletoe, have green leaves and can thus photosynthesise. Others, such as the dodder, have scale leaves which lack chloroplasts. They rely entirely on the host for organic nutrients.

5 Respiration

Parasites which live where there is a high oxygen tension release energy by **aerobic respiration**. They include ectoparasites and endoparasites which inhabit well-oxygenated tissues such as the mesophyll of leaves and the blood of vertebrates. The intestines of animals have a low oxygen tension, so gut parasites respire anaerobically. The relative inefficiency of **anaerobic respiration** in releasing energy (Chapter 5) is compensated for by the abundant supply of respiratory substrates in the host's food.

6 Reproduction and transmission

The chance of most parasites finding a new host is maximised because they have a prolific rate of reproduction. Both asexual and sexual reproductive methods are used.

Endoparasitic fungi release huge numbers of asexual spores from hyphae on the outside of the host. They are carried by air currents over vast areas (Fig 31.24). This is why many diseases of crop plants spread so rapidly.

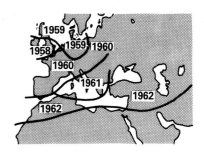

Fig 31.24 Spread of *Peronospora tabacina* through European tobacco crops (from Deverall, *Fungal Parasitism*, Inst. Biol. 'Studies in Biology', Edward Arnold)

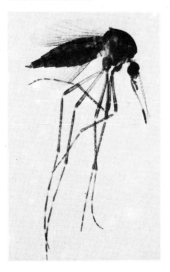

Fig 31.25 Anopheline mosquito, the vector of malaria

Asexual spores produced by protozoan parasites which inhabit blood are carried to new hosts by blood-sucking insect **vectors**. This is how the malarial parasite is transmitted by the mosquito, *Anopheles* (Fig 31.25). Extremely large numbers of sexually-produced cysts and eggs of gut parasites pass to the outside world in the faeces of their host (Table 31.5).

Table 31.5 Number of eggs produced by selected parasites (after Marshall 1978)

Species	No. produced
Taenia solium (pork tapeworm)	800×10^6 (total production)
Ascaris (roundworm)	200×10^3 per day
Ankylostoma (hookworm)	35×10^3 per day
Diphyllobothrium (fish tapeworm)	1×10^6 per day

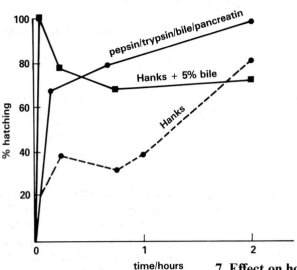

Fig 31.26 Effect of intestinal enzymes on the hatching of cysts of the dog tapeworm (after Smyth 1967). Hanks is a physiological solution with a composition similar to intestinal fluid but it contains no enzymes

They are able to resist desiccation and can survive for years. A fresh host usually becomes parasitised when consuming infected food or water. Cysts and eggs of gut parasites are often stimulated to hatch by intestinal enzymes (Fig 31.26). In this way they remain dormant until they reach an appropriate host.

Some parasites have **secondary hosts** in which they spend part of their life cycle. The rabbit is the secondary host of the dog tapeworm. Others have **alternative hosts**. Many game animals in Africa are alternative hosts to humans for the malarial parasite. Additional hosts of these kinds improve the survival chances of such parasites.

7 Effect on host

Parasites sometimes have no detectable effect on their hosts. Humans can be **carriers** of parasites, such as the bacterium which causes typhoid fever, without any ill-effects. However, there are many instances when the harmful effects of parasites are all too apparent.

Physical damage to important organs such as the brain and liver is caused by the hydatid cysts of flatworm parasites. Gut parasites sometimes damage the intestinal mucosa, causing haemorrhage. **Enzymic damage** of host tissues by invasive bacteria and fungi is quite common. Bacterial and protozoan parasites of the bloodstream frequently bring about haemolysis of the host's red blood cells. Substances which are toxic to the host are secreted by many parasites. Probably the best-known of these are bacterial **toxins**. Tumorous growth of the host is a feature of some parasitic diseases. Fungal parasites of plants are known to produce plant hormones, which are thought to be the cause of swollen organs in diseases such as club-root of brassicas (Fig 31.37). Gut parasites often absorb sufficient food for the host to be undernourished. Blood-suckers may take enough blood for the host to become anaemic.

The general effect of parasites is to bring about a loss in fitness of the host. Even though the host may survive, its inability to thrive when parasitised may mean that it competes less favourably or that it is more easily taken by predators.

31.2.2 Mutualism

Mutualism is an association between organisms of different species in which each partner benefits. In the narrow sense, mutualism is often referred to as symbiosis. The partners in mutualistic associations are often so dependent on each other that they are unable to live independently. One or both partners are sometimes found as free-living organisms.

1 Mycorrhizae

A **mycorrhiza** consists of a fungus and the root of a higher plant. In **ectotrophic** mycorrhizae, the fungus forms a sheath which covers the lateral roots of forest trees such as oak, beech and conifers. The fungus is

Fig 31.27 Base of cabbage stalk with club root (Biophoto Associates)

648

often a basidiomycete unable to hydrolyse lignin and cellulose, the most abundant energy-rich substrates in soil. It depends on photosynthesis by the tree to provide organic materials. Experiments in which $^{14}CO_2$ was fed to mycrorrhizal pine seedlings have shown that up to 10 % of the products of photosynthesis pass into the fungus. Even so, mycorrhizal plants grow more rapidly than those without the fungal partner (Table 31.6).

Table 31.6 Comparison of growth of mycorrhizal and non-mycorrhizal plants (after Harley 1971)

	Mean dry mass (g) of seedlings or cuttings	
	non-mycorrhizal	mycorrhizal
Pinus strobus	0.303	0.405
Quercus robur	1.14	1.69
Eucalyptus macrorhiza	7.8	11.3

Fig 31.28 Transverse section of an endotrophic mycorrhiza

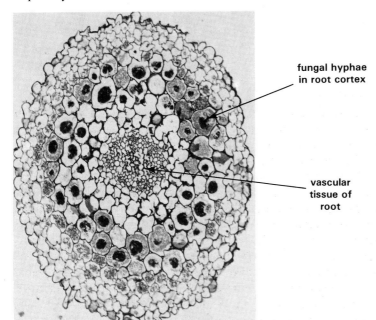

fungal hyphae in root cortex

vascular tissue of root

Endotrophic mycorrhizae have most of the fungus inside the root (Fig 31.28). Phycomycete, basidiomycete and imperfect fungi have been found in this type of mycorrhiza. The partner includes trees and orchids, as well as a wide range of important crop plants. Some of the fungi rely on photosynthesis by the partner to provide organic materials. Others can digest lignin and cellulose in the soil and pass the end-products into the roots of the plant. In this way growth of the plant is stimulated. Some orchids are unable to photosynthesise and rely entirely on the fungus for organic materials.

The more vigorous growth of most mycorrhizal plants is thought to be due to their efficiency at absorbing minerals even from mineral-deficient soils (Table 31.7). It enables them to compete more effectively with other plants in places where soils are infertile and allows them to become established in poor soils. The benefit to the fungus is apparent where it uses the plant's photosynthetic products. The advantage is less obvious where the fungus is not reliant on its partner for such materials.

Table 31.7 Growth of mycorrhizal and non-mycorrhizal plants in soil with high (H) and low (L) phosphate content (after Harley 1971)

		Mean dry mass (g)	
		non-mycorrhizal	mycorrhizal
strawberry	L	6.69	16.04
	H	18.83	17.04
tomato	L	0.09	0.39
	H	0.70	0.74

2 Lichens

A **lichen** is an association between an **alga** and a **fungus**. Lichens are often found as crust-like, leaf-like or shrub-like growths on tree bark, roof tiles and stones (Fig 31.29). More than 90 % of the dry mass of most lichens is made up of the fungus. The algal cells are scattered near the surface of the lichen, where light can penetrate. The unicellular green *Trebouxia* is most commonly the algal partner; an ascomycete fungus is nearly always the fungal partner. Neither have been known to live independently. Some algal partners such as *Nostoc* are, however, free-living. It is able to use nitrogen gas in the same way as nitrogen-fixing bacteria.

Fig 31.29 (a) Crustose lichens growing on bare rock, (b) transverse section of a lichen *Xanthoria*, (c) *Physcia*, TS (Biophoto Associates)

(a)

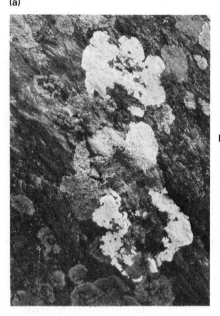

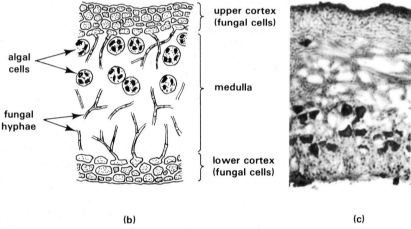

(b)

(c)

Experiments with $^{14}CO_2$ show that between 40 and 70 % of the photosynthetic products of the algal partner pass into the fungus. Where the alga is a nitrogen-fixer, most of the nitrogenous compounds it makes are also taken up by its partner. It is therefore quite obvious how the fungus gains from the association. The advantages to the alga are less apparent. It may be protected by the fungus from intense sunlight and from desiccation. It has also been suggested that minerals absorbed by the fungus are transferred to the alga, but there is no experimental proof that this occurs.

Fig 31.30 (a) Root system of broad bean plant, $\times \frac{3}{4}$

(a)

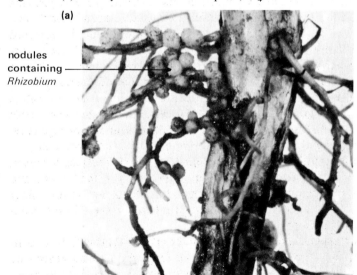

nodules containing *Rhizobium*

3 Root nodules and nitrogen-fixing bacteria

Probably the best known of this type of association is between nitrogen-fixing bacteria of the genus *Rhizobium* and the roots of leguminous plants such as clover. The bacteria get into the roots through root hairs. They then cause the cortex cells to divide, forming **nodules** in which the bacteria multiply (Fig 31.30). *Rhizobium* reduces nitrogen gas to form ammonia.

Fig 31.30 (b) Section through a root nodule, × 30

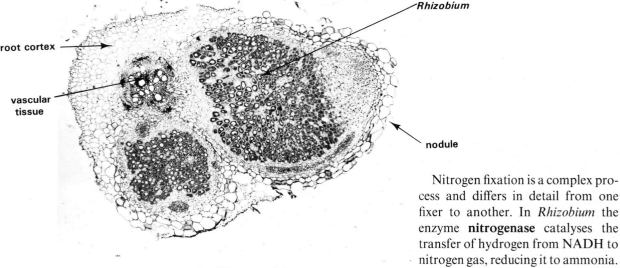

root cortex

vascular tissue

cells containing *Rhizobium*

nodule

Nitrogen fixation is a complex process and differs in detail from one fixer to another. In *Rhizobium* the enzyme **nitrogenase** catalyses the transfer of hydrogen from NADH to nitrogen gas, reducing it to ammonia. The ammonia combines with carboxylic acids formed in the Krebs cycle to make a range of amino acids. Any surplus to requirements are used by the host for protein and nucleic acid synthesis, thus stimulating its growth.

$$N_2 \xrightarrow[\text{NADH} \quad \text{NAD}^+]{\text{nitrogenase}} NH_3 \rightarrow \text{amino acids} \rightarrow \text{proteins}$$

In return, the bacteria are supplied with sugars and vitamins and have a sheltered habitat.

Apart from growing more vigorously in partnership with nitrogen-fixing bacteria, plants with root nodules are able to grow in soils deficient in ammonia and nitrates. This makes them very competitive in poor soils.

4 Herbivorous animals and gut microbes

Ruminant mammals such as cattle, goats and sheep have complex stomachs (Chapter 15). One of the chambers is the **rumen**. It is here that a variety of bacteria and protozoa live in millions. They are anaerobic and obtain energy by fermenting carbohydrates and proteins in herbage. About 60–90 % of the carbohydrates in herbage are fermented in the rumen. The main products are volatile fatty acids such as ethanoic, lactic, butyric and propionic acids, as well as carbon dioxide and methane gas. The gases are belched out, but the acids are absorbed into the ruminant's bloodstream. Up to 70 % of the animal's energy requirements come from the absorbed acids. What is even more significant is that without the rumen bacteria very little of the carbohydrate in herbage could be used by the ruminant. This is because no mammal produces an enzyme capable of breaking down cellulose, the main ingredient of plant cell walls. Unless the cell wall is destroyed, starch grains, sugars and proteins inside the cells are inaccessible to enzyme attack. Rumen protozoa are mainly ciliates and a few flagellates. They ferment starch, forming the same end-products as the bacteria.

Between 40 and 50 % of the proteins in herbage are hydrolysed by rumen bacteria into amino acids. A small proportion is used by the bacteria and protozoa for protein synthesis. The remainder is fermented to form fatty acids and ammonia. Most of the ammonia is taken into the bloodstream and used for making ruminant proteins.

The environment in the rumen is ideal for microbial growth. It is warm and moist, and there is an abundance of food. As a consequence, the microbes multiply quickly. Their numbers are kept in check, because every time food passes from the rumen into the other chambers of the stomach some of the microbes are taken along too. This is an example of a natural continuous fermentation system. In the abomasum the microbes are digested and are a significant part of the ruminant's source of protein, B-vitamins and vitamin K. Similar activities take place in the caecum and colon of ruminants and also in non-ruminants.

5 Cleaner animals

Many species of African game, such as antelopes, buffaloes and zebras, are groomed by **oxpecker birds** which remove ticks and blood-sucking flies from their hides (Fig 31.31). When predators come near the birds alert the game by calling and flying off.

Egyptian plovers and sandpipers have been observed taking leeches from inside the open mouths of basking crocodiles on the River Nile. The birds are safe because crocodiles do not feed while they are lying out in the sun.

Barber fish or **cleaner fish**, as they are alternatively called, play an important role in keeping many species of large fish in good health. They do this by picking off lice, parasitic fungi and bits of dead skin. The fish which is being cleaned allows the barber fish to go into its mouth and even into its gill chamber to do its job.

In all the cleaning associations described above the groomer obtains a supply of food. The groomed animal is relieved of harmful parasites and is thus more likely to thrive.

31.2.3 Commensalism

Commensalism is the association between organisms of different species in which one species benefits but does no apparent harm to its partner.

Living on the skin, in the mouth and respiratory passages of mammals are millions of coccoid bacteria. The large intestine teams with bacilli such as *Escherichia coli*. The bacteria have a permanent source of food and are sheltered from the outside world. Laboratory investigations have shown that some species of commensal skin bacteria can produce antibiotics which are effective against pathogenic bacteria. Whether they do this while living on the skin is yet to be determined. A common feature of all commensal bacteria is that they take up sites in the mammalian body which disease-causing microbes might otherwise occupy. In this way they may help to reduce the incidence of microbial diseases among mammals.

Another commensal partnership is seen between the hermit crab *Eupagurus bernhardus* and the sea anemone *Adamsia* (Fig 31.32). The crab is a scavenger and lives in the abandoned shells of gastropod molluscs such as whelks. As it tears up its food, small pieces float away and are picked up by the anemone. The tentacles of the anemone have a sharp sting and protect the crab from would-be predators.

Attempts to define associations sometimes fall down. This is because there is a very delicate balance in many partnerships. Given appropriate circumstances, an association can change from one kind to another. Hospital staff frequently carry pathogenic staphylococci in their nasal passages without coming to any harm. Does it mean that in these people the bacteria are commensal? Conversely, commensal bacteria can become pathogenic if given the opportunity. *E. coli* is quite safe in our bowel, but if it gets into the urinary tract it causes a severe infection.

The nitrogen-fixing bacterium *Rhizobium* is mutualistic so long as its host plant thrives. However, when its supply of carbohydrate slows down,

Fig 31.31 Oxpeckers on African buffaloes (Bruce Coleman: Gunter Ziesler)

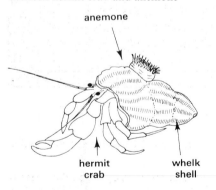

Fig 31.32 (a) Commensal partnership between hermit crab and anemone

anemone

hermit crab

whelk shell

Fig 31.32 (b) Hermit crab *Dardanus* in whelk shell with anemone *Calliactis* attached to shell (Geoscience Features)

it starts attacking the root cells of the legume. In contrast, root cells in mycorrhizal associations sometimes appear to digest some of the hyphae of the fungal partner. This has led some biologists to believe that such relationships may in fact be examples of **controlled parasitism**.

SUMMARY

Ecology is the study of the interaction between living organisms and between living organisms and their environment. It seeks to account for the fluctuations in numbers of species and offers reasons for the distribution of species.

Biotic factors affect, to some extent, the changes in size of populations of species and their distribution. Competition for the same resources may exclude a species from a habitat. Grazing may exclude some species but encourage others. Animals are only likely to be found where their source of food is. Man in his capacity as farmer and landscaper clearly influences species distribution. The widespread use of pesticides and the release of toxic wastes of industry into the environment are other ways in which man has a potent influence in this context.

Edaphic factors determine the properties of soil and have a direct bearing on the distribution of plant species. In turn this influences the distribution of animals which rely on plants as a food source. Climatic factors such as temperature, solar radiation, movement and humidity of the air play their part in this respect too.

It is unusual for a single factor to determine species fluctuation and distribution. Normally it is a complex of interacting factors. The factor complex explains the zonation of marine life on rocky shores around our coast. Tolerance to dehydration when the tide is out, ability to survive buffeting by waves, competition and predation are just a few of the many factors which determine which zone is inhabited by different species of seaweed, crustaceans and molluscs.

Organisms which frequent the same general habitat often occupy only a very small part of it, a microhabitat. There the environmental conditions are often rather different. Micro-ecology is the study of such habitats and their inhabitants. It provides explanations for the distribution of many creatures. The woodlouse has been extensively studied in this respect.

The reason for the distribution of some species is to be found in their dependence on other species for support, often in terms of food provision. There are several kinds of such symbiotic associations which range from parasitism where the relationship can be fleeting to mutualism where the partnership is permanent. In such associations there may be no clear advantage to either partner, both partners may benefit or one partner may benefit at the expense of the other.

QUESTIONS

1 The graph below shows variations in light intensity, nitrate concentration, numbers of phytoplankton and numbers of zooplankton, in a fresh water lake, over a period of twelve months.

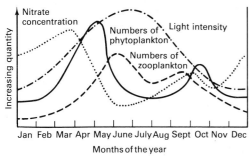

Nitrate concentration ······
Light intensity —·—
Numbers of phytoplankton ——
Numbers of zooplankton – – –

(a) Using only the data given, suggest possible reasons for:
(i) the increase in numbers of phytoplankton from March to early May,
(ii) the sharp decline in numbers of phytoplankton during mid-May to mid-June,
(iii) the low numbers of phytoplankton in July and August.
(b) (i) Suggest a possible explanation for the second peak in numbers of phytoplankton in October.
(ii) Suggest reasons why this peak is lower than the peak in May.
(c) State **two** physical factors, other than those given, which might limit the numbers of phytoplankton from mid-November to early February.　　　　　　(C)

2 Duckweed grows on or near the surface of ponds. Its growth can be measured by counting the number of fronds. Two species of duckweed, *Lemna trisulca* and *Lemna minor* were grown separately, and together, in identical beakers in the laboratory. The following results were obtained:

	Total number of fronds			
	Species grown separately		Species grown together	
Days	*L. trisulca*	*L. minor*	*L. trisulca*	*L. minor*
0	30	30	30	30
16	63	78	48	105
36	126	142	84	234
46	177	225	84	324
54	165	276	48	360
60	129	219	45	354

(a) Draw graphs to compare the rates of growth of the two species when grown separately and when grown together.
(b) (i) What do the graphs suggest about the growth of the two species when grown separately?
(ii) Offer an explanation for this difference.
(c) Offer an explanation for the interaction of the two species when grown together.
(d) Account for the changes in growth rate between 46 and 60 days for *Lemna trisulca* grown separately.

(WJEC)

3 The diagrams show the main paths of ion and water movement in freshwater and marine bony fish.

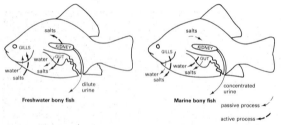

(a) What are the problems of osmoregulation and how are they overcome in
(i) a freshwater bony fish?
(ii) a marine bony fish?
(b) Many marine species of bony fish have kidneys that lack glomeruli. Suggest how this helps in water conservation.
(c) Desert rats live in a very dry climate. They excrete very small quantities of urine with a salt concentration almost twice that of sea water. They produce large quantities of antidiuretic hormone (ADH) and aldosterone.
(i) From which glands are these hormones secreted?
(ii) Suggest **three** problems associated with osmoregulation in desert rats and suggest how they may be overcome.
(d) Terrestrial animals gain water by drinking and eating. Explain **one** other method by which terrestrial animals gain water. (AEB 1984)

4 A small parasitic roundworm. *Litomoisoides carinii*, inhabits the pleural cavity of rats. The larvae enter the blood stream and are transmitted to other rats by a mite which feeds on rats' blood.

Graph 1 shows by dotted lines the mean growth rate of young rats infected with this roundworm and fed on diets with protein values of 2.5%, 5%, 10% and 15% but with

the same energy value. The mean growth rates of young rats kept uninfected and similarly fed are shown as solid lines on the graph.

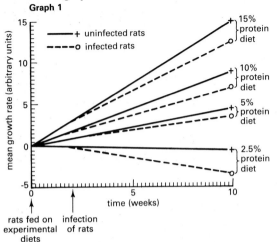

Graph 2 shows the mean number of larvae produced by adult parasites in rats fed two of the diets. The table shows the mean number and mean length of adult male and female parasites found in the pleural cavity eight weeks after infection.

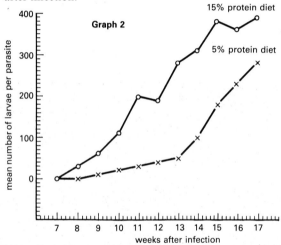

Diets (% protein)	Number of parasites in pleural cavity	Length of adult parasites (mm)	
		Male	Female
2.5	45	17.0	52.5
5.0	30	18.0	60.0
10.0	31	20.5	71.5
15.0	29	20.0	71.0

(a) Compare the growth rates of infected and uninfected rats on the different diets.
(b) Consider the effect of the rats' diets on the parasites'
(i) number,
(ii) growth,
(iii) reproductive capacity,
(iv) potential for transmission to new hosts. (AEB 1985)

5 (a) Explain what is meant by the terms *parasitism, mutualism* and *predation*, indicating, with the help of suitable examples, how they differ from one another.
(b) To what extent is competition important in regulating the size of a population? (C)

32 Ecology 2: Populations, communities and ecosystems

32.1 Populations   656

    32.1.1 Population growth   656
    32.1.2 Growth of human
        populations   657
    32.1.3 Population dynamics   660

32.2 Communities   663

    32.2.1 Succession   663

32.3 Ecosystems   665

    32.3.1 Trophic relationships   665
    32.3.2 Energy flow   666
    32.3.3 Ecological pyramids   671
    32.3.4 Cycling of matter   672
    32.3.5 Soil as an ecosystem   675

Summary   684

Questions   685

32 Ecology 2: Populations, communities and ecosystems

An **ecosystem** is part of a biome such as a pond, stream, woodland or seashore, each inhabited by a typical **community** of living organisms. A **population** is a group of individuals of the same species living in the same place at the same time. The place in which they live is called a habitat. An organism's **niche** is the role it plays in a community.

32.1 Populations

The number of organisms of the same species in a habitat at any time is determined by innumerable factors. Many are those which affect the distribution of the species (Chapter 31). One of the factors, called the **key factor**, is more important than any other. **Density-independent factors** affect a population, whether it is dense or sparse. Those which have a greater effect on a dense population are known as **density-dependent factors**. To find out how such factors regulate the size of a population, it is useful to look at a model of population growth.

32.1.1 Population growth

Growth of a population can be studied by inoculating a fixed volume of nutrient broth with bacteria, incubating the culture at the optimum temperature and counting the number of bacteria at known time-intervals. Four phases of population growth are usually seen (Fig 32.1).

1 Lag phase

In this phase the population grows rather slowly. There are several reasons for the slow growth. The bacteria may have been in a dormant state, and time is required before their metabolism begins to work efficiently. They may have been growing previously in a different medium and have to adjust to the new medium.

2 Log (exponential) phase

Here the population increases at its most rapid rate. It doubles its number in a given time. There are no factors limiting growth. Nutrients and oxygen are in plentiful supply and there is ample space. It is called the log phase because a straight line is obtained if the logarithm of the number of bacteria is plotted against time during this phase of the growth cycle.

3 Stationary phase

By this time the population size is stable and does not increase further. There is a limit to which any habitat can support a population. It is called the habitat's **carrying capacity**. The key factors here are density-dependent factors such as the amount of nutrients and supply of oxygen. Shortage of nutrients and oxygen prevents any further increase in the density of the population. This is known as **environmental resistance**.

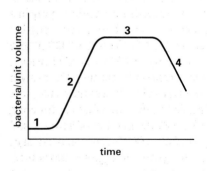

Fig 32.1 Growth of a population of bacteria

4 Death phase

At this stage the carrying capacity of the environment has declined. It is unable to support such a high density of bacteria and they begin to die. They may die of starvation, a shortage of oxygen, or waste products may be present in toxic amounts.

If the investigation is repeated at 10 °C lower, the growth rate would be about half of what was seen at the optimum temperature. The change in temperature has this effect whatever the density of the population (Fig 32.2). It is a density-independent factor. Remember that the laboratory model does not necessarily indicate what happens in populations outdoors. It has not allowed for the possibility of resource renewal, competition with other species, disease, predation and climatic changes. Nevertheless, there are two very important conclusions to be drawn from such studies. The first is that the density of a population will increase up to the carrying capacity of its environment. Secondly, as the carrying capacity alters, so does the population density (Fig 32.3).

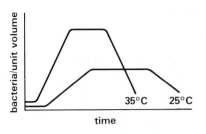

Fig 32.2 Effect of temperature on the growth of a population of bacteria

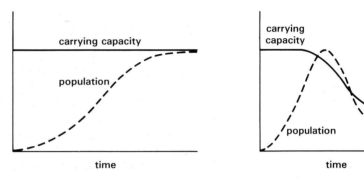

Fig 32.3 Effect of population on the carrying capacity of the environment (from Owen, *What is Ecology?*, OUP) (a) stability resulting from pressures exerted by carrying capacity (b) decrease in population caused by reduction of carrying capacity

32.1.2 Growth of human populations

How far can the principles outlined above be applied to human populations? We first need to know the size of the human population over a long period of time. In many countries, population censuses have been carried out only in the last 150 years or so. Hence the figures for earlier times are estimates based on burial sites, parish registers and other documents. Populations increase when the number born exceeds the number dying in a given time. The human population appears to have had a very long lag period which ended about 300 years ago (Fig 32.4). Between 4000 BC and about 1659 AD the population size doubled every 2000 years. However, from the middle of the seventeenth century the growth rate began to increase considerably. In less than 200 years the population had doubled again and a further doubling had occurred by 1930. In 1987 the number of humans had risen to a staggering 5 000 000 000. What has been happening for the past 300 years is **hyper-exponential growth**. The interval required for the population to double its size is almost halved as time goes on. This contrasts with exponential growth where the interval of time required for the population to double remains fixed.

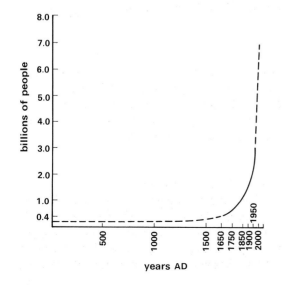

Fig 32.4 Growth of the human population (from J Z Young 1976 after Dorn 1966)

The explosive increase in the numbers of people has not occurred evenly throughout the world. It first began mainly in western countries and subsequently spread to most other parts of the globe. Another important feature is that growth of the human population in every country has occurred in spurts (Fig 32.5). Famine, disease and wars have accounted for the intermittent drastic reductions.

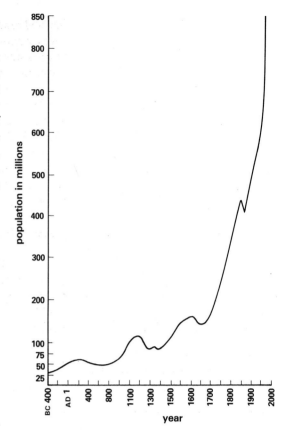

Fig 32.5 Growth of the population of China (from McEvely and Jones 1978)

1 Birth rate

The **fertility** of a woman is the number of live children she bears by the time her reproductive life ends. Each British mother gave birth on average to 7.6 children in 1890. In 1975, it was down to 2.3. The downward trend is seen in many countries. Better education, more reliable, acceptable and readily available methods of **contraception** have been mainly responsible for the declining birth rate. Even so, in many countries more children are born each year than people dying (Table 32.1).

Table 32.1 Population growth in selected countries for 1978 (World Population Data Sheet 1979)

Country	Annual birth rate	Annual death rate	Doubling time (y)
	(per 1000 population)		
Kenya	51	12	18
India	34	15	36
China	20	8	58
USSR	18	10	87
USA	15	9	116
Sweden	12	11	693
United Kingdom	12	12	—

2 Death rate

Improved hygiene, sanitation, diet, disease control and medical care are some of the more important reasons why the human death rate has declined in many parts of the world. In the 1940s, about 3 million people died of malaria each year. Now it is about 2 million. There were 162 infants among every 1000 deaths in Britain in 1900. Today it is 29 per thousand. Modern medical practice has greatly increased the chance of diagnosing and curing many human diseases. Disease prevention by vaccination and immunisation is now widely practised. Such measures have been so successful that some killer diseases, such as smallpox, have been eradicated.

3 Life expectancy

A century ago in Britain, a woman could expect to reach the age of about 45 years on average, a man 40 years. Today it is roughly 75 and 70 years respectively. Hence more people now live their full reproductive span and are potentially capable of adding to the population. One of the reasons for the greatly increased life expectancy is the reduced mortality rate among children. These days 29 infants out of every 1000 born in Britain die before their first birthday. A hundred years ago the figure was nearly 400. Remarkable advances, in medicine in particular, have contributed enormously to the increased rate of survival. Many previously fatal infectious diseases, for which there was no cure in the past, can now be cured, or prevented by vaccination. We now have a much better understanding of how infectious diseases are spread. People can be advised to avoid passing on hereditary diseases, and dietary diseases can be avoided as our knowledge of human nutrition has improved.

How successful such advances have been can be judged from **survivorship curves**. They show that life expectancy is increasing in many countries (Fig 32.6). It means that more and more people of all ages survive longer, thus adding to the existing population and contributing to a population explosion.

What are the probable reasons for the change in survivorship of the people of Sweden shown in Fig 32.6(b)?

Fig 32.6 (a) Survivorship curves (from J Z Young 1976 after Clark 1967)

Fig 32.6 (b) Change in survivorship curves with time for Sweden (from J Z Young 1976, after Clark 1967)

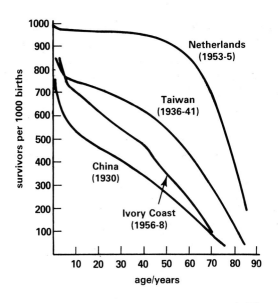

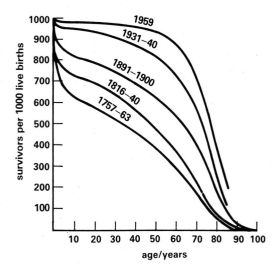

4 The outlook

The earth has finite resources and there is a limited amount of space. If it continues to grow at its present rate, the human population could outgrow the earth's carrying capacity. Humans are adaptable and intelligent creatures. They have the ability to avert the dire consequences of unlimited population growth. However, there is a limited number of options. The most obvious of these is to reduce the birth rate. Increased food production is another possibility, but it will only delay the inevitable if growth of the human population is not kept in check.

32.1.3 Population dynamics

Long-term studies of wild populations show successive growth curves of the kind we have just studied (Fig 32.7). There are many possible causes of such fluctuations. Spells of very cold weather kill many individuals (Fig 32.8). Epidemics of disease and fluctuations in food supply have comparable effects. Predation is a key factor in the ups and downs observed in some animal populations.

Fig 32.7 Fluctuations in the numbers of the water flea *Daphnia* with time (from Sands 1978)

Fig 32.8 Numbers of breeding pairs of grey heron in two areas of England (from Owen, *What is Ecology?*, OUP)

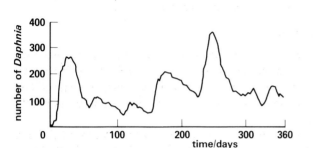

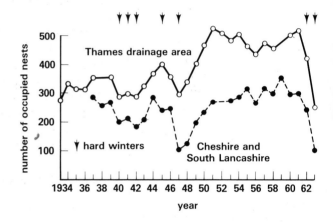

1 Predation

A **predator** is an animal which captures and rapidly kills live animals, called **prey**, for its food. Repeated fluctuations in numbers of predators and their prey have been seen over a fairly short time in laboratory studies. The density of predators is slightly out of phase with the number of prey. A peak in the number of prey is followed by an increased density of predators. When the numbers of predators increase, there is a fall in the density of the prey (Fig 32.9).

Similar fluctuations are thought to occur in wild populations. Perhaps the best documented outdoor study is of the snowshoe hare and its predator the lynx in Canada between 1845 and 1935 (Fig 32.10). The density of the hare and lynx populations was estimated from the numbers of their pelts received by the Hudson Bay Company, from information provided by trappers and from field observations. Fluctuations in each of the populations occurred every 9–10 years. In other predator–prey relationships, the interval may be larger or smaller.

Fig 32.9 Relationship between the density of a predator *Hydra* and its prey *Daphnia* (from Brown, *Ecology of Fresh Water*, Heinemann)

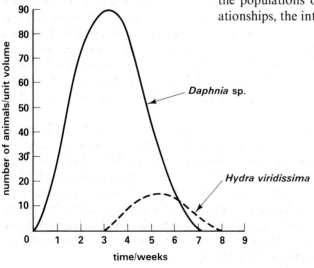

Fig 32.10 Fluctuation in abundance of the lynx and snowshoe hare in Canada (from Odum, *Fundamentals of Ecology*, Saunders)

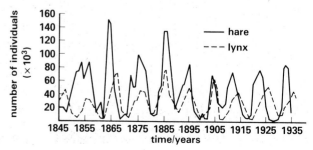

Predation is a **density-dependent factor**. Changes in prey density usually have a delayed effect on the density of predators. Food supply, weather, the incidence of disease and the predation rate are among the factors which determine the numbers of prey. Specialist predators which eat a limited variety of prey either starve or are forced to migrate when prey is scarce. General predators can turn to other kinds of prey.

The relationship between predators and prey helps to maintain the biological fitness of animal populations. Young, old or infirm prey are more easily captured by predators. When prey is in short supply, predators which are low in the pecking order are likely to go without food. In these ways the fittest of the prey and predators survive.

2 Humans as predators

Humans kill many species of animals as part of their omnivorous diet. Domesticated stock such as pigs, sheep, cattle and fowl are farmed to maintain a stable supply for market. Wild prey also make up part of our food supply. Marine animals, notably fish and whales, figure prominently in the diets of many humans.

Improved fishing methods have meant that more and more fish have been taken in recent years from the world's seas and oceans (Table 32.2).

Table 32.2 Amounts of fish taken from the world's oceans (after Lucas and Critch 1974)

Year	Global catch (tonnes)
1940	20×10^6
1963	46×10^6
1970	70×10^6

Fishing vessels now travel over a thousand miles from their home ports. Factory ships stay at sea for months, processing the catch from fishing fleets. Intensive fishing sooner or later results in a fall in the catch. The fleet then moves to other fishing grounds. It often takes many years for **overfished** areas to recover. There is the fear that, in the not-too-distant future, fish stocks generally will become so depleted that the oceans will be unable to meet the demand placed on them.

Fluctuations in fish numbers are sometimes due to natural causes. Many fish eat plankton. Hence their numbers depend to some extent on the amount of plankton available to them. The decline of the herring shoals in the North Sea during the 1930s may have been caused partly by overfishing and partly by a fall in phosphate concentration. Phosphate stimulates growth of plankton. Recovery of the herring population in the 1960s was probably because the fish had overcome the intensive fishing of earlier days and also because of increased phosphate concentration. The phosphate came from domestic detergents which became popular at that time.

There is no doubt that predation by humans has been the main cause of the dramatic fall in numbers of many species of whale this century. Following the severe decline in the numbers of blue whales in the 1940s, by which time they had been hunted almost to the point of extinction, whaling fleets switched to catching fin whales. When the fin whale population slumped in the 1950s from overhunting, whalers took to capturing sei whales. In a little more than a decade the sei whale population fell drastically (Fig 32.11).

Fig 32.11 (a) Numbers of five species of whale caught over a 25 year period (from Lucas and Critch 1974) (b) Fin whales (Eric and David Hoskins)

(a)

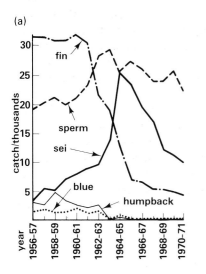

(b)

661

3 Regulation of population size

Studies on wild populations have made it possible to find out how the sizes of natural populations are regulated. The findings should enable us to control the size of wild populations to our advantage. To keep the population size stable, the rate at which mature stock is harvested must be equalled by replacement from immature stock (Fig 32.12). In this way the **maximum sustainable yield** is maintained and the population is kept in the log phase of growth. If the population reaches the stationary phase then young fish grow less quickly as they have to compete with a larger number of adults. Further, adult fish do not grow any more even though they continue to feed. Hence removing mature fish relieves the competition, enabling the young to increase the biomass of the population for the same amount of food consumed.

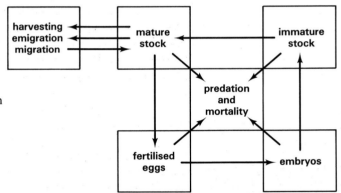

Fig 32.12 Model of population regulation

Managing this in a massive habitat such as the sea is easier said than done. Shoals are usually a mixture of mature and young stock. If a net with a small mesh is used, many young fish are caught, leaving too few for populations to recover rapidly.

Various measures have been taken to try to conserve whale and fish stocks. Worldwide bans on the hunting of humpback whales came into being in 1966. A similar measure was taken to protect the blue whale in 1967. A ten-year ban on the capture of any whales was proposed at the International Whaling Commission (I.W.C.) in 1972 but was turned down. In 1986 the I.W.C. agreed to a cessation of whaling but several countries continue to catch whales for what they describe as scientific purposes. Catch quotas of over-exploited species of fish have been set by international agreement; much still needs to be done however to rectify the mistakes of the past. Had the blue whale been maintained at its optimum population density, a sustained catch of 6000 individuals a year could have been achieved. The amount of meat alone would have provided three million people with a 200 g steak every day!

4 Biological control

The dispersal of plants and animals to new habitats often upsets the equilibrium of the community to which they have moved. Introduced species often increase in numbers very quickly. It happens because the factors which checked their numbers in the previous habitat are absent or less strong in their new environment. In other words, there is less **environmental resistance**. They key factor is often the absence of any natural enemies.

When the prickly-pear cactus, *Opuntia stricta*, was taken to Australia from North America in 1900, it soon spread to grassland. Within 25 years it covered millions of acres and was a serious nuisance to farmers. In its natural habitat the cactus population was kept down by the browsing larvae of a moth, *Cactoblastis*. When the moth was introduced into Australia from Argentina, its caterpillars soon stopped the spread of the cactus. Within a few years most of the lost farmland was reclaimed.

In recent years, upland farmers in Britain have been unable to afford to use chemicals to kill bracken. As a result it is spreading at a rate of over 100 km a year. A comparable solution to controlling cactus may be used to deal with the problem in the near future, using the larvae of a South African moth.

Another method of biological control has been used with some success in keeping down populations of insects such as locusts and mosquitoes. It involves releasing males which have been made sterile by irradiation. A proportion of subsequent matings thus fail and fewer offspring are produced in the subsequent generation.

There are many other instances of the successful **biological control** of pests. The advantage of regulating populations of pests in this way is that the control is very specific, so that useful organisms are unaffected. In contrast, pesticides are often unable to distinguish between the target organism and other closely related but desirable species (Chapter 33).

Fig 32.13 Oak forest, the climax vegetation of much of Britain (Biofotos)

32.2 Communities

Only a few hundred years ago, much of the British countryside was covered with deciduous forest inhabited by wild boar, fallow deer and woodland birds. Deciduous forest is the **climax community** for most low-lying areas of Britain (Fig 32.13). Today vast tracts of land are intensively cultivated and trees are confined to hedgerows and copses. However, if arable land is neglected, it eventually reverts to forest. The sequence of changes which occur in arriving at a climax community is called **succession**. It occurs wherever plants have an opportunity to colonise bare ground.

32.2.1 Succession

Primary successions begin where vegetation has never grown previously. Bare rock, sand and mud are examples. **Secondary successions** occur where vegetation has grown but has since been destroyed by fire, farming or by flood.

1 Primary successions

The three successions described here are seen on bare rock, sandy beaches and in the mud of river estuaries around Britain. Such habitats are fairly accessible to most students.

i. Bare rock   Crustose lichens such as *Xanthoria parietina* cling by their rhizoids to the surface of bare rock. Though they grow rather slowly, lichens are very hardy organisms. They endure the wide fluctuations of temperature and dryness of bare rock. Lichens overcome the lack of available nutrients on the rock face by absorbing minerals dissolved in moisture from the air. When lichens die, they add organic matter to the particles of rock eroded by weather. Tufts of small mosses such as *Tortula muralis* are now able to grow on the rock. They can tolerate long spells of drought and retain moisture like a sponge when rain falls. The organic matter they add to the weathered rock attracts saprophytic microbes and detritivores such as nematodes, molluscs and arthropods.

At this stage the rock is covered with a thin layer of soil in which seedlings of herbaceous flowering plants become established. Their roots penetrate

Fig 32.14 Succession on a sandy shore at Ynyslas, Wales (a) embryo dune colonised by prickly saltwort (left) and sea rocket (b) yellow dune colonised by marram grass

(a)

(b)

Fig 32.15 Plants growing in saline mud at Gibraltar Point, Lincs: (a) glasswort (b) sea lavender (c) cord grass

fissures in the rock and prise it apart into pieces. As the developing soil deepens, large-rooted shrubs and eventually trees are able to colonise the area. This is the primary succession which occurs in the development and colonisation of sedentary soils. The process may take thousands of years. Climate mainly determines the climax community which eventually appears.

ii. Sand Sandy beaches are found in many places around the coast of Britain. When exposed to the sun and wind, the sand dries and is blown onshore. On the upper part of the beach, just beyond the reach of the highest tides, annual flowering plants, notably the prickly saltwort *Salsola kali* and the sea rocket *Cakile maritima*, grow (Fig 32.14). They can withstand abrasion by blown sand, the salinity and dryness of the beach and wide fluctuations in temperature. Around them, blown sand forms small hummocks which become colonised by sand couch grass, *Agropyron junceiforme*. The spear-like leaves of the couch grass gather more sand, which is consolidated by the plant's extensive fibrous roots. It prepares the way for the xerophyte marram grass, *Ammophila arenaria*, whose spiky shoots trap large amounts of blown sand. The sand builds up quickly to form **yellow dunes** which are stabilised by the long rhizomes of the marram grass. They are called yellow dunes because fresh sand is continually deposited between the tufts of marram grass. As the marram grass spreads, the spaces between the tufts become smaller and more stable. They become colonised by ground-hugging angiosperms such as sea holly *Eryngium maritimum* and sea bindweed *Calystegia soldanella*. The organic matter in the top few centimetres of sand increases as the dunes become clad in vegetation. It encourages the growth of lichens such as *Cladonia*, mosses such as *Tortula ruraliformis* and flowering plants, the sand sedge *Carex arenaria* and red fescue grass *Festuca rubra*. As the dune vegetation develops insects, birds and small mammals come in to feed and shelter thus diversifying the community.

In winter, the maturing dunes look grey when the shoots of the vegetation die back. For this reason they are called **grey dunes**. By now a crust of peat-like organic matter lies on the surface of the sand. Rain leaches the organic crust of lime, and acid-loving plants such as heathers are able to compete successfully with the dune plants. The **heathland** so formed eventually gives way to scrub vegetation and finally to **forest**.

iii. Mud Clay and silt carried down by rivers often settle as mud in estuaries. When the mud is sufficiently thick, it is exposed to the air at low tide and colonisation by plants begins. In Britain, the first flowering plants to grow in the saline mud are halophytes such as the glasswort *Salicornia europaea* and cord grass *Spartina anglica* (Fig 32.15). When the tide comes

(a)

(b)

(c)

in, their shoots are immersed in brackish water and speed up the sedimentation of mud. Sea aster *Aster tripolium* and sea poa grass *Puccinellia maritima* now begin to colonise the mud. The fibrous roots of the grass help to stabilise the mud. As the mud thickens, it is covered less frequently by tides and begins to dry out.

The drier conditions enable red fescue, *Festuca rubra*, to compete with *Puccinellia*. In many parts of Britain the vegetation is grazed by sheep and cattle, causing the succession to stop at this stage. Where the natural sequence is allowed to go on, the mud continues to thicken and the salt marsh vegetation is now covered only by high spring tides. Organic matter accumulates and sea pink *Armeria maritima*, sea lavender *Limonium vulgare*, and sea plantain *Plantago maritima* thrive. They are finally supplanted by the sea rush, *Juncus maritimus*, which in turn may give way to **heathland** and **forest**.

An interesting feature of primary successions is that at each stage the inhabitants change the habitat, making it amenable to more competitive organisms which eventually oust the previous community. The process is repeated over and over until the climax community develops.

Fig 32.16 Heather moor: (a) during burning (Biofotos: Brian Rogers) and (b) strips of heather moor burnt in rotation show new growth in paler areas (Biofotos)

(a)

(b)

2 Secondary successions

Burning of heather, *Calluna vulgaris*, is part of the management of grouse moors. It causes the heather to produce young, tender shoots which make up the bulk of the diet of grouse. Burning also prevents trees from growing among the heather. Strip-burning is the normal practice, leaving stands of old, tall heather for the birds to nest in and take cover from predators.

After a strip of heather has been burned, charred heather stems stick out of the bare peat. The heather roots are unaffected, and buried in the peat at the bases of the charred stems are reserve buds which soon spurt into growth. Gelatinous green algae and lichens colonise patches of bare peat. They are followed by other lichens, especially *Cladonia*, mosses, notably *Polytrichum*, and heather seedlings. Wavy hair grass, *Deschampsia caespitosa*, and crowberry, *Empetrum nigrum*, may later dominate the vegetation, but the heather plants soon outgrow the other species and within a year or two the community is back to normal (Fig 32.16).

Secondary successions are often more rapid than primary successions. This is because the soil is already developed and contains the spores, seeds and vegetative organs of propagation of the colonising plants.

32.3 Ecosystems

An **ecosystem** is a more or less self-sufficient community comprised of populations of organisms in equilibrium with each other and with their environment. Energy and matter are exchanged between ecosystems and their surroundings. The activities of the community cause **energy to flow** through the ecosystem. The community consists essentially of producers and consumers, which remove matter from the environment, and decomposers, which return it. The way in which producers, consumers and decomposers interact ensures that **matter is cycled** in ecosystems.

32.3.1 Trophic relationships

Green plants are the lifeblood of every ecosystem. They make organic molecules by photosynthesis from simple inorganic molecules. For this reason they are called **producers**. Organic substances are eaten mainly by animals which are the **consumers** in ecosystems. **Primary consumers** are mainly herbivorous animals which feed on green plants. Predators,

Fig 32.17 Trophic levels in a food chain

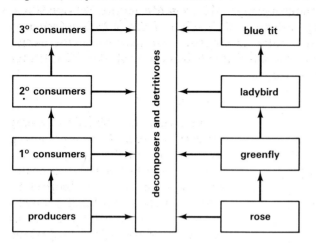

scavengers and many parasites are **secondary consumers**. They feed on primary consumers. **Tertiary consumers** feed on secondary consumers, and so on. When producers and consumers die, their bodies are broken down by **decomposers** such as bacteria and fungi, which also dispose of animal faeces. Dead organic matter is also consumed by animals called **detritivores**. Millipedes, woodlice and earthworms feed mainly on dead vegetation in soil ecosystems. In aquatic ecosystems, particles of dead organic matter suspended in the water are eaten by detritivores such as bivalve molluscs and lugworms.

It is convenient to think of the feeding relationships in ecosystems as a **food chain** with several **trophic levels** (Fig 32.17). In reality, feeding relationships are rarely as simple as this. Insectivorous plants, for example, are producers and also consumers. Omnivores can be primary, secondary and tertiary consumers, so too can decomposers and detritivores. It is therefore more informative to show feeding relationships as **food webs** in which food chains are interlinked (Fig 32.18). This gives a much clearer picture of what happens to food in a community.

Fig 32.18 Food web in a freshwater ecosystem (after Owen 1974)

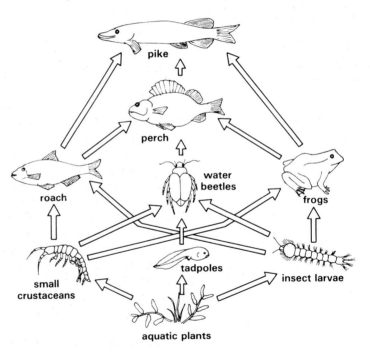

32.3.2 Energy flow

Energy can be transformed from one form to another. Green plants, for example, convert the energy in sunlight into the chemical bond energy of photosynthetic products. When energy is converted from one form to another, some energy always appears as heat. A proportion of the energy in respiratory substrates is released as heat energy.

The energy source for all ecosystems is the sun. About 641×10^4 kJ m^{-2} y^{-1} of solar energy reaches the earth's atmosphere. A good deal of solar energy does not, however, penetrate the atmosphere. It is reflected or absorbed and radiated back into space by the ozone layer, dust particles and clouds. The precise amount reaching the earth's surface depends on geographical location. Britain gets about $105 \times 10^4\,kJ\,m^{-2}\,y^{-1}$, which is roughly a third of the energy received in tropical countries. Between 90 and 95 % of the energy getting to the surface of the earth is reflected by vegetation, soil and water or absorbed and radiated to the earth's atmosphere as heat. It means that only between 10 % and 5 % is left for producers to make use of.

1 Producers

Let us assume that 100 units of energy per unit time reach the leaves of a crop plant. What exactly happens to the energy? About half is not of an appropriate wavelength to be absorbed by chloroplast pigments. Roughly a quarter of what is absorbed ends up as chemical bond energy in photosynthetic products such as starch. The rate at which the products are formed is called **gross primary productivity** (GPP). A substantial amount of the **gross production** is respired by the plant. What is left, in this example just 5.5 % of the energy reaching the leaf, is **net production** (Fig 32.19). The rate at which the products of photosynthesis accumulate is known as **net primary productivity** (NPP). The factors which affect NPP in crop plants are discussed in Chapter 13.

Fig 32.19 Fate of solar energy reaching the leaf of a crop plant (data from Hall 1979)

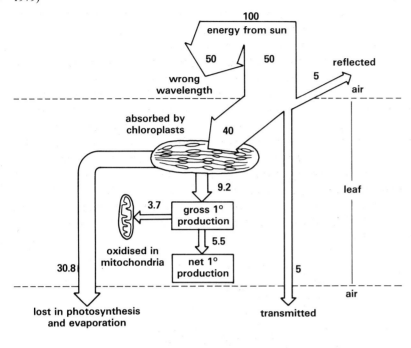

Because they photorespire more, plants which use the C_3 photosynthetic pathways have lower NPP values than C_4 species (Chapter 5). Productivity in temperate areas could in theory be improved by growing C_4 crops. Unfortunately, many C_4 species are native to tropical climates and do not grow well in temperate countries. Those which can be cultivated, such as maize, are less productive because of the lower temperature and less intense light (Table 32.3). One approach to improve the productivity of C_3 species such as wheat has been the developing of early maturing varieties. This ensures the crop is ready for harvesting before the hot summer weather begins to stimulate photorespiration. An alternative strategy, which is being actively researched, is to spray C_3 crops with a chemical to block photorespiration. Another possibility is to breed out the biochemical pathway of photorespiration.

Table 32.3 The growth of maize under different conditions (from Bland and Bland 1983)

Plant	Region	Net primary production $(g\,m^{-2}\,day^{-1})$	Irradiance $(J\,cm^{-2}\,day^{-1})$	Conversion of photosynthetically active radiation (%)
maize	California	52	2000	9.8
maize	Europe	17	1200	4.6

In winter, fields often have no crops growing in them. During the early growth of a crop, a good deal of sunlight is absorbed or reflected by the soil. It is only when plants are fully grown that their leaves absorb the maximum amount of sunlight. When the NPP is averaged for the whole year, it is therefore much less than in the growing season. In natural ecosystems where the climate allows plants to grow all year, the NPP is higher. In others, climatic factors limit the rate of photosynthesis, so the NPP is lower (Table 32.4).

Table 32.4 Net primary productivity of some of the world's ecosystems (after Whittaker 1975)

| | NPP $(g\,m^{-2}\,y^{-1})$ | |
Ecosystem	Range	Mean
open ocean	2– 400	125
lakes and streams	100–1500	250
continental shelf	200–1000	360
temperate grassland	200–1500	600
temperate forest	600–2500	1250
tropical forest	1000–3500	2000

2 Consumers

Let us consider the transfer of energy from producers to primary consumers. **Secondary productivity** is the rate at which consumers accumulate energy in the form of cells and tissues. The first point to bear in mind is that **herbivores** do not eat all the vegetation in an ecosystem. Cattle grazing a pasture will eat the grasses and palatable weeds such as buttercups. They leave inedible weeds such as nettles and thistles. Even the plants that are grazed are not eaten entirely. The roots are inaccessible and part of the shoot system is also left uneaten. It means that only part of the NPP of the ecosystem is transferred to the primary consumers. The amount depends to some extent on the variety and density of primary consumers. In pasture grazed by cattle it is about 30–45 %.

Fig 32.10 shows the fate of energy consumed by a bullock grazing on pasture. Nearly two-thirds of this energy, 1909/3056 kJ, is excreted in urine and egested in its faeces. The remaining third, 1147/3056 kJ, is absorbed into the circulatory system. Roughly 90 % of the absorbed energy, 1022/1147 × 100, is lost as heat or is belched out as methane gas. Only about 10 % of the absorbed energy, 125/1147 × 100, is used to synthesise new cells and tissues. The secondary production of the bullock is thus only 0.6%, 125/21 436 × 100, of the NPP of the pasture.

Fig 32.20 Fate of energy in a year's growth of grass from $1\,m^2$ of pasture (data from *Nuffield Study Guide in Advanced Biology* 1974)

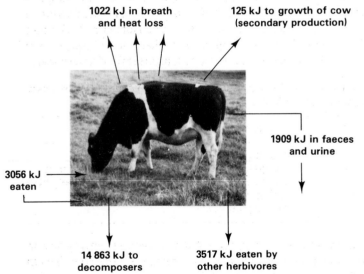

1022 kJ in breath and heat loss

125 kJ to growth of cow (secondary production)

3056 kJ eaten

1909 kJ in faeces and urine

14 863 kJ to decomposers

3517 kJ eaten by other herbivores

Carnivores have a much higher secondary productivity. This is because a protein-rich diet is more readily and efficiently digested. They do not have energy-consuming symbiotic microbes in their digestive tracts and their faeces contain much less undigested matter. Only about 20 % of the energy intake is lost in their faeces and urine. They absorb almost twice as much energy per unit mass of food compared with herbivores.

The proportion of absorbed energy released in respiration is determined by the activity of the consumer and whether or not it maintains a constant body temperature. **Endotherms** have a stable temperature. They use some of the energy in absorbed food to make good the energy lost as heat to their surroundings. In cold weather, more absorbed energy is used for this purpose. **Ectotherms** do not use energy in this way and direct a higher proportion of absorbed energy into secondary production.

Of the consumers eaten by humans, fish are probably the most productive (Table 32.5). This is one reason why there has been increased interest in fish-farming in recent years. Changes have also been made in the rearing of farm animals. They involve intensive methods of farming such as raising poultry by the broiler method. Keeping livestock indoors reduces the amount of energy they use for movement and minimises heat loss through the skin.

Table 32.5 Production of a range of animals (mainly after Open University 1974)

			% Absorbed	% Respired	% Produced
Endotherms	herbivores	rabbit	50.0	47.4	2.6
		cow	37.5	33.4	4.1
Ectotherms	carnivores	trout	86.0	55.9	30.1
		spider	91.0	62.0	29.0
	herbivores	grasshopper	37.0	24.0	13.0
		caterpillar	41.0	17.6	23.4

3 Decomposers and detritivores

The proportion of the primary production used by herbivores varies from one ecosystem to another. Up to 90 % of the phytoplankton in the oceans is eaten by microphagous animals. About 50 % of the primary production of pastureland is grazed, whereas as little as 10 % is consumed by herbivores in forests. The faeces of herbivores and most of what is left of the primary production are used as an energy source by **decomposers** and **detritivores**.

Table 32.6 Effect of mesh size of nylon bags on access to leaf discs by soil organisms (after Phillipson 1966)

Mesh size (mm)	Organisms with free access
7.000	Decomposers, detritivores
1.00	Decomposers and detritivores except earthworms
0.500	Decomposers and small detritivores such as mites, springtails and enchytraeids
0.003	Decomposers only

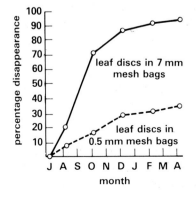

Fig 32.21 Effect of detritivores on the disappearance of buried oak leaf discs (from Phillipson, *Ecological Energetics*, Inst. Biol. Studies in Biology, Edward Arnold)

Some idea of the roles of decomposers and detritivores in energy flow in terrestrial ecosystems has been obtained by burying leaf discs contained in nylon bags of various mesh sizes in soil. The mesh size determined which decomposers and detritivores had access to the leaf discs (Table 32.6). The bags were dug up at regular intervals to measure the rate at which the discs were broken down. In the large mesh bags, which earthworms could get into, the discs were broken down three times faster than in any of the others. Where only microbes could enter, the discs showed little signs of deterioration over the nine months the investigation lasted (Fig 32.21). The investigation fails to show what proportion of the energy in the leaf discs is used by the detritivores and how much is used by the decomposers. Detritivores are herbivorous animals which egest much of the energy they consume in their faeces. A millipede feeding on dead hazel leaves, for example, absorbs only 10 % of the energy in the leaf material. Evidence of this kind points to the decomposers as the main users of energy in dead organic matter. So why are the detritivores apparently so important? Dead plant material is often deficient in nitrogen which limits microbial activity. The faeces of detritivores contain more nitrogen and moisture than the dead vegetation they eat. When altered in these ways, the plant material is used much more readily by decomposer bacteria and fungi.

One ecosystem in which the energy flow through decomposers has been determined is a salt marsh in Georgia, USA (Table 32.7). The primary production available to the animals and decomposers living in the marsh is $34\,354 - 15\,370 = 18\,984\,kJ\,m^{-2}\,y^{-1}$ of which the decomposers use $16\,287\,kJ$. The proportion used by the decomposers is $16\,287/18\,984 \times 100 = 85.8\,\%$. It is thought that this is a reflection of the energy flow through the decomposers in most ecosystems.

Table 32.7 Energy flow in a Georgia salt-marsh (after Teal 1962)

Process	Energy $(kJ\,m^{-2}\,y^{-1})$
gross production	152 323
producer respiration	117 969
net production	34 354
decomposer respiration	16 287
export from marsh	15 370
primary consumer respiration	2 495
secondary consumer respiration	201

Fig 32.22 Flow of energy in Silver Springs, Florida (after Odum 1957)

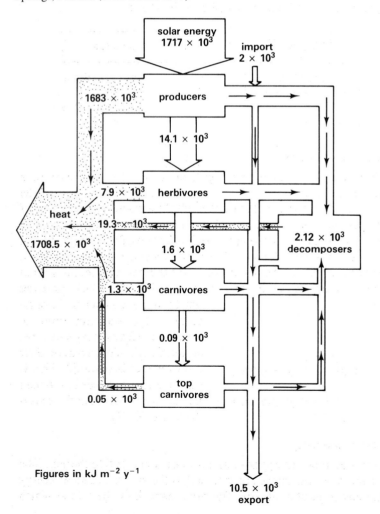

Figures in $kJ\,m^{-2}\,y^{-1}$

4 Whole ecosystems

In most ecosystems there are many species, and feeding relationships are often poorly known. The few ecosystems where the flow of energy is known in detail are fairly simple. A freshwater spring is a simple ecosystem. The producers are diatoms, green filamentous algae and small flowering plants such as duckweed. When they die, the producers become detritus and sink to the water bed. Some of the detritus is carried away by the current. Detritus also enters the spring with the incoming water. The detritivores include freshwater lice, bivalve and gastropod molluscs and small crustaceans. Larvae of the caddis fly feed on larger pieces of dead vegetation. Predatory turbellarians and midge larvae are the consumers.

The flow of energy between the various groups of organisms in such an ecosystem is shown in Fig 32.22. The first important thing to notice is that the energy leaving the system mainly as heat is the same as the amount of solar energy coming in. The second is that the rate at which energy passes into each trophic level is about 10 % of that entering the previous level. For example, the secondary consumers in the spring receive $1.6 \times 10^3\,kJ\,m^{-2}\,y^{-1}$ from the primary consumers which get $14.1 \times 10^3\,kJ\,m^{-2}\,y^{-1}$

from the producers. The rate at which energy passes to the secondary consumers compared with the rate it enters the primary consumers is thus $1.6/14.1 \times 100 = 11.4\%$. This is called the **gross ecological efficiency**. Because only about 10% of the energy passes from one level of consumers to the next, the number of levels is limited to two or three in most ecosystems.

32.3.3 Ecological pyramids

The food and energy relationships in ecosystems can be displayed as bar diagrams. The producers are placed at the bottom, the primary and secondary consumers in the middle and tertiary consumers at the top. Diagrams of this kind often have the shape of a pyramid. Decomposers and detritivores are not often shown on **ecological pyramids**.

1. Pyramids of numbers

These show the numbers of organisms at each trophic level. The length of each bar is proportional to the number of organisms. A major drawback of number pyramids is that they fail to distinguish between the sizes of organisms. A mature tree in a forest ecosystem is counted the same as a diatom in a lake. A cow in a meadow is equated with a leaf-eating caterpillar in a woodland. For this reason, number pyramids are bulged at the middle when the producers are large and few in number as in forest ecosystems. They are inverted when the consumers carry parasites. For example, a single rose bush may support thousands of aphids, which in turn may be the hosts for millions of parasites. (Fig 32.23).

Fig 32.23 Pyramids of numbers

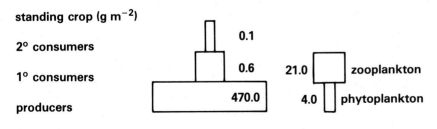

2 Pyramids of biomass

These show the dry mass of the organisms in the trophic levels at a particular point in time. The length of the bars is proportional to the biomass in each level. They tell us what is called the **standing crop**. One of the problems with this approach is that it does not allow for changes in biomass at different times of the year. Deciduous trees, for example, have a larger biomass in summer than in winter when they have shed their leaves. Another difficulty is that the rate at which the biomass accumulates is not taken into account. A mature tree has a large biomass which increases slowly over many years. In comparison, the standing crop of diatoms in a lake is very small, yet production is high because their turnover rate is so rapid. This is why inverted pyramids of biomass are obtained for aquatic ecosystems (Fig 32.24).

Fig 32.24 Pyramids of biomass (a) a field in Georgia, U.S.A. and (b) the English Channel

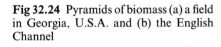

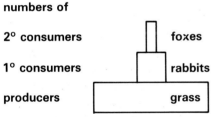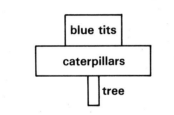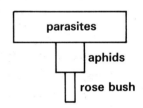

3 Pyramids of energy

These show how much energy passes from one trophic level to the next. The length of the producer bar is proportional to the amount of solar energy used annually in photosynthesis. The other bars show the rate at which

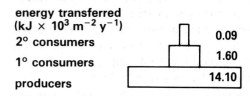

Fig 32.25 Pyramid of energy in Silver Springs, Florida

energy transferred
$(kJ \times 10^3\ m^{-2}\ y^{-1})$

2° consumers	0.09
1° consumers	1.60
producers	14.10

energy passes along the food chain (Fig 32.25). Note that the transfer of energy from producers to primary consumers is less efficient than between the other trophic levels. Energy pyramids are more informative than pyramids of numbers and biomass. They tell us how much energy is required to support each trophic level. Because only a proportion of energy in a level is transferred to the next, energy pyramids are never inverted nor do they have a central bulge.

32.3.4 Cycling of matter

Energy comes into ecosystems from the sun and is lost as heat to the surroundings. In contrast, matter is cycled between organisms and their environment. The availability of minerals, especially nitrogen, carbon dioxide and water affects the distribution and abundance of plants and hence animals.

1 The nitrogen cycle

Nitrogen is vital to plants and animals for the synthesis of amino acids, proteins and nucleic acids. The earth's atmosphere contains over 78 % by volume of nitrogen gas, yet in this form it is useless to most living organisms. Plants take in nitrogen as ammonium and nitrate ions. On a global scale, farmers add about 75×10^6 tonnes of ammonium and nitrate fertilisers to cultivated soils each year. Treating fields with nitrogen fertilisers is essential, because most of the nitrogen previously in the soil is removed in the crop at harvesting. In uncultivated soils, the crop is returned each year as dead leaves, roots, animal carcasses and faeces. Nitrogen in the detritus is converted to ammonium and nitrate ions and is absorbed once more by the vegetation. The cycling of nitrogen takes place in several steps (Fig 32.26).

Fig 32.26 The nitrogen cycle

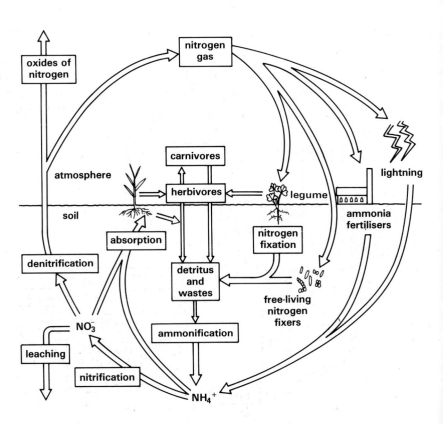

i. Ammonification   A wide range of **saprophytic soil bacteria** and **fungi** decompose detritus. Peptidase enzymes released by the decomposers hydrolyse proteins to amino acids, some of which are absorbed and used to synthesise microbial protein. Unwanted amino acids are deaminated in a way comparable to that which occurs in the mammalian liver and ammonia is released. As much as $7000–10\,000 \times 10^6$ tonnes of nitrogen enter soil annually in this way. Some ammonia escapes into the atmosphere, but much of it is converted to nitrate ions by nitrifying bacteria.

ii. Nitrification   Well aerated soil contains chemo-autotrophic **nitrifying bacteria**. They oxidise ammonia to nitrate ions in two stages. The first stage is the conversion of ammonia to nitrite ions, mainly by bacteria of the genus *Nitrosomonas*. Nitrite ions are toxic to plants, but they are oxidised as quickly as they are formed into nitrate ions by bacteria of the genus *Nitrobacter*.

Plants absorb nitrate ions, but these are also rapidly leached from soil by rain water. In anaerobic conditions as in waterlogged soils nitrate ions are reduced to nitrite, ammonia, nitrogen and oxides of nitrogen by **denitrifying bacteria** such as *Pseudomonas denitrificans*.

iii. Nitrogen fixation   Something like $200–300 \times 10^6$ tonnes of soil nitrogen are lost each year by leaching, denitrification and ammonification. The loss is made good by **nitrogen-fixing bacteria** and **blue-green algae**. They are able to reduce nitrogen gas to ammonia which they use to form amino acids. Free-living bacteria such as *Azotobacter* and *Clostridium* and blue-green algae such as *Nostoc* account for $170–270 \times 10^6$ tonnes of nitrogen fixed annually. The rest is fixed by symbiotic bacteria, notably *Rhizobium* which lives in the nodules of legume roots (Chapter 31). Before artificial fertilisers were generally available, farmers and gardeners boosted soil fertility by growing legumes in rotation with other crops.

2 The carbon cycle

The earth's atmosphere contains 2.5×10^6 million tonnes of carbon dioxide. The oceans contain 150 times more. Terrestrial and aquatic organisms exchange about $100–200 \times 10^6$ tonnes of carbon dioxide annually with their surroundings. The exchange is mainly brought about by three activities (Fig 32.27).

Fig 32.27   The carbon cycle

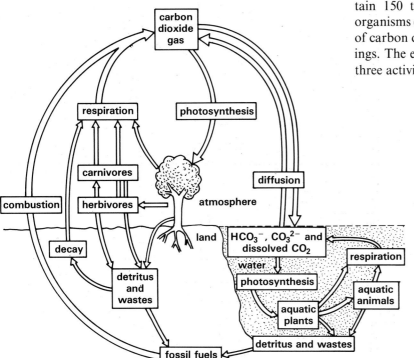

i. Photosynthesis Producers use carbon dioxide in photosynthesis. Each year, terrestrial plants remove between 50 and 100×10^6 tonnes of carbon dioxide from the atmosphere. Marine algae use about the same amount from the oceans. The standing crop of plants is an enormous reservoir of carbon. They hold about twice as much carbon as is contained in the atmosphere. Carbon fixed in photosynthesis millions of years ago, especially during the Carboniferous period, is preserved in vast reserves of fossil fuels such as coal, oil and natural gas. Peat is a recently-formed fossil fuel.

ii. Respiration Living organs make carbon dioxide as a respiratory product. The amount of respiratory carbon dioxide released by producers is over half the amount they fix in photosynthesis. Consumers contribute about 10 % and decomposers and detritivores roughly 40 % of the carbon dioxide returned to the environment as a product of respiration.

iii. Combustion Carbon dioxide is released when fossil fuels are burned. It is estimated that about $1-10 \times 10^6$ tonnes of carbon dioxide enter the atmosphere annually in this way. The amount of fossil fuels burned each year has increased over the past century. It is thought to be one of the reasons why the mean concentration of carbon dioxide in the earth's atmosphere is slowly rising (Chapter 33). The cutting-down of vast tracts of tropical forest may be another cause, because there is now less vegetation to remove atmospheric carbon dioxide in photosynthesis.

Ecologists have speculated on the possible effects of the increased carbon dioxide in the air. One possible consquence is what is called the **greenhouse effect** (Chapter 33). An alternative possibility is that a rise in temperature would cause increased evaporation, making the earth's atmosphere more cloudy. Less heat from the sun could then reach the earth's surface, so compensating for the greenhouse effect. The fact that the outcome is the subject of great controversy tells us how little is known of the rates of the processes which bring about the cycling of carbon.

3 The water cycle

Water has many roles in living organisms and life is impossible without it (Chapter 1). The total mass of water vapour in the atmosphere is equivalent to a mean rainfall of 2.5 cm a year over the earth's surface. The average rainfall is 90 cm a year, so water in the atmosphere is cycled $90/2.5 = 36$ times a year. Rain does not fall evenly throughout the world. The amount falling on land determines to a large extent the abundance and distribution of terrestrial plants. Two processes are the cause of the cycling of water (Fig 32.28).

Fig 32.28 The water cycle

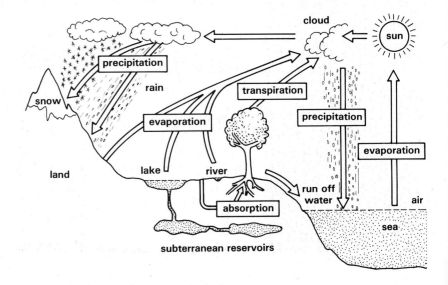

i. Evaporation The earth's atmosphere has a lower water potential than living organisms, aquatic habitats and all but the driest of environments. Consequently, water constantly evaporates into the air. Most **evaporation** occurs over the oceans, from which about $1500 \times 10^6 \, m^3$ of water vapour pass into the atmosphere each day. This compares with approximately $200 \times 10^6 \, m^3$ each day from the land, roughly half from the soil and half from vegetation. Water vapour is less dense than air, so it rises to the upper atmosphere where the temperature is lower. Here it condenses, forming clouds of water droplets. Most clouds form over the oceans, where evaporation is greatest. Clouds sometimes evaporate when they absorb heat from the sun. Alternatively they are blown elsewhere, often over land where water precipitates from them.

ii. Precipitation Water precipitates from clouds as rain, hail or snow. The mean daily **precipitation** over land is about $100 \times 10^6 \, m^3$ more than that lost by evaporation. The difference is because this amount of water runs off the land into the oceans in streams and rivers every day. The total precipitation of water over the earth is, however, the same as the amount evaporated, so the mean water vapour content of the atmosphere remains constant. It is important to realise that evaporation and precipitation are uneven in different parts of the world. It explains why some terrestrial habitats are constantly wet, whereas others are arid. In many areas, precipitation is also seasonal and wet spells alternate with dry ones.

The rate at which water is cycled causes the rapid removal of a variety of atmospheric pollutants. The clearing of a haze after a shower is because dust, soot, smoke and other particles suspended in the air have been washed out. Rain also dissolves other atmospheric pollutants, such as sulphur dioxide (Chapter 33).

The cloud cover in the atmosphere affects the earth's **heat balance**. Rain clouds reflect or absorb about 20 % of the radiation from the sun, mainly infra-red rays. Clouds also reflect heat radiating from the earth back to the earth's surface. The greenhouse effect is comparable to what we saw earlier with atmospheric carbon dioxide. It helps to prevent the loss of too much heat to outer space.

Fig 32.29 (a) Soil profile under heather moorland (b) Profiles of (i) podsol, and (ii) brown earth

(a)

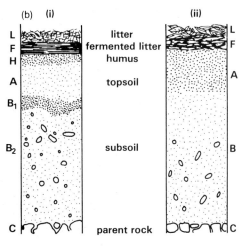

32.3.5 Soil as an ecosystem

Soil is a mixture of mineral particles and organic matter. It is the habitat for hordes of decomposers, detritivores and carnivores, as well as the roots of green plants which are the main producers in the soil ecosystem. In addition to providing nutrients for its inhabitants, soil is their source of air, heat and water. Many of the stages in the cycling of matter also take place in soil.

1 Soil profiles

When a deep trench is dug in soil, it is usual to see the soil in horizontal layers called **horizons**. The layering, which varies from one type of soil to another, is known as a **soil profile** (Fig 32.29). Freshly deposited **detritus**, mainly leaves, twigs and fruits, make up the topmost horizon. Beneath it is a layer of partly decomposed detritus, then a well-rotted horizon of **humus**. In soils where earthworms are plentiful, humus is thoroughly mixed with the underlying horizon of mineral particles. Horizons L–A are the **topsoil**. It lies on the lighter-coloured B horizons which make up the **subsoil**. The roots of plants grow mainly in the topsoil.

2 Mineral particles

Sand, silt and **clay** are the mineral particles in soil. They are derived from rock by weathering and are of different sizes (Table 32.8). The mineral particles of **sedentary soils** come from the underlying rock. The parent rock of **sedimentary soils** may be a long way off, the particles having been carried by water before settling as a sediment. Sand and silt are made of silica, a material which is physically and chemically inert. Clay is composed of layers of aluminosilicates which can attract and hold mineral ions and water. The ions can be displaced by others in a process called **ion exchange**. They are then available for absorption by soil inhabitants and plant roots.

The proportion of sand, silt and clay determines the feel or **texture** of a soil and also its **structure**. The mineral fraction of sandy soils contains more than 80 % sand particles by weight. When rubbed between the fingers, sandy soils have a friable gritty texture. Clay soils contain more than 40 % clay particles and have a clinging soapy texture. The structure of a soil is the way in which the particles bind into larger units called **aggregates**. A high proportion of clay usually causes the mineral particles to stick together in large **clods**. They make it difficult to obtain a fine tilth which is desirable for seedling growth. Loam soils contain small aggregates called **crumbs** in which clay, silt and sand are bound by humus and fungal hyphae. Adding lime to soils encourages crumbs to form by causing the **flocculation** of clay particles. Good crumb structure improves drainage and aeration of soil.

3 Pore space and air capacity

Between the mineral particles are pores filled with air or water. The pores are the spaces in which the soil inhabitants live and plant roots grow. Clay soils have a large **pore space** but a low capacity for retaining air, because many of the pores are extremely fine and hold water even when drainage has occurred. In contrast, the **air capacity** of sandy soils is higher, despite the fact that the pore space is smaller than in clay soils. This is because water drains more easily through the coarser pores, which then become filled with air.

4 Soil air

The air in the pore space is continuous with the earth's atmosphere, and gas exchange occurs between the two. **Soil air** in the uppermost horizons is little different in composition from the atmosphere. Deeper down, soil air generally contains more carbon dioxide and less oxygen (Table 32.9). The

Table 32.8 Sizes of soil particles (International Society of Soil Science scale)

Particle	Diameter (mm)
coarse sand	2.00–0.200
fine sand	0.20–0.020
silt	0.02–0.002
clay	< 0.002

Table 32.9 Comparison of composition of the atmosphere and soil air (after Russell 1961)

	% Vol oxygen	% Vol carbon dioxide
arable land (fallow)	20.4–21.1	0.02–0.38
grassland	16.7–20.5	0.30–3.30
atmosphere	20.97	0.03

differences come about because of respiration by plant roots and the soil inhabitants. The extra carbon dioxide encourages autotrophic micro-organisms such as nitrifying bacteria. Aerobic micro-organisms use oxygen gas at such a rate that air-filled pores may become anaerobic. This explains why anaerobic nitrogen-fixing bacteria such as *Clostridium* thrive even in well-aerated soils.

Soil air is usually saturated with water vapour which helps to prevent desiccation of soil inhabitants.

5 Soil moisture

The moisture content of well-drained soils fluctuates mainly according to rainfall. After prolonged heavy rain, the pore space is filled with water and the soil is waterlogged. If a soil is **waterlogged** for a long time, anaerobic bacteria flourish. They include denitrifying and sulphate-reducing bacteria. Sulphate-reducers produce hydrogen sulphide which is toxic to roots, as is nitrite formed by denitrifiers. Root growth and mineral absorption, both dependent on a supply of ATP from aerobic respiration, are also inhibited in waterlogged soils. A few species of plants, such as the common rush, *Juncus communis*, can tolerate such conditions (Fig 32.30). Waterlogging also slows down the rate at which organic matter is decomposed. This is why thick layers of partly decayed detritus accumulate in bogs and marshes as peat.

Fig 32.30 Common rush *Juncus communis* growing in poorly drained pasture

In places which are well-drained, gravity causes water to percolate through soil. The amount of water retained when **gravitational water** has drained away is the **field capacity** of the soil. A soil's ability to retain water draining through it depends on its structure and organic content. The smaller pores in clay soils hold moisture better than the coarser pores of sandy soils. Organic matter holds moisture like a sponge. The water-holding capacity of sandy soils can be improved by digging in organic matter in the form of peat or compost.

Water is held in the soil pores by capillarity. It is **capillary water** that is mainly accessible to plant roots and the inhabitants of soil. Capillary water is lost by evaporation and root absorption. If it is not replaced by rainfall or irrigation, a point is reached when the water potential of the soil is so low that roots are unable to absorb water. At this point plants begin to wilt. There is still **hygroscopic water** in the soil, but it is strongly held by clay particles and organic matter.

Soil water contains dissolved mineral ions and gases which soil organisms and plant roots may absorb.

6 Organic matter

Small amounts of organic matter are made in soils by autotrophic micro-organisms. However, the quantities are insignificant compared with **detritus** in the form of dead leaves, twigs, fruits and roots from vegetation. In well-aerated soils, such detritus is acted on by decomposers and detritivores and converted to black, amorphous **humus**. The type of humus which forms depends on the nature of the detritus and the soil inhabitants which break it down. **Mull humus** develops where deciduous trees grow in soils with a high calcium content. Earthworms, which abound in calcium-rich soils, pull enormous numbers of dead leaves into the soil. The humus which forms is thus thoroughly mixed with mineral particles in the topsoil, where a good crumb structure develops. **Mor humus** is acidic and is formed where coniferous trees or heathers grow in sandy soils with a low calcium content. Earthworms are scarce in such soils, so little detritus is mixed with the mineral particles. Most of the detritus stays on the surface of the soil, where it is humified.

Very thick deposits of **peat** accumulate in places where water stands permanently, such as bogs and fens. Peat is partly decayed detritus. Bog peat is very acidic, fen peat less so.

7 Soil pH

The pH of a soil is governed primarily by the amount of calcium it contains. Calcium-rich mineral particles are formed when calcareous rocks such as limestone are weathered. Silaceous rocks such as sandstone and granite are weathered to form particles lacking calcium.

Some plant species, for example beech, *Fagus sylvaticus*, are tolerant of a wide range of soil pH. It grows in British soils with a pH ranging from 3.8 to 8.8. However, soil pH is an important factor governing the distribution of many species of plants. **Calcifuge** species such as heather, *Calluna vulgaris*, thrive in acidic, calcium-deficient soils. Calcifuges can be grown in calcareous soils, provided plenty of acid peat is dug in first. **Calcicole** species such as dog's mercury, *Mercurialis perennis*, do best in alkaline soils where calcium is plentiful. Some calcicoles fail to thrive in acid soils because there is not enough calcium present. Others do not do well because they are poisoned by the high concentrations of available aluminium and manganese in acid soils. Dressing acid soils with lime enables calcicoles to be grown.

The pH of soil also affects the distribution of soil organisms. Fungi are more tolerant of acid conditions than bacteria. A few species of earthworm are acid-tolerant but most prefer neutral or alkaline soils (Fig 32.31).

Fig 32.31 Effect of soil pH on earthworm distribution (after Satchell 1955)

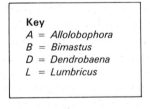

Key
A = *Allolobophora*
B = *Bimastus*
D = *Dendrobaena*
L = *Lumbricus*

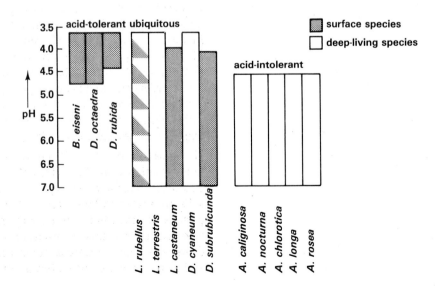

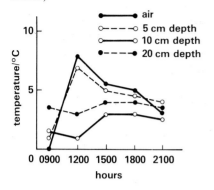

Fig 32.32 Temperature fluctuations in air and soil (from Brown, *Ecology of Soil Organisms*, Heinemann Educational Books)

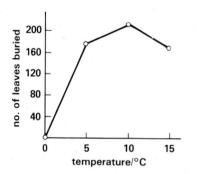

Fig 32.33 Effect of temperature on leaf burial by *Lumbricus terrestris* (data from Edwards and Lofty 1972)

8 Soil temperature

The **temperature** of soil depends mainly on how much solar radiation reaches the soil's surface and how well the radiation is absorbed and conducted. Where vegetation forms a dense canopy, little radiation reaches the soil. Dark soils absorb heat better than light-coloured soils. However, a thick layer of dark detritus on the surface may absorb heat so well that little is transferred to the underlying horizons. Soils with a high moisture content absorb heat well, but do not warm up a great deal because water has a high specific heat capacity (Chapter 1). Nevertheless, water conducts heat better than air. Soils which are dry at the surface may therefore be cold deeper down.

Soil temperatures vary from season to season. In the upper horizons there are daily fluctuations in temperature too, whereas the temperature of the hower horizons is fairly stable (Fig 32.32). Seeds fail to germinate and roots stop absorbing water and minerals if the temperature falls below 5 °C. The activity of earthworms is also affected by the temperature (Fig 32.33). Much is still to be learned about the effect of temperatures on many of the soil's inhabitants. Most are thought to be dormant at low temperatures.

9 Soil microflora

Soil is the natural habitat for representatives of all the main groups of **micro-organisms**. Their activities are vital to the cycling of matter.

i. Bacteria are the most numerous of soil microbes. A gram of soil may contain as many as 1×10^9 bacteria. They have a live mass of up to 90 g in a square metre.

Bacteria which use carbon dioxide as a source of carbon and solar radiation as an energy source are **photo-autotrophs**. They live near the soil surface where sunlight is available. **Chemo-autotrophs** include nitrifying, denitrifying and sulphate-reducing bacteria. The quantity of organic material made by autotrophic bacteria is very small in comparison with the amount of detritus entering the soil. Their activities in the cycling of matter are much more important. The most abundant of heterotrophic bacteria in soil are the **decomposers**. They feed saprophytically, using a wide range of organic substances in detritus as a source of energy and nutrients. Other heterotrophic soil bacteria are primary consumers. They feed as **parasites** on plant roots or as **symbionts**. The best known of the symbionts are the nitrogen-fixing bacteria.

ii. Fungi are just as abundant as bacteria in neutral and alkaline soils. In acid soils, fungi are more plentiful. All soil fungi are heterotrophs. The majority are **decomposers** feeding saprophytically on detritus. As a source of energy they use a wide variety of organic compounds including lignin and cellulose which few bacteria can break down. Saprophytic soil fungi are therefore effective in decomposing woody detritus. Others are primary consumers living as symbionts in mycorrhizae or as root **parasites**. A few soil fungi are **predators** catching protozoans with sticky discs on their hyphae or lassooing nematodes in hyphal nooses.

iii. Algae are found in all soils, but they are not very abundant. They are most plentiful near the surface of moist, undisturbed soils. In acid soils, unicellular and filamentous **green algae** are common. Neutral and alkaline soils are preferred by **diatoms**. Soils which are waterlogged for lengthy periods of time often contain large numbers of **blue-green algae**. Their nitrogen-fixing ability is important in maintaining the fertility of paddy fields. Algae are producers, but as they are relatively few in number they add little to the organic matter in soil.

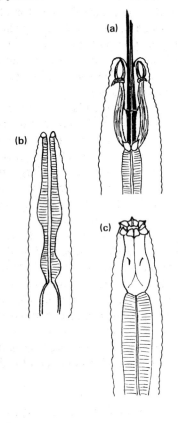

iv. Viruses are parasites. Those found in soil are mainly plant pathogens and bacteriophages which lyse bacteria.

10 Soil fauna

There are many animals which are found nowhere else but in soil. It is convenient to group them according to size.

i. Microfauna are only visible with a microscope. Flagellate, ciliate and amoeboid **protozoans** are common in most soils. A few, such as *Euglena*, can photosynthesise and are producers. Some are detritivores consuming small fragments of detritus. The majority are primary consumers feeding on the plentiful supply of soil bacteria.

ii. Mesofauna are soil animals between 0.2 and 10 mm long. A hand lens is required for the smallest of them to be seen clearly. The most abundant of the mesofauna are **nematodes**. Some are detritivores using their muscular pharynx to suck in small particles of detritus suspended in water. Many are primary consumers feeding as parasites on the juices of plant roots, which they pierce using sharp stylets. They often live permanently in roots and frequently are the vectors of viral diseases of plants. A few soil nematodes are predators on small soil animals, which they grasp using the tooth-like structures around their mouths (Fig 32.34).

Arthropods are well represented in the soil mesofauna (Fig 32.35). Collembolans, commonly called springtails, are small wingless insects. They are called springtails because of the springing organ which is folded under the abdomen while the animal is at rest. When extended, it acts as a lever causing the insect to spring into the air. Many species of springtails are detritivores. Predatory spiders live near the surface of soil where they catch insect larvae, small centipedes and millipedes. The most ferocious of soil arthropods are the false scorpions, which are found mainly in detritus where they prey on spiders and springtails. Mites are the most abundant of soil arthropods. They have a range of feeding habits. Some are detritivores using their blunt mandibles to chew dung and detritus. Others have a preference for fungal mycelia. Predatory mites capture nematodes and springtails, while spiders are the hosts for parasitic species.

Fig 32.35 Soil mesofauna

(a) springtail

(ARDEA: J L Mason)

(b) predatory spider

(ARDEA: Jonathan Player)

(c) false scorpion

(ARDEA: John Mason)

(d) mite

(Eric & David Hosking)

iii. Macrofauna are soil animals bigger than 10 mm in length. The arthropods are again well represented (Fig 32.36). Many are **insects** or insect larvae. They include predators such as ground beetles which prey on mites, detritivores such as rove beetles which feed on carrion and dor beetles which eat cowpats. Primary consumers such as wireworms (larvae of the click beetle) and leatherjackets (crane fly larvae) which browse on plant roots are also present. **Crustaceans** are relatively few in soil; the most frequent are woodlice which are detritivores. Chilopods are represented by

Fig 32.36 Soil macrofauna

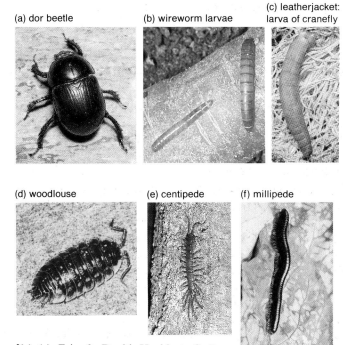

(a) dor beetle (b) wireworm larvae (c) leatherjacket: larva of cranefly

(d) woodlouse (e) centipede (f) millipede

[(a)–(c) Eric & David Hosking: G E Hyde; (d) ARDEA; (e) ARDEA: J L Mason; (f) Science Photo Library: Sinclair Stammers]

carnivorous centipedes which prey on mites and springtails. Millipedes are burrowing diplopods which feed mainly on leaf-litter.

Molluscs of the macrofauna are slugs and snails. They are best known as consumers of living plant material which they rasp away using their file-like radula. However, many are omnivorous, eating detritus and carrion, while a few species of slugs are carnivorous, preying on annelids.

The best-known of soil **annelids** are the earthworms. There are twenty-five species of earthworm in Britain. Ten are common in farmland and in gardens. Earthworms are detritivores feeding mainly on partly decayed leaves which they drag into their burrows. In this way they help to mix the mineral particles of soil with organic matter. Their burrows also help to aerate and drain the soil. Earthworms eat their way through soil as they burrow. The soil they egest is deposited underground or as casts on the soil surface. Casts are small piles of very fine soil rich in ammonia which improve the tilth and fertility of soil. Charles Darwin estimated that casting produced a top layer of soil 5 mm thick each year in his garden. In cultivated soils earthworms are fewer because many are killed by ploughing and are eaten by birds which follow the plough.

Table 32.10 Effect of soil type on earthworm populations (after Guild 1951)

Soil type	Worms m^{-2}	No. of species
light sandy	57	10
medium loam	56	9
clay	40	9
peaty soil	14	6

Fig 32.37 Vertical distribution of some soil arthropods (data from Jackson and Raw 1966)

11 Distribution of soil organisms

The types and numbers of soil inhabitants vary from one soil type to another (Table 32.10). As we have seen, soil moisture content, aeration and pH have an important bearing on the organisms present. Within the same soil, however, there is usually a vertical distribution of soil organisms. More of the inhabitants are found in the upper layers than in the deeper horizons (Fig 32.37). Little has been done to find out the reasons for vertical distribution. However, it is possible to speculate on what the causes may be.

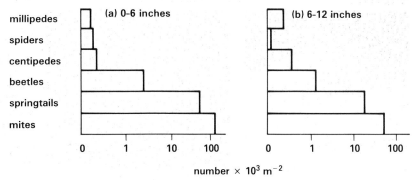

(a) 0-6 inches   (b) 6-12 inches

millipedes, spiders, centipedes, beetles, springtails, mites

number × 10^3 m^{-2}

Algae and photo-autotrophic bacteria can only flourish where **light** is available. This is probably why they are active only at the surface of soil. The majority of soil organisms respire aerobically. It is possibly for this reason that many of them inhabit the upper horizons where **oxygen** is more abundant.

Many of the larger soil animals are unable to burrow, so are confined to the upper horizons. Smaller species may use the **burrows** of earthworms to get to the deeper layers, or they may be washed downwards by water

681

percolating through the soil. Burrowing enables earthworms to inhabit the moister lower horizons where desiccation is less likely.

There is some evidence that availability of suitable **food** may account for the vertical distribution of different species of springtails. **Temperature** preference may be another factor influencing vertical distribution. Some species of mite move up and down in the soil in order, it is thought, to keep near to their optimum temperature. **Predation** and **competition** must also have a bearing on the vertical distribution of all soil organisms, but so far these factors have been little investigated.

One aspect about which rather more is known is the **rhizosphere** effect. As roots grow, surface cells are rubbed off by rough mineral particles. Young roots exude sugars and amino acids into the soil. The effect is to encourage the activity of soil micro-organisms in the immediate vicinity of growing roots. The rhizosphere is a zone extending several millimetres from the root surface. In it soil microbes are many times more numerous than elsewhere. They compete with plant roots for nutrients, but are generally beneficial to plants. Some of them produce auxins which stimulate root growth, others are symbionts which may form a mutualistic association with the roots. It is also possible that they keep parasites at bay.

12 Decomposition

Precisely what happens when detritus is broken down varies from one type of soil to another. It depends mainly on the nature of the detritus, decomposers and detritivores. The detritus in a meadow is mainly the soft leaves, stems and roots of grasses. Earthworms are usually abundant. In a coniferous forest, woody twigs and cones make up part of the detritus which consists mainly of tough, spiky leaves. Earthworms are scarce.

Very few comprehensive studies have been made of the decomposition process. Detritus usually contains a large proportion of cellulose and hemicelluloses. In woody detritus the cellulose is lignified. The amounts of starch, sugars and proteins are usually small, because these materials are withdrawn from leaves before they fall and are stored elsewhere in plants. However, minerals are usually present, often in substantial amounts.

The decomposition of pine needles occurs in a number of stages:

Stage 1 Colonisation by fungi; needles dark brown but still intact.

Stage 2 Mesophyll collapsed, containing fungal hyphae and faeces of detritivores; needles greyish but still intact.

Stage 3 Needles fragmented and mixed with hyphal fragments of detritivores.

Stage 4 Needles humidified and mixed with mineral particles.

Detritus usually has a high carbon to nitrogen ratio. As it is consumed by soil organisms the ratio falls. This is because the decomposers and detritivores use some of the carbon compounds for respiration, and carbon dioxide is released to the air. While the detritus is attacked by soil organisms it is leached of minerals by falling rain. The minerals are then available for absorption by plant roots. Eventually the only materials left are those resistant to decomposition or which cannot be digested by detritivores. Humus is an amorphous mixture of such compounds. Clearly the composition varies according to the type of detritus and the way in which it has been decomposed.

13 Food webs in soil

The **producers** of soil ecosystems are the vegetation, micro-organisms are the **decomposers**, and the soil fauna are the **consumers** (Fig 32.38). As in most ecosystems, the **food web** is imperfectly known, mainly because much has still to be learned about the food preferences of the soil fauna.

Fig 32.38 Food web in soil (simplified)

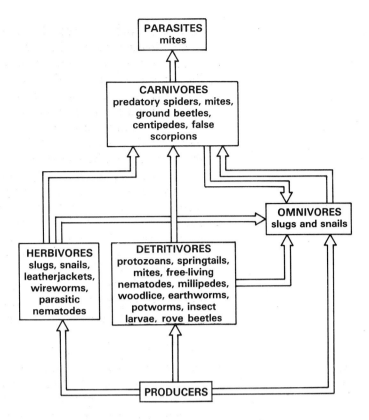

14 Nutrient cycling and energy flow

Many of the stages in the cycling of **carbon** and **nitrogen** take place in soil. Soil micro-organisms also play vital roles in the cycling of **phosphorus** and **sulphur** (Fig 32.39). The repeating cycles of growth, death and decay ensure that minerals are constantly circulated among the soil community.

Fig 32.39 The sulphur cycle and the phosphorus cycle

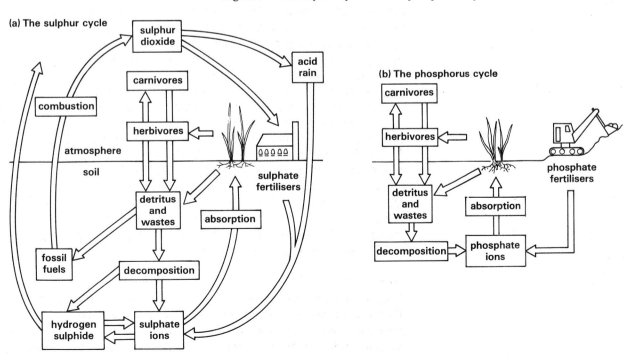

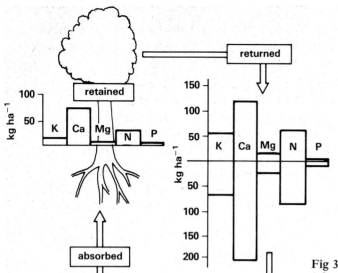

Soil usually contains a much larger reservoir of minerals than the vegetation growing in it (Fig 32.40).

In the past, ecologists concentrated mainly on counting the numbers of various kinds of organisms in soil samples. Today, interest is centred around their activities and in particular the **flow of energy** through the soil community. The few investigations which have been completed show that many of the detritivores are inefficient converters of energy (Table 32.11).

Fig 32.40 Annual cycling of five nutrients in a Belgian forest (after Mason 1976)

Table 32.11 Assimilation efficiencies of some detritivores (after Mason 1976)

Detritivore	Efficiency (%)	
earthworms	1.00	The assimilation efficiency is $\dfrac{C-F}{C} \times 100$
millipedes	6–15	
woodlice	15–30	
mites	10–65	where C = mass of food consumed
molluscs	50–75	and F = mass of faeces produced

This is because they are unable to digest much of the plant material they consume. Symbiotic microbes in the intestines of some detritivores may assist digestion in much the same way as in herbivorous mammals. A few detritivores, notably molluscs, produce cellulase enzymes in their digestive juices which considerably boost their efficiency of energy conversion. Nevertheless, the faeces of most detritivores contain much undigested plant material which decomposers can break down. Decomposers are also usually active in detritus before it is eaten by detritivores. It is for these reasons that it is presently thought that decomposers account for over 80 % of the energy entering the soil ecosystem.

SUMMARY

Ponds, streams, rivers, woods and forests are examples of ecosystems. They are inhabited by balanced communities of plant and animal populations which are in equilibrium with their environment.

The size of a population is determined by a complex of factors, some dependent on the density of the population, others not. Typically, populations fluctuate in size, having cycles of growth and decline. The phase of most rapid growth of a population is the log (exponential) phase. At the present time the human population is in a phase of hyperexponential growth. The carrying capacity of the environment determines the maximum population size it can support. Should we outgrow the resources available to us, the human population will eventually decline in numbers.

Spells of severe weather, epidemics of disease and fluctuations in food supply account for the decline in wild populations. Predation is another key factor in the ups and downs observed in many wild ▶

populations. Humans have preyed so intensively on some wild animals that they have become extinct or are on the verge of being wiped out. Various species of whale have suffered a serious decline in numbers in the second half of this century as a direct result of human predation.

Biological control of populations is a strategy which has been used successfully in several countries to check the number of some important pests which were introduced from other parts of the world. In a new environment populations often grow at an alarming rate because the factors which kept them in check in their native country are less strong or even absent. In Britain the rabbit population was checked by the myxomatosis virus; in Australia the prickly pear cactus by a moth larva.

Deciduous forest is the natural vegetation of much of lowland Britain. It developed on the bare sediments left behind by the melting glaciers and ice sheet which covered most of the country 10 000 years ago. The first colonisers of bare land are lichens and mosses. The changes they bring about encourage herbaceous flowering plants then shrubs and trees to take over. Successions of this kind are typical whenever new habitat is created as on sandy shores and estuarine mud flats. Climate is the most important factor in determining the resultant climax community.

Ecosystems exchange energy and matter with their surroundings. Within ecosystems energy flows from one group of organisms to another. For most ecosystems the sun is the source of energy. Solar energy trapped in producers (green plants) passes on to the consumers, decomposers and detritivores which form an intricate food web in any ecosystem. Only about 10 % of the energy flows from one feeding group to the next. This limits the number of trophic levels in a food web to three or four at the most.

Three kinds of ecological pyramid are used in describing ecosystems. Pyramids of numbers tell us the numbers of organisms at each trophic level; pyramids of biomass show the standing crop at each level whilst pyramids of energy show how much energy flows from one trophic level to the next.

Nitrogen and carbon are two essential elements for all living organisms. The various stages in the cycling of such elements helps maintain an equilibrium in any ecosystem. Mineral cycles keep the elements in various forms which different groups of organisms can exploit. Ammonification, nitrification and nitrogen fixation are the main stages of the nitrogen cycle. The most important processes which contribute to the cycling of carbon are respiration, photosynthesis and combustion. Water is vital to all forms of life. It is constantly recycled by the processes of precipitation and evaporation.

Soil is an ecosystem which is accessible to most students of biology. Many of the factors such as pH, moisture, air, organic and mineral particle content which affect such an ecosystem can be readily measured. Populations of soil organisms can also be investigated using simple, inexpensive equipment.

QUESTIONS

1 The diagram shows part of a food web.

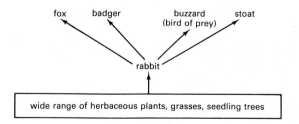

(a) At each trophic level in this food web
 (i) what is the main way in which energy is gained?
(ii) in what ways is energy lost?
(b) Approximately what proportion of the energy in this food web is lost in each way at each trophic level?
(c) With the help of information in the diagram, suggest **one** explanation of each of the following:

 (i) the disappearance of rabbits from many areas where they were common has been accompanied by an increase in the growth of scrub;
(ii) analyses of residues of chlorinated hydrocarbon insecticides in body fat show higher concentrations in the tissues of foxes and buzzards than in the tissues of rabbits.
(d) In Scotland the recent spread of the rabbit as a pest can be linked with the development of huge areas of agricultural monoculture. Define *plant monoculture* and explain how it leads to increases in the population of one or two animal species. (AEB 1986)

2 (a) With reference to a **named** ecosystem, what is meant by the following terms:
 (i) energy flow,
 (ii) trophic levels,
(iii) food web?
(b) Discuss the interactions between the living and non-living components of such an ecosystem. (C)

3 Here are three pyramids of numbers from different food chains.

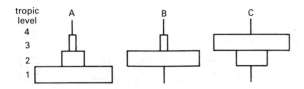

(a) (i) Name the four trophic levels in pyramid A and give a typical example of an organism found at each level. (The organisms you suggest should all come from the same food chain.)
(ii) Account for the shapes of pyramids B and C.

(b) The following pyramids were estimated for a single food chain in a terrestrial ecosystem.

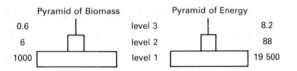

(i) In what units might the figures for each pyramid be expressed?
(ii) Calculate the energy content per unit biomass of organisms in levels 1 and 3.
(iii) Calculate the percentage energy flow from level 2 to level 3 in this ecosystem. Give **one** reason why this value is not 100 %.

(c) The table in the second column relates the number of fish per unit area to the annual rate of population growth. An annual rate of population growth of 1.0 signifies no net change in population.

Number of fish per unit area in year 1	10	15	20	25	30	35	40	45
Annual rate of population growth	0.5	0.75	1.0	4.3	5.0	3.9	1.5	0.1
Number of fish per unit area in year 2								

(i) Calculate and enter the missing values in the table.
(ii) Consider the column in the table with the highest number of fish per unit area in Year 2. How many fish should be harvested in Year 2 in order to maintain the same annual rate of population growth in Year 3? Explain your answer.
(iii) Suppose that there are 40 fish per unit area, of which 25 are harvested prior to reproduction. What would be the annual rate of population growth as a result of this cropping activity?
What would be the likely effect of continuing this level of cropping activity on the future of the population?

(JMB)

4 (a) What is an ecosystem?
(b) Describe the flow of energy and the cycling of carbon and nitrogen in any **named** ecosystem.
(c) Suggest reasons why felling and removal of forest trees result in changes in the levels of nutrients in the soil.

(C)

5 In an investigation of a freshwater pond, 35 water bugs (**Notonecta**) were caught, marked and released. Three days later 35 water bugs were caught and 7 were found to be marked.
(a) What is the approximate size of population of water bugs in the pond? Show your working.
(b) Give **three** reasons why capture-recapture is unlikely to be an accurate way of assessing the size of a water-bug population.

(AEB 1985)

6 Detritivores are animals which feed on particles of dead and decomposing plant and animal material. A study was made of interactions between river detritivores kept in plastic chambers and fed on leaves of one species. The temperature in all chambers was kept at 5 °C. Each chamber contained 9.7 g (dry mass) leaves in 1.5 dm³ aerated river water. All the chambers were inoculated with saprophytic bacteria and fungi; one week later detritivores were added to all the chambers except the controls. Two types of detritivore (both 8–10 mm in length) were used: 'shredders' (crane-fly larvae) which feed directly on leaves; 'collectors' (may-fly nymphs) which feed on faeces and fine leaf fragments.

The table summarises the results of the study. Each line in the table gives the mean values of results from three chambers.

(a) Using the information provided,
 (i) comment on the effect of the shredders on the loss of leaf tissue;
(ii) comment on the effect of the collectors on the loss of leaf tissue.
(b) Consider some of the weaknesses in the method used in this study and suggest how the study might have been improved.

(AEB 1986)

Duration of experiments	Initial number of individuals per chamber		Survivors at the end of the experiments (%)		Growth per day during the period of the experiments (%)		Loss in dry mass of leaves at the end of the experiments (%)
(days)	shredders	collectors	shredders	collectors	shredders	collectors	
110	0	0	—	—	—	—	26.1
105	10	0	66.7	—	1.1	—	63.9
51	30	0	56.6	—	0.75	—	68.6
57	0	10	—	70.0	—	0.13	18.9
86	0	30	—	35.5	—	0.5	24.1
110	10	10	43.3	50.0	1.53	0.2	49.8
60	30	30	56.7	58.3	0.66	0.4	67.3

(after Cummins et al., 1973)

33 Pollution and conservation

33.1 Pollution of the atmosphere 688

    33.1.1 Carbon dioxide 688
    33.1.2 Sulphur dioxide and smoke 689
    33.1.3 Carbon monoxide 691
    33.1.4 Oxides of nitrogen 691
    33.1.5 Chlorofluorocarbons 692

33.2 Pollution by heavy metals 692

    33.2.1 Lead 692
    33.2.2 Mercury 694
    33.2.3 Other heavy metals 695

33.3 Pesticides 696

    33.3.1 Herbicides 696
    33.3.2 Fungicides 697
    33.3.3 Insecticides 698

33.4 Radiation 701

    33.4.1 Forms of radiation 701
    33.4.2 Sources of radiation 702
    33.4.3 Biological effects of radiation 703

33.5 Pollution of aquatic habitats 705

    33.5.1 Freshwater pollution 705
    33.5.2 Marine pollution 708

33.6 Conservation 709

    33.6.1 The case for conservation 709
    33.6.2 Conservation in action 710

Summary 713

Questions 714

Bruce Coleman: WWF/Kojo Tanaka

33 Pollution and conservation

There are many toxic substances in the biosphere. They are part of the natural cycling of matter (Chapter 32). For example, when volcanoes erupt they discharge poisonous gases and ash is scattered for miles around. However, the amounts of toxic materials arising from natural events are often small in comparison with those released into the environment by man. A good many toxic substances are man-made and are of recent origin. Some are the products of industry and man's attempt to eradicate or control pests on a large scale. Others are the results of people living close to each other in towns and cities. In such places, the sheer mass of domestic waste poses disposal problems.

Substances introduced into the environment by man which adversely affect living organisms, man included, are called **pollutants**. Growth of the human population, industry and agriculture are at the heart of many aspects of pollution.

33.1 Pollution of the atmosphere

The earth's atmosphere is a mixture of gases which is about 2000 km thick. The lowermost layer, about 10 km thick, is called the **troposphere.** It is with this layer that gas exchange occurs between the atmosphere and living organisms. Above it is the **stratosphere**, which contains ozone, a gas which absorbs ultraviolet and infra-red radiation from the sun. In this way terrestrial organisms are protected from excessive exposure to such forms of radiation. Ultraviolet radiation is one of the causes of mutations, while infra-red radiation raises the temperature of living organisms and can be the cause of sunstroke and sunburn.

Table 33.1 shows the composition of the troposphere at sea level in a clean-air area. Variable amounts of other gases such as neon, helium, krypton, xenon, methane, carbon monoxide, hydrogen, ammonia, sulphur dioxide, hydrogen sulphide and nitrogen oxides are also present. They are there as part of the natural cycling of matter. Man's industrial activities have altered the amounts of the atmosphere's natural constituents. In recent times, man's technology has also caused new ingredients to enter the atmosphere. There is growing evidence that such changes are harmful to life.

Table 33.1 Composition of the troposphere at sea level

Gas	Volume (%)
nitrogen	78.08
oxygen	20.96
argon	0.93
carbon dioxide	0.03

The amount of water vapour varies from 0 to 4 %.

Fig 33.1 Annual mean concentration of CO_2 in the air at three widely spaced sites (from Spedding, *Air Pollution*, OUP)

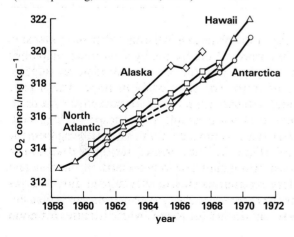

33.1.1 Carbon dioxide

The burning of fossil fuels such as coal, oil and natural gas provides us with most of the energy we require for domestic, industrial and agricultural purposes. When such fuels are burned, **carbon dioxide** is released into the air. Data obtained from widely-spaced sampling points indicate that the amount of carbon dioxide in the air is slowly increasing (Fig 33.1). The outcome of this is far from clear. However, two major possible effects of increasing atmospheric carbon dioxide are:

i. It may stimulate photosynthesis by plants. This is unlikely, as other factors such as scarcity of minerals and light are more likely to limit photosynthesis in plants growing outdoors. The results of recent field studies show that C3 plants are stimulated more by the extra carbon dioxide than C4 species. One effect of this could be that weeds would compete better than at present among C4 crops such as sugar cane.

ii. The earth's surface reflects some of the heat from the sun, thus warming the atmosphere. When air is enriched with carbon dioxide it reduces the amount of heat escaping into space just as the glass in a greenhouse does. For this reason it is called the **greenhouse effect**.

Fig 33.2 Temperature changes, measured and estimated in the northern hemisphere (from Spedding as above)

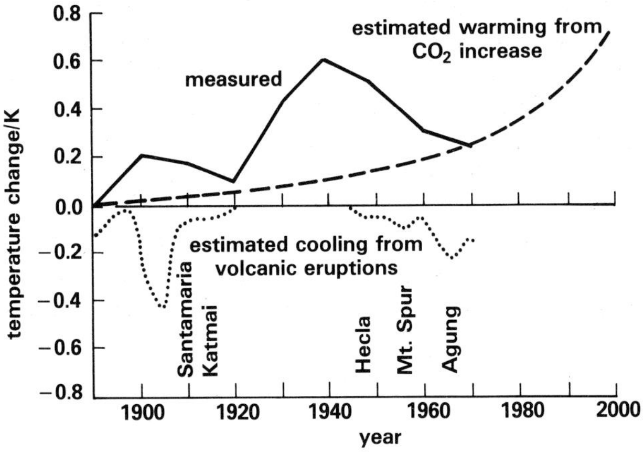

Fig 33.2 shows the predicted change in world temperature up to the year A.D. 2000, based on present emissions of carbon dioxide. Of course, the use of alternative sources of energy such as nuclear fuels may in the near future bring about a reduction in the emission rate. The falls in measured temperature are thought to be due to increases in particulate matter in the atmosphere from industry and volcanic eruptions. Modern industries release less dust, soot and smoke, so any future falls in the measured temperature are more likely to be caused by natural events. If, however, the overall result is a gradual increase in temperature, it will bring about melting of the polar ice-caps, thus raising the level of the oceans. One of the consequences of this would be a flooding of large areas of low land on which many modern cities such as London have been built as well as vast tracts of farmland.

33.1.2 Sulphur dioxide and smoke

Sulphur dioxide is released into the air when fossil fuels are burned and in the smelting of ores (Table 33.2). The background concentration is between 0.3 and 1.0 $\mu g\, m^{-3}$, but in areas near heavy industry the concentration can be as high as 3000 $\mu g\, m^{-3}$. The effects can be beneficial or harmful.

Table 33.2 Annual sulphur dioxide emissions for Britain in 1965 (after Robinson and Robbins 1970)

Source	Emission × 10^6 (tonnes)
coal	102.0
petroleum	28.5
copper smelting	12.9
lead smelting	1.5
zinc smelting	1.3

Sulphur dioxide is oxidised to sulphate ions in the air. When precipitated in rainfall, sulphate can be absorbed from the soil by plant roots. Sulphur is an essential nutrient for plant growth, so its absorption as sulphate stimulates the growth of crops. The effect could be important in poor countries where chemical fertilizers are not used for economic reasons.

Human life is endangered when people are exposed to high concentrations of sulphur dioxide in combination with smoke arising from the combustion of fossil fuels (Fig 33.3). The increased number of deaths when these factors were present was mainly due to bronchitis, pneumonia and heart failure. Laboratory experiments indicate that sulphur dioxide slows down ciliary activity in the respiratory tract. Particulate matter is then not cleared in the usual way and reaches the alveoli, where it causes irritation

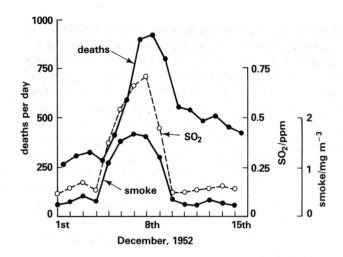

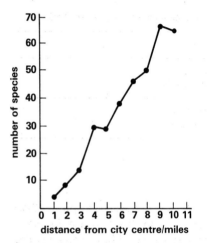

and interferes with gas exchange. Bronchitis is one of the main causes of death in British men over 45 years of age. About 30 million working days a year are lost in Britain by people who suffer from the condition.

Fig 33.3 Human death rates and pollution of the air with soot and SO_2 in London (from Mellanby, *The Biology of Pollution*, Inst. Biol. 'Studies in Biology', Edward Arnold)

Fig 33.4 Distribution of lichens in the Newcastle-upon-Tyne area (after Gilbert 1970)

The most obvious effect of the presence of particles in the air is a reduction in visibility. The **smogs** of the 1950s contained large amounts of smoke particles. Smoke, soot and ash particles are released into the air when fossil fuels are burned and ores refined. Ash and soot particles are heavy and are deposited near to their source of release. The smaller smoke particles remain suspended in the air for much longer. They are breathed in and may permanently blacken the alveoli, where they cause irritation and interfere with gas exchange. It was this which contributed to the increase in human deaths in the London smogs of the 1950s. Smoke particles may also contribute to the various forms of pulmonary ailments which are common among smokers.

When smoke, soot and other particles land on the surface of leaves, light is unable to penetrate and photosynthesis is inhibited. In this way the vigour of the plant is sapped. Plants absorb sulphur dioxide through their stomata. The leaves of some important crop plants such as barley, wheat and cotton are damaged by high concentrations of sulphur dioxide in the air. Others such as potatoes, maize and onions are resistant. Lichens are especially sensitive and have been used as indicators of aerial pollution (Fig 33.4).

The implementation of the Clean Air Act 1956 has done much to reduce the amounts of sulphur dioxide and smoke in the air around and in British towns and cities in recent years (Fig 33.5). The Act prohibits the emission of

Fig 33.5 Soot and SO_2 emissions in London air after the passing of the Clean Air Act (from Mellanby, *The Biology of Pollution*, Inst. Biol. 'Studies in Biology', Edward Arnold)

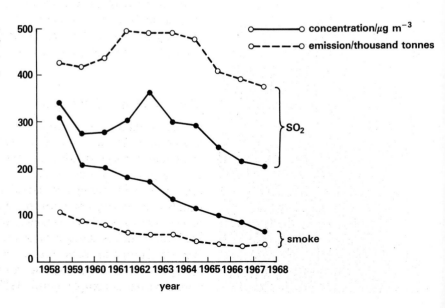

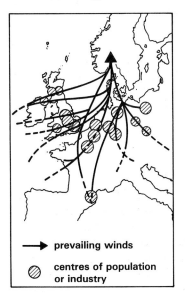

Fig 33.6 Accumulation of acid rain over Southern Norway (from Ottaway, *The Biochemistry of Pollution* Inst. Biol. 'Studies in Biology', Edward Arnold)

→ prevailing winds

⊘ centres of population or industry

Fig 33.7 Stonework at Canterbury Cathedral eroded by acid rain (Geoscience)

dark smoke from any trade or industrial premises. It also enables local authorities to designate clean air areas in which smokeless fuels have to be used in domestic fireplaces. In 1968 the Act was amended to include grit and dust. The Control of Pollution Act 1974 is aimed at reducing the sulphur dioxide content of the air by controlling the sulphur content of fuel oil. Even so, something like 30 million tonnes of sulphur dioxide are still released into the air in western Europe each year. It is blown by the prevailing winds to Scandinavia, where it falls as sulphuric acid in **acid rain** (Fig 33.6). Acid rain erodes and disfigures our buildings (Fig 33.7). It also lowers the pH of the soil, bringing toxic elements such as aluminium into solution. The extensive damage to plantations of conifers in Sweden and Germany in recent years is probably due to such factors. The high concentration of aluminium leached from soils receiving acid rain is thought to be the reason for the massive decline in fish populations in Norwegian lakes in the past decade.

So serious is the threat from acid rain that it was the sole topic at the 1982 Conference on the Environment held in Stockholm. To tackle the problem member countries of the European Economic Community (EEC) have set themselves targets aimed at minimising pollution of the air with sulphur dioxide and smoke (Table 33.3).

Table 33.3 EEC limit values ($\mu g\,m^{-3}$) for SO_2 and smoke to be met by 1993

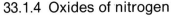

Period	Smoke	SO_2
Year	80	120 if smoke less than 40 80 if smoke more than 40
Winter	130	180 if smoke less than 60 130 if smoke more than 60

33.1.3 Carbon monoxide

Haemoglobin has a much higher affinity for **carbon monoxide** than oxygen. Breathing in high concentrations of carbon monoxide reduces the blood's ability to carry oxygen to respiring tissues. Many people suffer from oxygen deficiency in their tissues when the carbon monoxide haemoglobin content in their blood reaches 5%. Cigarette smokers often have up to 10% carbon monoxide haemoglobin in their blood. This is probably why babies born to mothers who smoke heavily are on average smaller than those of non-smokers.

33.1.4 Oxides of nitrogen

Nitrogen oxides play an important role in the formation of **photochemical smog** which was first recognised in Los Angeles, USA. Both nitric oxide and nitrogen dioxide, as well as hydrocarbons, are present in the exhaust gases of motor vehicles. In sunny weather they become trapped near the ground causing a severe reduction in visibility. Nitrogen dioxide is mainly a secondary pollutant formed from oxidisation of nitric oxide by ozone and hydrocarbons. Oxides of nitrogen are important contributors to **acid rain** which is thought to be the main cause of damage to forests in Western Europe in recent years. An EEC Directive sets a limit value of $200\,\mu g\,m^{-3}$ of nitrogen dioxide in the air with which member countries were expected to comply by July 1987.

33.1.5 Chlorofluorocarbons

Chlorofluorocarbons (CFCs) are gases used in refrigeration and air-conditioning units and in making polyurethane foams. Their best known use is as propellants in aerosol spray cans. Research suggests that CFCs can destroy ozone in the stratosphere. The effect would be to allow more UV light to reach the earth which could result in an increased incidence of skin cancer. It could also destroy plankton, the basis of food chains in all aquatic ecosystems. CFCs also contribute to the greenhouse effect. Concern for this form of aerial pollution has been worldwide.

By 1981 EEC countries had agreed to a 30% reduction in CFC production compared with 1976. The Vienna Convention of 1985 signed by the EEC, Canada, the USA and Scandinavia calls for a further reduction leading to an eventual ban on the use of CFCs.

33.2 Pollution by heavy metals

33.2.1 Lead

Heavy metals are those with relative atomic masses above 100. **Lead, mercury** and **cadmium** are in this category. They accumulate in living organisms, often reaching concentrations which inhibit enzyme activity.

Not all environmentalists agree that the present concentration of lead in our surroundings is harmful to life in general. However, there are many pieces of evidence which show that the amount of lead in the environment is increasing. There is lead in the air we breathe, in the water we drink and in the food we eat.

1 Lead in the air

Combustion of petrol in motor vehicles accounts for most of the lead in air. Lead is added to petrol as **tetraethyl lead** (TEL) which serves as an **anti-knock compound**. It improves the efficiency with which petrol is burned in internal combustion engines. Most of the lead comes out in the exhaust gases as lead chloro-bromide (PbClBr). In rural areas, about 20% of the lead taken in by humans comes from the air. It is over 50% in towns and cities. People who work close to motor vehicles have higher concentrations of lead in their blood than those who do not (Table 33.4).

The first EEC Directive on the lead content of petrol sought member countries to set a maximum permitted level of $0.4\,g\,dm^{-3}$ in 1978 and to work towards reducing it to $0.15\,g\,dm^{-3}$ as soon as possible. Britain did so by December 1985. The effects have been significant; the lead content of our air falling from 0.56 to $0.29\,\mu g\,m^{-3}$ in the first quarter of 1986 compared with the same period in 1985. A second Directive asked that unleaded petrol be available from October 1989 (Fig 33.8). These measures have resulted in only a marginal increase in the price of petrol to the consumer. From December 1987 the annual mean concentration of lead over EEC countries must not exceed $2\,\mu g\,m^{-3}$. Air sampling stations are to be set up by December 1989 to monitor this form of air pollution.

2 Lead in water

Tributaries of some Welsh rivers such as the Rheidol and Ystwyth are devoid of fish life. They receive water from disused **lead mines**. In places, the concentration of lead is more than 0.5 ppm, which is toxic to freshwater fish. Only a few species of algae and insects survive.

Lead poisoning was common among Ancient Romans. It happened because **lead pipes** were used to supply drinking water. Of the 18.5 million houses in Britain, something like 7–10 million have lead pipework between

Table 33.4 Concentrations of lead in the blood of people in Cincinnati, USA (after Lagerwerff 1972)

Occupation	Lead concentration (μg per $100\,cm^3$ in blood)
office worker	19
policeman	21
postman	23
parking attendant	34
garage mechanic	38

Fig 33.8 Unleaded petrol on sale at a British garage (an Esso photograph)

the mains supply and the taps. In soft water areas especially this contributes significantly to lead intake. The cost of replacement on a national scale has been estimated at £1000 million at present-day prices.

3 Lead in food

Lead in the air is washed out by falling rain. Soil near to busy roads is more likely to be polluted in this way than elsewhere (Fig 33.9). Sewage sludge may also contain relatively high concentrations of lead. Plant roots absorb lead from polluted soil. Foliage may also take in lead from the air. Either way, the amount of lead in plants may impair human health if the contaminated plants are eaten. Another source of lead which plants may absorb is sewage sludge applied to land to maintain soil fertility. To safeguard the environment from such contamination the EEC has produced a Directive which bans application of sewage sludge whenever the concentration of lead in the soil exceeds a limit set by each member country.

Fig 33.9 Lead in surface soil as a function of distance from a highway (after Harrison and Laxen 1981)

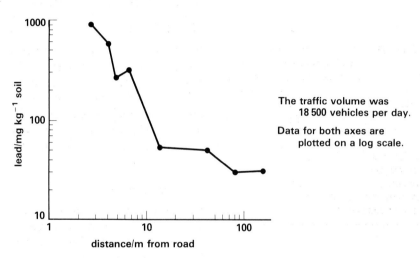

The traffic volume was 18 500 vehicles per day.

Data for both axes are plotted on a log scale.

As a safeguard it is also recommended that there is a 10 month gap between applying sewage sludge and the harvesting of fruit and vegetables which grow in contact with soil and are eaten raw. Member countries are expected to comply by June 1989.

The amount of lead in canned foods is higher than in fresh foods. This is because lead solders are used to seal cans.

4 Lead in humans

Whether inhaled or ingested, lead is absorbed into the blood and **accumulates** in the liver, kidneys and bones. About 90% of lead in the body is in the bones and teeth. The amounts of lead absorbed from air, food and water have been measured in the USA (Table 33.5). Food is the main source, wherever people live. In urban areas, lead from air makes up a large proportion of absorbed lead. Most of it is thought to have come from car

Table 33.5 Average daily intake of lead by people in USA (after Walker 1975)

Source	Lead absorbed per day (µg)
food	17.0
water	1.0
urban air	10.4
rural air	0.4
tobacco smoke	9.6

exhausts. There is evidence that the **mental development** of children in urban areas may be affected by the lead they breathe in.

Tobacco smokers take in more lead than non-smokers, because nicotine is thought to stimulate the absorption of lead by the lungs.

People whose work exposes them to lead contaminated air are protected by an EEC Directive with which member countries had to comply in January 1986. It requires such workers to have regular medical tests and sets limits of $150 \, \mu g \, m^{-3}$ for lead in the air and $70 \, \mu g$ per $100 \, cm^3$ in the workers' blood.

33.2.2 Mercury

An estimated 230 tonnes of mercury reach the oceans from natural sources, mainly the weathering of rocks. Worldwide production of mercury for industrial and agricultural uses amounts to more than 7000 tonnes annually, of which 25 to 50 % is believed to be discharged into the environment. Mercury is used extensively as a floating electrode in the production of chlorine and caustic soda from brine. It is also used as a catalyst in the manufacture of vinyl plastics and batteries. Organic mercury fungicides are used to preserve wood pulp and as seed dressings to ensure good germination.

1 Mercury in the air

When coal and oil are burned, mercury is released into the air. Ore-refining is another source of aerial mercury. Mercury also evaporates from the earth's crust. High concentrations of atmospheric mercury detected by sniffer aircraft are valuable indicators of where ores of mercury can be found.

2 Mercury in water

Rain picks up mercury as it falls through the air. Waste water from paper mills and factories making vinyl plastics also contains mercury. Most of the mercury accumulates in mud at the bottom of rivers, lakes and seas. There it is converted to **methyl mercury** by methane-producing bacteria. Methyl mercury is volatile and very toxic. It is absorbed by aquatic organisms which may be part of a food web involving man.

3 Mercury in food

In the recent past, **sprays** and **seed dressings** containing mercury were used a great deal to prevent and to control many plant diseases caused by pathogenic fungi. The annual usage of mercury worldwide for this purpose was 140 tonnes in 1967. It fell to about half this amount by 1970 and the trend is still downward.

Mercury is retained in the upper layers of well-drained soils. From here it is absorbed by the roots of plants. Mercury in sprays is taken in through the leaves. Translocation to the fruit and other edible parts of plants ensues. Mercury poisoning has occurred in wood pigeons consuming dressed seed. In Sweden, Canada and the USA freshwater fish have been shown to contain much higher concentrations of mercury than marine fish (Table 33.6). It is thought to have entered lakes in the waste water from **wood pulp mills**. The pollution is so bad in parts of North America that commercial fishing in some lakes has been banned in the past few years.

Table 33.6 Concentrations of mercury in Swedish freshwater and marine fish (after Walker 1975)

Species	Habitat	Mercury conc. (ppm)
mackerel	marine	0.06–0.08
herring		0.03–0.11
cod		0·15–0·19
pike	freshwater	5.00

4 Mercury in humans

The appalling events at Minamata and Niigata, Japan, in the 1950s underlined the danger to humans of pollution by mercury. At Minamata 111 people were seriously affected by mercury poisoning and 46 of them died. Shellfish caught in Minamata Bay made up a large proportion of their diet. It was later discovered that a vinyl plastics factory had for some time discharged waste water into the bay. Table 33.7 shows the concentrations of mercury in the seawater and in marine organisms from Minamata Bay in 1960. Note how the mercury had **accumulated** higher up the food chain.

Table 33.7 Concentrations of mercury in seawater and marine organisms from Minamata Bay in 1960

Source	Mercury concentration
seawater	1.6– 3.6 ppb
plankton	3.5–19.0 ppm
shellfish	30–102 ppm

Post-mortem examinations revealed high concentrations of mercury in the kidneys, liver and brain of those poisoned (Table 33.8). At Niigata, 120 people were poisoned by mercury in similar circumstances. The victims had eaten up to three fish meals a day. At both places the survivors were partially or totally paralysed and had lost most forms of sensation. Many were blinded. Mothers who three years earlier had shown no signs of mercury poisoning bore infants who suffered from mental retardation and cerebral palsy. The developing foetuses had evidently accumulated mercury from their mother's blood.

Table 33.8 Concentrations of mercury in body organs of Minamata victims (1) and a control group (2) of humans

Organs	Mercury concentration	
	1	2
kidneys	106 ppm	1.0 ppb
liver	42 ppm	3.0 ppb
brain	21 ppm	0.1 ppb

In Europe, protection of the environment against pollution with mercury has been tackled by the development of two EEC Directives based on **limit values** and **quality objectives**. The former sets the maximum concentration of mercury permitted in waste water from industry at 0.05 ppm and states limits for the total amount of mercury discharged. The latter indicate the highest concentrations allowed in the environment (Table 33.9).

Table 33.9 EEC quality objectives for mercury in the environment to be achieved by July 1989

fish flesh	0.3 ppm
inland water	0.0001 ppm
estuaries	0.00005 ppm
sea	0.00003 ppm

33.2.3 Other heavy metals

Copper enters streams and rivers in drainage water from copper mines and in the liquid effluent from copper-plating works. It is a very toxic metal. Freshwater plants and fish are killed by less than 1 ppm of copper. Fish absorb it through their gills. The rate at which water passes over the gills mainly determines how much copper is absorbed in a given time. In water containing a low concentration of dissolved oxygen, gill ventilation is rapid. In such conditions fish are more quickly killed by copper polluting the water.

Copper also pollutes the spoil heaps of copper mines. Its presence limits the vegetation to copper-tolerant metallophytes (Chapter 14). Pollution of soil by copper has occurred in some English orchards. It gets there from copper-containing fungicides, such as Bordeaux mixture, which have been sprayed on the trees for over a hundred years. The polluted soil is devoid of most kinds of animal life. The absence of earthworms is thought to be the reason why a thick layer of poorly decayed detritus has accumulated on the soil surface.

Zinc appears in streams and rivers in the effluent from zinc mines. Water boatmen and the larvae of caddis and stone flies tolerate up to 500 ppm. Freshwater fish are killed by less than 1 ppm. As with copper, fish are more quickly killed in poorly aerated water. Most terrestrial plants can tolerate up to 7 ppm of zinc in soil. The spoil heaps hear zinc mines contain much more than this and are inhabited by zinc-tolerant metallophytes.

Cadmium is released into the environment in the refining of lead and zinc, in the manufacture of batteries and from the electroplating industry. The air in the vicinity of zinc smelters is polluted with oxides of zinc and cadmium. Humans can absorb cadmium from the air through their lungs and there is some evidence that people working and living near smelters suffer from acute cadmium poisoning, which causes renal damage and high blood pressure. We can also consume cadmium in our food and water. The drainage water from zinc mines also contains cadmium. In the Jintsu Basin of Japan, where riverwater containing such effluent is used to irrigate paddy fields, people suffer from a chronic form of cadmium poisoning called *itai-itai*. Minerals are withdrawn from the bones and there is severe pain. Cadmium is taken up by the rice crop, which is a staple part of the diet of the Japanese. They therefore ingest much more cadmium than people living in unpolluted areas.

In dealing with pollution by cadmium the EEC has taken a similar approach to the one it took with mercury. A quality objective of 0.0005 ppm has been set for inland waters receiving effluent from industries which generate cadmium.

33.3 Pesticides

A wide range of toxic chemicals called **pesticides** is used by man to control or eradicate pests.

33.3.1 Herbicides

Herbicides make up about 40 % of the world's production of pesticides. They are used to control or eliminate weeds. Gardeners use herbicides to get rid of daisies, dandelions, plantains and other weeds growing in lawns. Farmers use them to eliminate weeds such as thistles, docks and poppies, which grow among cereal crops and in pastures. Herbicides are also used in forestry to kill forest weeds such as rhododendrons and gorse. Railways are kept free of weeds by spraying herbicides on the track.

Many herbicides are similar to auxins (Chapter 21). Auxin-like herbicides are biodegradable and are soon broken down to harmless products by soil bacteria. For this reason they seldom cause environmental problems. However, an impurity which is present in one group of herbicides has recently given cause for concern. Several widely-used herbicides are derivatives of trinitrophenol, for example 2:4:5 trichlorophenoxyacetic acid (2:4:5T). In manufacturing trinitrophenol, a harmful impurity called **dioxin** is formed and this is usually present in 2:4:5T. Dioxin is so potent that half a gram of it can kill 3000 people. Although it is normally present in only small amounts in 2:4:5T, dioxin is a potential threat because a good deal of the herbicide is released into the environment. In the USA, over 3000 tonnes of 2:4:5T are used annually, mainly in forestry. In the UK, about 3 tonnes are used each year, chiefly by British Rail and the Forestry Commission.

The effects of dioxin on humans and its potential effects have been studied in three main ways:

1 Laboratory studies

Table 33.10 Minimum dose of dioxin required to kill 50 % of a batch of animals (LD 50 values)

Species	LD 50 (ppm)
rat	0.0005
mouse	0.1000
guinea pig	0.0006

When small amounts of dioxin are painted on to the ears of rabbits, the treated area develops a very severe form of acne known as **chloracne**. When given to animals in their food and water, dioxin has been shown to be one of the most potent of poisons (Table 33.10). The animals which died in these experiments showed general wastage of body tissues and organs. Tumours and cancer were frequent. A number of serious **teratological defects** (defects of the foetuses) were observed when dioxin was given to pregnant animals. They included cleft palate, kidney malfunction, enlarged livers and internal haemorrhage. In monkeys, whose anatomy and physiology are very similar to those of humans, spontaneous abortion, haemorrhage, difficulty in conception, gangrene of the fingers and toes and loss of hair occurred even at the lowest dosage tested.

2 Vietnamese and American forests

Eleven million gallons of 'Agent Orange' (50 % 2:4:5T) were sprayed by the Americans onto the forests of Vietnam in the 1970s. The herbicide defoliated the trees, making it easier to observe enemy movements. It has since been shown that defects among children born in the contaminated areas are much higher than normal. USA servicemen working in the area later fathered children with physical and mental handicaps. Dioxin was found in the bodies of the fathers. More recently, miscarriages in women living in the forests of Oregon, USA, where 2:4:5T has been used for several years, have been greater than expected. It has led to a temporary ban on the use of the herbicide in this part of the USA. For these reasons there has also been a campaign to stop the use of 2:4:5T in Britain.

3 Factory accidents

There have been several major explosions at factories where trinitrophenol is produced. One of the most recent was at Seveso, Italy, in 1976. Cats and dogs died in large numbers. There were 400 cases of chloracne in children alone. Miscarriages and cancer rates were higher than normal. Chloracne among workers and their families, abnormal liver function and higher incidences of skin and heart disease followed an explosion at Bolsover, England in 1968. The effects of an accident at Nitro, West Virginia, USA in 1949 are still unknown. Recently, a comprehensive survey of the health of the affected people has been carried out. The results are awaited and could be vitally important because some conditions such as cancer often have a long latent period. There is little doubt that some dioxin was released in the explosions and was taken into the bodies of workers and people living nearby by absorption through the skin and lungs and by swallowing.

33.3.2 Fungicides

Compounds containing mercury and copper are the most serious pollutants of the **fungicides**. Their environmental effects are discussed in sections 33.2.2 and 33.2.3.

33.3.3 Insecticides

Chemicals which kill insects have been known for a long time. Before 1940, the **insecticides** most widely used were natural products such as pyrethrum and nicotine. Since then, two groups of synthetic insecticides, **organochlorine** and **organophosphorus compounds**, have replaced the older insecticides. Like pyrethrum and nicotine, organophosphorus insecticides are biodegradable and are quickly broken down into harmless products in the environment. Although very toxic, they are not pollutants if used in a responsible way. In contrast, organochlorine insecticides persist for many years in the environment. Their persistence has been the cause of pollution.

The best known of the organochlorine insecticides is DDT (dichlorodiphenyl trichloroethane). It was first made nearly a century ago, but its insecticidal powers were not realised until the Second World War. Its non-toxicity to man and its effectiveness in killing insects led people to believe that DDT would soon eradicate insect pests. Malaria, yellow fever and other insect-borne diseases would become a thing of the past. Other organochlorine insecticides such as aldrin, dieldrin, heptachlor and lindane were just as effective in eradicating insect pests of crops. Hopes were high that the use of such compounds would eliminate starvation in developing countries. Today the widespread use of DDT and some of the other organochlorine insecticides has been banned in many developed countries. People die in their millions each year of insect-borne diseases and crops are still ravaged by insect pests. What has gone wrong?

1 Overuse

Rachel Carson's aptly titled book, *Silent Spring* (1963) publicised the ecological effects of the **overuse** of organochlorine insecticides in the USA. In places where farmers were grossly exceeding the recommended dose rate, there were hardly any songbirds. Some had been killed by breathing in air containing large amounts of organochlorine insecticides or by eating contaminated food. The massive reduction of the songbird population was, however, mainly caused by the insecticides reducing their fertility. Many of the eggs which were laid failed to hatch. Enough of the insecticides had also been leached from the land to kill fish in nearby rivers and lakes.

2 Persistence

Organochlorine insecticides are very stable substances. They persist unchanged for a long time in the environment (Table 33.11). **Persistence**

Table 33.11 Persistence of some organochlorine insecticides (after Walker 1975)

Compound	Time for 95 % loss (y)
Dieldrin	4–30
DDT	5–25
Lindane	3–10

increases the chances of poisoning non-target organisms. In the 1950s, thousands of seed-eating birds were poisoned after eating cereal grain dressed with aldrin and dieldrin. Endotherms convert DDT to less toxic DDE and aldrin to dieldrin. The amounts of DDE and dieldrin in the flesh and eggs of birds provide an estimate of the extent of pollution of the environment by organochlorine insecticides. Such a survey was carried out between 1960 and 1965 in Britain after reports of a severe decline in the numbers of predatory birds such as the golden eagle and peregrine falcon (Fig 33.10). The survey showed that predatory birds had higher concentrations of DDE and dieldrin in their bodies than other species (Fig 33.11). The insecticides had probably entered the flesh-eating species along a food chain.

Fig 33.10 (a) Distribution of peregrine falcons in Britain in 1961 (from Mellanby, *The Biology of Pollution*, Inst. Biol. 'Studies in Biology', Edward Arnold (b) Peregrine falcon (RSPB, William S. Paton)

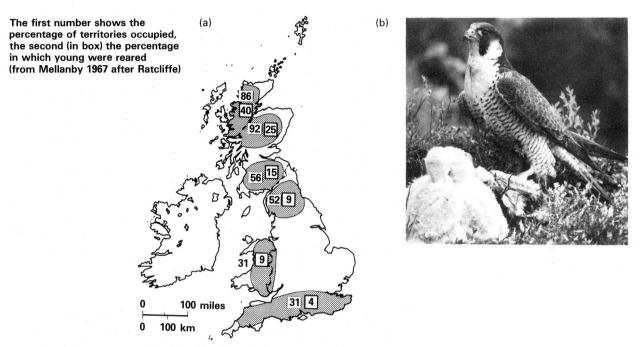

The first number shows the percentage of territories occupied, the second (in box) the percentage in which young were reared (from Mellanby 1967 after Ratcliffe)

Fig 33.11 Residues of organochlorine insecticides in British birds (from Walker 1975)

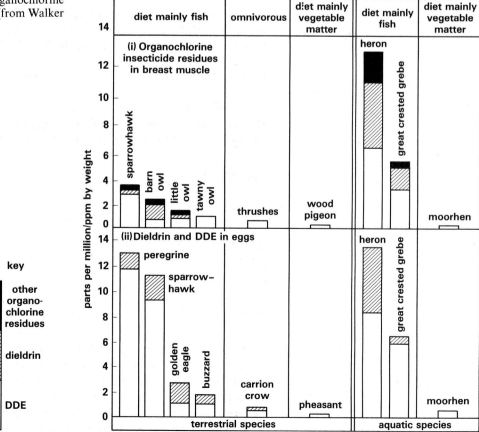

699

Because they are insoluble in water, the organochlorine compounds are stored in fatty tissues instead of being excreted. In this way they **accumulate** at each trophic level. Toxic amounts of DDE and dieldrin are probably released into the bloodstream when fat reserves are used at times of food shortage and at egg-laying. Laboratory experiments show that sublethal amounts of organochlorine insecticides cause birds to lay eggs with thin, fragile shells. When incubated by the parents, they are likely to be broken. A study of the eggshells of peregrine falcons showed that they had become significantly thinner after organochlorine insecticides came into general use in Britain in the 1940s (Fig 33.12). Circumstantial evidence of this kind led to a restriction of the use of organochlorine insecticides for agricultural purposes in the USA. Britain and other western countries adopted similar measures in the late 1960s. Subsequently there has been a marked improvement in the breeding success of predatory birds.

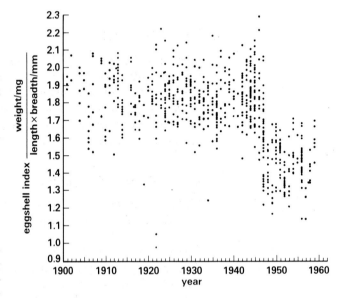

Fig 33.12 Change in time of the eggshell index of the peregrine falcon in Britain (from Ratcliffe, *Journal of Applied Ecology*, Vol. 7, Blackwell Scientific Publications Ltd)

Fig 33.13 The number of insects of public health importance which have developed resistance to insecticides (from Ottaway, *The Biochemistry of Pollution*, Inst. Biol. 'Studies in Biology', Edward Arnold)

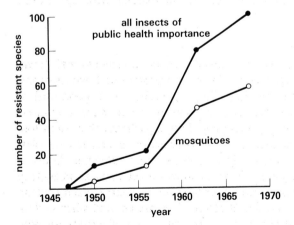

Table 33.12 EEC quality objectives for lindane and DDT

lindane	inland waters	$100\,\mu g\,dm^{-3}$
lindane	estuaries and sea water	$20\,\mu g\,dm^{-3}$
DDT	all waters	$25\,\mu g\,dm^{-3}$

3 Resistance

Soon after DDT was introduced, houseflies in Italy were found to be unaffected by it. They had acquired a means of converting DDT to less toxic DDE. It is now known that over 100 species of insect vectors have developed a **resistance** to organochlorine insecticides (Fig 33.13). It is for this reason that any hope of eradicating malaria in the near future has subsided. More than 200 species of insects which are pests of crops are also resistant. Whilst this is not a pollution problem, it does give cause for concern to those who have to control insect pests and insect-borne diseases.

Environmental protection against pollution by two insecticides, lindane and DDT, is embodied in two EEC Directives with which member countries have recently agreed to comply. They specify limit values for the amounts of lindane and DDT to be released into watercourses at sites where they are manufactured. In addition they lay down quality objectives indicating the maximum concentrations permitted in the environment (Table 33.12).

The adverse environmental effects of pesticides have often presented the latter in a rather bad light. However, there is little doubt that they have saved millions of people from disease and starvation. Much of the publicity centred around them has stimulated research into producing safer pesticides and alternative ways of pest control. There is now a move towards **integrated control** of pests. It involves using natural methods of control where possible and resorting to the use of pesticides only when necessary. A typical example is the control of the red spider mite, a pest of greenhouse plants such as tomatoes and cucumbers. Its numbers can be kept in check biologically by introducing another species of mite which preys on red spider mites. Unfortunately, the predator does not do well near the roof where the temperature is high. Any red mites here can be killed with a pesticide spray. The amount of pesticide used is much less than required if it was the only control available.

33.4 Radiation

Ever since life evolved, living organisms have been subjected to various forms of **radiation**. The most obvious is visible light from the sun. The sun also radiates ultraviolet and infra-red waves. Radiation comes from outer space too as cosmic rays, and from radioactive minerals in rocks. Natural forms of radiation are not thought of as pollutants. But radiation introduced into the environment by human activity is a form of pollution.

33.4.1 Forms of radiation

The forms of radiation which pose a threat to the well-being of living organisms fall into two categories:

1 Electromagnetic waves

Gamma rays (γ-rays) are electromagnetic waves with a wavelength of about 10^{-12} m. Their very short wavelength gives them considerably more energy than visible light and consequently much better powers of penetration. Gamma rays are emitted when radioactive atoms (radionuclides) disintegrate or when atomic nuclei are bombarded with sub-atomic particles.

2 Sub-atomic particles

Alpha particles (α-particles) are given off when radionuclides break down. They are positively charged, consisting of two protons and two neutrons. When they collide with anything solid, α-particles lose a good deal of energy. **Beta-particles** (β-particles) are electrons and have a negative charge. They are not very penetrating. Like α-particles, they soon lose energy when they hit something solid. Because of this energy transfer, α- and β-particles are biologically among the most damaging forms of radiation. However, they are unlikely to do very much damage unless they are inside a living organism. **Neutrons** have no net charge. Each neutron consists of one proton and one electron. They are emitted when certain elements are bombarded with α-particles or with γ-rays. Neutrons are more penetrating than α- and β-particles. Like γ-rays, they can do a great deal of harm, even when originating outside living organisms.

Some radionuclides give off radiation for only a short time. Others do so for many years. The time required for a radionuclide to emit half its radiation is called its **half-life** (Table 33.13). The half-life and type of radiation emitted determine the degree of danger.

Table 33.13 Half-lives of some radionuclides (after Ottaway 1980)

Nuclide	Name	Radiation	Half-life
^{90}Sr	Strontium	β-particles	28 y
^{129}I	Iodine	β-particles γ-rays	16×10^6 y
^{131}I	Iodine	β-particles γ-rays	8 days
^{137}Cs	Caesium	β-particles γ-rays	30 y
^{237}Np	Neptunium	α-particles	2.2×10^6 y
^{239}Pu	Plutonium	α-particles γ-rays	2.4×10^4 y

33.4.2 Sources of radiation

Natural sources account for more than 80 % of the radiation absorbed by humans (Table 33.14). There is little that can be done about natural radiation. Living organisms have had to cope with it since life began. What is of concern is the radiation released by human activities.

Table 33.14 Radiation absorbed by people in Britain (after Mellanby 1980)

Source	Annual dose (μGy)	Percentage
natural background	870	83.8
medical	140	13.5
nuclear fallout	21	2.02
miscellaneous	7	0.67
radioactive waste	0.1	0.01

The unit of radiation most meaningful to biologists is the gray (Gy). 1 Gy is equal to 1 J of radiation energy absorbed by 1 kg of body mass.

Though non-natural radiation presently accounts for a relatively small amount of the total, it is on the increase and adds to the risks created by natural sources. Two human activities which release radiation into the environment are the use and testing of **nuclear weapons** and the use of **nuclear fuels**. Both are emotive subjects to people the world over.

In the early 1960s there was a dramatic increase in the amount of ^{90}Sr in the milk consumed in Britain. It followed the testing of nuclear weapons by the USA and USSR. Soon after the Nuclear Test Ban Treaty, contamination by ^{90}Sr reverted to normal and has stayed there since (Fig 33.14).

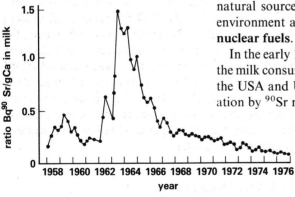

Fig 33.14 Average ratio of ^{90}Sr to Ca in milk in Britain (from Mellanby, *The Biology of Pollution*)

Thermal nuclear reactors use natural uranium, a mixture of ^{235}U and ^{238}U. When bombarded with neutrons, uranium decays into a variety of **fission products**, including ^{90}Sr, ^{129}I, ^{131}I, ^{137}Cs and ^{239}Pu. The fission products which are potentially most dangerous to living organisms are ^{90}Sr, ^{129}I, ^{131}I and ^{137}Cs. Radioactive iodine absorbed by grasses and eaten by cows appears in their milk. Humans drinking contaminated milk concentrate radio-iodine in the thyroid gland. In this way the thyroid can be exposed to high doses of radiation. ^{90}Sr accumulates in bone adjacent to the marrow. Its long half-life means that it irradiates dividing bone marrow cells for years. Caesium is taken up by soft tissues, especially muscle.

Accidents have happened at nuclear power plants, spilling fission products into the surrounding countryside. The chief risk is failure of the cooling system in a reactor. The build-up of heat could be enough to melt a reactor. Partial failure of the cooling system in a fast reactor caused an explosion at Three Mile Island near Pittsburgh, USA, in 1979. The effects are still being assessed. The worst nuclear accident in Britain happened at Sellafield, Cumbria, in 1957 when ^{131}I and ^{137}Cs were released. As a precautionary measure, milk produced for several weeks within 5 km^2 of the nuclear station was declared unfit for consumption.

Without doubt the most serious accident in a nuclear power plant occurred in 1986 at Chernobyl in the Ukraine, USSR. On 26 April one of the four reactors exploded, releasing a huge cloud of radioactive dust. Within days, over 100 000 inhabitants were evacuated from Chernobyl and the surrounding area. Crops in a zone of diameter 29 km were immediately designated as unfit for harvesting. A reserve supply of drinking water was hurriedly arranged for Kiev, the Ukrainian capital. Before the accident, water for about half of Kiev's population came from the River Dnieper which has tributaries in the Chernobyl area.

The radioactive cloud drifted across Europe, and in less than a week radiation levels in Scandinavia had increased five-fold. In the UK the highest fallout occurred in the wettest areas, the uplands of Scotland, Cumbria and Wales. Within a month the concentration of ^{137}Cs in the muscles of sheep was so high that the animals were declared unfit for human consumption. Slaughter of some sheep was permitted about two months later when the amount of radioactivity in their bodies had fallen to acceptable levels. However at least 250 000 animals are still too contaminated to be moved from their farms.

The full consequences of the Chernobyl catastrophe may never be known. Medical experts have predicted an inevitable increase in the incidence of cancers such as leukaemia in Europe as a whole as well as the USSR. At a news conference in August 1986 the Soviet government disclosed that 31 Russians had been killed in the accident and that 203 were hospitalised with acute radiation sickness.

Another problem faced by the nuclear fuel industry is what to do with spent reactor cores which are still radioactive. In Britain, high and medium radioactive waste is sealed in concrete containers. If stored indefinitely, the fission products decay and become harmless. The difficulty is in deciding where they should be kept. Storage in underground caverns is the disposal method used mainly at present. Prior to 1983, Britain dumped hundreds of tonnes of radioactive waste at sea. In 1984 the London Dumping Convention stopped the disposal of such waste in the Atlantic Ocean. Hence the government agency for the disposal of radioactive waste is presently seeking dumping sites on land.

33.4.3 Biological effects of radiation

The effects of radiation on living organisms are two-fold.

1 Genetic effects

Living cells are very sensitive to radiation at interphase. It is at this stage in the cell cycle that replication of DNA occurs. Radiation alters the genetic code by breaking hydrogen bonds between the pairs of nitrogenous bases in the DNA molecules. Bases may be lost or reassembled in a way different

Fig 33.15 Possible behaviour of fragmented chromosomes during nuclear division

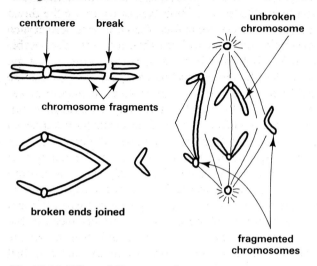

centromere   break

unbroken
chromosome

chromosome fragments

broken ends joined

fragmented
chromosomes

Fig 33.16 Effect of X-rays on the mutation rate of *Drosophila*

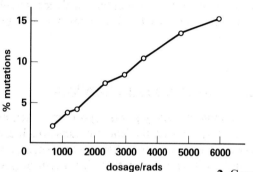

from that in the original DNA. Such changes are called **gene mutations** (Chapter 4). The faulty DNA then codes the synthesis of abnormal proteins which may be defective in function. Gene mutations may be passed to future generations in gametes. Intense doses of radiation cause chromosomes to break. Chromosome fragments do not behave as normal chromosomes in nuclear division. They often fail to become attached to the spindle and are lost when the cytoplasm cleaves. The ends of large chromosome fragments may join up. It is then impossible for the chromatids to separate at anaphase (Fig 33.15). Such changes are chromosome mutations.

Muller, in 1927, using the fruit-fly *Drosophila*, was the first to demonstrate that radiation was a cause of gene mutations. He also showed that the mutation rate was proportional to the dosage of radiation (Fig 33.16).

The probability of a mutation occurring increases with each exposure to radiation. The critical exposure for any organism is when it is producing offspring. It is then that mutations may occur in cells which give rise to progeny formed asexually or in gamete-forming cells. For women the critical period extends over thirty-child bearing years. It has been estimated that the rate of mutation among humans would double if women absorbed an additional 0.5 Gy over their reproductive life span. If the increase was only 0.01 Gy, in each generation the mutation rate would increase by 1.5 %.

2 Somatic effects

Table 33.15 summarises the observations made on Japanese people

Table 33.15 Somatic effects of radiation

Dosage (Gy)	Death within	General effect	Specific effects
>100	1 day	CNS syndrome	Vomiting, diarrhoea, convulsions, brain damage, coma
10–100	2 weeks	GI syndrome	Gut lining not replaced, inability to digest food, vomiting, diarrhoea, emaciation
2–10	3–4 weeks	BM syndrome	Blood cells not replaced, anaemia, fever, vomiting, haemorrhages, loss of hair
<2	unknown	life-shortening	Inability to make antibodies, small wounds remain septic for long periods

CNS = central nervous system. GI = gastro-intestinal. BM = bone marrow.

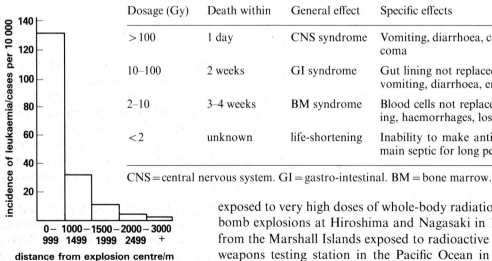

Fig 33.17 Relationship between the incidence of leukaemia and distance from centre of the atomic explosion at Hiroshima, Japan

exposed to very high doses of whole-body radiation following the nuclear bomb explosions at Hiroshima and Nagasaki in 1945, and on fishermen from the Marshall Islands exposed to radioactive fall-out near a nuclear-weapons testing station in the Pacific Ocean in 1954. The children of pregnant women exposed to radiation at Hiroshima and Nagasaki weighed much less than normal, had smaller brains, and many suffered from severe mental and physical handicaps. Survivors at Hiroshima were more likely to have **leukaemia** if they were near the centre of the explosion (Fig 33.17).

Fall-out of ^{90}Sr from the testing of nuclear weapons may well have been the cause of the increased number of cases of leukaemia in Britain in the 1960s.

As with pollution by pesticides, it is important to try to see the threat from radiation in perspective. Reserves of fossil fuels are being used at such a rate that some experts predict they will be exhausted in the foreseeable future. At present, nuclear fuel offers the main alternative source of energy. Another advantage is that the nuclear fuel industry does not pollute the atmosphere with acid rain and particulate matter. As more nuclear power plants are built, the risk of radiation pollution is likely to increase. So far, rigorous safeguards have been taken. Although accidents have happened and some radiation has escaped from nuclear power stations, there has been no widespread danger to human life. However, the amount or radioactive waste to be disposed of is growing rapidly. There is concern about the safety of dumping at sea where the containers may eventually corrode. Nuclear waste buried underground may escape if earth tremors occur. The very long half-lives of some fission products increase the possibility of contamination. Future generations could then inherit the hazard of pollution by the radioactive wastes of today's nuclear fuel industry.

A worse hazard would come from the wide-scale use of nuclear weapons. The radiation released by explosions of nuclear bombs worldwide would kill most of us immediately and many parts of the earth could be uninhabitable for a very long time.

Fig 33.18 Effects of organic effluent on (a) physical changes, (b) chemical changes, (c) flora, (d) fauna of a river (from Hynes *The Biology of Polluted Waters*, Liverpool University Press, 6th imp. 1978)

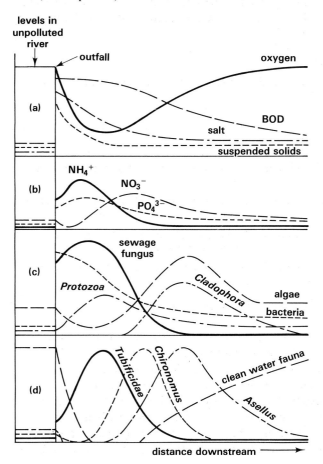

33.5 Pollution of aquatic habitats

The more obvious of society's waste products, such as plastic containers, glass bottles and tin cans which are often dumped in streams, rivers, lakes and the oceans, are eyesores. There is little evidence that they harm organisms. Agricultural, domestic and industrial wastes are the main pollutants of aquatic habitats.

33.5.1 Freshwater pollution

1 Rivers

The pollution of British rivers by heavy metals is described in section 33.2. Other industrial wastes such as ammonia and cyanide are also toxic to aquatic life. Particulate matter released by industry can pollute rivers too. Coal washings and china clay settle on the beds of rivers, fouling the spawning places of salmon and trout. When suspended in water, the particles reduce penetration of light for aquatic plants and asphyxiate fish by blocking their gills

Sewage is the biggest pollutant of fresh water. It is the waterborne waste of society. The effects of discharging untreated sewage into a river have been studied by Hynes (Fig 33.18). One of the most striking features is a substantial and immediate drop in the amount of oxygen dissolved in the water. It happens because organic matter stimulates decomposer organisms, especially bacteria, which break down suspended solids in the sewage. As they respire, the decomposers use up dissolved oxygen. The amount of oxygen used in breaking down the organic matter into simple soluble products, such as ammonia, sulphate and phosphate, is the **biochemical oxygen demand (BOD)** of the sewage. The ammonia is quickly oxidised to nitrate by nitrifying bacteria. Some of the phosphate may have come from detergents in the sewage. Downstream from the sewage outfall, the dissolved oxygen concentration returns to normal as oxygen is absorbed by the river from the air. A gradual fall in BOD occurs simultaneously as the decomposers are dispersed by the flowing water.

Changes in the flora and fauna of the river accompany the chemical changes. Filamentous bacteria called sewage fungus thrive near the outfall. They are anaerobic and tolerate the high concentration of ammonia which makes the water alkaline. Algae, *Cladophora* in particular, are sensitive to pollution by sewage. Downstream they flourish, stimulated by the minerals released from the decomposed organic matter. Their numbers then return to normal as the minerals are used up or diluted by water from tributaries. The fauna most tolerant of sewage are annelids of the family *Tubificidae* (tubifex worms). Their blood contains a form of haemoglobin with an exceptional affinity for oxygen. It enables them to extract oxygen even when the concentration of dissolved oxygen in water is very low. Larvae of the midge *Chironomus* are the next most tolerant. They too have a type of haemoglobin which binds very efficiently with dissolved oxygen. The water louse *Asellus* begins to appear when the dissolved oxygen begins to recover. Clean-water fauna such as the freshwater shrimp *Gammarus* and larvae of the caddis fly, stonefly and mayfly are wiped out by sewage. They reappear when the amount of dissolved oxygen is back to normal (Fig 33.19).

Fig 33.19 Changes in the fauna of a polluted river (from Sands 1978)

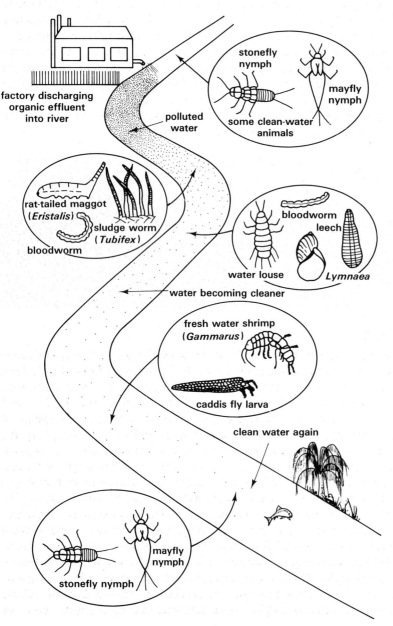

The numbers and types of animals at different sampling points in a river provide a **biotic index** to the extent and type of pollution. Given time, a river cleans itself in the way described. However, if it receives successive outfalls of sewage close to each other, the stretches of water between the outfalls do not have time to recover. In these circumstances the water becomes deoxygenated. Anaerobic bacteria partly decompose the sewage, releasing toxic ammonia and hydrogen sulphide. Rivers as badly polluted as this have an obnoxious smell and contain little flora or fauna.

What is popularly called **thermal pollution** has occurred in some of our rivers. A good deal of water is taken from rivers to cool industrial machinery. The biggest user of water for cooling purposes in Britain is the Central Electricity Generating Board. When returned, the water is hotter than when it was removed. In an unpolluted river, the increase in temperature lowers the amount of dissolved oxygen (Table 31.2). At the same time, it increases the metabolic rate of aquatic organisms and therefore raises their oxygen demand. Up to now no gross changes in the flora and fauna have been observed in Britain, because the increase in temperature is slight. In other parts of the world, salmon and trout have disappeared from stretches of clean rivers which have been warmed to a greater extent. Ironically, water from cooling towers that is returned to polluted rivers accelerates their return to normal. While cascading down cooling towers (Fig 33.20). water takes up a good deal of oxygen from the air, so the water returned to polluted rivers often contains more oxygen than the water extracted. The increase in temperature and oxygen also encourages decomposers to break down organic matter more quickly. In this way the BOD returns to normal sooner.

Fig 33.20 Cooling towers of the electricity generation station at Averham, Notts. with River Trent in foreground.

2 Lakes

Lakes are fairly still bodies of water often with no outlet. Any pollutants they receive will thus accumulate gradually. Reference has already been made to pollution of lakes by acid water draining from land polluted with sulphur dioxide and by mercury in industrial effluent (sections 33.1 and 33.2). In unpolluted lakes, the water contains only small amounts of dissolved minerals. The low mineral content limits the growth of aquatic plants. In such **oligotrophic** lakes there is consequently little organic matter to be decayed. For this reason, the water has a low BOD and remains well oxygenated. When polluted with sewage or with drainage water from nearby arable land, the water is enriched with minerals. The water is now **eutrophic** and growth of phytoplankton is stimulated. Blooms of algae, blue-greens especially, quickly appear. Some of them release toxins which kill fish. When the algae die, they contribute substantially to the dead organic matter which sinks to the bed of the lake. Here it creates a high BOD, and what oxygen there is in the water is soon used by decomposers. When this point is reached, anaerobic bacteria release ammonia, methane and hydrogen sulphide, making the water unfit for most forms of aquatic life. If there is a large amount of organic matter to be decayed, anaerobism may extend to the upper layers of water.

There are two ways in which the effects of such pollution can be reversed. One is to dredge the lake bed. Clearly this is too large a task for all but the smallest lakes. The other is to oxygenate the water so that decomposers break down the organic matter. This expensive approach is presently being tried in an attempt to clean up Lake Geneva in Switzerland. For some of the world's largest lakes it seems that the effects of eutrophication are irreversible. The flora and fauna of large parts of Lake Erie in Canada are pollution-tolerant species. Masses of decaying algae and dead fish are regularly washed up on the lake's shores. The water is unfit for bathing and the local fishing industry has virtually collapsed.

33.5.2 Marine pollution

Unlike the water in a lake, sea water moves continuously. In some seas the movement brings about a rapid replacement of water. Turnover of all the water in the North Sea takes about two years. For this reason, pollutants are soon carried away and dispersed. The effects of pollution are more likely to be serious in shallow, enclosed seas. Here the turnover may take hundreds of years and marine organisms soon concentrate toxic pollutants. This was a major contributory factor in the poisoning of people with **mercury** at Minamata, Japan (section 33.2.2).

Eutrophication has been another cause of pollution in enclosed seas. Within the past 30 years, parts of the southern Baltic Sea have been made anaerobic in the same way as an enriched lake. Enrichment with phosphate fertiliser coming from surrounding agricultural land is mainly to blame. Eutrophication also occurs when raw or partly-treated sewage is discharged at sea, a common feature of many coastal towns on the Mediterranean Sea. Bathing in sewage-tainted sea water is unpleasant and likely to be a health risk. Humans have also contracted typhoid fever after eating shellfish from sewage-contaminated water.

Shellfish also concentrate **organochlorine insecticides** in their tissues. Oysters, for example, accumulate DDT to a concentration 70 000 times that of DDT in seawater, and as little as 0.001 ppm reduces their growth. Organochlorine insecticides pass from one trophic level to another in the sea, as in other ecosystems, and with comparable effects. The decline of predatory marine birds such as the herring gull, the osprey and brown pelican in various parts of North America in the 1960s is thought to have occurred because they accumulated high concentrations of DDT residues in their bodies.

Marine organisms concentrate **radionuclides** in their tissues too. They are often used as a measure of the amounts of radionuclides in the environment near coastal nuclear power plants. So far the evidence indicates little danger to marine life or to humans eating seafood.

Oil has captured the headlines as one of the most serious forms of marine pollution. When the *Torrey Canyon* went aground off Land's End in 1967, about 60 000 tonnes of crude oil were spilled. Much of it was washed up on Cornish beaches. The rest was blown across the English Channel and ended up on the French coast. The foundering of the *Amoco Cadiz* off the coast of France in 1978 caused 200 000 tonnes of crude oil to be released. Much of the oil came up on the beaches of Brittany. There have been many others similar incidents elsewhere. Large spillages have also occurred more recently from oil rigs in the North Sea and at oil terminals in the Shetland Islands. It has been estimated that up to 10 million tonnes of crude oil are released annually into the world's seas and oceans.

Fig 33.21 Guillemot polluted with oil (RSPB/Michael W. Richards)

The worst casualties of oil pollution are fish-eating birds. From the air, oil slicks resemble patches of plankton where fish are likely to be feeding. Attracted by what looks like a good source of food, the birds settle on the slick or dive through it. Their feathers become clogged with oil, stopping the birds from flying (Fig 33.21). Oil-soaked feathers are poor insulators of body heat, so many of the birds die of hypothermia. The birds attempt to rid themselves of the oil by preening, but in doing so they swallow crude oil and are poisoned. Something like 100 000 sea birds, especially guillemots were killed by the oil spillage from the *Torrey Canyon* alone.

Crude oil also kills seaweeds, molluscs and crustaceans when it is washed up on rocky shores. The seaweeds grow again quite quickly, often better than before. Browsing molluscs which keep them in check are slower to return, but within two years the shore is more or less back to normal. The oil disappears because it is broken down by bacteria. Concerned by the possible harm to their tourist industries, British and French authorities tried various measures to deal with the oil from the sunken tankers. In Britain the polluted shoreline was sprayed with detergents to disperse the oil. Because detergents are toxic to marine life, recovery of populations of tidal organisms was slower when this was done. The French sprinkled chalk and sawdust on the slicks to sink the oil before it could be washed up on the shore. Molluscs on the seabed were poisoned by the oil and by the mid 1980s their numbers had still not returned to normal.

The world's oceans contain over $14 \times 10^{18} \, m^3$ of water. The earth's atmosphere extends up to 2000 km. It is tempting to believe that such huge volumes can absorb all the wastes of human activities without any ill effects. Indeed, there is a catchphrase which says 'dilution is the solution to pollution'. Unfortunately, aerial pollutants are concentrated by natural forces such as rain or snowfall. Marine pollutants do not become evenly dispersed. Some of the world's largest freshwater lakes have been unable to take the wastes of human activities without irreversible and undesirable effects. Living organisms have the knack of accumulating toxic materials and passing them on to others. The unwanted products of the activities of some people are the problems of others.

33.6 Conservation

Early man lived by hunting, fishing and gathering food. His numbers were few and he led a nomadic existence. The resources he took from his surroundings had time to recover as he moved from place to place. It was when he settled as a farmer in Neolithic times that man's impact on the environment was first felt. He cut down forest to provide land on which to grow crops. As the human population grew, more forest was cleared and marshland was drained to be used for growing crops and rearing livestock. Habitat destruction has since gone on to such an extent that the wildlife in many areas is now impoverished and some species of plants and animals are near to extinction. Added to this are the pressures put on wildlife by man's pollution of the environment. It has led to the **conservation** of habitats in which living organisms are managed and protected from man's excesses.

33.6.1 The case for conservation

During the past 300 years, animal species have disappeared at an alarming rate (Fig 33.22). It has been estimated that about 25 000 species of plants are now near to extinction. **Habitat destruction** and **over-exploitation** by man are the main causes of extermination. Natural forest and woodland

Fig 33.22 The number of exterminated mammal and bird forms over the past 300 years (from Spellerberg, *Ecological Evaluation for Conservation*, Inst. Biol. 'Studies in Biology', Edward Arnold)

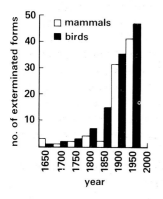

have been chopped down, wetlands have been drained and heathland burned to provide more land for growing crops. Large areas have been flooded to provide water for industrial and domestic use. Birds and mammals have been mercilessly hunted. Wild grasslands have been overgrazed and turned into desert.

Environmentalists argue that it is morally wrong for humans to condemn other species to extinction. Many people derive pleasure from observing wildlife. There are also good practical reasons for conserving all existing forms of life. Many species of wild plants are the sources of medicinal drugs, but so far less than 2 % have been screened for the useful products they may yield. Many wild species may be hardier and more resistant to disease than their domesticated relatives. It may be desirable in future to transfer the genes for these characteristics to domesticated stock. Many pests are kept in check by wild animals which are their natural enemies. If their enemies are eradicated, we could be overrun by pests. Other species are useful as monitors of pollution and alert us to potential hazards to man's health. Many species of wild animals are more productive than domesticated livestock in natural ecosystems. They can be managed as a resource for humans to utilise. Such a strategy has conserved the Tsaiga antelopes in the steppes of Russia, where it yields about half a million carcases annually for meat and hides.

Conserving wildlife is also important for present-day economic reasons. The big game reserves of African countries attract many tourists. Fishing and shooting are important sources of income locally in many developed countries.

33.6.2 Conservation in action

The aim of conservation is to manage wildlife to the benefit of mankind. It requires formulating policies and regulations to protect and maintain populations of wild plants and animals, identifying and preserving habitats in which wildlife can flourish, controlling pollution of the environment and setting up agencies to promote and monitor conservation strategies.

1 Conservation agencies

The need for conservation of wildlife was realised nearly a century ago in a number of countries. National Parks were established in various parts of the world where wild species could exist without man's interference. The first to be created was the Yellowstone National Park, USA, in 1876. Protection societies were founded and legislation was passed to help preserve wildlife. In Britain, the Royal Society for the Protection of Birds was founded in 1889, the National Trust in 1895. After the passing of the National Parks and Countryside Act 1949, the Nature Conservancy was set up to establish and manage British nature reserves. In 1973, the Nature Conservancy became the Nature Conservancy Council (NCC). County trusts for Nature Conservation have done much to foster an interest in conservation at a local level.

International bodies have also helped to promote and co-ordinate conservation projects. The International Union for Conservation of Nature (IUCN) was founded by UNESCO in 1956. More recently, the World Wildlife Fund has done much to conserve endangered species in developing countries. The giant panda (Fig 33.23), a threatened species, was symbolically chosen for the fund's logo. Countries have also worked together on conservation issues in the International Whaling Commission, whilst the EEC has issued a number of Directives and Regulations aimed at member countries co-operating in tackling environmental problems.

Fig 33.23 Giant Panda (Bruce Coleman: WWF/Kojo Tanaka)

2 Habitat preservation

In Britain the NCC manages over 150 **Nature Reserves**, representative of all our major ecosystems. It has identified more than 200 sites which are regarded as important to birds. The RSPB, which controls over 60 reserves, considers the NCC figure to be inadequate for conservation purposes. The NCC has the power to designate **Sites of Scientific Interest** (SSIs) which are special by virtue of their flora, fauna or geological features. So far over 4000 SSIs have been identified. The concept of **Environmentally Sensitive Areas** (ESAs) was established by the EEC in an attempt to integrate environmental considerations into the Common Market's Agricultural Policy (CAP). Member countries were expected to comply with the regulations from September 1985.

Evaluating the worth of an ecosystem for conservation purposes is an elaborate exercise. The NCC has taken particular interest in the conservation of old woodlands in Britain. Deciduous woodland is the natural climax vegetation for much of this country and contains a wide variety of native flora and fauna. Today, deciduous woodland covers about 10 % of the land area of Britain, mainly in fairly small patches surrounded by farmland (Fig 33.24). Small woods are constantly being chopped down and

Fig 33.24 English landscape showing pockets of woodland (Bruce Coleman: Robert Burton)

the land used for other purposes. The NCC often has to decide which woods are worth conserving, so that measures can be taken to protect them. The first thing to be done is to make a list of all the species. The next is to identify **indicator species**. According to Peterken (1977), a typical indicator species of old woodland helps create the canopy of the wood, can tolerate the shade of a closed woodland canopy or requires woodland conditions to survive. On this basis the species indicative of old British woodland include wood sorrel, wood millet, sedges and columbine. If they are present, the flora of the wood and probably the fauna too is likely to be diverse. Conserving such a wood maintains viable populations of a wide range of plant and animal species. Other criteria of importance in evaluating a site are:

i. Area Every species requires a minimum area for its survival. If the aim is to conserve a population of large herbivores, a large area of habitat is required. In general, the larger the area the more diverse is the flora and fauna. Nevertheless, small areas may be worth maintaining if the aims of conservation are more modest.

ii. Structural diversity An area with a diversity of habitats is more valuable for conservation purposes than a homogeneous area. Banks, paths, rides and streams in a wood provide places where a wide range of species can survive.

iii. Rarity Some sites are the last refuges for rare species. Although the site may lack other desirable features, it is worth conserving. Otherwise the species could be lost for ever.

iv. Vulnerability Some sites may be near to urban areas and are likely to be interfered with by trespassers. Others may be close to industry and there is a risk of pollution.

v. Cost Money has to be found to purchase sites. Some are more expensive than others. Conserving a site also means that it has to be **managed**. Access to visitors may have to be controlled. If the vegetation is not the natural climax, succession will occur if the site is left to its own devices and the flora and fauna will change. At many sites it may be desirable to prevent succession. In this way diverse sites can be maintained supporting viable populations of a large range of species. This is a key distinction between preservation and conservation. However, it does mean that some sites are more costly than others to conserve.

3 Conservation policies and regulations

Protective legislation aimed at conserving endangered species in Britain includes the Protection of Birds Acts 1954 and 1967, Badgers Act 1973, Conservation of Wild Creatures and Plants Act 1975 and Wildlife and Countryside Act 1981. Every five years the NCC is required to review lists of species under threat.

Perhaps the biggest problem confronting conservationists is in identifying which species need protection. One way of doing so is to determine the population sizes of threatened species. Measures which can be taken to sustain viable populations in their natural environments can then be suggested. This approach was initiated by Sir Peter Scott in the mid-1960s. The information is collected from many parts of the world and documented in **Red Data Books** kept at the IUCN headquarters at Morges, Switzerland.

In the 1970s there was a considerable public disquiet at the annual slaughter of millions of migratory birds in Southern Europe. As a counter to this the EEC issued a Directive with which member states agreed to comply from April 1987. It seeks to control the hunting and killing of wild birds and to protect their nests and eggs. It also requires EEC countries to provide a diversity of habitats so as to maintain populations of all indigenous species of birds. A ban on the importation of whale products and skins of harp and hooded seal pups is now in operation in the EEC as a contribution to marine conservation. Another measure recently agreed to by EEC states is to introduce farming practices compatible with conserving natural habitats. It means that national governments can pay farmers not to develop some or all of their land in such a way that habitats for wildlife are lost.

SUMMARY

Pollution is the introduction into the environment of substances harmful to living organisms, man included. Growth of the human population, industry and agriculture are the causes of the worst aspects of pollution.

Pollution of the atmosphere is of particular concern to all terrestrial life forms. Mammals such as ourselves constantly breathe in air. Plants also continuously exchange gases with the atmosphere. In this way toxic substances such as lead and sulphur dioxide enter the bodies of living organisms directly. Sulphur dioxide is also the main contributor to acid rain which corrodes our buildings. Acid rain is blamed by many ecologists for the deaths of large tracts of coniferous forest in western Europe and for the disappearance of fish from Scandinavian lakes. The gradual increase in carbon dioxide in the air is a potential problem because of the attendant greenhouse effect. Another group of aerial pollutants giving cause for concern are the chlorofluorocarbons. They are thought to be destroying the ozone layer of the stratosphere which protects us from too much ultraviolet light from the sun.

Heavy metals such as lead, mercury and cadmium pollute our drinking water and food as well as the air. Zinc and copper in industrial effluent kill off aquatic life in canals and rivers. Pesticides are toxic to other forms of life as well as the target organisms they are intended to kill. This is vividly illustrated in the dwindling numbers of predatory birds such as the peregrine falcon when organochlorine insecticides were widely used. Following a ban on the use of such pesticides there has been a revival of the falcon population in Britain.

Radiation is a form of pollution which people probably fear most of all. Widespread contamination by radionuclides has occurred from testing nuclear weapons and as a result of accidents at nuclear power plants. Whereas human life has not been threatened on a large scale there is concern that such pollution will increase the incidence of gene mutations in populations, humans included. Another potential source of radiation pollution is the waste products of the nuclear industry.

Ever since man became a settled agriculturalist he has used streams and rivers to carry away sewage. Aquatic pollution became worse during the Industrial Revolution when people crowded into towns and cities to work and industry also poured its wastes into nearby water courses. In some parts of the world the effect has been to irreversibly change aquatic ecosystems, lakes in particular. Elsewhere successful attempts have been made to clean up rivers. The building of sewage treatment works has been very effective in this context. A major pollutant of the sea has been oil released in large quantities from foundered oil tankers. Hundreds of thousands of marine birds have met their deaths as a direct result of such contamination.

There are many agencies worldwide which are concerned with monitoring pollution and are responsible for establishing policies for dealing with it. In the past 20 years the EEC has been fairly active in producing Directives on pollution control with which member states are expected to comply. Such co-operation is clearly required as pollutants cross international boundaries. However, individual countries can do much for themselves in environmental protection as has been portrayed in the upturn in numbers of predatory birds in Britain following the banning of DDT in the late 1960s. River Authorities have also proved reasonably effective watchdogs in dealing with aquatic pollution.

The enormous pressure put on wildlife by the growing human population has caused the extinction of many species. Habitat destruction, over-exploitation and environmental pollution have been the root causes. There is a clear need for protecting and managing wildlife if such trends are to be halted. Conservation has many practical outcomes from which humans will benefit in future years. Legislation, the setting up of conservation agencies and the establishment of nature reserves in individual countries has done much to protect wildlife. However there is a need for international cooperation on many conservation issues such as the protection of whales and fish stocks. Although much has yet to be achieved there is some evidence that recent agreements between countries, such as the member states of the EEC, provide a basis on which a sensible international conservation policy can be founded.

QUESTIONS

1 A factory emitting smoke containing sulphur dioxide was sited in a rural district. The table below gives distances and directions of (i) number of lichen species and (ii) sulphur dioxide concentration in the atmosphere at different directions from the factory chimney.

Distance in miles from factory in a south-south-west direction	1	2	4	8	12	16	28
Number of lichen species	0	1	2	3	7	9	14
Sulphur dioxide concentration in parts per million	28	27	26	23	19.5	16	2
Distance in miles from factory in a north-north-east direction	1	2	4	8	12	16	28
Number of lichen species	1	2	3	4	4	5	5
Sulphur dioxide concentration in parts per million	27	26.5	25	24	23	22	19

(a) Plot the information to show the relationships between the lichen distribution and the sulphur dioxide concentration using the same X axis and two Y axes, the one on the right for sulphur dioxide concentration and the one on the left hand for number of species.

(b) Explain the difference in the results between those obtained for the south, south west direction and those obtained for the north, north east direction.

(c) Fully describe one example of evolution in action e.g. industrial melanism and its significance in the study of natural selection. (O)

2 The diagram shows some of the effects of discharging sewage and copper waste into a river.

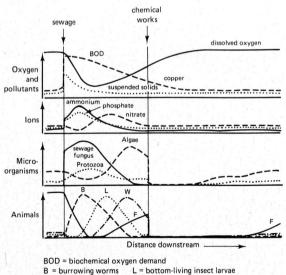

BOD = biochemical oxygen demand
B = burrowing worms   L = bottom-living insect larvae
W = water lice on plant and stones
F = clean-water fish   (*modified after Mellanby*)

(a) What is meant by the term *biochemical oxygen demand* (BOD)? Explain the changes in BOD shown in the diagram.

(b) Explain the changes in nitrate level shown in the diagram.

(c) Compare and comment on the curves for the sewage fungus and the algae in the diagram.

(d) Using evidence from the diagram, suggest a method by which an organism might be used as a pollution indicator. Your answer should include practical details of your method.

(e) Suppose that the chemical works also discharged thermal pollution. Suggest **one** possible effect on the river's chemical content and **one** possible effect on its biological content. (AEB 1984)

3 During the past thirty years many soft-water lakes in Sorlandet (Southern Norway) have lost their fish populations and have become more acid in their surface waters. At the same time there has been a marked increase in the combustion of fossil fuels in industrial Europe.

(a) What are the possible links between these observations?

(b) The diagrams below show the relationship between fish stocks and (i) pH, and (ii) excess sulphate concentration in the lake water.

(i) The relationship between pH and fish stocks of lakes

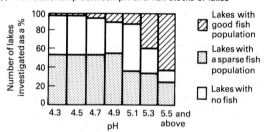

(ii) The relationship between excess sulphate concentration and fish stocks of lakes

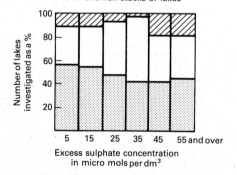

Which of these two factors, pH or excess sulphate concentration, had the greater influence on fish stocks? Give evidence from the diagrams in support of your answer.

(c) What would be the likely effect on fish stocks of increasing by 0.2 units the pH of lakes which currently have a pH range of (i) 4.3–4.5 and (ii) 5.1–5.3? (L)

4 (a) Describe the factors which influence the concentration of dissolved oxygen in a river.

(b) Explain the presence or absence of those organisms which may be used as indicators of the concentration of dissolved oxygen. (Nuffield/JMB)

34 Biotechnology

34.1 Commercial production of
enzymes   716

  34.1.1 Sources of enzymes   716
  34.1.2 Enzyme manufacture   717
  34.1.3 Extraction and purification
         of enzymes   717

34.2 Immobilised enzyme
technology   718

34.3 Manufacture of alcoholic
beverages   718

34.4 Manufacture of dairy
products   720

34.5 Manufacture of amino acids,
proteins and vitamins   720

34.6 Manufacture of antibiotics,
vaccines, antibodies and
hormones   721

34.7 Production of fuels   723

34.8 Extraction of oil and metals   724

34.9 Improvement of crop plants   725

34.10 Making of silage   725

34.11 Environmental manage-
ment   725

Summary   730

Questions   730

Milk Marketing Board

34 Biotechnology

Micro-organisms have long been used to manufacture alcoholic beverages, bread and dairy products as well as to dispose of sewage. **Biotechnology** is the use of organisms, microbes especially but not exclusively, to make useful products and to provide useful services. In recent years biotechnology has evolved into an integrated applied science drawing on the expertise of biochemists, chemical engineers, geneticists, microbiologists and even computing scientists. The modern approach to biotechnology is best appreciated by examining in detail an example such as the commercial production of enzymes. Following this, a wide range of biotechnological activities will be reviewed. Some of these are well established processes. Others are at the forefront of the latest research and promise to herald a new golden era in scientific advances from which people throughout the world will benefit.

34.1 Commercial production of enzymes

Over 2000 enzymes have been isolated to date, yet less than 20 are produced on a large scale. Relatively crude preparations used primarily in the food processing industries and in the manufacture of 'biological' detergents make up the bulk of the output. Many commercially produced enzymes are hydrolases such as amylases, cellulase, pectinases and peptidases. Nevertheless there is a growing demand for purified enzymes such as glucose oxidase for biochemical analysis of blood glucose concentration and asparaginase for treatment of a form of leukaemia.

34.1.1 Sources of enzymes

The traditional sources of many enzymes have been plants and animals. Amylase from germinating seeds and rennet from the stomachs of slaughtered animals are typical examples. Such sources are no longer adequate to meet the growing demand for enzymes. Consequently manufacturers are increasingly turning to **micro-organisms** as sources. Up to now most of the microbes used have been bacteria, moulds and yeasts which have long been acceptable in the food and beverage producing industries (Table 34.1). Mutant strains have been selected which maximise

Table 34.1 Examples of microbial sources of commercially important enzymes (from Smith 1985)

Enzyme	Source	Uses
Alcohol dehydrogenase	*Saccharomyces cerevisiae*	Assay of ethanol
Asparaginase	*Penicillium camemberti*; *Aspergillus niger*	Treatment of acute lymphatic leukaemia
Cellulase	*Trichoderma viride*	Preparation of dehydrated vegetables
Glucose isomerase	*Bacillus coagulans*	Production of fructose from glucose
Lipase	*Rhizopus arrhizus*	Flavour improvement in ice-cream
Pectinase	*Aspergillus niger*	Clarification of fruit juices
Penicillin acylase	*Escherichia coli*	Production of semi-synthetic penicillins
Peptidases	*Bacillus subtilis*	Biological detergents, meat tenderisers
Rennin	*Mucor* sp	Cheese production

enzyme production. **Genetic engineering** will come to the fore in the near future as a means of tailoring producer organisms for enzyme synthesis (Chapter 4).

34.1.2. Enzyme manufacture

Many kinds of enzymes derived from microbes are **extracellular**. They are secreted to break down insoluble substrates to form simple products which the microbes can absorb. In a pure culture, such enzymes appear in virtually pure form in the growth medium. In comparison, **intracellular enzymes** have to be obtained from the cells in which they are produced alongside many others. They require expensive extraction and separation procedures.

In most countries the producer organisms are grown in large stainless steel tanks called **bioreactors** of volume 10–50 m^3 (Fig 34.1). The tank is steam-sterilised before a sterile liquid nutrient medium is introduced. An inoculum of producer cells is then transferred to the medium which is maintained at a temperature optimal for enzyme production. The medium is constantly aerated to provide an aerobic environment thus encouraging rapid growth of the producer. A stirrer agitates the medium to prevent stale pockets developing. Within 30–50 hours the culture will have reached a stationary phase of growth (Chapter 32) when the secretion of extracellular enzymes reaches a peak. Chemicals which induce enzyme secretion may be added to the culture medium. The composition of the medium is also such that it does not contain enzyme repressors.

In Japan, moulds used in the commercial production of enzymes are grown in trays of rotating drums on solid media. Whatever the culture medium it has to be sterile and must satisfy the producer organism with regard to pH, temperature, water potential, nutritional and gaseous requirements. Some bioreactors allow continuous growth. Others are based on batch culture whereby growth of the producer organism is stopped after a given time to extract the product.

34.1.3 Extraction and purification of enzymes

A major part of the overall cost of producing enzymes commercially is spent on extracting and purifying them, a process called **downstreaming**. If the enzymes are intracellular the producer organisms have to be harvested from the culture medium, then subjected to **cell disruption**. For enzyme extraction, lysis of the cells with a weak solution of alkali is often used. The remains of the cells are next removed by **filtration** or **centrifugation**. The enzymes are then **precipitated** from the cell extract. For many years ammonium sulphate has been used for this purpose. The salt is gradually added until a sediment containing the enzyme is formed, a process called salting-out. Alternatively organic solvents such as isopropanol which are effective enzyme precipitants may be employed (Fig 34.2). Purification of enzymes can be achieved by **absorption chromatography** in columns of sepharose or agarose (Chapter 2). If required in powder form the solution of pure enzymes is spray-dried in a film evaporator at a temperature below 30°C. Enzymes in solution are usually stabilized by the addition of calcium ions.

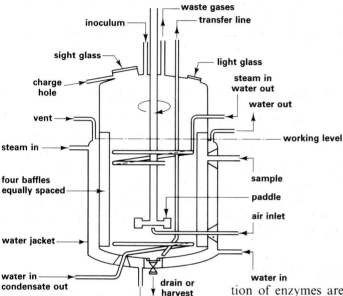

Fig 34.1 An industrial bioreactor (after Atkinson 1974)

inoculum — waste gases — transfer line
sight glass
charge hole
light glass
steam in / water out
water out
vent
steam in
working level
four baffles equally spaced
sample
paddle
air inlet
water jacket
water in / condensate out
water in
drain or harvest
condensate

Fig 34.2 Flowchart for extraction of an intracellular enzyme

Medium containing producer cells
|
CENTRIFUGATION
↓
Slurry of cells
|
BEAD MILL
↓
Disrupted cells
|
CROSS-FLOW FILTER
↓
Filtrate
|
PRECIPITATION
with isopropanol
↓
Intracellular enzyme

Fig 34.3 Ways of immobilising enzymes (after Smith 1985)

(i) **adsorption**

(ii) **cross-linking**

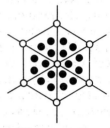

(iii) **entrapment**

(iv) **encapsulation**

● = enzyme molecules

Fig 34.4 Yeast bloom on grape skins (Food & Wine From France Ltd)

34.2. Immobilised enzyme technology

When product molecules are released from enzyme-substrate complexes, the enzyme is free to catalyse the same reaction again (section 3.2). On a commercial scale it is usually too expensive to try to recover the enzyme from a product. **Immobilised enzyme technology** has been developed to enable the same enzyme to be used over and over again. It involves converting the enzyme from a water-soluble, mobile state to a water-insoluble, immobile condition. There are several ways in which enzymes can be immobilised (Fig. 34.3).

Although well-perfected in the laboratory for many enzymes, only two examples have so far been scaled up for industrial use. They involve the use of:

(i) penicillin acylase from *E. coli* which catalyses changes in the penicillin molecule to form semi-synthetic derivatives such as ampicillin.
(ii) glucose isomerase which catalyses the conversion of glucose to fructose; the latter is used as an alternative sweetener to sucrose in many foods these days.

A major drawback to further commercial exploitation of immobilised enzymes is the fact that many enzymes can be produced cheaply and are thus replaced at no great cost. In addition the capital outlay required to establish large scale operation of the technology is enormous.

34.3 Manufacture of alcoholic beverages

The fermentation of plant materials by yeast to produce ethanol was the earliest use man made of micro-organisms. As long ago as 3000 BC the Ancient Egyptians could brew alcohol. Today considerable quantities of alcohol are used in the manufacture of such commodities as paints, plastics and textiles. However most of the ethanol produced at the present time is consumed in the form of alcoholic drinks. Two of the principal fermented beverages are wine and beer.

1 Wine production

Ripe grapes are picked and crushed to produce a **must** which contains juice, pulp, skins and pips. The must is then treated with about 100 p.p.m. of sulphur dioxide to inhibit undesirable bacteria and yeasts present on the skins. Fermentation of the juice is brought about by wine yeast, *Saccharomyces cerevisiae var. ellipsoideus*, which also grows on the skins of the fruit (Fig 34.4). Some wine producers add starter cultures of known strains of the fungus. As fermentation proceeds alcohol is formed which extracts the red pigments located in the skins of red grapes. If the skins are removed before the process is too advanced a pink or rosé wine is produced. White wine is produced if the skins are removed early or if the must is from green grapes. Complete breakdown of the sugar in the must produces a dry wine with an alcohol content of up to 15 % at which concentration the yeast is killed. Sweet wines are made from musts with very high sugar content or by stopping fermentation before all the sugar is used up. The fermented liquor is then aged in vats, when sediment is deposited, then filtered before it is bottled.

Sparkling wines such as Champagne are made by adding sugar to the fermented must. Wine yeast is then added and the mixture bottled. Champagne bottle corks have to be held in place with wire to withstand the pressure of the carbon dioxide formed inside the bottle by further fermentation.

The characteristics of wine depend not only on the way in which the must is treated but also on the composition of grape juice. This will depend on the type of grape and the environmental conditions under which the fruit is grown. Each major wine-producing region uses the variety of vine most suited to its climate and soil.

2 Beer

Its production depends on yeast fermentation of an infusion made from grain, mostly barley, and compared to wine production is microbiologically a more controlled process.

The barley grains are germinated under carefully controlled conditions so that grain enzymes, amylases and peptidases, convert endospermic starch to maltose and protein to amino acids. The **malting** is brought to an end by raising the temperature to kill the barley embryos. The addition of warm water to the malt forms a liquor, the **wort**. Dried hops are then added to give flavour, and to release chemicals which have antimicrobial properties. The wort is boiled for several hours.

Following cooling and filtering, the wort is **pitched**, that is inoculated with brewers' yeast. *Saccharomyces cerevisiae* is used for most beers but if a light beer or lager is being produced *S. carlsbergensis* is added instead. In the large fermentation vats the yeast converts soluble sugars in the wort into ethanol and carbon dioxide over a period of two to five days. A thick yeast head forms on the surface of the vat, and is periodically skimmed off.

Fig 34.5 Flowchart for beer production (after Williams and Shaw 1982)

When the beer has reached the desired alcohol content, between 4–8 %, it is racked off into casks or storage tanks, where a continued, secondary fermentation may occur at a slow rate. The beer is clarified, frequently enriched with carbon dioxide and is then ready for the market (Fig 34.5 and 34.6).

Industrial alcohol is used as a solvent in the chemical industry. Some of the requirement is met by alcohol of microbiological origin. Any waste substance rich in carbohydrate, such as molasses from sugar refining or spent brewery grains, can be fermented with selected strains of *S. cerevisiae*. The demand for industrial alcohol is so great, however, that the catalytic hydration of ethene using sulphuric acid is now a more rapid and economic process for the production of ethanol.

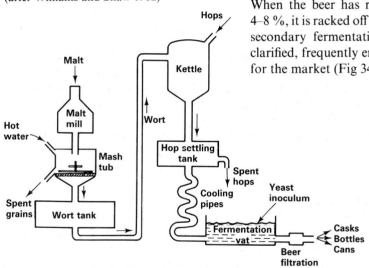

Fig 34.6 Production of beer (Heritage Brewery Museum, Burton Upon Trent)
(a) Hop kettle for the preparation of wort

(b) Open tank fermentation. Note yeast floating on surface of the wort

34.4 Manufacture of dairy products

i. Butter is prepared by churning cream, which causes fat globules in the cream to coalesce into granules. The liquid part of the cream, called buttermilk, is then drained off and the granules are compressed to make butter. In Britain nearly all butter is made from pasteurised milk in which the natural microflora is killed. However, the cream is sometimes soured by inoculating it with starter cultures of *Streptococcus lactis* and *Leuconostoc citrovorum*. The streptococci ferment milk sugar, lactose to lactic acid, while the leuconostocs produce volatile substances which give the butter flavour and aroma.

ii. Cheese Pasteurised milk is generally used in the commercial production of cheese. Starter cultures of *Streptococcus lactis* or *S. cremoris* are added to the milk, and changes in pH and texture are monitored as the milk is curdled by the bacteria. The enzyme rennin, in rennet prepared from calves' stomachs, is usually added. This speeds up curdling by coagulating the milk protein casein. Soft cheeses are made by allowing the liquid called whey to drain from the curd. In making hard cheeses, the curd is compressed. Common salt is mixed with the curd or applied to its surface to prevent the growth of undesirable micro-organisms and to add flavour. (Fig 34.7).

A great variety of cheeses is ripened by moulds and sometimes yeasts. Such cheeses differ from non-ripened cheeses in digestibility, vitamin content, appearance and palatability. (Fig 34.8).

iii. Yoghurt was originally prepared from buttermilk produced in butter-making. In Britain it is now manufactured from pasteurised milk. In Eastern Europe and Asia, it is frequently made from the milk of goats, sheep and mares. The milk is inoculated with a mixed starter culture containing *Streptococcus thermophilus* and *Lactobacillus bulgaricus* and kept at 42–45°C. The bacteria ferment lactose to lactic acid, so curdling the milk and giving it a sour taste. The flavour of natural yoghurt made in this way can be altered by adding fruit pulp and fruit juices.

Fig 34.7 Flow diagram for cheese manufacture

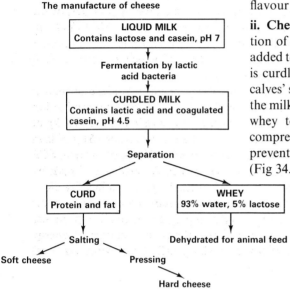

The manufacture of cheese

Fig 34.8 A blue-vein cheese (Milk Marketing Board)

Fig 34.9 *Spirulina maxima*

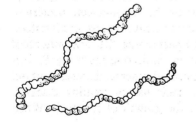

34.5 Manufacture of amino acids, proteins and vitamins

1 Amino acids

In recent years, amino acids have increasingly been produced commercially with the aid of bacteria. It is possible to make amino acids by hydrolysing proteins or by chemical synthesis. However, microbial production is more economical and the product is of better quality. When *Micrococcus glutamicus* is grown in a liquid medium containing glucose, minerals, vitamins and peptone, about $15\,g\,dm^{-3}$ of the amino acid lysine appear in the medium within 3–4 days. Substantial amounts of other amino acids such as glutamic acid, valine, tryptophan and alanine appear when other species of bacteria are used. Some of the acids are 'essential' amino acids and are used to supplement animal feedstuffs containing protein of low biological value.

2 Proteins

As long ago as the early 1500s the Aztecs were eating cakes made from the blue green alga *Spirulina maxima* (Fig 34.9) which grows in warm, shallow,

Fig 34.10 Pruteen manufacturing plant (ICI)

freshwater lakes. This tradition is still carried on in Mexico and in Chad where the alga contributes significantly to the protein consumption by the natives. It has a protein content as high as 65 % of its fresh mass, about 35 times more than that of maize. Its growth rate is also very rapid. Trials are now underway to assess the possibility of growing *Spirulina* commercially as a source of edible protein.

Early this century, yeasts were cultured in oil-based media to provide supplementary protein for sausages. The experiment was not a success as the **single cell protein** (SCP) as it is called was not very palatable. In the past decade or so much greater success in producing SCP has come from growing the producer organisms in media based on the by-products of the petroleum and natural gas industries. Bacteria such as *Methylophilus*, yeasts such as *Candida* and moulds such as *Fusarium* are currently employed in making SCP (Fig 34.10) some of which is used to feed livestock, the rest for human consumption.

The efficiency with which some micro-organisms synthesise protein is remarkable compared with traditional methods of protein production using livestock. For example, one hectare of maize yields about 4000 kg of grain on well farmed land. When fed to cattle this produces 85 kg of lean meat. If instead it is used to make SCP the yield is 360 kg of protein. SCP contains a good range of amino acids, is free of cholesterol, and if made by moulds has a high fibre content. As such it can be part of a healthy diet.

Production of SCP could become an important contribution to the present shortage of edible protein in developing countries. The producer organisms will grow in media containing almost any source of energy thereby increasing the feasibility of making SCP from whatever local resources are available.

3 Vitamins

Brewer's yeast, *Saccharomyces cerevisiae*, synthesises a range of **vitamins** of the B-group. For this reason yeast products such as brewer's yeast tablets are taken by humans to supplement diets deficient in these vitamins. Animal feedstuffs are supplemented by riboflavin (vitamin B_2) which is made by the yeast *Ashbya gossypii*.

34.6 Manufacture of antibiotics, vaccines, antibodies and hormones

1 Production of antibiotics

Antibiotics are chemicals produced by micro-organisms which, at very low concentrations, have the capacity to inhibit or destroy other micro-organisms. They are of medical and veterinary importance, because they can be used to kill pathogenic micro-organisms in humans and livestock without adversely affecting the patient. Antibiotics are made by moulds and by bacteria. Some are effective against several pathogens (broad-spectrum), others act on only one or two (narrow-spectrum). **Streptomycin** is a broad-spectrum antibiotic made by the actinomycete bacterium *Streptomyces griseus*. It is used extensively to treat humans and livestock suffering from tuberculosis and bacterial pneumonia. **Chloramphenicol** (chloromycetin) is another broad-spectrum antibiotic made by *S. venuzulae*. This antibiotic is used in the treatment of venereal diseases, typhoid fever and skin infections. Apart from their use in treating diseases, antibiotics are widely used to stimulate the growth of livestock. When

Fig 34.11 Fermentation tanks for antibiotic production (Science Photo Library)

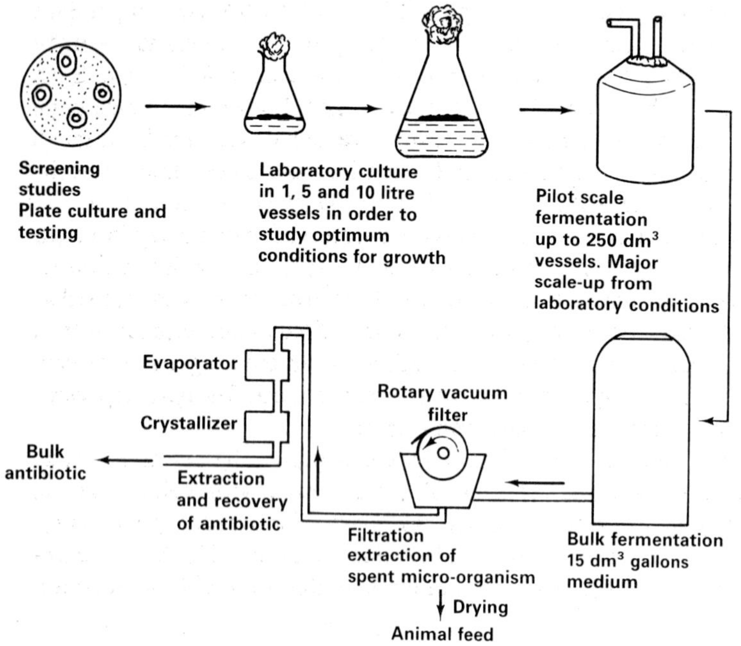

Fig 34.12 Flowchart for production of antibiotics

Screening studies
Plate culture and testing

Laboratory culture in 1, 5 and 10 litre vessels in order to study optimum conditions for growth

Pilot scale fermentation up to 250 dm³ vessels. Major scale-up from laboratory conditions

Evaporator

Rotary vacuum filter

Crystallizer

Bulk antibiotic

Extraction and recovery of antibiotic

Filtration extraction of spent micro-organism
↓ Drying
Animal feed

Bulk fermentation 15 dm³ gallons medium

added to animal feedstuffs, antibiotics kill micro-organisms including pathogens in the gut and promote protein synthesis in the body.

Many **antibiotics** made by fungi have been isolated and tested, but medical and veterinary use is confined so far to **penicillin, cephalosporin** and **griseofulvin**.

Penicillin was discovered almost accidently by Alexander Fleming in 1928 when he was working at St Mary's Hospital, Paddington. A plate culture of staphylococci, one of the bacteria which infect wounds, had become contaminated in his laboratory with a mould. He noticed that there were areas where the bacteria had apparently been destroyed by some unknown substance produced by the mould. Later the mould was identified as *Penicillium notatum*. Fleming published his observations, but nobody took much notice until the Second World War. The large number of injuries gave impetus to a search for drugs which would help heal infected wounds. Howard Florey and Ernst Chain succeeded in isolating the anti-bacterial substance, now called penicillin. It was found to be effective against many different species of bacteria.

Some of the earliest produced penicillin eventually proved to be less effective, mainly because resistant strains of bacteria evolved. However, modern production methods such as immobilised enzyme technology have made it possible to make modifications to the original penicillin molecule (section 34.2). The newer penicillins are more effective. Today the main producer mould is *P. chrysogenum* which is grown aerobically in a liquid medium in huge tanks of several hundred thousand litres' capacity. (Fig 34.11 and 34.12). By altering the composition of the medium, the mould can be made to release greater amounts of penicillin, which can be extracted from the medium and prepared for medicinal use. For example, if lactose is used as an energy source and the penicillin precursor phenylethanoic acid is also included the yield of penicillin can be as much as $20\,\mathrm{g\,dm^{-3}}$ of culture medium. This contrasts with a few mg obtained by Florey and Chain in their early work. Genetic engineering may further enhance the output in the near future.

2 Production of vaccines

Vaccines have contributed enormously to the fight against infectious diseases (Chapter 10). Some idea of their impact can be judged from the fact that in the mid 1960s over 10 million people suffered from smallpox. Today, thanks to an effective vaccination programme, the disease has been eradicated. Comparable successes have been achieved with poliomyelitis, german measles, yellow fever and rabies, though cases of these diseases continue to crop up.

Vaccines which give immunity against viral diseases are usually produced from viruses raised in cell cultures or in laboratory animals, both expensive procedures. It is the protein coats of viruses that are antigenic, stimulating us to produce antibodies. If antigenic viral proteins could be made in another cheaper way, vaccines would be affordable to developing countries. An encouraging step forward in this direction has been made

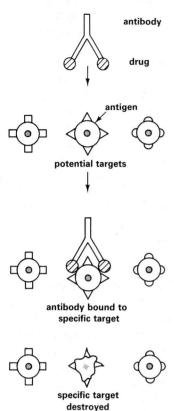

Fig 34.13 Drug delivery using monoclonal antibodies

antibody

drug

antigen

potential targets

antibody bound to
specific target

specific target
destroyed

recently by biotechnologists. They have succeeded in genetically engineering the bacterium *Escherichia coli* so that is synthesises proteins of the kind found in the coat of the hepatitis virus. This has resulted in the production of a cheap vaccine against hepatitis. Similar approaches may provide a vaccine against the Aids virus.

3 Production of monoclonal antibodies

In 1975 Köhler and Milstein showed that monoclonal antibodies could be made by hybridomas, hybrid cells produced by fusing lymphocytes with repeatedly dividing tumour cells (Chapter 10). Drugs can be bound to such antibodies. When administered to a patient the drug–antibody complex finds its way to selected target cells to which the antibodies specifically bind (Fig 34.13). In this way diseased tissues and cells can be eliminated with very low dosages of the drug. Another use of monoclonal antibodies is in tissue typing prior to organ transplantation.

4 Production of hormones

The way in which genetically engineered *E. coli* is now used to produce human insulin is described in detail in Chapter 4. Mammalian growth hormone and calcitonin are also made in a comparable way. Such hormones have both clinical and veterinary uses. For example injecting a cow with bovine growth hormone can increase milk production by as much as 25 %, whereas beef cattle show a 10–15 % increase in body mass. Results of this kind may help alleviate a world shortage of first class protein.

Pharmaceutical interest in **steroids** has increased enormously in recent years. This results from the discovery of the value of **cortisone** and its derivatives in the treatment of arthritis. Other steroid hormones are used as **oral contraceptives**. When plant steroids are added to a culture medium in which fungi can grow, high yields of the finished steroid drugs are achieved. Among the fungi which are used for this purpose are *Rhizopus nigricans*, *Fusarium culmorum* and *Trichoderma viride*.

Plant hormones too have commercial applications. **Gibberellin** is made by the fungus *Gibberella fujikuroi*. The main use of gibberellin is in the production of malt from germinated barley grain for the brewing industry. Barley grains make their own gibberellin (Chapter 23) but if extra gibberellin is given, the production of maltose is considerably speeded up.

34.7 Production of fuels

Fossil fuels such as coal, oil and natural gas are used at such a rate that reserves will be exhausted within the foreseeable future. At present rates of consumption supplies of oil will dry up in about 50 years and coal in 100 years unless additional supplies are found. Nuclear energy is an option which governments are turning to more and more. However people are concerned about the dangers of environmental contamination with nuclear waste (Chapter 33). Hydro-electricity is a renewable form of energy but requires massive capital investment. On a smaller scale windmills and water mills can generate power. Provision of a cheap supply of energy which will not run out and is non-polluting would do much to improve the quality of life for many people the world over. Biotechnology may provide the answer.

1 Gasohol

In Brazil the world's largest project aimed at producing renewable supplies of energy has been underway since 1975. Brazil's climate is suited to growing sugar cane on a large scale. The sugar is fermented by yeasts and currently yields about 10 billion litres of fuel alcohol called **gasohol**. Car engines have been adapted at modest cost to run on gasohol. In this way the country's dependence on petroleum has been halved. The National Alcohol Programme is not without teething problems. Fermentation waste poses a disposal problem because of its high biochemical oxygen demand (section 34.11). Taking over vast tracts of land for monoculture of sugar cane has also presented undesirable ecological changes.

One of the difficulties in repeating such a project elsewhere is that sugar cane grows well only in tropical and sub-tropical climates. In many parts of the world the main store of energy in crops is starch. For this reason attempts are presently being made to genetically engineer yeasts to produce the enzyme amylase which converts starch to fermentable sugar. Gasohol programmes could then become universal.

2 Biogas

Massive quantities of domestic, industrial and agricultural waste are generated each year. The energy it contains can be converted to forms which are usable and renewable. Whenever such waste has been disposed of at landfill sites and covered with earth, methane gas is released by bacteria which degrade organic molecules. In some places it has been worthwhile to tap the trapped methane and burn it as a source of energy to heat small factories and housing estates. Biotechnologists are now looking at ways in which such **biogas** can be generated on a larger scale.

34.8 Extraction of oil and metals

A substantial amount of oil clings to underground rocks in the vicinity of oil wells, hence oil extraction is less efficient than it might be. Biotechnology has come up with an answer to this difficulty. Polysaccharide gums such as xanthan synthesised by the bacterium *Xanthomonas* are pumped into the ground near to oil wells. The gum oozes through the rocks, pushing the oil adhering to the rocks towards the well where it is pumped out.

The bacterium *Thiobacillus ferro-oxidans* obtains energy by oxidising sulphides to sulphates and sulphuric acid. Such an approach is applied to the extraction of metals from low grade ores which are not exploited by traditional methods of ore refining for economic reasons. However as the supply of high grade ores is quickly being used up, cheap ways of extracting metals from low grade ores will become more important. At present, copper and uranium are microbiologically extracted on a commercial basis.

Copper is leached from the spoil heaps of ore-refining plants. *Thiobacillus* generates acid from a sulphide ore called chalcopyrites. The acid maintains the very acid environment in which the bacterium thrives. Copper sulphate solution released from the ore is drained from the spoil and collected in tanks containing scrap iron. The sulphate reacts with the iron coating it with copper which is then removed for further refinement (Fig 34.14).

Fig 34.14 Leaching bed for bacterial mining of copper

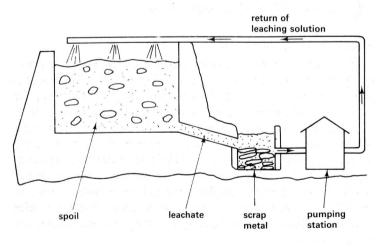

return of leaching solution

spoil   leachate   scrap metal   pumping station

Pilot studies have shown that cobalt, lead and nickel could also be extracted in this way.

34.9 Improvement of crop plants

Huge amounts of expensive fertilisers are applied to the land each year by farmers and gardeners (Chapter 14). Yet bacteria which live symbiotically with some important crops, notably legumes, can fix gaseous nitrogen free (Chapter 31). Over many years unsuccessful attempts have been made to get nitrogen-fixing bacteria to associate with non-legumes such as cereals. Much more needs be known before compatibility between 'strange' partners can be achieved. Genetic engineering has as yet also failed to provide a solution, direct attempts at transferring the nitrogen fixing genes proving impossible. Nevertheless a breakthrough may be around the corner, using other bacteria which commonly associate with cereal plants, as recipients of the genes.

Greater success in redesigning crop plants by genetic engineering has been achieved in protecting crops from insect damage. A polypeptide produced by the bacterium *Bacillus thuringiensis* is toxic to caterpillars but not to humans and livestock. The gene which regulates synthesis of the peptide has been cloned and transferred to tobacco plants. The success of the venture offers hope that many edible crops can be improved in this way and opens up possibilities that almost any species can be redesigned to maximise yield in the habitat to which it is best suited.

34.10 Making of silage

Silage-making is important to British agriculture, because it enables farmers to store grass over winter without having to dry it. In many summers it is virtually impossible to make hay, and drying grass indoors is expensive. In making silage, fresh-cut grass is compressed in a large silo or a pit. Bacteria such as *Lactobacillus* and *Streptococcus* on the grass ferment sugars in the plant juices and release fatty acids such as butyric, propionic an ethanoic acid. These give the silage its characteristic flavour and smell.

34.11 Environmental management

1 Treatment of industrial effluent

In Finland a system has been developed using the mould *Paecilomyces* to detoxify sulphite liquor from the paper producing industry. The liquor is normally discharged into rivers and lakes. Without such treatment the environmental consequences are catastrophic. Aerial contamination with sulphur in the effluent discharged by coal and oil burning power stations is also avoidable. Pilot studies have shown that some species of bacteria can 'scour' sulphur from aerial effluent at half the cost of chemical methods. Liquid effluent of high biochemical oxygen demand can also be turned to good use. Sugar rich effluent from the confectionery industry is used to grow yeasts. The carbon dioxide and alcohol so generated is sold at a price which more than offsets the capital cost of installing and maintaining the required process plant.

2 Treatment of oil spillages

Marine pollution with oil discharged from foundered and war damaged tankers has occurred on a wide scale. Early attempts at dealing with the problem involved the use of detergents, chalk dust and sawdust (Chapter 33). More recently oil slicks have been dealt with by spraying them with a suspension of the bacterium *Pseudomonas*. It can degrade hydrocarbons in oil, thus causing the slick to disperse.

3 Sewage disposal

Sewage is the water-borne waste of a community. It has three main components: domestic waste, industrial waste and surface water. Domestic waste includes tea leaves from our kitchen sinks, detergents from washing machines, soaps in bathwater, faeces and urine from lavatories. Industrial waste contains animal waste such as blood from slaughterhouses, plant waste such as cellulose from paper mills, pharmaceuticals, clay from potteries, oil from garages and engineering works, heavy metals such as lead, zinc and mercury. Surface water includes rain water from paved areas, roads and roofs. Many of the older towns in Britain are sewered on the combined system in which surface water drains are connected to foul sewers carrying domestic and industrial waste. To prevent the sewers from flooding when there is heavy rainfall, overflows release excess sewage into natural watercourses such as streams, rivers and sea. Because this arrangement results in pollution of watercourses, newly installed sewers have a separate system for collecting surface water and discharging it into natural watercourses. Domestic and industrial wastes are channelled to sewage treatment plants.

(i) Composition of sewage

The composition of sewage varies from place to place and from time to time. An approximate analysis of 'typical' sewage shows it to contain 99.9 % water, 0.02–0.03 % suspended solids, and 0.07–0.08 % dissolved organic and inorganic compounds. The amount of organic material determines the **Biochemical Oxygen Demand, BOD**, of the sewage. The larger the amount of organic material, the higher is the BOD. The BOD is the quantity of oxygen required, mainly by decomposer organisms, to break down the organic matter into simple, stable products. Sewage also contains micro-organisms. Among the bacteria present are relatively harmless species, for example *Escherichia coli* which comes from the intestines of healthy people, and pathogens which come from the bowels of people suffering from enteric diseases such as typhoid fever, cholera and dysentery. Viruses in sewage include harmless bacteriophages as well as pathogens such as the viruses which cause poliomyelitis and infectious hepatitis.

(ii) Aims of sewage treatment

There are two main aims of sewage treatment.
(a) To lower the BOD sufficiently so that the effluent from the sewage works can be discharged into a natural watercourse without grossly upsetting its ecological stability (Chapter 33).
(b) As far as possible to destroy or eliminate pathogens which may endanger wildlife such as shellfish and birds living in the water. Humans may also eat shellfish from, or bathe in natural watercourses into which treated sewage is discharged.

Both these aims can be achieved by a combination of physical, chemical and microbiological processes.

(iii) Stages of sewage treatment

There are many different ways in which sewage can be treated. The choice depends on the amount of sewage to be dealt with, the fate of the effluent and its possible ecological effects and the cost of treatment. In general there are three main stages (Fig 34.15).

Fig 34.15 Flow diagram of the trickling filter system of sewage treatment

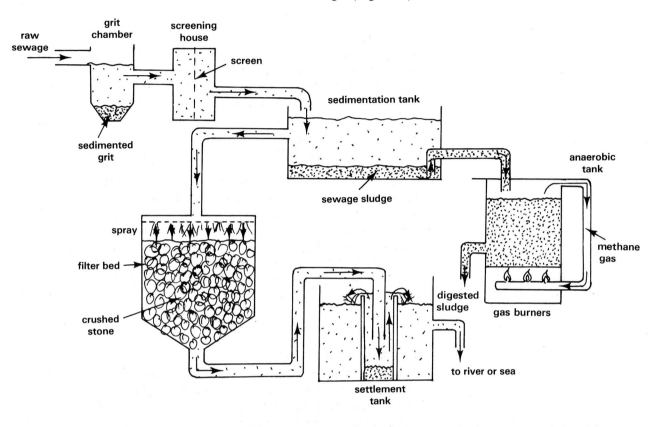

Fig 34.16 Trickling filter system: circular tanks in foreground (Water Authorities Association)

Primary treatment   physically removes the larger suspended and floating solids by **sedimentation** and **screening**. Raw sewage is first channelled into small detritus tanks where soil and grit form a sediment. Grit is removed to prevent excessive wear on the pumps which move the sewage from one part of the treatment plant to another. The supernatant liquid then passes through a screen house where floating debris such as paper, rags and wood are filtered out using coarse metal screens. Screening is carried out to minimise the risk of pump and pipe blockages in the later stages of treatment. The screened sewage enters large sedimentation tanks where after several weeks many suspended solids are precipitated as crude **sewage sludge**. It contains about 4% of the total sewage solids. The supernatant liquid and sludge are then subjected separately to secondary treatment which is entirely microbiological.

Secondary treatment   There are two main ways of dealing with the supernatant liquid from the sedimentation tanks the trickling filter system or the activated sludge process.
Trickling Filter System. The filter is a bed of crushed stone onto which the liquid sewage is sprayed through rotating pipes (Fig 34.16). Spraying enriches the liquid with oxygen and is carried out intermittently to maintain aerobic conditions in the filter. The stone becomes coated with a thick film of slime containing bacteria, fungi, protozoans and algae. Aerobic saprophytic bacteria such as *Zooglea ramigera* and fungi such as *Fusarium* and *Geotrichum* inhabit the surface of the slime where they decompose organic substances in the sewage as follows:

$$\text{Carbohydrates} \quad \rightarrow CO_2 + H_2O$$
$$\text{Proteins} \qquad\quad \rightarrow NH_3 + CO_2 + H_2S$$
$$\text{Lipids and soaps} \ \rightarrow CO_2 + H_2O$$
$$\text{Urea} \qquad\qquad \rightarrow NH_3 + CO_2$$

Aerobic chemo-autotrophic bacteria such as *Nitrosomonas* and *Nitrobacter* also live near the surface of the slime. They carry out important oxidations such as the conversion of ammonia to nitrate (nitrification).

$$2NH_4^+ + 3O_2 \rightarrow 2NO_2^- + 4H^+ + 2H_2O$$
$$2NO_2^- + \ O_2 \rightarrow 2NO_3^-$$

Others such as *Thiobacillus* oxidise hydrogen sulphide to sulphate:

$$2H_2S + 5O_2 \rightarrow 2SO_4^{2-} + 2H_2O$$

As a result of these activities the BOD of the sewage is greatly reduced and the products are simple, relatively harmless substances. With little further treatment the liquid leaving the filter can be released into a nearby watercourse. Many motile protozoans are found in the upper part of the filter, stalked forms lower down. They consume the bacteria and maintain a suitable balance of organisms in the filter.

Activated Sludge Process. In recent years the trickling filter system has been superseded by the activated sludge process. About half of the sewage in Britain is now treated using the activated method. The supernatant liquid from the primary treatment is pumped into **aeration tanks** (Fig 34.17). The liquid is constantly stirred and vigorously aerated, resulting in the formation of small aggregates of finely suspended colloidal organic matter called **floc**. The formation of floc is not fully understood. It has a gelatinous texture and contains numerous slime-forming bacteria, notably *Zooglea ramigera*. Nitrifying bacteria are also present as well as protozoans. Stalked protozoans are more common than motile species. Fungi are normally

Fig 34.17 Flow diagram of the activated sludge process of sewage treatment

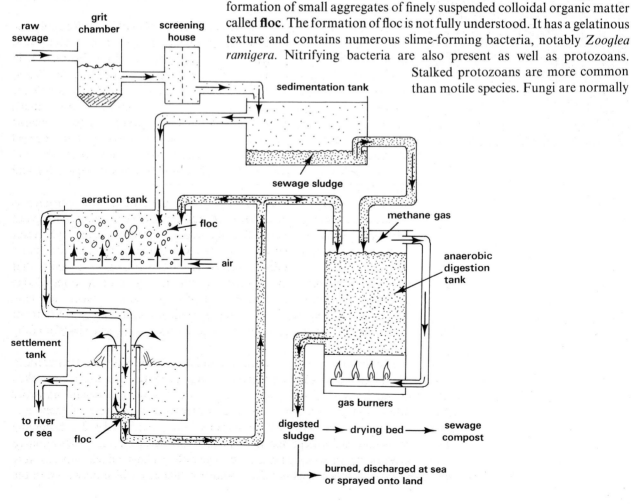

728

Fig 34.18 (a) Activated sludge aeration tanks (Water Authorities Association)

Fig 34.18 (b) Anaerobic digestion tank (Water Authorities Association)

absent. The biochemical changes which take place in the floc are similar to those which occur in the bed of a trickling filter system, but because the sewage is better aerated the breakdown of organic matter is much more rapid. Within 5 to 15 hours the BOD of the liquid is lowered by 85–95 %. The contents of the tank are then pumped into settlement tanks for tertiary treatment.

Several methods are used in Britain for the treatment of raw sewage sludge. One of the most important is **anaerobic digestion**. About half of the sewage sludge in the country is now treated in this way. The digester is a large closed tank which is almost filled with sludge so little air is present and thus very little oxygen (Fig 34.18). Facultative bacteria such as *E. coli* multiply and break down carbohydrates, lipids and proteins into a variety of products including fatty acids, alcohols, hydrogen sulphide, ammonia and carbon dioxide. Methane-producing bacteria such as *Methanobacillus* then metabolise some of the fatty acids and carbon dioxide as follows:

$$CH_3CH_2CH_2COOH + H_2O + CO_2 \rightarrow CH_4 + CH_3COOH$$

$$\text{butanoic} \quad + \text{water} + \text{carbon} \rightarrow \text{methane} + \text{ethanoic}$$
$$\text{acid} \qquad\qquad\qquad \text{dioxide} \qquad\qquad \text{acid}$$

The digestion process is speeded up by constantly stirring the sludge and heating it to 50–55 °C. At this temperature many pathogens in the sludge are killed. Within 2–4 weeks about 50 % of the organic matter is broken down, 70 % to methane and 30 % to carbon dioxide. The methane is burned and the heat generated is used to warm the digestion tanks or converted to electrical energy which drives the pumps in the sewage works. About two thirds of the works' energy requirements comes from sewage in this way.

In Britain the digested sludge is dealt with in several ways. Roughly one fifth of the treated sludge produced at inland sewage works is dumped at sea, two fifths applied to agricultural land and two fifths disposed of in other ways such as burning or burying. Wet sludge is sometimes applied to land and ploughed in after it has dried sufficiently. Alternatively, the sludge is dried at the sewage treatment plant and sold as 'compost'. These methods make use of the limited mineral constituents in the sludge and its organic matter contributes to humus formation in the soil. However, land-disposal is not a feasible long-term proposition because the sludge is of variable and limited quality and may contain high concentrations of pesticides and heavy metals.

Another process which has recently been introduced into this country is to incinerate the sludge. In 1976 the largest British sludge incineration plant was opened at Coleshill, near Birmingham. The plant occupies 2.9 hectares compared with 54.6 hectares which would have been required for drying the sludge in the conventional way. An additional important benefit is that the mass of ash to be disposed of is only one eighth of air-dried sludge prepared from the same volume of sewage. The sewage works at many coastal towns discharge untreated sludge directly into the sea. The fate of sludge dumped at sea and its effects on sea-life including fisheries is not fully known.

The time taken for the microbiologically-based secondary treatment to occur depends mainly on how much organic matter there is to be broken down. Effluents from abattoirs, dairies and food processing industries often pose problems to sewage treatment works because of their high organic content. For similar reasons Water Authorities are becoming concerned at the increased use of kitchen sink homogenisers. They grind up most forms of soft waste such as vegetable peelings which hitherto were disposed of in domestic refuse. Another difficulty which may arise in the secondary stage of treatment is the presence in sewage of wastes such as phenols which are toxic to the decomposer micro-organisms. Phenol is a byproduct of the coal tar industry which produces antiseptics and disinfectants. This problem can be overcome by introducing phenol-oxidising bacteria to activated sludge tanks.

Tertiary treatment Where a high standard of liquid effluent is required, the liquid from the secondary stage is treated further. At many sewage works the liquid is simply pumped into settlement tanks where the floc forms a sediment. About 10 % of the floc is used to inoculate fresh batches of sewage entering the aeration tanks. The remainder is pumped to the anaerobic tanks for digestion. The supernatant liquid is then run into a watercourse. Liquid from the settlement tanks may be disinfected with chlorine or with ozone to kill pathogenic bacteria. However, this is only done if the water is required for immediate re-use. Chlorine is toxic to fish and other aquatic organisms and would have a catastrophic effect on the ecological balance of rivers and streams. Other components in the liquid effluent such as organic matter, phosphate, nitrate and other inorganic ions, can be removed by filtration though activated carbon and ion-exchange resins. These methods are however comparatively expensive and at the moment are not widely practised.

SUMMARY

Biotechnology is a multidisciplinary applied science which uses organisms, microbes especially to make useful products and provide useful services.

Some aspects of biotechnology such as wine making go back to ancient times. However in the last decade, major advances in the biological sciences in particular have opened up a new and exciting future for biotechnology. Genetic engineering promises to provide solutions to hitherto intractable problems in the fields of crop improvement, antibiotic and vaccine production. Microbial synthesis of single cell protein offers hope to the malnourished. Microbial production of renewable sources of energy will help extend the life of fossil fuels.

Running alongside those remarkable developments are traditional but nevertheless equally important biotechnology-based industries such as the manufacture of dairy products and the treatment of sewage. Biotechnology will undoubtedly figure even more prominently in improving the health and quality of life of people the world over in the years ahead.

QUESTIONS

1 (a) Give an account of the principal physical and chemical changes which occur in the 'natural' souring of milk, stating clearly the role played by named microorganisms.
(b) Explain how this basic souring process may be modified during the commercial production of any two dairy products.   (JMB)

2 (a) Describe how microorganisms play a part in the synthesis of a range of products useful to man.
(b) Give an account of the commercial preparation and extraction of any one such product.   (JMB)

3 Describe, with full emphasis on the biological principles, the practical application of microorganisms to any two organic industrial processes.   (O)

4 Write an essay on the economic importance of micro-organisms.   (AEB 1985)

5 Advances in biology are essentially dependent on technical advances.' Discuss this statement.   (C)

6 Write an essay on alcohol production.   (WJEC)

7 Describe the importance of micro-organisms in the food (not drink) industry.   (WJEC)

8 (a) What is meant by sewage?
(b) Describe the sequence of stages used in the treatment of sewage produced by an urban community.
(c) What are the problems associated with the disposal of large amounts of sewage?   (London)

Appendix 1 Units, Symbols and Quantities

SI and SI derived units, symbols and quantities

Système International (SI) and SI-derived units, symbols and quantities have been used throughout the text. Non-SI units have been given in addition where it is anticipated that the SI system may not be adopted in the near future.

Non-SI units and their conversion to SI units are given in brackets under each section.

Length

1 metre (m) = 1000 millimetres (mm)

1 mm $(10^{-3}$m) = 1000 micrometres (μm)

1 μm $(10^{-6}$m) = 1000 nanometres (nm)

(1 in = 25.4 mm)

Energy

1 kilojoule (kJ) = 1000 joules (J)

(1 calorie = 4.187 J)

Pressure

1 kilopascal (kPa) = 1000 pascals (Pa)

(1 atm = 760 mm Hg = 101.325 kPa)

Mass

1 kilogram (kg) = 1000 grams (g)

1 g $(10^{-3}$kg) = 1000 milligrams (mg)

1 mg $(10^{-6}$kg) = 1000 nanograms (ng)

1 ng $(10^{-9}$kg) = 1000 picograms (pg)

(1 lb = 0.454 kg)

Volume

1 cubic decimetre (dm^3) = 1000 cubic centimetres (cm^3)

1 cubic centimetre $(10^{-3}$ dm^3) = 1000 cubic millimetres (mm^3)

(1 in^3 = 16.38 cm^3)

Electrical potential difference

1 volt (V) = 1000 millivolts (mV)

Time

1 second (s) = 1000 milliseconds (ms)

Temperature

Thermodynamic temperature   kelvin (K)

degree Celsius = °C

(t Fahrenheit = $\frac{5}{9}$(t − 32)°C)

Appendix 2 Chemical Nomenclature

In recent years the International Union of Physics and Chemistry (IUPAC) and the International Union of Biochemistry (IUB) have reached agreement on the rules of nomenclature for many compounds of biological interest. The systematic names of most biochemicals are extremely cumbersome and require a knowledge of chemistry beyond what is expected of GCE Advanced Level Biology students. We have used systematic names where students should reasonably be able to cope with them or where they present little or no more difficulty than the trivial names. Trivial names have been given where it is not anticipated that the systematic names will be used at this level of study in the near future. Because some students may wish to know the systematic names of some of the biochemicals given trivial names in the text the following list has been drawn up. The list may also be useful for those who are not familiar with the systematic names used in the text.

The names of mammalian hormones are those recommended by the Institute of Biology

Trivial name	Systematic name
acetaldehyde	ethanal
acetic acid	ethanoic acid
acetone	propanone
alanine	2-aminopropanoic acid
aniline	phenylamine
benzoic acid	benzenecarboxylic acid
carbon tetrachloride	tetrachloromethane
chloroform	trichloromethane
citric acid	2-hydroxypropane, 1,2,3-tricarboxylic acid
ethylene	ethene
formaldehyde	methanal
fumaric acid	trans-butenedioic acid
glyceraldehyde	2,3-dihydroxypropanal

Trivial name	Systematic name
glycerol	propane-1,2,3-triol
glycine	aminoethanoic acid
glycollic acid	hydroxyethanoic acid
hippuric acid	N-benzoylglycine
lactic acid	2-hydroxypropanoic acid
malic acid	2-hydroxybutanedioic acid
malonic acid	propanedioic acid
ornithine	2,5-diaminovaleric acid
phosphoenolpyruvic acid	2-hydroxy-2-propanoic acid
pyruvic acid	2-oxopropanoic acid
succinic acid	butanedioic acid
urea	carbamide
xylene	dimethylbenzene

Appendix 3 Practical work

Introduction

Most Examination Boards require candidates to undertake laboratory exercises and fieldwork studies as part of their A-level course in Biology. Reports of practical work are assessed and contribute to the final grading, as do observations made by teachers of the performance of candidates whilst performing selected exercises. Some Boards also include a practical examination. The following general notes may prove helpful to candidates when considering how practical work should be recorded for assessment purposes.

Preparing reports of practical work

A. Laboratory notebooks

1. Drawings should be in pencil and labelled or annotated where appropriate in pencil. This makes it easier for mistakes to be corrected. Annotations should indicate features of biological interest displayed by a specimen. Each drawing should have an underlined heading. The size of the drawing relative to actual size of the specimen should be given. Shading should be avoided as should the use of colouring. Stippling may be used to indicate dense or dark features. Drawings must be as life-like as possible! Plan diagrams are 'contour maps' indicating the distribution of tissues in a section of an organ. They should not include cells. When microscopic drawings of cells are requested it is best to draw a few cells accurately, as you see them. Quality is far preferable to quantity!

B. Field reports

These should be structured as follows:
1. Introduction – which provides background information of the habitat and field investigations
2. Aims—what the study intends to achieve.
3. Methods—the procedures used for analysing populations of plants and animals and factors affecting their distribution.

2. Experimental investigations should include:

(a) an introduction which explains the basis on which the experiment is founded.
(b) an account of the method you used written in the past imperfect tense.
(c) a report of the observations (results), and where appropriate
(d) a discussion of the findings; this could include an explanation of why the results were not as anticipated and suggestions as to how the method could be improved.

Diagrams of apparatus should be neat (use a ruler to draw straight lines!), drawn in pencil and labelled or annotated where relevant.

4. Results presented in tabular, graphical and pictorial forms.
5. Discussion of results.
6. Bibliography—list of references used.

SUGGESTED PRACTICAL EXERCISES

In the space available it is impossible to do other than offer a limited range of suggestions. Many other practicals are usually included in a complete 'A' level course. Nevertheless many of the exercises that have been chosen are similar to those which often appear in practical examinations. They are vehicles for developing the range of practical skills which are commonly assessed and are designed to be thought-provoking. Ideas for extension work are offered in many instances. The exercises centre around important biological themes, frequently drawing on material from more than one chapter. Several examples are presented in detail to offer an insight into the degree of practical skill development expected.

The resources for carrying out the practicals are typical of those expected in school and college laboratories.

1 CELLS AND TISSUES

a. Microscopic observation of the structure of plant and animal cells

1. Using a pair of forceps carefully remove a leaf from near the tip of a shoot of the pondweed *Elodea*. Transfer the leaf on a brush to a drop of glycerine on a microscope slide. Lower a cover slip onto the leaf, taking care not to trap any air bubbles. Examine the preparation under the low power (× 10) objective of a compound microscope. Find a few cells near the edge of the leaf and switch to high power. Make a labelled drawing of three adjacent cells. Remember to include an appropriate heading and the scale of magnification (this is the magnifying power of the eyepiece lens multiplied by that of the objective).

2. Examine a commercially prepared, stained section of mammalian liver, first under low power then high power. Make a labelled drawing of three adjacent cells. Include a heading and scale of magnification.

3. Tabulate the observed differences in structure of the two types of cell.

4. Use an eyepiece graticule to measure the length and width of an *Elodea* leaf cell.

b. Electronmicrograph study

Examine Fig A3.1, an electronmicrograph of a thin section through a mammalian liver cell.

Fig A3.1

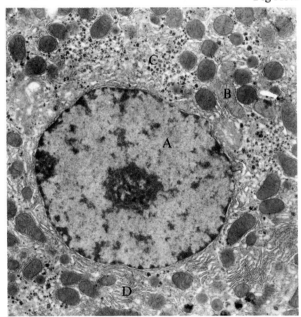

1. Identify the structures labelled **A**, **B**, **C** and **D**.

2. Comment on the functional relationship between these structures.

3. Given that the magnification of the electronmicrograph is 30 000, calculate the actual length of **D**.

c. Nuclear division—chromosome squashes

Excellent commercially prepared microscope slides of mitosis in root tips and meiosis in anthers and invertebrate testes are available for studying the various stages of nuclear division. Nevertheless students can be successful in making their own, at the same time acquiring valuable practical experience at making temporary, stained microscopic preparations. For mitosis a supply of young, actively growing root tips is required. These can be obtained by standing an onion bulb over water, such that the base of the bulb is about half an inch above the water surface. Small bulbs used for growing onions are best. Large stored bulbs are often treated chemically to inhibit root growth.

1. Allow the roots to reach 3–6 cm long. Remove several and cut off the tip, the first 5 mm from the apex.

2. Fix the tips overnight in glacial acetic acid and absolute alcohol (1:3 by volume).

3. Immerse the root tips in 1.0 M HCl and aceto-orcein stain (1:10 by volume) for about 5 minutes at 60°C in a water bath.

4. Transfer the stained tips to a few drops of the stain on a microscope slide. Carefully lower a coverslip over the preparation and cover it with a piece of folded filter paper. Gently press down on the coverslip to squash the tips.

5. If necessary, irrigate the squash with a little more stain to eliminate any trapped air bubbles. The same technique can be used to prepare squashes of locust testes and pollen sacs from lily flower buds.

6. Make a labelled drawing of as many stages as you can see in both forms of nuclear division. Place the diagrams in an appropriate sequence. Fill in any missing stages from commercial preparations.

7. Tabulate the main observed differences in mitosis and meiosis.

d. Osmotic gain and loss of water by plant cells

1. Weigh out and dissolve 34.2 g of sucrose in 100 cm³ distilled water. Place 20 cm³ of this 1.0 M solution in a boiling tube. Use the remainder to make 20 cm³ samples of 0.8, 0.6, 0.4 and 0.2 M solutions by dilution with distilled water.

2. Into each tube place three strips cut from the same potato tuber. Each strip must be exactly 5.0 cm long and approximately 2.0 mm wide and thick. Leave for 20 minutes, then remove the strips and remeasure their lengths.

3. Calculate the mean gain or loss in length for each solution. Plot a graph of molar concentration (horizontal axis) against mean change in length. From the graph calculate the molar concentration of sucrose which would cause no net osmotic loss or gain of water.

Questions
1. What volumes of 1.0 M sucrose and distilled water did you use to make the 0.4 M solution?
2. What is the reason for having three strips taken from the same potato in each solution?
3. Why were the strips left for 20 minutes before their lengths were remeasured?

Extension work
This exercise can be made open ended, for example by investigating the effect of a range of solutes – some electrolytes, others non-electrolytes; some organic, others inorganic – on the osmotic uptake or loss of water by a variety of plant tissues.

e. Determination of the water potential of plant tissue

1. Dissolve 34.2 g sucrose in 100 cm³ water to make a 1.0 M solution.

2. From the 1.0 M sucrose solution prepare 0.8, 0.6, 0.4 and 0.2 M solutions by dilution with water.

3. Place 3 or 4 leaves of *Elodea* or pieces of coloured epidermis of rhubarb petiole in 10 cm³ of each solution and leave for 20 minutes. Make sure the tissue is covered by the solution.

4. Transfer the tissue together with a few drops of the solution in which it was immersed onto a microscope slide. Carefully apply a cover slip so as to avoid trapping any air bubbles.

5. Observe 50 cells from each preparation under the microscope and record the number showing evidence of plasmolysis.

6. Plot a graph of % plasmolysed cells (vertical axis) against molarity of solution. From the graph read off the concentration at which 50 % of the cells are plasmolysed. At this concentration the water potential of the tissue is the same as the solution.

7. Knowing that a 1.0 M solution of sucrose has a water potential of -22.4×101.32 kPa at STP, calculate the water potential of the tissue.

f. Investigation of the effect of temperature on the permeability of the cell membranes of beetroot

1. Cut a raw beetroot into slices measuring 5 cm long by 1 cm $\times$ 1 cm. Wash in running water overnight until no more red pigment escapes from the damaged cells.

2. Place one slice in 15 cm^3 of water at 75 °C for 5 minutes. Remove the tissue and measure the absorbance of the solution in a colorimeter using a blue filter. Zero the instrument with water.

3. Repeat the procedure at 5 °C temperature intervals down to 30 °C. If a colorimeter is not available a subjective visual estimate of the intensity of the red pigment in each solution can be made.

4. Draw a graph of absorbance (vertical axis) against temperature.

5. Discuss your observations.

Extension work
The procedure can be modified to investigate the effect of different ions on membrane permeability. For example, 0.1 M solutions of NaCl and CaCl$_2$ separately and in combination can be used at a temperature of 20 °C. Alternatively the effect of different solvents such as ethanol, glycerol and propanone can be investigated.

2 BIOCHEMISTRY

a. Testing for reducing sugars

1. Place 5cm^3 of 1 % glucose solution (aqueous) in a test tube and add a similar volume of Benedict's reagent. Set up a control containing 5 cm^3 water instead of glucose.

2. Heat both tubes for 5 minutes in a boiling water bath. Observe and record any colour changes.

3. Repeat with 1 % aqueous solutions of fructose, lactose, maltose and sucrose. Explain any observed differences. What are the limitations of the test in investigating the presence of reducing sugars in cell extracts?

4. Incubate 5 cm^3 of 1 % aqueous sucrose solution with 5 cm^3 0.1 % invertase solution for 20 minutes at 37°C. Test the reaction mixture as above and explain any observed difference. What is an appropriate control for this part of the procedure? In what other way could the sucrose have been treated to achieve a similar result?

5. Mix equal quantities of 1 % glucose and sucrose solutions. Test as above and filter off the precipitate. Repeat the test on the filtrate until a precipitate fails to form. Boil the filtrate for 10 minutes with dilute HCl and neutralise with solid NaHCO$_3$. Once again test with Benedict's reagent. Explain your observations. Why is it necessary to ensure that the initial filtrate does not give a precipitate before boiling with HCl?

b. Digestion of starch by amylase

1. Mix 5 cm^3 of 1.0 % starch suspension in phosphate buffer pH 7.0 with a similar volume of 0.1 % amylase also made up in the same buffer. Set up a control containing 5 cm^3 of buffer solution instead of amylase. A second control containing the same volume of boiled amylase can also be included.

2. Incubate the tubes at 37 °C. At 30 second intervals remove drops of the reaction mixtures and add to separate drops of dilute iodine in potassium iodide solution on a white tile. Note the time required for the blue-black coloration failing to appear in the test and control mixtures.

Questions
1. What is the reason for using a buffer of pH 7.0 and for incubating the mixturest at 37 °C?
2. What other factors may have affected the activity of amylase?

Extension work
The exercise can be repeated at a range of temperatures and pH to ascertain the optimum conditions for amylase activity. Design and carry out one of these experiments showing clearly how one factor at a time can be varied whilst maintaining all other conditions at their optima.

c. Separation and detection of amino acids by paper chromatography

Throughout this exercise the chromatogram should be handled at the corners only. Otherwise the amino acids in sweat will show up.

1. Place enough developing solvent in the tank to give a depth of 1 cm and replace the lid.

2. Draw a starting line in pencil 2 cm from the bottom of a large piece of chromatography paper and on it mark four origins 3cm apart.

3. To the first origin apply a drop of aspartic acid using a capillary pipette. Apply leucine to the second origin, lysine to the third and a mixture of all three amino acids to the fourth, using a new pipette in each case. Keep the spots no more than 5 mm in diameter.

4. Place the chromatogram in the tank, origins lowermost, ensuring that the developing solvent is just below the starting line. Some tanks have a rack to support the paper, in others the paper is folded into a cylinder held by paper clips.

5. Allow the solvent to rise to about three quarters of the way up the paper. Remove the chromatogram and mark the position of the solvent front.

6. After allowing the chromatogram to dry, place it in a flat tray containing the locating agent.

7. Place the chromatogram in a hot-air oven at 100 °C for about 2 minutes until the purple coloured spots appear.

8. Measure the distance moved by each amino acid and calculate the R_f values.

Extension work
The method can be used to detect amino acids in cell extracts such as orange, lemon or tomato juice. Such extracts may need treatment before use to remove proteins and cell debris.

This can be achieved by adding absolute ethanol to the juice (3:1 vol./vol.) then centrifuging the mixture to sediment the precipitated components. The supernatant liquid will contain the amino acids. Changes in the amino acids as fruits ripen could be investigated in this way.

Developing solvent: ammonia : ethanol : water
(10:10:80 by vol.)

Locating agent: 200 mg ninhydrin in 100 cm³ propanone (acetone)
Amino acids: 0.1 % solutions in the developing solvent.

d. Protein assay by the biuret method

1. Prepare a standard protein suspension by dissolving 0.5 g albumen in 100 cm³ water. A few drops of 1 M KOH will help the albumen dissolve.

2. Use the standard to make albumen solutions containing 0.4, 0.3, 0.2 and 0.1 g in 100 cm³ water.

3. Place 2 cm³ of each concentration into a test tube, add 3 cm³ of quantitative biuret reagent, mix and warm at 37°C for 10 minutes.

4. Allow the mixture to cool before transferring each to a separate cuvette. Measure the absorbance of each protein concentration at 540 nm in a colorimeter after zeroing the instrument with a blank.

5. Draw a calibration graph of absorbance (vertical axis) against protein concentration.

6. Repeat the procedure using raw egg white diluted 1 in 100 parts of water. From the calibration graph read off the protein concentration in the diluted egg white. Calculate the actual concentration of protein in the egg white before dilution.

Questions
1. What is an appropriate blank for this procedure?
2. What is the main limitation of this method for assaying a specific protein in a biological fluid, e.g. albumen in blood serum?

e. Investigation of dehydrogenase activity in yeast cells

Various indicator dyes can be used to investigate redox reactions catalysed by dehydrogenase enzymes. During the reactions the dyes become reduced and change colour. Among the dyes commonly used for this purpose are:
(i) dichlorophenol indophenol (DCPIP) and methylene blue which change from blue to colourless,
(ii) tetrazolium chloride (TTC) which changes from colourless to pink.

1. Incubate 10 g of dried bakers' yeast in 100 cm³ of 1 % glucose solution at 30 °C overnight.

2. Place 10 cm³ of the yeast suspension in a clean test tube and add 1 cm³ of 0.5% TTC.

3. Incubate in a water bath at 25 °C and note the time required for a pink colour to develop.

4. Repeat with a sample of boiled yeast suspension. Explain your observations. What is the purpose of boiling the yeast in the second tube?

Extension work
Design procedures for investigating the effect of temperature and pH on the activity of yeast dehydrogenase. The procedure can also be used to investigate dehydrogenase activity in germinating seeds. Sites of dehydrogenase activity can be detected by cutting the seeds lengthwise and applying the redox dye. Germinated maize and pea seeds are suitable subjects. 1 % solutions of DCPIP and methylene blue are recommended.

f. Investigation of catalase activity in plant and animal tissues

Catalase breaks down hydrogen peroxide to water and oxygen. The evolved oxygen causes effervescence of the reaction mixture.

1. Place 5 cm³ of 2 % aqueous hydrogen peroxide in a clean test tube. Add a cube of fresh liver measuring approx. 1 cm × 1 cm × 1 cm. Note the amount of effervescence and place a glowing splint into the mouth of the tube. Record your observations.

2. Repeat using the same amount of liver which has been finely chopped and a piece of boiled liver. Explain the differences in your observations.

Extension work
The same procedure can be used to compare catalase activity in pieces of storage organs such as potato and carrot. Differences in catalase activity of radicles and plumules of germinating seeds can also be investigated in this way.

3 GAS EXCHANGE AND BREATHING

a. Measurement of the rate of oxygen uptake by germinating seeds

1. A simple respirometer (Fig 5.24) can be used for this exercise. Weigh about 10 g of pea seeds which have been soaked for 24 h in water, then spread out on moist cotton wool for 2 to 3 days. Record the exact weight of seeds used. Place the seeds in the gauze basket which fits into the respiration chamber. Immerse the basket and seeds in 100 cm³ water in a 250 cm³ measuring cylinder. The increase in volume is the volume of basket and seeds (A).

2. Place 10 cm³ (B) of 15 % aqueous KOH solution in the respiration chamber. Insert a filter paper wick in the solution as shown in the diagram. Lower the basket containing the seeds into the chamber, taking care to avoid them touching the KOH solution.

3. Add A + B cm³ of KOH solution to the compensation tube, together with a filter paper wick of the same size as used in the respiration chamber.

4. With the screw-clip valves open, join the tubes to the manometer. The index fluid in the manometer should be of the same height in both arms. Note the position of the plunger in the syringe. Immerse the tubes in water at 25 °C.

5. After 10 minutes, close both clips simultaneously. When the index fluid is near the top of the scale, depress the plunger to return it to the starting point. The change in volume recorded on the syringe is the volume of oxygen taken up by the seeds. Calculate the volume of oxygen absorbed in mm³ g⁻¹ tissue h⁻¹. Alternatively the change in height of the manometer fluid (h) can be recorded and the volume of oxygen absorbed calculated, using the formula $\pi r^2 h$, where r = radius of the bore of manometer.

Questions
1. What is the function of the compensation tube?
2. What is the purpose of immersing the tubes in water at 25 °C and waiting for 10 minutes before starting to take readings?
3. How could the experiment be modified to determine the respiratory quotient (RQ) of the seeds?

b. Measurement of the rate of oxygen production by photosynthesising tissue

1. The equipment shown in Fig 5.16 can be used to determine the number of bubbles of oxygen evolved in a given time. This is a measure of the relative rate of photosynthesis and provides a rapid method of comparing the effects of various factors on the photosynthetic rate. To determine the volume of oxygen evolved a photosynthometer (Fig A3.2) is required.

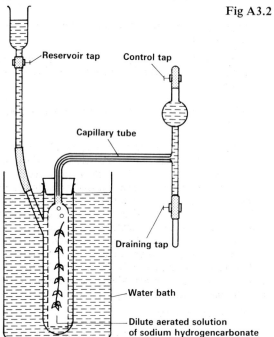

Fig A3.2

2. Using whichever method is available, place a 0.1% aqueous solution of sodium hydrogencarbonate ($NaHCO_3$) in the tube. It is preferable to aerate the solution in advance, so that it is saturated with oxygen. This ensures that the oxygen produced by the photosynthesising subject is evolved instead of dissolving in the solution.

3. Using a sharp razor make a clean cut across the lower end of a fresh green shoot of the pondweed *Elodea*. Immediately immerse the shoot upside down in the tube of $NaHCO_3$ solution.

4. Place the tube in a glass water bath at 25°C and position a lamp close to the latter. Constantly check the temperature of the water, and if necessary add cold water to keep it at 25°C.

5. If you are using the bubbling method, record the number of bubbles evolved each minute over a 10 minute period and calculate the mean relative rate of oxygen production. If using the photosynthometer, allow the evolved oxygen to collect at the top of the tube for about 10 minutes then slowly open the reservoir and drain taps to push the gas into the capillary tube. Measure the length of the gas column (l) and calculate the volume using the formula $\pi r^2 l$ where r = radius of the bore of the tube. The shoot can be weighed after removing water from its surface by gently dabbing it with absorbent paper. The rate of photosynthesis can then be calculated and expressed as mm^3 oxygen evolved g^{-1} tissue h^{-1}.

Questions
1. What measures are taken to ensure that no factor is limiting the photosynthetic rate?
2. What is the difference between gross and net photosynthesis?
3. Does either method measure gross or net photosynthesis? Explain.

Extension work
Either method can be extended into an open-ended exercise whereby the effect of temperature, light intensity, wavelength of light and concentration of CO_2 on the rate of oxygen evolution can be investigated. The exercise would be a useful test of experimental design, in this case of how to vary one factor at a time whilst keeping all other factors optional. It would also provide a useful insight into the law of limiting factors.

c. Investigation of some factors affecting stomatal opening and closing

1. Immerse a piece of a lettuce leaf in neutral distilled water in a petri dish and place under a bench lamp. Repeat with another lettuce leaf but this time place the dish in a dark cupboard. A crispy variety of lettuce such as 'Iceberg' is best suited for this experiment as its epidermis can be easily peeled off.

2. After 20 minutes, remove a piece of the epidermis from each and immediately plunge it into absolute alcohol. This treatment fixes the stomata in the same state as they were in life—either open or closed.

3. Observe 25 stomata on each epidermal strip and record the number open and closed. Calculate the percentage open and closed in each case.

Questions
1. What factor is most likely to have affected the opening or closing of the stomata in this experiment?
2. How could the design of the experiment have been improved to ensure that no other factor influenced stomatal movement?
3. How could the method be modified to investigate the effect of pH on stomatal opening and closing?

d. Investigation of the release of carbon dioxide by respiring organisms using bicarbonate indicator

Bicarbonate indicator consists of an aqueous solution of sodium hydrogencarbonate (0.001 M) containing two pH indicator dyes—cresol red and thymol blue. Carbon dioxide evolved by organisms immersed in the solution, dissolves in the surrounding water to form carbonic acid, causing a fall in the pH of the solution. The increase in acidity results in a colour change of: purple→red→orange→yellow.

1. Place 10 cm^3 of bicarbonate indicator solution in a boiling tube together with 5 pond snails. Stopper the tube and stand it near a bench lamp. Observe any colour change every 5 minutes for half an hour.

2. Repeat using 2 or 3 shoots of *Elodea* and a mixture of pond snails and *Elodea*.

3. At the same time keep a control tube without living organisms.

4. Record your observations and attempt to explain them.

Extension work
The exercise can be made open-ended by performing it in the absence of light and at different temperatures. The compensation point of *Elodea* can be investigated using different intensities of light, and by noting the minimum intensity to bring about a change of yellow→purple of indicator in which shoots, previously kept in the dark overnight, are immersed.

e. Locust ventilation rate

Insects ventilate their tracheal systems by means of body movements. Air is drawn in and out of the tracheae through the spiracles along the sides of the abdomen. The most rapid ventilation is caused by flight movements. The ventilation rate can be determined by counting the number of abdominal movements in an insect such as a locust.

Carefully place a healthy locust into a glass tube just big enough to accommodate it without squeezing its body.

Fig A3.3

1. Design an experiment to determine the effect of carbon dioxide on the locust's ventilation rate. You can expose the locust to carbon dioxide by gently breathing into its glass tube.

2. What controls are necessary to ensure that any observed effect is due to carbon dioxide and not just to blowing air through the locust's tube?

3. Use the same apparatus to investigate the effects of temperature on the locust's respiratory ventilation rate. Again, incorporate suitable controls into the experimental design.

4. Discuss the significance of the results of your experiment to the life of locusts in the wild.

4 TRANSPORT SYSTEMS

a Investigation of some factors affecting the rate of water absorption by a leafy shoot

1. A potometer (Fig 12.10) can be used for this investigation. The rate at which the shoot absorbs water is usually proportional to the rate of transpiration. A shoot which was cut from a woody plant such as *Rhododendron* is preferred. The stem of the shoot should be bent under water while it is cut and the cut end kept in water before use.

2. Fill the potometer by immersing it in a sink of water. Make sure there are no air locks anywhere in the apparatus.

3. Lower the cut end of the shoot under the surface of the water and cut off the last few inches using a pair of sharp secateurs so that the stem just fits into the rubber sleeve. Insert the stem into the sleeve and close the screw clip.

4. Remove the potometer from the sink and mop away any water using absorbent paper. Allow the apparatus to stand for 15–20 minutes before taking any readings. Transpiration and hence water absorption can be stimulated by placing a fan 18 inches from the shoot to create a current of cool air.

5. Measure the rate of water uptake by recording the distance moved by the air–water meniscus in the capillary tube about every 2 minutes. When the capillary tube is filled with air open the screw clip to allow water from the reservoir to fill the tube. Close the screw clip and continue to take readings. Plot a graph of distance moved by the meniscus (vertical scale) against time. Not the air temperature and humidity in the vicinity of the shoot using a wet and dry bulb hygrometer. The speed of the air current can be measured using an anemometer.

Questions
1. What is the reason for cutting the shoot from the parent plant under water?
2. What method would be suitable for determining the rate of water loss by an intact plant? Would the method also be appropriate for a cut shoot?

Extension work

The effect of various factors on the rate of water absorption can be investigated by altering such parameters as:

1. Air movement—the same procedure could be followed with the fan switched off, or with the speed of air current increased or decreased.

2. Air humidity—the shoot could be enclosed in a polythene bag or a fine mist of water could be sprayed over it.

3. Air temperature—a fan which blows warm air could be used.

4. Blocking the stomata—by smearing petroleum jelly onto the leaf surface.

Consideration should be given to altering only one parameter at a time if possible if the effects of individual factors are to be determined.

b. The effects of various factors on the heart rate of *Daphnia*

Use a Pasteur pipette to carefully place a healthy specimen of *Daphnia* into the depression of a cavity slide. Make sure the animal has enough water in which to swim freely when a cover slip is placed over the cavity. Do not harm the animal; it can be returned to its habitat after the experiment.

Use the low power of a microscope or a binocular dissecting microscope to locate the *Daphnia*. After a while the animal should slow down enough for you to be able to see the heart through its exoskeleton. (Look at figure 29.51)

The heart rate can easily be determined by using a stop clock to time 10 or 20 beats. A simple calculation can then be used to find the rate in 'beats per minute'.

A. Temperature effects

1. Devise a method by which you can obtain a fairly constant temperature in the cavity of the slide. You could do this by placing the slide on a thermostatically-controlled hotplate. Alternatively you could place the slide in a petri plate and float the plate on heated (or cooled) water in a beaker.

2. Measure the *Daphnia*'s heart rate at four or five different temperatures within the range 5 °C–35 °C. Measure the heart rate three times at each temperature in order to check reliability. You should obtain more or less the same rate each time. If the results are very different at the same temperature then something is wrong and you will need to look for sources of error.

3. Record the average heart rate at each temperature and plot this as a graph of rate against temperature.

Questions
1. Are there any trends which are displayed by the shape of the curve?
2. What is the relationship between increase in rate and each 10 °C rise in temperature? This is called Q10.
3. Is the Q10 for *Daphnia* heart rate similar to that for enzymes (Chapter 3)? Explain any similarities.
4. Write brief notes on the significance of the effects of temperature on heart rate in *Daphnia*.

B. Effects of various substances

1. Add to the cavity containing the *Daphnia* a drop of the following substances one at a time:

  70 % ethanol
  dissolved aspirin, weak solution
  dissolved aspirin, strong solution

Allow each substance to flow underneath the coverslip. Replace the water in the cavity between each addition so as to assess the effect of each substance separately.

2. Report the effects which each addition has on the heart rate.

3. List the main sources of error in the experimental design. Suggest ways in which the experiment can be improved, especially in terms of reliability. What controls could you incorporate into the method?

c. Dissection of mammalian heart

Use the fresh heart from a pig or sheep. This can be obtained from a local butcher.

1. Distinguish between the dorsal and ventral surfaces of the heart. Identify:
   (a) right and left atria
   (b) right and left ventricles.

2. Find the following vessels:
   (a) pulmonary arteries
   (b) aorta
   (c) pulmonary veins
   (d) anterior and posterior venae cavae
   (e) coronary artery and vein.
   (Look at Fig. 9.3(a).)

3. Make a labelled drawing of a ventral view of the heart showing the features referred to in 1 and 2.

4. Remove the ventral walls of the atria and ventricles. Identify
   (a) atrio-ventricular valves
   (b) semi-lunar valves
   (c) chordae tendineae
   (d) papillary muscles.
   (Look at Fig. 9.3(b).)

5. Note the difference in thickness between the walls of the:
   (a) atria and ventricles
   (b) left and right ventricles.

6. Make a labelled drawing to show the internal features referred to in 4 and 5. On your drawing use arrows to indicate the direction of blood flow.

7. Fill the ventricles with water and gently turn the heart upside down. Does water flow from the ventricles into the atria?

8. Which chambers carry
   (a) oxygenated blood
   (b) deoxygenated blood?

9. What is the significance of the difference in thickness of the walls of
   (a) atria and ventricles
   (b) right and left ventricles?

10. What are the functions of the:
    (a) chordae tendineae
    (b) coronary blood vessels?

11. List the anatomical features of the heart which ensure one-way blood flow.

d. Comparison of artery and vein

Examine a prepared slide of mammalian artery and vein in transverse section.

1. Make a labelled plan drawing of the walls of the vessels to show
   (a) the thin inner tunica intima layer.
   Include the collagen and elastic fibres covered with a layer of squamous epithelium, the endothelium.
   (b) the thicker middle tunica media layer.
   Include the circular elastic fibres interspersed with collagen and smooth muscle fibres. Arteries have more smooth muscle in this layer than veins.

(c) the outer tunica adventitia layer.
This contains mainly collagen fibres.
(Look at Fig. 9.14.)

2. Use an eyepiece graticule to measure the relative thickness of each layer in both vessels.

3. What is the significance of the smooth layer of squamous epithelium lining the blood vessels?

4. (a) What are the main characteristics of elastin, collagen and smooth muscle fibres?
   (b) Use this information and your measurements referred to in 2 to discuss the significance of the differences in wall structure of arteries and veins.

5. Semi-lunar valves occur at regular intervals in veins but not in arteries. What is the significance of this difference? (Look at Fig 9.17)

6. The walls of capillaries consist of only an endothelium made up of a single layer of flat cells. How does this structure relate to the functions of capillaries?
(Look at Fig 9.16)

5 HOMEOSTASIS

a. Contractile vacuole activity in Protozoa

Fresh-water *Protozoa* absorb water from their surroundings by osmosis. They collect the water into vacuoles which swell as the water accumulates. See Chapter 28. When the water vacuole reaches a critical size it suddenly empties its contents to the surrounding environment. For this reason they are called contractile vacuoles. The whole process begins again and in this way fresh-water *Protozoa* regulate the water content of their bodies.

It is possible to observe the actions of the contractile vacuoles of a variety of *Protozoa* such as *Paramecium* and *Amoeba*. (Look at Figs 1.13, and 28.14). A convenient species is the sessile ciliate called *Carchesium*. This can be found in colonies attached to the leg joints of the fresh-water shrimp *Gammarus*.

Carefully place a healthy specimen of *Gammarus* into the depression of a cavity slide. Make sure the cavity is filled with water and place a cover slip over it. Handle the *Gammarus* gently; it can be returned to its habitat after the experiment.

1. Locate a good specimen of *Carchesium*. Observe its contractile vacuoles and practice a technique for determining the rate of activity. This can be done by timing a fixed number of contractions. A simple calculation can then be used to find out the rate in 'contractions per minute'.

2. Design an experiment to determine the effect of water potential on the rate of vacuole contraction. You could expose the *Protozoa* to a series of different concentrations of sodium chloride solution.

   (a) You will have to fill the cavity on the slide with a new solution before determining the contraction rate.

   (b) Determine the contraction rate at four or five different concentrations between 0.1 % and 0.9 %.

   (c) Plot the rate of contraction against sodium chloride concentration.

   (d) Explain the shape of your graph.

   (e) What is the significance of your results to the life of *Carchesium* in its natural habitat?

b. Effects of temperature on water loss in woodlice

Terrestrial arthropods lose water by evaporation through their exoskeletons. The epicuticle contains wax or grease and minimises water loss but is not perfect in this respect.

You can measure the amount of water loss from terrestrial arthropods by determining the loss of mass of their bodies. To study the effects of temperature on water loss a constant relative humidity must be maintained. This is because humidity can also affect water loss by evaporation.

A constant relative humidity can be created in a experimental chamber by placing a saturated solution of a particular salt in the bottom of the chamber. A boiling tube is an ideal chamber.

A change in temperature also changes the humidity in the chamber, so a saturated solution must be selected which creates the required humidity at the temperature used.

Table A3.1 **Some saturated salt solutions and the effects of temperature on the % relative humidity above their surfaces** (from G. W. C. Kaye and T. H. Laby, Tables of physical and chemical constants, 13th edn., Longman, 1966 after Hickman)

Saturated salt solution	Temperature/°C										
	0	5	10	15	20	25	30	35	40	50	60
K_2SO_4	99	98	98	97	97	97	96	96	96	96	96
KNO_3	97	96	95	94	93	92	91	89	88	85	82
KCl	—	88	88	87	86	85	85	84	82	81	80
$(NH_4)_2SO_4$	83	82	82	81	81	80	80	80	79	79	78
$NaCl$	76	76	76	76	76	75	75	75	75	75	75
$NaNO_2$	—	—	—	66	65	63	62	62	59	59	
NH_4NO_3	77	—	72	69	65	62	59	55	53	47	42
$Mg(NO_3)_2$	60	58	57	56	55	53	52	50	49	46	—
K_2CO_3	—	—	47	44	44	43	43	43	42	—	—
$MgCl_2$	35	34	34	34	33	33	33	32	32	31	30
CH_3COOK	—	—	21	21	22	22	22	21	20	—	—
$LiCl_2$	15	14	14	13	12	12	12	12	11	11	11
KOH	—	14	13	10	9	8	7	6	6	6	—

(% R.H.)

1. Use the table above to select saturated solutions which create a fixed humidity over a temperature range of 5°C–35°C. Pick four or five different temperatures.

2. Place about five woodlice into each of the experimental chambers, one for each temperature. Weigh the woodlice (all together) to the nearest mg.

3. Allow the chambers to stand in incubators or water baths at controlled temperatures for 24 hours. Then weigh the woodlice again.

4. Express the mass loss as a percentage of the original mass. Plot a graph of mass loss against temperature.

5. Explain the pattern of your results.

6. Discuss the significance of your results to woodlice living in their natural habitat.

7. In what ways could the design of the experiment give misleading results? How could you improve the experiment in order to obtain more reliable results?

c. Structural modifications of hydrophyte and xerophyte leaves

You are provided with prepared slides of transverse sections through the leaves of water lily (a hydrophyte) and marram grass (a xerophyte)

1. Examine each section under the microscope. Note in particular the following points:
   (a) the relative amounts of supporting and transporting tissues
   (b) presence or absence of external protective tissues
   (c) relative numbers of stomata and their distribution
   (d) differences in leaf shape
   (e) presence or absence of aerating tissue.

2. Make labelled low power plan drawings to show the distribution of tissues in each leaf.

3. Using the observations made in (a), annotate each drawing to show clearly how the structure of the leaf is related to the habitat of the plant.

4. Describe a non-structural modification which would prevent excessive water loss from a xerophyte.

Questions
Some plants—called halophytes inhabit regions where the environmental water potential is very low (e.g. mud flats, salt marshes).
1. Outline the problems facing these plants in terms of water uptake.
2. Suggest modifications which would allow these plants a continuous water supply.

6 NUTRITION

a. Dissection of the mammalian digestive system

Use a good dissection guide to help you to dissect and display the digestive system of a preserved mammal such as a rat or mouse.

Dissect as much of the system as you can. Clear away any fat covering the various organs. Leave major blood vessels intact and all glands and organs which are attached to the digestive system. (See Chapter 15.)

1. Make a labelled drawing of your dissection. (Look at Fig 15.14.)

2. On the drawing name and label the following:
   (a) the organ in which hydrochloric acid is secreted
   (b) the gland that secretes enzymes which hydrolyse starch and polypeptides
   (c) the blood vessel which transports absorbed nutrients from the intestine to the liver
   (d) the organ whose internal surface area is greatly enlarged to facilitate absorption of nutrients
   (e) the tissue which holds the gut in position in the abdomen
   (f) the organ which stores carbohydrate and fat-soluble vitamins
   (g) the organ in which microbial digestion of cellulose occurs
   (h) the organ in which absorption of water from undigested food occurs
   (i) the organ which prepares undigested food for passage to the exterior.

b. The use of a model gut to investigate the action of amylase on starch

1. Prepare a dilute solution of amylase as follows: rinse out your mouth with clean water. Chew for a few minutes on a clean rubber band. Place a couple of 'dribbles' of saliva into a measuring cylinder. Dilute to $10\,cm^3$ with distilled water.

2. Take a 25 cm length of Visking tubing and tie a tight knot at one end. Fill the tubing to about three quarters of its capacity with 1% starch solution. Mark this level by loosely tying a piece of cotton around the outside of the tubing.

3. Top up the tubing so it is full of saliva solution. Be careful NOT to mix the two solutions. Fasten the open end of the tubing with a paper clip. Thoroughly rinse the outside of the tubing with water.

4. Place the filled Visking tubing into a test tube of distilled water. Make sure the end which is fastened with a paper clip is above the water level. Incubate the test tube at 37°C for at least 30 minutes.

5. Repeat the entire procedure for another piece of Visking tubing, but **mix** the saliva and starch solutions thoroughly before incubating the tubing.

6. When the mixtures have been incubated for the same period of time remove a sample of water from each tube and test it with Benedict's reagent and, separately, iodine solution. Record your results. Explain any differences in the results for the two tubes.

7. Add iodine solution to the water surrounding the Visking tubing and leave for 10 minutes. Record your results. Explain any differences in the results for the two tubes.

8. Write a brief discussion about the deductions which can be made from the experiment as a whole.

c. Investigation of the process of digestion in small invertebrates

It is possible to follow the course of digestion in small invertebrates such as *Paramecium* (Fig 28.16) or *Daphnia* (Fig 29.34).

Both animals will feed on yeast. The yeast can be stained with a dye such as Congo red by placing the yeast culture in a tube containing the diluted dye. This enables you to see the yeast clearly, even when it has been taken into the bodies of the experimental animals. In addition, the dye is an indicator which changes colour when the pH of the surroundings changes. As digestion proceeds the pH changes. Congo red changes from red at pH 5 to violet at pH 3 and yellow at pH 8.

1. Place a drop of water containing *Paramecium* or a few *Daphnia* into the depression of a cavity slide. Use a microscope to locate a good specimen. You can usually slow down *Paramecium* by placing some strands of cotton wool on the slide. They become trapped in the strands long enough to make your observations.

2. Note the mode of feeding displayed by the selected animal. The ciliary feeding currents of *Paramecium* are usually very clear.

3. Notice the gradual accumulation of yeast inside the animal's body. Record any changes in colour of the dye. Note the time taken for these changes to take place and, if possible, the position in the animal's body where the changes occurred.

4. Explain your observations.

d. Comparison of the skulls and dentition of a carnivore, a herbivore and an omnivore

Obtain the skulls of a carnivore such as a dog (Fig 15.4) and a herbivore such as a sheep (Fig 15.6). Make sure that the teeth are all in place.

1. Make labelled drawings of each skull as viewed from the side. Pay particular attention to the structure of the teeth.

2. Write brief annotations to go with your labels regarding the following:
(a) gripping and piercing food
(b) cutting and chewing food
(c) grinding vegetation
(d) the relationship between the type of food and the articulation of the lower jaw (look at Figs 15.5 and 15.8)
(e) the dental formulae of the two animals.

3. Make a labelled drawing of a human skull as viewed from the side (Fig 15.9).

4. Write brief notes comparing the dentition of the human, as an omnivore, with that of the carnivore and the herbivore.

Include in your account observations you have made from the skulls about points of attachment for muscles used to move the lower jaw during feeding.

e. Chromatographic separation of chloroplast pigments

This can readily be achieved on a small scale using the apparatus shown in Fig 5.5.

1. Pour 3 cm³ of the developing solvent propane (acetone) and petroleum ether (1:9 by volume) into a stoppered boiling tube.

2. Finely chop about 1 g of dark green leaf tissue into a mortar and add a pinch of sand. Finely grind the tissue as quickly as possible.

3. Transfer the ground tissue to a stoppered boiling tube, add 4 cm³ of propanone and shake vigorously for about 10 seconds. Stand for 10 minutes then add 4 cm³ water and shake again.

4. Add 3 cm³ petroleum ether (40–60°C) and shake vigorously once more. Stand to allow the pigments to separate into the upper layer. Remove this layer to a clean tube using a teat pipette.

5. Cut a piece of chromatography paper just big enough to fit the boiling tube. Using a pencil draw a starting line about 1 cm above the surface of the developing solvent and mark an origin at its centre.

6. Using a capillary pipette apply a sample of the extracted pigments to the origin, keeping the spot no bigger than 5 mm in diameter. Allow to dry at room temperature before adding further samples until the spot is dark green.

7. Place the chromatogram into the boiling tube, origin lowermost and replace the stopper. Allow to stand until the solvent front is near the top of the paper.

8. Observe and record the positions and colours of the various pigments.

f. Investigation of the Hill reaction

The reducing power developed in the light-dependent reactions of photosynthesis can be investigated by observing the conditions required for a chloroplast suspension to reduce the dye dichlorophenol indophenol (DCPIP) causing it to change from a blue to a colourless state.

1. Weigh out 50 g of fresh spinach or dark green cabbage leaves and chop into small pieces.

2. Place 100 cm³ of ice-cold phosphate buffer solution containing 6 g sucrose and 0.01 g KCl in a blender. Switch on the blender at half speed and add the chopped tissue over a period of 20 seconds. Now switch to full speed for 5 minutes.

3. Remove cell debris from the homogenate by filtering it through several layers of muslin into a beaker standing in ice.

4. Centrifuge the filtrate at high speed for 2 minutes to sediment the chloroplasts. Pour off the supernatant liquid and resuspend the chloroplasts in 10–15 cm³ of ice-cold buffer.

5. Transfer 0.5 cm³ of the chloroplast suspension to a clean test tube. Quickly add 8 cm³ of buffer solution and 0.2 cm³ of a 0.1% solution of DCPIP. Place near a bench lamp and note the time required for the mixture to become colourless. At the same time set up a control with an additional 0.5 cm³ buffer instead of choroplast suspension.

6. Repeat stage 5 in a dark cupboard.

7. Record your observations and attempt to explain any differences.

Question

What is the purpose of (i) including sucrose in the phosphate buffer, and (ii) keeping the homogenate ice-cold during the extraction process?

7 GERMINATION, GROWTH AND DEVELOPMENT

a. Investigation of the hydrolysis of food reserves in germinating barley seeds

1. Surface sterilise 20–30 barley seeds by immersion in 1% sodium hypochlorite solution for 5 minutes. Rinse the seeds in two changes of distilled water.

2. Soak the seeds for 24 hours, then place them between sheets of moist absorbent paper for 24–48 hours until the radicle begins to emerge.

3. Cut the seeds lengthwise in half and carefully dissect out the embryo and part of the aleurone layer from several seeds.

4. Place the embryos and pieces of aleurone layer on plates of starch agar (2% starch in 2% plain agar) as shown in Fig 23.5. Set up controls with boiled samples.

5. After 2–3 days, flood the plates with dilute iodine solution and record the colour reaction beneath each sample. Explain your observations.

Questions
1. What is the reason for surface sterilising the seeds?
2. Which hydrolytic enzyme is under investigation?

Extension work
The procedure can be repeated using milk agar (5% sterile milk in 2% plain agar) to investigate the hydrolysis of protein reserves.

b. Investigation of the effect of auxin concentration on the growth of cress roots

1. Thickly sow about 200 cress seeds on wet absorbent paper and leave for 24 hours at 25°C.

2. Make up 500 cm³ of an aqueous stock solution of indole-acetic acid (IAA) of concentration 1 mg dm⁻³. From this prepare a range of concentrations by dilution with distilled water. Concentrations of 0.1, 0.01, 0.001 and 0.00001 mg dm⁻³ should be suitable.

3. Place a filter paper in each of fifteen petri dishes. Add 5 cm³ of distilled water to three of the dishes. On each dish space out equidistantly 10 germinated seeds. Repeat with each of the concentrations of IAA.

4. Incubate at 25°C in the dark for 24 hours and measure the length of root of each seedling. Calculate the mean root

length and standard deviation. Plot a graph of the data (vertical axis) against the logarithm of the concentration of IAA.

Extension work
The procedure can be adapted to investigate the effect of auxin concentration on the growth of oat or wheat coleoptiles. The range of IAA concentration will need to be between 1.0 and 0.1 mg dm⁻³. The seeds are germinated by covering them with moist vermiculite for about 3 days at 25°C. The coleoptiles are then cut off at their base and the top 2 mm removed with a sharp razor. The next 10 mm is then cut off and ten such lengths are placed in 10 cm³ of each concentration of auxin, using distilled water as a control. After 24 hours at 25°C the lengths of the sections are again measured. Mean lengths are calculated and plotted as above. How can the method be used to determine an unknown concentration of auxin?

c. Measuring the rate of growth of the shoot of a potted plant

One method of measuring the growth rate of an organism is to record increases in its height against time. With the shoot of a potted plant this can be achieved using an auxanometer (Fig. A 3.4). The free end of the lever is longer than the part attached to the shoot, consequently it exaggerates actual increases in height, making it possible to record growth even though it may be occurring slowly.

Fig A3.4

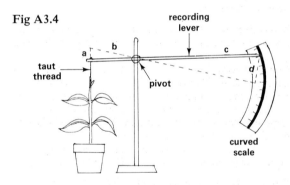

1. Carefully tie a piece of non-stretchable thread to the tip of the shoot, making sure it is not pulled too tight so as to cause any damage to the growth region.

2. Attach the free end of the thread to the short arm of the lever. Ensure that the thread is vertical.

3. Set up the vertical scale at the free end of the recording lever. It is not necessary for the pointer to start at the top of the scale.

4. Take readings every 24 hours. The actual increase in height after a given time (a) can be calculated from the formula (b × d) ÷ c.

5. Express your results as mm increase in height h⁻¹.

Questions
1. What are the advantages and limitations of this method as a means of measuring growth?
2. What other methods are available and how do they compare with the above in terms of reliability and utility?

d. Measuring the rate of growth of a yeast population

1. Dispense 15 cm³ of sterile 2% sucrose solution into a sterile test tube.

2. Add one drop of a 1 % suspension of fresh yeast in the same sucrose solution. If fresh yeast is not available it can be substituted by dry yeast which has been suspended overnight in the solution.

3. Stopper the tube with a cotton wool bung and incubate at 25°C.

4. Every 24 hours determine the total number of cells using a counting chamber (haemocytometer). If this is not possible the number of viable cells can be determined by preparing dilution plates with yeast extract agar and counting the number of colonies after 48 hours' incubation at 25 °C.

5. Plot a graph of number of cells cm^{-3} of sucrose solution against time. Use your graph to calculate the maximum rate of growth of the population. Explain any observed changes in the growth rate with time.

Extension work

The method can be easily modified to investigate the effect of environmental factors such as temperature, pH and different sugars on the growth rate. The effect of daily removal of $1 cm^3$ of the suspension and its replacement with a similar volume of sucrose solution can also be investigated.

e. Comparison of the functional histology of the stem and root of buttercup at the primary stage of growth

For this exercise you require permanent preparations of stained transverse sections of both organs.

1. Make plan drawings of both sections to show accurately the distribution of tissues as seen under low power of a microscope.

2. The drawings should be to the same scale.

3. Examine tissues under high power to assist identification but do not show individual cells in your drawings.

4. Annotate your drawings using the answers to the following questions.

Questions

1. What differences can be seen in the structure of the epidermis of the two organs? How do these differences relate to the environment and functions of the stem and root?

2. Compare the distribution of xylem and phloem in both organs. What are the main functions of these tissues? Comment on the relative amounts of xylem and phloem in each section.

3. Are there any observed differences in the presence of supporting tissues in the sections? Comment on the distribution of supporting tissues in the stem.

8 REPRODUCTION

a. A comparative study of mammalian ovary and testis

For this exercise you require prepared sections of a mammalian ovary and testis.

1. Make labelled drawings of the structures at:
   (a) low power (a plan diagram)
   (b) high power.
(Look at Figs 25.3 and 25.9.)

2. Write annotations to accompany the labels regarding the following:
   (a) the germinal epithelium from which the gametes originate
   (b) the developing gametes
   (c) mature gametes
   (d) means of release of gametes from the gonads
   (e) tissue which secretes sex hormones

3. (a) Tabulate the similarities and differences between ovaries and testes which you have been able to see in the slide preparations
   (b) Explain the significance of these similarities and differences

b. Dissection of the mammalian reproductive system

Use a good dissection guide to help you to dissect and display the reproductive system of a preserved mammal such as a rat or mouse. If you can only dissect a specimen of one sex make sure you study someone else's dissection of the other sex.

Dissect as much of the system as you can. Clear away any fat covering the various organs. Leave major blood vessels intact.

1. Make a labelled drawing of your dissection. (Look at Figs 25.1 and 25.8.)

2. On the drawing (if it is a male) name and label the following:
   (a) the sites of gametogenesis
   (b) the structures which ensure gametogenesis occurs at a temperature lower than that in the abdomen
   (c) urinary bladder
   (d) structures where sperm are stored before passing to the outside
   (e) site of semen ejaculation
   (f) gland which secretes mucus that helps copulation
   (g) the tube which delivers sperm into the female during copulation
   (h) the penis.

3. On the drawing (if it is a female) name and label the following:
   (a) the sites of gametogenesis
   (b) site where fertilization usually takes place
   (c) structure where the embryos develop
   (d) site where sperm are deposited during copulation
   (e) blood vessel which carries FSH to the ovaries
   (f) blood vessel which carries sex hormones from the ovaries to the rest of the body.

c. Investigation of flower structure and pollination mechanism of white dead nettle

1. Remove a mature, open flower from the inflorescence. Note the numbers of sepals. Are they free or fused? Is the calyx symmetrical or asymmetrical? Repeat these observations on the corolla. What is the colour of the petals? Does the flower have any scent?

2. Using a sharp safety razor cut the flower lengthways into two identical halves starting at the receptacle. Note the number of stamens in each half flower and the positions of the anthers. How are the stamens attached to the flower? Place an anther on a microscope slide and observe how it dehisces to release pollen. How many compartments are visible in the ovary? What do they contain? Note the length of the style and the position of the stigma.

3. Find out the kinds of insects that pollinate this species. What features attract insects to this flower? How is the flower structured such that cross-pollination will occur when such insects visit it?

4. Make an accurate drawing of one half flower and annotate it with the observations and information obtained in response to the above questions.

5. The white dead nettle belongs to the family *Labiatae*. Practise using the keys to this family in the *Excursion Flora* by Clapham, Tutin and Warburg (Cambridge University Press) to identify the genus and species to which it belongs.

Extension work

This procedure can be followed with any insect-pollinated flower. It can also be modified for use with wind-pollinated species.

d. Investigation of some factors affecting germination of pollen

1. Place a drop of 0.3 M sucrose solution on a cover slip and onto it brush a few pollen grains from a freshly dehisced anther. Invert the cover slip and lower it over the hollow in a cavity slide.

2. Place the slide in a petri dish lined with moist filter paper and incubate at 20 °C. Observe under the microscope every 20 minutes and note any signs of the emergence of a pollen tube. A drop of aceto-carmine may then be added to stain the pollen tube nucleus and male gametes.

3. Repeat with 0.1, 0.5 and 1.0 M sucrose.

Extension work

The procedure can be modified to include 0.001 % boric acid and yeast extract in the sucrose solution, separately or in combination. The percentage germination and growth rate of the pollen tubes can be compared in the various solutions.

9 SENSITIVITY

a. Investigation of kineses in woodlice

Choice-chambers are large circular dishes containing a number of compartments (Fig 31.22(a)). Different environmental conditions such as light and dark or different humidities can be established in the compartments.

Small invertebrates such as woodlice are placed in the centre of the chamber. They are allowed to move freely into and between any of the compartments.

1. Design an experiment to investigate the humidity preferences of woodlice. Water in the bottom of one compartment provides a humid choice; silica gel in another provides a dry choice.

Take account of the following points:

(a) one or two specimens may act in a peculiar way and not represent normal behaviour

(b) be very careful that you only give the animals one choice at a time

(c) make sure other environmental factors do not confuse your experiment—humidity must be the only variable

(d) devise controls to ensure your results are meaningful

(e) check the validity of your findings

2. Design and perform another choice-chamber experiment to investigate the light–dark preferences of woodlice.

3. Discuss the significance of your results regarding the lives of woodlice in their natural habitat.

b. Examination of the structure of the mammalian eye

Obtain a fresh sheep's or ox's eye from a local butcher or abattoir.

1. Make a labelled drawing of the external appearance of the eye. (See Chapter 19.)
(a) Why is the cornea transparent?
(b) Why is the sclera thick and tough?

2. Remove fat from the eyeball and expose the extrinsic eye muscles and the optic nerve.
What is the function of these muscles?

3. Make a circular cut around the edge of the cornea just in front of where it joins the sclera. (Look at Fig 19.1.)
(a) What is the watery fluid that emerges from the cut?
(b) What is the function of this fluid?

4. Deflect the cornea forwards. Identify the iris, the pupil and the front of the lens. Use a blunt needle to push the lens backwards and forwards and from side to side.
What restricts the movement of the lens?

5. Make a circular cut all round the eyeball so as to cut the eye into anterior and posterior halves. Remove the gelatinous vitreous humour.
What is the function of the vitreous humour?

6. In the posterior half of the eyeball identify the retina, the choroid and the sclera. Note blood vessels radiating from the point where the optic nerve leaves the retina.
(a) What are the functions of the choroid and sclera?
(b) What is the region called where the optic nerve leaves the retina?
(c) Just to one side of this region the fovea centralis should be visible. What is its function?

7. Look at the front half of the eyeball. The retina joins the ciliary body which forms a circular band around the lens. Gently move the lens and observe that it is attached to the ciliary body by delicate suspensory ligaments.
What part do these structures play in the accommodation of the eye for near and distant vision? Use sketches to illustrate your answer.

c. Investigation of phototaxis in *Euglena*

1. Prepare a light-proof cover to fit around a fresh culture of *Euglena* as shown in Fig A3.5.

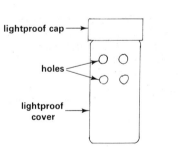

Fig A3.5

lightproof cap

holes

lightproof cover

2. Leave the first hole uncovered, stick a piece of tracing paper over the second hole, two pieces over the third and three pieces over the fourth.

3. Place the culture in a dark place for 24 hours before fitting the cover.

4. Stand the culture on a windowsill for a day with the holes facing the light. Notice the distribution of the *Euglena* on removing the cover.

Question

What is the purpose of applying different numbers of layers of tracing paper over the holes?

Extension work

The holes in the cover can be covered instead with coloured cellophane to investigate the phototactic response of *Euglena* to different wavelengths of light. The intensity of light passing through each colour of cellophane must be the same.

d. Investigation of phototropism in oat coleoptiles

1. Thinly sow some oat seeds in pots of moist seed compost. When the coleoptiles begin to emerge place one pot in a box which allows light to enter from one side only and another on a rotating clinostat (Fig A 3.6). Place both near a source of natural light and keep the compost moist by regular watering.

Fig A3.6

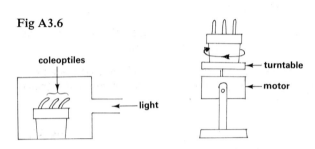

2. After 3 or 4 days note the direction in which coleoptiles are pointing. Explain your observations.

Questions
1. The box should be painted inside with matt black paint. Why is this so?
2. What is the purpose of rotating one of the batches of seedlings on the clinostat?
3. How can the method be adapted to investigate the effect of gravity on the direction of growth of the coleptiles?

Extension work

The procedure can be modified to investigate the effect of covering the coleoptile tips with light proof caps or of removing the tips. The coleoptiles can also be marked at 2 mm intervals from the tip backwards to determine which part enlarges most in the phototropic response.

10 MOVEMENT AND SUPPORT

a. Making a permanent stained microscopic preparation of striated muscle

Place a small piece of muscle from the thigh of a mouse onto a microscope slide. Add a drop of saline and tease the fibres apart using mounted needles. Now stain the fibres as follows:

1. Add a few drops of haematoxylin; leave for 2 minutes.
2. Wash in tap water.
3. If the preparation is very heavily stained, differentiate quickly in acid–alcohol.
4. Wash in tap water.
5. Counterstain with eosin for 1–2 minutes.
6. Differentiate in tap water.
7. Dehydrate by adding 30, 50, 70 % and absolute alcohol (2 minutes for each).
8. Clear in xylene to remove the alcohol.

9. Mount in Canada balsam and apply a coverslip. Label your preparation.
10. Make a labelled drawing of a few fibres from your preparation and state the colour each cell component is stained.

Questions
1. What is the purpose of:
   (a) differentiation
   (b) dehydration?
2. In what ways is the structure of striated muscle suited to its functions?

b. Investigation of supporting tissues in the stem of a flowering plant

For this exercise you will require permanent microscopic preparations of transverse and longitudinal sections of a dicotyledonous stem such as *Lamium* (dead nettle) or *Ranunculus* (buttercup).

1. Make a lower power plan diagram of the transverse section showing the distribution of collenchyma and sclerenchyma.
2. Draw three adjacent cells of both tissues as seen in transverse and longitudinal section.
3. Annotate your diagrams using the answers to the questions below.

Questions
1. Comment on the distribution of the sclerenchyma and collenchyma as seen in the transverse section.
2. In what ways are the tissues suited to their supporting role? Use evidence from your observations of both transverse and longitudinal sections.
3. What other tissues in the sections help to support the stem? In what ways is the supporting function of these tissues comparable to collenchyma and sclerenchyma?

c. Vertebrate skeletons

1. Obtain specimens of the following mammalian vertebrae: atlas, axis, cervical, thoracic and lumbar (Fig 16.5).
   (a) Make a large, labelled drawing of each vertebra.
   (b) Write annotations on your drawings which highlight the functions of the different parts in the vertebral column. Make particular reference to:
   (i) points of articulation with other vertebrae
   (ii) points of attachment for muscles
   (iii) load-bearing function
   (iv) flexibility of movement.

2. Obtain specimens of the fore- and hind-limb skeletons of several different mammals, e.g. dog, rabbit, mouse, human (Figs 16.9 and 16.10)
   (a) Make a large labelled drawing of each specimen.
   (b) Write annotations on your drawings which highlight the functions of the limbs in the lives of the animals from which they came. Make particular reference to:
   (i) strength and load-bearing function,
   (ii) flexibility of movement during locomotion
   (iii) points of attachment of muscles,
   (iv) relative proportions of the different components of the limbs
   (v) fusion of bones.

3. Examine a microscopical preparation of a section through compact bone (Fig 16.13).
   (a) Make a labelled drawing of the preparation.
   (b) List ways in which the histological structure of bone enables it to carry out its skeletal function.

744

11 INHERITANCE

Practical studies in genetics at this level are limited by time and the expertise required to handle and maintain cultures of the test organisms. The following suggestions which can help overcome such difficulties are based on the use of tomato seeds which can be easily and quickly germinated. The ratios of seedlings with specific traits are determined and provide a basis for statistical analysis. The seeds are obtainable in packaged form from biological suppliers.

a. Monohybrid inheritance – stem colour

The seeds are in two batches. The first is from self-pollinated F_1 plants which were obtained by crossing a pure-breeding purple stemmed variety with a green-stemmed variety.

1. Sow the seeds in moist seedling compost.

2. When the seedlings have reached the two-leaf stage count the number with purple stems and the number having green stems.

3. Use your data to calculate the F_2 monohybrid ratio.

4. Perform a χ^2 test on the results to determine whether or not the observed ratio is significantly different from that expected.

5. Repeat the exercise on the second batch of seeds which was taken from F_1 plants which were back-crossed to the pure-breeding green stemmed variety.

b. Dihybrid inheritance – stem colour and leaf shape

Again, material provided consists of separate packs of seeds obtained from self-pollinated F_1 plants and from F_1 plants back-crossed to the double homozygous parent. The same procedure is followed but this time a dihybrid ratio is calculated and statistically analysed.

c. Incomplete dominance – colour of stem and cotyledons

Separate packs of F_2 seeds and those obtained from a backcross are provided to investigate the inheritance of alleles which show incomplete dominance. Give a genetic explanation of your results.

d. Linkage

Loose linkage can be detected among three traits – stem and leaf colour and leaf shape. Tight linkage can be investigated using the traits of stem colour and seedling vigour. Use your results to prepare chromosome maps of the alleles in question.

As an alternative, prepared cobs of corn (maize) are available for investigating mono- and dihybrid patterns of inheritance, including back-crosses. Colour of the aleurone and seed texture are two traits which can be readily followed. Segregation of the trait for waxy pollen in maize can be seen by observing microscopically and counting the number of stained pollen grains after treatment with an aqueous solution of iodine. Grains without a waxy coat absorb the iodine solution and starch inside them is stained blue. Waxy pollen repels aqueous solutions and remains unstained.

12 ECOLOGY

a. Field-study techniques

The aims of the study must be determined as this will determine the techniques employed. The amount and types of data collected will influence the ability or otherwise to draw valid conclusions. The following techniques are typical of those used in terrestrial habitats.

A. Mapping the area

1. Use an ordnance survey map (large scale) to locate the site and record the grid reference. Instructions on how to do this are given on the map.

2. Place several vertical poles around the site. Lie a long piece of rope in a square around the site.

3. Measure and record the distances between the poles and the rope. Use the data to outline the site on squared paper (Fig A3.7)

Fig A3.7

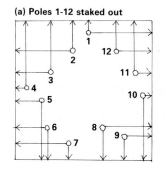
(a) Poles 1-12 staked out

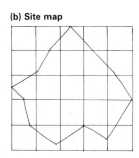
(b) Site map

B. Sampling the area

As many areas are too large to study everything within them it is necessary to select sample sites. Transects and quadrats are most frequently used for this purpose.

The following methods are appropriate for studying most populations of plants and of fixed animals such as barnacles and limpets on rocky shores.

1. **Line transects** are especially useful in areas where there is zonation as on rocky sea shores and where there are obvious changes in vegetation or animal species. A tape is laid on the ground across the area to be sampled and the species touching the tape are recorded in sequence from one end of the line to the other. Where dense populations occur and the line is long it is expedient to record species every metre or less along the transect.

2. **Belt transects** are formed by having two parallel line transects 1.0 m apart. The belt is then divided into squares of area 1 m^2 and the species in each square recorded. If the belt is long, recording say every fifth square would be time saving.

3. **Quadrats** are square frames usually ranging in size from 0.25 to 1.0 m^2. They are laid on the ground and the species inside the square are noted. Enough quadrats should be used to ensure that all species in the area are recorded. This can be

determined by plotting a graph of no. of species against no. of quadrats (Fig A3.8).

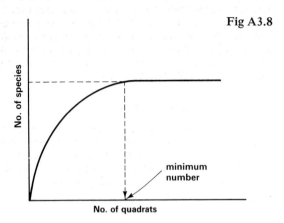

Fig A3.8

No. of species

minimum number

No. of quadrats

Small quadrats are suitable for sampling lichens on a wall or tree trunk. For most sites the larger quadrats are appropriate. The optimum size of quadrat can be determined by trial and error (Fig A3.9). In woodland and forest very large quadrats are required for recording the distribution of shrubs and trees.

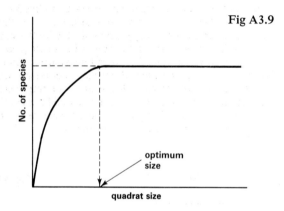

Fig A3.9

No. of species

optimum size

quadrat size

It is desirable that **random sampling** is carried out when using quadrats so as to avoid bias in the results which could arise from subjective choice of sites. A simple way of doing this is to divide the area into numbered squares (Fig. A3.10) and pick several numbers out of a hat. Sampling is then performed in the squares chosen. In practice this is time consuming and a reasonable alternative is to blind-fold a colleague, turn him/her around so he/she is unaware of his/her bearings, and get him/her to throw a stick over his/her shoulder. The place where the stick lands is the first sampling point. The procedure is repeated for subsequent samples.

Fig A3.10

			29	39				
3	11	20	30	40				
4	12	21	39	41	49	56		
1	5	13	22	32	42	50	57	62
2	6	14	23	33	43	51	58	63	66
	7	15	24	34	44	52	59	64	67
	8	16	25	35	45	53	60	65
	9	17	26	36	46	54	61
	10	18	27	37	47	55
		19	28	38	48

4. Pin frames are alternatively called **point quadrats** (Fig A.3.11). They are used to determine the percentage of ground covered by the shoots of plants when viewed from above. The frame has a number of holes through which knitting needles are pushed. Each species touched by the lower end of the needle is recorded. Some needles will touch tall species first then low-grading species when lowered further. Where organisms are widely spaced, as in early stages of colonisation of bare ground, some of the needles will touch the soil. This technique cannot cannot be used if species are tall, e.g. trees in a wood. Random sampling as described above should be used.

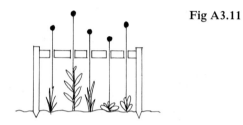

Fig A3.11

C. Recording and analysis of results

Subjective and objective methods are used for recording purposes.

1. Subjective methods are based on visual estimates of the abundance of species. The Braun–Blanquet is one of several scales used for recordings of this kind. It has seven categories of **abundance**:

$+ = <1\%$ $1 = 1–5\%$ $2 = 6–10\%$ $3 = 11–25\%$
$4 = 26–50\%$ $5 = 51–75\%$ $6 = >75\%$

Others, such as the Domin scale have more divisions, thus offering greater refinement. Such methods are best used for rough indications of changes in abundance of species along belt transects. Beginners have difficulties in judging abundance and will often produce somewhat different estimates.

2. Objective methods give the same result irrespective of the experience or otherwise of the worker. The **frequency** of a species is the number of times it is recorded in a given number of quadrats. For example, if it is listed 5 times in 20 quadrats it has a frequency of 25 in 100 or 25 %. Some workers go to the trouble of counting the number of individuals of each species in a given area. This gives an objective picture of abundance but is very time consuming where large numbers of organisms grow or live very close to each other. **Cover** is the number of times a species is recorded using a given number of point quadrats. For example, if a species is recorded 50 times when a quadrat with 5 pins is used on 30 occasions, the % cover is 50 out of 150 = 33 %. For very abundant species it is possible to have cover values in excess of 100 %.

D. Presentation of results

Differences in abundance, frequency and % cover can be presented in **tabular** form. However, such data has much more impact if converted to some form of **graph**. For this purpose **histograms** and **kite diagrams** are useful (Fig A3.12(a) and (b)), especially to display changes along a belt transect. They clearly show the differences in distribution of species as would be observed in the zonation of seaweeds and invertebrate animals on a rocky shore. A kite diagram is constructed by plotting each of the bars of a histogram such that one half of each is above the horizontal axis and one below, then joining up the tops and bottoms of each bar.

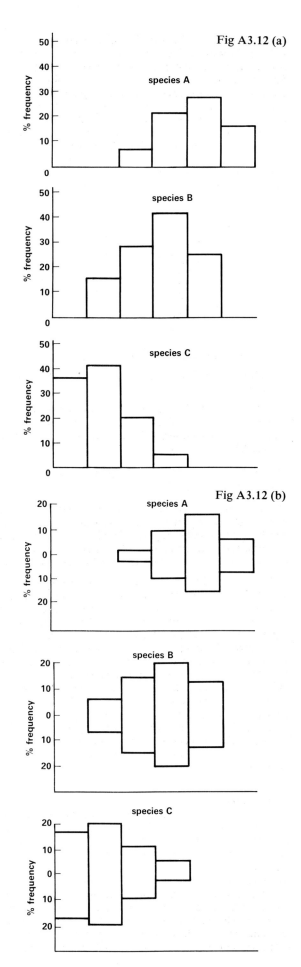

Fig A3.12 (a)

Fig A3.12 (b)

b. Population studies of soil animals

Many populations of animals are difficult to observe in the field as they are constantly on the move and often take cover when humans intrude on their habitat. Soil contains many kinds of invertebrate animals however, which are relatively easy to collect and count. The following is a selection of the many methods available for estimating the sizes of populations of selected groups of soil animals.

A. Capture–recapture technique

Some soil animals such as snails and beetles can be easily captured, marked with spots of non-toxic paint and released. The number of marked animals in a later sample caught at random provides the basis on which the size of the population of the species is calculated using the formula:

$$N = \frac{N_1 \times N_2}{N_3}$$

where N = estimated population size
N_1 = no. of individuals marked and released
N_2 = total no. of individuals in second catch
N_3 = no. of marked individuals in second catch

B. Pitfall traps

Invertebrates such as predatory beetles, millipedes and centipedes which are active at the soil surface can be collected using pitfall traps. An empty jar sunk flush with the surface of the soil will do. Baiting the trap with small pieces of catfood or with jam spread on bread helps attract victims. However, some of the trapped animals may escape whilst others may be taken by birds. More elaborate traps such as that shown in Fig A3.13 are designed to avoid losses of these kinds. The number of animals captured depends on how active they are as well as their abundance.

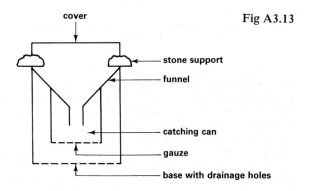

Fig A3.13

C. Chemical methods

Applying solutions of irritant chemicals to the soil surface will flood the burrows of earthworms such as *Lumbricus terrestris* driving them to the surface. The technique works well when the soil temperature is about $10\,^{\circ}C$ and is best done at sites where the soil has been left undisturbed for a long time. The site is marked into squares measuring $25\,cm \times 25\,cm$ and $25\,cm^3$ of 2 % formalin or 0.25 % potassium permanganate in water evenly applied to two squares. Within a few minutes worms begin to appear at the soil surface. They are picked up and immediately rinsed in water. When no more worms appear a second application is made. The collected worms are counted and then returned to an untreated plot.

D. Heating methods

The Tullgren funnel (Fig A3.14) is based on the principle that many active soil animals, arthropods especially, will leave a soil sample as it dries out on heating. Heat from the lamp dries the soil from the top causing the animals to move down through the sieve into the funnel. From there they slip into the preservative where they are killed and can be counted. The amount of heat applied to the soil sample affects the numbers of animals extracted. For this reason it is recommended that the temperature at the soil surface should be about 30°C. Complete extraction usually takes several days.

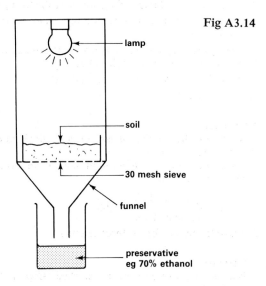

Fig A3.14

lamp

soil

30 mesh sieve

funnel

preservative
eg 70% ethanol

E. Immersion methods

One of the methods most frequently used for extracting nematodes from soil involves using a Baermann funnel (Fig. A3.15). Nematodes are resistant to dessication and will not leave heated, drying soil. However, they will move out of flooded soil. Being denser than water they sink to the bottom of the stem of the funnel from where they can be run into the collection vessel by opening the screw clip. Sampling must be carried out every 24 hours or the nematodes will die of lack of oxygen.

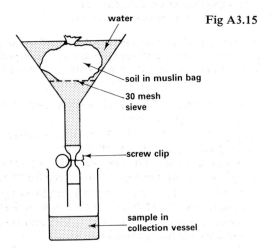

water **Fig A3.15**

soil in muslin bag

30 mesh sieve

screw clip

sample in collection vessel

F. Flotation methods

Resting stages such as eggs and pupae of soil arthropods cannot move and leave soil treated in the methods described previously. Their numbers can be determined by getting

748

them to float on a medium of greater density than themselves. Flotation methods are lengthy and comprise several stages. The soil sample is pre-treated by covering it with water then freezing it for several days before allowing it to thaw. This helps to break up the soil structure, enabling the organisms to be more readily extracted. The thawed soil is then washed in a jet of water to further break up soil crumbs and to free roots. Small stones are removed using a 2 mm mesh sieve. In the next stage the organic material, including soil animals is separated from the mineral particles by flotation in a solution of magnesium sulphate. Finally, the animals are freed from the organic matter by partitioning in benzene or xylene. The arthropod cuticle is wetted by such organic solvents causing the arthropods to separate from the aqueous phase which retains the soil debris.

c. Edaphic factors

The following methods of soil analysis give a measure of some of the edaphic factors operating in a terrestrial habitat.

A. Determination of the percentage of water

The water in soil evaporates when it is heated to 100 °C. The loss in weight of a soil sample treated in this way expressed as a percentage of the weight of the sample before treatment is the percentage of water in the soil. The water content of soil at any point in time depends largely on the rainfall immediately before the time of sampling and the ability of soil to retain water. Fluctuations in soil water content are best measured using soil tensiometers. Basically a soil tensiometer consists of an unglazed porcelain pot containing water which is attached to a manometer. As water enters or leaves the pot a change in height of the manometer fluid is registered. These are bulky instruments and must be left undisturbed for long periods of time. The following method is thus preferred for routine analysis.

1. Spread the freshly collected soil in a layer 1–2 cm deep on a clean tray. Leave at room temperature until air dry.
2. Weigh in an evaporating basin 10 g of air-dried soil which has been passed through a 2 mm sieve to remove stones.
3. Place the container in an oven at 100 °C for 24 hours.
4. Remove the container from the oven and allow it to cool in a desiccator.
5. Weight the basin and sample.
6. Repeat the procedure until the sample has a constant weight.
7. Carry out the entire procedure on two further samples.
8. Keep the dried samples of soil for further analysis.

Calculation
1. Determine the weight of the oven-dried sample in grammes.
2. Determine the loss in weight of the sample in grammes (10 g − weight of oven-dried soil). This is the weight of water in the sample in grammes.

3. Calculate $\% \text{ water} = \dfrac{\text{weight of water}}{\text{weight of oven-dried soil}} \times 100$

4. Average the results for the three samples.

Questions
1. What is the reason for allowing the heated samples to dry in a desiccator?
2. Why is it necessary to remove stones from the soil?
3. Why is it desirable to determine the water content of three samples?

B. Determination of the percentage of organic matter

Organic matter in the soil is oxidized largely to carbon dioxide and water when strongly heated in air. The oxidation products are driven off and the resultant loss in weight is taken as a measure of the organic material in the sample. The sample of soil used must be oven-dried. In chalky soils some of the loss in weight is due to combustion of calcium carbonate.

1. Weigh out exactly 5 g of oven-dried soil and place in a crucible which has previously been heated at $100\,°C$ to drive off any moisture and cooled in a desiccator.
2. With the crucible lid in place, heat to red-heat for at least one hour.
3. Cool in a desiccator and weigh again.
4. To check that all organic matter has been fully oxidised heat again for 10 minutes and re-weigh. Continue until there is no further loss in weight.
5. Repeat the procedure on two further samples.

Calculation

Weight of soil before oxidising $= p$ grams
Weight of soil after oxidising $= q$ grams

$$\% \text{ organic matter} = \frac{p - q}{p} \times 100$$

Average the percentages for the three samples.

Questions
1. Why is it necessary to do the analysis on oven-dried soil?
2. What is the purpose of reheating the sample until it has a constant weight?

C. Determination of the pH

The hydrogen ions in soil come largely from organic acids released when organic matter is decomposed. The pH of soil depends partly on the rate at which organic acids are formed and retained in the soil and on the mineral content of the soil. Chalky soil, for example, contains lime which neutralises organic acids.

1. Weigh in a $100\,cm^3$ beaker 10 g of air-dried soil which has been sieved through a 2 mm sieve to remove stones.
2. Add $10\,cm^3$ neutral distilled water and stir thoroughly.
3. Set up and calibrate a pH meter as follows:
   (a) Wash the electrode by swirling it in distilled water.
   (b) Plug the electrode into the meter and plug the meter into the mains.
   (c) Place $20-30\,cm^3$ of buffer solution pH 7.0 in a $100\,cm^3$ beaker and place the electrode in the solution.
   (d) Set the temperature dial so that it is at the same temperature as the solution.
   (e) If a pH of 7.0 is not shown on the meter, adjust the buffer dial until this reading is obtained.
   (f) Wash the electrode in distilled water and repeat the procedure with a buffer solution of pH 4.0.
4. Now measure the pH of the soil sample by placing the electrode in the soil-water mixture.
5. Wash the electrode once more in distilled water and repeat with two further samples of soil.
6. Average the figures for the three samples.

Questions
1. Why is it necessary to wash the electrode in distilled water?
2. For what reason is neutral distilled water used to prepare the soil for analysis?

D. Determination of the percentage of air

In a well-drained soil, air fills the spaces between the soil particles. To determine the air content accurately, the soil must not be disturbed when it is collected. This can be achieved using a soil corer which is used to cut cylinders of soil as they exist in the ground. The air content of the soil core can then be determined by displacing it with water. The following method is suitable for a 3 cm diameter soil core.

1. Snip off any aerial growth of plants in the core.
2. Measure accurately in mm the diameter and length (l) of the core. Trim the core if necessary to make it cylindrical.
3. Calculate the volume of the core using the formula $\pi r^2 l$.
4. Place the soil core in a measuring cylinder containing $50\,cm^3$ of water.
5. Gently break up the core with a glass rod to ensure all the air in the soil is displaced with water.
6. Note the final volume of water and soil.
7. Repeat with two other samples.

Calculation

1. Calculate the volume of soil in cm^3 (final volume of water + soil − $50\,cm^3$).
2. Calculate the volume of air in the core in cm^3 (volume of core − volume of soil).

3. $$\% \text{ air} = \frac{\text{volume of air}}{\text{volume of core}} \times 100$$

Average the results for the three samples.

Question
Why is it necessary to use undisturbed samples of soil for this determination?

E. Mineral particle (mechanical) analysis

Mineral particles of many sizes are present in soil. To compare the amounts of particles of different sizes the following scale is recommended by the International Society of Soil Science.

2.000–0.200 mm mean diameter—coarse sand
0.200–0.020 mm mean diameter—fine sand
0.020–0.002 mm mean diameter—silt
0.002 and less mm mean diameter—clay

A rapid method for mechanical analysis involves using a soil hydrometer to measure the relative density of a suspension of soil in water. The organic matter in the soil is first destroyed by mild oxidation with hydrogen peroxide.

1. Weigh out 40 g of 2 mm sieved air-dry soil. Add $250\,cm^3$ of water and boil to half volume.
2. Add $250\,cm^3$ of 6% H_2O_2 in small volumes. When frothing has stopped, boil to half volume.
3. Wash into a $800\,cm^3$ beaker, add $5\,cm^3$ of 1.0 M sodium pyrophosphate and make up to $500\,cm^3$. Stir for 20 minutes.
4. Transfer to a $1\,dm^3$ measuring cylinder. Make up to the mark with water. Stand the cylinder in a water bath at $20°C$.
5. Invert the cylinder several times to evenly disperse the soil then quickly add the hydrometer.
6. Take readings for silt and clay at 4 minutes and for clay at 2 hours.
7. Pour off the supernatant liquid. Oven-dry the sediment and pass through a 0.2 mm sieve to determine the amount of fine and coarse sand.

d. Climatic factors

Some measure of several of the factors which contribute towards the general climate of a terrestrial habitat is readily attainable using simple instrumentation. It must be remembered that successive recordings of such factors need be made over lengthy periods of time if the data are to be meaningful in assisting any interpretation of species distribution. One-off measurements made at the time of a visit, give no idea of the fluctuations of climate which can occur in a habitat. Weather stations will provide long term recordings but if there is not one nearby the necessary instruments can be installed near the site of study and regular readings must be taken. Another point to note is that the methods may give little idea of the climatic factors operating at a micro-ecological level.

1. Temperature. An ordinary mercury in glass thermometer can be used to measure air, soil or water temperature.

2. Humidity. A paper or hair hygrometer gives a direct reading of the % relative humidity of the air. Continuous readings of air temperature and humidity can be recorded with a thermohygrograph.

3. Light. A light meter will record the intensity of light but gives no indication of its wavelength or its direction.

4. Wind. An anemometer can be used to measure wind speed but will not indicate how long or how often it has been blowing. Wind direction is determined with a weather vane.

5. Rainfall. The amount of rain falling at a site can be measured by means of a rain gauge.

In aquatic habitats, other environmental factors such as the oxygen content of the water and the speed of the water current if any can provide useful data.

e. Relationships

Obtain good specimens or photographs of the following organisms:

pond weed, freshwater snail, *Paramecium*, diatoms, stickleback, freshwater algae, perch, pike, heron.

1. Make labelled drawing of the specimens. On your drawing write annotations which indicate the ways in which the organisms are adapted to their way of life.

2. Classify the organisms into one major group and subgroup. In each case give reasons for your classification based on external features.

3. Indicate the possible ecological relationships between the organisms.

f. Comparison of the features of different animals

Select pairs of animals from the suggested examples listed below.
- (a) locust and crayfish
- (b) lizard and mouse
- (c) bird and butterfly
- (d) housefly and honeybee.

1. Make labelled drawings of the external features of the two animals you have selected.

2. List the differences and similarities between the two animals.

3. For each animal write brief notes on those features of their bodies which are adaptations to their
- (a) habitat, and
- (b) way of life

4. (a) Classify the two animals as far as you can.
   (b) List reasons for your classification.

g. Autecology and synecology

The techniques described are equally applicable to practical studies of communities of different species in the same habitat (synecology) and to studies of the distribution of a given species in different habitats (autecology). Autecological studies are often supplemented with observations on patterns of behaviour, reproduction, life cycles, physiological and structural features, all of which contribute to an explanation of the niche of the selected species.

h. Code of conduct for fieldwork

Whenever working in natural habitats it is necessary to follow a strict code of conduct which is basically aimed at leaving the habitat unspoiled and the wildlife unharmed so that others can continue to use and enjoy it in the future. The more important things to bear in mind are:

1. Familiarise yourself with and observe the Countryside Code.

2. If the land is privately owned, always obtain the permission of the landowner.

3. Study the various Acts which are aimed at protecting and conserving wildlife (p 710). They will tell you what you can and cannot collect.

4. Where specimens can be collected always take as few as is necessary and leave the habitat as undisturbed as possible.

Index

Numbers prefixed by the letter 'F' refer to Figures, Numbers prefixed by the letter 'T' refer to Tables.

A-antigen 194
A-bands 334
abdomen 598, 602
abiotic factors 637–42
ABO blood groups, inheritance 196
ABO system 194–5, F10.17, F10.18, T10.6, T10.7, 503–4
abomasum 291, F15.21, 558
aboral surface 609
abscissic acid 469
  effect on seed germination 445, 447, 448
  and flowering 431
  in leaf abscission 474–5
abscission 475
abscission zone F24.30(a)
absorption chromatography 717
absorption spectra F19.11
  chloroplast pigments 5.8
  in photosynthesis 77–9
  in phytochromes F23.11
acatalasia 66, 517
accelerator nerve F9.9
accessory sex organs 483,487
accommodation 360–1, F19.5
acellular organisms 559
acetabulum 311, F16.7
acetyl coenzyme A 87, 280
acetylcholine 166, 331, 332, F17.18, 350, F18.14
acetylcholinesterase 331, F17.18
achene 440
acid–base balance 212–13
acid–base pair 7
acid rain 691
acidic solution 7
acidosis 213
acids, bases and buffers 7
acoelomates 589
acoustic nerve 372, F19.17
Acquired Immune Deficiency Syndrome see AIDS
acquired immunity 185, 186–94
acromegaly 409, F21.30
acrosome F25.5(a)
ACTH 29, 404, 408, F21.22, F21.27
actin 113, 335, F17.26-8
  binding sites F17.27
Actinia 587
Actinophrys 559
Actinozoa 587–8
action potential 323, F17.7, F17.8, F17.11
action spectrum
  photosynthesis 77–9, F5.4
  phototropism 472
activated sludge process 728–30
activation energy 41
activation, nucleic acids 61–2
activator, enzyme 43
active absorption, gut 403
active centres, enzymes 42
active immunisation 193
active immunity 185
active transport, minerals in plants 103, 272
active uptake
  in kidney 206
  of minerals by plants 207–1
acylglycerol 23, F2.15
Adamsia 652
adaptations to high altitude 147–8
adaptive immunity 183, 186–94
adaptive radiation 539
adaxial meristem 458
adenine 54, 433
adenosine diphosphate see ADP
adenosine monophosphate see AMP
adenosine triphosphate see ATP
ADH 29, 210, 211, F11.5, 410
adhesion 228
adipose tissue 385, 386, F20.7(a)
adolescence 483
ADP 85
adrenal bodies F11.1, 404, F21.1, F21.20
adrenal cortex 404, F21.22, F21.25
adrenal medulla 404, 406–7
adrenaline 166, 350, 384, 406, 407, F21.26, F21.27
adrenergic neurons F18.14
adrenocorticotrophic hormone see ACTH
adsorption chromatography 34
adventitious roots 441
Aedes aegypti 637
Aegilops 546
aerating tissue 468, 640
aerobes 84

aerobic respiration 109, 647
aerofoil 621
aestivation 390
afferent arterioles, kidney F11.4(b)
afferent neurons 345
affinities between organisms 529, 539–41
  anatomical 539
  behavioural 541
  biochemical 540
  embryological 540
  immunological 540
afterbirth 492
aftershaft 621
'Agent Orange' 697
agglutination 190, 191
agglutination inhibition test 191, F10.16
agglutinin 190, 194
agglutinogens 190, 194
Agropyron 546, 664
Agrostis canina 267
AIDS 193
air bladder 640
air capacity, soil 676
air movement
  and distribution of organisms 642
  effect on transpiration of 223–4
air sac 623
akinetes 559
alanine F2.22, F2.23
  albinism 499
albumen 184, T10.5
alcoholic beverages, manufacture of 718–19
alcoholic fermentation 85
aldoses 16
aldosterone 405, 406, F21.23, F21.25
aleurone layer 444
algae 531, 650, 679
algin 565
alimentary system 278–304
alkaline solution 6
alkalosis 213
alkapton 66
alkaptonuria 66, 518
all-or-nothing law 323, F17.7
allanto-chorionic placenta villi F25.17
allantois 489, F25.17, F25.18, 620
alleles 496
allergen 192
allergic response 182
allergy 192
alligators 542
alternation of generations 423
alternative hosts 648
altitude, adaptation to 147–8, T8.6
alveolar ventilation 140
alveoli 136, F8.4
amber 530
ambulacrum 609
ameloblast cells 282, F15.1
amides 443
amino acid pool 300
amino acids 27–8, F2.21, F2.23, T2.3, 278
  amphoteric 28
  biochemical tests 33
  essential 278, T15.1
  ionisation of F2.24
  isoelectric point 28, F2.34
  manufacture of 720
  non-essential 278
  in proteins 443
  structure F2.21, F2.23
amino group 27
aminotransferases 50
ammonification 673
Ammophila arenaria 664
amniocentesis 518
amnion 489, F25.17–19, 620
amniote egg 620, 624, 625
amniotic cavity F9.23, 490
amniotic fluid F25.18
Amoeba 12, 559, 560
amoeboid movement 181, 559
AMP 84
  cyclic 395, F21.3
amphibians 531, 618–19
amphion 28
amphoteric amino acids 28
ampulla 374, F19.23, F19.25(a)(b)
amylase 19
  in germinating seeds 442
amylopectin 20, F2.8
amylose 20, F2.7
anaemia 181
anaerobes 84

anaerobic digestion of sewage 729
anaerobic respiration 151, 647
analogous structures 539, 553
anaphase
  meiosis 128
  mitosis 123
androecium 414
androgens 483
angiosperms 532, 577, 580–1
  reproduction, physiology, of 430–4
angiotensin 353, 405, 406, F21.25
angle of aperture F6.25
annelids 531, 594–8, 681
annual rings 465
annuals 581
annulus 575
Anopheles 562–3, 647
ant 602
antagonistic muscles 316, F18.11, 584–5
antenna 601
antennule 601
anterior pituitary 408–10, F21.31, 483, F25.14
anther 414, F22.1(a)
  dehiscence F22.3(b)
antheridium 565, 568, 569, 571, 572
antherozoid 565, 574
Anthozoa 587–8
Anthropoidea 534, 628
anti-A 194
anti-B 195
anti-knock compound 692
anti-mitotic drugs 125
antibiosis 636
antibiotics 636
  production of 721–2
antibodies 103, 174, F9.20, 182, 185, F10.8, F10.12, F10.15, F10.16, 540
  monoclonal 190, 723
anticodon 57, 62
antidiuresis 211
antidiuretic hormone 29, 210, 211, 410
antigen–antibody complex F10.15
antigen–antibody reactions 190–1
antigens 103, F10.8, F10.10, F10.12, F10.15, F10.16, 185
  binding sites F10.13
anti-immunoglobulin 195
antipodal cells F22.5(d)
antiserum 196
anus 298, F15.14, F15.26(b)
aorta F9.2, F9.3(a), F9.10, F11.1
aortic valve 7, F9.3(b), F9.6
apex of cochlea 372
aphids 251, F13.14, F13.22, 602, 603
apical dominance 470
apical meristem 452, F24.1(a)
  root 460
  shoot 452–4
aplanospore 567
apneustic area 140, F8.10
apnoea 141
appendicular skeleton 306
appendix F15.14, F15.26(b)
apomixis 429
aqueous humour 360, F19.1(b)
aqueous solutions 3, 7–8
Arachnida 606
archegonium 571, 572, 579
Arenicola 596
Aristotle's lantern 610
arm F16.9
Armeria maritima 665
arteries 156, 168, F9.2, F9.3(a), F9.18, F9.20
arterioles 156, 169, F9.11, F9.12, F9.18, F9.19, F11.4(b), F20.7(b)
arthropods 531, 598–600, 680
articular process 16.4
artificial classification 552
artificial selection 545, 546–7
artiodactyls 625
Ascaris 593
ascending chromatography 35
asci 569
ascocarp 569
ascogonium 569
Ascomycetes 568, 569
ascorbic acid 280
ascospores 569
ascus 569
asexual reproduction, flowering plants 428–9
aseptate hyphae 566
association area of cerebral cortex 354
associations between organisms 645
Aster tripolium 665
Asterias 610
Asteroidea 610
asthma 192
astigmatism 361, 362
atheromatous plaques 168, F9.15

atlas 309, F16.5(a), F16.17
atmospheric humidity and transpiration 224
atmospheric pollution 688–92
atmospheric pressure 639–40
atoll 588
ATP 76
  in chloroplasts 112
  in fruit ripening 434
  in growth and development 467
  in guard cells 245
  in mitochondria 109
  in muscle 337–8
  in photosynthesis 245
  in protein synthesis 61
  in respiration 84–5
  and seed germination 446
  structure F5.16
ATP-ase 337, F17.27
atretic follicles 484
atrial systole F9.5(a)
atrio-ventricular node 159, F9.4, F9.5(b)
atrio-ventricular valves 159
atrium 159, F9.3(a)
auditory area, primary 372, F19.21
auditory nerve F18.16, 372, F19.17, F19.18(a)
auditory tube 371
Aurelia 588
Australopithecus 534–5, 536
autogamy 561
autoimmune responses 192
autolysis 108, 404
autonomic nervous system 342, 348–50, F18.1, F18.13
autophagosomes 107
autoradiography 82
autosomes 496
  abnormalities of 509–60
  inheritance of 496–510
auxins 432, 433, 469–73
Aves 532, 620–4
axial skeleton 306
axillary bud primordia 453–4
axis 309, F16.5(b), F16.17
axon 321, F17.3(a), F17.5, F17.9(d), F17.12, F17.14, F17.15
  myelinated F17.12(b)
Azotobacter 673

B-antigen 194
bacillus 556
Bacillus Calmette Guérin (BCG) 193
back-cross ratio 498
bacteria 555–9, 679
  cellulolytic 291, 295
  chemo-autotrophic 556, 679
  nitrogen-fixing 558, 673
  phagocytosis of F10.3(a)
  photo-autotrophic 679, 681
bacteriochlorophyll 556
bacteriophage 553
balance 373–4
Balanus 601, 643
baleen plates F15.12(b), 627
barber fish 652
barbs 621
barbules 621
barnacle 600, 601
baroreceptors 8.11, F8.12(b), 165, F9.10, F9.12, 141
Barr body 520
basal bodies 110, 561
basal ganglia 354
basal metabolic rate (BMR) 92, F5.25, 380, 383, 384, 396, 398
basement membrane 137, F11.6(a), 204
bases 7
basidia 569
basidiocarp 569
Basidiomycetes 568, 569
basidiospore 569
basilar membrane 371, F19.18(a),(c), F19.20, F19.22
basophils 181
bat 627
BCG vaccine 193
beak 623
bee 602
beer production 719
beetle 602
behaviour 541
  control of 354
bends 152
Benedict's reagent 16
beri-beri 280
β-cells F10.12, 186, 187, 192
β-particles 701
β-carotene F5.7
biceps 316, F16.19, F16.22, F18.11
bilateral symmetry 589, 607, 614
bile 292, 299, F15.21, F15.25
bile duct 299, F15.22, F15.29(a)
bile pigments 292

751

bile salts 292, 297
bilharzia 592
bilirubin 301, F15.21
bilirubin diglucuronide 301, F15.21
binary fission 557, 560, 561
binocular vision 366
binomial system 552
biochemical oxygen demand (BOD) 706, 707, 726
biochemical techniques 34–7
biochemical tests
  amino acids 33
  amylopectin 20
  cellulose 22
  glycogen 21
  proteins 33
  starch 20
  sugars 16
biogas 724
biological control 554, 662–3
biological effects of radiation 703–5
biomes 632
bioreactors 717
biosphere 632
biotechnology 716–30
biotic factors 633–6
bipedal gait 619
birds 532, 620–4
birth 491–3, F25.21
birth rate 658
Biston betularia 545
biuret test 33
bivalents 126
Bivalvia 607–8
bladder sphincter 201
bladder wrack 565
bladderworm 593
blastostyle 587
blind spot 363, F19.1(b)
blood
  capillary 156, 168, 169, F9.16, F9.19–20
  cells 180–5
  circulation 156–78, F9.1
  clot F10.6, 183
  coagulation 182–4, F10.5
  factors 183
  glucose, effect of medullary hormones on 407
  groups 194–6, T10.5
  lymph 156–78, F9.1
  pH 146
  pressure 163, 167, F9.8(a), F9.12, F9.18
    in dialysis 215
    hydrostatic 170
  vessels, cutaneous 386
  volume T9.1
blood fluke 592
blow-hole 627
blubber 627
blue cones 366
blue–green algae 558, 673, 679
blue weakness 366
body temperature F20.1, F20.11, F20.12
body volume 382
Bohr effect 146
bolus 288
bone 313–14, F16.15, F21.13, F21.14
  compact 314, F16.12, F16.13
  spongy 314, F16.12
bone marrow 147, F8.19, 186, 313, F16.14, F16.15
  cavity F16.12
bony fish 616
boundary layer F12.4, 223, 226
boutons 329, F17.3(a)
Bowman's capsule 202
brachial artery F9.8(b)
brachydactyly 498
bract scale 578
brain 350–5, F18.1, F18.13, F18.15–17, 590, 624
  lobes 354
brainstem 354, F18.15
'breaking of the waters' 492
breast-feeding 491
breathing 136–42, F8.6, 493
  control 139–42
brewing 718–19
bristleworms 595
bronchi 137, F8.3, 192
bronchioles 137, F8.3
brown algae 565–6
brown fat 386
Brunner's glands 293, F15.28
Bryophyta 531, 570, 570–3
bud 459
  winter 468, 475
budding
  of *Hydra* 586
  of yeast 568
buffer 7, 212
buffering F2.33
buffering capacity 7
  of proteins 32

bulb F22.17, 428, 468, 580
bulbo-urethral gland 488, F25.1
bundle of His 159, 162, F9.4, F9.5(b)
buoyancy 623
butterfly 602
byssus threads 608

C-cells 400
C_3 plants 689
C_4 plants 689
C_4 pathway 84
Cactoblastis 663
cadmium 696
caeca 603
caecum 295, F15.14, F15.26(a), 558
Cakile maritima 664
calcareous shell 607, 624
calcicole species 678
calcifuge species 678
calcitonin 400, F21.14
calcium 280, F21.12, F21.15
  absorption from gut 400
  in blood coagulation 183
  in muscle 338
  in plant growth 266
  reabsorption from urine 400
  release from bones 399
  role of thyroid gland in control of 400
callose 255
Calluna vulgaris 665
callus 474
calorimeter 89, F5.22
Calvin cycle 83, F5.14(a), 249
calyptra 571, 572
calyptrogen 460
Calystegia soldanella 664
calyx 414
cambium
  cork F24.12, 463, 465–6
  vascular 463–5, F24.9
Cambrian period 531
camel 211–12
canaliculus 299, F15.29(a)–(c), 314
canines 283
cankers 558
capillaries F9.18
  blood 156, 168, 169, F9.16, F9.19, F9.20
  lymphatic F9.20, 172, 297
capillarity 228, 229, 233, 236, 677
capillary bed 169
capsid 553
capsomeres 553
capsule 556, 571, 572
  Bowman's 202
  fibrous F16.15
carapace 600
carbohydrates 16–22, 278, 297, 442
  metabolism of 300
carbon cycle 673–4, 683
carbon dioxide 240
  diffusion in plants 3
  effect on photosynthesis 639
  effect on plant growth 468
  fixation 82–3, 84
  pollution 688–9
  tension 146, F8.17, 166
  transport in blood T8.4
  uptake by leaves 244–5
carbon monoxide
  poisoning 144
  pollution 691
carbon monoxide haemoglobin 144
carbonic acid 145
carbonic acid–hydrogencarbonate 212
carbonic anhydrase 144
  in leaves 243, 244
  in mammals 212
Carboniferous period 531
carboxyl group 27
cardiac centre F8.12(b), F9.13
cardiac cycle 159–61, F9.5(a)
cardiac muscle 159
cardiac nerve 164
cardiac output 159, 164
cardiac reserve 165
cardiac sphincter F15.17(a)
cardioacceleratory centre 164, 9.9
cardioinhibitory centre 164, 9.9
cardiovascular system
  effect of medullary hormones on 407
Carex arenaria 664
carnassial teeth 283, F15.4
carnivores 283, F15.5, 636 668
carotene 434, 558, 563
  β-carotene F5.7
carotenoids 78
carotid artery, common F9.2, F9.3(a), F9.10
carpals 312, F16.1, F16.8, F16.9, F27.19
carpels F22.1(a), 414, 580
carrier molecules 271, F14.11
carrier proteins, in cell membrane 103
carriers 512, 517–18, 648
carrying capacity 656

cartilage 313, F16.14
  articular F16.12, F16.15
  elastic 313
  epiphysial F16.12
  fibro-cartilage 313
  hyaline 313, F16.11
cartilaginous fish 614, 617
caryopsis 441
Casparian band (strip) F12.17(a), 272, 461
cast of fossils 529, F27.3
castration 483
catalase 109
catecholamines F21.26
cat's eyes 363
cell
  coat 103
  cycle 124–5, 7.6
  disruption 717
  division 120–32, 473
    meiosis 126–31
    mitosis 120–6
  expansion 442
  extension 469
  fractionation 116–17
  growth 442
  membranes 9, 99, 102–4
  plate 124
  structure 96–118
  theory 96
  wall 99
    fungi 566
    plant 9, 113, F6.23, 442
cell-mediated response 187
cellular immune response 187, 216
cellulolytic bacteria 295
cellulose 22, F2.12(a), 113, F6.22, 278, 434, 563
cellulotoxic substances 187
centipede 598, 602
central canal F18.8
central nervous system (CNS) 342
central sulcus F18.18
centrifugation, differential 117, F6.28
centrifuge 116, 717
centrioles 110–11, F6.20
centromere 123
centrum 308, F16.4
cephalisation 589, 594
Cephalopoda 608
cephalosporin 722
cephalothorax 600
cercaria 591
cerebellum 351, F18.15, F18.16
cerebral cortex 17.3(c), 17.3(d), 353, T18.4
  functional areas of F18.19
cerebral ganglion 603
cerebral hemispheres 353, F18.18, 629
cerebrum 353, F18.15
ceruminous glands 375
cervical vertebrae F16.1
cervix 292, F25.8, F25.21
Cestoda 592–3
cetaceans 627
chaetae 595
chelicera 606
chemoautotrophs 556, 679
chemoreceptors 141, F8.12(b), 590
chemotaxis 182, 571, 576
chemotropism of pollen grains 432
chewing 282
Chi-square test 501
chiasmata 127, F7.9(b)
chilling
  breaking corm dormancy 468
  of seed 445
chin 308
chiropterans 626–7
chitin 595
chloracne 697
chloramphenicol 721
chloride shift 145
chlorocruorin 595
chlorofluorocarbons 692
chlorophyll 78, 112, 558
  a 563, 565
  b 563
  c 565
  structure F5.6
chlorophyta 563–4
chloroplasts 99, 111–12, 240
  pigments of F5.8
  ultrastructure F6.21(a)
chlorosis 264, 266
cholesterol 26, F2.20(b), 102, 297, 301
choline 25
cholinergic neurons 350, F18.14
cholinesterase 45
Chondrichthyes 614–15
chondrin 313
chondroblast 313
Chondrus 566
chordae tendineae F9.3(b), F9.6
choriogonadotropin (human) HGC 490
chorion 489, F25.17–19

chorionic villi 490, F25.17
  sampling 518
choroid layer 358, F19.1(b), F19.2
chromatids 123
chromatin 110
chromatography 34–6
  absorption 717
  adsorption 34
  ascending 35
  column 35, F2.36
  descending 35
  paper 34–5, F2.35
  partition 34
  thin-layer 36
  two-dimensional 35
chromatophore 556
chromosome maps 507–8
chromosomes 99, 110, 120, F7.4(b)
  deletion of 544
  duplication of 544
  homologous pairs of 120, 126
  human F7.1
  inversion of 544
  mutation 543
  random orientation of 130–1, F7.14
  translocation of 544
Chthamalus 643
chyle 295
chylomicrons 297
chyme 290, 294, F15.25
chymotrypsin 293, F15.24
chymotrypsinogen 40, 293
  structure F3.1
cilia 110–11, F6.20, 137, F19.10, 560–1
ciliary body 360, F19.1(b)
Ciliata 560–1
Cimex 603
circulation
  of blood and lymph 156–78, F9.1
  minerals in plants 468
cisternae 105, 106
civilisation 629
Cladonia 664, 665
Cladophora 77
clasper 615
classes 552
classification 552–3
cleaner animals 652
cleaner fish 652
cleavage, cytoplasmic 123–4, 7.5
cleidoic eggs 624
climate, effect on transpiration 222–4
climatic factors 637
climax community 663
clitellum 597
clitoris 487
cloaca 625
clones 429, 515
Clostridium 557, 673
  C. pasteurianum 556
clot 183, F10.6
cloven hoof 626
clubmoss 573
Cnidaria 584–9
cnidoblasts 585
co-dominance 503
co-enzymes 43
co-factors, enzyme 43
co-ordination in homeostasis 320
coacervates 528
coagulation, blood 182–4, 10.5
cobalt 280
coccyx 309, F16.5(f)
cochlea 371, F19.17, F19.23
coconut milk 433
cocoon 598, 605
codon 57, 62
codon–anticodon binding 62
coelenterates 584
coelom 594
  extra-embryonic F25.17
coenocytic 566
cohesion, water 228
colchicine 519
cold centre 384, F20.13(a)
cold receptors F20.7(b)
Coleoptera 603
coleoptile 441, F23.3(a), (b)
coleorhiza 441, F23.3(a)
collagen F2.29, 168, F9.14, 313
collecting duct 203, F11.4(b)
collenchyma 457
colloid osmotic pressure 170, 184
colloidal state 3–4
colon 298, F15.14, F15.26(b)
colostrum 491
colour blindness 366
colour vision 365
columella 572
combustion 674
commensalism 652
communities 656, 663–5
compact bone F16.12, F16.13

companion cells 256, F13.18(b), 580
compensation period 248, F13.12
  point 248, F13.12, 467
competition 682
  interspecific 634, T31.1
  intraspecific, 633–4
competitive exclusion 635
competitive inhibitors F3.5(a), F3.6
complement 190, 197
compound eye 599, 601
compound microscope 96, F6.2
composts 268
compression of air 368
concentration gradient 207
conceptacle 565
condenser 114
conditioned reflex 348
conditioning 348
conduction 382, F20.3
conductive deafness 375
cone cells 362–4, F19.8, F19.9, F19.11
cone pigments 365
conidiophores 568
conidium 568
Coniferales 578–9
conifers 577
conjugate acid–base pair 7
conjugated proteins 40
conjugation 561, 564, 569
conjugation tubes 564
conjunctiva 358, F19.1(b)
connective tissue 168, 182
conservation 709–12
consumers 528, 665, 668–9, 682
  secondary 666
  tertiary 666
continuous flow ventilation 139, F8.9
continuous variation 512
contour feathers 621
contraception 658
contraceptives, oral 723
contractile vacuole 12, F1.13, 560, 561
contralateral reflex 347
controlled parasitism 653
converging pathways 344
copper 280, 695–6, 724–5
copulation 487–93
corals 584, 585, 587
cork cambium 463, 465–6, F24.12
cork cells 466, F24.12
corm 428, F22.17, 468, 580
cornea 358, F19.1(b), F19.3
corolla 414
corona radiata 488
coronary artery F9.3(a)
coronary sinus F9.3(b)
coronary thrombosis 184
corpora quadrigemina 353
corpus F24.1(b)
corpus allatum 603
corpus callosum 354
corpus luteum 485, F25.14
cortex
  adrenal 404, F21.22, F21.25
  cerebral F17.3(c), F17.3(d), 353, F18.19
  root 232
  shoot 453
Corti, organ of 372 F19.20
cortical nephron 203, 207, F11.4(b), F11.10,
corticotrophin (ACTH) 29, 404, 408, F21.22, F21.27
cortisol 404, F21.21, F21.22, F21.27
cortisone 723
costal cartilage 310, F16.6
cotyledon F22.8(a), 438, 580
counter-current multiplication system 208
counterflow gas exchange 617
crab 600
cranial nerves 351, F18.16, T18.2
cranium 307, 614
crassulacean acid metabolism (CAM) 227
crayfish 600
creatine 338, F17.29
creation 11, F1.11
Cretaceous period 532
cretinism 398
cristae 108, F19.23
crop improvement 72, 725
cross-fertilisation 424, 480, 590, 591
cross-pollination 424
crossed extensor reflex 347, F18.10
crossing-over 130, F7.13
crown of tooth 282, F15.2
Crustacea 598, 600–2, 680–1
crypt of Lieberkühn 293, F15.28
cud 291
cudding F15.21
cupula 374, F19.23, F19.25(a) (b)
current, electric 324
cuticle 26, 227, 573, 577
cutin 113, 457

cuttings, striking of 476
cyanocobalamin 280
cycads 577
cycles per second, of sound 369
cyclic AMP 395, F21.3
cyclic photophosphorylation 81 F5.11
cycling of matter 672–5
cysteine 2.23
cysticercus 593
cytochrome 88
  c 540
  oxidase 50
cytogenetic laboratories 518
cytokinesis 210
cytokinins 432, 433, 444, 469, 474
cytoplasm 99
cytoplasmic cleavage 123–4, F7.5
cytoproct 561
cytosine 58
cytoskeleton 112–13
cytostome 561

2,4D 475, 476
D-agglutinogen 195
dairy products, manufacture of 720
damping off 645
Danielli-Davson membrane model F6.8
dark-adapted eye 365
dark period 431
Darwin, Charles 542, 547, 549
Darwinism 542–3
Darwin's finches 547
day-length, effect of
  on flowering 430–1
  on mammals 485
  on plant growth 467
day-neutral plants 430
DDE 698–9
DDT 563, 698, 700, 708
dead space 138, 144
deafness 375
deaminases 50
deamination 300, F15.21
death phase 657
death rate 658–9
decibels 370
decomposer organisms 268, 558, 666, 669–71, 679, 682
decomposition, soil 682
decompression sickness 152
degenerate organisms 645
dehiscence 418, F22.3(b), 575
dehydration in mammals 211–12
dehydrogenases 50
delayed implantation 485
delivery 292, F25.22
demes 547–9
denaturation
  enzymes 46
  proteins 33
dendrites 321, F17.3(a)
Dendrocoelum 589
density-dependent factor 656, 661
density-independent factor 656
dental formula see dentition
dentine 282, F15.1, F15.2
dentition 282–6
  dog F15.4
  elephant F15.13
  hedgehog F15.10
  herbivores F15.8
  man F15.9
  rabbit F15.11(b)
  rat F15.11(a)
  rodents 285
  sheep F15.6
  toothed whale F15.12(a)
deoxygenation of haemoglobin F8.14
deoxyribonucleic acid see DNA
deoxyribose 54, F4.1
depolarisation 323, F17.6, F17.8, F17.9(b)
dermis 385, F20.7(b)
Deschampsia caespitosa 665
desensitisation 192
detection of stimuli 320
detoxification 302
detritus 675, 677
detritivores 66, 669–71
Devonian period 531
diabetes 404, 21.19
  diabetes insipidus 403
  diabetes mellitus 192, 403
diagnostic features 522
dialysis 214, F11.20
diaphragm 136, 149, F8.3, F8.6, F8.12(b), F15.14, 624
diaphysis 313, F16.12
diastema 284, F15.6
diastole 163
diatoms 679
dichlorophenoxyacetic acid see 2,4D
Dicotyledones 581
diencephalon 353, F18.15
diet 281, 581, 658

dietary fibre 278
differential centrifugation 117, F6.28
Difflugia 559
diffusion 13, 103
  of carbon dioxide, plants 3
  of minerals, plants 270
  of respiratory gases, mammals 137
  of water vapour, plants 222
diffusion shells 225
digestion 287–95
digestive enzymes 287, 293, T15.3
digestive system F15.14
digits F16.8
dihybrid inheritance 500–2
dihybrid ratio 500
dinosaurs 532, 619
dioecious species 594, 596
dioxin 696–7
dipeptide 29, F2.25
diploblastic organisms 584
diploid 130
diploid parthenogenesis 428
diploid zygote 420, 488
Dipodomys 135, 212
Diptera 603–4
direct secretion 205, F11.7
disaccharide sugars 18–19
discharge zones 343, F18.3
discontinuous variation 512
disease control 658
dispersal 642
disperse phase 3, 588
dispersion medium 3
dissociation curves
  haemoglobin 148, F8.15
  myoglobin F8.23
dissociation of water 6–7
distal convolution 202
distribution curve, normal 513
disulphide bonds 32
disulphide bridges 45
disulphide cross linkages in proteins 31
diuresis 211, 403
diverging pathways 343–4
diving
  and mammals T8.7
  and man 152
dizygotic twins 489
DNA 110, 540
  changes in mitotic division 125–6
  copy (cDNA) 70–1, F4.25
  effects of radiation on 703–4
  and gene mutations 64
  as hereditary material 57–8
  hydrogen bonds in 55
  and insulin production 70–2
  and protein synthesis 60–1
  recombinant 68–9, F4.23
  replication of 58–9, F4.11, F4.12
  structure 54–6
  Watson–Crick model of F4.5
DNA ligase 58, 68
DNA polymerase enzymes 70
DNA probe 73
dominant characteristics 497
donor 216
Doppler effect 369, F19.15
dormancy of seeds 448–9
dormant winter buds 468
dorsal nerve ganglia F18.8, 594
dorsal root 346, F18.8
dorso-ventral flattening 615
double fertilisation 420, 580
down feather 620
Down's syndrome 509, 544
downstreaming 717
drone 605
Drosophilia 508, 633, 704
dry fruits 422
Dryopteris 575
ductless glands 394
ductus arteriosus 176, F9.24
ductus venosus 176, F9.24
duodenum 293, F15.14, F15.21, F15.22
dwarfism 409, F21.29

ear 367–75, F19.17
  defects of 375
eardrum 371
earthworm 596
ecdysis 599
ecdysone 599, 603
Echinodermata 609–10
Echinoidea 610
Echinostoma 590
Echinus 610
echolocation 627
ecological pyramids 671–2
ecosystems 656, 665–84
  soil as 675–84
  whole 670–1
ectoderm 584
ectoparasites 645
ectotherms 380, F20.1, 668

eczema 192
edaphic factors 637
efferent arterioles F11.4(b)
efferent neurons F17.3(b), 345, F18.8–10, F18.14
egg cell, flowering plant 420
egg-laying mammals 532
eggs 281
ejaculation 488
elastic arteries 168, F9.14, F9.18
elastic fibres F9.14
elastic lamina F9.14
elastic recoil
  of arteries 571
  of lungs 137
elastin 31, F2.28
elater 571
elbow 311, F16.17, F16.18
electro-osmosis 258, F13.25
electrocardiogram (ECG) 162, F9.7(a)
electrochemical basis of nervous activity 321–32
electroencephalogram (EEG) 355
electromagnetic waves 701
electron microscope 99–113, 114–16, F6.5, F6.24
electron transfer in chloroplasts 79–80
electron transport chain 88
electronmicrographs 114
electrophoresis 36–7, F2.37, F2.38, F10.7
electrostatic interaction 34
electrovalent bonds 32
elephantiasis 174. F9.22
Elodea 246
elution 35
elytra 603
embolism 184
embolus 184
embryo
  in conifers 579
  flowering plant 421, 438
  human F25.15
  immaturity in plante. 448
  sac 418, F22.5(b) (d), 580
embryonic cells, flowering plant 421
embryonic membranes F25.17–18
Empetrum nigrum 665
emphyra larvae 589
emulsification 292, 297
enamel 282, F15.1, F15.2
enamel organ 282, F15.1
endergonic reactions 76
endochondral ossification 314
endocrine glands 394, F21.1, F21.2
endocrine system 320, 394–412
  interrelationship with nervous system F17.2
endoderm 584
endodermis 232, F12.17(a), 461
endogenous roots 461
endolymph 19.20
endometrium 485, F25.14
endoparasites 590, 645
endoplasmic reticulum 105–6, 566
  rough 105, F6.12(a)
  smooth 105–6
endosperm 421, F22.8(a), 438, F23.3(c)
endospermic seeds 421
endospores 557
endothelium F9.14, 668
endotherms 380, F20.1, 622, 668
endothermy 624
energy conversion 76–94
energy flow 666–71
  in soil 684
Entamoeba histolytica 560
enterocrinin 294, F15.25
enterokinase 293, F15.24
enteron 584
environment 452
environmental resistance 656, 662
enzymes 40–52, 646
  activators 43
  catalysis 41, F3.8, F3.9, F3.11, F3.13
  classification 49–51
  co-factors 43
  commercial production of 716–17
  concentration 47, F3.11
  control 43
  denaturation 46
  effect of gibberellic acid on 444
  embryo production of 445
  extraction and purification of 717
  extracellular 717
  factors affecting 43–8
  induction 67
  inhibitors 43
  intracellular 717
  lock and key mechanism of action F3.3
  manufacture 717
  and pH F3.9
  repression 67
  secreted by small intestine T15.3
  sources of 716–17

specificity 42, F3.4
structure 40
and substrate concentration 3.13
enzyme active centre 42
enzyme-secreting cells 105
enzyme–substrate complex 42
eosinophils 181, 184, 192, F10.1
epicotyl 439, F23.1(b)
epidermis
root 230
skin 385, F20.7(b)
stem 453, 457
epididymis 481, F25.1, F25.2
epigeal germination 440
epiglottis 288, F15.14, F15.15
epimysium 333, F17.21
epiphysis 313, F16.12
epistasis 504–5
epistatic genes 504
epithelium 388
germinal 481
pseudostratified ciliated columnar
F8.6(b)
squamous F8.4(b)
Equisetales 574
Equisetum 574
erectile tissue 487, F25.1
erection 487
erector-pili muscles 385, F20.7(b),
F20.13(b)
ergocalciferol 279
erythrocytes 142, 180
erythropoiesis 301, F8.19
impaired 181
erythropoietin 148, F8.19, 215
Escherichia coli
and commensalism 652
DNA replication in 59
genetic engineering 68–9
protein synthesis in 61, 66–7
essential amino acids 278, T15.1
essential fatty acids 279
ester linkages 23
esterification 23
estradiol 485, 491, F25.11, F25.12, F25.14
ethanol 85–6
etiolation 467
Euglena 563, 680
euglenoid movement 563
Euglenoidea 563
eukaryotes 101, 563, 565
euryhaline 641
Eustachian tube 371
Eutheria 625–9
eutrophic lakes 707
eutrophication 708
evaporation 13, 134, 382, F20.3, 640, 642,
675
evolution 526–50
convergent 539
divergent 539
evidence of 529–33
explanations for 542–9
excitable cells 321
excitatory postsynaptic potential (EPSP)
331
excretion 200
excretory canal 590
exergonic reactions 76
exocrine glands 287, F17.2, F21.2
exodermis 231
exophthalmos 398
exophthalmos-producing substance 398
exoskeleton 598
expiratory reserve volume F8.8
explants 474
exponential phase 656
extension F16.19
external auditory meatus F16.2, 371, F19.17
external factors in plants 452
extrinsic thromboplastin 183, F10.5
eye 358–67, F19.1(a) (b), F19.3, F19.12
defects of 361–2
eyepiece lens 114
eyespot 590

facets 309
facial bones 308
facilitated transport 103
facilitated zones 344, F18.3
factor complex 643
facultative anaerobes 84, 556
FAD 43
faeces 298
families 552
fanworms 595
fasciculi 333, F17.21
Fasciola 590, 592
F. hepatica 591
fat-soluble vitamins 279, 297
fats 22, 23–4, 279, 281, 443
brown 386
fatty acids 22–4, F2.14, 297, 443

essential 279
non-essential 279
saturated 23
unsaturated 23
feather 620–1
feedback
in breathing 142
in nervous control 320, F17.1, F17.2
feeding centre 353
Fehling's test 16
female cone 578
femur 311, F16.1, F16.7, F16.8, F16.10
fenestration F11.6(a)
fern 573, 575–6
ferritin 301, F15.21
fertilisation
flowering plants 420–2
implantation of the embryo 488–91
internal 620
fertilisers 262
Festuca ovina 267
Festuca rubra 664, 665
fetus F9.23, F9.24, 489, 25.16
fetal haemoglobin 68, 149, F8.21, 176
fetal membranes 489
fibres, muscle 333
fibrin 183, F10.5, 10.6
fibrinogen 183, F10.5, 10.6
fibrinolysin 183, F10.5
fibrocystic disease of the pancreas (FCD)
499, 521
fibrous layer 184
fibrous protein 418
fibula 311, F16.1, F16.8, F16.10
filaments 336
of blue–green algae 558
of flower 414, F22.1(a)
protein 336, F17.26(a), F17.27
filtering sensory signals 352
filtrate, glomerular 205
filtration pressure 171
fins 617
fish 281
fission 559
fissures, brain 353
flaccid cells 9
flagella 11, 556, 560
Flagellata 560
flagellin 556
flame cell 590
flatworm 589
flavin-adenine dinucleotide see FAD
flexion 316, F16.18
flight 620–2
flight or fight hormones 406
flipper 627
flowering
effect of day-length 430–1
effect of temperature 432
physiology 430–2
flowering plants 580–1
flowers 580
insect-pollinated 426–7
structure of 414
fluid-mosaic model of membrane structure
103 F6.10
fluke 627
blood 592
fluorescence 79
fluorescent microscopical techniques 520
focusing 360–1, F19.5
fogging 82
folia 351
folic acid 280
follicle
hair 385, F20.7(b)
thyroid 395
follicle cells 483
follicle stimulating hormone (FSH) 408,
483, 485, 491, F25.11, F25.14
fontanelles 308, F16.3
food chain 666
food web 666, 682
'foolish seedling' 473
foot of liverwort 571
of moss 572
'foot projection' of renal capsule F11.6(a)
foramen magnum 308
foramen ovale 176, F9.24
Foraminifera 559
forebrain 353–5
fossils 529–33
founder principle 517
fovea 363–4, F19.1(b)
fractionation of cells 116–17
free energy 76, F5.1
frequency distribution curve 512
frequency of sound 368
freshwater pollution 705–8
frond 575
frontal bone 307, F16.2, F16.3
frontal lobes of brain 354, F18.18
frostbite 386

fructose 18, F2.3, F2.4, F2.5
fruit 281
formation 422
physiology of 433–4
ripening 434
fucoxanthin 565
Fucus serratus 643
Fucus spiralis 643
Fucus vesiculosus 565, 643
fulcrum 317
functional residual capacity F8.8
fungi 566–70, 650, 679
parasitic 570
saprophytic 570
fungicides 690, 697
funicle 418, 438
fusiform initials 463, F24.9

galactocerebroside 26
galactosaemia 66, 517
galactose F2.3, 66
Galapagos finches 547
gall bladder 294, 299, F15.14, F15.22
galls 557
gallstones 301
gametangia 568
gametes
female 420
flowering plants 414, 416
male 420
mammalian 480
gametogenesis 480
gametophyte 423, 564, 565, 570, 575, 576
female 423, 579
male 423, 579
gamma-globulin 188
gamma rays 701
ganglion cell F19.8
basal 354
gas exchange
and evaporation 134–5
in mammals 134–54
in the tissues 144–7
gasohol 723–4
gastric cavity 589
gastric glands F15.8(a) (b)
gastric juice 288, F15.19
gastric mucosa 288
gastric pit 288, F15.18(b)
gastric secretion 294, F15.25
gastrin 290, F15.19, F15.25
Gastropoda 607
gastrovascular cavity 585
gemma cups 570
gemmae 570
gene
cloning 69, F4.24
flow 516
interaction 504–5
mutations 64–6, F4.19, F4.20, 543–4,
704
regulator 67
structural 67
therapy 66
genes 63, 496
general sensory area of brain 355
generations, alternation of 423
generative cell 416, F22.4(d)
generator potential 358
genetic code 60, T4.1
genetic counselling 520–1
genetic effects of radiation 703–5
genetic engineering 68, 717
genetic fingerprinting 73, F4.28
genetic polymorphism 517
genetic variation 130, 424
genetics 496–524
genome 126, 496
genotype 496
genus 552
geographical distribution 529, 541
geographical isolation 547–8
geographical zones 632
geotropism 473
germinal epithelium 481
germinal layer 385
gestation 492
germination
epigeal 440
factors affecting 444–8
hypogeal 439, 441
inhibitors 448–9
physiology of 442–9
of seeds 438–50
broad bean 438–9
maize 441
sunflower 440
gestation 149, 485, 492
gibberellic acid 469, 473
effect on flowering 431, 432
effect on seed germination 444, 445, 447
structure F24.26

gibberellins 473, 723
gigantism 409, F21.29
gill cleft 614, 615
gill slits 615
gills 139, 617, 618
glands see under names
glenoid cavity F16.6, 311
gliding joint F16.17
globin 143
globular protein 40
globulin 184
glomerulus F11.4(b), 202, F21.24
glucagon 402, F21.17
glucocorticoid hormones 404
glucose 17–18, F2.3, F2.4, 442, F21.18
control of blood 401–2
effect of medullary hormones on 407
tolerance 403, F21.19
glutamic acid F2.23, F2.24
glutamic–oxaloacetic transaminase (GOT)
305
glyceraldehyde 17, F2.2
glycerol 22, F2.13, 443
glycerophosphocholine see lecithin
glycine F2.23, 2.24
glycogen F2.10, 99, 278, 300, 338, 403
biochemical tests 21
granules F2.11
glycolipid 26, F2.19, 103
glycolysis 85, 86, F5.17(a)
glycoproteins 34, 103
glycosidic linkages 19, 288
glycosuria 403
goblet cells F15.28, 290
goitre 280, 398, F21.10
golgi body 99, 106, F6.13
ultrastructure F6.14
gonadotrophins 408
gonads 480
goose-pimples 387
Graafian follicle 484
grafting 476
grana 111
gravity
effects on the ear 374
effect on plant growth 468
grazing 635–6
Great Barrier Reef 588
green algae 563, 679
greenhouse effect 674, 689
grey matter 346, F18.8
Griffith's experiment F4.10
griseofulvin 722
growth inhibitors 469
growth promoters 469
gross ecological efficiency 671
gross primary productivity (GPP) 667
gross productivity 250, 667
ground meristem 460
growth
and development of flowering plants
452–78
physiology of 467–76
of human populations 657–9
primary 452–62
secondary 463–6, 580, 581
guanine 55
guard cells 242, F13.3
ATP in 245
mineral concentration in 245
photosynthesis by 244
of stomata 457
gut
absorption in 295–8, 400, 403
effect of medullary hormones 407
guttation 233
Gymnospermae 532, 573, 577–9
gynoecium 414
gyri 353

H-bands 334
habitats 632
destruction of 709
haem 143
haemocoel 600, 606
haemocyanin 607
haemoglobin 142–4, 149, 180–1, F15.21,
301, 540, 594
abnormal 65
adult 149, F8.21
breakdown F15.21
deoxygenation of F8.14
fetal 68, 149, F8.21, 176
haemoglobin S 150
oxygen dissociation curve of 144, 146,
F8.15, F817–19, F8.23
oxygenation of F8.14
quaternary structure F2.32
structure 29, 32, F2.32, F8.13
haemolysis 11, F1.11, 181
haemolytic disease of the newborn
(HDNB) 195–6
haemophilia 512, 517

haemorrhage 181
haemosiderin 301, F15.21
haemostasis 183
hair 385, 387, 624
  skin F20.9
hair cells in ear 327, F19.18(a), F19.20, F19.24, F19.25(a)
hair follicle 385, F20.7(b)
half-life 530, 701
hallux 312, F16.10
halteres 603
haploid 130
Hardy–Weinberg equation 515–17
Hardy–Weinberg equilibrium 516
Hatch–Slack pathway 84, F5.15, 249
haustoria 645
Haversian canal 314
hay fever 192
hearing 371–2
hearing aid 375, F19.26
hearing threshold level (HTL) 370
heart 156, 158–67, F9.3(a)
  murmurs 163
  pressure changes in 163
  rate 164, 166–7
  sounds 163
  valves F9.6
heat see oestrus
heat centre 384
heat distribution F20.2
heat exchange mechanism 135
heat loss 381–3, 385–9, F20.3, F201.10
heat production 380–1, 384
heavy metal pollution 692–6
heel 312
helicotrema 372
helium 152
α-helix 30, F2.27
hemicelluloses 113
Hemiptera 603
hepatic artery 9.2, 299, F15.29(a)
hepatic portal vein 299, F15.29(a)
hepatic vein 9.2, 299, F15.29(a)
Hepaticae 571
herbaceous plants 580
herbicides 696–7
herbivores 284, 668
heredity 452
Hering–Breuer reflex 141
hermaphrodite species 414, 590
Hertz (Hz) 369
heterocercal tail 615
heterocysts 558
heterodont 283, 624
heterogametic sex 510
heteromorphic alternation of generations 570
heterospores 574
heterosporous species 578
heterothallic fungi 568
heterotrophs 556, 567
heterozygote 496
hexosephosphate isomerase 51
hibernation 390
high energy bonds 85
Hill reaction 80
hilum 438, F23.1(a)
hindbrain 351
hip F16.17
hirudin 598
Hirudinea 598
histamine 192
histocompatibility antigens 197
hive 605
HLA 197
holdfast 565, 642
hollow nerve cord 614
homeostasis 200, 298
  cellular 66
  coordination in 320
  mammalian 320
Hominidae 628
Homo 535–8
  H. erectus 563
  H. sapiens 549, 552, 636
    H. s. neanderthalensis 536–71
    H. s. sapiens 537–8, 549
homocercal tail 616
homodont 620
homogametic sex 510
homogentisic acid 66
homologous structures 539, 553
homothallic fungi 568
homozygous 496
hooks 592
horizons 675
hormone receptor complex F21.3
hormones
  flowering plant 496–76
  glucocorticoid 404
  mammalian F17.2
  plant, practical uses 475–6
  production 723
horsetail 573, 574

hot centre F20.13(a)
house flies 602
human chorionic gonadotrophic hormone (HCG) 191, 490
Human Immunodeficiency Virus (HIV) 193–4
humerus 311, F16.1, F16.6, F16.8, F16.9, F16.18, F27.19
humoral immune response 188–90, F10.12
humus 570, 675, 678
Huntington's chorea (HC) 498
hyaluronic acid 103
hyaluronidase 489
hybridomas 190
hydathodes 233
Hydra 584, 585, 586
hydranths 586
hydrocarbons 79
hydrochloric acid 288
hydrogen bonds, in water 4
  in cellulose 22
  in DNA 55
  in protein 30, 32
hydrogencarbonate 145
  concentration in blood 141
  ion 6–7, 213, 215
hydrogen sulphide 529
hydroid 584, 586
hydrolases 49
hydrolysis 49, F3.15
hydrolytic enzymes 287, 442
hydrophilic colloid 3–4
hydrophobic bonds 32
hydrophytes 226
hydroskeleton 596
hydrostatic pressure of blood 170
hydrotropism 473
hydroxyl ion 6
Hydrozoa 586–7
hygiene 658
Hymenoptera 605
hyper-exponential growth 657
hyperactivity 192
hypercalcaemia 401
hypercapnia 141
hyperglycaemia 403
hypermetropia 361, 362, F19.7
hyperparathyroidism 401
hyperpolarisation 332, F17.8, F17.19, 365
hypersensitivity 192
hyperthyroidism 398, F21.10
hyphae 566
hypocalcaemia 401
hypocotyl 440, F23.2(b)
hypogeal 439, 441
hypoparathyroidism 401
hypophysial stalk F21.28
hypophysis 408
hypothalamic–hypophysial neurons F21.32
  portal veins F21.31
hypothalamus 210, 353, F18.15, F20.6, F20.10, F20.13(a), 396, 408, F21.31, F21.32
hypothermia 389
hypothyroidism 398

IAA 433, 469–73, 474
  effect on root and shoot growth 444
  role in apical dominance 470
  role in tropisms 470–3
  structure F24.17
  transport F21.18
I-bands 334
ICSH F25.6
IgA 188
IgD 188
IgE 188, 192
IgG 188, 195, F10.13
IgM 188, 196, F10.13
ileum 292, F15.14, F15.26(b), 15.27
iliac artery, common 9·2
iliac vein, common 9·2
ilium 311, F16.7
image formation in the eye 360
imago 603
imbibition 445
immune antibodies 194
immune rejection 216
immune system 185–98, 646
immunity 185, 186–94
immunobilised enzyme technology 718
immunoglobulins 37, 184, 188, F10.13, T10.3
immunosuppression 197, 216
implantation 486, 489
  delayed 485
impression, fossils 529, F27.3
impulse 323, 324, F17.8, F17.17
  conduction 324–5
  rate of 326–8
  transmission F17.9
inbreeding 424, 429, 514–15
incipient plasmolysis 10, F1.10

incisors 283
incompatible blood 195
incomplete antibody 196
incomplete dominance 503
incus 371, F19.17
index finger 312
indicator species 711
indole-3-acetic acid see IAA
induced fit mechanism 42, F34.(b)
induced ovulation 485
inductive photoperiod 430
indusium 575
industrial effluent 725
industrial melanism 546
inflammation 192
inflorescence 414
infra-red radiation 381
inheritance 496
  autosomal 496–510
  and natural selection 543
  non-autosomal 510–12
  quantitative 512–14
  of sex 510–11
inhibitors
  competitive 44–5
  effect on enzymes 43
  germination 448–9
  growth 469
  muscle F18.11
  non-competitive 45
  non-reversible 45
  reversible 44–5
inhibitory postsynaptic potential (IPSP) 332
inhibitory synapses 331–2, F17.19
innate immunity 185
innate reflexes 348
inner ear 370
innominate artery F9.3(a)
insect-pollinated flowers 426–7
Insecta 602–5, 680
insecticides 698–701
insectivores 625
inspiratory reverse volume F8.8
insulin 300
  and glucagon 402–4, F21.17, F21.18
  production of, by genetic engineering 70–1
  structure 29, F2.26
integuments 418, 579
intelligence 624, 628
intercalary meristems 466
intercalated discs 335
intercostal muscles 137, F8.3, F8.12(b), 210, 619, F25.26
intercostal space F16.6
interferon 71, 187
intermediate neurone F18.8
internal factors in plants 452
internal fertilisation 620
internode elongation 459, 473
interphase 120–1, F7.2
interstitial cell stimulating hormone (ICSH) 483
interstitial cells (Leydig) 481
interventricular septum F9.3(b)
intervertebral discs 308, F16.4
intestinal fluid 293–4
intestinal microorganisms 294–5
intramembranous ossification 314
intravascular thrombus 184
intrinsic factor (IF) 181
intrinsic thromboplastin 183, F10.5
invertebrates 584
inverted image 360
inverted retina 363
iodine 280, 396, F21.5, F21.7, 21.8
iodopsin 364, F19.11
ion exchange in soil 268, 676
ion pump 245
ion uptake, plants 264–7
ionic product, water 6
ipsilateral reflex arc 347
iris, eye 359, F19.1(b)
iron
  in mammals 280
  metabolism 301–2
  in plant growth 266
ischium 311, F16.7
islets of Langerhans 192, 300, 402, 403, F21.2, F21.16, F21.17
isoantibodies 194
isoelectric point
  of amino acids 28, F2.34
  of proteins 32–3
isogamous species 564
isolating mechanisms 547–9
isomerases 51

Jacob–Monod hypothesis 67, F4.22
jaundice 301
jawed fishes 531
jaws 308
jejunum 292

jellyfish 531, 584, 588
Jenner, Edward 192–3
joints 315–16, F16.15
  ball and socket 315, F16.17
  gliding F16.17
  hinged 315, F16.17
  pivot 315, F16.17
jugular veins 9.2
Juncus communis 677
Juncus maritimus 665
Jurassic period 532
juvenile hormone 603
juxtaglomerular cells F21.24
juxtamedullary nephron 203, 208, F11.4(b)

kangaroo rat 135, F8.1, F8.2, 212
karyotype 519
karyotyping 518–19
keel 622
keratin 30, 385
ketose sugars 16
key factor 656
kidney 200–22, F11.1, F11.3(a)
  medulla 202
kidney failure 192
kidney machines 214–15, F11.19(b)
kidney transplantation 214, 216
kieselguhr 35
kinesis 644
kinetic energy 76, 223
kinetin 433, 474
kinetodesmata 561
kinetosome 561
Klinefelter's syndrome 511, 520
klinokinesis 644
knee 311
knee-jerk reflex 346
Krebs cycle 86–8, F5.20
Küpffer cells 301

labour 492
lachrymal glands 358
lactation 302, 491
lacteals F15.28
lactic acid 85, 151
lactic fermentation 85
lactiferous duct 25.20(a)
lactose 19, F2.6
lacunae in bones 314
lag phase 656
Lagomorpha 285
Lamarckism 542
lamellae 569
  in bone 314
  in cartilaginous fish 617
  in chloroplasts 111
  in rods 364, F19.10
Lamellibranchi 607–8
large intestine 298
larvae 603
latent heat
  of melting 5
  of vaporisation 5, 382
latent period 554
lateral canal 616
lateral line system 615
lateral meristems 463
lateral root development 461–2
lateral shoot development 458–9
lattice precipitate 190
law of independent assortment 502
law of segregation 499
lead, pollution by 692–4
leaf
  development 458–9
  mosaic F13.2
  of moss 572
  penetration by light 241
  as a photosynthetic organ 240
  primordia 453, F24.1(a)
  shade 241, F13.12
  structure 225–8
  suction 229
    unit rate 249, 250
  sun 241, F13.12
  water supply 241
leaf area index 249, 250
leafy liverwort 571
lecithin 25
leech 598
lens 360, F19.1(b), F19.3, F19.4
  eyepiece 114
  objective 114
lenticels 466, F24.13
lenticular transpiration 220
Leodice viridis 596
Lepidoptera 604
leucocytes 181–4
leukaemia 52
leverage 317
levers 317, F16.20–2
Leydig cells 481
lichens 650
life cycle, flowering plant 423

life expectancy 659
ligaments
  in joints 316, F16.18
  suspensory 360, F19.1(b)
ligases 51
light
  effect on plant growth 467–8
  effect on seed germination 447–8
  penetration of water 6
  penetration through leaves 241
  refraction of 360
light-adapted eye 365
light-dependent reactions 77–81
light-independent reactions 81–4
light intensity
  and compensation point 467
  effect on eye 362
  effect on photosynthesis 248
  effect on transpiration 222
light microscope 97–9, 114–16, F6.24
lignin 22, 113
limbic system 354
limbs 311–12
limiting factors, law of 247
Limnaea truncatula 591
Limonium vulgare 665
linkage 505–7
Linnaeus 552
lipase enzymes F15.23
  in germinating seed 443
  in mammals 293, 297, T15.3
lipids
  bilayer 102
  in mammals 279, 297
  metabolism 301
  in seeds 442
  simple 22–4
lipoids 22–6
lipoproteins 34
littoral belt 643
Littorina saxatilis 643
liver 298–302, F15.14, F15.22, 403
  lobule F15.29(a)(b)(c)
liver fluke 591
liverwort 570, 571
Ilama F8.20(b)
lobes, brain 354
lobster 600
lobules of liver F15.29(a)(b)(c)
locating agents 35
lock and key mechanism 42
locust 602
log phase 656
long-day plants 430
long-sightedness 361
loop of Henle 200, F11.4(b), F11.10
lugworm 596
lumbar vertebrae F16.1, F16.5(e), F16.7
Lumbricus 596
lumen 169
lung books 606
lungfish 616
lungs 136, F8.3, F8.4, F8.5, 158, 619
Lupinus arcticus 438
luteinising hormone (LH) 408, 483, 485,
  F25.13, F25.14
lyases 51
Lycopsida 573–4
lymph 172–4, F9.20
  circulation of 156–78, F9.1
lymphatic capillaries 172, F9.20, 297
  nodes 174, F9.20, F9.21
  system 172, F9.20, F9.21
  vessels F9.21
lymphatic duct F9.21
lymphocytes 174, F9.20, 181, 186,
  F10.1
lymphokines 187
lysine F2.23, F2.24
lysis 190
lysosomes 107–8
  structure F6.16
lytic cycle 554

macrofauna, soil 680
macronutrients
  in mammals 280
  plant 263, 264–6
macrophages 182, 187, 190
maculae 374, F19.23, F19.24
madreporite 609
magnesium in plant growth 263, 266
malaria 150, 562–3
male cone 578
malic acid 576
malleus 371, F19.17
Malpighian body 202
Malpighian tubule 602, 603, 606
maltase
  in germinating seeds 442
  in mammals T15.3
maltose 19, F2.6, 288, 442
Mammalia 534, 624–9

mammary ducts 491, F25.20
mammary glands 491, F25.20(a)(b), 624,
  625
mandible 308, F16.2, 601
manganese deficiency in plants 266
manometer, Warburg 90, F5.23
mantle 607
mantle cavity 607
manubrium 310
manures 268
marginal meristem 458
marine pollution 708–9
marrow cavity F16.12
Marsupella 571
marsupials 625
mass-flow hypothesis 257–8, F13.23,
  F13.24
mast cells 182, 192
matrix 108
matrix potential 13
matter, cycling of 672–5
maxilla F16.2, 601
maxilliped 601
maxillule 601
mean 513
median canal 371, F19.18(a)
medical care 658
medulla
  adrenal 404, 406–7
  kidney 202
medulla oblongata F8.10, 164, 166, 351,
  F18.15
medullary gradient F11.10
medullary hormones 406–7
medullary rhythmicity area 140, F8.10
medusa 584
medusoid 588
megakaryocytes 182, F10.4
meganucleus 560
megasporangia 423, 574, 579
megaspore mother cell 418
megaspores 418, 423, 574, 579
meiosis 126–31, F7.9(a), F7.10
  significance of 130–1
melanin 363, 513–14, F19.8
melanism, industrial 546
membranes 9, 99, 102–4
  basement 137, 204, F11.6(a)
  bones 313
  cell 9, 99, 102–4
memory 354, 367, 624
  B-cells 190
Mendel, Gregor 496
Mendel's First Law 499
Mendel's Second Law 502
menopause 484, 487
menstrual cycle 485
menstrual flow 487
menstruation 487
mercury pollution 694–5, 708
meristems
  adaxial 458
  apical 452, F24.1(a)
  ground 460
  intercalary 466
  lateral 463
  marginal 458
Merulius lachrymans 570
mesentery 587
mesocotyl 441, F23.3(b)
mesoderm 589
mesofauna, soil 680
mesogloea 584
mesophyll F24.5(c)
mesophytes 226
mesosome 556
messenger RNA 56, 57, F4.8, 110
metabolic activity in seeds 445
metabolic pathways 76
metabolic rate 139, 380, F20.4
metacarpals 312, F27.19
metacarpals F16.1, F16.8, F16.9
metachronal rhythm 561
metallophytes 267
metameric segmentation 594
metamorphosis 603, 619
metaphase
  meiosis 128
  mitosis 122–3
metaphloem 455
metatarsals 312, F16.1, F16.8, F16.10
Metatheria 625
metaxylem 232, 456, F24.2(d)
micelles 297
Michaelis constant 48
micro-ecology 644
microbodies 109
microfauna, soil 680
microfilaments 113, 123
microflora, soil 679–80
microhabitat 644
micronucleus 560
micronutrients, plant 263, 267, T14.6

micropyle 418, F22.5(d), 438, F23.1(a), 579
microscope
  compound 96, F6.2
  electron 99–113, 114–16, F6.5, F6.24
  light 97–9, 114–16, F6.24
  phase-contrast 97, 115–15, F6.26
  simple 96, F6.1
  Van Leeuwenhoek's 96
microsporangia 423, 574, 578
microspores 416, 423, 574, 578
microsporophyll 578
microtome 98
microtubules 110, 113
micturition 201
midbrain 352–3, F18.15
middle ear 370
middle lamella 99, F14.6
milk 281, 491
milk expulsion reflex F25.20(b)
milk teeth 283, F15.3
millipede 598, 602
Millon's reagent 33
mineralocorticoid hormones 405
minerals 280
  absorption
    mechanisms of 269–71
    in plants 268–71
  circulation of, in plants 273
  deficiency, effect on plant growth 264
  elements, effect on plant growth 468
  fertilisers 268
  functions in plants 264–7
  ion concentration, guard cells 245
  movement of, into shoot system 272
  nutrition of plants 262–76, 468
  plant 262
  requirements, plants 262–4
  toxic 267
  transport in plants 272–3
  uptake, roots 272
minisatellites 73
minute volume 140
miracidium larva 591
mites 606
mitochondria 87, 99, 108–9, 566
  ultrastructure F6.17(a)
mitosis 120–6, F7.4(a)
  significance of 125–6
mitral valve F9.3(b)
molars 283, F15.2
molecular clusters, in water 4
Mollusca 531, 606–8, 681
moments, principle of 317, F16.23
Monera 552, 555
mongolism see Down's syndrome
monoclonal antibodies 190, 723
Monocotyledons 580
monocytes 181, F10.1
monoecious species 578
monohybrid inheritance 497–9
monohybrid ratio 497
monomers 526–7
monosaccharides 16–18, T2.1, 297
monotremes 625
monozygotic twins 489
mosquito 602, 637
moss 570, 572–3
moth 602
motor end plates 332
motor region, of forebrain 354
mould, fossils 529
moulting hormone 603
mouth 288
movement 316–17
moving parallax 367
Mucor 567, 568
mucin 290, F15.18(a)
mucopolysaccharides 103
mucosa, gastric 288
mucus 106, 137, 290, F15.28, 597
multiple alleles 503–4
murmurs, heart 163
Musci 572–3
muscle 333–8, F17.2
  antagonistic 318, 18.11, 584–5
  calcium in 338
  cardiac 333, 335, F17.24
  contraction 336–8
  fibres F17.16(e), F17.21
  inhibitors F18.11
  involuntary 335
  skeletal 333–4, F17.21, F17.22
  smooth F9.14
  striated 333
  striped 333
  visceral 333, F17.25
  voluntary 333
muscular arteries 168, F9.14, F9.18
mutagenic substances 543
mutation
  chromosome 543, 544
  gene 543–4
mutualism 648–62
mycelium 566

mycorrhiza 269, 648–9
mycotoxins 570
myelin sheath 326, F17.12(a)(b), F17.13,
  F17.16(c)(f)
myocardium 159
myofibrils 333–4, 336, F17.21
myogenic activity 159, 164, 335
myoglobin 150–1, 338
  oxygen dissociation curve F8.23
  structure 31, F2.30
myopia 361, F19.6
myosin 113, 335, F17.26–28
myotomal muscles 615
Myriapoda 602
Mytilus 607–8
myxoedema 192, 398
myxomatosis 554, 635

NAD 43, 50, 86
NADH 86, 528
NADP 43, 50, 80
NADP$^+$ 528
NADPH 112
nail-patella syndrome 505
nails 628
natural classification 552
natural selection 517, 545–6
Nauplius 601
Nautilus 608
Neanderthal man 536–7
nectar 426
negative feedback F8.12(a), 183, 396, 404,
  483, 485
  inhibition 43
nematoblast 585
nematocyst 585
Nematoda 593–4, 680
Neo-Darwinism 543
neotenin 603
neoteny 619
nephridiopores 594
nephridium 594, 596
nephrons 201
  cortical 203, 208, F11.4(b), F11.10
  functions of 203–12
nephrostome 594
Nereis 595
nerve cord 590
nerve impulse see impulse
nerve net 585
nerves see neurons
nervous control and coordination in
  mammals 320–40, 342–56
nervous integration 342–56
nervous system 320, 321 F18.1
  effect of medullary hormones on 407
net primary productivity 250, 667
neural spine F16.4
neuro-effector junctions 339–32
neuro-muscular junction F17.16(f)
neurogenic activity 334
neuronal pathways 343–5
neurons 320–40, F17.3(a)
  afferent 345
  bipolar F17.3(b), F19.8
  converging F18.4
  efferent F17.3(b), 345, F18.8–10, F18.14
  intermediate F18.8
  interrelationship with endocrine system
    F17.2
  multipolar F17.3(b)
  neural canal F16.4, F16.5(d)
  neuronal pathway, diverging F18.2–3
  non-myelinated F17.14
  pseudo-unipolar F17.3(b)
  receptor F17.3(b), F18.8–10, F19.8
  reverberating F18.6
neutral solution 6
neutralisation 190
neutrons 701
neutrophil 181, 190, F10.1, F10.3(a)
niche 634, 656
nicotinamide 280
nicotinamide-adenine dinucleotide see
  NAD
nicotinamide-adenine dinucleotide
  phosphate see NADP
nicotinic acid 280
nictitating membrane 620
ninhydrin reaction 33, 35, 37
nipple F25.20(a)
Nissl bodies F17.3(a)
nitrifiers 558
nitrification 673
Nitrobacter 673
nitrogen
  deficiency in plants 264
  narcosis 152
  in plant growth 264
  in soil 683
nitrogen cycle 672–3
nitrogen-fixing bacteria 558, 673
nitrogen oxide pollution 691

nitrogenous bases in nucleic acids F4.2
Nitrosomonas 673
node of Ranvier 327, F17.12(c), F17.13
nodules
    root 650
    in thyroid 398
nomenclature and classification 552–3
non-autosomal inheritance 510–12
non-cyclic photophosphorylation 81
non-endospermic seeds 421
non-essential amino acids 278
non-essential fatty acids 279
non-myelinated axons 334
non-reducing sugars 19
non-self antigens 186, 216
non-vital centres 351
noradrenaline 166, 331, 332, 350, F18.14, 384, 406
noradrenergic nerves 350
normal distribution curve 513
Nostoc 650, 673
Nucella lapillus 643
nucellus 418, 579
nuclear division
    meiosis 126–9
    mitosis 121–3, F7.4(a)
nuclear fuels 702
nuclear membrane 99, 109, 6.19
nuclear reactors, thermal 702–3
nuclear weapons 702
nucleic acids 54–9, 60–3, 110, 544
nucleolus 99, 110
nucleoside F4.3
nucleotide 54, F4.3
nucleus 99, 109–10, F6.18, 566
    in brain 353
numerical aperture 115
nutrient-film technique 274, F14.14
nymphs 603

Obelia 584, 586–7
objective lens 114
occipital 307, F16.2, F16.3
occipital lobes 354, F18.18
odontoblasts 282, F15.1
odontoid process 309, F16:5, F16.17
oedema 174
oesophagus 15.14, F15.15, F15.17(a), F15.21, F15.22
oestrogen 485
oestrus cycle 485
oil
    extraction of 724–5
    pollution 708–9
    spillages 725–6
oils 22, 24, 279, 281, 443, 563
olecranon process 311
oleic acid 23, F2.14
Oligochaeta 596–8
oligotrophic lakes 707
omasum 291, F15.21
ommatidia 599
omnivores 284, 628
onchosphere 593
oocyte F25.10
oogamy 565, 570
oogenesis 483–4, F25.10
oogonia 483, 565, F25.10
oosphere 565, 576
oospore 568
oozes 559
open roots 285
operculum 573, 616
opisthosoma 606
opsin 365
optic chiasma F19.12
optic nerve F18.16, 360, F19.1(b), F19.8, F19.12
optic tract F19.12
optimum pH 47
optimum temperature 46, 445
Opuntia stricta 662–3
oral arms 589
oral groove 561
oral surface 609
orbit 308, F16.2
orders 552
Ordovician period 531
organ differentiation 474
organ, level of development 589
organ of Corti 372, F19.19, F19.20
organ transplantation 197
organic farming 268
organic matter in soil 677–8
organochlorine insecticides 698, 708
organophosphorus compounds 698
origin of life 526–50
ornithine cycle 300, F15.30
orthokinesis 644
os calcis 312, F16.10
osmometer F1.6
osmoregulation 12
    in fish 641, F31.15
osmoregulation centre 210

osmosis 8, F1.7, F1.9, 103, 206, 210, 229, 233, 446, 640
osmotic behaviour
    animal cells 11–12
    plant cells 9–11
osmotic fragility curves 12, F1.12
osmotic fragility test 12
osmotic properties 7
ossicles 371
ossification 314
    centres F16.14
Osteichthyes 616–18
osteoblast 314, F16.12
osteoclast 314, F16.12
osteocytes 314
ostia 600
ostiole 565
otoliths 374, F19.24
outbreeding 424, 514–15
    mechanisms 424–5
outer ear 370
oval window 371, F19.17, F19.18(c), F19.19
ovarian cycle F25.14
ovaries
    flower F22.1(a), 414
    in *Hydra* 586
    mammalian 480, 483–4, F25.8–9, F25.11, F25.13
over-exploitation 709
oviducts 483, 487, F25.8
ovulation 484, 485–7, F25.10, F25.14
ovule 414, 416, F22.1(a)
    formation 418
    in gymnosperms 578
ovuliferous scale 578
ovum 483, F25.10
oxidases 50
oxidation of respiratory substrates 85–8
oxidation water 13, 212
oxidative phosphorylation 88
oxido-reductases 50
oxpecker birds 652
oxygen 638–9
    availability in water 638–9
    carrying capacity T8.3
    debt 142, 151
    effect on plant growth 468
    effect on seed germination 446–7
    tension 141, 166
    uptake and carriage 142–51
    uptake by plant tissues 258, T13.6
oxygen dissociation curves
    of haemoglobin 144, 146, F8.15, F8.17–19, F8.23
    of myoglobin 150
oxygenation
    of blood 142–4
    of haemoglobin F8.14
oxyhaemoglobin 143, 145, 146, 338
oxysomes 108
oxyntic cells 288, F15.18(a)
oxytocin 29, 410, 491, 492
ozone layer 638

P-wave (ECG) 162
pacemaker, cardiac F9.9
palate 624
palisade mesophyll F11.2(a), F24.5(c)
palmitic acid 23, F2.14
palps 595, 608
pancreas 292, F15.14, F15.22, F15.23, F15.25
pancreatic amylase 293, F15.23
pancreatic duct F15.22, F15.23
pancreatic juice 292–3
pancreatic lipase 293
pancreozymin 294, F15.25
Paneth cells F15.28
panting 389
pantothenic acid 280
papillary muscle F9.3(b), F9.6
papullae 610
parallel-flow arrangement 615
Paramecium 12, 560, 561
parapodium 595
parasites 556, 567, 679
    facultative 645
    obligate 645
parasitism 645–8
    controlled 653
parasympathetic nerves 342, 348, F18.1, F18.13, F18.14
parathormone 399, 400, 401, F21.12, F21.14, 487
parathyroid F21.11
parathyroid glands 399–401, F21.1, F21.15
parathyroidectomy F21.15
parenchyma 457
parietal bone F16.2, F16.3
parietal cell 418
parietal lobes 354, F18.18
parotid gland F15.15
parthenocarpy 434

parthenogenesis 428
    diploid 428
    haploid 428
partition coefficient 34
parturition 492
passive immunity 185, 490, 491
passive transport in plants 272
pathogenic bacteria 557
patella 312, F16.1, F16.10
Patella vulgata 643
pathogen 570, 645
pathways, nervous 342
pavement epithelium 136
peat 678
pectic acid 113
pectin 434
pectoral girdle 310, 618
pedicellariae 609
pedipalp 606
pelagic 311, F16.1
Pellia 571
pellicle 563
Pelvetia canaliculata 643
pelvic bone 311, F16.7, F16.10
pelvic girdle 310–11, F16.7, 618
pelvis 311, F16.1
penicillin 722
Penicillium 544, 636
penis 483, 487, F25.1
pentadactyl limbs 311, F16.8, 539, 618
pentamerous species 609
peppered moth 545
pepsin 288
pepsinogen 288
peptic cells 288, F15.18(a)
peptidase enzymes 443
percentage relative humidity 224
perception 354
    of colour 365
    of distance and size 366–7
    of sound 373
perceptive deafness 375
perennials 581
pericarp 422, 434, F22.21, 438, 580
perichondrium 313
pericycle 232, F12.17(a)
periderm 231, 465, F24.12
perikaryon F17.3(a), F17.16(b)
perilymph F19.20
perimysium 333, F17.21
perinuclear space 109
peripheral nervous system 342, F18.1
periosteum 313, F16.12
perisarc 586
perissodactyls 626
peristalsis 258, 288, F15.16
peristome 572
permanent teeth 28, 30, F15.3
Permian period 532
pernicious anaemia 181
peroxisomes 109
persistence of insecticides 698
pesticides 696–701
petals 414, F22.1(a)
petrification 529
pH
    blood, and respiration 146
    and enzyme action 43, 46–7
    optimum 47
pH scale 6
Phaeophyta 565–6
phagocytes F9.20
phagocytosis 12, 103, 181, 187, 190, F10.3, 554, 559, 646
phalanges 312, F16.1, F16.8–10, 27.19
pharynx 288, 590
phelloderm 466, F24.12
phenotype 496
phenoxyacetic acids 475
phenylalanine hydroxylases 65
phenylketonuria 65–6, 517
pheromones 636
phloem 251, F13.18, F13.19
    protein 255
    role in translocation 251
    structure F13.18
    ultrastructure F13.19
phosphate 280
phosphocreatine 85, 338, F17.29
phosphoglyceraldehyde (PGAL) 86
phospholipids 25, 26, F2.17, F2.18, 102, 137, 297
phosphoric acid 25
phosphorus
    in plants 265
    in soil 683
phosphorylation, oxidative 88
phosphotransferases 50
photoautotrophism 556
photoautotrophs 679, 681
photochemical smog 691
photolysis 80
photomicrographs 114
photoperiod 430
photoperiodic response 430

photoperiodism, vegetative F22.18
photophosphorylation 80–1
    cyclic 81, F5.11
    non-cyclic F5.9, F5.10
photoreception 362–6
photoreceptors 362
photorespiration 84
photosensitive pigments 364–6
photosynthesis 2, 77–84, 240–60, 528–9, 674
    in chloroplasts 112
    effect of carbon dioxide concentration 249, 639
    effect of light intensity 248
    effect of temperature 247
    factors affecting rate of 246–9
    in guard cells 244
    light-independent 247
    pigments T5.1
phototaxis 563
phototropism
    action spectrum 472
    negative 472
phragmoplast 124
phycocyanin 558
phycoerythrin 558, 566
Phycomycetes 568
phyla 552
phyllotaxis 459
physiological immaturity 448
phytoalexins 557
phytochrome 431, 447, 467
    absorption spectra F23.11
phytohaemagglutinin 519
Phytophthora 567, 570
pileus 569
pinna
    of ear 371
    fern 575
pinnule 575
pinocytosis 12, 103, 554
Pinus sylvestris 578
pitch 368
pith, shoot 453
pituitary body 210, F11.15, F18.15, F20.13(a), 396, 407, 408–10, F21.2, F21.6, F21.22, F21.22, F21.28
pituitary gland F25.6, F25.11, F25.13
    anterior 408–10, F21.31, 438, F25.14
pituitary hormones 355
pivot 317
placcoid scale 615
placenta 149, F9.23, F9.24
    of ferns 575
    flowering plant 418
    mammalian 175, 486, 489–91, F25.16, F25.19, F25.21, 625
    villi F25.17
placental mammals 532, 625
Planaria 589–90
plankton 286
plant communities 249
plant tissue differentiation 452
Plantae 552, 570–81
Plantago maritima 665
planula larva 587
plasma
    blood 142
    cells 188, F10.10–12
    inorganic iodine 397
    proteins 184, F10.7, 205, F15.21
plasmagel 559
plasmasol 559
plasmids 69
plasmodesmata 256
Plasmodium 562
plasmolysis F1.9
    incipient F1.9
    total F1.9
plate tectonics 541
platelets F10.4
    and coagulation 182–4
    factors 183
Platyhelminthes 589–93
pleated sheets 30, F2.28
pleura F8.3
plumule 421, F22.8(a), 438
pluteus larva 609
pneumatic bone 621, 623
Pneumococcus 57, 58, 68
pneumotaxic area 140, F8.10
podocyte F11.6(a)(b)
polar body 484, F25.10
polar molecules 2
polarisation 322
pollen 577
    chemotropism of 432
    formation 416–18
    grains 414, 416, 432–3, F22.3(d), 530
    growth, physiology 432–3
pollen mother cells 416
pollen tube 420, 579
pollex F16.9, 312

pollinating insects 532
pollination 420, 424–7, 579
  mechanisms 426–7
pollinator 425
pollutants 688
pollution
  of aquatic habitats 705–9
  of atmosphere 688–92
  by heavy metals 692–6
  by pesticides 696–701
  radiation 701–5
Polychaeta 595–6
polydactyly 498
polygenic inheritance 513
polyhybrid inheritance 502–3
polymerisation 527
polymers 20
  isolation and replication 527–8
  synthesis 527
polymorphism, genetic 517
polynucleotide 54, F4.4
polyp 586
polypeptides 28–33, F4.17, F4.18
polyploidy 510, 544
polysaccharides 20–2
polysomes 63
polysomics 544
polysomy 509
Polytrichum 665
pons varolii 351, F18.15, F18.16
population 656
  dynamics 660–3
  genetics 514–17
  growth 656–7, T32.1
  regulation of 662
pores (stomatal)
  and carbon dioxide uptake 242–3
  effect on transpiration 225–6
porphyrins 78
positive chemotropism 432
positive feedback 389
positive phototropism 470
post-anal tail 614
posterior pituitary 408, 410, F21.32,
  F25.20(b)
potassium 280
  blood control, of 405–6
  deficiency in plants 265
potato blight 570
potential energy 76
potometer 228
power pack transformer 37
prawns 600
Precambrian period 530
precipitation 190, 675
precipitins 190
predation 660–1, 682
predators 660, 679
  humans as 661
preen gland 621
pregnancy 302
  energy requirements during 92
  and fetal haemoglobin 149
  glucose in 404
  test 191, F10.16
prehensile tail 628
premolars 283
pressure
  atmospheric 639–40
  blood 163, 167, F9.8(a), F9.12, F9.18,
    215
pressure potential 9
  root 233–4
  waves in ear 371
pressure receptor F20.7(b)
prey 660
primary auditory areas 372, F19.21
primary consumers 665
primary follicles 483
primary growth 452–62
primary oocyte 484
primary plant body 452
primary producers 528
primary productivity 250
primary root system 460–2
primary shoot system 452–9
primary spermatocytes 481
primates 534, 628–9
primordia
  axillary bud 453–4
  leaf 453, F24.1(a)
primordial germ cells 481, 483, F25.4,
  F25.10
principle of moments 317, F16.23
proangiotensin F21.25
proboscis 604
procambial strands 455, 460, F24.1(a)
producer organisms 665, 667–8, 682
productivity of plants 249–50
profibrinolysin 184
progesterone 485, 491, F25.12–14
proglottides 592
projector lens 114
prokaryotes 101, 555, 558

prolactin 491
prolactin inhibiting factor (PIF) 491
prolactin releasing factor (PRF) 491
promycelium 568
prophase
  meiosis 126–8
  mitosis 122
Prosimii 534, 628
prosoma 606
prostacyclin 183
prostate gland 488, F25.1
prosthetic group 34, 40, 43
protandry 424
protein-bound iodine 397
protein pool 300
proteins 27–34, 278–9, 297
  biochemical tests 33
  buffering capacity 32
  in cell membrane 102
  conjugated 40
  denaturation 33
  functions 33
  globular 40
  insolubility 32
  insoelectric point 32–3
  manufacture of 720–1
  metabolism 300
  properties 32–3
  repressor 67
  in seeds 442, 443
  solubility 32
  structure 30–2
  synthesis 60–8, 300
prothallus 576
prothrombin 34, 183, 184, F10.5
Protoctista 563–7
protoderm 460
protogyny 425
protonema 573
protophloem 455
protoplasm 99
protoplasmic streaming 258
Prototheria 625
protoxylem 232, F24.2(d)
  differentiation 455
Protozoa 559–63, 680
proximal convoluted tubules 202, F11.8,
  F11.9
pseudocoel 594
pseudoheart 597
Pseudomonas denitrificans 673
pseudopodia 559
pseudostratified ciliated columnar
  epithelium 137
psilophytes 531
Pteridophyta 531
pterodactyl F27.9
Pteropsida 575–6
pterosaurs 619
puberty 483, 484
pubic bone F25.1
pubis 311, F16.7
Puccinellia maritima 665
pulmonary artery F9.3(a), F9.10
pulmonary embolism 184
pulmonary valve F9.3(b)
pulmonary vein F9.3(a)
pulmonary ventilation 140
pulp cavity F15.2
pulses 281
pupa 603
pupil 359, F19.1(b)
purines 55
Purkinje fibres 159, 162, F9.4, F9.5(b)
pyloric caeca 610
pyloric sphincter 291, F15.17(a)
pyramid cell F17.3(c)
pyramids of biomass 671
pyramids, ecological 671–2
pyramids of energy 671–2
pyramids in kidney medulla 203, F11.3(a)
pyramids of numbers 671
pyrimidines 55
Pythium 645

Q10 (temperature coefficient) 46, 247
QRS-complex (ECG) 162
quadriceps extensor muscle 312
quaternary structure, proteins 32
queen bee 605
Quercus robur 466
quiescent centre 460, F24.7

race 549
rachis
  feather 621
  fern 575
radial symmetry 584, 609
radiation 701–5
  biological effects of 703–5
  forms of 701

heat 381, F20.3
  solar 222, 637–8
  sources of 702–3
radicle 421, F22.8(a), 438, F23.1(b)
radio-iodine 397
radioisotopes
  in carbon dioxide fixation 82
  in chromosome replication 126
  dating 530
  in DNA replication 59
  in oxygen evolution 80
  in thyroid measurement 397
  in translocation 253
Radiolaria 559
radionuclides 253, 397, 708
  see also radioisotopes
radius 311, F16.1, F16.8, F16.9, F16.18,
  F27.19
radula 607
ragworm 595
ragwort 635
Ramapithecus 534
random coil 31
random genetic drift 517
random orientation, chromosomes 130,
  F7.14
Ranvier, node of 327, F17.12(c), F17.13
ray initials 463, F24.9
reagins 192
receptacles 566
  flower 414, F22.1(a)
receptor neurons F17.3(b), 345, F18.8–10,
  F19.8
receptor sites 330
receptors of target cells 394
recessive characteristics 497
recipient 216
recombinant DNA technology 68–9
recombinant genes 506
rectum F15.14, F15.26(b)
red algae 566
red blood cells 142, 149, 180–1, F10.2
red–green colour blindness 366, 512
redia 591
reducing sugars 16
reduction division 128
reef 588
reflex arc 347
reflex centres 351
reflexes 316, 345–8
  conditioned 348
  contralateral 347
  infantile F18.12
  innate 348
refraction
  absolute and relative F17.11
  of light 360
refractory period 325
  absolute 326
  relative 326
regeneration 590
Reissner's membrane 371, F19.18(a)
relative density, water 5
renal artery F9.2, 201, F11.1
renal capsule F11.4(b), F11.6(a)
renal dialysis see dialysis
renal pelvis 203, F11.3(a)
renal system F11.1
renal vein F9.2, 201 F11.1
renin 211, 405, F21.25
repolarisation 325–6, F17.8
repressor, protein 67
reproduction
  asexual in flowering plants 414
  mammalian 480–94, F25.8
  physiology of, in angiosperms 430–4
  sexual in flowering plants 414–22
reproductive isolation 548–9
Reptilia 532, 619–20
residual volume F8.8
resistance
  environmental 656, 662
  to insecticides 700–1
  to movement 317
resolution 115
resolving power 115
respiration 84–92, 674
  aerobic 109, 647
  in germinating seeds 442, 446
  measuring rate of 90
respiratory centre 140, 151, F8.10,
  F8.12(b), F9.13
respiratory chain 88, F5.21
respiratory distress syndrome 26, 493
respiratory physiology of diving mammals
  151
respiratory quotient see RQ
respiratory substrates
  energy yields from 88–9
respiratory surface, properties of T8.1
respiratory system, effect of medullary
  hormones on 407
respirometer
  simple 90, F5.24

Warburg 90
response to stimulus 320
resting potential 322, F17.7, F17.8, F17.11
resting state 322
restriction endonucleases 68, 73
retardation factor see Rf value
reticular connective tissue 174
reticular formation 353
reticulin 174, 291
reticulo-endothelial system 182, 301
reticulum F15.21
retina 358, F19.1(b), F19.4, F19.8, F19.10,
  F19.12
  inverted 363
retinal 365
retinol see vitamin A
reverberating pathways 345
reverse transcriptase enzymes 70
Rf value 35
rhesus
  blood group 514
  factor 195
  negative 195
  positive 195
  system 195–6
Rhizobium 72, 650, 652, 673
rhizoids 570, 572, 575, 576
rhizome 428, F22.17, 468, 580
rhizosphere 682
Rhodnius 603
rhodopsin 364, 365
rib cage 310–11
riboflavin 43, 280
ribonuclease 29
ribonucleic acid see RNA
ribose F2.3, F4.7
ribosomes 556
ribs F8.3, F8.6, F16.1, F16.5(d), 619
  false 310
  floating 310
  true 310
ribulose T2.1
rickets 280, 538
rigid fins 615
ringing experiments 253, F13.17
RNA 56–7
  messenger 56, 57, F4.8, 110
  ribosomal 56, 57
  role in protein synthesis 54, 56–7
  structure F4.7
  transfer 56, 57, F4.9
RNA ligase 61
rod cells 362–4, F19.8, F19.9, F19.10
rodents 626
root
  adventitious 441
  cap 460
  cortex 461
  hairs 231–2, F12.15
  lateral 458–9
  mineral absorption 268–9
  nodules 650–1
  open 285
  pressure 233–4
  structure 460
  system, primary 460–2
  tissues, differentiation of 404
roots of teeth 268–9, F15.2
rough endoplasmic reticulum 105,
  F6.12(a)
round window 372, F19.17, F19.18(c),
  F19.19
roundworm see Nematoda
royal jelly 605
RQ 91, 446–7
rubella 521
rumen 291, F15.21, 558
  microbes 651
ruminants 291, F15.21, 558, 651
rumination 291
runner, strawberry F22.16

Sabella 596
Saccharomyces 566, 568
saccule 373, F19.23
sacrum F16.1, F16.5(f), F16.7
Salicornia europaea 664
saliva 288
salivary amylase 288
salivary glands F15.15, 620
Salsola kali 664
salt excretion glands 620, 624
sanitation 658
sap 9
Saprolegnia 567
saprophytes 556, 567
sarcolemma 333
sarcomere 334, 336, F17.21, F17.23,
  F17.26, F17.28
sarcoplasm 336
satiety centre 353
scapula 310, F16.6, F16.9
Schistosoma 590, 592, 646

Schmidt-Niesen 211
Schwann cell 326, 328, F17.12(a), F17.14
sclera 358, F19.1(b)
sclerenchyma fibres 457
scolex 592
scorpions 606
scrotal sac 481, F25.1
scurvy 280
scutellum 441
Scyphozoa 588–9
sea anemone 584, 587
sea urchin 609
seals 151
seasonal reproducers 485
sebaceous gland F20.7(b)
sebum 26
secondary consumers 666
secondary growth 463–6, 580, 581
secondary hosts 562, 648
secondary mycelium 569
secondary oocyte 484
secondary plant body 463
secondary productivity 668
secondary sexual characteristics 483
secondary spermatocytes 481
secondary structure, proteins 30
secretin 294, F15.25
sedentary soils 676
sedimentary rock 529
sedimentary soil 676
seeds 468, 577
  coat, impermeability of 448
  in conifers 579
  dispersal 422
  dormancy 448–9
  endospermic 421
  formation 421
  germination
    factors affecting 444–8
    morphology 438–41
  metabolic activity in 445
  non-endospermic 421
  physiology 433–4
  structural immaturity in 448
  structure 438
segmentation of the gut 295
Selaginella 573–4
selective reabsorption 205–7
self antigens 186
self-fertilisation 424
self-heating 446
self pollination 424
self-sterility 425
semen 488
semi-lunar valves 159, 169, 172, F9.17
semicircular canals 373, F19.17, F19.23, F19.25(a)(b)
seminal groove 598
seminal vesicles 481, F25.1
seminiferous tubules 481, F25.2, F25.3
sense organs 358, F17.2, 624
sensorineural deafness 375
sensory region of brain 354
sepals 414, F22.1(a)
sepia 608
septa
  fungal 566
  interventricular F9.3(b)
septate hyphae 566
Sequoia 230
serotonin 183
Sertoli cells 481
seta 571, 572
sewage
  anaerobic digestion of 729
  composition of 726
  disposal 726
  pollution 705–7
  treatment
    aims of 726
    stages of 727–30
sex chromosomes 510
sex hormones 483–4
sex linkage 511–12
sexual reproduction
  flowering plants 414–22
  mammalian 480–94, F25.8
shade leaves 241, F13.12
shell of egg F25.18
shivering 384, F20.13(a)(b)
shoot system, primary 452–9
shore crab 601
short-day plants 430
short-sightedness 361
shrimps 600
sickle cells F8.22
  anaemia 544
  disease 150, 181, 517
  trait 150, 503
sieve cells 573, 577
sieve plates 255, F13.18(b)
sieve-tubes 254–5, F13.18(b), F13.20, 580
silage 725
simian shelf 308

single cell protein 721
sinks 251
sinu-atrial node 159, F9.4
sinusoids 169, 299, F15.29(a)(c)
siphon 608
skeletal muscle 333–4, F17.21, F17.22
skeletal system 306–18
skeleton 16.1, 306–12
skin 285, F20.7(a)(b), 20.8(a)(b)
skin prick test 192
skull 307–8, F16.1
sleep 352
sliding filament model 338
slipped disc 308
small intestine 291–4, F15.22–3, 15.25, 15.27
smallpox 192–3
smog 690, 691
smooth muscle 9.14
snow blindness 365
sodium
  blood, control of 405–6
  gates 323
  ions 297
  in muscle 338
sodium–potassium exchange pump 322, F17.4
soil
  air 676–7
  air capacity 676
  decomposition of 682
  as an ecosystem 675–84
  fauna 680–1
  field capacity of 677
  food webs in 682
  microflora 679–80
  mineral particles in 676
  moisture 677
  nutrient cycling and energy flow in 683–4
  organic matter 677–8
  organisms, distribution of 681–2
  pH 678
  phosphorus in 683
  pore space 676
  profiles 675–6
  sedimentary 676
  as a source of minerals 267–8
  structure 676
  temperature 679
  texture 676
sol 559
solar radiation 637–8
  effect on transpiration 222
solitary flowers 414
solution, true 3
solvent
  developing 35
  water as a 2
solvent front 35
soma 321
somatic effects of 704–5
somatic nerves 342
somatomedin 408
somatotropin 408
somesthetic sensory area of brain 355
sorus 575
sound 368–70, F19.13(a)
  amplitude of 369
  intensity 369
  loudness 369
sound waves 368, F19.13(b), F19.14, F19.16
Spartina 544–5
  S. townsendii 664
species 549, 552
  formation of 549
specific heat capacity 5
spectrum, visible F5.2
sperm F25.5(a)
sperm ducts 597
spermathecae 598
spermatids 482, F25.4
spermatocytes 481, F25.4
spermatogenesis 481–2, F25.4
spermatogonia 481, F25.4
Spermatophyta 531
spermatozoa 480
Sphenopsida 574–5
sphincters 291, F11.2
  bladder 201
sphingolipids 26, F2.19(a)
sphingosine 26
sphygmomanometer 163, F9.8 (b)
spider 598, 606
spinal column 308
spinal cord F16.4, F18.1, F18.7, F18.11, F18.13, F18.15, F18.16
spinal nerve F18.8
spindle 122, 346
spindle fibres 110
spinneret 606
spiracle 602, 615
spirillum 556

Spirogyra 564
spirometer 138, F8.7
  tracing F8.8
spleen F15.14
sponges 531
spongy bone F3.14, F16.12
spongy mesophyll F11.2(a), F24.5(c)
sporangia 567, 574
sporangiophores 567, 574
spores 562, 567
sporocyst 591
sporogonium 571, 572
sporophyll 574
sporophyte 423, 570, 574, 575, 577
Sporozoa 562–3
sports 429
spot disease 558
spring wood 465
squamous epithelium 136
staining cells 98, 100
stamens 414, F22.1(a)
standard deviation 513
standing crop 671
stapes 371, F19.17, F19.18(c), F19.19
Staphylococcus 557
starch 563
  in germinating seeds 442
  grains F2.9
  in mammals 278
  in plant cells 99
  in potato tuber F2.9
  and salivary amylase 288
  test for 20
starfish 609
Starling's law of the heart 165
stationary phase 656
steady state 240
stearic acid 23, F2.14
stem cells 186
stem of moss 572
stenohaline 641
stenosis 168, F9.15
stereocilia 372, F19.20
stereoisomers 17–18
  alanine F2.22
  of glyceraldehyde F2.2
  sugars 18
stereoscopic vision 367, 628
sternal puncture 310
sternum 8.3, 310, F16.1, F16.6
  body of 310
steroid-secreting cells 106
steroids 26, F2.20(a), 483, 723
sterols 26
stethoscope 163, F9.8(b)
stigma 414, F22.1(a), 580
  euglenoid 563
stipe 569
stomach 288–9, F15.14, F15.20, F15.22, F15.25
stomata F12.9(a), 242, F13.3
  effect of light and carbon dioxide on F13.4
  opening and closing of 244–5
  stem 457
stomium 576
storage
  cells 99
  organs F13.13
strata 530
stratified epithelium 385
stratosphere 688
Strecker synthesis 527
Streptococcus 557
Streptomyces 558
streptomycin 721
stretch receptors 141, F8.12(b), F18.9
stretch reflex 346, F18.9
striated muscle 333
strict anaerobes 84, 556
striped muscle 333
strobilus 574
stroma 111–12
stroma of moss 572
stromatolites 530
strong acids 7
style 414, F22.1(a)
sub-species 549
subclavian artery F9.2, F9.3(a)
suclavian vein 174, F9.2, F9.21
subcutaneous layer F20.7(b)
suberin 113
suberisation 231–2
sublingual gland F15.15
submandibular gland F15.15
subsoil 676
substrate concentration, effect on
  enzymes 43, 48
succession 663–5
  primary 663–5
  secondary 665
succulent fruits 422
succulent plants 227
succus entericus 293
suckers 590, 592, 598

sucrose 18–19, F2.6, 442
sudorific glands 382, 385, F20.7(b)
sugars 278, 281
  non-reducing 19
  reducing 16
  test for 16
sulphur cycle 683
sulphur dioxide pollution 689–91
sulphur in plant growth 266
summer wood 465
sun leaves 241, F13.12
surface active agent 137, 493
surface area
  and body volume 382–3
  of red blood cells 180
  of skin F20.5
  of small intestine 295–6
  to volume ratio (SA/V) 382
surface tension 137
  of water 5–6
surfactant 137, 493
survival of the fittest 542
survivorship curves 659
suspensions 4
suspensors 421, F22.8(a)
  in fungi 568
suspensory ligaments 360, F19.1(b)
sutures 308, F16.2
sweat 382, 385, 387–9
sweat gland 624
sweating 212, F20.13(a)
swim bladder 616, 640
symbionts 567, 679
symbiosis 559, 585, 645, 648
sympathetic nerves 342, 348, F18.1, F18.13, F18.14
symphysis pubis 311, F16.7, F25.21
synapses 329–32, F17.16(a)(d), F18.14
  axodendritic F17.16(b)
  axosomatic 17F16(b)
  inhibitory 331–2, F17.19
synaptic cleft 329, F17.16(c)(d)
synaptic transmission F17.17
syncitium 594
synergids F22.5(d)
synovial capsule 315, F16.15
synovial fluid 315
synovial joint 315, F16.4, F16.17
synovial membrane 315, F16.15
syrinx 623
systole 163

T-cells 186, 187, F10.10
T-wave (ECG) 162
Taenia 592–3
tapetum 416, F22.3
  of eye 363
tapeworm 592–3
target cells 394, F21.3
target organs 394
tarsals 312, F16.1, F16.8, F16.10
taxa 552
taxonomy 552
tears 358
technetium 398
tectorial membrane 372, F19.18(a), F19.20
teeth 282–6, F15.1
telophase
  meiosis 128–9
  mitosis 123
temperature 637
  effect on enzyme action 43, 46, 432
  effect on flowering 432
  effect on haemoglobin 147
  effect on haemoglobin–oxygen
    dissociation curve F8.18
  effect on heart rate 247
  effect on photosynthesis 468
  effect on seed germination 445
  effect on transpiration 222
  mammalian body F20.1, F20.11, F20.12
  and soil organisms, distribution of 682
temperature coefficient 46, 247
temporal bone F16.2, F16.3, F18.18
temporal lobe 354
tendons 316, F17.21
tentacles 584, 595, 607, 608
teratogenic agents 521
teratological defects 697
tertiary consumers 666
Tertiary period 532
tertiary structure, proteins 31
test-cross 498
test-tube baby 488
testa 438, F22.8(a)
testes F21.1, 481, F25.1, F25.2, F25.6, 586
testosterone 483, F25.6
tetany 401, F21.15
tetraethyl lead (TEL) 692
tetraiodothyronine (T_4) 396, F21.5
thalamus 353, F18.15
thalassaemia 503
thalidomide 521
thallus 565, 571

theca externa 484
theca interna 484
Theria 625–9
thermal pollution 707
thermal properties of water 4–5
thermals 622
thermobarometer 90
thermodynamics, laws of 76
thermography 381
thermoneutral zone F20.12
thermoregulation 380–92, F20.13
  centre 353, 384, F20.6, F20.13(b)
thiamin 279
thin-layer chromatography 36
thirst centre 353, 406, F21.25
thoracic ducts 174, F9.21
thoracic vertebra F16.5(d), F16.6
thorax 136, 310
  of arthropod 598
  of insect 602
threshold, transmembrane potential 323, F17.11
thrombin 183, F10.5
thrombocytes *see* platelets
thromboplastin 183, F10.5
thrombosis 168, 184
thrombus, intravascular 184
thymine 55
thymus gland 186, 10.9, F21.4
thyroid follicle 395
thyroid gland 395–8, F21.1, F21.6–9, F21.11, F21.12
thyroid stimulating hormone (TSH) 396, 408, 410, F21.6
thyrotropin releasing hormone (TRH) 396, 410
thyroxin 384, 396, F21.5–7
tibia 311, F16.1, F16.8, F16.10
tick 606
tidal ventilation 139, F8.9
tidal volume 138, F8.8
tip propagation 429
tissue
  differentiation 473
  factors 183
  fluid 170–1, F9.19
  level of development 584
  root 460–2
  stem 455–8
  typing 197, 216
titre 10.8
tonoplast membrane 9
tooth bud 282
toothed whale 286
topsoil 676
torsion 607
total plasmolysis 10
touch receptor F20.70(b)
toxins 193, 648
toxoids 193
trabeculae in bones 314
trace elements
  in mammals 280
  in plants 267
trachea 137, F8.3, 602, F21.11
tracheids 234–7, 573, 577, 580
tracheole 602
Tracheophyta 573–81
transamination 301
transcellular strands 255
transcription 61, 4.14
transfer cells F13.21
  in leaf 256, F13.21
  in root 234
  in shoot 272
transfer RNA 56, 57, F4.9
transferases 50
transferrin 301, F15.21
transformation 58
transfusion reaction 195
transitional epithelium 201
translation, genetic code 62–3, F4.16
translocation
  effect of auxin on 470
  mechanisms 256–8
  pathway of 251–6
  of photosynthetic products 251–8
  rate of F13.5
transmembrane potential F17.4, F17.5, F17.8
transmission of information stimuli 320
transmitter substances 330, F17.18
transpiration 220–2
  effects of 228–33
  factors affecting 222–4
  lenticular 220
  and light intensity 222
  ratio 227, T12.3
  stream 229, 272
  and water absorption 228–30
transpiration–cohension–tension theory 230
transplantation of organs 197
transverse foramen 309, F16.5

transverse ligament 309, F16.5(a)
transverse process F16.4
Trebouxia 650
Trematoda 590–2
TRH F21.6
triacylglycerol F2.15
Triassic period 532
Tribolium 635
tricarboxylic acid cycle *see* Krebs cycle
triceps muscle 316, F16.19, F18.11
trichocyst 561
trichromatic theory of colour vision 365
trickling filter system 727–8
tricuspid valve F9.3(b)
triiodothyronine (T₃) 396, F21.5
triple-X syndrome 511
triplet code 61
triploblastic 589, 594
triploid fusion nucleus 420
Tritium 546–7
trochophore larva 596, 607
trophic hormones 396
trophic levels 666
trophic relationships 665–6
trophoblasts 489
tropins 396, 408
tropisms 470–3
tropomyosin 336, F17.27
troposphere 688
trypsin F15.24, 293
trypsinogen 293
TSH *see* thyroid stimulating hormone
tube feet 609
tube nucleus 416, F22.3(d)
tubers 428, 468
  potato F22.17
tubulin 113
tumour cells 190
tunica 452, F24.1(b)
tunica adventitia 168, F9.14
tunica intima 168, F9.14
tunica media 168, F9.14
Turbellaria 589–90
turgid cells 9
turgidity F1.9, 110, 468
Turner's syndrome 511, 520
turnover number 47
tusks F15.13
tympanic canal F19.18(a)
tympanic membrane 371, F19.17
tyrosine F2.23

ulcer 290
ulna 311, F16.1, F16.8, F16.9, F16.18, F27.19
ultracentrifuge 116–17, F6.27
ultrafiltration 204–5, F11.6(a), F11.7, F11.8
ultramicrotome 100
ultraviolet radiation 544
umbilical artery F9.24, F25.19
umbilical blood vessels 175
umbilical cord 175, F9.23, F25.19, F25.21
umbilical vein F9.24, F25.19
ungulates 626
unicellular organisms 528, 558, 559, 566
unicellular rhizoids 571
unisexuality 480
unit leaf rate 249, 250
unit membrane 102, F6.9
  hypothesis 102
universal donors 195
universal recipients 195
unspecialised dentition 628
uracil 56
urea 300, F15.21
ureter 201, F11.1–3, F25.1
urethra 201, F11.1, F11.2, F25.1
uric acid 603, 620, 624
urinary bladder 201, F11.1, F11.2, F25.1, F25.21
urine 201
uterus F9.23, 483, 487, F25.8, F25.21
utricle 373, F19.23, F19.24

vaccines 192–3, T10.4, 722–3
vacuolation of cells 445, 455, 468
vacuoles 9, 99
  contractile 12, F1.13, 560, 561
vagina 483, 487, F25.8, F25.21
vagus nerve 164, F9.9, F15.19, F18.13, F18.16
Van Leeuwenhoek's microscope 96
variation 543–5
variety 549
vasa deferentia 481, F25.1, F25.2
vasa efferentia 481, F25.2
vasa recta 208
vasa vasorum 168
vascular bundles 453, F22.4(d)
vascular cambium 463–5
vascular rays 464
vascular tissue 232, 573, 577
  differentiation of 469

vasoconstriction 166, 183, 386, F20.8(a), F20.13(a)
vasodilation 166, 386, F20.8(b). F20.13(a)
vasomotor centre F8.12(b), 166, F9.12, F9.13
vasomotor control 166–7
vector 554, 557, 562, 646, 647
vegetables 281
vegetative photoperiodism 467
vegetative propagation 428–9
veins 156, 168, 169, 170, F9.2, F9.18, F9.20
veliger larva 607
vena cava
  inferior F9.2, F9.3(a), F11.1
  superior 9.3(a)
venation 12.1
ventral gland 603
ventricles F9.3(a)
  of heart 159
ventricular diastole 163
ventricular systole F9.5(b)
venules 156, 168, 169–70, F9.18, F9.19
vertebrae 308–9, F16.1, F16.4, F16.5, F16.17
vertebral arch 309, F16.5(d)
vertebral column 614
vertebrates 614
venter 576
ventral root 346, F18.8
vernalisation 432
vesicles 106, F17.16(d)
vessels in plants 234–5, 580
vestibular apparatus 373, F19.17, F19.23
vestibular canal 371, F19.18(a)
vestibular nerve 374, F19.17, F19.25(b)
vestigial structures 539
villus F15.28
  chorionic 490, 518, F25.17
  intestinal 296
  placental 25.17
virion 553
viruses 553–4, 680
visceral hump of molluscs 607
viscosity, water 5
visual acuity 364
visual field 366
vital capacity 138, F8.8
vital centres 351
vitamins 279–80
  A 279
  B 279
  B₂ 280
  B₁₂ 280
  C 280
  D 279, 538
  E 279
  K 279, 558
  fat-soluble 279, 297
  manufacture of 721
  water-soluble 279–80
vitelline gland 538
vitreous humour 360, F19.1(b)
Vorticella 560, 561
vulva 483, F25.8

Wallace, Alfred Russel 542
Warburg manometer 90, F5.23
water
  absorption
    by gut 298
    by roots 230–2
  and aqueous solutions 2–14
  availability, effect on plant growth 468
  balance T1.6
  capillarity 228, 229, 233, 236, 677
  cohesion 288
  density 4–5
  dissociation of 6–7
  effect on seed germination 445–6
  gradient 8, 233, 237
  gravitational 677
  hydrogen bonding in 4
  hygroscopic 677
  molecule 2, F1.1, F1.2
  movement 642
    through shoot system 228–30
  oxygen in 638–9
  penetration by light 6
  polarity of 2
  potential 8, T12.1, T12.2, 640–2
  as a solvent 2–4
  supply to leaves 241
  surface tension 5–6
  thermal properties of 4–5, T1.4
  uptake and loss by cells 8
  uptake and transport in plants 220–38
  vapour pressure T12.1, T12.2
  vascular system 609
  viscosity 5
water cycle 674–5
water culture experiments 263–4
water flea 600
water potential gradient 8, 233, 237

water-soluble vitamins 279–80
Watson–Crick model F4.5
wavelength of light
  effect on eye 362
  effect on growth 467
wavelength of sound 368, F19.14
  effects on ear 373
waxes 26, 598
weak acids 7
web, spider's 606
weedkillers 476
weeping lubrication theory 315, F16.16
whales 151, 286, 627, 661, 662
white blood cells *see* leucocytes
white fat 386
white matter 346, F18.8
wilts 558
wind-pollinated flowers 426
winter buds 468, 475
winter sleep 390
wisdom teeth 283
woodlouse 600
wood tissue 577
worker bee 605

X-ray diffraction 30
Xanthomonas 558
xanthophyll 434, 558, 563, 565
Xanthoria parietina 663
xerophytic plants 226–7
xiphoid process 310, F16.6
xylem 241, 573
yeast 568
yolk sac 489, 490, F25.17, F25.18, 620

Z-line 334, 17.26, 17.28
zeatin 474
zinc 280, 696
zonation 643
zoospores 563, 565, 567
zygomatic bone F16.2
zygospores 564, 568
zygote, diploid 420, 421, 488